# Elasto-Plasticity of Frame Structure Elements

Andreas Öchsner

# Elasto-Plasticity of Frame Structure Elements

## Modeling and Simulation of Rods and Beams

Springer

Andreas Öchsner
Griffith University
Southport
Australia

and

The University of Newcastle
Callaghan
Australia

ISBN 978-3-662-44224-1 ISBN 978-3-662-44225-8 (eBook)
DOI 10.1007/978-3-662-44225-8

Library of Congress Control Number: 2014945136

Springer Heidelberg New York Dordrecht London



Printed on acid-free paper

Springer is part of Springer Science+Business Media (www.springer.com)

*Honest disagreement is often a good sign of progress*

Mahatma Gandhi (1869–1948)

# Preface

The scientific literature offers many excellent textbooks on the elasto-plastic behavior of structural members. However, many students who are entering this research area find it difficult to master this topic based on classical textbooks. To facilitate the understanding of the theory, concepts, and involved algorithms, this textbook offers for students and researchers who are new in this topic a simpler introduction to elasto-plastic material behavior and its computational treatment.

In this intention, the focus of this textbook is on two simple one-dimensional structural members, i.e., the rod and the beam. A classical approach is presented where first the governing differential equations are derived based on the basic equations of continuum mechanics. The corresponding analytical solutions serve many times to judge the accuracy of derived approximate solutions. The focus in regard to approximate solutions is on the finite element method. This method can be regarded nowadays as the standard tool to solve engineering problems in the industrial context. The treatment of one-dimensional members allows to keep the mathematical notation and treatment relatively simple. Nevertheless, the derivations are presented in such a way that a transfer to higher dimensions can be achieved. Thus, the considered one-dimensional elements are from an educational point of view a good demonstrator to introduce basic concepts of continuum mechanics and numerical strategies for approximate solutions. Pure elasticity is treated first and then the important case of nonlinear material behavior in the elasto-plastic range is considered.

Furthermore, rod and beam elements have structural analogies in higher dimensions. The procedures applied to the rod element can be transferred to three-dimensional solid elements. Knowing the theory of one-dimensional beam elements allows the understanding of two-dimensional plate elements. Thus, a solid understanding of the basic equations and procedures for one-dimensional elements allows an easier access to elements of the two- and three-dimensional space.

Chapter 1 illustrates a few technical applications, which can be modeled in a simplified approach based on one-dimensional elements. Chapter 2 introduces the basic equations of plasticity theory. The yield condition, the flow rule, and the hardening rule are introduced. After presenting the equations for the one-dimensional stress and strain state, the equations are generalized for a two-component case. Looking at the two-component case, the notation stays relatively simple and the major differences between the one-dimensional case and a higher dimensionality can easily be shown. Chapter 3 covers the analytical description of rod or bar members. Based on the three basic equations, i.e., the kinematics relationship, the constitutive law, and the equilibrium equation, the partial differential equation which describes the problem is derived. Analytical solutions in the elastic and elasto-plastic range for different loading and boundary conditions are derived and discussed. Chapter 4 follows the approach of the previous chapter and introduces beam elements. Three different types of theories are treated, i.e., the EULER–BERNOULLI theory, the TIMOSHENKO theory and higher-order beam theories. Chapter 5 introduces the finite element theory for elastic problems based on rod, beam, and generalized beam elements. First, each element type is considered separately, then in pure one-dimensional structures and finally the case of plane frame structures which allows to arrange and combine different elements. Chapter 6 covers elasto-plastic finite element simulations, an important type of nonlinear problems. The solution strategy is derived for pure one-dimensional problems. However, the derivations are presented in such a way that a transfer to higher dimensions is easily possible. Chapter 7 is devoted to an alternative approach to derive approximate solutions. Based on one-dimensional problems, the finite difference method is introduced for elastic and elasto-plastic problems. Chapter 8 summarizes the basic equations for a three-dimensional continuum. The derived partial differential equations serve to derive finite element procedure for solid elements. The chapter concludes with an introduction to elasto-plastic finite element simulations.

In order to deepen the derived equations and theories, each technical chapter collects at its end supplementary problems. In total over 120 of such additional problems are provided, and a short solution for each problem is included in this book. It should be noted that these short solutions contain major steps for the solution of the problem and not only, for example, a numerical value for the final result. This should ensure that students are able to successfully master these problems. I hope that students find this book a useful complement to many classical textbooks. I look forward to receive their comments and suggestions.

Gold Coast, Australia, May 2014 Andreas Öchsner

# Acknowledgments

It is important to highlight the contribution of many undergraduate and postgraduate students, which helped to finalize the content of this book. Their questions and comments during different lectures and their work in the scope of final year projects helped to compile this book. Furthermore, I would like to express my sincere appreciation to Springer-Verlag, especially to Dr. Christoph Baumann, for giving me the opportunity to realize this book. A professional publishing company with the right understanding was the prerequisite to complete this long-term project. Corrections, comments, and suggestions by Prof. Graeme E. Murch and Prof. Holm Altenbach are greatly appreciated. Finally, I like to thank my family for the understanding and patience during the preparation of this book.

# Contents

# Symbols and Abbreviations

## Latin Symbols (Capital Letters)

| | |
|---|---|
| $A$ | Area, cross-sectional area |
| $\bar{A}$ | Effective resisting area |
| $A_{\mathrm{D}}$ | Total area of microcracks and cavities |
| $\boldsymbol{A}$ | Matrix, cf. derivation of shape functions |
| $\boldsymbol{B}$ | Matrix which contains derivatives of shape functions |
| $C_{ijkl}$ | Fourth-order elasticity tensor |
| $\boldsymbol{C}$ | Elasticity matrix |
| $C^{\mathrm{elpl}}$ | Elasto-plastic modulus matrix |
| $D$ | Damage variable |
| $D^{\mathrm{pl}}$ | Generalized plastic modulus |
| $D_{\mathrm{c}}$ | Critical damage at crack initiation in pure tension |
| $\boldsymbol{D}$ | Compliance matrix |
| $D^{\mathrm{pl}}$ | Matrix of generalized plastic moduli |
| $\boldsymbol{E}$ | YOUNG's modulus |
| $\bar{E}$ | Elastic modulus of damaged material |
| $E^{\mathrm{elpl}}$ | Elastic-plastic modulus |
| $E^{\mathrm{pl}}$ | Plastic modulus (isotropic hardening) |
| $\boldsymbol{EA}$ | Axial tensile stiffness |
| $EI$ | Bending stiffness |
| $F$ | Force, yield condition |
| $F^{\mathrm{el}}_{\mathrm{lim}}$ | Maximum elastic force |
| $G$ | Shear modulus |
| $GA$ | Shear stiffness |
| $H$ | Kinematic hardening modulus |
| $I$ | Second moment of area, abbreviation for an integral statement |
| $\mathbf{I}$ | Identity matrix (diagonal matrix), $\mathbf{I} = \lceil 1\ 1\ 1 \ldots \rfloor$ |
| $J$ | Jacobian, cf. coordinate transformation |

| | |
|---|---|
| $\boldsymbol{K}$ | Global stiffness matrix |
| $\boldsymbol{K}^{\mathrm{e}}$ | Elemental stiffness matrix |
| $\boldsymbol{K}_{\mathrm{T}}$ | Tangent stiffness matrix |
| $L$ | Element length |
| $M$ | Moment |
| $M_{\mathrm{lim}}^{\mathrm{el}}$ | Maximum elastic bending moment |
| $M_{\mathrm{lim}}^{\mathrm{pl}}$ | Limiting bending moment |
| $M_{\mathrm{s,\,lim}}^{\mathrm{pl}}$ | Limiting bending moment under shear influence |
| $N$ | Normal force (internal), shape function |
| $\boldsymbol{N}$ | Column matrix of shape functions |
| $\boldsymbol{N}_i$ | $3 \times 3$ Matrix of shape functions for node $i$ |
| $P$ | LEGENDRE polynomial, point |
| $Q$ | Shear force (internal), plastic potential function (plastic potential) |
| $\dot{Q}$ | Heat transfer rate |
| $Q_{\mathrm{lim}}^{\mathrm{pl}}$ | Plastic shear force |
| $R$ | Equivalent nodal force, radius of curvature of a curve, stress ratio |
| $\boldsymbol{T}$ | Transformation matrix |
| $U$ | Perimeter |
| $V$ | Volume |
| $V_{\mathrm{D}}$ | Total volume of microcracks and cavities |
| $W$ | Weight function |
| $W^*$ | Fundamental solution |
| $\boldsymbol{W}$ | Column matrix of weight functions |
| $X$ | Global Cartesian coordinate |
| $Y$ | Global Cartesian coordinate, damage energy release rate |
| $Z$ | Global Cartesian coordinate |

## Latin Symbols (Small Letters)

| | |
|---|---|
| $a$ | Basis coefficient, geometric dimension |
| $\boldsymbol{a}$ | Column matrix of basis coefficients |
| $b$ | Coefficient, function, geometric dimension |
| $\boldsymbol{b}$ | Column matrix of body forces acting per unit volume |
| $c$ | Constant of integration, coefficient, geometric dimension |
| $d$ | Coefficient |
| $f$ | Body force, scalar function |
| $\boldsymbol{f}$ | Column matrix of loads |
| $g$ | Scalar function, standard gravity |
| $h$ | Evolution function of hardening parameter, geometric dimension |
| $\boldsymbol{h}$ | Evolution function of hardening parameters |
| $i$ | Iteration index, node number |
| $j$ | Iteration index |

| | |
|---|---|
| $k$ | Auxiliary function, elastic embedding modulus, elastic foundation modulus, layer number, stiffness, spring constant, thermal conductivity, yield stress |
| $k_{\mathrm{s}}$ | Shear correction factor, shear yield stress |
| $k_{\mathrm{t}}$ | Tensile yield stress |
| $k_{\mathrm{t}}^{\mathrm{r}}$ | Uniaxial stress at rupture |
| $m$ | Distributed moment, element number, mass |
| $\boldsymbol{m}$ | Matrix function |
| $n$ | Node number, increment number |
| $n_j$ | Components of the normal vector |
| $\boldsymbol{n}$ | Normal vector |
| $p$ | Distributed load in $x$-direction |
| $q$ | Distributed load in $y$-direction, internal variable (hardening) |
| $\boldsymbol{q}$ | Column matrix of hardening variables |
| $\dot{q}$ | Heat flux |
| $r$ | Damage evolution parameter, residual |
| $\boldsymbol{r}$ | Plastic flow direction, residual column matrix |
| $s$ | Damage evolution parameter |
| $\boldsymbol{s}$ | Column matrix of stress deviator components |
| $t$ | Time, traction force |
| $t_i$ | Components of the traction force vector |
| $t_{\mathrm{end}}$ | Convergence value |
| $\boldsymbol{t}$ | Column matrix of traction forces |
| $u$ | Displacement |
| $u^0$ | Exact solution |
| $\boldsymbol{u}$ | Column matrix of displacements, column matrix of nodal unknowns |
| $v$ | Auxiliary function |
| $\boldsymbol{v}$ | Variable matrix |
| $w$ | Weight for numerical integration |
| $x$ | Cartesian coordinate |
| $\boldsymbol{x}$ | Column matrix of Cartesian coordinates |
| $y$ | Cartesian coordinate |
| $z$ | Cartesian coordinate |

## Greek Symbols (Capital Letters)

| | |
|---|---|
| $\Gamma$ | Boundary |
| $\Lambda$ | Factor |
| $\Omega$ | Domain |

## Greek Symbols (Small Letters)

| | |
|---|---|
| $\alpha$ | Back stress/kinematic hardening parameter, parameter, rotation angle |
| $\boldsymbol{\alpha}$ | Column matrix of kinematic hardening parameter |

| | |
|---|---|
| $\beta$ | Parameter |
| $\gamma$ | Shear strain (engineering definition), load per unit volume, specific weight per unit volume, $\gamma = \varrho g$ |
| $\delta$ | Dirac delta function, geometric dimension |
| $\varepsilon$ | Strain |
| $\bar{\varepsilon}$ | Effective strain |
| $\varepsilon^{\mathrm{el}}$ | Elastic strain |
| $\varepsilon^{\mathrm{pl}}$ | Plastic strain |
| $\varepsilon^{\mathrm{pl}}_{\mathrm{D}}$ | Damage plastic strain threshold in pure tension |
| $\varepsilon^{\mathrm{pl}}_{\mathrm{eff}}$ | Effective plastic strain (equivalent plastic strain) |
| $\varepsilon^{\mathrm{pl}}_{\mathrm{r}}$ | Plastic strain at rupture |
| $\varepsilon^{\mathrm{pl}}_{\mathrm{V}}$ | Volumetric plastic strain |
| $\varepsilon_{ij}$ | Second-order strain tensor |
| $\varepsilon_{\mathrm{c}}$ | Elastic limit strain in compression |
| $\varepsilon_{\mathrm{t}}$ | Elastic limit strain in tension |
| $\boldsymbol{\varepsilon}$ | Column matrix of strain components |
| $\zeta$ | Natural coordinate |
| $\eta$ | Natural coordinate |
| $\dot{\eta}$ | Rate of energy generation per unit volume |
| $\kappa$ | Curvature, isotropic hardening parameter |
| $\kappa^{\mathrm{el}}_{\mathrm{lim}}$ | Maximum elastic curvature |
| $\lambda$ | Lamé's constant, consistency parameter (cf. plasticity) |
| $\mu$ | Lamé's constant |
| $\nu$ | Poisson's ratio |
| $\xi$ | Natural coordinate, relative stress ($\xi = \sigma - \alpha$) |
| $\dot{\emptyset}$ | Rate of energy generation per unit length |
| $\varrho$ | Mass density |
| $\sigma$ | Stress, normal stress |
| $\bar{\sigma}$ | Effective stress (damage mechanics) |
| $\sigma_{\mathrm{eff}}$ | Effective stress (yield condition) |
| $\sigma_{\mathrm{m}}$ | Mean stress (hydrostatic stress) |
| $\sigma_{ij}$ | Second-order stress tensor |
| $\boldsymbol{\sigma}$ | Column matrix of stress components |
| $\varsigma$ | Coordinate where elasto-plastic interface reaches beam surface |
| $\tau$ | Shear stress |
| $\phi$ | Rotation (Timoshenko beam) |
| $\varphi$ | Basis function, rotation (Bernoulli beam) |
| $\chi$ | Column matrix of basis functions |
| $\psi$ | Basis function |
| $\omega$ | Material integrity |

## Mathematical Symbols

| | |
|---|---|
| $\times$ | Multiplication sign (used where essential) |
| $[\ldots]$ | Matrix |
| $\lceil\ldots\rfloor$ | Diagonal matrix |
| $[\ldots]^{\mathrm{T}}$ | Transpose |
| $\langle\ldots\rangle$ | MACAULAY's bracket |
| $\langle\ldots,\ldots\rangle$ | Inner product |
| $\mathcal{L}\{\cdots\}$ | Differential operator |
| $\mathcal{L}$ | Matrix of differential operators |
| $\mathrm{sgn}(\ldots)$ | Signum (sign) function |
| $\partial$ | Partial derivative symbol (rounded *d*) |
| $\mathbb{R}$ | Set of real numbers |
| $\delta$ | DIRAC delta function |
| $\mathbf{1}$ | Identity column matrix, $\mathbf{1} = [111000]^{\mathrm{T}}$ |
| $\mathbf{L}$ | Diagonal scaling matrix, $\mathbf{L} = \lceil 111000 \rfloor$ |

## Indices, Superscripted

| | |
|---|---|
| $\ldots^{\mathrm{e}}$ | Element |
| $\ldots^{\mathrm{el}}$ | Elastic |
| $\ldots^{\mathrm{init}}$ | Initial |
| $\ldots^{\mathrm{pl}}$ | Plastic |
| $\ldots^{\mathrm{trial}}$ | Trial state (return mapping) |

## Indices, Subscripted

| | |
|---|---|
| $\cdots_{\mathrm{b}}$ | Bending |
| $\cdots_{\mathrm{c}}$ | Center, compression |
| $\cdots_{\mathrm{lim}}$ | Limit |
| $\cdots_{\mathrm{p}}$ | Nodal value ('point') |
| $\cdots_{\mathrm{s}}$ | Shear |
| $\cdots_{\mathrm{t}}$ | Tensile |

## Abbreviations

| | |
|---|---|
| 1D | One-dimensional |
| 2D | Two-dimensional |
| 3D | Three-dimensional |
| a.u. | Arbitrary unit |
| BC | Boundary condition |
| BEM | Boundary element method |
| const. | Constant |

| | |
|---|---|
| dim. | Dimension |
| DOF | Degree(s) of freedom |
| EBT | EULER–BERNOULLI beam theory (elementary beam theory) |
| FD | Finite difference |
| FDM | Finite difference method |
| FEM | Finite element method |
| FGM | Functionally graded material |
| FSDT | First-order shear deformation theory |
| inc | Increment |
| max | Maximum |
| PDE | Partial differential equation |
| RVE | Representative volume element |
| SI | International system of units |
| SSDT | Second-order shear deformation theory |
| sym. | Symmetric |
| TBT | TIMOSHENKO beam theory |
| TSDT | Third-order shear deformation theory |
| WRM | Weighted residual method |

## Some Standard Abbreviations

| | |
|---|---|
| ca. | About, approximately (from Latin 'circa') |
| cf. | Compare (from Latin 'confer') |
| ead. | The same (woman) (from Latin 'eadem') |
| e.g. | For example (from Latin 'exempli gratia') |
| et al. | And others (from Latin 'et alii') |
| et seq. | And what follows (from Latin 'et sequens') |
| etc. | And others (from Latin 'et cetera') |
| i.a. | Among other things (from Latin 'inter alia') |
| ibid. | In the same place (the same), used in citations (from Latin 'ibidem') |
| id. | The same (man) (from Latin 'idem') |
| i.e. | That is (from Latin 'id est') |
| loc. cit. | In the place cited (from Latin 'loco citato') |
| N.N. | Unknown name, used as a placeholder for unknown names (from Latin 'nomen nescio') |
| op. cit. | In the work cited (from Latin 'opere citato') |
| pp. | Pages |
| q.e.d. | Which had to be demonstrated (from Latin 'quod erat demonstrandum') |
| viz. | Namely, precisely (from Latin 'videlicet') |
| vs. | Against (from Latin 'versus') |

# Chapter 1
# Introduction

**Abstract** The first chapter classifies the content as well as the focus of this textbook. Based on common applications taken from structural engineering, the investigation of rod and beam elements in spatial arrangements is motivated.

Frame structures are spatial arrangements or skeletons composed of beams and rods. The design principle of arranging one-dimensional members in a spatial framework can be found in many areas of civil, mechanical and structural engineering. A major idea of these structures is to optimize the use of a material such as steel, wood or concrete and to arrange it in such a way to obtain maximum stiffness and strength at a lower weight and costs than solid structures. Typical examples are some of the constructions of Gustave EIFFEL,[1] such as the railway arch bridge Garabit Viaduct which spans the river La Truyère in Southern France (Auvergne), see Fig. 1.1.

Some other typical examples from civil engineering are:

- Eiffel Tower, Paris, France (construction period 1887–1889),
- Maria Pia Bridge (Douro Viaduct), Porto, Portugal (construction period 1876–1877),
- internal frame structure of the Statue of Liberty,[2] New York, United States (construction period 1875–1886).

In the same context, one may mention, for example, tower structures for power transmission, cranes, and wind energy converters. Within this textbook, we distinguish two different types of basic members of such spatial structures, i.e. rods and beams. A rod is understood as a prismatic body where loads act only in the direction of the principal axis of the body. This member is sometimes called a bar and in some applications a rotational deformation around the principal axis is also considered.

[1] Alexandre Gustave EIFFEL (1832–1923), French civil engineer and architect.

[2] The Statue of Liberty was realized by the French sculptor Frédéric Auguste BARTHOLDI (1834–1904). A bronze model of the statue stands in the Jardin du Luxembourg (Paris, France) and was used by Bartholdi as part of the preparatory work for the New York statue. The second Statue of Liberty in Paris is near the Grenelle Bridge on the Île aux Cygnes.

 
A. Öchsner, *Elasto-Plasticity of Frame Structure Elements*,
DOI 10.1007/978-3-662-44225-8_1

**Fig. 1.1** Railway arch bridge (Garabit Viaduct), France (construction period 1884–1884)

However, we are going to consider in the following only the elongation and compression of rods. If rods are spatially arranged, a so-called truss structure is obtained. The common characteristics and simplifications for truss structures are as follows [59]:

- two- or three-dimensional framework of straight rod members,
- the ends are connected by frictionless hinged joints (not subjected to moments, rotations are zero),
- the loads and reactions act only at the joints,
- the rods are only subjected to axial compressive or tensile forces.

The second basic member is a beam element which is understood as a prismatic body where loads act only perpendicular to the principal axis of the body. This member can be generalized by superimposing a rod so that a member which can bend and elongate is obtained (generalized beam). As a matter of course, torsion around the principle axis could be also superimposed. The spatial arrangement of generalized beams is called a frame structure and has the following common characteristics and simplifications [59]:

- two- or three-dimensional framework of straight generalized beam members,
- the ends are connected by rigid (and/or hinged) connections (rigid connections allow rotations),
- the members are subjected to axial forces, shear forces, and bending moments.

Naturally we can combine rods and beams as separated members in a spatial arrangement and we simply call such a structure a frame structure. As an example,

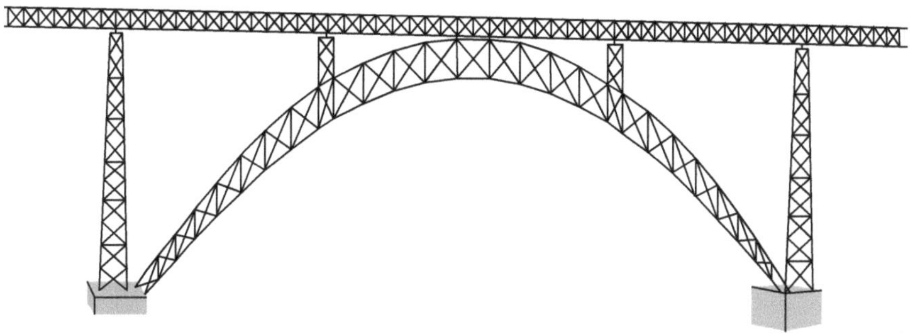

**Fig. 1.2** Simplified two-dimensional model of the railway arch bridge (Garabit Viaduct in France, see Fig. 1.1) based on rods and beams

**Table 1.1** Some steps in the early historical development of the finite element method

| Year | Author | Comment | Ref. |
|---|---|---|---|
| ... | ... | ... | ... |
| 1909 | W. Ritz | First discrete variational method. Formulation of the approximate solution within the *whole* domain | [89] |
| 1915 | B.G. Galerkin | Orthogonality of the residuum and the weight function. Formulation of the approximate solution and the weight function within the *whole* domain | [42] |
| 1943 | R. Courant | Subdivision of the whole domain into triangular elements for solving the torsion problem in a variational approach. Consideration of a very small number of elements | [27] |
| 1956 | M.J. Turner, R.W. Clough, H.C. Martin, L.J. Topp | First engineering formulation of the FEM based on the principle of virtual work. Derivation of a triangular plane stress element with six degrees of freedom. Deduction of a quatrilateral and a rectangular based triangular element | [108] |
| ... | ... | ... | ... |
| 1960 | R.W. Clough | This conference manuscript introduces for the first time the expression 'finite element' | [24] |
| ... | ... | ... | ... |

these basic elements can be used to create simple models of complex structures as shown in Fig. 1.2 where a two-dimensional model of the Garabit Viaduct (see Fig. 1.1) based on straight frame elements is schematically indicated.

The focus concerning approximation methods within this book is on the finite element method. Chapter 5 introduces the finite element method for linear-elastic material behavior while Chap. 6 extends the procedures to the treatment of elasto-plastic material response. A few steps in the very early historical development of this widespread numerical tool are collected in Table 1.1, see [40, 91, 97, 119, 121].

# Chapter 2
# Continuum Mechanics of Plasticity

**Abstract** This chapter introduces first the basic equations of plasticity theory. The yield condition, the flow rule, and the hardening rule are introduced. After deriving and presenting these equations for the one-dimensional stress and strain state, the equations are generalized for a two-component $\sigma$-$\tau$ stress state. Based on these basic equations, the concept of effective stress and strain as well as the elasto-plastic modulus is introduced. The chapter finishes with an introduction to two different damage concepts, i.e. the LEMAITRE and GURSON damage model, which are derived for the one-dimensional case.

## 2.1 General Comments and Observations

Let us consider in the following a uniaxial tensile test whose idealized specimen is schematically shown in Fig. 2.1. The original dimensions of, for example, the cylindrical specimen are characterized by the cross-sectional area $A$ and length $L$. This specimen is now elongated in a universal testing machine and its length increases to $L+\Delta L$. In the case of a real specimen made of a common engineering material, the cross-sectional area would reduce to $A-\Delta A$. This phenomenon could be described based on POISSSON's ratio.[1] However, if we assume an idealized uniaxial state, i.e. a uniaxial stress *and* strain state, the contraction is disregarded and the cross-sectional area is assumed to remain constant, see Fig. 2.1b. During the tensile test, the force is normally recorded by a load cell attached to the movable or fixed cross-head of the machine. If this force is divided by the (initial) cross-sectional area, the engineering stress is obtained as:

$$\sigma = \frac{F}{A}. \tag{2.1}$$

The deformation or elongation of the specimen can be measured, for example, by an external extensometer which should be directly attached to the specimen.

[1] Siméon Denis POISSON (1781–1840), French mathematician, geometer, and physicist.



A. Öchsner, *Elasto-Plasticity of Frame Structure Elements*,
DOI 10.1007/978-3-662-44225-8_2

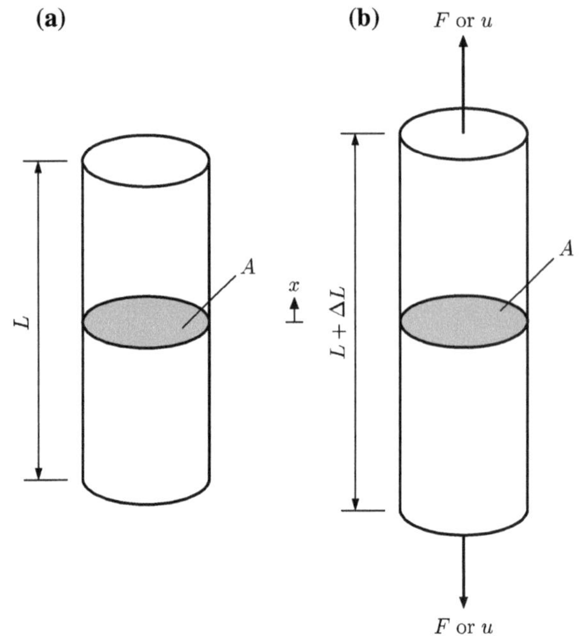


**Fig. 2.1** Schematic representation of **a** an unloaded and **b** an idealized uniaxial tensile specimen loaded by a force $F$ or a displacement $u$

These devices are either realized as strain gauge or inductive extensometers.[2] Any measurement based on the movement of the cross-head must be avoided since it does not guarantee an accurate determination of the specimen's behavior. The definition of strain is given in its simplest form as elongation over initial length and the engineering strain can be calculated as:

$$\varepsilon = \frac{\Delta L}{L}. \tag{2.2}$$

Relating the stress to its corresponding strain, the engineering stress-strain diagram can be plotted as schematically shown in Fig. 2.2. In the pure elastic range, a linear line is observed and its slope is equal to YOUNG's[3] modulus:

$$E = \frac{\Delta\sigma}{\Delta\varepsilon} \quad \text{or} \quad E = \frac{\mathrm{d}\sigma}{\mathrm{d}\varepsilon}. \tag{2.3}$$

[2] More modern devises are contactless laser or video based extensometers.

[3] Thomas YOUNG (1773–1829), English polymath.

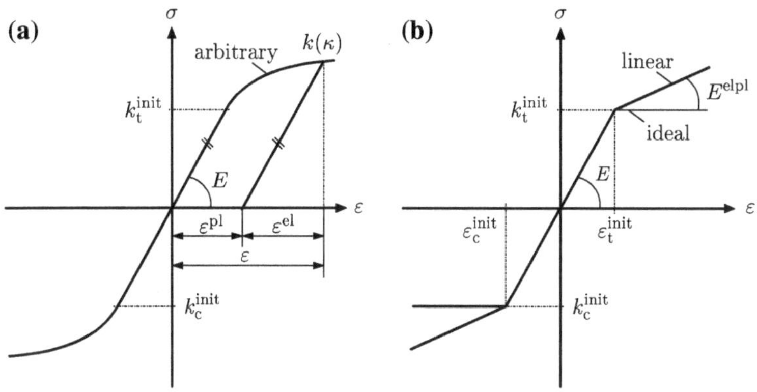


**Fig. 2.2** Uniaxial stress-strain diagrams for different isotropic hardening laws: **a** arbitrary hardening; **b** linear hardening and ideal plasticity

The last relation is also known as HOOKE's[4] law and often written in the following form for linear elastic behavior:

$$\sigma = E\varepsilon \,. \tag{2.4}$$

As soon as the initial yield stress $k_t^{\text{init}}$ is reached, plastic material behavior occurs and the slope of the stress-strain diagram changes.

The characteristic feature of plastic material behavior is that a remaining strain $\varepsilon^{\text{pl}}$ occurs after complete unloading, see Fig. 2.2a. Only the elastic strains $\varepsilon^{\text{el}}$ returns to zero at complete unloading. An additive composition of the strains by their elastic and plastic parts

$$\varepsilon = \varepsilon^{\text{el}} + \varepsilon^{\text{pl}} \tag{2.5}$$

is permitted at restrictions to small strains. The elastic strains $\varepsilon^{\text{el}}$ can hereby be determined via HOOKE's law, whereby $\varepsilon$ in Eq. (2.3) has to be substituted by $\varepsilon^{\text{el}}$.

Furthermore, no explicit correlation is given anymore for plastic material behavior in general between stress and strain, since the strain state is also dependent on the loading history. Due to this, rate equations are necessary and need to be integrated throughout the entire load history. Within the framework of the time-independent plasticity investigated here, the rate equations can be simplified to incremental relations. From Eq. (2.5) the additive composition of the strain increments results in:

$$\text{d}\varepsilon = \text{d}\varepsilon^{\text{el}} + \text{d}\varepsilon^{\text{pl}} \,. \tag{2.6}$$

[4] Robert HOOKE (1635–1703), English natural philosopher, architect and polymath.

The constitutive description of plastic material behavior includes

- a yield condition,
- a flow rule and
- a hardening law.

In the following, the case of the monotonic loading[5] is considered first, so that isotropic hardening is explained first in the case of material hardening. This important case, for example, occurs in the experimental mechanics at the uniaxial tensile test with monotonic loading. Furthermore, it is assumed that the yield stress is identical in the tensile and compressive regime: $k_t = k_c = k$.

## 2.2 Yield Condition

The yield condition enables one to determine whether the relevant material suffers only elastic or also plastic strains at a certain stress state at a point of the relevant body. In the uniaxial tensile test, plastic flow begins when reaching the initial yield stress $k^{\text{init}}$, see Fig. 2.2. The yield condition in its general one-dimensional form can be set as follows ($\mathbb{R} \times \mathbb{R} \rightarrow \mathbb{R}$):

$$F = F(\sigma, \kappa), \tag{2.7}$$

where $\kappa$ represents the inner variable of isotropic hardening. In the case of ideal plasticity, see Fig. 2.2b, the following is valid: $F = F(\sigma)$. The values of $F$ have the following mechanical meaning, see Fig. 2.3a:

$$F(\sigma, \kappa) < 0 \quad \rightarrow \quad \text{elastic material behavior,} \tag{2.8}$$

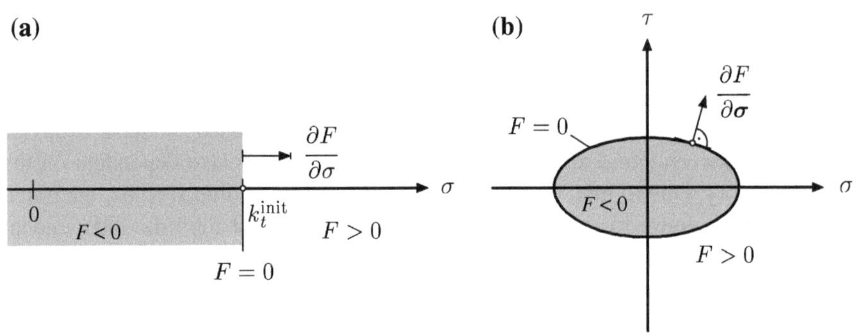


**Fig. 2.3** Schematic representation of the values of the yield condition and the direction of the stress gradient in the **a** uniaxial and **b** two-component $\sigma$-$\tau$ stress space (in the $\sigma$-$\tau$ space, the shown normal vector is the projection of this vector onto this plan)

[5] The case of unloading or alternatively load reversal will be treated separately in Sect. 2.7.

$$F(\sigma, \kappa) = 0 \quad \rightarrow \quad \text{plastic material behavior,} \tag{2.9}$$

$$F(\sigma, \kappa) > 0 \quad \rightarrow \quad \text{invalid.} \tag{2.10}$$

A further simplification results under the assumption that the yield condition can be split into a pure stress fraction $f(\sigma)$, the so-called yield criterion,[6] and into an experimental material parameter $k(\kappa)$, the so-called flow stress:

$$F(\sigma, \kappa) = f(\sigma) - k(\kappa)\,. \tag{2.11}$$

For a uniaxial tensile test (see Fig. 2.2) the yield condition can be noted in the following form:

$$F(\sigma, \kappa) = |\sigma| - k(\kappa) \leq 0\,. \tag{2.12}$$

If one considers the idealized case of the linear hardening (see Fig. 2.2b), Eq. (2.12) can be written as (Fig. 2.4)

$$F(\sigma, \kappa) = |\sigma| - (k^{\text{init}} + E^{\text{pl}}\kappa) \leq 0\,. \tag{2.13}$$

The parameter $E^{\text{pl}}$ is the plastic modulus (see Fig. 2.5), which becomes zero in the case of ideal plasticity:

$$F(\sigma, \kappa) = |\sigma| - k^{\text{init}} \leq 0\,. \tag{2.14}$$

In the following, we are going to extend this concept to a stress state where in addition to a normal stress ($\sigma$) a shear stress component ($\tau$) is acting. This two-component stress state occurs, for example, in the case of general beam formulations, see Sects. 4.3 and 4.4. Two classical representatives of yield conditions for solid dense

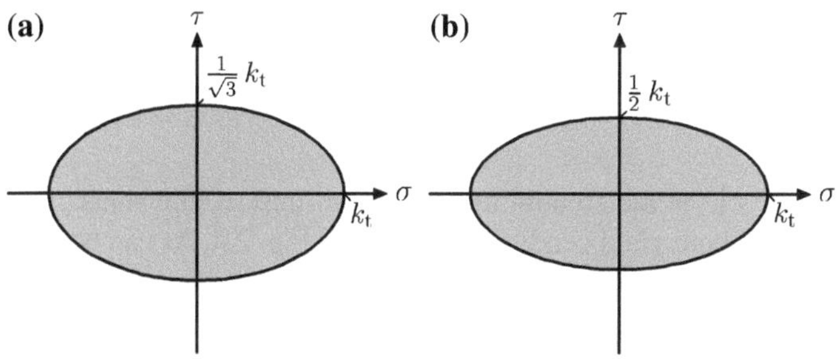


**Fig. 2.4** Graphical representation of the yield condition according to **a** VON MISES and **b** TRESCA in the two-component $\sigma$-$\tau$ space

[6] If the unit of the yield criterion equals the stress, $f(\sigma)$ represents the equivalent stress or effective stress. In the general three-dimensional case the following is valid under consideration of the symmetry of the stress tensor $\sigma_{\text{eff}} : (\mathbb{R}^6 \rightarrow \mathbb{R}_+)$.

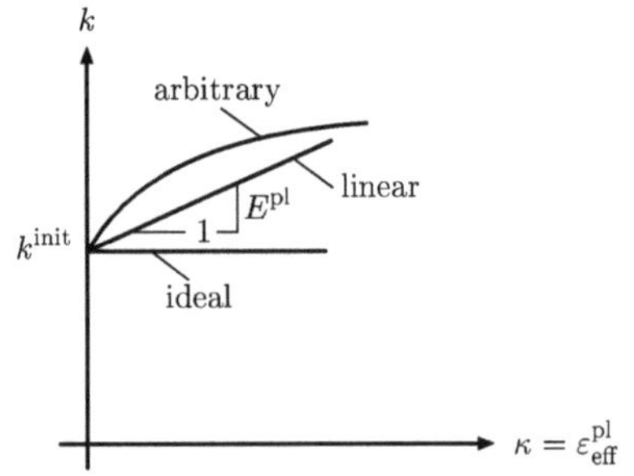


**Fig. 2.5** Flow curve for different isotropic hardening laws

materials (classical metals) will be introduced for this special two-component stress state[7] in the following [53]. The TRESCA[8] yield condition [107] is based on the assumption that yielding begins when the maximum shear stress reaches a certain value and can be written, for example, in its mathematical form based on the two acting stress components as

$$F_{\sigma-\tau} = \sqrt{\sigma^2 + 4\tau^2} - k_{\text{t}} = 0 \quad (\text{TRESCA}), \tag{2.15}$$

where the relationship between shear ($k_{\text{s}}$) and tensile yield stress ($k_{\text{t}}$) is given by $k_{\text{s}} = \frac{k_{\text{t}}}{2}$, see [120].

The VON MISES[9] yield condition [69] is based on the assumption that yielding begins when the elastic energy of distortion reaches a critical value [50]. The mathematical formulation of this condition reads in the $\sigma$-$\tau$ space as

$$F_{\sigma-\tau} = \sqrt{\sigma^2 + 3\tau^2} - k_{\text{t}} = 0 \quad (\text{VON MISES}), \tag{2.16}$$

where the relationship between shear and tensile yield stress is given by $k_{\text{s}} = \frac{k_{\text{t}}}{\sqrt{3}}$, see [20]. Both yield conditions can be written in a more general form as

$$F = F(\boldsymbol{\sigma}) = 0, \tag{2.17}$$

where $\boldsymbol{\sigma} = \begin{bmatrix}\sigma & \tau\end{bmatrix}^{\text{T}}$ is the column matrix of the stress components. The graphical representation of both yield conditions is given in Fig. 2.4. As can be seen, centered ellipses are obtained where the major and minor axes of the ellipses are parallel to the stress axes of the coordinate system. Comparing both shapes, it can be concluded that the TRESCA yield condition gives a more conservative prediction of the yield stress since the minor axis is smaller compared to the VON MISES condition.

---

[7] A two-dimensional stress state would comprise in its general form, for example, the three components $\sigma_x$, $\sigma_y$ and $\tau_{xy}$.

[8] Henri Édouard TRESCA (1814–1885), French mechanical engineer.

[9] Richard Edler VON MISES (1883–1953), Austrian scientist and mathematician.

**Table 2.1** Different formulations of the VON MISES and TRESCA yield condition in the two-component $\sigma$-$\tau$ space

| VON MISES | TRESCA |
|---|---|
| *Based on tensile yield stress* | |
| $\sqrt{\sigma^2 + 3\tau^2} - k_t = 0$ | $\sqrt{\sigma^2 + 4\tau^2} - k_t = 0$ |
| $\sigma^2 + 3\tau^2 - k_t^2 = 0$ | $\sigma^2 + 4\tau^2 - k_t^2 = 0$ |
| *Based on shear yield stress* | |
| $\frac{\sqrt{3}}{3}\sqrt{\sigma^2 + 3\tau^2} - k_s = 0$ | $\frac{1}{2}\sqrt{\sigma^2 + 4\tau^2} - k_s = 0$ |
| $\frac{1}{3}\left(\sigma^2 + 3\tau^2\right) - k_s^2 = 0$ | $\frac{1}{4}\left(\sigma^2 + 4\tau^2\right) - k_s^2 = 0$ |

It should be noted here that both conditions given in Eqs. (2.15) and (2.16) reduce for a uniaxial stress state where only a normal stress $\sigma$ is acting to the formulation presented in Eq. (2.12). Finally, Table 2.1 summarizes different formulations of the considered yield conditions in the two-component $\sigma$-$\tau$ space.

## 2.3 Flow Rule

The flow rule serves as a mathematical description of the evolution of the infinitesimal increments of the plastic strain $d\varepsilon^{pl}$ in the course of the load history of the body. In its most general one-dimensional form, the flow rule can be set up as follows [96]:

$$d\varepsilon^{pl} = d\lambda\, r(\sigma, \kappa)\,, \tag{2.18}$$

whereupon the factor $d\lambda$ is described as the consistency parameter ($d\lambda \geq 0$) and $r : (\mathbb{R} \times \mathbb{R} \rightarrow \mathbb{R})$ as the function of the flow direction.[10] One considers that solely for $d\varepsilon^{pl} = 0$ then $d\lambda = 0$ results. Based on the stability postulate of DRUCKER[11] [33] the following flow rule can be derived[12]:

$$d\varepsilon^{pl} = d\lambda \frac{\partial F(\sigma, \kappa)}{\partial \sigma}\,. \tag{2.19}$$

Such a flow rule is referred to as the normal rule[13] (see Fig. 2.3a) or due to $r = \partial F(\sigma, \kappa)/\partial \sigma$ as the *associated* flow rule.

---

[10] In the general three-dimensional case $\boldsymbol{r}$ hereby defines the direction of the vector $d\varepsilon^{pl}$, while the scalar factor defines the absolute value.

[11] Daniel Charles DRUCKER (1918–2001), US engineer.

[12] A formal alternative derivation of the associated flow rule can occur via the LAGRANGE multiplier method as extreme value with side-conditions from the principle of maximum plastic work [12].

[13] In the general three-dimensional case the image vector of the plastic strain increment has to be positioned upright and outside oriented to the yield surface, see Fig. 2.3b.

Experimental results, among other things from the area of granular materials [13] can however be approximated better if the stress gradient is substituted through a different function, the so-called plastic potential $Q$. The resulting flow rule is then referred to as the *non-associated* flow rule:

$$\mathrm{d}\varepsilon^{\mathrm{pl}} = \mathrm{d}\lambda \frac{\partial Q(\sigma, \kappa)}{\partial \sigma} . \tag{2.20}$$

In quite complicated yield conditions often the case occurs that a more simple yield condition is used for $Q$ in the first approximation, for which the gradient can easily be determined.

The application of the associated flow rule (2.19) to the yield conditions according to Eqs. (2.12)–(2.14) yields for all three types of yield conditions (meaning arbitrary hardening, linear hardening and ideal plasticity):

$$\mathrm{d}\varepsilon^{\mathrm{pl}} = \mathrm{d}\lambda \, \mathrm{sgn}(\sigma) , \tag{2.21}$$

where $\mathrm{sgn}(\sigma)$ represents the so-called sign function,[14] which can adopt the following values:

$$\mathrm{sgn}(\sigma) = \begin{cases} -1 \text{ for } \sigma < 0 \\ \;\;0 \;\text{ for } \sigma = 0 \\ +1 \text{ for } \sigma > 0 \end{cases} . \tag{2.22}$$

For the two-component $\sigma$-$\tau$ stress space, the associated flow rule (2.19) can be written as

$$\mathrm{d}\boldsymbol{\varepsilon}^{\mathrm{pl}} = \mathrm{d}\lambda \frac{\partial F(\boldsymbol{\sigma})}{\partial \boldsymbol{\sigma}} , \tag{2.23}$$

where $\mathrm{d}\boldsymbol{\varepsilon}^{\mathrm{pl}} = \left[\mathrm{d}\varepsilon^{\mathrm{pl}} \; \mathrm{d}\gamma^{\mathrm{pl}}\right]^{\mathrm{T}}$ is the column matrix of the plastic strain increments. Application of this definition to the yield conditions given in Eqs. (2.15) and (2.16) gives finally:

$$\mathrm{d}\boldsymbol{\varepsilon}^{\mathrm{pl}} = \begin{bmatrix} \mathrm{d}\varepsilon^{\mathrm{pl}} \\ \mathrm{d}\gamma^{\mathrm{pl}} \end{bmatrix} = \frac{\mathrm{d}\lambda}{\sqrt{\sigma^2 + 4\tau^2}} \begin{bmatrix} \sigma \\ 4\tau \end{bmatrix} \quad (\text{TRESCA}) , \tag{2.24}$$

$$\mathrm{d}\boldsymbol{\varepsilon}^{\mathrm{pl}} = \begin{bmatrix} \mathrm{d}\varepsilon^{\mathrm{pl}} \\ \mathrm{d}\gamma^{\mathrm{pl}} \end{bmatrix} = \frac{\mathrm{d}\lambda}{\sqrt{\sigma^2 + 3\tau^2}} \begin{bmatrix} \sigma \\ 3\tau \end{bmatrix} \quad (\text{VON MISES}) . \tag{2.25}$$

These two equations can be generally expressed in the manner of Eq. (2.18) as:

$$\mathrm{d}\boldsymbol{\varepsilon}^{\mathrm{pl}} = \mathrm{d}\lambda \, \boldsymbol{r}(\boldsymbol{\sigma}, \kappa) . \tag{2.26}$$

[14] Also signum function; from the Latin 'signum' for 'sign'.

## 2.4 Hardening Rule

The hardening law allows the consideration of the influence of material hardening on the yield condition and the flow rule.

### 2.4.1 *Isotropic Hardening*

In the case of isotropic hardening, the yield stress is expressed as being dependent on an inner variable $\kappa$:

$$k = k(\kappa). \tag{2.27}$$

If the equivalent plastic strain[15] is used for the hardening variable ($\kappa = |\varepsilon^{\text{pl}}|$), then one talks about strain hardening.

Another possibility is to describe the hardening being dependent on the specific[16] plastic work ($\kappa = w^{\text{pl}} = \int \sigma \text{d}\varepsilon^{\text{pl}}$). Then one talks about work hardening. If Eq. (2.27) is combined with the flow rule according to (2.21), the evolution equation for the isotropic hardening variable results in:

$$\text{d}\kappa = \text{d}|\varepsilon^{\text{pl}}| = \text{d}\lambda\,. \tag{2.28}$$

Figure 2.5 shows the flow curve, meaning the graphical illustration of the yield stress being dependent on the inner variable for different hardening approaches.

The yield condition which was expressed in Eq. (2.17) for the case of ideal plasticity can now be expanded to the formulation

$$F = F(\boldsymbol{\sigma}, q) = 0\,, \tag{2.29}$$

where the internal variable $q$ considers the influence of the material hardening on the yield condition. The evolution equation for this internal variable can be stated in its most general form based on Eq. (2.29) as

$$\text{d}q = \text{d}\lambda \times h(\boldsymbol{\sigma}, q)\,, \tag{2.30}$$

---

[15] The effective plastic strain is in the general three-dimensional case the function $\varepsilon^{\text{pl}}_{\text{eff}} : (\mathbb{R}^6 \rightarrow \mathbb{R}_+)$. In the one-dimensional case, the following is valid: $\varepsilon^{\text{pl}}_{\text{eff}} = \sqrt{\varepsilon^{\text{pl}}\varepsilon^{\text{pl}}} = |\varepsilon^{\text{pl}}|$. Attention: Finite element programs optionally use the more general definition for the illustration in the post processor, this means $\varepsilon^{\text{pl}}_{\text{eff}} = \sqrt{\frac{2}{3}\sum \Delta\varepsilon^{\text{pl}}_{ij} \sum \Delta\varepsilon^{\text{pl}}_{ij}}$, which considers the lateral contraction at uniaxial stress problems in the plastic area via the factor $\frac{2}{3}$. However in pure one-dimensional problems *without* lateral contraction, this formula leads to an illustration of the effective plastic strain, which is reduced by the factor $\sqrt{\frac{2}{3}} \approx 0.816$.

[16] This is the volume-specific definition, meaning $\left[w^{\text{pl}}\right] = \frac{\text{N}}{\text{m}^2}\frac{\text{m}}{\text{m}} = \frac{\text{kg}\,\text{m}}{\text{s}^2\text{m}^2}\frac{\text{m}}{\text{m}} = \frac{\text{kg}\,\text{m}^2}{\text{s}^2\text{m}^3} = \frac{\text{J}}{\text{m}^3}$.

where the function $h$ defines the evolution of the hardening parameter. Assigning for the internal variable $q = \kappa$ (in the case that $\kappa$ equals the effective plastic strains, one talks about a strain space formulation) and considering the case of associated plasticity, a more specific rule for the evolution of the internal variable is given as

$$\begin{aligned} \mathrm{d}\kappa &= -\mathrm{d}\lambda \times \left(D^{\mathrm{pl}}\right)^{-1} \frac{\partial F(\boldsymbol{\sigma}, \kappa)}{\partial \kappa} = -\mathrm{d}\lambda \times \frac{\partial \kappa}{\partial k(\kappa)} \frac{\partial F(\boldsymbol{\sigma}, \kappa)}{\partial \kappa} \\ &= -\mathrm{d}\lambda \times \frac{1}{E^{\mathrm{pl}}} \frac{\partial F(\boldsymbol{\sigma}, \kappa)}{\partial \kappa}, \end{aligned} \tag{2.31}$$

where $D^{\mathrm{pl}}$ is the generalized plastic modulus. Considering the yield stress $k$ as the internal variable, one obtains a stress space formulation as $F = F(\boldsymbol{\sigma}, k)$ and the corresponding evolution equation for the internal variable is given by:

$$\mathrm{d}k = -\mathrm{d}\lambda \times D^{\mathrm{pl}} \frac{\partial F(\boldsymbol{\sigma}, k)}{\partial k} = -\mathrm{d}\lambda \times E^{\mathrm{pl}} \frac{\partial F(\boldsymbol{\sigma}, k)}{\partial k}, \tag{2.32}$$

where $\mathrm{d}k$ can be written as $E^{\mathrm{pl}}\mathrm{d}\kappa$. Thus, one may alternatively formulate:

$$\mathrm{d}\kappa = -\mathrm{d}\lambda \times \frac{\partial F(\boldsymbol{\sigma}, k)}{\partial k}. \tag{2.33}$$

Application of the instruction for the evolution of the internal variable according to Eq. (2.31) or (2.33) to the TRESCA or VON MISES yield condition with $k_{\mathrm{t}} = k_{\mathrm{t}}(\kappa)$ (cf. Eqs. (2.15) and (2.16)) gives:

$$\mathrm{d}\kappa = \mathrm{d}\lambda\,. \tag{2.34}$$

Thus, it turned out that $h$, i.e. the evolution equation for the hardening parameter in Eq. (2.30), simplified to $h = 1$. However, in the case of more complex yield conditions, the function $h$ may take a more complex form. This will be shown in Sect. 2.8.2 where the GURSON's damage model is introduced. The graphical representation of the initial and subsequent yield surface for isotropic hardening in the two-component $\sigma$-$\tau$ space is shown in Fig. 2.6.

### *2.4.2 Kinematic Hardening*

In the case of a pure monotonic loading, i.e. pure tensile or pure compression, it is not possible to distinguish from the stress-strain diagram the cases of isotropic or kinematic hardening. Let us look in the following on a uniaxial test with plastic deformation and stress reversal as schematically shown in Fig. 2.7. The test starts without any initial stress or strain in the origin of the stress-strain diagram (point '0') and a tensile load is continuously increased. The first part of the path, i.e. as long as

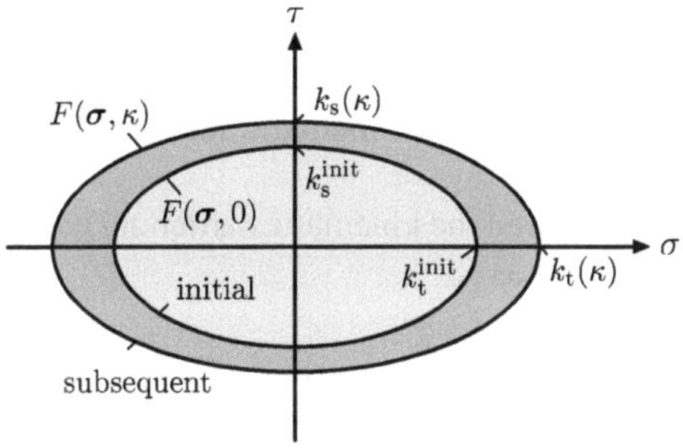


**Fig. 2.6** Initial and subsequent yield surface for isotropic hardening in the two-component $\sigma$-$\tau$ space

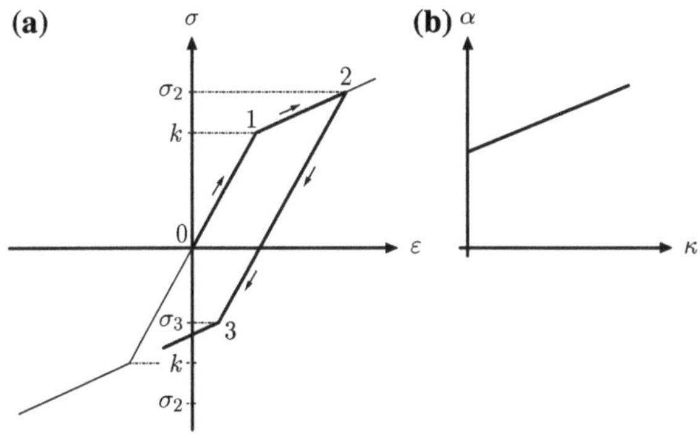


**Fig. 2.7** Uniaxial kinematic hardening: **a** idealized stress-strain curve with BAUSCHINGER effect and **b** kinematic hardening parameter as a function of the internal variable (linear hardening)

the stress is below the yield stress $k$, is in the pure elastic range and HOOKE's law describes the stress-strain behavior. Reaching the yield stress $k$ (point '1'), the slope of the diagram changes and plastic deformation occurs. With ongoing increasing load, the plastic deformation and the plastic strain increases in this part of the diagram. Let us assume now that the load is reversed at point '2'. The unloading is completely elastic and compressive stress develops as soon as the load path passes the strain axis. The interesting question is now when the subsequent plastic deformations starts in the compressive regime. This plastic deformation occurs now in the case of kinematic hardening at a stress level $\sigma_3$ which is lower than the initial yield stress $k$ or the subsequent stress $\sigma_2$. This behavior is known as the BAUSCHINGER[17] effect [8] and requires plastic pre-straining with subsequent load reversal.

The behavior shown in Fig. 2.7a can be described based on the following yield condition

$$F = |\sigma - \alpha(\kappa)| - k = 0\,, \tag{2.35}$$

---

[17] Johann BAUSCHINGER (1834–1893), German mathematician and engineer.

where the initial yield stress $k$ is constant and the kinematic hardening parameter[18] $\alpha$ is a function of an internal variable $\kappa$. Figure 2.7b shows the case of linear hardening where a linear relationship between kinematic hardening parameter and internal variable is obtained.

The simplest relation between the kinematic hardening parameter and the internal variable was proposed in [68] as

$$\alpha = H\varepsilon^{\text{pl}} \quad \text{or} \quad \text{d}\alpha = H\text{d}\varepsilon^{\text{pl}}\,, \tag{2.36}$$

where $H$ is a constant called the kinematic hardening modulus and the plastic strain is assigned as the internal variable. Thus, Eq. (2.36) describes the case of linear hardening. A more general formulation of Eq. (2.36) is known as PRAGER's[19] hardening rule [80, 81]:

$$\text{d}\alpha = H(\sigma, \kappa_i)\text{d}\varepsilon^{\text{pl}}\,, \tag{2.37}$$

where the kinematic hardening modulus is now a scalar function which depends on the state variables $(\sigma, \kappa_i)$. One suggestion is to use the effective plastic strain $\varepsilon^{\text{pl}}_{\text{eff}}$ as internal variable [6]. A further extension is proposed in [60] where the hardening modulus is formulated as a tensor.

Another formulation was proposed by ZIEGLER[20] [94, 118] as

$$\text{d}\alpha = \text{d}\mu(\sigma - \alpha)\,, \tag{2.38}$$

where the proportionality factor $\text{d}\mu$ can be expressed as:

$$\text{d}\mu = a\text{d}\varepsilon^{\text{pl}}\,, \tag{2.39}$$

or in a more general way as $a = a(\sigma, \kappa_i)$. The rule given in Eq. (2.38) is known in the literature as ZIEGLER's hardening rule. It should be noted here that the plastic strain increments in Eqs. (2.39) and (2.37) can be calculated based on the flow rules given in Sect. 2.3. Thus, the kinematic hardening rules can be expressed in a more general way as:

$$\text{d}\alpha = \text{d}\lambda h(\sigma, \alpha). \tag{2.40}$$

For the two-component $\sigma$-$\tau$/$\varepsilon$-$\gamma$ space, the kinematic hardening rules according to PRAGER and ZIEGLER can be written as

$$\text{d}\boldsymbol{\alpha} = H(\boldsymbol{\sigma}, \boldsymbol{\kappa})\text{d}\boldsymbol{\varepsilon}^{\text{pl}} \quad (\text{PRAGER})\,, \tag{2.41}$$

$$\text{d}\boldsymbol{\alpha} = \text{d}\mu(\boldsymbol{\sigma} - \boldsymbol{\alpha}) \quad (\text{ZIEGLER})\,, \tag{2.42}$$

---

[18] An alternative expression for the kinematic hardening parameter is back-stress.

[19] William PRAGER (1903–1980), German engineer and applied mathematician.

[20] Hans ZIEGLER (1910–1985), Swiss scientist.

or in components as

$$\begin{bmatrix} d\alpha_\sigma \\ d\alpha_\tau \end{bmatrix} = H(\boldsymbol{\sigma}, \boldsymbol{\kappa}) \begin{bmatrix} d\varepsilon^{pl} \\ d\gamma^{pl} \end{bmatrix} \quad (\text{PRAGER}), \tag{2.43}$$

$$\begin{bmatrix} d\alpha_\sigma \\ d\alpha_\tau \end{bmatrix} = d\mu \begin{bmatrix} \sigma - \alpha_\sigma \\ \tau - \alpha_\tau \end{bmatrix} \quad (\text{ZIEGLER}). \tag{2.44}$$

Thus, the generalization of the one-dimensional yield condition given in Eq. (2.35) can be written in the case of TRESCA and VON MISES as:

$$F_{\sigma-\tau} = \sqrt{(\sigma - \alpha_\sigma)^2 + 4(\tau - \alpha_\tau)^2} - k_t = 0 \quad (\text{TRESCA}), \tag{2.45}$$

$$F_{\sigma-\tau} = \sqrt{(\sigma - \alpha_\sigma)^2 + 3(\tau - \alpha_\tau)^2} - k_t = 0 \quad (\text{VON MISES}). \tag{2.46}$$

The graphical representation of Eqs. (2.45) and (2.46) is schematically shown in Fig. (2.8) where it can be seen that the center of the subsequent yield surface is described by the back-stress vector $\boldsymbol{\alpha}^T = \left[\alpha_\sigma \; \alpha_\tau\right]^T$.

### 2.4.3 Combined Hardening

The isotropic and kinematic hardening rules presented in Sects. 2.4.1 and 2.4.2 can simply be joined together to obtain a combined hardening rule for the one-dimensional yield condition as:

$$F(\sigma, \boldsymbol{q}) = |\sigma - \alpha| - k(\kappa), \tag{2.47}$$

or for the special case of isotropic linear hardening as

$$F = |\sigma - \alpha| - (k^{init} + E^{pl}\varepsilon^{pl}_{eff}), \tag{2.48}$$

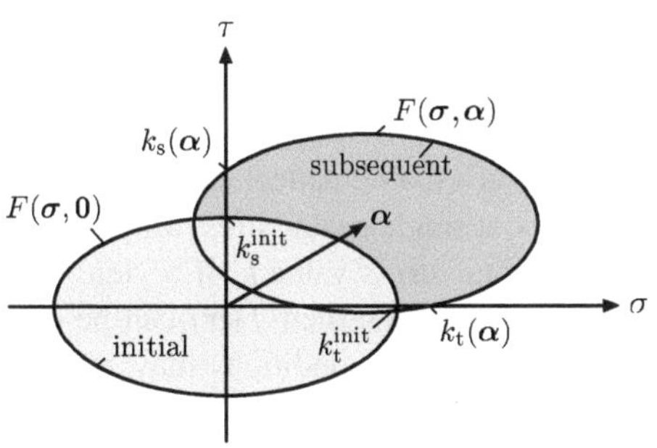


**Fig. 2.8** Initial and subsequent yield surface for kinematic hardening in the two-component $\sigma$-$\tau$ space

where the back-stress $\alpha$ can be a function as indicated in Eqs. (2.36)–(2.38). The associated flow rule is then obtained according to Eq. (2.19) as:

$$\mathrm{d}\varepsilon^{\mathrm{pl}} = \mathrm{d}\lambda \frac{\partial F}{\partial \sigma} = \mathrm{d}\lambda \, \mathrm{sgn}(\sigma - \alpha) \tag{2.49}$$

and the isotropic and kinematic hardening (PRAGER) laws can be written as:

$$\mathrm{d}\kappa = \mathrm{d}|\varepsilon^{\mathrm{pl}}| = |\mathrm{d}\lambda \, \mathrm{sgn}(\sigma - \alpha)| = \mathrm{d}\lambda \,, \tag{2.50}$$

$$\mathrm{d}\alpha = H \mathrm{d}\varepsilon^{\mathrm{pl}} = H \, \mathrm{d}\lambda \, \mathrm{sgn}(\sigma - \alpha) \,. \tag{2.51}$$

The last two equations can be combined and generally expressed as:

$$\mathrm{d}\boldsymbol{q} = \mathrm{d}\lambda \boldsymbol{h}(\sigma, \boldsymbol{q}) \,, \tag{2.52}$$

or

$$\begin{bmatrix} \mathrm{d}\kappa \\ \mathrm{d}\alpha \end{bmatrix} = \mathrm{d}\lambda \begin{bmatrix} 1 \\ H \, \mathrm{sgn}(\sigma - \alpha) \end{bmatrix} . \tag{2.53}$$

For the two-component $\sigma$-$\tau$/$\varepsilon$-$\gamma$ space, Eqs. (2.45), (2.46) and (2.27) can be combined to obtain the yield conditions for combined hardening. In this case, the evolution equations for the isotropic and kinematic hardening parameters can be generally expressed as:

$$\mathrm{d}\boldsymbol{q} = \mathrm{d}\lambda \boldsymbol{h}(\boldsymbol{\sigma}, \boldsymbol{q}) \,. \tag{2.54}$$

## 2.5 Effective Stress and Effective Plastic Strain

The previous sections introduced a two-component stress state $\boldsymbol{\sigma} = \begin{bmatrix} \sigma \; \tau \end{bmatrix}^{\mathrm{T}}$. If a material is subjected to such a multi-axial stress state, it is difficult to directly compare these different stress components with the experimental uniaxial stress-strain as schematically shown in Fig. 2.2. How to decide if the state is still in the elastic or already in the plastic range? To overcome this problem, one can define the so-called effective stress.[21] This is a mathematical equation which calculates, based on the acting stress components, a single scalar stress value $\sigma_{\mathrm{eff}}$ which can be directly compared with the experimental stress value $k$. If a yield condition is defined, for example, as $F(\boldsymbol{\sigma}, \kappa) = f(\boldsymbol{\sigma}) - k(\kappa)$, then $f(\boldsymbol{\sigma})$ can be considered as the effective stress definition. However, yield conditions may have different formulations

[21] Sometimes called the equivalent stress.

(see Table 2.1) and we may indicate the yield condition in a more general form as

$$F(\boldsymbol{\sigma}, \kappa) = f(\boldsymbol{\sigma}) - g(k) = 0\,. \tag{2.55}$$

Thus, the yield condition is split into a fraction $f(\boldsymbol{\sigma})$ which depends only on the acting stress state and a fraction which involves the experimental stress-strain response of the material. Then, it follows that the function $f(\boldsymbol{\sigma})$ must be some constant $c$ times the effective stress to some power $d$ [22]:

$$f(\boldsymbol{\sigma}) = c \times \sigma_{\text{eff}}^{d}\,. \tag{2.56}$$

Let us assume, for example, a yield condition of the form $\sigma^2 + 3\tau^2 - k_{\text{t}}^2 = 0$, see Table 2.1. Since the effective stress should reduce to the normal stress component $\sigma$ in the uniaxial tensile test ($\sigma \neq 0 \wedge \tau = 0$), we can write that

$$\sigma^2 = c \times \sigma_{\text{eff}}^{d}\,, \tag{2.57}$$

and comparing coefficients gives $c = 1$ and $d = 2$. Thus, the effective stress in this specific case is given as:

$$\sigma_{\text{eff}} = 1 \times \left(\sigma^2 + 3\tau^2\right)^{\frac{1}{2}}\,. \tag{2.58}$$

For the definition of the effective plastic strain $\varepsilon_{\text{eff}}^{\text{pl}}$, different approaches can be found in the literature [22]. We will use in the following a definition which is based on the volume-specific plastic work which can be expressed either by the stress and plastic strain components or the corresponding effective values as:

$$\mathrm{d}w^{\text{pl}} = \boldsymbol{\sigma}^{\text{T}}\mathrm{d}\boldsymbol{\varepsilon}^{\text{pl}} \stackrel{!}{=} \sigma_{\text{eff}}\,\mathrm{d}\varepsilon_{\text{eff}}^{\text{pl}}\,, \tag{2.59}$$

or rearranged for the effective plastic strain increment as:

$$\mathrm{d}\varepsilon_{\text{eff}}^{\text{pl}} = \frac{\boldsymbol{\sigma}^{\text{T}}\mathrm{d}\boldsymbol{\varepsilon}^{\text{pl}}}{\sigma_{\text{eff}}}\,. \tag{2.60}$$

Under the assumption of an associated flow rule, the last equation can be expressed as:

$$\mathrm{d}\varepsilon_{\text{eff}}^{\text{pl}} = \mathrm{d}\lambda \underbrace{\frac{\boldsymbol{\sigma}^{\text{T}}\frac{\partial F}{\partial \boldsymbol{\sigma}}}{\sigma_{\text{eff}}}}_{h}\,. \tag{2.61}$$

In the following let us look at the VON MISES yield condition as given in Eq. (2.16), i.e.

$$F_{\sigma-\tau} = \underbrace{\sqrt{\sigma^2 + 3\tau^2}}_{\sigma_{\text{eff}}} - k_{\text{t}} = 0\,, \tag{2.62}$$

where the effective stress is easily identified. The derivative of the yield condition with respect to the stress components can be obtained from Eq. (2.25) and the effective plastic strain increment can be written as:

$$\mathrm{d}\varepsilon_{\mathrm{eff}}^{\mathrm{pl}} = \mathrm{d}\lambda \frac{\begin{bmatrix}\sigma & \tau\end{bmatrix}^{\mathrm{T}} \frac{1}{\sqrt{\sigma^2+3\tau^2}} \begin{bmatrix}\sigma \\ 3\tau\end{bmatrix}}{\sqrt{\sigma^2+3\tau^2}} = \mathrm{d}\lambda \times 1 . \tag{2.63}$$

## 2.6 Elasto-Plastic Modulus

The stiffness of a material changes during plastic deformation and the strain state is dependent on the loading history. Therefore, HOOKE's law which is valid for the linear-elastic material behavior according to Eq. (2.3) must be replaced by the following infinitesimal incremental relation:

$$\mathrm{d}\sigma = E^{\mathrm{elpl}} \mathrm{d}\varepsilon , \tag{2.64}$$

where $E^{\mathrm{elpl}}$ is the elasto-plastic modulus. The algebraic expression for this modulus can be obtained in the following manner. The total differential of a yield condition $F = F(\sigma, \boldsymbol{q})$, see Eq. (2.47), is given by:

$$\mathrm{d}F(\sigma, \boldsymbol{q}) = \left(\frac{\partial F}{\partial \sigma}\right) \mathrm{d}\sigma + \left(\frac{\partial F}{\partial \boldsymbol{q}}\right)^{\mathrm{T}} \mathrm{d}\boldsymbol{q} = 0 . \tag{2.65}$$

If HOOKE's law (2.3) and the flow rule (2.18) are introduced in the relation for the additive composition of the elastic and plastic strain according to Eq. (2.6), one obtains:

$$\mathrm{d}\varepsilon = \frac{1}{E} \mathrm{d}\sigma + \mathrm{d}\lambda r . \tag{2.66}$$

Multiplication of Eq. (2.66) from the left-hand side with $\left(\frac{\partial F}{\partial \sigma}\right) E$ and inserting in Eq. (2.65) gives, under the consideration of the evolution equation of the hardening variables (2.52), the consistence parameter as:

$$\mathrm{d}\lambda = \frac{\left(\frac{\partial F}{\partial \sigma}\right) E}{\left(\frac{\partial F}{\partial \sigma}\right) E r - \left(\frac{\partial F}{\partial \boldsymbol{q}}\right)^{\mathrm{T}} \boldsymbol{h}} \mathrm{d}\varepsilon . \tag{2.67}$$

This equation for the consistency parameter can be inserted into Eq. (2.66) and solving for $\frac{\mathrm{d}\sigma}{\mathrm{d}\varepsilon}$ gives the elasto-plastic modulus as:

$$E^{\mathrm{elpl}} = E - \frac{E\left(\frac{\partial F}{\partial \sigma}\right) E r}{\left(\frac{\partial F}{\partial \sigma}\right) E r - \left(\frac{\partial F}{\partial \boldsymbol{q}}\right)^{\mathrm{T}} \boldsymbol{h}}. \tag{2.68}$$

Let us consider now the case of combined linear kinematic and isotropic hardening [see Eq. (2.48)] where the kinematic hardening modulus $H$ (PRAGER) and the plastic modulus $E^{\mathrm{pl}}$ are constant. Furthermore, the flow rule is assumed to be associated. We assume in the following that the yield condition is a function of the following internal variables: $F = F(\sigma, \boldsymbol{q}) = F(\sigma, \varepsilon^{\mathrm{pl}}, \varepsilon^{\mathrm{pl}}_{\mathrm{eff}})$. The corresponding terms in the expression for the elasto-plastic modulus are as follows:

$$\left(\frac{\partial F}{\partial \sigma}\right) = \operatorname{sgn}\left|\sigma - H\varepsilon^{\mathrm{pl}}\right|, \tag{2.69}$$

$$r = \frac{\partial F}{\partial \sigma} = \operatorname{sgn}\left|\sigma - H\varepsilon^{\mathrm{pl}}\right|, \tag{2.70}$$

$$\left(\frac{\partial F}{\partial \boldsymbol{q}}\right) = \begin{bmatrix} -E^{\mathrm{pl}} \\ -\operatorname{sgn}\left(\sigma - H\varepsilon^{\mathrm{pl}}\right) H \end{bmatrix}, \tag{2.71}$$

$$\boldsymbol{h} = \begin{bmatrix} 1 \\ \operatorname{sgn}\left(\sigma - H\varepsilon^{\mathrm{pl}}\right) \end{bmatrix}. \tag{2.72}$$

Introducing these four expressions in Eq. (2.68) gives finally:

$$E^{\mathrm{elpl}} = \frac{\mathrm{d}\sigma}{\mathrm{d}\varepsilon} = \frac{E(H + E^{\mathrm{pl}})}{E + (H + E^{\mathrm{pl}})}. \tag{2.73}$$

A slightly different derivation is obtained by considering the yield condition depending on the following internal variables: $F = F(\sigma, \boldsymbol{q}) = F(\sigma, \alpha, \varepsilon^{\mathrm{pl}}_{\mathrm{eff}})$. Then, the following different expressions are obtained:

$$\left(\frac{\partial F}{\partial \boldsymbol{q}}\right) = \begin{bmatrix} -E^{\mathrm{pl}} \\ -\operatorname{sgn}\left(\sigma - H\varepsilon^{\mathrm{pl}}\right) \end{bmatrix}, \tag{2.74}$$

$$\boldsymbol{h} = \begin{bmatrix} 1 \\ H\operatorname{sgn}\left(\sigma - H\varepsilon^{\mathrm{pl}}\right) \end{bmatrix}, \tag{2.75}$$

which result again in Eq. (2.73). The different general definitions of the moduli used in this derivation are summarized in Table 2.2.

**Table 2.2** Comparison of the different definitions of the stress-strain characteristics (moduli) in the case of the one-dimensional $\sigma$-$\varepsilon$ space

| Range | Definition | Graphical representation |
|---|---|---|
| Elastic | $E = \dfrac{\mathrm{d}\sigma}{\mathrm{d}\varepsilon^{\mathrm{el}}}$ | Figure 2.2 |
| Plastic | $E^{\mathrm{elpl}} = \dfrac{\mathrm{d}\sigma}{\mathrm{d}\varepsilon}$ for $\varepsilon > \varepsilon^{\mathrm{init}}$ | Figure 2.2b |
| | $E^{\mathrm{pl}} = \dfrac{\mathrm{d}k}{\mathrm{d}\lvert\varepsilon^{\mathrm{pl}}\rvert}$ | Figure 2.5 |

For the two-component $\sigma$-$\tau$/$\varepsilon$-$\gamma$ space, the incremental relation between the stresses and strains reads $\mathrm{d}\boldsymbol{\sigma} = \boldsymbol{C}^{\mathrm{elpl}}\mathrm{d}\boldsymbol{\varepsilon}$ where $\boldsymbol{C}^{\mathrm{elpl}}$ is the elasto-plastic modulus matrix. The yield condition can be generally stated as $F = F(\boldsymbol{\sigma}, \boldsymbol{q})$ and the total differential of such a yield condition is given as:

$$\mathrm{d}F(\boldsymbol{\sigma}, \boldsymbol{q}) = \left(\frac{\partial F}{\partial \boldsymbol{\sigma}}\right)^{\mathrm{T}} \mathrm{d}\boldsymbol{\sigma} + \left(\frac{\partial F}{\partial \boldsymbol{q}}\right)^{\mathrm{T}} \mathrm{d}\boldsymbol{q} = 0\,. \tag{2.76}$$

Following the same line of reasoning as in the one-dimensional case, the consistence parameter is obtained as

$$\mathrm{d}\lambda = \frac{\left(\frac{\partial F}{\partial \boldsymbol{\sigma}}\right)^{\mathrm{T}} \boldsymbol{C}}{\left(\frac{\partial F}{\partial \boldsymbol{\sigma}}\right)^{\mathrm{T}} \boldsymbol{C}\boldsymbol{r} - \left(\frac{\partial F}{\partial \boldsymbol{q}}\right)^{\mathrm{T}} \boldsymbol{h}} \mathrm{d}\boldsymbol{\varepsilon}\,, \tag{2.77}$$

and finally the elasto-plastic modulus matrix as:

$$\boldsymbol{C}^{\mathrm{elpl}} = \boldsymbol{C} - \frac{\left(\boldsymbol{C}\frac{\partial F}{\partial \boldsymbol{\sigma}}\right) \times (\boldsymbol{C}\boldsymbol{r})^{\mathrm{T}}}{\left(\frac{\partial F}{\partial \boldsymbol{\sigma}}\right)^{\mathrm{T}} \boldsymbol{C}\boldsymbol{r} - \left(\frac{\partial F}{\partial \boldsymbol{q}}\right)^{\mathrm{T}} \boldsymbol{h}}\,, \tag{2.78}$$

or under consideration of the dyadic product '$\otimes$' (see Sect. A.14.3):

$$\boldsymbol{C}^{\mathrm{elpl}} = \boldsymbol{C} - \frac{\left(\boldsymbol{C}\frac{\partial F}{\partial \boldsymbol{\sigma}}\right) \otimes (\boldsymbol{C}\boldsymbol{r})}{\left(\frac{\partial F}{\partial \boldsymbol{\sigma}}\right)^{\mathrm{T}} \boldsymbol{C}\boldsymbol{r} - \left(\frac{\partial F}{\partial \boldsymbol{q}}\right)^{\mathrm{T}} \boldsymbol{h}}\,. \tag{2.79}$$

Let us consider now the case of combined linear kinematic and isotropic hardening (see Sect. 2.4.3) where the kinematic hardening modulus $H$ (PRAGER) and the plastic modulus $E^{\mathrm{pl}}$ are constant. Furthermore, the flow rule is assumed to be associated. We assume in the following that the yield condition is a function of the following internal variables: $F = F(\boldsymbol{\sigma}, \boldsymbol{q}) = F(\sigma, \varepsilon^{\text{pl}}_{\text{eff}}, \alpha_\sigma, \alpha_\tau)$. The corresponding terms in the expression for the elasto-plastic modulus matrix for the two-component $\sigma$-$\tau$/$\varepsilon$-$\gamma$ space are as follows:

$$\boldsymbol{C} = \begin{bmatrix} E & 0 \\ 0 & G \end{bmatrix}, \tag{2.80}$$

$$\left(\frac{\partial F}{\partial \boldsymbol{\sigma}}\right) = \frac{1}{\sqrt{(\sigma - \alpha_\sigma)^2 + 3\,(\tau - \alpha_\tau)^2}} \begin{bmatrix} \sigma - \alpha_\sigma \\ 3(\tau - \alpha_\tau) \end{bmatrix}, \tag{2.81}$$

$$\boldsymbol{r} = \frac{1}{\sqrt{(\sigma - \alpha_\sigma)^2 + 3\,(\tau - \alpha_\tau)^2}} \begin{bmatrix} \sigma - \alpha_\sigma \\ 3(\tau - \alpha_\tau) \end{bmatrix}, \tag{2.82}$$

$$\left(\frac{\partial F}{\partial \boldsymbol{q}}\right) = \begin{bmatrix} -E^{\text{pl}} \\ -\frac{\sigma - \alpha_\sigma}{\sqrt{(\sigma-\alpha_\sigma)^2 + 3(\tau-\alpha_\tau)^2}} \\ -\frac{3(\tau-\alpha_\tau)}{\sqrt{(\sigma-\alpha_\sigma)^2 + 3(\tau-\alpha_\tau)^2}} \end{bmatrix}, \tag{2.83}$$

$$\boldsymbol{h} = \begin{bmatrix} 1 \\ \frac{H(\sigma-\alpha_\sigma)}{\sqrt{(\sigma-\alpha_\sigma)^2 + 3(\tau-\alpha_\tau)^2}} \\ \frac{H3(\tau-\alpha_\tau)}{\sqrt{(\sigma-\alpha_\sigma)^2 + 3(\tau-\alpha_\tau)^2}} \end{bmatrix}. \tag{2.84}$$

Introducing these five relationships in Eq. (2.78) gives finally the following specific expression for the elasto-plastic modulus matrix:

$$\boldsymbol{C}^{\text{elpl}} = \begin{bmatrix} E & 0 \\ 0 & G \end{bmatrix} - \frac{1}{(E + H + E^{\text{pl}})(\sigma - \alpha_\sigma)^2 + (9G + 9H + 3E^{\text{pl}})(\tau - \alpha_\tau)^2} \times \begin{bmatrix} E^2(\sigma - \alpha_\sigma)^2 & 3EG(\sigma - \alpha_\sigma)(\tau - \alpha_\tau) \\ 3EG(\sigma - \alpha_\sigma)(\tau - \alpha_\tau) & 9G^2(\tau - \alpha_\tau)^2 \end{bmatrix}. \tag{2.85}$$

The last equation can be simplified to the special case of a one-dimensional stress and and strain state by assigning $G = 0$ and $\tau = 0$. Then one obtains again Eq. (2.73).

At the end of this section, Table 2.3 compares the different equations and formulations of one-dimensional plasticity with the general two-dimensional representations (see for example [10, 96]).

## 2.7 Consideration of Unloading, Reversed Loading and Cyclic Loading

The previous sections considered only monotonic loading either in the tensile or compressive regimes. We will now briefly look at the cases where the loading direction can change. Figure 2.9a shows the case of loading in the elastic ($0 \rightarrow 1$) and

**Table 2.3** Comparison between general 2D plasticity and 1D plasticity with combined hardening ($E^{pl}$ and $H$ are assumed constant)

| General 2D plasticity | 1D *Linear* hardening plasticity |
|---|---|
| Yield condition<br>$F(\boldsymbol{\sigma}, \boldsymbol{q}) \leq 0$ | $F = \vert\sigma - \alpha\vert - (k^{\text{init}} + E^{\text{pl}}\varepsilon^{\text{pl}}_{\text{eff}}) \leq 0$ |
| Flow rule<br>$\varepsilon^{\text{pl}} = \text{d}\lambda \times \boldsymbol{r}(\boldsymbol{\sigma}, \boldsymbol{q})$ | $\text{d}\varepsilon^{\text{pl}} = \text{d}\lambda \times \text{sgn}(\sigma - \alpha)$ |
| Hardening law<br>$\boldsymbol{q} = [\kappa, \boldsymbol{\alpha}]^{\text{T}}$ | $\boldsymbol{q} = [\text{d}\varepsilon^{\text{pl}}_{\text{eff}}, \alpha]^{\text{T}}$ |
| $\text{d}\boldsymbol{q} = \text{d}\lambda \times \boldsymbol{h}(\boldsymbol{\sigma}, \boldsymbol{q})$ | $\text{d}\varepsilon^{\text{pl}}_{\text{eff}} = \text{d}\lambda,\ \text{d}\alpha = \text{d}\lambda H \text{sgn}(\sigma - \alpha)$ |
| Elasto-plastic modulus matrix<br>$\boldsymbol{C}^{\text{elpl}} = \left(\boldsymbol{C} - \dfrac{\left(\boldsymbol{C}\frac{\partial F}{\partial \boldsymbol{\sigma}}\right) \times (\boldsymbol{C}\boldsymbol{r})^{\text{T}}}{\left(\frac{\partial F}{\partial \boldsymbol{\sigma}}\right)^{\text{T}} \boldsymbol{C}\boldsymbol{r} - \left(\frac{\partial F}{\partial \boldsymbol{q}}\right)^{\text{T}} \boldsymbol{h}}\right)$ | $E^{\text{elpl}} = \dfrac{E \times (H + E^{\text{pl}})}{E + (H + E^{\text{pl}})}$ |

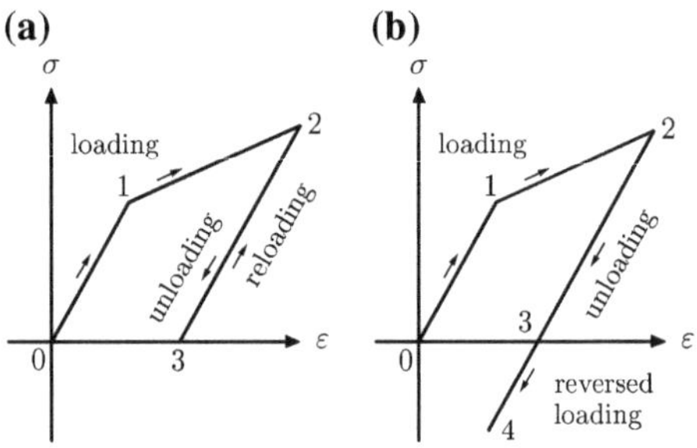


**Fig. 2.9** Idealized stress-strain curve with **a** loading—unloading—reloading and **b** loading—unloading—reversed loading

elasto-plastic ($1 \rightarrow 2$) range, followed by elastic unloading ($2 \rightarrow 3$) and elastic reloading ($3 \rightarrow 2$).

In the case of Fig. 2.9b, the elastic unloading ($2 \rightarrow 3$) is followed by reversed loading ($3 \rightarrow 4$). The important feature which should be highlighted here is that the unloading phase ($2 \rightarrow 3$) can be described based on HOOKE's law, cf. Eq. (2.3). Figure 2.10 shows the case of cyclic loading where a specimen is exposed to fluctuating loads $F(t)$.

Some characteristic stress quantities are indicated in Fig. 2.10b: The stress range $\Delta\sigma$ is the difference between the maximum and minimum stress:

$$\Delta\sigma = \sigma_{\text{max}} - \sigma_{\text{min}}\,. \tag{2.86}$$

The stress amplitude $\sigma_{\text{a}}$ is half the value of the stress range:

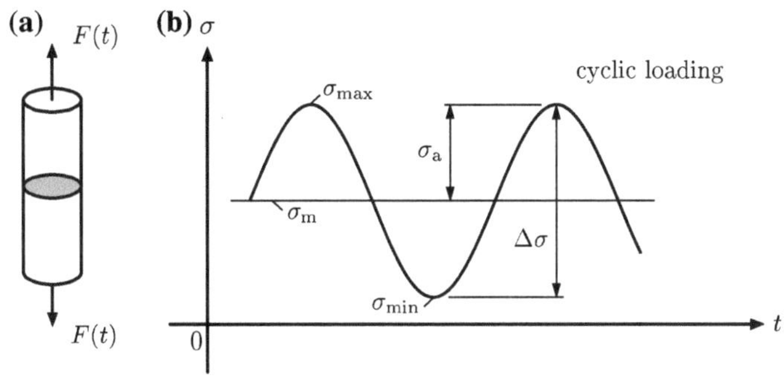


**Fig. 2.10** Cyclic loading: **a** idealized specimen and **b** stress-time curve at constant amplitude

$$\sigma_{\mathrm{a}} = \frac{\Delta\sigma}{2} = \frac{\sigma_{\mathrm{max}} - \sigma_{\mathrm{min}}}{2}. \tag{2.87}$$

The so-called stress ratio $R$ is often used to characterize the stress level in cyclic tests:

$$R = \frac{\sigma_{\mathrm{min}}}{\sigma_{\mathrm{max}}}, \tag{2.88}$$

where a value $R = -1$ characterizes a fully-reversed load cycle, $R = 1$ stands for static loading and $R = 0$ refers to the case where the mean stress is positive and equal to the stress amplitude. In materials testing cyclic tests are performed to determine the fatigue life of components and structures. Further details can be found in [78, 93].

## 2.8 Consideration of Damage

### *2.8.1 Lemaitre's Damage Model*

The following section summarizes briefly the major ideas of ductile damage based on the concept given by LEMAITRE in [61, 62]. Let us consider an idealized uniaxial tensile sample as shown in Fig. 2.11 where the undamaged or initial tensile specimen is shown on the left-hand side and the damaged or deformed specimen on the right-hand side. A pure uniaxial tensile sample is assumed in the following which means in this context that the specimen deforms only in its longitudinal direction (i.e. the $x$-direction as shown in Fig. 2.11) and does not show any deformation perpendicular, i.e. contraction, to the loading direction. This corresponds to the assumption that Poisson's ratio is equal to zero.

It should be noted here that the size of the specimen must be in such a range that the considered volume represents a representative volume element (RVE) for the considered material. Some estimates for the minimum size of RVEs for different

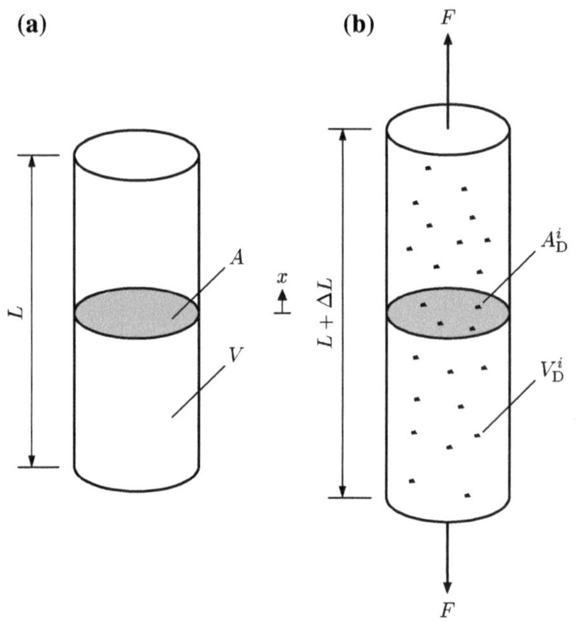


**Fig. 2.11** Schematic representation of **a** an undamaged and **b** a damaged tensile specimen

materials are given in [63]. Let $A$ be the overall cross-sectional area of the specimen (marked in grey in Fig. 2.11a) and $A_D$ be the total area of the micro-cracks and voids in the considered area which is in Fig. 2.11b marked in black. The effective resisting area is denoted by $\bar{A}$. Based on these quantities, the damage variable $D$ can be introduced as:

$$D = \frac{\sum_i A_D^i}{A} = \frac{A_D}{A} = \frac{A - \bar{A}}{A}. \tag{2.89}$$

If this definition is based on a RVE, then the same damage variable is obtained based on the volume of the micro-cracks and voids [111]: $D = V_D/V$. A state $D = 0$ corresponds to the undamaged state, $D = 1$ represents the rupture of the specimen into two parts and $0 < D < 1$ characterizes the damaged state. In the scope of this chapter, an isotropic damage variable is assumed. This means that the defects are equally distributed in all directions of the specimen. Thus, a scalar description of the damage is sufficient under the *hypothesis of isotropy*. If the resisting area in Eq. (2.89) is used to calculate the stress in the specimens, the *concept of effective stress* is obtained which states that the effective stress in the specimen is given by:

$$\bar{\sigma} = \frac{\sigma}{1 - D}. \tag{2.90}$$

It must be stated here that this definition of the effective stress holds only in the tensile regime. Under compression, some defects may close again or in the limiting case, all defects can be closed again so that the effective stress is again equal to the usual stress $\sigma$. However, this effect in the compressive regime will be not considered within this chapter. In the case of the strain, the *hypothesis of strain equivalence* is applied which states that the strain behavior of a damaged material is represented by the virgin material:

$$\bar{\varepsilon} = \varepsilon \,. \tag{2.91}$$

Based on these assumptions and simplifications, HOOKE's law can be written with the effective stress $\bar{\sigma}$ and elastic strain $\varepsilon^{\text{el}}$ as:

$$\bar{\sigma} = E\varepsilon^{\text{el}} \,, \tag{2.92}$$

which can be expressed with the definition of the effective stress given in Eq. (2.90) as:

$$\sigma = \underbrace{(1 - D)E}_{\bar{E}}\,\varepsilon \,, \tag{2.93}$$

where $E$ is the elastic modulus of the undamaged material (initial modulus) and $\bar{E}$ is the modulus of the damaged material. The last equation offers an elegant way to experimentally determine the evolution of the damage variable $D$. Measuring during tests with reversed loading stress and strain based on the usual engineering definitions, i.e. $\sigma = F/A$ and $\varepsilon = \Delta L/L$, the damage variable can be indirectly obtained from the variation of the elasticity modulus as:

$$D = 1 - \frac{\bar{E}}{E} \,. \tag{2.94}$$

The classical continuum theory of plasticity is based on three equations, i.e. the yield condition, the flow rule and the hardening law. For a one-dimensional stress state, the yield condition reads under consideration of the damage effects as

$$F = \frac{|\sigma|}{1 - D} - k(\kappa) \,, \tag{2.95}$$

where $|\sigma|/(1 - D)$ is the equivalent stress which is compared to the experimental value $k$. The flow rule and the evolution equation for the internal variable $\kappa$ do not change and are given by Eqs. (2.19) and (2.28).

In the case of damage mechanics, there is in addition the evolution equation required for the damage variable. Following the notation in [74] and considering a one-dimensional stress state, the model for the ductile damage evolution can be expressed as

**Table 2.4** Basic equations of the LEMAITRE model (isotropic hardening) in the case of a uniaxial stress state with $\sigma$ as acting stress

| 1D LEMAITRE model |
|---|
| HOOKE's law<br>$\sigma = (1 - D)E\varepsilon$ |
| Yield condition<br>$F = \dfrac{\lvert\sigma\rvert}{1 - D} - k(\kappa)$ |
| Flow rule<br>$\mathrm{d}\varepsilon^{\mathrm{pl}} = \mathrm{d}\lambda \dfrac{\mathrm{sgn}(\sigma)}{1 - D}$ |
| Evolution of hardening variable<br>$\mathrm{d}\kappa = \mathrm{d}\lambda$ |
| Evolution of damage variable<br>$\mathrm{d}D = \dfrac{\mathrm{d}\lambda}{1 - D}\left(\dfrac{-Y}{r}\right)^s = \mathrm{d}\lvert\varepsilon^{\mathrm{pl}}\rvert\left(\dfrac{-Y}{r}\right)^s$<br>with $Y = -\dfrac{\sigma^2}{2E(1 - D)^2}$ |

$$\mathrm{d}D = \frac{\mathrm{d}\lambda}{1 - D}\left(\frac{-Y}{r}\right)^s = \mathrm{d}\lvert\varepsilon^{\mathrm{pl}}\rvert\left(\frac{-Y}{r}\right)^s, \tag{2.96}$$

where $Y$ is the so-called damage energy release rate which corresponds to the variation of internal energy density due to damage growth at constant stress, and $r$ and $s$ are damage evolution material parameters. For a one-dimensional stress state, $Y$ takes the form:

$$Y = -\frac{\sigma^2}{2E(1 - D)^2} \overset{(2.90)}{=} -\frac{\overline{\sigma}^2}{2E}. \tag{2.97}$$

The basic equations for the one-dimensional model according to LEMAITRE are collected in the following Table 2.4.

An experimental strategy based on a simple tensile test and a fatigue test (WÖHLER curve) for the determination of the parameters $r$ and $s$ is given in [64]. A simpler form of Eq. (2.96) is given in [63] as:

$$\mathrm{d}D = \mathrm{d}\lvert\varepsilon^{\mathrm{pl}}\rvert \times \frac{-Y}{r}, \tag{2.98}$$

or under consideration of Eq. (2.97) as:

$$dD = d|\varepsilon^{pl}| \times \frac{\sigma^2}{2Er(1-D)^2}. \quad (2.99)$$

The last equation can be rearranged to give

$$\frac{dD}{d|\varepsilon^{pl}|} = \frac{\sigma^2}{2Er(1-D)^2}, \quad (2.100)$$

or

$$r = \frac{\sigma^2}{2E(1-D)^2 \frac{dD}{d|\varepsilon^{pl}|}}, \quad (2.101)$$

which allows to determine the material parameter $r$ from a tensile test with reversed loading in the following way:

- Perform a uniaxial tensile test with reversed loading as shown in Fig. 2.12a. Determine the elastic modulus $E$ of the undamaged material (initial modulus) and for each unloading-loading cycle $n$ the modulus of the damaged material $\bar{E}$, the corresponding stress $\sigma$ where the unloading starts and the plastic strain at $\sigma = 0$.
- Construct from the values of stress and corresponding plastic strain the flow curve as shown in Fig. 2.12b.
- Calculate from the modulus of the damaged material the damage variable $D$ as given in Eq. (2.94). Plot the damage variable over the plastic strain as shown in Fig. 2.12c.
- Calculate the damage material parameter $r$ according to Eq. (2.101) for given values of stress $\sigma$, elastic modulus $E$, damage variable $D$ and slope $\frac{dD}{d|\varepsilon^{pl}|}$.
- Calculate an average (or interpolated) value of $r$ by considering several evaluations as described in the previous step.

## 2.8.2 Gurson's Damage Model

To model the arbitrary distribution of voids in a matrix (cf. Fig. 2.13a), GURSON introduced the idealized models of two void geometries [48]. For the first model, a single long circular cylindrical void is considered in a similarly shaped matrix (cf. Fig. 2.13b) whereas the second model considers a spherical shape of void and matrix (cf. Fig. 2.13c). The matrix material is considered as a homogeneous, isotropic, rigid-plastic VON MISES material. It should be highlighted here that the approach neglected the elastic material response and assumed the matrix material to be imcompressible.

Based on these assumptions, the following yield conditions could be derived

$$F = \left(\frac{\sigma_{eff}}{k_t}\right)^2 + 2D\cosh\left(\frac{3\sqrt{3}}{2}\frac{\sigma_m}{k_t}\right) - (1 + D^2) = 0\,, \text{(cylindrical)} \quad (2.102)$$

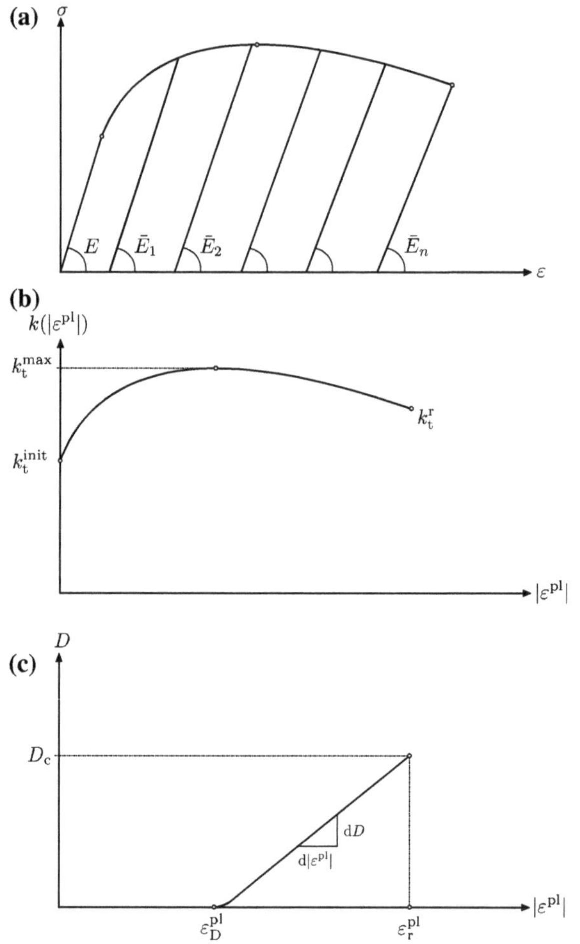


**Fig. 2.12** **a** Uniaxial stress-strain diagram with reversed loading; **b** flow curve; **c** damage variable as a function of plastic strain

$$F = \left(\frac{\sigma_{eff}}{k_t}\right)^2 + 2D\cosh\left(\frac{3}{2}\frac{\sigma_m}{k_t}\right) - (1 + D^2) = 0\,, \qquad \text{(spherical)} \qquad (2.103)$$

where $\sigma_{eff}$ is the effective stress based on the VON MISES definition and $\sigma_m$ is the mean stress (hydrostatic stress). The yield stress $k_t$ is used to normalize the effective and mean stress. The damage variable $D$ in this model is based on the void volume fraction, i.e. $D = V_D/V$ where $V_D$ is the volume of the void and $V$ is the volume of the matrix. If the calculation of the void volume fraction is based on real distributions as schematically shown in Fig. 2.13, then some account is taken of the interaction of neighboring voids [48].

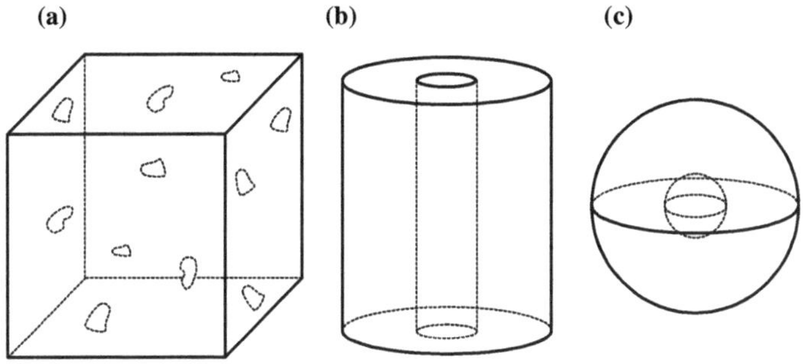


**Fig. 2.13** **a** Voids of random shape and orientation distributed in a matrix; **b** long circular cylindrical void; **c** spherical void. Adapted from [48]

**Table 2.5** Intersections of the GURSON flow curves with the coordinate axes

| $D$ | $\left(\frac{\tau}{k_t}\right)_{x=0}$ | $\left(\frac{\sigma}{k_t}\right)_{y=0}$ |
|---|---|---|
| 0.00 | ±0.5774 | ±1.000000 |
| 0.01 | ±0.5715 | ±0.988639 |
| 0.10 | ±0.5132 | ±0.877740 |
| 0.30 | ±0.3215 | ±0.536890 |

Let us consider in the following the two-component $\sigma$-$\tau$ stress space. Thus, the yield conditions given in Eqs. (2.102) and (2.103) take the following form:

$$F = \left(\frac{\sqrt{\sigma^2 + 2\tau^2}}{k_t}\right)^2 + 2D\cosh\left(\frac{\sqrt{3}}{2}\frac{\sigma}{k_t}\right) - (1 + D^2) = 0\,, \text{ (cylindrical)} \tag{2.104}$$

$$F = \left(\frac{\sqrt{\sigma^2 + 2\tau^2}}{k_t}\right)^2 + 2D\cosh\left(\frac{1}{2}\frac{\sigma}{k_t}\right) - (1 + D^2) = 0\,. \quad \text{(spherical)} \tag{2.105}$$

A graphical representation of the GURSON yield condition based on Eq. (2.105) is shown in Fig. 2.14 where different values of the damage parameter have been assigned. A value of $D = 0$ results in the classical VON MISES ellipse and values $D > 0$ result in this graphical representation in similar looking shapes[22] but with a smaller elastic region ($F < 0$), i.e. plastic yielding will start earlier compared to the classical VON MISES condition.

Abscissae and ordinates of the intersection of the flow curve with the coordinate axes can be obtained from Eq. (2.105) for $\sigma = 0$ or $\tau = 0$ and are summarized in Table 2.5.

[22] The yield surfaces for $D > 0$ look like ellipses but because of the cosh function, they are not from a mathematical point of view classified as ellipses.

The derivative of the yield condition or the plastic strain increments with respect to the stresses are obtained in the case of a general stress state from Eq. (2.103)

$$\mathrm{d}\boldsymbol{\varepsilon}^{\mathrm{pl}} = \mathrm{d}\lambda \frac{\partial F}{\partial \boldsymbol{\sigma}} = \mathrm{d}\lambda \left( \frac{3\mathbf{L}\boldsymbol{s}}{k_{\mathrm{t}}^2} + \frac{D}{k_{\mathrm{t}}} \sinh\left( \frac{3\sigma_{\mathrm{m}}}{2k_{\mathrm{t}}} \right) \mathbf{1} \right), \tag{2.106}$$

where $\boldsymbol{s} = \left[ s_x \; s_y \; s_z \; \tau_{xy} \; \tau_{yz} \; \tau_{xz} \right]^{\mathrm{T}}$ is the column matrix of the stress deviator components,[23] $\mathbf{L} = \lceil 1\,1\,1\,0\,0\,0 \rfloor$ is a diagonal scaling matrix and $\mathbf{1} = [1\,1\,1\,0\,0\,0]^{\mathrm{T}}$ is an identity column matrix. In the case of a $\sigma$-$\tau$ stress space, Eq. (2.105) gives under consideration of $\frac{\mathrm{d}}{\mathrm{d}x}|x|^2 = 2|\sigma|\mathrm{sgn}(\sigma) = 2\sigma$ the following expression for the plastic strain increments:

$$\mathrm{d}\boldsymbol{\varepsilon}^{\mathrm{pl}} = \begin{bmatrix} \mathrm{d}\varepsilon^{\mathrm{pl}} \\ \mathrm{d}\gamma^{\mathrm{pl}} \end{bmatrix} = \mathrm{d}\lambda \begin{bmatrix} \frac{2\sigma}{k_{\mathrm{t}}^2} + \frac{D}{k_{\mathrm{t}}} \sinh\left( \frac{\sigma}{2k_{\mathrm{t}}} \right) \\ \frac{6\tau}{k_{\mathrm{t}}^2} \end{bmatrix}. \tag{2.107}$$

At the end of this section, the evolution equation for the ductile damage should be given. According to [74], the evolution equation is given as

$$\mathrm{d}D = (1 - D)(\mathrm{d}\varepsilon_x^{\mathrm{pl}} + \mathrm{d}\varepsilon_y^{\mathrm{pl}} + \mathrm{d}\varepsilon_z^{\mathrm{pl}}) = (1 - D)\mathrm{d}\varepsilon_{\mathrm{V}}^{\mathrm{pl}}, \tag{2.108}$$

where $\varepsilon_{\mathrm{V}}^{\mathrm{pl}}$ is the volumetric plastic strain. Consideration of Eq. (2.106) and the fact that $s_x + s_y + s_z = 0$ allows to reformulate the last equation to obtain:

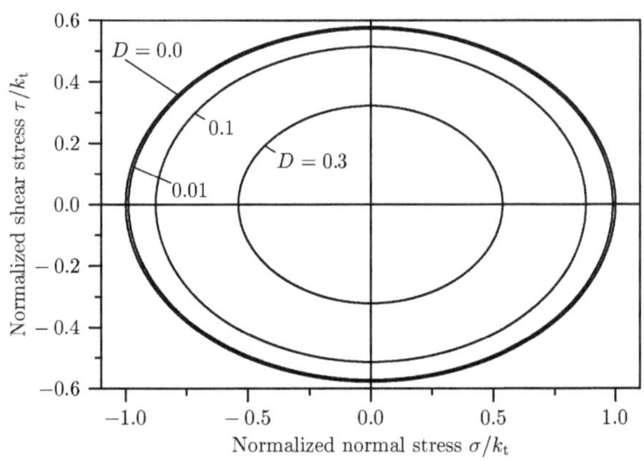


**Fig. 2.14** Graphical representation of the yield condition according to GURSON in the two-component $\sigma$-$\tau$ space for different values of the damage variable (spherical voids assumed)

[23] The first component of the stress deviator is given by $s_x = \frac{2}{3}\sigma_x - \frac{1}{3}\left(\sigma_y + \sigma_z\right)$.

**Table 2.6** Basic equations of the original GURSON model (spherical inclusions, rigid-plastic material) in the case of a uniaxial stress state with $\sigma$ as acting stress

| 1D GURSON model |
| --- |
| Yield condition |
| $F = \left(\frac{\lvert\sigma\rvert}{k_\mathrm{t}}\right)^2 + 2D\cosh\left(\frac{\sigma}{2k_\mathrm{t}}\right) - (1+D^2) = 0$ |
| Flow rule |
| $\mathrm{d}\varepsilon^\mathrm{pl} = \mathrm{d}\lambda\left(\frac{2\sigma}{k_\mathrm{t}^2} + \frac{D}{k_\mathrm{t}}\sinh\left(\frac{\sigma}{2k_\mathrm{t}}\right)\right)$ |
| Damage evolution equation |
| $\mathrm{d}D = \mathrm{d}\lambda\frac{3(D-D^2)}{k_\mathrm{t}}\sinh\left(\frac{\sigma}{2k_\mathrm{t}}\right)$ |

$$\mathrm{d}D = \mathrm{d}\lambda\frac{3(D-D^2)}{k_\mathrm{t}}\sinh\left(\frac{3\sigma_\mathrm{m}}{2k_\mathrm{t}}\right). \tag{2.109}$$

Considering a pure one-dimensional stress state where only the normal stress $\sigma$ is acting, Eqs. (2.105), (2.107) and (2.109) can be simplified to the forms given in Table 2.6.

In many practical applications, the GURSON yield condition is applied to *elasto*-plastic material behavior, even under consideration of isotropic hardening ($k_\mathrm{t} = k_t(\kappa)$). In such a case it is necessary to indicate the evolution equation for the internal variable $\kappa$. Assuming that the effective plastic strain is assigned as the internal variable, i.e. $\varepsilon^\mathrm{pl}_\mathrm{eff} = \kappa$, and furthermore assuming that the increment of equivalent plastic work in the matrix material equals the macroscopic increment of plastic work [10, 109], i.e.

$$\sigma\mathrm{d}\varepsilon^\mathrm{pl} = (1-D)k_\mathrm{t}\mathrm{d}\varepsilon^\mathrm{pl}_\mathrm{eff}, \tag{2.110}$$

the evolution of the internal variable is given in the one-dimensional case as:

$$\mathrm{d}\kappa = \frac{\mathrm{d}\lambda}{1-D}\left(\frac{2\sigma^2}{k_\mathrm{t}^3} + \frac{D\sigma}{k_\mathrm{t}^2}\sinh\left(\frac{\sigma}{2k_\mathrm{t}}\right)\right). \tag{2.111}$$

Thus, the equations presented in Table 2.6 can be extended to the case of an *elasto*-plastic material with isotropic hardening as summarized in Table 2.7.

**Table 2.7** Basic equations of the extended GURSON model (elastic range, isotropic hardening) in the case of a uniaxial stress state with $\sigma$ as acting stress

| 1D GURSON model |
|---|
| HOOKE's law<br>$\sigma = E\varepsilon$ |
| Yield condition<br>$F = \left(\frac{\lvert\sigma\rvert}{k_t(\kappa)}\right)^2 + 2D\cosh\left(\frac{\sigma}{2k_t(\kappa)}\right) - (1 + D^2) = 0$ |
| Flow rule<br>$\mathrm{d}\varepsilon^{\mathrm{pl}} = \mathrm{d}\lambda\left(\frac{2\sigma}{(k_t(\kappa))^2} + \frac{D}{k_t(\kappa)}\sinh\left(\frac{\sigma}{2k_t(\kappa)}\right)\right)$ |
| Evolution of hardening variable<br>$\mathrm{d}\kappa = \frac{\mathrm{d}\lambda}{1-D}\left(\frac{2\sigma^2}{(k_t(\kappa))^3} + \frac{D\sigma}{(k_t(\kappa))^2}\sinh\left(\frac{\sigma}{2k_t(\kappa)}\right)\right)$ |
| Evolution of damage variable<br>$\mathrm{d}D = \mathrm{d}\lambda\frac{3(D-D^2)}{k_t(\kappa)}\sinh\left(\frac{\sigma}{2k_t(\kappa)}\right)$ |

## 2.9 Supplementary Problems

**2.1 Derivation of Basic Equations for von Mises' Yield Condition**

Given are the following two forms of a one-dimensional VON MISES (or TRESCA) yield condition:

(a) $F = \left(\frac{|\sigma|}{k_t}\right)^2 - 1 = 0,$

(b) $F = \frac{|\sigma|}{k_t} - 1 = 0.$

Derive the expressions for the associated flow rule and the evolution equation for the internal hardening variable in the case of strain gardening.

**2.2 Derivation of the Elasto-plastic Modulus for the One-Dimensional Case with Isotropic Hardening**

The general expression for the elasto-plastic modulus for a one-dimensional stress and strain state is generally given in Eq. (2.68) as:

$$E^{\mathrm{elpl}} = E - \frac{E\left(\frac{\partial F}{\partial \sigma}\right)Er}{\left(\frac{\partial F}{\partial \sigma}\right)Er - \left(\frac{\partial F}{\partial \boldsymbol{q}}\right)^{\mathrm{T}}\boldsymbol{h}}. \tag{2.112}$$

Simplify this equation for the case of pure isotropic hardening.

**2.3 Derivation of the Elasto-plastic Modulus Matrix for Tresca Yield Condition**
Given is the two-dimensional TRESCA yield condition in the following form

$$F_{\sigma-\tau} = \sqrt{(\sigma - \alpha_\sigma)^2 + 4(\tau - \alpha_\tau)^2} - (k^{\text{init}} + E^{\text{pl}}\varepsilon^{\text{pl}}_{\text{eff}}) = 0\,, \tag{2.113}$$

which considers combined linear kinematic and isotropic hardening where the kinematic hardening modulus $H$ (PRAGER) and the plastic modulus $E^{\text{pl}}$ are constant. Furthermore, the flow rule is assumed to be associated. Assume in the following that the yield condition is a function of the following internal variables: $F = F(\boldsymbol{\sigma}, \boldsymbol{q}) = F(\sigma, \varepsilon^{\text{pl}}_{\text{eff}}, \alpha_\sigma, \alpha_\tau)$. Derive the expression for the elasto-plastic modulus matrix $\boldsymbol{C}^{\text{elpl}}$.

**2.4 Derivation of Basic Equations for Gurson's Yield Condition**
Given is the following form of a one-dimensional GURSON's yield condition:

$$F = |\sigma|^2 + 2D\cosh\left(\frac{1}{2}\frac{\sigma}{k_{\text{t}}(\kappa)}\right)k_{\text{t}}(\kappa)^2 - \left(1 + D^2\right)k_{\text{t}}(\kappa)^2 = 0\,. \tag{2.114}$$

Derive the expressions for the flow rule, damage evolution equation, and evolution equation for the internal hardening variable. Assume for the derivation of the internal hardening variable that the increment of equivalent plastic work in the matrix material equals the macroscopic increment of plastic work.

# Chapter 3
# Axial Loading of Rods

**Abstract** This chapter covers the analytical description of rod or bar members. Based on the three basic equations of continuum mechanics, i.e. the kinematics relationship, the constitutive law and the equilibrium equation, the partial differential equation which describes the problem is derived. Analytical solutions in the pure elastic and elasto-plastic range for different loading and boundary conditions are derived and discussed.

## 3.1 Kinematics

To derive the strain-displacement relation (kinematics relation), an axially loaded road is considered as shown in Fig. 3.1. The length of the member is equal to $L$ and the constant axial tensile stiffness is equal to $EA$. The load is either given as a single force $F_x$ and/or as a distributed load $p_x(x)$.

This distributed load has the units force per unit length. In the case of a body force $f_x$ (unit: force per unit volume), the distributed load takes the form $p_x(x) = f_x(x)A(x)$ where $A$ is the cross-sectional area of the rod. A typical example for a body force would be the dead weight, i.e. the mass under the influence of gravity. In the case of a traction force $t_x$ (unit: force per unit area), the distributed load can be written as $p_x(x) = t_x(x)U(x)$ where $U(x)$ is the perimeter of the cross section. Typical examples are frictional resistance, viscous drag and surface shear.

Let us now consider a differential element $\mathrm{d}x$ of such a rod as shown in Fig. 3.2. Under an acting load, this element deforms as indicated in Fig. 3.2b where the initial point at the position $x$ is displaced by $u_x$ and the end point at the position $x + \mathrm{d}x$ is displaced by $u_x + \mathrm{d}u_x$. Thus, the differential element which has in the unloaded state a length of $\mathrm{d}x$ elongates to a length of $\mathrm{d}x + (u_x + \mathrm{d}u_x) - u_x$.

 
A. Öchsner, *Elasto-Plasticity of Frame Structure Elements*,
DOI 10.1007/978-3-662-44225-8_3

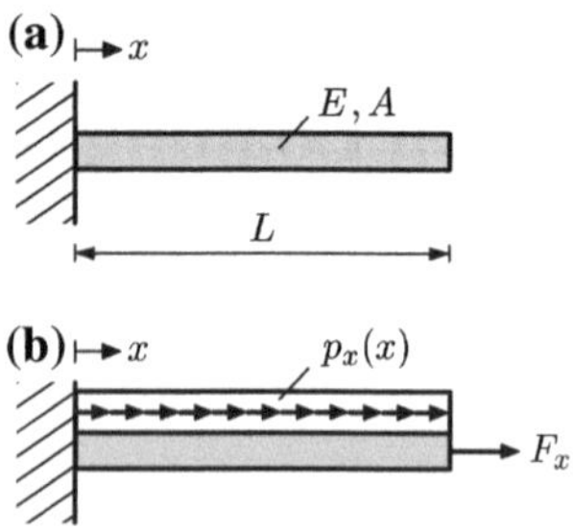


**Fig. 3.1** General configuration of an axially loaded rod: **a** geometry and material property; **b** prescribed loads

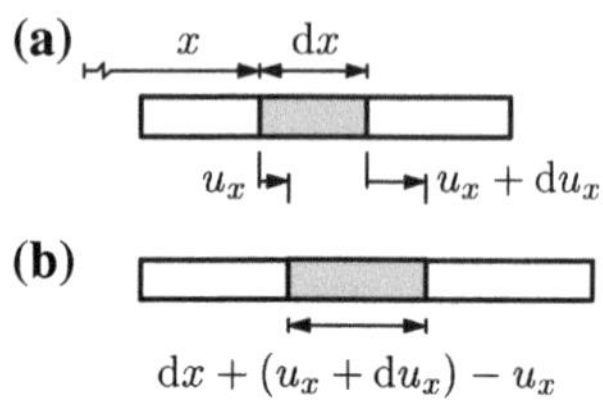


**Fig. 3.2** Elongation of a differential element of length d$x$: **a** undeformed configuration; **b** deformed configuration

The engineering strain, i.e. the increase in length related to the original length, can be expressed as

$$\varepsilon_x = \frac{(\mathrm{d}x + (u_x + \mathrm{d}u_x) - u_x) - (\mathrm{d}x)}{\mathrm{d}x}, \tag{3.1}$$

or finally as:

$$\varepsilon_x(x) = \frac{\mathrm{d}u_x(x)}{\mathrm{d}x}. \tag{3.2}$$

The last equation is often expressed in a less mathematical way (non-differential) as $\varepsilon_x = \frac{\Delta L}{L}$ where $\Delta L$ is the change in length of the entire rod element.

## 3.2 Constitutive Equation

The constitutive equation, i.e. the relation between stress $\sigma_x$ and strain $\varepsilon_x$, is given in its simplest form as HOOKE's law

$$\sigma_x(x) = E\varepsilon_x(x), \tag{3.3}$$

where the YOUNG's modulus $E$ is in the case of linear elasticity a material constant. For the considered rod element, the normal stress and strain is constant over the cross section as shown in Fig. 3.3.

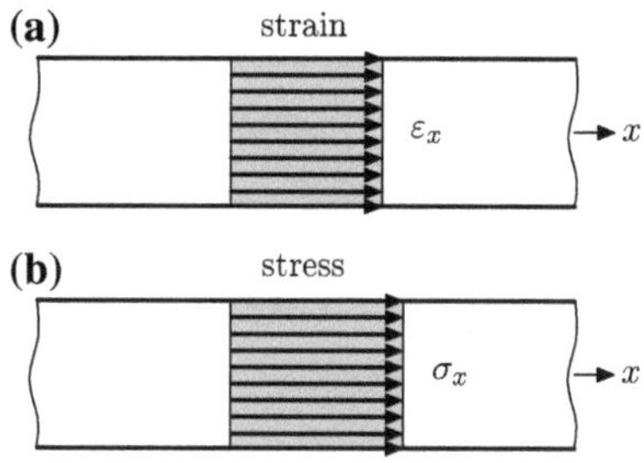


**Fig. 3.3** Axially loaded rod: **a** strain and **b** stress distribution

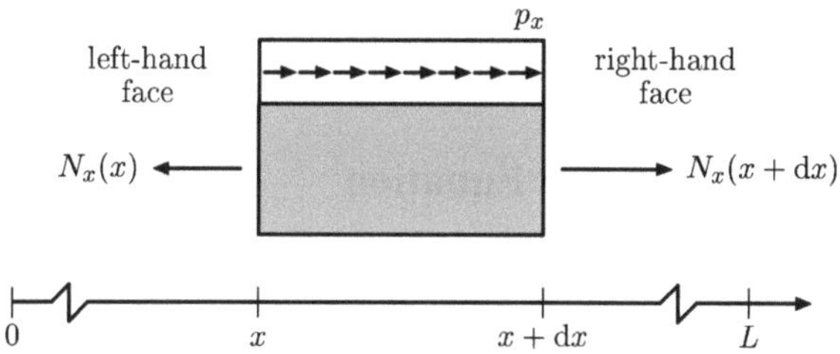


**Fig. 3.4** Differential element of a rod with internal reactions and constant external distributed load

## 3.3 Equilibrium

The equilibrium equation between the external forces and internal reactions can be derived on a differential element of length d$x$ as shown in Fig. 3.4. It is assumed for simplicity in this figure that the distributed load $p_x$ and the cross-sectional area $A$ are constant. The internal reactions $N_x$ are drawn in their positive directions, i.e. at the left-hand face in the negative and at the right-hand face in the positive $x$-direction.

The force equilibrium in the $x$-direction requires for a static configuration that

$$-N_x(x) + p_x\mathrm{d}x + N_x(x + \mathrm{d}x) = 0 \tag{3.4}$$

holds. A first order TAYLOR's[1] series expansion (cf. Appendix A.12) of $N_x(x + \mathrm{d}x)$ around point $x$, i.e.

$$N_x(x + \mathrm{d}x) \approx N_x(x) + \left.\frac{\mathrm{d}N_x}{\mathrm{d}x}\right|_x \mathrm{d}x\,, \tag{3.5}$$

allows to express Eq. (3.4) finally as:

$$\frac{\mathrm{d}N_x(x)}{\mathrm{d}x} = -p_x(x)\,. \tag{3.6}$$

[1] Brook TAYLOR (1685–1731), English mathematician.

**Table 3.1** Fundamental governing equations of a rod for deformation along the $x$-axis

| Expression | Equation |
|---|---|
| Kinematics | $\varepsilon_x(x) = \dfrac{\mathrm{d}u_x(x)}{\mathrm{d}x}$ |
| Equilibrium | $\dfrac{\mathrm{d}N_x(x)}{\mathrm{d}x} = -p_x(x)$ |
| Constitution | $\sigma_x(x) = E\varepsilon_x(x)$ |

The three fundamental equations to describe the behavior of a rod element are summarized in Table 3.1.

## 3.4 Governing Differential Equation

To derive the governing partial differential equation, the three fundamental equations given in Table 3.1 must be combined. Introducing the kinematics relation (3.2) into HOOKE's law (3.3) gives:

$$\sigma_x(x) = E\frac{\mathrm{d}u_x}{\mathrm{d}x}. \tag{3.7}$$

Considering in the last equation that a normal stress is defined as an acting force $N_x$ over a cross-sectional area $A$:

$$\frac{N_x}{A} = E\frac{\mathrm{d}u_x}{\mathrm{d}x}. \tag{3.8}$$

The last equation can be differentiated with respect to the $x$-coordinate to give:

$$\frac{\mathrm{d}N_x}{\mathrm{d}x} = \frac{\mathrm{d}}{\mathrm{d}x}\left(EA\frac{\mathrm{d}u_x}{\mathrm{d}x}\right), \tag{3.9}$$

where the derivative of the normal force can be replaced by the equilibrium Eq. (3.6) to obtain in the general case:

$$\frac{\mathrm{d}}{\mathrm{d}x}\left(E(x)A(x)\frac{\mathrm{d}u_x(x)}{\mathrm{d}x}\right) = -p_x(x). \tag{3.10}$$

If the axial tensile stiffness *EA* is constant, the last formulation can be simplified to:

$$EA\frac{\mathrm{d}^2u_x(x)}{\mathrm{d}x^2} = -p_x(x). \tag{3.11}$$

**Table 3.2** Different formulations of the partial differential equation for a rod ($x$-axis: right facing)

| Configuration | Partial differential equation |
|---|---|
|   | $EA\frac{\mathrm{d}^2u_x}{\mathrm{d}x^2}=0$ |
| 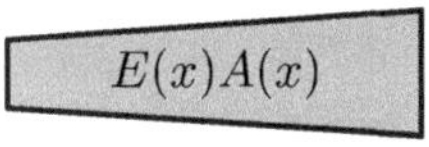  | $\frac{\mathrm{d}}{\mathrm{d}x}\left(E(x)A(x)\frac{\mathrm{d}u_x}{\mathrm{d}x}\right)=0$ |
| 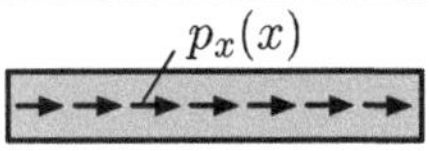  | $EA\frac{\mathrm{d}^2u_x}{\mathrm{d}x^2}=-p_x(x)$ |
| 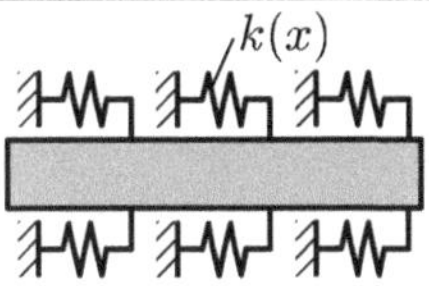  | $EA\frac{\mathrm{d}^2u_x}{\mathrm{d}x^2}=k(x)u_x$ |

Some common formulations of the governing partial differential equation are collected in Table 3.2. It should be noted here that some of the different cases given in Table 3.1 can be combined. The last case in Table 3.1 refers to the case of elastic embedding of a rod where the embedding modulus $k$ has the units force per unit area.

## 3.5 Analytical Solutions

### *3.5.1 Elastic Range*

**3.1 Example: Elongation of a rod due to a single force**
Given is a rod of length $L$ and constant axial tensile stiffness $EA$ as shown in Fig. 3.5. At the left-hand side there is a fixed support and the right-hand side is loaded by a single force $F$. Determine the elongation $u_x(x)$ along the rod axis.

**3.1 Solution** The describing partial differential equation can be taken from Table 3.2 as:

$$EA\frac{\mathrm{d}^2u_x(x)}{\mathrm{d}x^2}=0\,. \tag{3.12}$$

Integrating twice gives

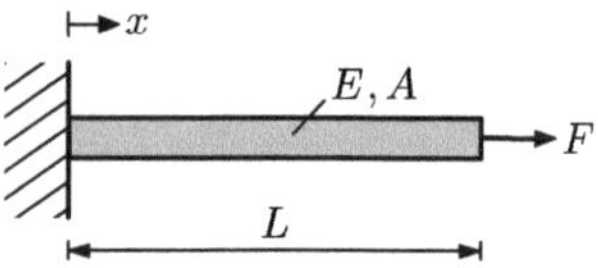


Fig. 3.5 Analytical solution for the elongation of a rod due to a single force

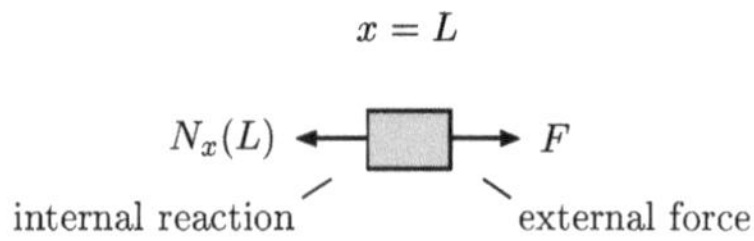


Fig. 3.6 Equilibrium between internal normal force $N_x$ and external load $F$

$$EA\frac{\mathrm{d}u_x(x)}{\mathrm{d}x}=c_1 \tag{3.13}$$

and

$$u_x(x)=\frac{1}{EA}(c_1x+c_2)\ , \tag{3.14}$$

where the two constants of integration $c_i(i=1,2)$ must be determined based on the boundary conditions. The boundary condition at the fixed support, i.e. $u_x(x=0)=0$, gives immediately from Eq. (3.14) the second constant of integration as $c_2=0$. To determine the first constant of integration $c_1$, the equilibrium at the point of load application, i.e. at $x=L$, must be considered, cf. Fig. 3.6.

Equilibrium between the internal force $N_x$ and the external force $F$ gives

$$-N_x(L)+F=0 \quad \text{or} \quad N_x(L)=F\,. \tag{3.15}$$

This result can be introduced into Eq. (3.8) to obtain

$$EA\frac{\mathrm{d}u_x(L)}{\mathrm{d}x}=F\,. \tag{3.16}$$

Comparison with Eq. (3.13) gives that the first constant of integration is equal to $c_1=F$. Thus, the elongation along the rod axis is given as:

$$u_x(x)=\frac{F}{EA}\times x\,. \tag{3.17}$$

**3.2 Example: Elongation of a rod due to dead weight**

A rod is vertically oriented as shown in Fig. 3.7 and only loaded by its dead weight $p_0$. The length is equal to $L$ and the constant axial tensile stiffness is $EA$. The standard

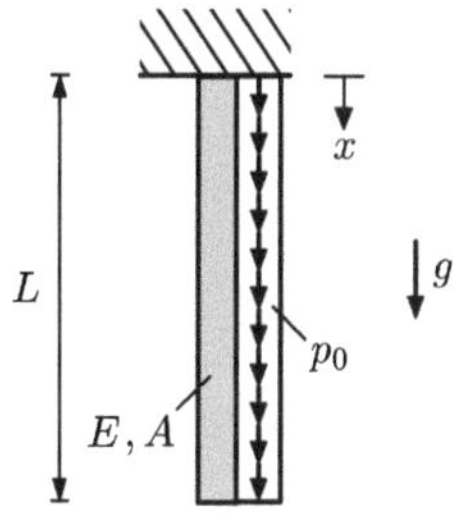


**Fig. 3.7** Analytical solution for the elongation of a rod due to the dead weight

gravity[2] is given as $g$ and the mass density as $\varrho$. Determine the elongation $u_x(x)$ along the rod axis and the value $u_x(L)$ at its free end.

**3.2 Solution** The describing partial differential equation can be taken from Table 3.2 as:

$$EA\frac{\mathrm{d}^2u_x(x)}{\mathrm{d}x^2} = -p_0\,. \tag{3.18}$$

Integrating twice gives

$$u_x(x) = \frac{1}{EA}\left(-\frac{1}{2}p_0x^2 + c_1x + c_2\right). \tag{3.19}$$

In a similar way as in the previous example, the constants of integration can be determined based on the boundary conditions $u_x(0) = 0$ and $N_x(L) = 0$ as $c_2 = 0$ and $c_1 = p_0L$. Thus, the elongation along the rod axis as a function of the distributed load $p_0$ is given as:

$$u_x(x) = \frac{p_0}{EA}\left(xL - \frac{x^2}{2}\right). \tag{3.20}$$

We may express the distributed load $p_0$ which has the units of force per unit length based on the standard gravity and mass density. The resulting force due to the dead weight is given as the product $mg$ where $m$ is the entire mass of the rod. Expressing the mass by the volume, i.e. $m = \varrho V = \varrho AL$, the distributed load can be expressed as $p_x = \varrho Ag$ and the displacement at the end of the rod is obtained as:

$$u_x(L) = \frac{p_0}{EA}\frac{L^2}{2} = \frac{1}{2}\frac{\varrho gL^2}{E}. \tag{3.21}$$

---

[2] The standard gravity or standard acceleration due to free fall is equal to 9.80665 $\frac{\mathrm{m}}{\mathrm{s}^2}$.

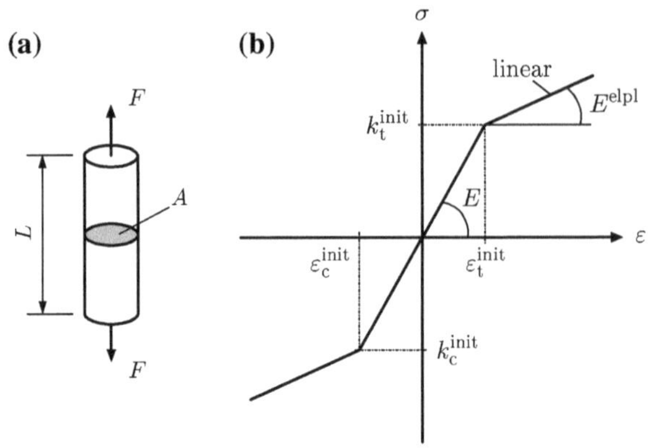


**Fig. 3.8** Tensile test of a rod: **a** idealized specimen and **b** uniaxial stress-strain diagram for linear hardening

Table 3.3 summarizes some standard solutions for cantilevered rods subjected to different types of simple loadings and Table 3.4 gives a similar compilation for fixed supported rods.

### 3.5.2 Elasto-Plastic Range

To derive an analytical solution for the behavior of rods in the elasto-plastic range, we consider again an idealized tensile specimen without contraction, see Fig. 3.8a. Furthermore, we assume a bilinear material behavior as shown in Fig. 3.8b, i.e. linear elasticity which is characterized by YOUNG's modulus $E$ and plasticity with isotropic linear harding which is characterized by the elasto-plastic modulus $E^{\mathrm{elpl}}$. A further simplification is that the behavior in tension and compression should be the same, i.e. $k_{\mathrm{c}}^{\mathrm{init}} = k_{\mathrm{t}}^{\mathrm{init}}$.

The stress-strain behavior in Fig. 3.8b can be described in the elasto-plastic range ($\varepsilon \geq \varepsilon_{\mathrm{t}}^{\mathrm{init}}$) by the following relationship

$$\sigma(\varepsilon) = \left(E - E^{\mathrm{elpl}}\right) \varepsilon_{\mathrm{t}}^{\mathrm{init}} + E^{\mathrm{elpl}} \varepsilon \tag{3.22}$$

$$= \left(1 - \frac{E^{\mathrm{elpl}}}{E}\right) k_{\mathrm{t}}^{\mathrm{init}} + E^{\mathrm{elpl}} \varepsilon \,. \tag{3.23}$$

The last equation describes also the homogeneous stress and strain state in the idealized specimen in the case that only edge loads are applied in the form of a given force or displacement. Equation (3.22) allows also to determine the elongation of a specimen as shown in Fig. 3.8a in the elasto-plastic range. Considering the definitions for stress and strain, we obtain

**Table 3.3** Elongation of cantilevered rods subjected to simple load cases along the $x$-axis. In addition, the maximum elongation is given

| Load case | Elongation |
|---|---|
| 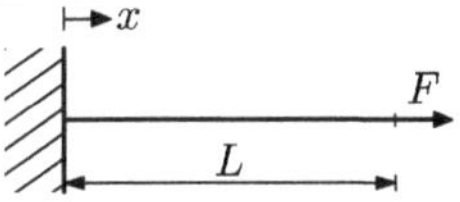 | $u_x(x) = \frac{Fx}{EA}$<br>$u_x(L) = \frac{FL}{EA}$ |
| 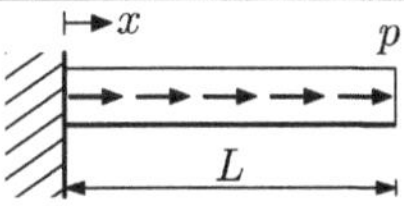 | $u_x(x) = \frac{p}{EA}\left(xL - \frac{x^2}{2}\right)$<br>$u_x(L) = \frac{pL^2}{2EA}$ |
| 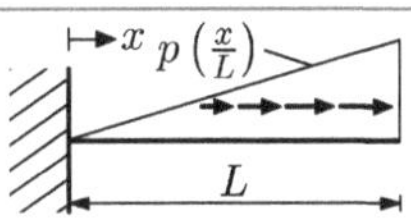 | $u_x(x) = \frac{p}{EA}\left(\frac{xL}{2} - \frac{x^3}{6L}\right)$<br>$u_x(L) = \frac{pL^2}{3EA}$ |
| 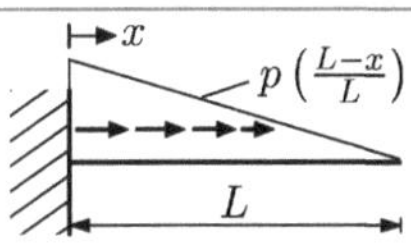 | $u_x(x) = \frac{p}{EA}\left(\frac{xL}{2} - \frac{x^2}{2} + \frac{x^3}{6L}\right)$<br>$u_x(L) = \frac{pL^2}{6EA}$ |
| 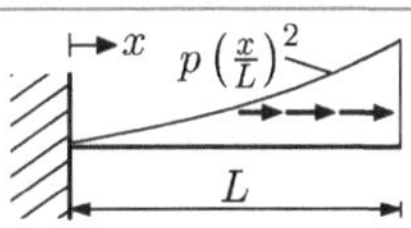 | $u_x(x) = \frac{p}{EA}\left(\frac{xL}{3} - \frac{x^4}{12L^2}\right)$<br>$u_x(L) = \frac{pL^2}{4EA}$ |
| 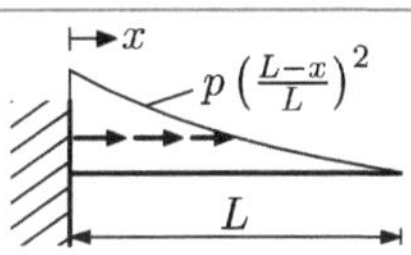 | $u_x(x) = \frac{p}{EA}\left(\frac{xL}{3} - \frac{x^2}{2} + \frac{x^3}{3L} - \frac{x^4}{12L^2}\right)$<br>$u_x(L) = \frac{pL^2}{12EA}$ |

**Table 3.4** Elongation of fixed supported rods subjected to simple load cases along the $x$-axis

| Load case | Elongation |
|---|---|
| 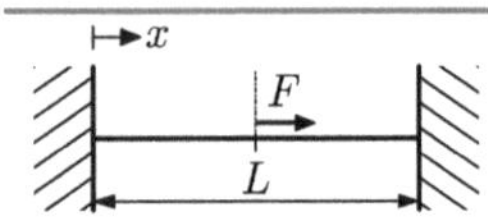  | $u_x(x) = \frac{Fx}{2EA}$, for $x \leq \frac{L}{2}$ <br> $u_x(x) = \frac{F}{EA}\left(\frac{L}{2} - \frac{x}{2}\right)$, for $x \geq \frac{L}{2}$ |
| 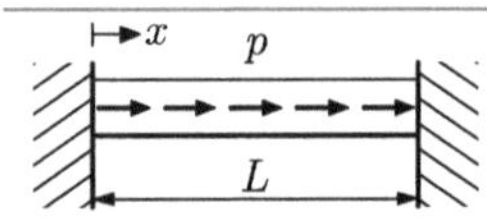  | $u_x(x) = \frac{p}{EA}\left(\frac{xL}{2} - \frac{x^2}{2}\right)$ |
| 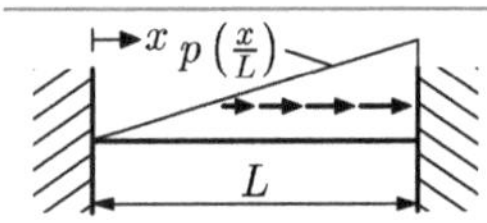  | $u_x(x) = \frac{p}{EA}\left(\frac{xL}{6} - \frac{x^3}{6L}\right)$ |
| 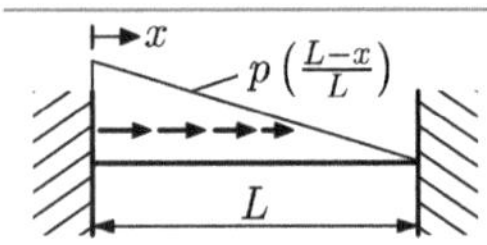  | $u_x(x) = \frac{p}{EA}\left(\frac{xL}{3} - \frac{x^2}{2} + \frac{x^3}{6L}\right)$ |
| 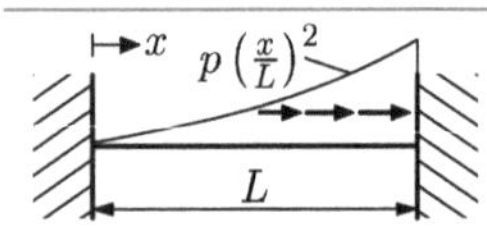  | $u_x(x) = \frac{p}{EA}\left(\frac{xL}{12} - \frac{x^4}{12L^2}\right)$ |
| 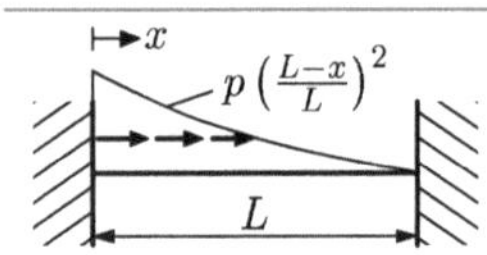  | $u_x(x) = \frac{p}{EA}\left(\frac{xL}{4} - \frac{x^2}{2} + \frac{x^3}{3L} - \frac{x^4}{12L^2}\right)$ |

$$\frac{F}{A} = \left(E - E^{\text{elpl}}\right)\varepsilon_{\text{t}}^{\text{init}} + E^{\text{elpl}}\frac{u(L)}{L}, \tag{3.24}$$

which gives the elongation $\Delta L$ (the expression $u(L)$ would refer to a specimen which is fixed at one side ($x = 0$) and the other side ($x = L$) is loaded by $F$) as:

$$\Delta L = \frac{FL}{AE^{\text{elpl}}} - \frac{L}{E^{\text{elpl}}}\left(E - E^{\text{elpl}}\right)\varepsilon_{\text{t}}^{\text{init}}. \tag{3.25}$$

If the intention is to determine the displacement field $u(x)$, the approach presented in Sects. 3.1–3.4 should be followed. Introducing the kinematics relation (3.2) into the stress-strain relation (3.23) gives:

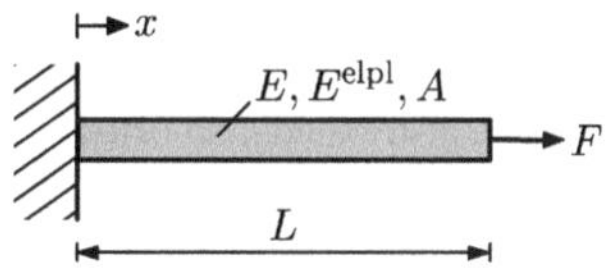


**Fig. 3.9** Analytical solution for the elongation of a rod due to a single force in the elasto-plastic range

$$\sigma(\varepsilon) = \left(1 - \frac{E^{\text{elpl}}}{E}\right) k_{\text{t}}^{\text{init}} + E^{\text{elpl}} \frac{\text{d}u_x}{\text{d}x} . \tag{3.26}$$

Considering in the last equation that a normal stress is defined as an acting force $N_x$ over a cross-sectional area $A$ and differentiating with respect to the $x$-coordinate gives:

$$\frac{\text{d}N_x}{\text{d}x} = \frac{\text{d}}{\text{d}x}\left[ A \left(1 - \frac{E^{\text{elpl}}}{E}\right) k_{\text{t}}^{\text{init}} + A E^{\text{elpl}} \frac{\text{d}u_x}{\text{d}x} \right] , \tag{3.27}$$

or under consideration of the equilibrium Eq. (3.6) and rearranging:

$$\frac{\text{d}}{\text{d}x}\left[ A \left(1 - \frac{E^{\text{elpl}}}{E}\right) k_{\text{t}}^{\text{init}} + A E^{\text{elpl}} \frac{\text{d}u_x}{\text{d}x} \right] = -p_x(x) . \tag{3.28}$$

The last equation is the governing partial differential equation which describes a rod in the elasto-plastic range ($\varepsilon \geq \varepsilon_{\text{t}}^{\text{init}}$). Equation (3.28) can be simplified to the pure elastic range by the transition $E^{\text{elpl}} \to E$ and Eq. (3.10) is obtained.

**3.3 Example: Elongation of a rod due to a single force in the elasto-plastic range** Given is a rod of length $L$ and constant cross-sectional area $A$ as shown in Fig. 3.9. At the left-hand side there is a fixed support and the right-hand side is loaded by a single force $F$. The stress-strain behavior is given by Eq. (3.23) and Fig. 3.8b and it is assumed that the material is in the elasto-plastic range, i.e. ($\varepsilon \geq \varepsilon_{\text{t}}^{\text{init}}$). Determine the elongation $u_x(x)$ along the rod axis.

**3.3 Solution** The describing partial differential equation is given by Eq. (3.6) as:

$$\frac{\text{d}}{\text{d}x}\left[ A \left(1 - \frac{E^{\text{elpl}}}{E}\right) k_{\text{t}}^{\text{init}} + A E^{\text{elpl}} \frac{\text{d}u_x}{\text{d}x} \right] = 0 . \tag{3.29}$$

Integrating twice gives

$$A \left(1 - \frac{E^{\text{elpl}}}{E}\right) k_{\text{t}}^{\text{init}} + A E^{\text{elpl}} \frac{\text{d}u_x}{\text{d}x} = c_1 = N_x(x) \tag{3.30}$$

and

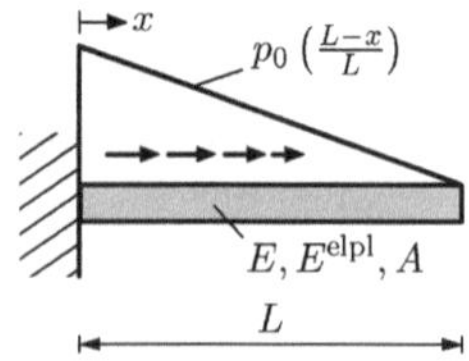


**Fig. 3.10** Analytical solution for the elongation of a rod due to a distributed load in the elasto-plastic range

$$AE^{\mathrm{elpl}}u_x(x) = \left(c_1 - A\left(1 - \frac{E^{\mathrm{elpl}}}{E}\right)k_{\mathrm{t}}^{\mathrm{init}}\right)x + c_2\,, \tag{3.31}$$

where the two constants of integration $c_i (i = 1, 2)$ must be determined based on the boundary conditions. The boundary condition at the fixed support, i.e. $u_x(x = 0) = 0$, gives immediately from Eq. (3.31) the second constant of integration as $c_2 = 0$. To determine the first constant of integration $c_1$, the equilibrium at the point of load application, i.e. at $x = L$, must be considered, cf. Fig. 3.6: $N_x(L) = F \Rightarrow c_1 = F$. Thus, the elongation along the rod axis is given as:

$$u_x(x) = \frac{x}{AE^{\mathrm{elpl}}}\left[F - A\left(1 - \frac{E^{\mathrm{elpl}}}{E}\right)k_{\mathrm{t}}^{\mathrm{init}}\right]. \tag{3.32}$$

This gives for $x = L$ the same result as in Eq. (3.25).

### 3.4 Example: Elongation of a rod due to a distributed load in the elasto-plastic range

Given is a rod of length $L$ and constant cross-sectional area $A$ as shown in Fig. 3.10. At the left-hand side there is a fixed support and the rod is loaded by a distributed load $p_x(x)$. The stress-strain behavior is given by Fig. 3.8b and it is assumed that a certain section $0 \leq x \leq \varsigma$ of the material is in the elasto-plastic range while the other part is still in the elastic range. Determine the elongation $u_x(x)$ along the rod axis.

**3.4 Solution** The distribution of the normal force $N_x(x)$ results from the distributed load as:

$$N_x(x) = -\int p_x(x)\,\mathrm{d}x = -\frac{p_0}{L}\left(Lx - \frac{x^2}{2} - \frac{L^2}{2}\right). \tag{3.33}$$

As soon as this force reaches the yield limit, i.e. $k_{\mathrm{t}}^{\mathrm{init}}A$, the material undergoes plastic deformation. Thus, the location of the interface $x = \varsigma$ which separates the elasto-plastic from the pure elastic part (see Fig. 3.11) is obtained from the following condition:

$$k_{\mathrm{t}}^{\mathrm{init}} = -\int p_x(x)\,\mathrm{d}x = -\frac{p_0}{AL}\left(L\varsigma - \frac{\varsigma^2}{2} - \frac{L^2}{2}\right), \tag{3.34}$$

or solved for the location of the interface:

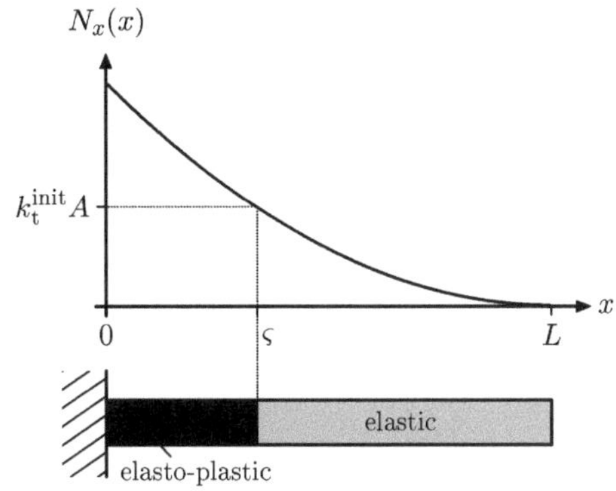


**Fig. 3.11** Normal force distribution and location of plastic and elastic zones

$$\varsigma = L - \sqrt{\frac{2k_t^{init}AL}{p_0}} \quad \text{or} \quad \frac{\varsigma}{L} = 1 - \sqrt{\frac{k_t^{init}A}{\frac{1}{2}p_0L}}. \tag{3.35}$$

Let us look first on the elasto-plastic region $0 \leq x \leq \varsigma$. The describing partial differential equation can be stated as

$$\frac{d}{dx}\left[A\left(1 - \frac{E^{elpl}}{E}\right)k_t^{init} + AE^{elpl}\frac{du_x}{dx}\right] = -\frac{p_0}{L}(L - x)\,. \tag{3.36}$$

Integrating twice gives

$$A\left(1 - \frac{E^{elpl}}{E}\right)k_t^{init} + AE^{elpl}\frac{du_x}{dx} = -\frac{p_0}{L}\left(Lx - \frac{x^2}{2}\right) + c_1 = N_x(x)\,, \tag{3.37}$$

and

$$AE^{elpl}u_x = -\frac{p_0}{L}\left(\frac{Lx^2}{2} - \frac{x^3}{6}\right) + \left[c_1 - A\left(1 - \frac{E^{elpl}}{E}\right)k_t^{init}\right]x + c_2. \tag{3.38}$$

Let us look now on the pure plastic region $\varsigma \leq x \leq L$. The describing partial differential equation can be stated as

$$EA\frac{d^2u_x}{dx^2} = -\frac{p_0}{L}(L - x). \tag{3.39}$$

Integrating twice gives

$$EA\frac{du_x}{dx} = -\frac{p_0}{L}\left(Lx - \frac{x^2}{2}\right) + c_3 = N_x(x), \tag{3.40}$$

and

$$EAu_x = -\frac{p_0}{L}\left(\frac{Lx^2}{2} - \frac{x^3}{6}\right) + c_3 x + c_4. \tag{3.41}$$

The constants of integration $c_2$ and $c_3$ can be obtained from the boundary conditions at $x = 0$, i.e. $u_x(0) = 0$, and at $x = L$, i.e. $N_x(L) = 0$, as:

$$c_2 = 0\,, \tag{3.42}$$

$$c_3 = \frac{1}{2}p_0 L. \tag{3.43}$$

The constants $c_1$ and $c_4$ can be obtained from the continuity conditions at $x = \varsigma$, i.e. $N_x(\varsigma^-) = N_x(\varsigma^+)$ and $u_x(\varsigma^-) = u_x(\varsigma^+)$, as:

$$c_1 = c_3 = \frac{1}{2}p_0 L\,, \tag{3.44}$$

$$c_4 = \left(\frac{E}{E^{\text{elpl}}} - 1\right)\left[-\frac{p_0}{L}\left(\frac{L\varsigma^2}{2} - \frac{\varsigma^3}{6}\right) + \frac{p_0 L}{2}\varsigma - Ak_{\text{t}}^{\text{init}}\varsigma\right]. \tag{3.45}$$

Alternatively, the constant $c_4$ may be written in the following forms:

$$c_4 = \left(\frac{E}{E^{\text{elpl}}} - 1\right)\left[p_0 L^2\left(\frac{1}{2}\left(\frac{\varsigma}{L}\right) - \frac{1}{2}\left(\frac{\varsigma}{L}\right)^2 + \frac{1}{6}\left(\frac{\varsigma}{L}\right)^3\right) - ALk_{\text{t}}^{\text{init}}\left(\frac{\varsigma}{L}\right)\right],$$

$$= \frac{p_0 L^2}{6}\left(\frac{E}{E^{\text{elpl}}} - 1\right)\left(1 - \sqrt{\frac{k_{\text{t}}^{\text{init}}A}{\frac{1}{2}p_0 L}}\right)\left(1 + \sqrt{\frac{k_{\text{t}}^{\text{init}}A}{\frac{1}{2}p_0 L}} - 2\frac{k_{\text{t}}^{\text{init}}A}{\frac{1}{2}p_0 L}\right). \tag{3.46}$$

Thus, the displacement in the entire range is given by the following two equations:

$$\frac{u_x(x)}{L} = \frac{p_0 L}{E^{\text{elpl}}A}\left[\frac{1}{2}\left(\frac{x}{L}\right) - \frac{1}{2}\left(\frac{x}{L}\right)^2 + \frac{1}{6}\left(\frac{x}{L}\right)^3 - \frac{k_{\text{t}}^{\text{init}}A}{p_0 L}\left(1 - \frac{E^{\text{elpl}}}{E}\right)\left(\frac{x}{L}\right)\right], \tag{3.47}$$

which is valid in the range $0 \leq x \leq \varsigma$, i.e. in the zone where the material is elasto-plastic, and

$$\frac{u_x(x)}{L} = \frac{p_0 L}{EA}\left[\frac{1}{2}\left(\frac{x}{L}\right) - \frac{1}{2}\left(\frac{x}{L}\right)^2 + \frac{1}{6}\left(\frac{x}{L}\right)^3 + \frac{1}{6}\left(\frac{E}{E^{\text{elpl}}} - 1\right)\right.$$
$$\left. \times\left(1 - \sqrt{\frac{k_{\text{t}}^{\text{init}}A}{\frac{1}{2}p_0 L}}\right)\left(1 + \sqrt{\frac{k_{\text{t}}^{\text{init}}A}{\frac{1}{2}p_0 L}} - 2\frac{k_{\text{t}}^{\text{init}}A}{\frac{1}{2}p_0 L}\right)\right], \tag{3.48}$$

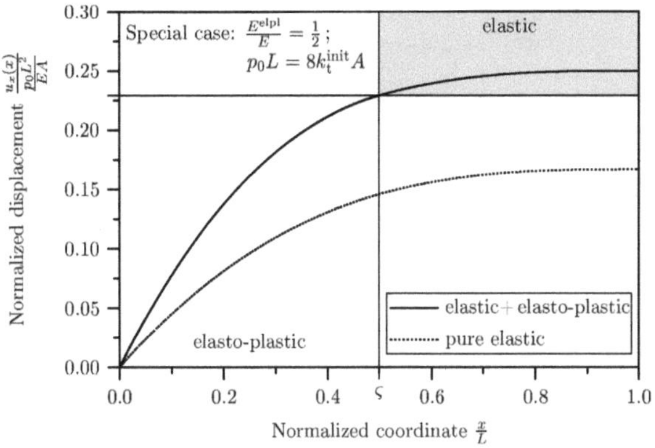


**Fig. 3.12** Normalized deformed shape of a rod due to distributed load in the elasto-plastic range

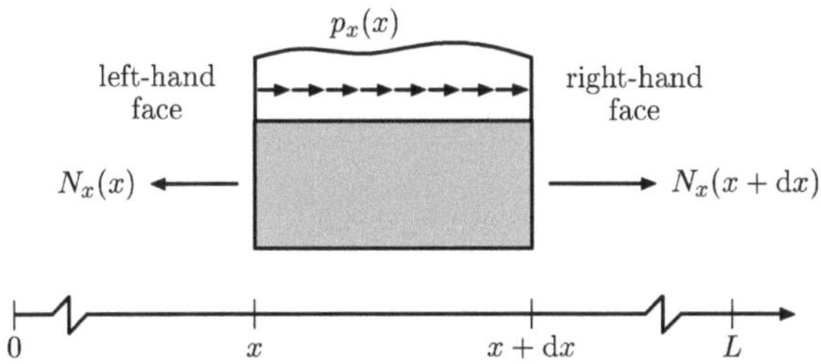


**Fig. 3.13** Differential element of a rod with internal reactions and external variable distributed load

which is valid in the elastic range, i.e. $\varsigma \leq x \leq L$. A graphical representation of the deformed rod shape based on Eqs. (3.47) and (3.48) and the comparison with the pure elastic solution (see Table 3.3) is given in Fig. 3.12 for the special case of $\frac{E^{\text{elpl}}}{E} = \frac{1}{2}$ and $p_0 L = 8k_t^{\text{init}} A$ ($\rightarrow \frac{\varsigma}{L} = \frac{1}{2}$). It can be seen from this figure that the consideration of plastic material behavior results in remarkably larger deformations.

## 3.6 Supplementary Problems

**3.5 Differential equation for a rod with variable distributed load**
Derive the governing differential equation based on a differential element as shown in Fig. 3.13 where the distributed load is a function of the $x$-coordinate. The cross-sectional area and the YOUNG's modulus is constant for this problem.

**3.6 Differential equation for a rod with variable cross-sectional area**
Derive the governing differential equation based on a differential element as shown in Fig. 3.14 where the cross-sectional area is changing with the $x$-coordinate. The

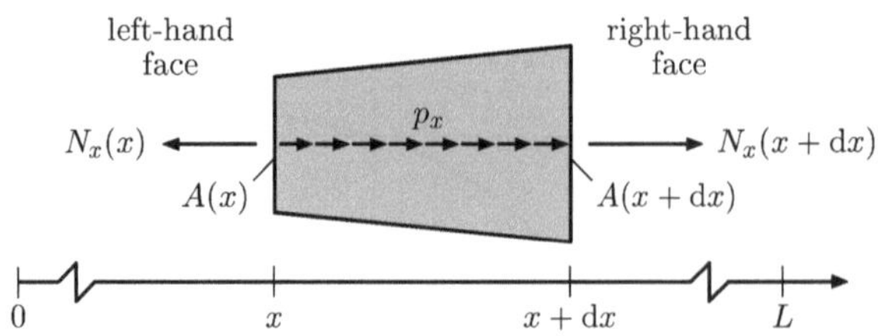


**Fig. 3.14** Differential element of a rod with internal reactions and variable cross-sectional area

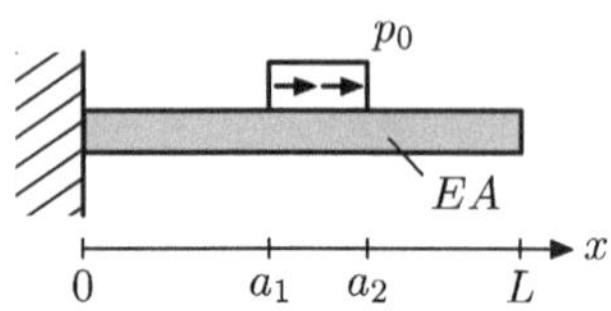


**Fig. 3.15** Elongation of a rod due to a distributed load in the segment $a_1 \leq x \leq a_2$

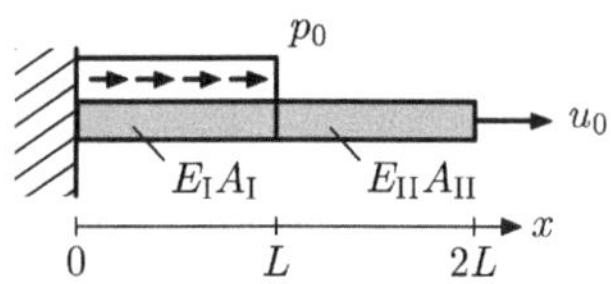


**Fig. 3.16** Elongation of a bi-material rod due to a distributed load $p_0$ and a displacement $u_0$

distributed load $p_x$ and the YOUNG's modulus can be assumed as constant in this problem.

**3.7 Elongation of a rod due to distributed load**
Calculate the end elongation $u(L)$ of the rod shown in Fig. 3.15. The rod is loaded in the section $a_1 \leq x \leq a_2$ by a constant distributed load $p_0$.

**3.8 Deformation of a bi-material rod**
Given is a rod as shown in Fig. 3.16 which is made of two different sections with axial stiffness $k_I = E_I A_I$ and $k_{II} = E_{II} A_{II}$. Each section is of length $L$ and in the left-hand section, i.e. $0 \leq x \leq L$, is a constant distributed load $p_0$ acting while the right-hand end is elongated by $u_0$.

(a) Calculate the general analytical expressions for the elongation $u(x)$, strain distribution $\varepsilon(x)$ and stress distribution $\sigma(x)$.
(b) Sketch the distributions $u(x)$, $\varepsilon(x)$ and $\sigma(x)$ for the case $k_I = 2k_{II} = 1$, $L_I = L_{II} = 1$, $p_0 = 1$, $u_0 = 1$ and $E_I = 2E_{II} = 1$.
(c) Redraw Fig. 3.16 for the special case $u_0 = 0$.

**3.9 Elongation of a rod made of a functionally graded material**
Calculate the elongation $u(L)$ of the rod shown in Fig. 3.17. The rod material is a functionally graded material (FGM) where the YOUNG's modulus is described as:

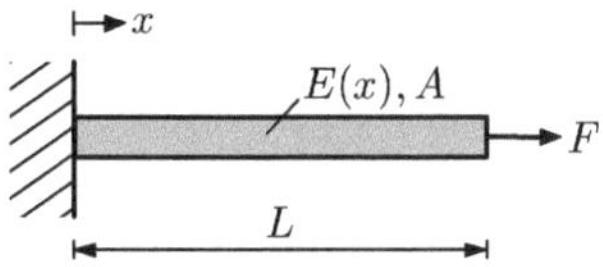


**Fig. 3.17** Elongation of a rod made of a functionally graded material

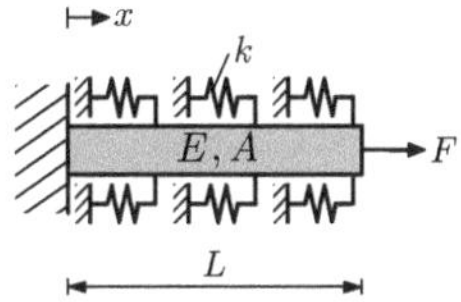


**Fig. 3.18** Rod with elastic embedding loaded by a single force

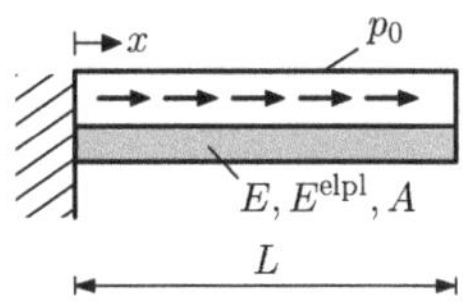


**Fig. 3.19** Analytical solution for the elongation of a rod due to constant distributed load in the elasto-plastic range

$$E(x) = E_0 \left( 1 - x \times \frac{1 - E_1/E_0}{L} \right) \tag{3.49}$$

$$= E_0 \left( 1 - x \times \alpha_{01} \right) . \tag{3.50}$$

**3.10 Elongation of a rod with elastic embedding**
Calculate the elongation $u(x)$ and $u(L)$ of the rod shown in Fig. 3.18. It can be assumed for this exercise that the embedding modulus $k$ and the axial tensile stiffness $EA$ are constant. Calculate in addition the limiting cases $u(x)$ for $\beta \to 0$ and $\beta \to \infty$, where $\beta = \sqrt{\frac{k}{EA}}$.

**3.11 Elongation of a rod due to constant distributed load in the elasto-plastic range**
Given is a rod of length $L$ and constant cross-sectional area $A$ as shown in Fig. 3.19. At the left-hand side there is a fixed support and the rod is loaded by a constant distributed load $p_0$. The stress-strain behavior is given by Fig. 3.8b and it is assumed that a certain section $0 \leq x \leq \varsigma$ the material is in the elasto-plastic range while the other part is still in the elastic range. Determine the elongation $u_x(x)$ along the rod axis.

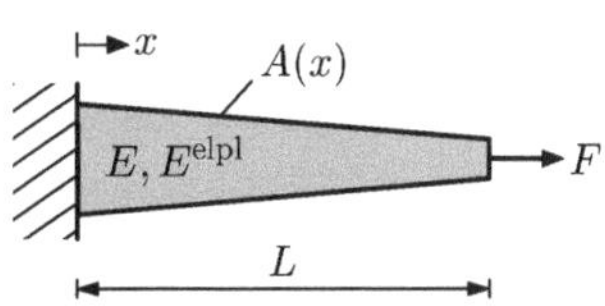


**Fig. 3.20** Analytical solution for the elongation of a rod with variable cross section in the elasto-plastic range

**3.12 Elongation of a rod with changing cross section in the elasto-plastic range**
Given is a rod of length $L$ and variable cross-sectional area $A(x)$ as shown in Fig. 3.20. At the left-hand side there is a fixed support and the rod is loaded by single force $F$. The stress-strain behavior is given by Fig. 3.8b and it is assumed that a certain section $0 \leq x \leq \varsigma$ the material is in the elasto-plastic range while the other part is still in the elastic range. Determine the elongation $u_x(x)$ along the rod axis for a linear function $A(x) = A_1 + \frac{x}{L}(A_2 - A_1)$ where $A_1 = A(x = 0)$ and $A_2 = A(x = L)$.

# Chapter 4
# Bending of Beams

**Abstract** This chapter covers the analytical description of beam members. Based on the three basic equations of continuum mechanics, i.e. the kinematics relationship, the constitutive law and the equilibrium equation, the partial differential equations which describe the problem are derived. Three different types of theories are treated, i.e. the EULER- BERNOULLI theory, the TIMOSHENKO theory and higher-order beam theories. Analytical solutions in the pure elastic range are presented for all three types of theories. The elasto-plastic range is covered for the EULER- BERNOULLI and a higher-order beam theory.

## 4.1 Introduction

A beam is defined as a long prismatic body as schematically shown in Fig. 4.1a. The following derivations are restricted to some simplifications:

- straight beams,
- no elongation along the $x$-axis,
- no torsion around the $x$-axis,
- deformations in a single plane, i.e. symmetrical bending,
- small deformations,
- simple cross sections.

The external loads, which are considered within this chapter, are single forces $F_y$, single moments $M_z$, distributed loads $q_y(x)$, and distributed moments $m_z(x)$. These loads have in common that their line of action (force) or the direction of the momentum vector are orthogonal to the center line of the beam and cause its bending. This is a different type of deformation compared to the rod element from Chap. 3, see Table 4.1. It should be noted here that these basic types of deformation can be superposed to account for more complex loading conditions [14].

 
A. Öchsner, *Elasto-Plasticity of Frame Structure Elements*,
DOI 10.1007/978-3-662-44225-8_4

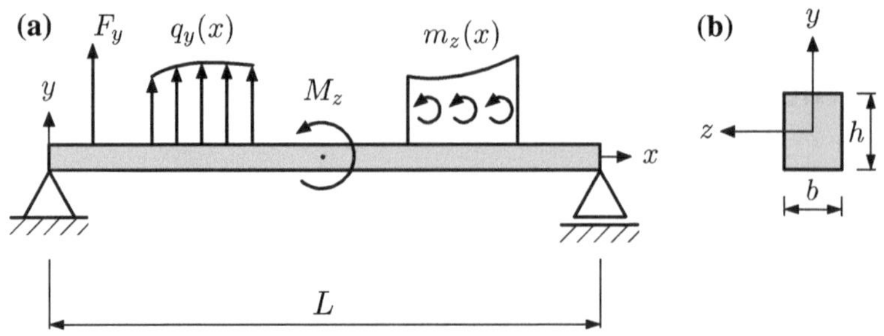


**Fig. 4.1** General configuration for beam problems: **a** example of boundary conditions and external loads, **b** cross-sectional area

**Table 4.1** Differentiation between rod and beam element; center line parallel to the $x$-axis

| | Rod | Beam |
|---|---|---|
| Force | Along the rod axis | Perpendicular to the beam axis |
| Unknown | Displacement in or<br>Rotation around rod axis | Displacement perpendicular to and<br>Rotation perpendicular to the beam axis |

The classic theories of beam bending distinguish between shear-rigid and shear-flexible models. The shear rigid-beam, also called the BERNOULLI[1] beam,[2] neglects the shear deformation from the shear forces. This theory implies that a cross-sectional plane which was before the deformation perpendicular to the beam axis remains in the deformed state perpendicular to the beam axis, see Fig. 4.2a. Furthermore, it is assumed that a cross-sectional plane stays plane and unwrapped in the deformed state. These two assumptions are also known as BERNOULLI's hypothesis. Altogether one imagines the cross-sectional planes are rigidly fixed to the center line of the beam[3] so that a change of the center line affects the entire deformation. Consequently, it is also assumed that the geometric dimensions[4] of the cross-sectional planes do not change.

In the case of a shear-flexible beam, also called the TIMOSHENKO[5] beam, the shear deformation is considered in addition to the bending deformation and cross-sectional planes are rotated by an angle $\gamma$ compared to the perpendicular line, see Fig. 4.2b. For beams for which the length is 10–20 times larger than a characteristic dimension of the cross section, the shear fraction is usually disregarded in the first approximation.

[1] Jakob I. BERNOULLI (1655–1705), Swiss mathematician and physicist.

[2] More precisely, this beam is known as the EULER-BERNOULLI beam. A historical analysis of the development of the classical beam theory and the contribution of different scientists can be found in [51].

[3] More precisely, this is the neutral fibre or the bending line.

[4] Consequently, the width $b$ and the height $h$ of a, for example, rectangular cross-section remain the same, see Fig. 4.1b.

[5] Stepan Prokopovych TYMOSHENKO (1878–1972), Ukrainian/US engineer.

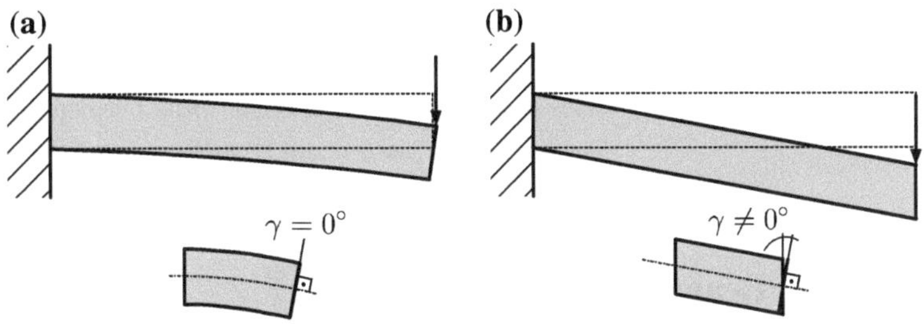


**Fig. 4.2** Different deformation modes of a bending beam: **a** shear-rigid, **b** shear-flexible. Adapted from [49]

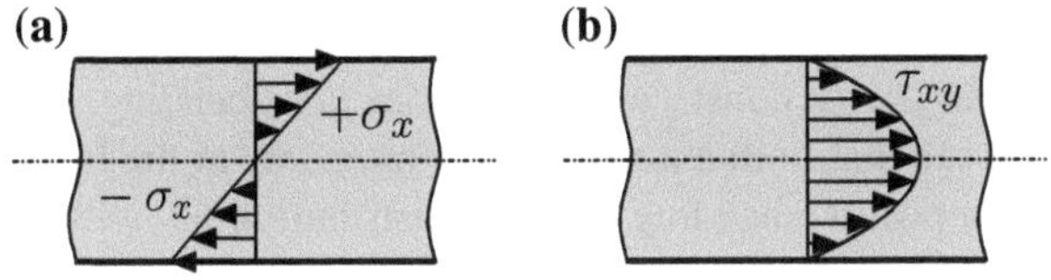


**Fig. 4.3** Different stress distributions of a beam with rectangular cross-section and linear-elastic material behavior: **a** normal stress (shear-rigid and flexible) and **b** shear stress (shear-flexible)

The different load types, meaning pure bending moment loading or shear due to shear force, lead to different stress fractions in a beam. In the case of a BERNOULLI beam, loading occurs solely through normal forces, which are linearly distributed over the cross section. Consequently, a tension—alternatively a compression maximum on the bottom face—maximum on the top face occurs, see Fig. 4.3a. In the case of symmetric cross sections, the zero crossing[6] occurs in the middle of the cross-section. The shear stress distribution for a rectangular cross-section is parabolic at which the maximum occurs at the neutral axis and is zero at both the top and bottom surface, see Fig. 4.3b.

Finally, it needs to be noted that the one-dimensional beam theories have corresponding counterparts in the two-dimensional space, see Table 4.2. In plate theories, the BERNOULLI beam corresponds to the shear-rigid KIRCHHOFF[7] plate and the TIMOSHENKO beam corresponds to the shear-flexible REISSNER[8]-MINDLIN[9] plate, [4, 45, 105].

Further details regarding the beam theory and the corresponding basic definitions and assumptions can be found in references [18, 46, 52, 101]. In the following Sect. 4.2, only the BERNOULLI beam is considered. Consideration of the shear part takes place in Sects. 4.3 and 4.4.

---

[6] The sum of all points with $\sigma = 0$ along the beam axis is called the neutral fiber.

[7] Gustav Robert KIRCHHOFF (1824–1887), German physicist.

[8] Eric REISSNER (1913–1996), German/US engineer.

[9] Raymond David MINDLIN (1906–1987), US engineer.

Table 4.2 Analogies between the beam and plate theories

| | Beam theory | Plate theory |
|---|---|---|
| Dimensionality | 1D | 2D |
| Shear-rigid | BERNOULLI beam | KIRCHHOFF plate |
| Shear-flexible | TIMOSHENKO beam | REISSNER-MINDLIN plate |

## 4.2 Bernoulli Beam Theory

### *4.2.1 Kinematics*

For the derivation of the kinematics relation, a beam with length $L$ under constant moment loading $M_z(x) = \text{const.}$, meaning under *pure* bending, is considered, see Fig. 4.4. One can see that both external single moments at the left- and right-hand boundary lead to a positive bending moment distribution $M_z$ within the beam. The vertical position of a point with respect to the center line of the beam *without action* of an external load is described through the $y$-coordinate. The vertical *displacement* of a point on the center line of the beam, meaning for a point with $y = 0$, under action of the external load is indicated with $u_y$. The deformed center line is represented by the sum of these points with $y = 0$ and is referred to as the bending line $u_y(x)$.

Only the center line of the deformed beam is considered in the following. Through the relation for an arbitrary point $(x, u_y)$ on a circle with radius $R$ around the center point $(x_0, y_0)$, meaning

$$(x - x_0)^2 + (u_y(x) - y_0)^2 = R^2, \tag{4.1}$$

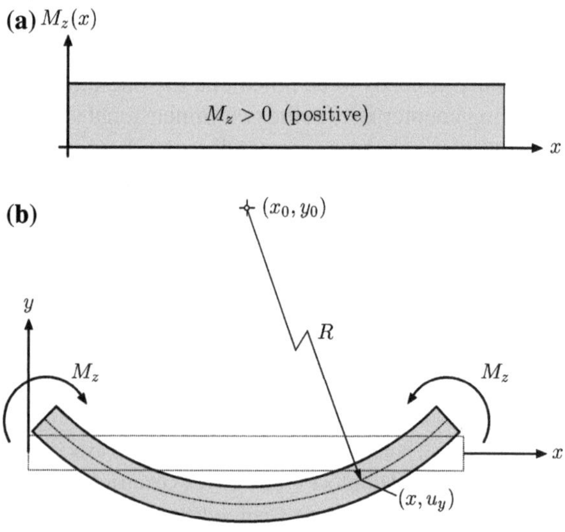


Fig. 4.4 Beam under pure bending in the $x$-$y$ plane: **a** moment distribution, **b** deformed beam. Note that the deformation is overdrawn for better illustration. For the deformations considered in this chapter the following applies: $R \gg L$

through differentiation with respect to the $x$-coordinate, one obtains

$$2(x - x_0) + 2(u_y(x) - y_0)\frac{\mathrm{d}u_y}{\mathrm{d}x} = 0, \tag{4.2}$$

alternatively after another differentiation:

$$2 + 2\frac{\mathrm{d}u_y}{\mathrm{d}x}\frac{\mathrm{d}u_y}{\mathrm{d}x} + 2(u_y(x) - y_0)\frac{\mathrm{d}^2u_y}{\mathrm{d}x^2} = 0. \tag{4.3}$$

Equation (4.3) provides the vertical distance between an arbitrary point on the center line of the beam and the center point of a circle as

$$(u_y - y_0) = -\frac{1 + \left(\frac{\mathrm{d}u_y}{\mathrm{d}x}\right)^2}{\frac{\mathrm{d}^2u_y}{\mathrm{d}x^2}}, \tag{4.4}$$

while the difference between the $x$-coordinates results from Eq. (4.2):

$$(x - x_0) = -(u_y - y_0)\frac{\mathrm{d}u_y}{\mathrm{d}x}. \tag{4.5}$$

If the expression according to Eq. (4.4) is used in Eq. (4.5) the following results:

$$(x - x_0) = \frac{\mathrm{d}u_y}{\mathrm{d}x}\frac{1 + \left(\frac{\mathrm{d}u_y}{\mathrm{d}x}\right)^2}{\frac{\mathrm{d}^2u_y}{\mathrm{d}x^2}}. \tag{4.6}$$

Inserting both expressions for the $x$- and $y$-coordinate differences according to Eqs. (4.4) and (4.6) in the circle equation according to (4.1) leads to:

$$\begin{aligned} R^2 &= (x - x_0)^2 + (u_y - y_0)^2 \\ &= \left(\frac{\mathrm{d}u_y}{\mathrm{d}x}\right)^2 \frac{\left(1 + \left(\frac{\mathrm{d}u_y}{\mathrm{d}x}\right)^2\right)^2}{\left(\frac{\mathrm{d}^2u_y}{\mathrm{d}x^2}\right)^2} + \frac{\left(1 + \left(\frac{\mathrm{d}u_y}{\mathrm{d}x}\right)^2\right)^2}{\left(\frac{\mathrm{d}^2u_y}{\mathrm{d}x^2}\right)^2} \end{aligned} \tag{4.7}$$

$$= \left( \left( \frac{\mathrm{d}^2 u_y}{\mathrm{d}x^2} \right)^2 + 1 \right) \frac{\left( 1 + \left( \frac{\mathrm{d}u_y}{\mathrm{d}x} \right)^2 \right)^2}{\left( \frac{\mathrm{d}^2 u_y}{\mathrm{d}x^2} \right)^2}$$

$$= \frac{\left( 1 + \left( \frac{\mathrm{d}u_y}{\mathrm{d}x} \right)^2 \right)^3}{\left( \frac{\mathrm{d}^2 u_y}{\mathrm{d}x^2} \right)^2}. \tag{4.8}$$

Since the center line shown in Fig. 4.4 is a concave-upward curve[10] $\left( \frac{\mathrm{d}^2 u_y}{\mathrm{d}x^2} > 0 \right)$, the radius of curvature $R$ results in:

$$R = \overset{+}{(-)} \frac{\left( 1 + \left( \frac{\mathrm{d}u_y}{\mathrm{d}x} \right)^2 \right)^{3/2}}{\left( \frac{\mathrm{d}^2 u_y}{\mathrm{d}x^2} \right)}. \tag{4.9}$$

Note that the expression curvature, which results as a reciprocal value from the curvature radius, $\kappa = \frac{1}{R}$, is used as well.

For small bending deflections, meaning $u_y \ll L$, $\frac{\mathrm{d}u_y}{\mathrm{d}x} \ll 1$ results and Eq. (4.9) simplifies to:

$$R = \frac{1}{\frac{\mathrm{d}^2 u_y}{\mathrm{d}x^2}} \quad \text{or} \quad \kappa = \frac{1}{R} = \frac{\mathrm{d}^2 u_y}{\mathrm{d}x^2}. \tag{4.10}$$

For the determination of the strain, one refers to its general definition, meaning elongation referred to initial length. Referring to the configuration shown in Fig. 4.5, the longitudinal elongation of a fibre at distance $y$ to the neutral fibre allows to express the strain as:

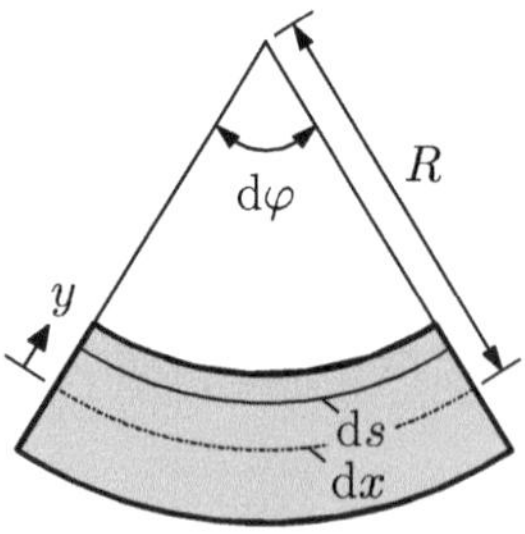


**Fig. 4.5** Beam element under pure bending in the $x$-$y$ plane. Note that the deformation is overdrawn for better illustration

[10] Sometimes referred to as 'convex' or a 'left-handed curve'.

$$\varepsilon_x = \frac{\mathrm{d}s - \mathrm{d}x}{\mathrm{d}x}. \tag{4.11}$$

The lengths of the circular arcs d$s$ and d$x$ result from the corresponding radii and the enclosed angles in radian measure as:

$$\mathrm{d}x = R\mathrm{d}\varphi_z\,, \tag{4.12}$$

$$\mathrm{d}s = (R - y)\mathrm{d}\varphi_z. \tag{4.13}$$

If these relations for the circular arcs are used in Eq. (4.11), the following results:

$$\varepsilon_x = \frac{(R - y)\mathrm{d}\varphi_z - R\mathrm{d}\varphi_z}{\mathrm{d}x} = -y\frac{\mathrm{d}\varphi_z}{\mathrm{d}x}. \tag{4.14}$$

From Eq. (4.12) $\frac{\mathrm{d}\varphi_z}{\mathrm{d}x} = \frac{1}{R}$ results and together with relation (4.10) the strain can finally be expressed as follows:

$$\varepsilon_x(x, y) = -y\frac{\mathrm{d}^2 u_y(x)}{\mathrm{d}x^2} \overset{(4.10)}{=} -y\kappa. \tag{4.15}$$

The definition of the positive and negative rotation angle, cf. Eq. (4.14), is illustrated in Fig. 4.6.

An alternative derivation of the kinematics relation results from consideration of Fig. 4.7. From the relation of the right-angled triangle $0'1'2'$, this means[11] $\sin\varphi_z = \frac{-u_x}{y}$, the following relation results for small angles ($\sin\varphi_z \approx \varphi_z$):

$$u_x = -y\varphi_z. \tag{4.16}$$

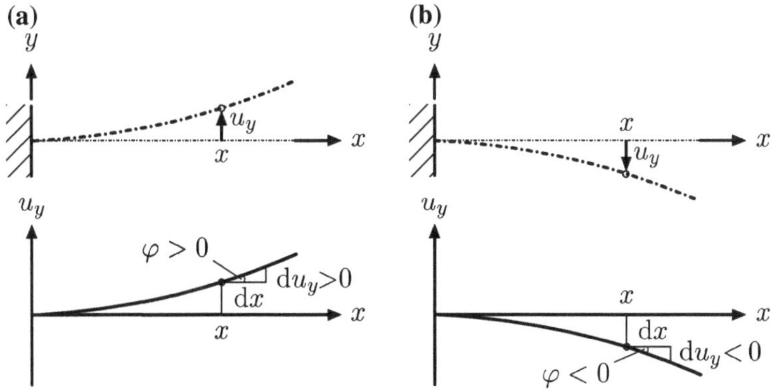


**Fig. 4.6** Definition of the rotation angle: **a** $\varphi_z = \frac{\mathrm{d}u_y}{\mathrm{d}x}$ positive, **b** $\varphi_z = \frac{\mathrm{d}u_y}{\mathrm{d}x}$ negative

[11] Note that according to the assumptions of the BERNOULLI beam the lengths 01 and $0'1'$ remain unchanged.

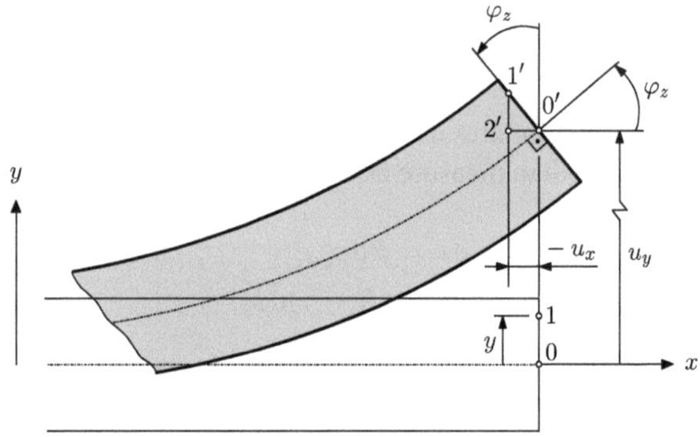


**Fig. 4.7** Alternative configuration for the derivation of the kinematics relation. Note that the deformation is overdrawn for better illustration

Furthermore, it holds that the rotation angle of the slope equals the center line for small angles:

$$\tan \varphi_z = \frac{\mathrm{d}u_y(x)}{\mathrm{d}x} \approx \varphi_z. \tag{4.17}$$

If Eqs. (4.16) and (4.17) are combined, the following results:

$$u_x = -y\frac{\mathrm{d}u_y(x)}{\mathrm{d}x}. \tag{4.18}$$

The last relation equals $(\mathrm{d}s - \mathrm{d}x)$ in Eq. (4.11) and differentiation with respect to the $x$-coordinate leads directly to Eq. (4.15).

### *4.2.2 Equilibrium*

The equilibrium conditions are derived from an infinitesimal beam element of length $\mathrm{d}x$, which is loaded by a constant distributed load $q_y$, see Fig. 4.8. The internal reactions are drawn on both cut faces, i.e. at location $x$ and $x + \mathrm{d}x$. One can see that a positive shear force is oriented in the positive $y$-direction at the right-hand face[12] and that a positive bending moment has the same rotational direction as the positive $z$-axis (right-hand grip rule[13]). The orientation of shear force and bending moment is reversed at the left-hand face in order to cancel in sum the effect of the internal

[12] A positive cut face is defined by the surface normal on the cut plane which has the same orientation as the positive $x$-axis. It should be regarded that the surface normal is always directed outward.

[13] If the axis is grasped with the right hand in a way so that the spread out thumb points in the direction of the positive axis, the bent fingers then show the direction of the positive rotational direction.

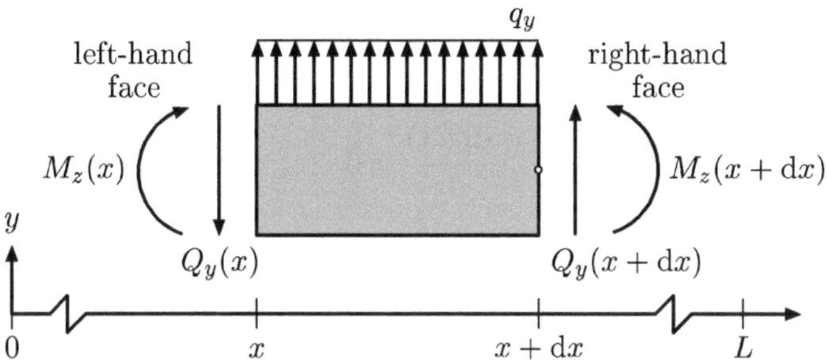


**Fig. 4.8** Infinitesimal beam element in the $x$-$y$ plane with internal reactions and constant distributed load

reactions at both faces. This convention for the direction of the internal directions is maintained in the following. Furthermore, it can be derived from Fig. 4.8 that an upwards directed *external* force or alternatively a mathematically positive oriented *external* moment at the right-hand face leads to a positive shear force or alternatively a positive internal moment. In a corresponding way, it results that a downwards directed *external* force or alternatively a mathematically negative oriented *external* moment at the left-hand face leads to a positive shear force or alternatively a positive internal moment.

The equilibrium condition will be determined in the following for the vertical forces. Assuming that forces in the direction of the positive $y$-axis are considered positive, the following results:

$$-Q(x) + Q(x + \mathrm{d}x) + q_y \mathrm{d}x = 0. \tag{4.19}$$

If the shear force on the right-hand face is expanded in a TAYLOR's series of first order, meaning

$$Q(x + \mathrm{d}x) \approx Q(x) + \frac{\mathrm{d}Q(x)}{\mathrm{d}x}\mathrm{d}x, \tag{4.20}$$

Equation (4.19) results in

$$-Q(x) + Q(x) + \frac{\mathrm{d}Q(x)}{\mathrm{d}x}\mathrm{d}x + q_y \mathrm{d}x = 0, \tag{4.21}$$

or alternatively after simplification finally to:

$$\frac{\mathrm{d}Q(x)}{\mathrm{d}x} = -q_y. \tag{4.22}$$

For the special case that no distributed load is acting ($q_y = 0$), Eq. (4.22) simplifies to:

$$\frac{dQ(x)}{dx} = 0. \tag{4.23}$$

The equilibrium of moments around the reference point at $x + dx$ gives:

$$M_z(x + dx) - M_z(x) + Q_y(x)dx - \frac{1}{2}q_y dx^2 = 0. \tag{4.24}$$

If the bending moment on the right-hand face is expanded into a TAYLOR's series of first order similar to Eq. (4.20) and consideration that the term $\frac{1}{2}\, q_y dx^2$ as infinitesimal small size of higher order can be disregarded, finally the following results:

$$\frac{dM_z(x)}{dx} = -Q_y(x). \tag{4.25}$$

The combination of Eqs. (4.22) and (4.25) leads to the relation between bending moment and distributed load:

$$\frac{d^2 M_z(x)}{dx^2} = -\frac{dQ_y(x)}{dx} = q_y. \tag{4.26}$$

### *4.2.3 Constitutive Equation*

The one-dimensional HOOKE's law according to Eq. (3.3) can also be assumed in the case of the bending beam, since according to the requirement that only normal stresses are regarded in this section:

$$\sigma_x = E\varepsilon_x. \tag{4.27}$$

Through the kinematics relation according to Eq. (4.15), the stress results as a function of deflection to:

$$\sigma_x(x, y) = -Ey\frac{d^2 u_y(x)}{dx^2}. \tag{4.28}$$

The stress distribution shown in Fig. 4.9a generates the internal moment, which acts in this cross section. To calculate this internal moment, the stress is multiplied by a surface, so that the resulting force is obtained. Multiplication with the corresponding lever arm then gives the internal moment. Since the stress is linearly distributed over the height, the evaluation is done for an infinitesimal small surface element:

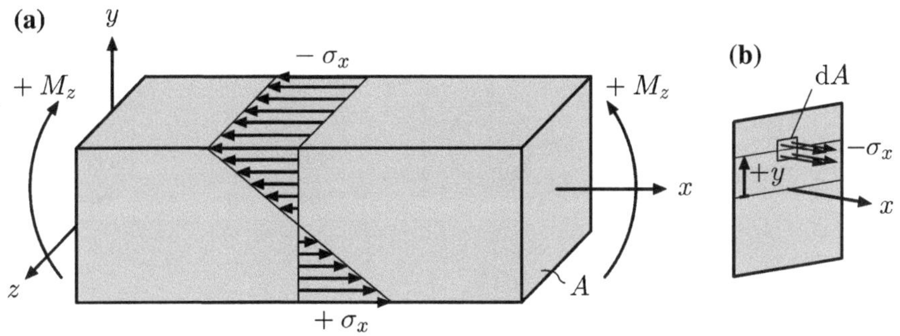


**Fig. 4.9** **a** Schematic representation of the normal stress distribution $\sigma_x = \sigma_x(y)$ of a bending beam, **b** Definition and position of an infinitesimal surface element for the derivation of the resulting moment action of the normal stress distribution

$$\mathrm{d}M_z = (+y)(-\sigma_x)\mathrm{d}A = -y\sigma_x\mathrm{d}A. \tag{4.29}$$

Therefore, the entire moment results via integration over the entire surface in:

$$M_z = -\int_A y\sigma_x \mathrm{d}A \overset{(4.28)}{=} +\int_A yEy\frac{\mathrm{d}^2u_y}{\mathrm{d}x^2}\mathrm{d}A. \tag{4.30}$$

Assuming that the YOUNG's modulus is constant, the internal moment around the $z$-axis results in:

$$M_z = E\frac{\mathrm{d}^2u_y}{\mathrm{d}x^2}\underbrace{\int_A y^2\mathrm{d}A}_{I_z}. \tag{4.31}$$

The integral in Eq. (4.31) is the so-called axial second moment of area or axial surface moment of 2nd order in the SI unit $\mathrm{m}^4$. This factor is only dependent on the geometry of the cross section and is also a measure of the stiffness of a plane cross section against bending. The values of the axial second moment of area for simple geometric cross sections are collected in Table 4.3.

Consequently the internal moment can also be expressed as

$$M_z = EI_z\frac{\mathrm{d}^2u_y}{\mathrm{d}x^2} \overset{(4.10)}{=} \frac{EI_z}{R} = EI_z\kappa. \tag{4.32}$$

Equation (4.32) describes the bending line $u_y(x)$ as a function of the bending moment and is therefore also referred to as the bending line-moment relation. The product $EI_z$ in Eq. (4.32) is also called the bending stiffness. If the result from

**Table 4.3** Axial second moment of area around the $z$- and $y$-axis

| Cross section | $I_z$ | $I_y$ |
|---|---|---|
| 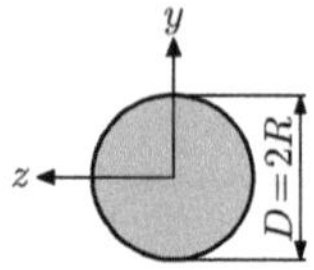 | $\frac{\pi D^4}{64} = \frac{\pi R^4}{4}$ | $\frac{\pi D^4}{64} = \frac{\pi R^4}{4}$ |
| 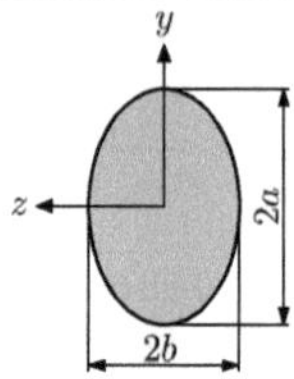 | $\frac{\pi b a^3}{4}$ | $\frac{\pi a b^3}{4}$ |
| 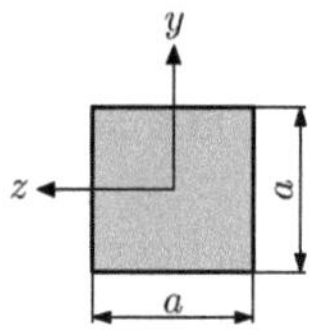 | $\frac{a^4}{12}$ | $\frac{a^4}{12}$ |
| 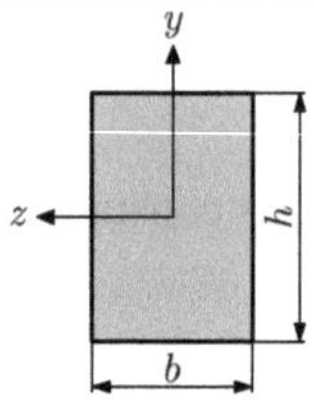 | $\frac{bh^3}{12}$ | $\frac{hb^3}{12}$ |
| 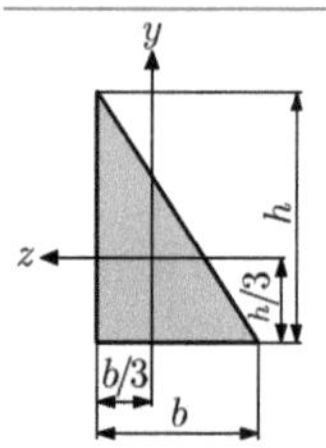 | $\frac{bh^3}{36}$ | $\frac{hb^3}{36}$ |
| 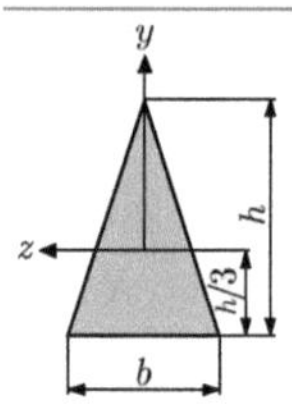 | $\frac{bh^3}{36}$ | $\frac{bh^3}{48}$ |

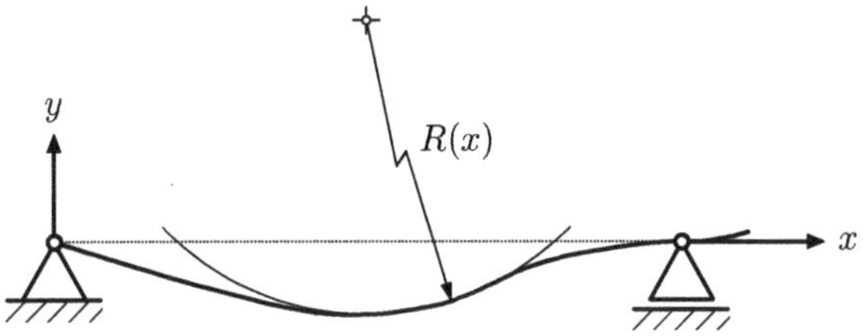


**Fig. 4.10** Deformation of a beam in the $x$-$y$ plane with $M_z(x) \neq$ const

Eq. (4.32) is used in the relation for the bending stress according to Eq. (4.28), the distribution of stress across the cross section results in:

$$\sigma_x(y) = -\frac{M_z}{I_z} y. \tag{4.33}$$

The minus sign in Eq. (4.33) causes that a positive bending moment (see Fig. 4.4) leads to a compressive stress in the upper beam half (meaning for $y > 0$). The corresponding equations for a deformation in the $x$-$z$ plane are collected in Appendix B.3.

In the case of plane bending with $M_z(x) \neq$ const., the bending line can be approximated in each case locally through a circle of curvature, see Fig. 4.10. Therefore, the result for *pure* bending according to Eq. (4.32) can be transferred to the case of plane bending as:

$$EI_z \frac{\mathrm{d}^2 u_y(x)}{\mathrm{d}x^2} = M_z(x). \tag{4.34}$$

Finally, the elementary basic equations for the bending of a beam in the $x$-$y$ plane for arbitrary moment loading $M_z(x)$ are summarized in Table 4.4.

### *4.2.4 Governing Differential Equation*

Two-time differentiation of Eq. (4.32) and consideration of the relation between bending moment and distributed load according to Eq. (4.25) leads to the classical type of differential equation of the bending line,

$$\frac{\mathrm{d}^2}{\mathrm{d}x^2}\left(EI_z \frac{\mathrm{d}^2 u_y}{\mathrm{d}x^2}\right) = q_y, \tag{4.35}$$

which is also referred to as the bending line-distributed load relation. For a beam with constant bending stiffness $EI_z$ along the beam axis, the following results:

**Table 4.4** Elementary basic equations for the bending of a BERNOULLI beam in the $x$-$y$ plane. The differential equations are given under the assumption of constant bending stiffness $EI_z$

| Relation | Equation |
|---|---|
| Kinematics | $\varepsilon_x(x, y) = -y \dfrac{\mathrm{d}^2 u_y(x)}{\mathrm{d}x^2}$ |
| Equilibrium | $\dfrac{\mathrm{d}Q_y(x)}{\mathrm{d}x} = -q_y(x)\,; \quad \dfrac{\mathrm{d}M_z(x)}{\mathrm{d}x} = -Q_y(x)$ |
| Constitution | $\sigma_x(x, y) = E\varepsilon_x(x, y)$ |
| Stress | $\sigma_x(x, y) = -\dfrac{M_z(x)}{I_z} y(x)$ |
| Differential equation | $EI_z \dfrac{\mathrm{d}^2 u_y(x)}{\mathrm{d}x^2} = M_z(x)$ |
| | $EI_z \dfrac{\mathrm{d}^3 u_y(x)}{\mathrm{d}x^3} = -Q_y(x)$ |
| | $EI_z \dfrac{\mathrm{d}^4 u_y(x)}{\mathrm{d}x^4} = q_y(x)$ |

$$EI_z \frac{\mathrm{d}^4 u_y}{\mathrm{d}x^4} = q_y. \tag{4.36}$$

The differential equation of the bending line can of course also be applied through the bending moment or the shear force as

$$EI_z \frac{\mathrm{d}^2 u_y}{\mathrm{d}x^2} = M_z \quad \text{or} \tag{4.37}$$

$$EI_z \frac{\mathrm{d}^3 u_y}{\mathrm{d}x^3} = -Q_y\,. \tag{4.38}$$

Different formulations of the fourth order differential equation are collected in Table 4.5 where different types of loading, geometry and bedding are differentiated. The last case in Table 4.5 refers to the elastic foundation of a beam which is also know in the literature as WINKLER[14] foundation [116]. The elastic foundation or WINKLER foundation modulus $k$ has in the case of beams[15] the units force per unit area.

---

[14] Emil WINKLER (1835–1888), German engineer.

[15] In the general case, the unit of the elastic foundation modulus is force per unit area per unit length, i.e. $\frac{\mathrm{N}}{\mathrm{m}^2}/\mathrm{m} = \frac{\mathrm{N}}{\mathrm{m}^3}$.

**Table 4.5** Different formulations of the partial differential equation for a BERNOULLI beam in the $x$-$y$ plane ($x$-axis: right facing; $y$-axis: upward facing)

| Configuration | Partial differential equation |
|---|---|
| 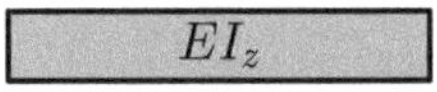 | $EI_z \frac{d^4 u_y}{dx^4} = 0$ |
| 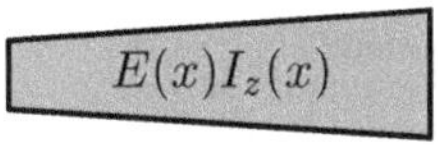 | $\frac{d^2}{dx^2}\left(E(x)I_z(x)\frac{d^2 u}{dx^2}\right) = 0$ |
| 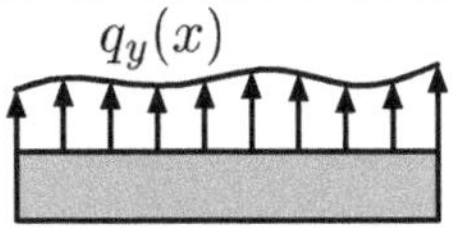 | $EI_z \frac{d^4 u_y}{dx^4} = q_y(x)$ |
| 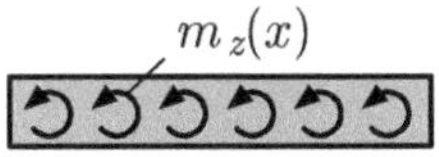 | $EI_z \frac{d^4 u_y}{dx^4} = -\frac{dm_z(x)}{dx}$ |
| 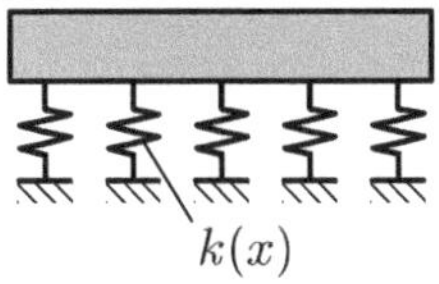 | $EI_z \frac{d^4 u_y}{dx^4} = -k(x)u_y$ |

### *4.2.5 Analytical Solutions in the Elastic Range*

In the following, the analytical calculation of the bending line for simple statically determinate cases will be considered. The differential equation of the bending line has to be integrated analytically according to Eqs. (4.36), (4.37) or (4.38). The constants of integration occurring in this integration can be determined based on the boundary conditions, see Table 4.6.

If the distributed load (or the moment or shear force distribution) cannot be represented in closed form for the entire beam because supports, pin-joints, effects of jumps or kinks in the load function occur, the integration has to be done in sections. Then, the additional constants of integration have to be defined through the transition conditions. The following transition conditions (condition of continuity) for the illustrated beam division in Fig. 4.11 can, for example, be named:

$$u_y^{\mathrm{I}}(a) = u_y^{\mathrm{II}}(a), \tag{4.39}$$

$$\frac{du_y^{\mathrm{I}}(a)}{dx} = \frac{du_y^{\mathrm{II}}(a)}{dx}. \tag{4.40}$$

The analytical calculation of the bending line is shown by example in the following for a beam which is loaded by a single force, see Fig. 4.12. The differential equation of the bending line in the form with the fourth order derivative according to Eq. (4.36) is chosen as an initial point. A four-time integration progressively leads to the following equations:

**Table 4.6** Boundary conditions for bending in the $x$-$y$ plane

| Symbol | Type of bearing | $u_y$ | $\frac{du_y}{dx}$ | $M$ | $Q$ |
|---|---|---|---|---|---|
| 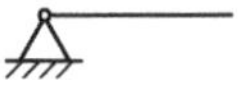 | Simply supported | 0 | – | 0 | – |
| 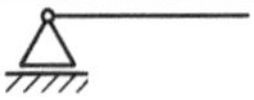 | Roller support | 0 | – | 0 | – |
| | Free end | – | – | 0 | 0 |
|  | Fixed support | 0 | 0 | – | – |
| 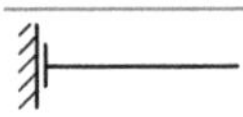 | Support with shear force link | – | 0 | – | 0 |
| 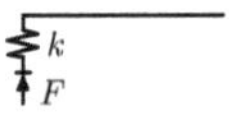 | Spring support | $\frac{F}{k}$ | – | 0 | – |

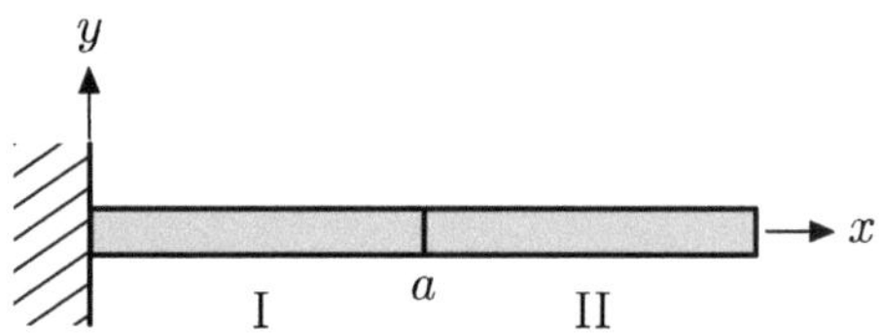


**Fig. 4.11** Beam subdivided into two different segments I and II. Interface location at $x = a$

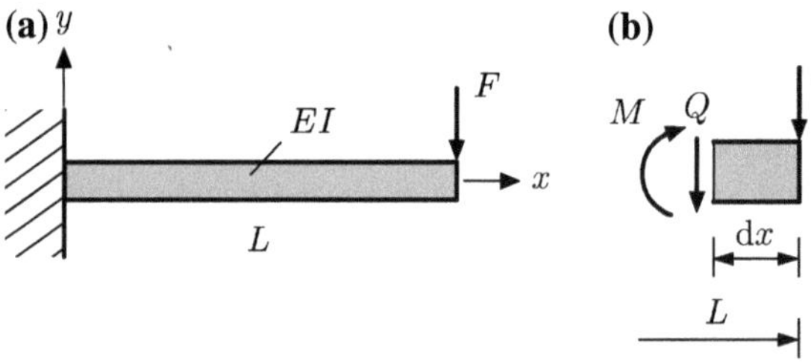


**Fig. 4.12** Calculation of the bending line for a beam loaded by a single force at the free end

$$EI_z \frac{d^3u_y}{dx^3} = c_1 \, (= -Q_y) \, , \tag{4.41}$$

$$EI_z \frac{d^2u_y}{dx^2} = c_1 x + c_2 \, (= M_z), \tag{4.42}$$

$$EI_z \frac{du_y}{dx} = \frac{1}{2} c_1 x^2 + c_2 x + c_3, \tag{4.43}$$

$$EI_z u_y = \frac{1}{6} c_1 x^3 + \frac{1}{2} c_2 x^2 + c_3 x + c_4. \tag{4.44}$$

Consequently, the general solution has to be adapted to the particular problem shown in Fig. 4.12a using the constants of integration $c_1, \ldots, c_4$.

The conditions $u_y(0) = 0$ and $\frac{\mathrm{d}u_y(0)}{\mathrm{d}x} = 0$ apply for the fixed support at the left-hand boundary ($x = 0$), see Table 4.6. It results immediately from Eqs. (4.43) and (4.44) together with these boundary conditions that $c_3 = c_4 = 0$ holds. For the determination of the remaining constants of integration one cannot use Table 4.6. In fact the external load has to be related to the internal reactions. To this end, the infinitesimal element shown in Fig. 4.12b, at which the external force $F$ acts has to be considered. The equilibrium between the external loads and the internal reactions has to be formulated at the position $x = L$, thus at the point of application of the external force. Therefore, the external force does not create a moment contribution since the case $\mathrm{d}x \to 0$ or in other words the position $x = L$ is being considered. The equilibrium of moments,[16] meaning $M_z(x = L) = 0$, together with Eq. (4.42) leads to the relation $c_2 = -c_1 L$. The vertical equilibrium of forces according to Fig. 4.12b leads to $Q_y(x = L) = -F$. It results through Eq. (4.41) that $c_1 = F$ holds. Consequently, the equation of the bending line can be formulated as

$$u_y(x) = \frac{1}{EI_z}\left(\frac{1}{6}Fx^3 - \frac{1}{2}FLx^2\right). \tag{4.45}$$

In particular, the maximum displacement at the right-hand boundary results in:

$$u_y(L) = -\frac{FL^3}{3EI_z}. \tag{4.46}$$

The calculation of the bending line can alternatively start, for example, from the moment distribution $M_z(x)$. To do so, the beam is 'cut' into two parts at an arbitrary location $x$, see Fig. 4.13. Subsequently, it is enough to consider only one of the two parts for the formulation of the moment equilibrium.

The moment equilibrium of the right-hand part around the reference point at the position $x$ delivers $+M_z(x) + (L - x)F = 0$, or alternatively solved for the moment distribution:

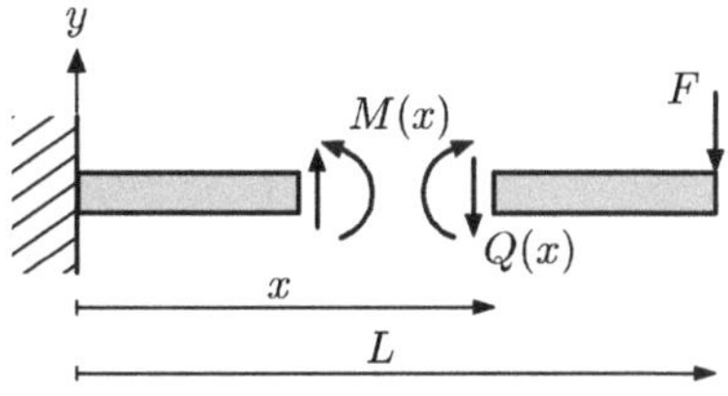


**Fig. 4.13** Calculation of the bending line for a beam under single force end load on the basis of the moment distribution

[16] Just for the case that an external moment $M^{\mathrm{ext}}$ would act at the position $x = L$, the internal moment would result in: $M_z(x = L) = M^{\mathrm{ext}}$. Hereby it was assumed that the external moment $M^{\mathrm{ext}}$ would be positive in a mathematical sense.

$$M_z(x) = (x - L)F. \tag{4.47}$$

The differential equation of the bending line in the form with the 2nd order derivative according to Eq. (4.37) is chosen as the starting point. Two-time integration progressively leads to the following equations:

$$EI_z \frac{\mathrm{d}^2 u_y}{\mathrm{d}x^2} = M_z(x) = (x - L)F, \tag{4.48}$$

$$EI_z \frac{\mathrm{d}u_y}{\mathrm{d}x} = \left(\frac{1}{2}x^2 - Lx\right)F + c_1, \tag{4.49}$$

$$EI_z u_y(x) = \frac{1}{6}x^3 F - \frac{1}{2}Lx^2 F + c_1 x + c_2. \tag{4.50}$$

Consideration of the boundary conditions on the fixed support, i.e. $u_y(0) = 0$ and $\frac{\mathrm{d}u_y(0)}{\mathrm{d}x} = 0$, finally leads to Eq. (4.45) and the maximum displacement according to Eq. (4.46).

To be able to realize a closed-form presentation with discontinuities, the so-called FÖPPL bracket can be used for the closed-form representation of bending lines. This mathematical notation, which was introduced by August Otto FÖPPL (1854–1942)[17] has the following meaning:

$$\langle x - a\rangle^n = \begin{cases} 0 & \text{for } x < a \\ (x - a)^n & \text{for } x \geq a. \end{cases} \tag{4.51}$$

In particular with the case $n = 0$

$$\langle x - a\rangle^0 = \begin{cases} 0 \text{ for } x < a \\ 1 \text{ for } x \geq a \end{cases} \tag{4.52}$$

the closed-form presentation of jumps can be realized. The first three discontinuous functions are shown in Fig. 4.14. Furthermore, derivations and integrals are defined by regarding the triangular bracket symbol as classical round brackets:

$$\frac{\mathrm{d}}{\mathrm{d}x}\langle x - a\rangle^n = n\langle x - a\rangle^{n-1}, \tag{4.53}$$

$$\int \langle x - a\rangle^n \mathrm{d}x = \frac{1}{n+1}\langle x - a\rangle^{n+1} + c. \tag{4.54}$$

Table 4.7 shows a few examples of discontinuous loads and their corresponding representations due to the discontinuous function given in Eq. (4.51).

[17] In English-speaking countries this mathematical notation is mostly named after the British mathematician and engineer W.H. MACAULAY (1853–1936). However this notation was originally proposed by the German mathematician Rudolf Friedrich Alfred CLEBSCH (1833–1872) [23].

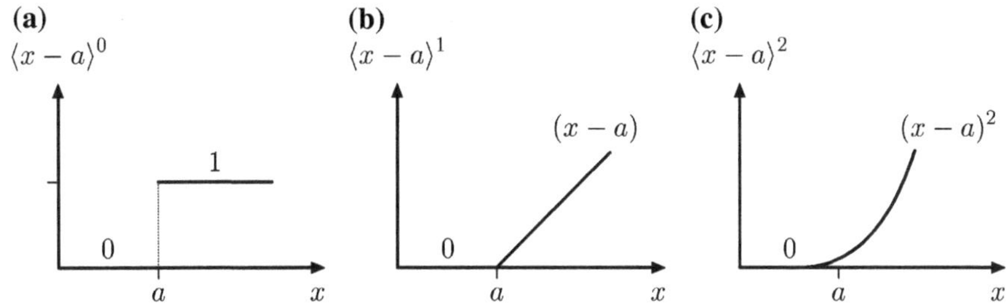


**Fig. 4.14** Graphical representation of the first three discontinuous functions according to Eq. (4.51): jump ($n = 0$); kink ($n = 1$); smooth transition ($n = 2$). Adapted from [9]

**Table 4.7** Discontinuous loads expressed due to discontinuous functions. Adapted from [9]

| Case | Load (discontinuity function) |
|---|---|
| 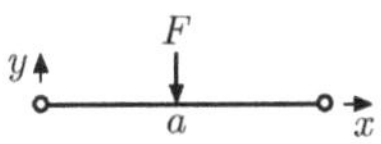  | $Q_y(x) = +F\langle x-a\rangle^0$ |
| 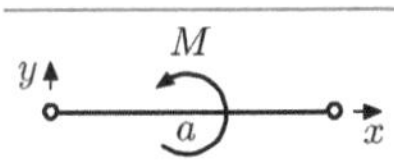  | $M_z(x) = -M\langle x-a\rangle^0$ |
| 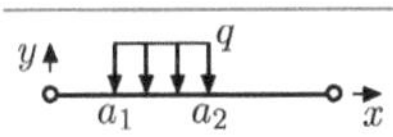  | $q_y(x) = -q\left(\langle x-a_1\rangle^0 - \langle x-a_2\rangle^0\right)$ |
| 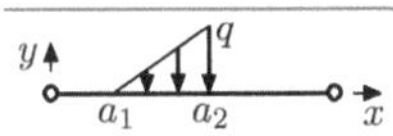  | $q_y(x) = \dfrac{-q}{a_2-a_1}(x-a_1)\left(\langle x-a_1\rangle^0 - \langle x-a_2\rangle^0\right)$ |
| 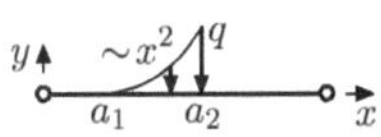  | $q_y(x) = \dfrac{-4q}{(a_2-a_1)^2}\left(x - \dfrac{a_1+a_2}{2}\right)^2 \left(\langle x-a_1\rangle^0 - \langle x-a_2\rangle^0\right)$ |

Closed-form solutions for the deflections of beams with different loading cases and boundary conditions are collected in Tables 4.8, 4.9, 4.10, 4.11 and 4.12 [31, 35].

Beams which are fixed at both ends are also known in the literature as built-in or encastré[18] beams.

**4.1 Example: Simply supported beam with a continuous triangular shaped load**
Given is a simply supported BERNOULLI beam as shown in Fig. 4.15 where a triangular shaped continuous load is acting in the negative $y$-direction. This continuous load acts in the range $a_1 \leq x \leq a_2$ and the maximum value is equal to $q$. Determine the deflection $u_y(x)$ by means of discontinuity functions.

---

[18] From French 'encastrer'.

**Table 4.8** Single equation solutions for the deflection of cantilevered and simply supported BERNOULLI beams subjected to loads for bending in the $x$-$y$ plane

| Load case | Deflection |
|---|---|
| 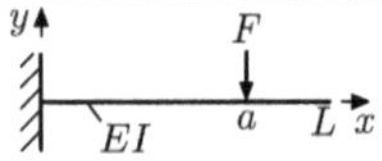 | $u_y(x) = \frac{-F}{6EI_z} \times \left[3ax^2 - x^3 + \langle x-a\rangle^3\right]$ |
| 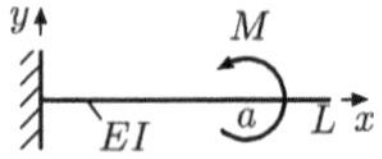 | $u_y(x) = \frac{-M}{2EI_z} \times \left[-x^2 + \langle x-a\rangle^2\right]$ |
| 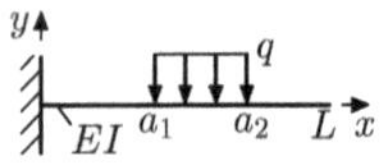 | $u_y(x) = \frac{-q}{24EI_z} \times \left[6(a_2^2 - a_1^2)x^2 - 4(a_2 - a_1)x^3 + \langle x-a_1\rangle^4 - \langle x-a_2\rangle^4\right]$ |
| 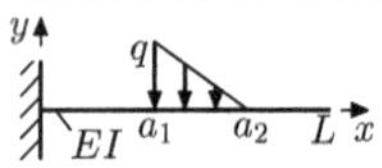 | $u_y(x) = \frac{-q}{(a_1-a_2)EI_z} \times \left[\frac{x-a_2}{24}\left(\langle x-a_1\rangle^4 - \langle x-a_2\rangle^4\right) - \frac{1}{30}\left(\langle x-a_1\rangle^5 - \langle x-a_2\rangle^5\right)\right] + \frac{q(a_2-a_1)}{12EI_z}(x^3 - 2a_1x^2 - a_2x^2)$ |
| 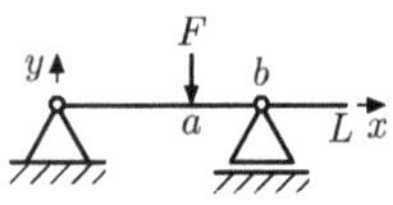 | $u_y(x) = \frac{-F}{6bEI_z} \times \left[(b-a)(b^2x - x^3) - x\langle b-a\rangle^3 + b\langle x-a\rangle^3 - a\langle x-b\rangle^3\right]$ |
| 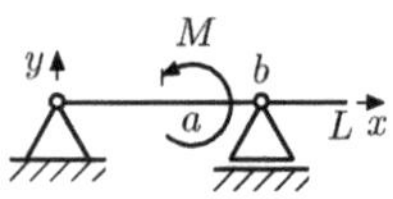 | $u_y(x) = \frac{-M}{6bEI_z} \times \left[b^2x - x^3 - 3x\langle b-a\rangle^2 + 3b\langle x-a\rangle^2 + \langle x-b\rangle^3\right]$ |
| 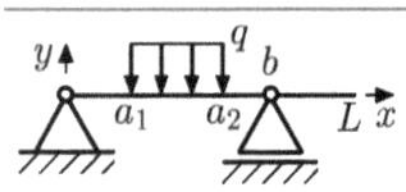 | $u_y(x) = \frac{-q}{24bEI_z} \times \left[2\left(a_2^2 - a_1^2 - 2b(a_2-a_1)\right)(x^3 - b^2x) - x\langle b-a_1\rangle^4 + x\langle b-a_2\rangle^4 + b\langle x-a_1\rangle^4 - b\langle x-a_2\rangle^4 - 2(a_2^2-a_1^2)\langle x-b\rangle^3\right]$ |
| 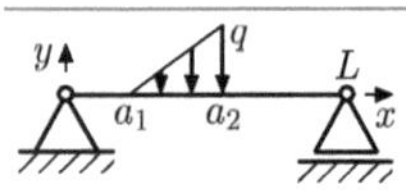 | $u_y(x) = \frac{-q}{(a_2-a_1)EI_z} \times \left[\frac{x-a_1}{24}\left(\langle x-a_1\rangle^4 - \langle x-a_2\rangle^4\right) - \frac{L-a_1}{24}\left((L-a_1)^4 - (L-a_2)^4\right)\frac{x}{L} - \frac{1}{30}\left(\langle x-a_1\rangle^5 - \langle x-a_2\rangle^5\right) + \frac{1}{30}\left((L-a_1)^5 - (L-a_2)^5\right)\frac{x}{L}\right] + \frac{q(a_2-a_2)}{12EI_z} \times \left(1 - \frac{1}{L}\left(\frac{1}{3}a_1 + \frac{2}{3}a_2\right)\right)(x^3 - xL^2)$ |

**Table 4.9** Single equation solutions for the deflection of fixed supported BERNOULLI beams subjected to loads for bending in the $x$-$y$ plane

| Load case | Deflection |
|---|---|
| 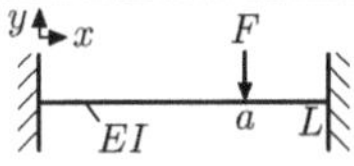 | $u_y(x) = \frac{-F}{EI_z} \times \left[\left(a + \frac{a^3}{L^2} - \frac{2a^2}{L}\right)\frac{x^2}{2} - \left(1 - \frac{3a^2}{L^2} + \frac{2a^3}{L^3}\right)\frac{x^3}{6} + \frac{1}{6}\langle x-a\rangle^3\right]$ |
| 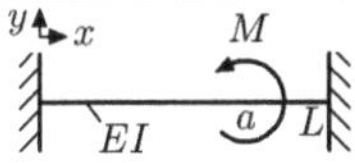 | $u_y(x) = \frac{-M}{2EI_z} \times \left[-\left(L^2 - 4aL + 3a^2\right)\left(\frac{x}{L}\right)^2 + \langle x-a\rangle^2 + 2\left(a^2 - aL\right)\left(\frac{x}{L}\right)^3\right]$ |
| 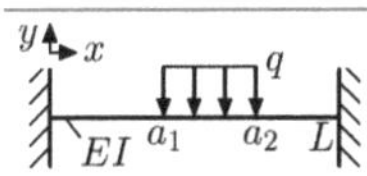 | $u_y(x) = \frac{-q}{EI_z} \times \left[\frac{1}{24}\left(\langle x-a_1\rangle^4 - \langle x-a_2\rangle^4\right) + \frac{x^3}{6}\left(\frac{1}{2L^3}\left((L-a_1)^4 - (L-a_2)^4\right) - \frac{1}{L^2}\left((L-a_1^3) - (L-a_2)^3\right)\right) + \frac{x^2}{2}\left(\frac{1}{3L}\left((L-a_1)^3 - (L-a_2)^3\right) - \frac{1}{4L^2}\left((L-a_1)^4 - (L-a_2)^4\right)\right)\right]$ |
| 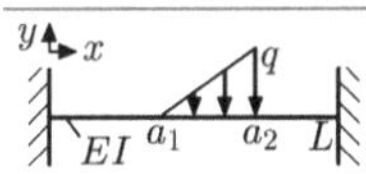 | $u_y(x) = \frac{-q}{(a_2-a_1)EI_z} \times \left[\frac{a_2 - x}{24}\left(\langle x-a_1\rangle^4 - \langle x-a_2\rangle^4\right) + \frac{1}{30}\left(\langle x-a_1\rangle^5 - \langle x-a_2\rangle\right)^5\right] - \frac{qx^3}{6(a_2-a_1)EI_zL^3} \times \left[\frac{2}{5}\left((L-a_1)^5 - (L-a_2)^5\right) + \frac{2a_2 - 5L}{4}\left((L-a_1)^4 - (L-a_2)^4\right) - L(a_2 - L)\left((L-a_1)^3 - (L-a_2)^3\right)\right] + \frac{qx^2}{2(a_2-a_1)EI_zL} \times \left[\frac{1}{5L}\left((L-a_1)^5 - (L-a_2)^5\right) + \frac{a_2 - 2L}{4L}\left((L-a_1)^4 - (L-a_2)^4\right) + \frac{L - a_2}{3}\left((L-a_1)^3 - (L-a_2)^3\right)\right]$ |

**Table 4.10** Deflection of cantilevered BERNOULLI beams subjected to simple load cases for bending in the $x$-$y$-plane. In addition, the maximum deflection and slope is given

| Load case | Deflection and slope | Maximum |
|---|---|---|
|   | $u_y(x) = \frac{-Fx^2}{6EI_z}(3L - x)$ ; | $u_{y,\max} = \frac{-FL^3}{3EI_z}$ |
| | $\varphi_z(x) = \frac{-Fx}{2EI_z}(2L - x)$ ; | $\varphi_{z,\max} = \frac{-FL^2}{2EI_z}$ |
|   | $u_y(x) = \frac{Mx^2}{2EI_z}$; | $u_{y,\max} = \frac{ML^2}{2EI_z}$ |
| | $\varphi_z(x) = \frac{Mx}{EI_z}$; | $\varphi_{z,\max} = \frac{ML}{EI_z}$ |
| 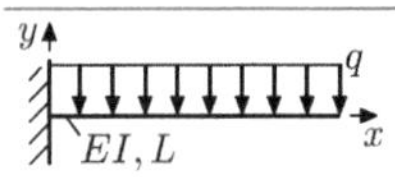  | $u_y(x) = \frac{-qx^2}{24EI_z}(x^2 - 4Lx + 6L^2)$ ; | $u_{y,\max} = \frac{-qL^4}{8EI_z}$ |
| | $\varphi_z(x) = \frac{-qx}{6EI_z}(x^2 - 3Lx + 3L^2)$ ; | $\varphi_{z,\max} = \frac{-qL^3}{6EI_z}$ |
| 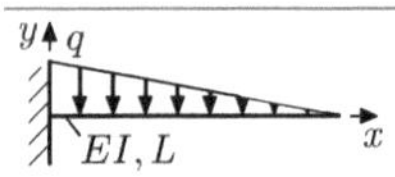  | $u_y(x) = \frac{-qx^2}{120EI_zL}(10L^3 - 10L^2x + 5Lx^2 - x^3)$ ; | $u_{y,\max} = \frac{-qL^4}{30EI_z}$ |
| | $\varphi_z(x) = \frac{-qx}{24EI_zL}(4L^3 - 6L^2x + 4Lx^2 - x^3)$ ; | $\varphi_{z,\max} = \frac{-qL^3}{24EI_z}$ |

**Solution 4.1 Solution**

The free body diagram of the loaded beam is shown in Fig. 4.16.

Let us first calculate the reactions at the supports. The resultant force of the continuous load is given by $R = -\frac{1}{2}q(a_2 - a_1)$. Based on this, the moment equilibrium about point 1 ($\sum M_1 = 0$) gives the reaction force acting at the right-hand support as:

$$R_2 = \frac{q}{2L}(a_2 - a_1)\left(\frac{1}{3}a_1 + \frac{2}{3}a_2\right). \tag{4.55}$$

The vertical force equilibrium ($\sum F_y = 0$) gives the reaction force at the left-hand support as:

$$R_1 = \frac{q}{2L}(a_2 - a_1)\left(L - \left(\frac{1}{3}a_1 + \frac{2}{3}a_2\right)\right). \tag{4.56}$$

**Table 4.11** Deflection of simply supported BERNOULLI beams subjected to simple load cases for bending in the $x$-$y$ plane. In addition, the maximum deflection and slope is given

| Load case | Deflection and slope | Maximum |
|---|---|---|
| 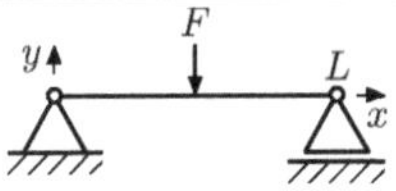 | $u_y(x) = \frac{-Fx}{48EI_z}(3L^2 - 4x^2), 0 \le x \le \frac{L}{2};$ | $u_{y,\max} = \frac{-FL^3}{48EI_z}$ |
| | $\varphi_z(x) = \frac{-F}{16EI_z}(L^2 - 4x^2), 0 \le x \le \frac{L}{2}$ | $\varphi_{z,\max} = \frac{\mp FL^2}{16EI_z}$ |
| 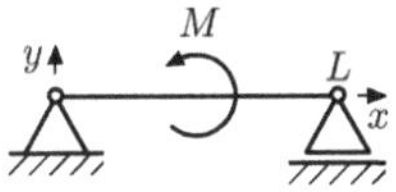 | $u_y(x) = \frac{Mx}{6EI_zL}\left(x^2 - \frac{1}{4}L^2\right), 0 \le x \le \frac{L}{2};$ | $u_{y,\max} = \mp\frac{ML^2}{128EI_z}$ |
| | $\varphi_z(x) = \frac{M}{6EI_zL}\left(3x^2 - \frac{1}{4}L^2\right), 0 \le x \le \frac{L}{2};$ | $\varphi_{z,\max} = \frac{ML}{12EI_z}$ |
| 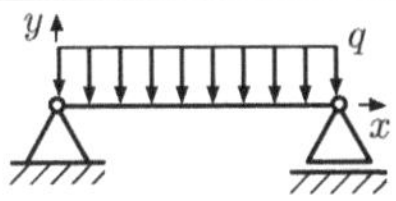 | $u_y(x) = \frac{-qx}{24EI_z}(x^3 - 2Lx^2 + L^3);$ | $u_{y,\max} = \frac{-5qL^4}{384EI_z}$ |
| | $\varphi_z(x) = \frac{-q}{24EI_z}(4x^3 - 6Lx^2 + L^3);$ | $\varphi_{z,\max} = \frac{\mp qL^3}{24EI_z}$ |
| 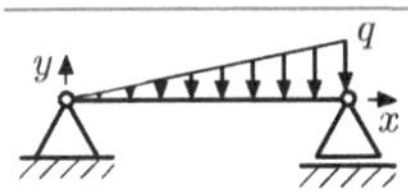 | $u_y(x) = \frac{-qx}{360EI_zL}(3x^4 - 10L^2x^2 + 7L^4);$ | $u_{y,\max} = -0.00652\frac{qL^4}{EI_z}$ |
| | $\varphi_z(x) = \frac{-q}{360EI_zL}(15x^4 - 30L^2x^2 + 7L^4);$ | $\varphi_{z,\max} = 0.02222\frac{qL^3}{EI_z}$ |

The boundary conditions of the problem can be stated as (cf. Fig. 4.17)

$$Q_y(x = 0) = -R_1, \qquad Q_y(x = L) = +R_2, \tag{4.57}$$

$$M_z(x = 0) = 0, \qquad M_z(x = L) = 0, \tag{4.58}$$

and

$$u_y(x = 0) = 0\,, \qquad u_y(x = L) = 0. \tag{4.59}$$

The next step is to derive the expression for the loading by means of discontinuous functions. Using a FÖPPL function to switch the load on at $a_1$ and to switch it off again at $a_2$, we can write the following expression for the triangular shaped load:

**Table 4.12** Deflection of fixed supported BERNOULLI beams subjected to simple load cases for bending in the $x$-$y$ plane. In addition, the maximum deflection and slope is given

| Load case | Deflection and slope | Maximum |
|---|---|---|
| 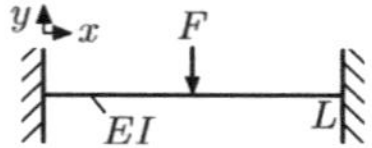  | $u_y(x) = \frac{-F}{EI_z}\left[\frac{Lx^2}{16} - \frac{x^3}{12} + \frac{1}{6}\left\langle x - \frac{L}{2}\right\rangle^3\right];$ | $u_{y,\max} = \frac{-FL^3}{192EI_z}$ |
| | $\varphi_z(x) = \frac{-F}{EI_z}\left[\frac{Lx}{8} - \frac{x^2}{4} + \frac{1}{2}\left\langle x - \frac{L}{2}\right\rangle^2\right];$ | $\varphi_{z,\max} = \frac{\mp FL^2}{64EI_z}$ |
| 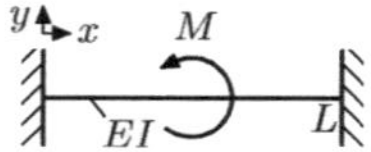  | $u_y(x) = \frac{-M}{2EI_z}\left(\frac{x^2}{4} + \left\langle x - \frac{L}{2}\right\rangle^2 - \frac{x^3}{2L}\right);$ | $u_{y,\max} = \frac{\mp ML^2}{216EI_z}$ |
| | $\varphi_z(x) = \frac{-M}{2EI_zL}\left(\frac{x}{2} + 2\left\langle x - \frac{L}{2}\right\rangle^1 - \frac{3x^2}{2L}\right);$ | $\varphi_{z,\max} = \frac{-ML}{48EI_z}$ |
| 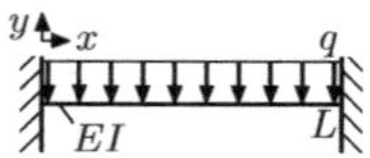  | $u_y(x) = \frac{-q}{24EI_z}(x^4 - 2Lx^3 + L^2x^2);$ | $u_{y,\max} = \frac{-qL^4}{384EI_z}$ |
| | $\varphi_z(x) = \frac{-q}{12EI_z}(2x^3 - 3Lx^2 + L^2x);$ | $\varphi_{z,\max} = \frac{\mp\sqrt{3}qL^3}{216EI_z}$ |
| 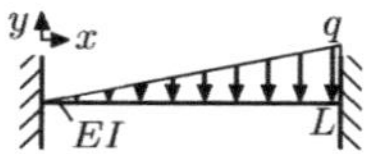  | $u_y(x) = \frac{-qL}{120EI_z}\left(-\frac{x^5}{L^2} + \frac{5x^4}{L} - 7x^3 + 3Lx^2\right);$ | $u_{y,\max} = \frac{-0.001309qL^4}{EI_y}$ |
| | $\varphi_z(x) = \frac{-qL}{120EI_z}\left(-\frac{5x^4}{L^2} + \frac{20x^3}{L} - 21x^2 + 6Lx\right);$ | $\varphi_{z,\max} = \frac{-0.512620qL^3}{EI_z}$ |

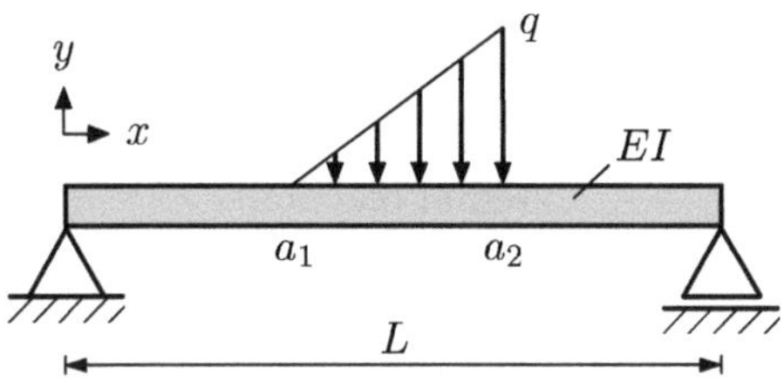


**Fig. 4.15** Simply supported beam with a continuous triangular shaped load acting in the range $a_1 \leq x \leq a_2$

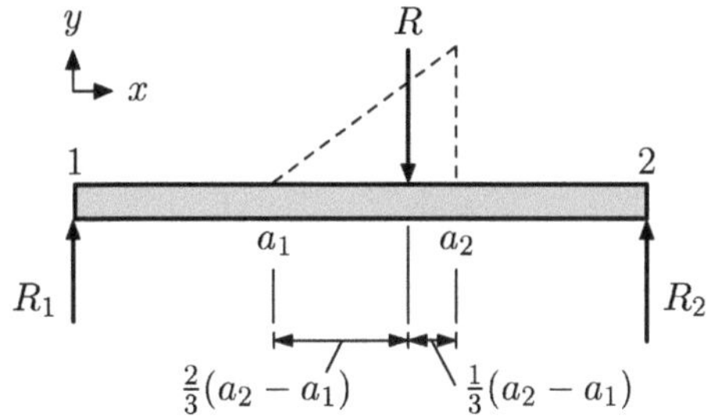


**Fig. 4.16** Free body diagram of the simply supported beam with a continuous triangular shaped load

$$q(x) = \frac{-q}{a_2 - a_1}(x - a_1)\Big(\langle x - a_1\rangle^0 - \langle x - a_2\rangle^0\Big). \tag{4.60}$$

This distribution can be inserted into the partial differential equation given in Table 4.5 and integrated four times to result in the deflection $u_y(x)$. The first integration, i.e.

$$EI_z \frac{\mathrm{d}^3 u_y}{\mathrm{d}x^3} = \int \Big( \underbrace{\frac{-q}{a_2 - a_1}(x - a_1)}_{f} \underbrace{\Big(\langle x - a_1\rangle^0 - \langle x - a_2\rangle^0\Big)}_{g'} \Big) \mathrm{d}x, \tag{4.61}$$

requires further efforts since an integration of the form $\int f(x)\langle \cdots \rangle \mathrm{d}x$ was not defined up to now. One way to overcome this problem is integration by parts as indicated by $f$ and $g'$ in Eq. (4.61). Thus, the right-hand side of Eq. (4.61) gives:

$$\begin{aligned} &\frac{-q}{a_2 - a_1}\left[(x - a_1)\Big(\langle x - a_1\rangle^1 - \langle x - a_2\rangle^1\Big) - \int 1 \times \Big(\langle x - a_1\rangle^1 - \langle x - a_2\rangle^1\Big)\mathrm{d}x\right] + c_1 \\ &\frac{-q}{a_2 - a_1}\left[(x - a_1)\Big(\langle x - a_1\rangle^1 - \langle x - a_2\rangle^1\Big) - \left(\frac{1}{2}\langle x - a_1\rangle^2 - \frac{1}{2}\langle x - a_2\rangle^2\right)\right] + c_1. \end{aligned} \tag{4.62}$$

The constant of integration $c_1$ can be determined based on the boundary condition given in Eq. $(4.57)_1$ as $c_1 = R_1$. Thus, the final result of the first integration is given by:

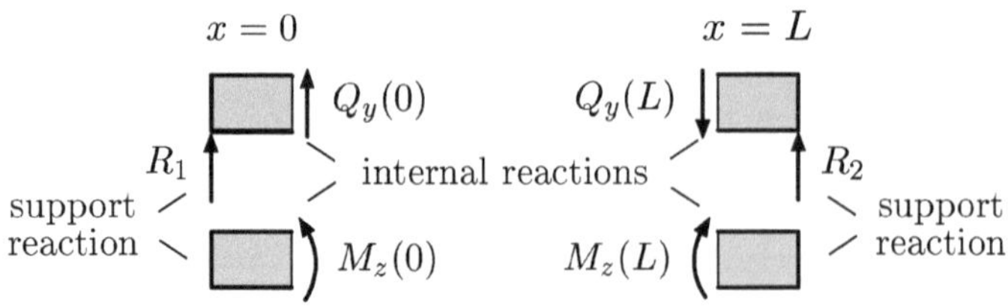


**Fig. 4.17** Equilibrium between support reactions and internal reactions at the ends of the simply supported beam

$$EI_z \frac{\mathrm{d}^3 u_y}{\mathrm{d}x^3} = \frac{-q}{a_2 - a_1}\Bigg[\underbrace{(x-a_1)}_{f}\underbrace{\Big(\langle x-a_1\rangle^1 - \langle x-a_2\rangle^1\Big)}_{g'} - \Big(\frac{1}{2}\langle x-a_1\rangle^2 - \frac{1}{2}\langle x-a_2\rangle^2\Big)\Bigg] + R_1. \tag{4.63}$$

Before performing the next integration, it may be useful to check the result obtained based on the boundary condition given in Eq. $(4.57)_2$ and the relation $EI_z \frac{\mathrm{d}^3 u_y}{\mathrm{d}x^3} = -Q_y$. The second integration gives under consideration of the boundary condition given in Eq. $(4.58)_1$ the following expression where again integration by parts was applied:

$$EI_z \frac{\mathrm{d}^2 u_y}{\mathrm{d}x^2} = \frac{-q}{a_2 - a_1}\Bigg[\underbrace{\frac{x-a_1}{2}}_{f}\underbrace{\Big(\langle x-a_1\rangle^2 - \langle x-a_2\rangle^2\Big)}_{g'} - \frac{1}{3}\Big(\langle x-a_1\rangle^3 - \langle x-a_2\rangle^3\Big)\Bigg] + R_1 x. \tag{4.64}$$

This result can be checked based on the boundary condition given in Eq. $(4.58)_2$ and the relation $EI_z \frac{\mathrm{d}^2 u_y}{\mathrm{d}x^2} = M_z$. The third and fourth integration give consecutively the following two expressions:

$$EI_z \frac{\mathrm{d}u_y}{\mathrm{d}x} = \frac{-q}{a_2 - a_1}\Bigg[\underbrace{\frac{x-a_1}{6}}_{f}\underbrace{\Big(\langle x-a_1\rangle^3 - \langle x-a_2\rangle^3\Big)}_{g'} - \frac{1}{8}\Big(\langle x-a_1\rangle^4 - \langle x-a_2\rangle^4\Big)\Bigg] + \frac{R_1}{2}x^2 + c_3, \tag{4.65}$$

$$
\begin{aligned}
EI_z u_y(x) = \frac{-q}{a_2 - a_1}\Bigg[\frac{x-a_1}{24}\Big(\langle x-a_1\rangle^4 - \langle x-a_2\rangle^4\Big) \\
- \frac{1}{30}\Big(\langle x-a_1\rangle^5 - \langle x-a_2\rangle^5\Big)\Bigg] + \frac{R_1}{6}x^3 + c_3 x + c_4 .
\end{aligned} \tag{4.66}
$$

The remaining two integration constants, i.e. $c_3$ and $c_4$, can be determined by means of the two displacement boundary conditions given in Eq. (4.59) as:

$$
c_4 = 0\,, \tag{4.67}
$$

$$
\begin{aligned}
c_3 = \frac{q}{(a_2 - a_1)L}\Bigg[\frac{L-a_1}{24}\Big((L-a_1)^4 - (L-a_2)^4\Big) \\
-\frac{1}{30}\Big((L-a_1)^5 - (L-a_2)^5\Big)\Bigg] - \frac{R_1 L^2}{6}.
\end{aligned} \tag{4.68}
$$

The final result for the deflection $u_y(x)$ reads as:

$$
\begin{aligned}
u_y(x) = \frac{-q}{(a_2 - a_1)EI_z} \times \Bigg[\frac{x-a_1}{24}\Big(\langle x-a_1\rangle^4 - \langle x-a_2\rangle^4\Big) \\
- \frac{L-a_1}{24}\Big((L-a_1)^4 - (L-a_2)^4\Big)\frac{x}{L} - \frac{1}{30}\Big(\langle x-a_1\rangle^5 \\
- \langle x-a_2\rangle^5\Big) + \frac{1}{30}\Big((L-a_1)^5 - (L-a_2)^5\Big)\frac{x}{L}\Bigg] + \frac{R_1}{6EI_z}(x^3 - xL^2).
\end{aligned} \tag{4.69}
$$

This result can be compared for the special case $a_1 = 0$ and $a_2 = L$ with the classical expression provided in Table 4.11.

### 4.2.6 Analytical Solutions in the Elasto-Plastic Range

Table 4.13 collects a few classical reference books[19] which cover the topic of analytical treatment of elasto-plastic beam bending.

[19] The original textbook by NÁDAI (Arpád Ludwig NÁDAI (1883–1963), Hungarian-American engineer) was the German monograph entitled 'Der bildsame Zustand der Werkstoffe' (171 pp.) which was published by Springer in 1927. This first German edition was revised and extended in 1931 under the assistance of Arthur M. WAHL and published as the English version by McGraw-Hill entitled 'Plasticity—a mechanics of the plastic state of matter' (349 pp.). The second edition was further enlarged and published in 1950 by McGraw-Hill as 'Theory of flow and fracture of solids' (572 pp.).

**Table 4.13** Classical reference books for the analytical treatment of elasto-plastic beam bending

| Year (1st ed.) | Author | Title | References |
|---|---|---|---|
| 1931 | A.L. NÁDAI | Theory of flow and fracture of solids | [72] |
| 1951 | W. PRAGER, P.G. HODGE | Theory of perfectly plastic solids | [79] |
| 1956 | B.G. NEAL | The plastic methods of structural analysis | [73] |
| 1959 | P.G. HODGE | Plastic analysis of structures | [55] |

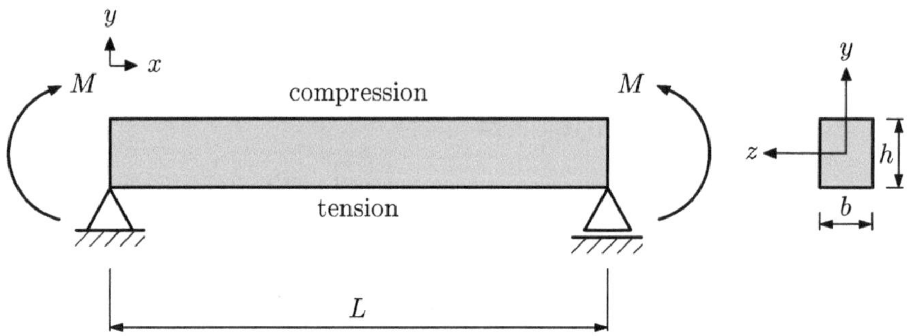


**Fig. 4.18** Simply supported BERNOULLI beam under pure positive bending load

### 4.2.6.1 Ideal Plasticity

The deformation of BERNOULLI beams in the elasto-plastic range will be investigated in the following. The derivations[20] will start with a very basic case of a simply supported beam of rectangular cross-section, cf. Fig. 4.18. Assuming that the beam is loaded by single moments at its ends as shown in Fig. 4.18, the bending moment distribution is constant and equal to the value of $+M$. This load case reveals the advantage that the stress distribution is the same at each position $x$ of the beam.

Furthermore, we assume that the beam material is linear-elastic ideal-plastic as shown in Fig. 4.19a and that the yield stress in tension and compression is the same: $k_t = k_c = k$.

The external moments are assumed to be linearly increased[21] from zero as shown in Fig. 4.19b. The first task is now to analyze the effect of different large values of the bending moment on the stress distribution over the height of the rectangular beam. At the beginning of this section, cf. Eq. (4.30), it was shown that the acting bending moment can be obtained by integration of the stress distribution according to

[20] The description of the basic relationships was presented in a similar way in [55].

[21] The moment is linearly increased over the time $t$. Since the material under consideration is time-independent, the time itself does not occur during the derivations and may be replaced by the expression 'increment'.

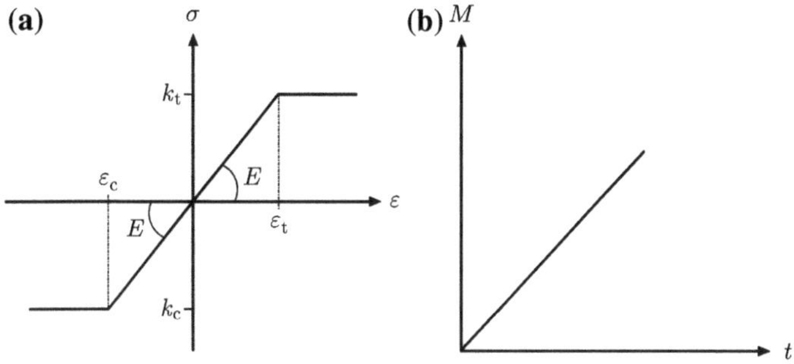


**Fig. 4.19** Material behavior and loading condition for the beam shown in Fig. 4.18: **a** uniaxial stress-strain diagram, **b** moment-time function

$$M_z = \int_A y\sigma_x \mathrm{d}A = -b \int_{-\frac{h}{2}}^{\frac{h}{2}} y\sigma_x(y)\mathrm{d}y \tag{4.70}$$

and that the axial strain distribution is given as being dependant on the curvature of the neutral fibre $\kappa$, cf. Eq. (4.15), as:

$$\varepsilon_x = -y\kappa. \tag{4.71}$$

Let us highlight here that the last two equations will be applied in the pure elastic and elasto-plastic deformation range. In the elastic range, HOOKE's law, i.e. $\sigma_x = E\varepsilon_x$, is valid and together with Eq. (4.71) the stress distribution—cf. Fig. 4.20a—can be expressed as $\sigma_x = -Ey\kappa$. Thus, the acting bending moment is obtained from Eq. (4.70) in the elastic range as:

$$M_z = -b \int_{-\frac{h}{2}}^{\frac{h}{2}} y(-Ey\kappa)\mathrm{d}y = bE\kappa \int_{-\frac{h}{2}}^{\frac{h}{2}} y^2\mathrm{d}y = bE\kappa \left[\frac{1}{3}y^3\right]_{-\frac{h}{2}}^{\frac{h}{2}}$$
$$= \frac{bE\kappa}{3}\left(\frac{h^3}{8} + \frac{h^3}{8}\right) = \frac{1}{12}Ebh^3\kappa. \tag{4.72}$$

The end of the elastic range is obtained as soon as the stress at the outer fiber reaches the yield stress, i.e. $|\sigma_x| = k$. At this point, the maximum elastic curvature, $\kappa^{\mathrm{el}}_{\mathrm{lim}}$, is given for $y = +\frac{h}{2}$ and $\sigma_x = -k$ as, cf. Eq. (4.71):

$$\kappa^{\mathrm{el}}_{\mathrm{lim}} = +\frac{2k}{Eh}. \tag{4.73}$$

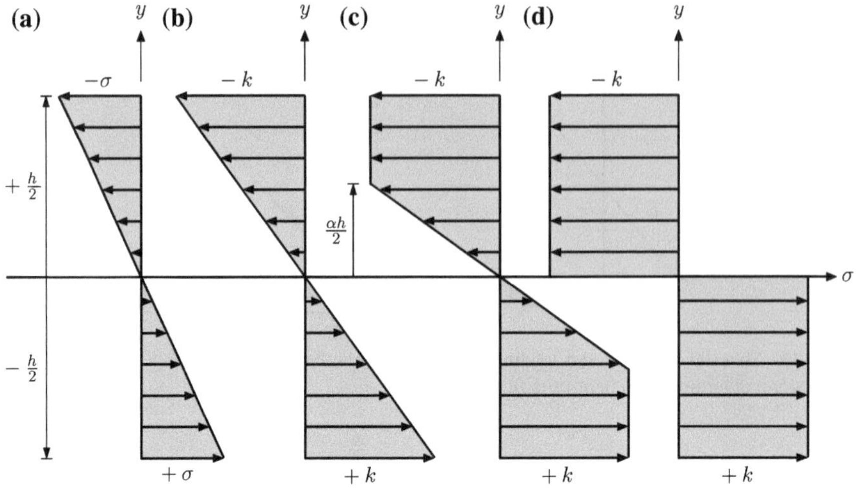


**Fig. 4.20** Stress distributions over the height of a rectangular BERNOULLI beam: **a** pure elastic, **b** maximum elastic (maximum elastic bending moment $M = M^{\text{el}}_{\text{lim}}$ acting), **c** elasto-plastic, **d** fully plastic (limiting yield moment $M = M^{\text{pl}}_{\text{lim}}$ acting). Adapted from [55]

The corresponding maximum elastic bending moment can be calculated based on this curvature due to Eq. (4.72) as:

$$M^{\text{el}}_{\text{lim}} = \frac{1}{12} E b h^3 \kappa^{\text{el}}_{\text{lim}} = +\frac{1}{6} b h^2 k = \frac{2k}{h} I_z . \tag{4.74}$$

The stress distribution for the elastic limit is indicated in Fig. 4.20b. Further increase of the bending moment beyond the maximum elastic bending moments results in an inward movement of the plastic zone and a stress distribution is obtained as shown in Fig. 4.20c. In the range $|y| \leq \frac{\alpha h}{2}$ the material is still pure elastic while the parts $|y| \geq \frac{\alpha h}{2}$ are in the elasto-plastic range.

The moment can be further increased until the so-called limiting or fully plastic moment $M^{\text{pl}}_{\text{lim}}$ is reached for $\alpha = 0$, cf. Fig. 4.20d. Looking at the elasto-plastic range as shown in Fig. 4.20c, the acting bending moment can be calculated as:

$$M_z = -2b \left( \int\limits_0^{\frac{\alpha h}{2}} y \left( -\frac{ky}{\alpha h/2} \right) \mathrm{d}y + \int\limits_{\frac{\alpha h}{2}}^{\frac{h}{2}} y(-k) \mathrm{d}y \right) \tag{4.75}$$

$$= 2bk \left( \left[ \frac{2y^3}{3\alpha h} \right]_0^{\frac{\alpha h}{2}} + \left[ \frac{y^2}{2} \right]_{\frac{\alpha h}{2}}^{\frac{h}{2}} \right) \tag{4.76}$$

or finally:

$$M_z(\alpha) = \frac{1}{12} bh^2 k \left(3 - \alpha^2\right), \tag{4.77}$$

where the yield stress is to be taken as positive, i.e. $k > 0$. The limiting bending moment can be obtained from the last equation for $\alpha = 0$ as:

$$M_{\text{lim}}^{\text{pl}} = \frac{1}{4} bh^2 k = \frac{3k}{h} I_z. \tag{4.78}$$

The ratio between the plastic and elastic limit moment results from Eqs. (4.74) and (4.78) as:

$$\frac{M_{\text{lim}}^{\text{pl}}}{M_{\text{lim}}^{\text{el}}} = \frac{\frac{1}{4} bh^2 k}{\frac{1}{6} bh^2 k} = \frac{3}{2}. \tag{4.79}$$

For the considered elastic ideal-plastic material (cf. Fig. 4.19a), this ratio is called the shape factor of the cross section. Further values of the shape factor for different types of cross sections can be taken from Refs. [55, 73, 117].

The relationship between the curvature and the moment can be expressed based on Eqs. (4.72), (4.73) and (4.78) in the elastic range as:

$$\frac{\kappa}{\kappa_{\text{lim}}^{\text{el}}} = \frac{\frac{12M_z}{Ebh^3}}{\frac{2k}{Eh}} = \frac{3M_z}{2M_{\text{lim}}^{\text{pl}}} \text{ (elastic)}. \tag{4.80}$$

At the transition between the pure elastic and the elasto-plastic range, i.e. at $|y| = \frac{\alpha h}{2}$, we can still use the relation $\sigma_x = -Ey\kappa$ which is based on HOOKE's law. For $y = +\frac{\alpha h}{2}$ and $\sigma_x = -k$ we can write that:

$$\kappa = \frac{2k}{Eh\alpha}. \tag{4.81}$$

The relationship between the parameter $\alpha$ and the acting bending moment can be obtained by balancing the external (this is $+M$ in the case of Fig. 4.18) and internal moment (obtained from the stress distribution shown in Fig. 4.20c), i.e.

$$+M = \frac{1}{12} bh^2 k (3 - \alpha^2) \tag{4.82}$$

as

$$\alpha = \sqrt{3}\sqrt{1 - \frac{4M}{bh^2 k}} = \sqrt{3}\sqrt{1 - \frac{M}{M_{\text{lim}}^{\text{pl}}}}. \tag{4.83}$$

Introducing the last relationship into Eq. (4.81) gives finally

$$\kappa = \frac{2k}{Eh\sqrt{3}\sqrt{1-\frac{M}{M_{\mathrm{lim}}^{\mathrm{pl}}}}} \text{ (elasto-plastic)}, \tag{4.84}$$

or with the definitions given in Eqs. (4.73) and (4.78):

$$\frac{\kappa}{\kappa_{\mathrm{lim}}^{\mathrm{el}}} = \left[3\left(1-\frac{M}{M_{\mathrm{lim}}^{\mathrm{pl}}}\right)\right]^{-\frac{1}{2}}, \tag{4.85}$$

$$\kappa = \frac{2M_{\mathrm{lim}}^{\mathrm{pl}}}{3EI_z}\left[3\left(1-\frac{M}{M_{\mathrm{lim}}^{\mathrm{pl}}}\right)\right]^{-\frac{1}{2}}. \tag{4.86}$$

Before going to the next question, let us highlight at this point the difference between the strain and stress distribution in the elasto-plastic range, cf. Fig. 4.21. Even in the elasto-plastic range, the strain distribution is assumed to be linear, cf. Fig. 4.21a, while the stress distribution has a kink at the transition between the pure elastic and elasto-plastic range, cf. Fig. 4.21b.

Let us concentrate in the following on the determination of the bending line, i.e. the vertical displacement $u_y(x)$ along the beam axis. In the pure elastic range, the second order differential equation $EI_z \frac{\mathrm{d}^2 u_y}{\mathrm{d}x^2} = M$ can be integrated twice to obtain under consideration of the boundary conditions $u_y(0) = u_y(L) = 0$ (cf. Fig. 4.18) the following solution:

$$u_y(x) = \frac{M}{2EI_z}\left(x^2 - Lx\right) = \frac{1}{2}\frac{M_{\mathrm{lim}}^{\mathrm{pl}}L^2}{EI_z}\frac{M}{M_{\mathrm{lim}}^{\mathrm{pl}}}\left[\left(\frac{x}{L}\right)^2 - \frac{x}{L}\right] \text{ (elastic)}, \tag{4.87}$$

or in a normalized form as

$$\frac{u_y(x)}{\frac{M_{\mathrm{lim}}^{\mathrm{pl}}L^2}{EI_z}} = \frac{1}{2}\frac{M}{M_{\mathrm{lim}}^{\mathrm{pl}}}\left[\left(\frac{x}{L}\right)^2 - \frac{x}{L}\right] \text{ (elastic)}, \tag{4.88}$$

which is valid in the range $0 \le M \le M_{\mathrm{lim}}^{\mathrm{el}}$. In order to calculate the vertical displacement in the elasto-plastic range, we make use of the assumption—cf. Eq. (4.15)—that the curvature can be approximated for small deformations by the second-order derivative of the displacement, i.e. $\kappa = \frac{\mathrm{d}^2 u_y(x)}{\mathrm{d}x^2}$. Combining this relationship with Eq. (4.84) gives:

$$\kappa = \frac{\mathrm{d}^2 u_y(x)}{\mathrm{d}x^2} = \frac{2k}{Eh\sqrt{3}\sqrt{1-\frac{M}{M_{\mathrm{lim}}^{\mathrm{pl}}}}} \text{ (elasto-plastic)}. \tag{4.89}$$

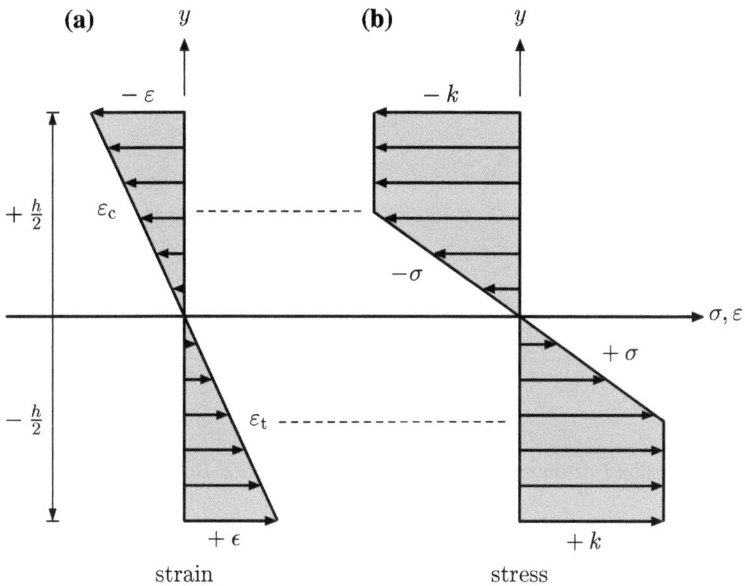


**Fig. 4.21** Difference between the **a** strain and **b** stress distribution in the elasto-plastic range for an ideal plastic material. The distributions are drawn for a positive bending moment which results in compression in the part $y \geq 0$

Integrating twice gives the following statement

$$u_y(x) = \frac{k}{Eh\sqrt{3}\sqrt{1-\frac{M}{M_{\text{lim}}^{\text{pl}}}}} x^2 + c_1 x + c_2, \tag{4.90}$$

from which under consideration of the boundary conditions, i.e. $u_y(0) = u_y(L) = 0$, the following equations results:

$$u_y(x) = \frac{1}{3\sqrt{3}} \frac{M_{\text{lim}}^{\text{pl}} L^2}{EI_z} \frac{1}{\sqrt{1-\frac{M}{M_{\text{lim}}^{\text{pl}}}}} \left[\left(\frac{x}{L}\right)^2 - \frac{x}{L}\right] \text{ (elasto-plastic)}, \tag{4.91}$$

or in a normalized form as:

$$\frac{u_y(x)}{\frac{M_{\text{lim}}^{\text{pl}} L^2}{EI_z}} = \frac{1}{3\sqrt{3}} \frac{1}{\sqrt{1-\frac{M}{M_{\text{lim}}^{\text{pl}}}}} \left[\left(\frac{x}{L}\right)^2 - \frac{x}{L}\right] \text{ (elasto-plastic)}. \tag{4.92}$$

The last equation is applicable in the range $M_{\text{lim}}^{\text{el}} \leq M \leq M_{\text{lim}}^{\text{pl}}$.

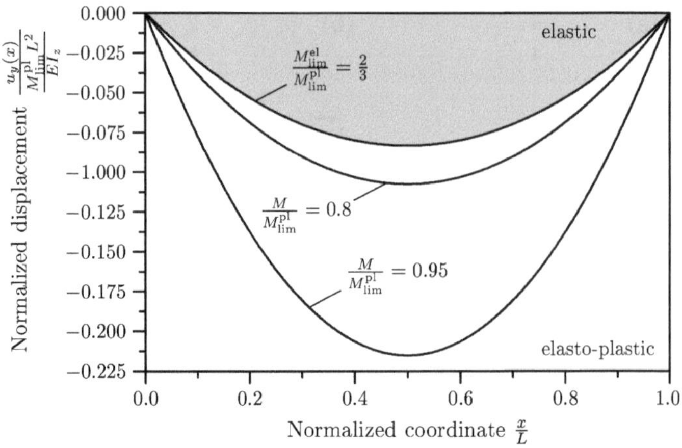


**Fig. 4.22** Normalized deformed shape of the simply supported BERNOULLI beam under pure bending load

A graphical representation of the deformed beam shape based on Eqs. (4.88) and (4.92) for different ratios of the acting bending moment is given in Fig. 4.22.

The following Table 4.14 collects a few functional evaluations in the pure elastic and elasto-plastic range which can be used in the following chapters on numerical methods.

A different graphical representation of the beam behavior can be obtained by plotting the bending moment over the displacement in the middle of the beam. Introducing $x = \frac{L}{2}$ in Eqs. (4.87) and (4.91) and solving for $\frac{M}{M^{pl}_{lim}}$ gives:

$$\frac{M}{M^{pl}_{lim}} = -\frac{3}{8} \times \frac{u_y\left(\frac{L}{2}\right)}{\frac{M^{pl}_{lim} L^2}{EI_z}} \quad \text{(elastic)}, \tag{4.93}$$

$$\frac{M}{M^{pl}_{lim}} = 1 - \frac{1}{48} \times \left(\frac{u_y\left(\frac{L}{2}\right)}{\frac{M^{pl}_{lim} L^2}{EI_z}}\right)^{-2} \quad \text{(elasto-plastic)}. \tag{4.94}$$

The graphical representation of these equations is given in Fig. 4.23. One should note that a positive bending moment results in a deformation in the negative $y$-direction. As the bending moment reaches the value of $M^{pl}_{lim}$, the deformation tends to infinity. Concluding, we are able to see that the derivation of the deformation in the elasto-plastic range requires the evaluation of Eq. (4.81) in the form $\kappa = \frac{d^2 u_y}{dx^2} = \frac{2k}{Eh\alpha}$ where the parameter $\alpha$ is obtained from balancing the external and internal bending moment.

**Table 4.14** Evaluation of Eqs. (4.88) and (4.92) at a few sampling points

| $\frac{M}{M_{\text{lim}}^{\text{pl}}} \setminus \frac{x}{L}$ | 0 | $\frac{1}{4}$ | $\frac{1}{2}$ | $\frac{3}{4}$ | $L$ |
|---|---|---|---|---|---|
| $\frac{1}{3}$ | 0 | $-\frac{1}{32}$ | $-\frac{1}{24}$ | $-\frac{1}{32}$ | 0 |
| $\frac{2}{3}$ | 0 | $-\frac{1}{16}$ | $-\frac{1}{12}$ | $-\frac{1}{16}$ | 0 |
| 0.8 | 0 | −0.0806872 | −0.107583 | −0.0806872 | 0 |
| 0.9 | 0 | −0.114109 | −0.152145 | −0.114109 | 0 |
| 0.95 | 0 | −0.161374 | −0.215166 | −0.161374 | 0 |

**4.2 Example: Cantilevered Bernoulli beam loaded by a single force at its end**
Given is a cantilevered BERNOULLI beam which is loaded by a single force $F$ as shown in Fig. 4.24. The cross section of the beam is rectangular with dimensions $b \times h$. Calculate the deflection $u_y(x)$ in the pure elastic and elasto-plastic range under the assumption that the material behavior is linear-elastic ideal-plastic with YOUNG's modulus $E$ and yield stress $k$.

**4.2 Solution**
The bending moment distribution can be derived as $M_z(x) = F(L-x)$ and the describing differential equation in the elastic range is given by:

$$EI_z \frac{\mathrm{d}^2 u_y(x)}{\mathrm{d}x^2} = M_z(x) = F(L-x). \tag{4.95}$$

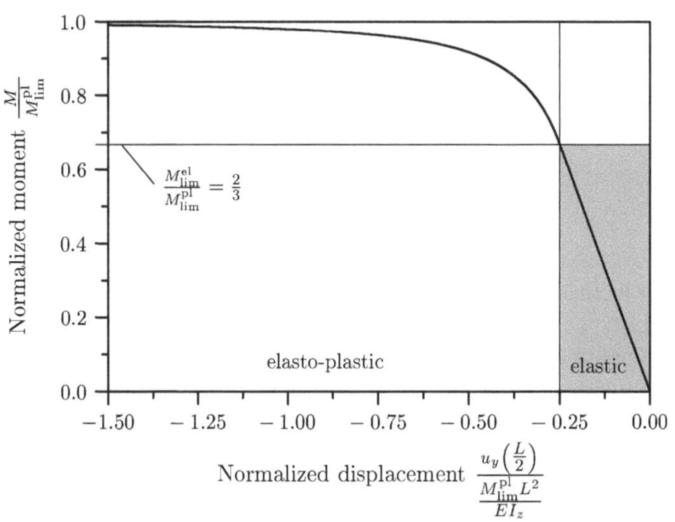


**Fig. 4.23** Normalized moment-displacement diagram of the simply supported BERNOULLI beam under pure bending load, cf. Fig. 4.18. The displacement is evaluated in the middle of the beam, i.e. $x = \frac{L}{2}$

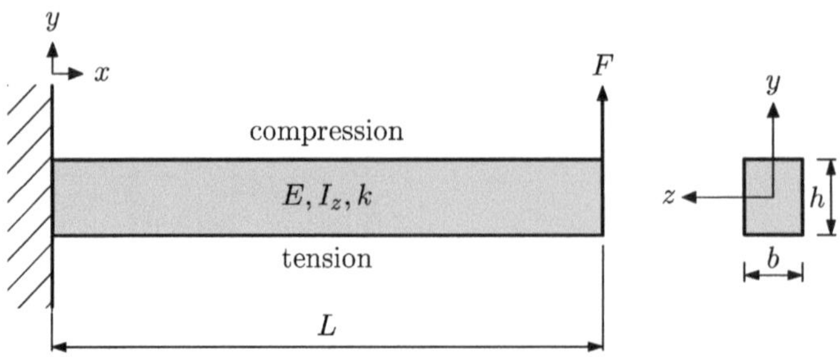


**Fig. 4.24** Cantilevered BERNOULLI beam loaded by a single force $F$ at its end

Integrating twice gives the following statements

$$EI_z \frac{\mathrm{d}u_y(x)}{\mathrm{d}x} = F\left(Lx - \frac{x^2}{2}\right) + c_1, \tag{4.96}$$

$$EI_z u_y(x) = F\left(\frac{Lx^2}{2} - \frac{x^3}{6}\right) + c_1 x + c_2, \tag{4.97}$$

from which under consideration of the boundary conditions, i.e. $u_y(0) = 0$ and $\frac{\mathrm{d}u_y(0)}{\mathrm{d}x} = 0$, the following expression in the pure elastic range can be derived:

$$u_y(x) = \frac{FL^3}{6EI_z}\left[3\left(\frac{x}{L}\right)^2 - \left(\frac{x}{L}\right)^3\right] = \frac{M_{\mathrm{lim}}^{\mathrm{pl}} L^2}{6EI_z}\frac{FL}{M_{\mathrm{lim}}^{\mathrm{pl}}}\left[3\left(\frac{x}{L}\right)^2 - \left(\frac{x}{L}\right)^3\right]. \tag{4.98}$$

The external force can be further increased up to the end of the pure elastic range. This is the case as soon as the moment at $x = 0$ reaches the maximum elastic bending moment and the outer fibres at $x = 0$ and $y = \pm\frac{h}{2}$ become plastic, cf. Fig. 4.25a:

$$F_{\mathrm{lim}}^{\mathrm{el}} L = M_{\mathrm{lim}}^{\mathrm{el}} = \frac{1}{6} bh^2 k. \tag{4.99}$$

A further increase of the external force beyond $F_{\mathrm{lim}}^{\mathrm{el}}$ results in the development of an elasto-plastic zone which spreads towards the positive $x$-direction. An interface separates the pure elastic zone (marked in light grey in Fig. 4.25b) and the elasto-plastic zone (marked in black in Fig. 4.25b). Thus, we must distinguish in the range $M_{\mathrm{lim}}^{\mathrm{el}} < M(0) < M_{\mathrm{lim}}^{\mathrm{pl}}$ two different zones along the principal axis of the beam. In the first zone where $0 \leq x \leq \varsigma$ holds, we have an elasto-plastic region which is separated by an interface from a pure elastic part, cf. Fig. 4.25b. The second zone, for $\varsigma \leq x \leq L$, is pure elastic and homogeneous, i.e. just a single phase.

Let us first have a look at the first zone where $0 \leq x \leq \varsigma$ holds. Balancing the external and internal bending moment (cf. Eq. (4.77)) gives:

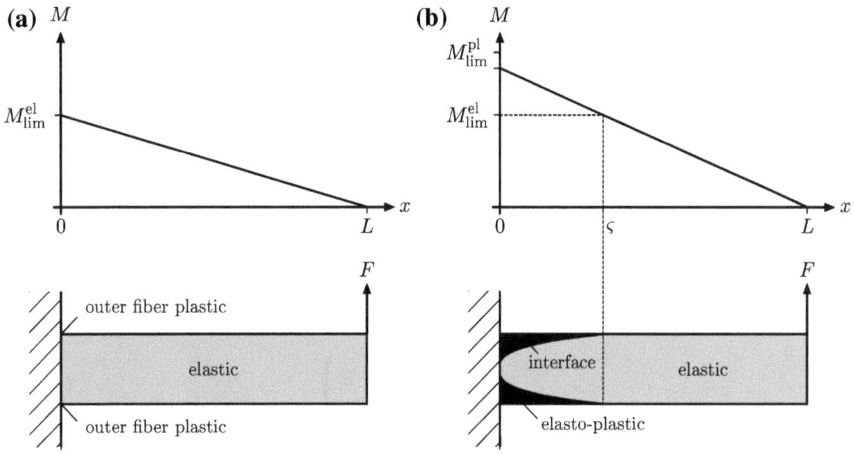


**Fig. 4.25** Bending moment distribution and location of elastic and plastic zones: **a** maximum elastic (maximum elastic bending moment $M = M_{\text{lim}}^{\text{el}}$ acting at $x = 0$), **b** elasto-plastic ($M_{\text{lim}}^{\text{el}} < M < M_{\text{lim}}^{\text{pl}}$ at $x = 0$)

$$F(L - x) = \frac{1}{12} bh^2 k (3 - \alpha^2), \tag{4.100}$$

which can be solved for $\alpha$ to obtain:

$$\alpha = \sqrt{3} \times \sqrt{1 - \frac{FL}{M_{\text{lim}}^{\text{pl}}} + \frac{FL}{M_{\text{lim}}^{\text{pl}}} \frac{x}{L}}. \tag{4.101}$$

Thus, the differential equation for the elasto-plastic range can be written as:

$$\kappa = \frac{\mathrm{d}^2 u_y(x)}{\mathrm{d}x^2} = \frac{2k}{Eh\sqrt{3}\sqrt{1 - \frac{FL}{M_{\text{lim}}^{\text{pl}}} + \frac{FL}{M_{\text{lim}}^{\text{pl}}} \frac{x}{L}}}. \tag{4.102}$$

Integrating twice gives the following statements:

$$\frac{\mathrm{d}u_y(x)}{\mathrm{d}x} = \frac{4\sqrt{3}}{3} \times \frac{(M_{\text{lim}}^{\text{pl}} - FL + Fx)k}{FEh\sqrt{\frac{M_{\text{lim}}^{\text{pl}} - FL + Fx}{M_{\text{lim}}^{\text{pl}}}}} + c_3, \tag{4.103}$$

$$u_y(x) = \frac{8\sqrt{3}}{9} \times \frac{(M_{\text{lim}}^{\text{pl}} - FL + Fx)^2 k}{F^2 Eh\sqrt{\frac{M_{\text{lim}}^{\text{pl}} - FL + Fx}{M_{\text{lim}}^{\text{pl}}}}} + c_3 x + c_4. \tag{4.104}$$

Before looking at the integration constants $c_i$, let us analyze the separation between the pure elastic and elasto-plastic zones which is indicated in Fig. 4.25b by the position $x = \varsigma$. The end of the zone $0 \leq x \leq \varsigma$ can be defined as soon as the interface reaches the free surface at $y = \pm\frac{h}{2}$. Thus, the position $\varsigma$ can be obtained from Eq. (4.101) by assigning $\alpha = 1$ and $x = \varsigma$ as:

$$1 = \sqrt{3} \times \sqrt{1 - \frac{FL}{M_{\text{lim}}^{\text{pl}}} + \frac{FL}{M_{\text{lim}}^{\text{pl}}} \frac{\varsigma}{L}}, \tag{4.105}$$

or

$$\varsigma = L\left(1 - \frac{2}{3}\frac{M_{\text{lim}}^{\text{pl}}}{FL}\right). \tag{4.106}$$

Let us return now to the set of equations which describe the deformation behavior $u_y(x)$ in the loading range $M_{\text{lim}}^{\text{el}} < M(0) < M_{\text{lim}}^{\text{pl}}$. In the spatial range $0 \leq x \leq \varsigma$, Eqs. (4.103) and (4.104) are valid and integration constants $c_3$ and $c_4$ must be determined whereas in the range $\varsigma \leq x \leq L$ the pure elastic equations (4.96) and (4.97) with its integrations constants $c_1$ and $c_2$ are valid. The constants $c_3$ and $c_4$ can be obtained from the boundary conditions at $x = 0$, i.e. $u_y(0) = 0$ and $\frac{\mathrm{d}u_y(0)}{\mathrm{d}x} = 0$, as:

$$c_3 = -\frac{4\sqrt{3}k}{3FEh}\left(M_{\text{lim}}^{\text{pl}} - FL\right)^{\frac{1}{2}}\left(M_{\text{lim}}^{\text{pl}}\right)^{\frac{1}{2}}, \tag{4.107}$$

$$c_4 = -\frac{8\sqrt{3}k}{9F^2Eh}\left(M_{\text{lim}}^{\text{pl}} - FL\right)^{\frac{3}{2}}\left(M_{\text{lim}}^{\text{pl}}\right)^{\frac{1}{2}}. \tag{4.108}$$

The constants $c_1$ and $c_2$ can be obtained from the continuity conditions at $x = \varsigma$, i.e. $\frac{\mathrm{d}u_y(\varsigma^-)}{\mathrm{d}x} = \frac{\mathrm{d}u_y(\varsigma^+)}{\mathrm{d}x}$ and $u_y(\varsigma^-) = u_y(\varsigma^+)$, as:

$$c_1 = \frac{M_{\text{lim}}^{\text{pl}}L}{6EI_z}\left(-\frac{3FL}{M_{\text{lim}}^{\text{pl}}} + \frac{4M_{\text{lim}}^{\text{pl}}}{FL} - \frac{8M_{\text{lim}}^{\text{pl}}}{\sqrt{3}FL}\left(1 - \frac{FL}{M_{\text{lim}}^{\text{pl}}}\right)^{\frac{1}{2}}\right), \tag{4.109}$$

$$c_2 = \frac{M_{\text{lim}}^{\text{pl}}L^2}{6EI_z}\left(\frac{FL}{M_{\text{lim}}^{\text{pl}}} - \frac{4M_{\text{lim}}^{\text{pl}}}{FL} + \frac{80}{27}\left(\frac{M_{\text{lim}}^{\text{pl}}}{FL}\right)^2 - \frac{48\sqrt{3}}{27}\left(\frac{M_{\text{lim}}^{\text{pl}}}{FL}\right)^2\left(1 - \frac{FL}{M_{\text{lim}}^{\text{pl}}}\right)^{\frac{3}{2}}\right). \tag{4.110}$$

Thus, the displacement in the elasto-plastic range is given by the following two equations:

$$u_y(x) = \frac{4M_{\text{lim}}^{\text{pl}}}{9\sqrt{3}FL}\frac{M_{\text{lim}}^{\text{pl}}L^2}{EI_z}\left(\frac{2M_{\text{lim}}^{\text{pl}}}{FL}\left(1-\frac{FL}{M_{\text{lim}}^{\text{pl}}}+\frac{FL}{M_{\text{lim}}^{\text{pl}}}\frac{x}{L}\right)^{\frac{3}{2}}\right.$$
$$\left.-\left(1-\frac{FL}{M_{\text{lim}}^{\text{pl}}}\right)^{\frac{1}{2}}\left(\frac{3x}{L}+\frac{2M_{\text{lim}}^{\text{pl}}}{FL}-2\right)\right), \quad (4.111)$$

which is valid in the range $0 \leq x \leq \varsigma$, i.e. the zone where elasto-plastic parts and pure elastic parts are separated by an interface and

$$u_y(x) = \frac{M_{\text{lim}}^{\text{pl}}L^2}{6EI_z}\frac{FL}{M_{\text{lim}}^{\text{pl}}}\left(3\left(\frac{x}{L}\right)^2-\left(\frac{x}{L}\right)^3\right)$$
$$+\frac{M_{\text{lim}}^{\text{pl}}L^2}{6EI_z}\left(-\frac{3FL}{M_{\text{lim}}^{\text{pl}}}+\frac{4M_{\text{lim}}^{\text{pl}}}{FL}-\frac{8M_{\text{lim}}^{\text{pl}}}{\sqrt{3}FL}\left(1-\frac{FL}{M_{\text{lim}}^{\text{pl}}}\right)^{\frac{1}{2}}\right)\frac{x}{L} \quad (4.112)$$
$$+\frac{M_{\text{lim}}^{\text{pl}}L^2}{6EI_z}\left(\frac{FL}{M_{\text{lim}}^{\text{pl}}}-\frac{4M_{\text{lim}}^{\text{pl}}}{FL}+\frac{80}{27}\left(\frac{M_{\text{lim}}^{\text{pl}}}{FL}\right)^2-\frac{48\sqrt{3}}{27}\left(\frac{M_{\text{lim}}^{\text{pl}}}{FL}\right)^2\left(1-\frac{FL}{M_{\text{lim}}^{\text{pl}}}\right)^{\frac{3}{2}}\right),$$

which is valid in the range $\varsigma \leq x \leq L$, i.e. the pure elastic zone. The limiting shape can be obtained from the last two equations for $FL = M_{\text{lim}}^{\text{pl}}$ as:

$$u_y(x)\big|_{\text{lim}} = \frac{16}{3\sqrt{3}}\frac{M_{\text{lim}}^{\text{pl}}L^2}{6EI_z}\left(\frac{x}{L}\right)^{\frac{3}{2}} \quad \text{for } 0 \leq x \leq \tfrac{L}{3}, \quad (4.113)$$

$$u_y(x)\big|_{\text{lim}} = \frac{M_{\text{lim}}^{\text{pl}}L^2}{6EI_z}\left(-\frac{1}{27}+\frac{x}{L}+3\left(\frac{x}{L}\right)^2-\left(\frac{x}{L}\right)^3\right) \quad \text{for } \tfrac{L}{3} \leq x \leq L. \quad (4.114)$$

A graphical representation of the deformed beam shape based on Eqs. (4.111) and (4.112) for different ratios of the maximum bending moment is given in Fig. 4.26.

A graphical representation of the maximum bending moment $M_z(x=0) = FL$ over the displacement of the force application point at $x = L$ can be obtained by introducing $x = L$ in Eqs. (4.98) and (4.112), cf. Fig. 4.27, as:

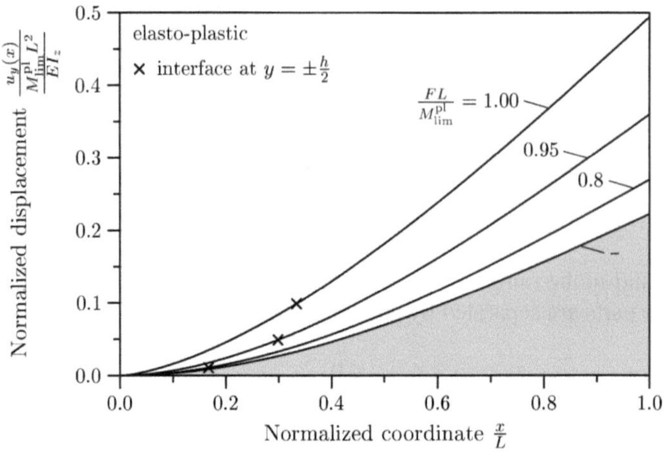


**Fig. 4.26** Normalized deformed shape of the cantilevered BERNOULLI beam loaded by a single force at its end

$$u_y(L) = \frac{1}{3}\frac{M^{pl}_{lim}L^2}{EI_z}\frac{FL}{M^{pl}_{lim}} \text{ (elastic)}, \tag{4.115}$$

$$u_y(L) = \frac{40}{81}\frac{M^{pl}_{lim}L^2}{EI_z}\left(\frac{M^{pl}_{lim}}{FL}\right)^2\left(1 - \frac{3\sqrt{3}}{10}\left(2 + \frac{FL}{M^{pl}_{lim}}\right)\left(1 - \frac{FL}{M^{pl}_{lim}}\right)^{\frac{1}{2}}\right). \tag{4.116}$$

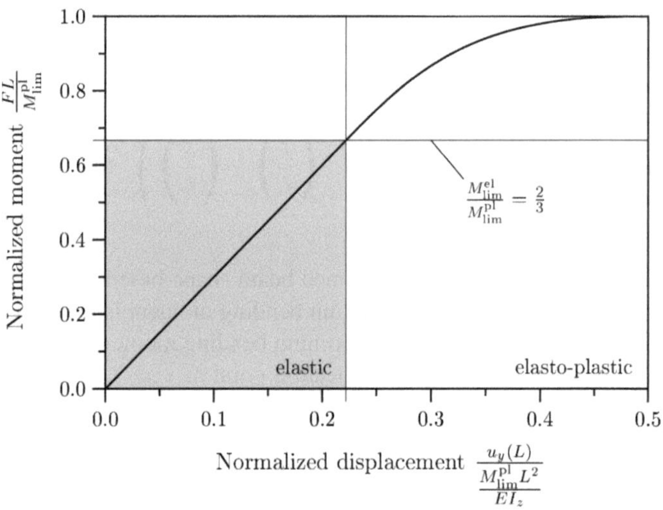


**Fig. 4.27** Normalized moment-displacement diagram of the cantilevered BERNOULLI beam loaded by a single force at its end. The displacement is evaluated in the *right-hand* end of the beam, i.e. at $x = L$, and the moment is the maximum moment at $x = 0$

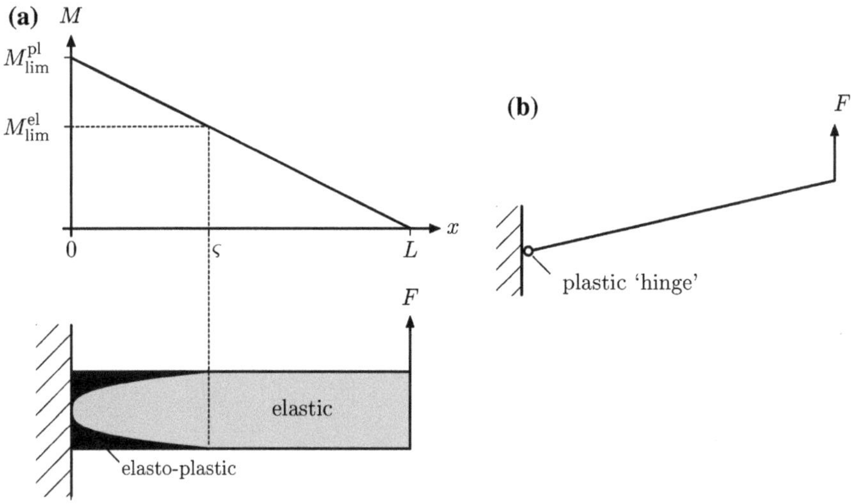


**Fig. 4.28** **a** Bending moment distribution and location of elastic and plastic zones in the limiting case $M = M_{\rm lim}^{\rm pl}$ at $x = 0$, **b** collapse mechanism

In the case that the limiting bending moment $M_{\rm lim}^{\rm pl}$ is reached at $x = 0$ as shown in Fig. 4.28a, the beam can no longer carry any load and the displacement tends to infinity. This situation is equivalent to the case where a hinge is introduced at the left support and thus, one can say that this limiting situation is caused by a plastic 'hinge' as indicated in Fig. 4.28b.

Let us compare at the end of this section the stress states in an elongated rod and a bent beam for a linear-elastic ideal-plastic material as shown in Fig. 4.29. Both structural elements are increasingly loaded from zero by an end displacement (in the case of the rod) or end rotations (in the case of the beam). The rod has a constant normal stress distribution—both in the pure elastic and the elasto-plastic range—over the height of the cross section. As soon as the yield stress $k$ is reached, the stress remains constant, even under increasing end displacement. The normal stress distribution in the case of the beam is in the elastic range linear increasing from the neural fibre, i.e. in the middle, to the maximum at the outer fiber.

As soon as the maximum stress at the outer fibers is equal to the yield stress, a constant stress $\sigma = k$ is developing inwards with increasing end rotation. The final stage is obtained as soon as both halves of the cross section are completely subjected to the constant yield stress $\pm k$.

#### 4.2.6.2 Consideration of Hardening

The following derivations are close to the line of reasoning as presented in Sect. 4.2.6.1. For further details, see also [58, 117]. The simplest case of a hardening material, i.e. the so-called linear hardening, is considered and schematically shown in Fig. 4.30a.

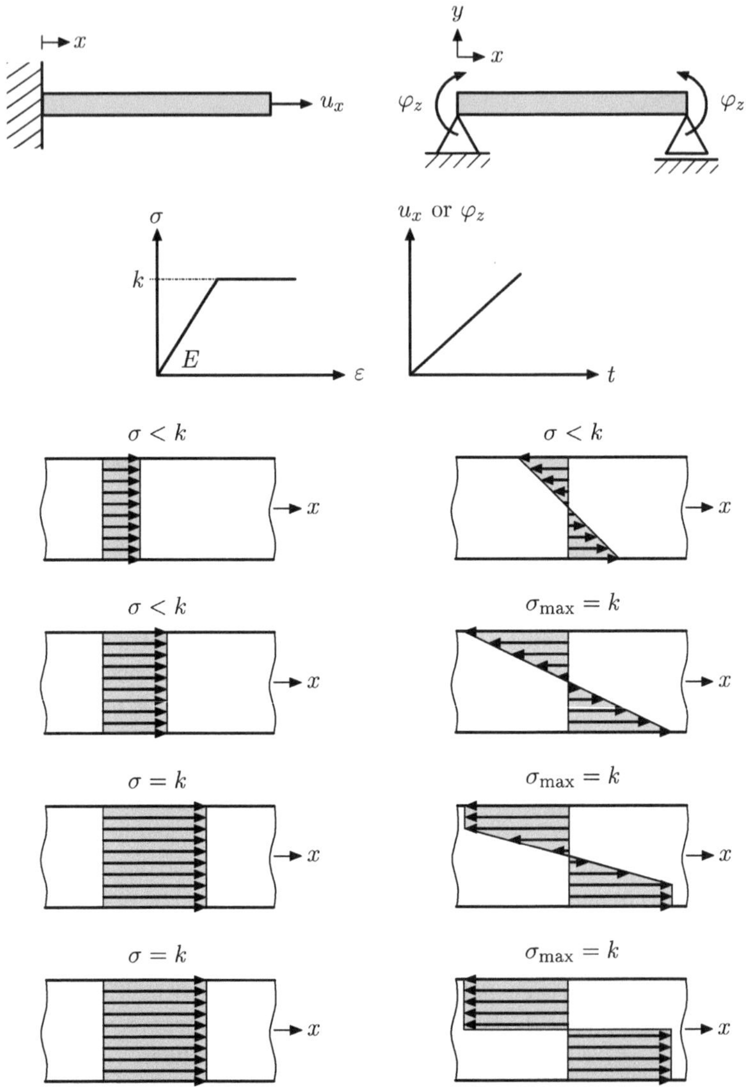


**Fig. 4.29** Comparison of stress states in an elongated rod (*left*) and a bent beam (*right*) for a linear-elastic ideal plastic material

As is shown in Fig. 4.30a, the material behavior is linear-elastic in the elastic range (characterized by Young's modulus $E$) and linear hardening in the elasto-plastic range (characterized by the elasto-plastic modulus $E^{\text{elpl}}$). For simplicity, we will additionally assume that the initial and subsequent yield stress is the same in tension and compression: $k_{\text{t}}^{\text{init}} = k_{\text{c}}^{\text{init}} = k^{\text{init}} \wedge k_{\text{t}} = k_{\text{c}} = k$. As in Sect. 4.2.6.1, the derivation will be demonstrated based on the example of a simply supported BERNOULLI beam (cf. Fig. 4.18) where the external moment is linearly increased

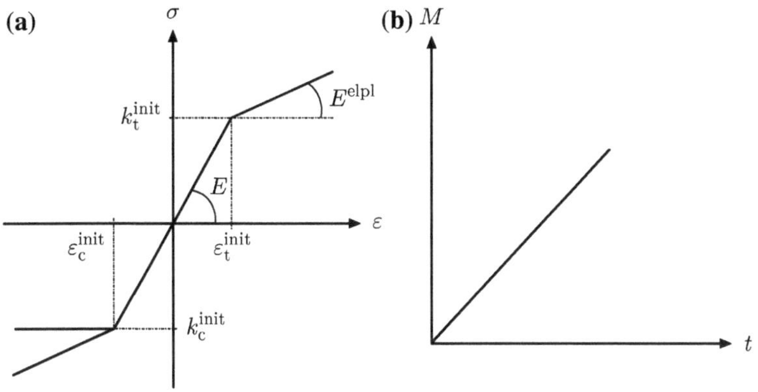


**Fig. 4.30** Material behavior and loading condition for the beam shown in Fig. 4.18: **a** uniaxial stress-strain diagram with linear hardening, **b** moment-time function

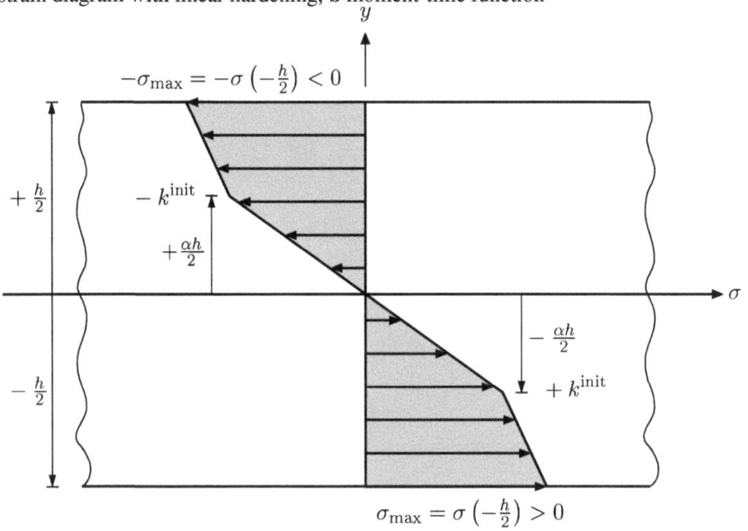


**Fig. 4.31** Stress distribution over the height of the cross section in the elasto-plastic range for a linear hardening material

from zero as shown in Fig. 4.30b. Let us consider now a typical elasto-plastic stress distribution over the height of a rectangular beam as shown in Fig. 4.31.

It might be advantageous for the following derivations to plot the stress in a $\sigma$–$y$ diagram for $y > 0$ as shown in Fig. 4.32a.

The slopes of the two line segments in Fig. 4.32a can be expressed as

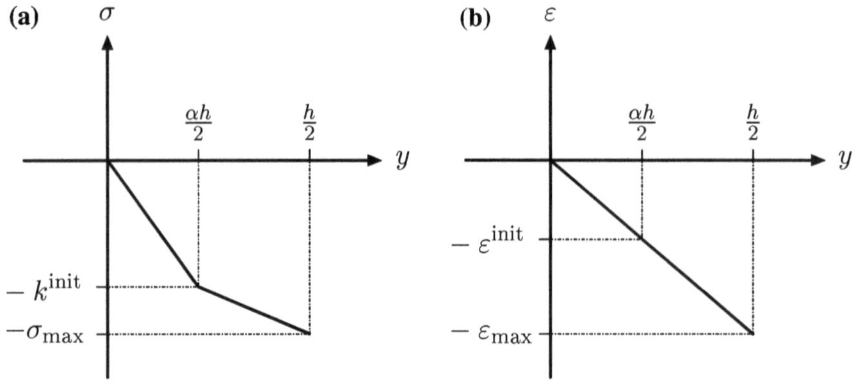


**Fig. 4.32** **a** Stress distribution and **b** strain distribution over the height of a beam in the elasto-plastic-range with linear hardening

$$\frac{\mathrm{d}\sigma}{\mathrm{d}y} = \begin{cases} -\dfrac{k^{\mathrm{init}}}{\frac{\alpha h}{2}} & \text{for } |y| \leq \dfrac{\alpha h}{2} \\ \dfrac{-\sigma_{\max} + k^{\mathrm{init}}}{\frac{h}{2} - \frac{\alpha h}{2}} & \text{for } |y| \geq \dfrac{\alpha h}{2} \end{cases}, \tag{4.117}$$

from which the stress course can be expressed as:

$$\sigma(y) = \begin{cases} -\dfrac{k^{\mathrm{init}}}{\frac{\alpha h}{2}} \times y & \text{for } |y| \leq \dfrac{\alpha h}{2} \\ -k^{\mathrm{init}} - \dfrac{\sigma_{\max} - k^{\mathrm{init}}}{\frac{h}{2} - \frac{\alpha h}{2}} \times \left(y - \dfrac{\alpha h}{2}\right) & \text{for } |y| \geq \dfrac{\alpha h}{2} \end{cases}. \tag{4.118}$$

The problem with the last equation is that the stress is still not expressed solely as a function of the $y$-coordinate, i.e. $\sigma = \sigma(y)$, since the maximum stress at the outer fiber ($\sigma_{\max}$) is not a constant. Thus, it would be more appropriate to write in Eq. (4.118): $\sigma = \sigma(y, \sigma_{\max})$. The application of the intercept theorem (see Appendix A.15) to the strain-coordinate curve shown in Fig. 4.32b gives the following relationship:

$$\frac{\varepsilon^{\mathrm{init}}}{\varepsilon_{\max}} = \frac{\frac{\alpha h}{2}}{\frac{h}{2}}. \tag{4.119}$$

Combining the last equation with Eq. (4.71) at the outer fiber gives:

$$\varepsilon_{\max} = -\frac{h}{2}\kappa \overset{(4.119)}{=} \varepsilon^{\mathrm{init}} \frac{\frac{h}{2}}{\frac{\alpha h}{2}} \overset{\text{HOOKE}}{=} \frac{k^{\mathrm{init}}}{E} \frac{\frac{h}{2}}{\frac{\alpha h}{2}}. \tag{4.120}$$

Looking at the stress-strain diagram shown in Fig. 4.30a, the maximum stress can be written as:

$$\sigma_{\text{max}} = k^{\text{init}} + E^{\text{elpl}}(\varepsilon_{\text{max}} - \varepsilon^{\text{init}}). \tag{4.121}$$

The maximum strain in the last equation can be replaced by the expression (4.120) to give:

$$\begin{aligned}\sigma_{\text{max}} &= k^{\text{init}} + E^{\text{elpl}}\left(\frac{k^{\text{init}}}{E}\frac{\frac{h}{2}}{\frac{\alpha h}{2}} - \frac{k^{\text{init}}}{E}\right)\\ &= k^{\text{init}} + \frac{E^{\text{elpl}}}{E}k^{\text{init}}\left(\frac{\frac{h}{2}}{\frac{\alpha h}{2}} - 1\right).\end{aligned} \tag{4.122}$$

The last expression for $\sigma_{\text{max}}$ can be introduced into Eq. (4.118) to give the stress distribution in the elasto-plastic range as:

$$\sigma = -k^{\text{init}} - \frac{\frac{E^{\text{elpl}}}{E}k^{\text{init}}\left(\frac{\frac{h}{2}-\frac{\alpha h}{2}}{\frac{\alpha h}{2}}\right)}{\frac{h}{2}-\frac{\alpha h}{2}} \times \left(y - \frac{\alpha h}{2}\right). \tag{4.123}$$

Thus, Eq. (4.118) can be finally written as:

$$\sigma(y) = \begin{cases} -\dfrac{k^{\text{init}}}{\frac{\alpha h}{2}} \times y & \text{for } |y| \leq \dfrac{\alpha h}{2} \\ -k^{\text{init}} - \dfrac{E^{\text{elpl}}}{E}k^{\text{init}}\left(\dfrac{y}{\frac{\alpha h}{2}} - 1\right) & \text{for } |y| \geq \dfrac{\alpha h}{2} \end{cases}. \tag{4.124}$$

Corresponding to Eqs. (4.70) and (4.75), the acting bending moment can be calculated as:

$$\begin{aligned}M_z(\alpha) &= -2b\left[\int_0^{\frac{\alpha h}{2}} y\left(-\frac{k^{\text{init}}}{\frac{\alpha h}{2}} \times y\right)\mathrm{d}y + \int_{\frac{\alpha h}{2}}^{\frac{h}{2}} y\left(-k^{\text{init}} - \frac{E^{\text{elpl}}}{E}k^{\text{init}}\left(\frac{y}{\frac{\alpha h}{2}} - 1\right)\right)\mathrm{d}y\right]\\ &= \underbrace{\frac{1}{3}k^{\text{init}}b\left(\frac{h}{2}\right)^2}_{\frac{1}{2}M_{\text{lim}}^{\text{el}}}\left[3 - \alpha^2 + \frac{E^{\text{elpl}}}{\alpha E} \times \left(2 - 3\alpha + \alpha^3\right)\right],\end{aligned} \tag{4.125}$$

or finally in a normalized form as:

$$\frac{M_z(\alpha)}{M_{\text{lim}}^{\text{el}}} = \frac{1}{2}\left[3 - \alpha^2 + \frac{E^{\text{elpl}}}{\alpha E} \times \left(2 - 3\alpha + \alpha^3\right)\right]. \tag{4.126}$$

The last equation must be solved for $\alpha$ in order to continue the derivation (again the internal moment $M_z(\alpha)$ is equalized with the constant external moment $M$). To this end, the last equation can be rearranged to:

$$\alpha^3 + \underbrace{\frac{2\frac{M_z}{M_{\text{lim}}^{\text{el}}} - 3 + 3\frac{E^{\text{elpl}}}{E}}{1 - \frac{E^{\text{elpl}}}{E}}}_{a}\alpha - \underbrace{\frac{2\frac{E^{\text{elpl}}}{E}}{1 - \frac{E^{\text{elpl}}}{E}}}_{b} = 0, \tag{4.127}$$

or in short as:

$$\alpha^3 + a\alpha + b = 0. \tag{4.128}$$

It should be noted here that Eq. (4.127) reduces for the special case of ideal plasticity with $\frac{E^{\text{elpl}}}{E} = 0$, $a = 2\frac{M_z}{M_{\text{lim}}^{\text{el}}} - 3$, $b = 0$ and $\frac{M_{\text{lim}}^{\text{pl}}}{M_{\text{lim}}^{\text{el}}} = \frac{3}{2}$ to the ideal plastic solution as given in Eq. (4.83).

The cubic equation (4.127) has for the given boundaries of the moment and stiffness ratio one real root and two imaginary roots. The real root is given by:

$$\alpha = \left(-\frac{b}{2} + \sqrt{\frac{a^3}{27} + \frac{b^2}{4}}\right)^{\frac{1}{3}} - \frac{a}{3}\left(-\frac{b}{2} + \sqrt{\frac{a^3}{27} + \frac{b^2}{4}}\right)^{-\frac{1}{3}}. \tag{4.129}$$

As in the case of Eqs. (4.81) and (4.89), one can write

$$\kappa = \frac{\mathrm{d}^2 u_y(x)}{\mathrm{d}x^2} = \frac{2k^{\text{init}}}{Eh\alpha}. \tag{4.130}$$

Integrating twice gives the following statement:

$$u_y(x) = \frac{k^{\text{init}}}{Eh\alpha}x^2 + c_1 x + c_2, \tag{4.131}$$

from which under consideration of the boundary conditions, i.e. $u_y(0) = 0$ and $u_y(L) = 0$, and the constants of integration, i.e. $c_2 = 0$ and $c_1 = -\frac{k^{\text{init}}L}{Eh\alpha}$, the following normalized displacement expression results:

$$\frac{u_y(x)}{\frac{M_{\text{lim}}^{\text{el}}L^2}{EI_z}} = \frac{1}{2\alpha} \times \left[\left(\frac{x}{L}\right)^2 - \frac{x}{L}\right], \tag{4.132}$$

where factor $\alpha$ is given by Eq. (4.129).

A few sampling points for a stiffness ratio of $\frac{E^{\text{elpl}}}{E} = 0.3$ are collected in Table 4.15. Assigning a stiffness ratio of $\frac{M_z}{M^{\text{el}}_{\text{lim}}} = 1$ to Eq. (4.132) means the pure elastic range and the same values as given in Table 4.14 are obtained. In order to correctly compare the numerical values in both tables, the values in Table 4.15 should be multiplied by $\frac{2}{3}$ since the normalization is done based on different limiting moments. Furthermore, it can be seen that the hardening material results in smaller displacements compared to the ideal plastic material since the bending stiffness is larger in the elasto-plastic range.

Let us consider in the following a similar problem as given in Example 4.2 where a cantilevered BERNOULLI beam under end load was considered, see Fig. 4.24. Now, the material is assumed as linear-elastic with linear hardening in the plastic range as shown in Fig. 4.33b. To simplify the expression for the moment distribution, the $x$-coordinate is directed from the right-hand end to left-hand side of the beam as shown in Fig. 4.33a.

Thus, the bending moment distribution can be written as $M_z(x) = Fx$ and the equalization between the internal moment due to the normal stress over the cross section and the moment distribution due to the external load gives (cf. Eq. (4.126)):

$$\frac{Fx}{M^{\text{el}}_{\text{lim}}} = \frac{1}{2}\left[3 - \alpha^2 + \frac{E^{\text{elpl}}}{\alpha E} \times \left(2 - 3\alpha + \alpha^3\right)\right], \tag{4.133}$$

or solved for $\alpha$ as (cf. Eq. (4.129))

**Table 4.15** Evaluation of Eq. (4.132) at a few sampling points for the stiffness ratio $\frac{E^{\text{elpl}}}{E} = 0.3$

| $\frac{M_z}{M^{\text{el}}_{\text{lim}}} \setminus \frac{x}{L}$ | 0 | $\frac{1}{4}$ | $\frac{1}{2}$ | $\frac{3}{4}$ | $L$ |
|---|---|---|---|---|---|
| 1.0 | 0 | −0.093750 | −0.125000 | −0.093750 | 0 |
| 1.2 | 0 | −0.117045 | −0.156060 | −0.117045 | 0 |
| 1.35 | 0 | −0.141656 | −0.188875 | −0.141656 | 0 |
| 1.425 | 0 | −0.156458 | −0.208611 | −0.156458 | 0 |

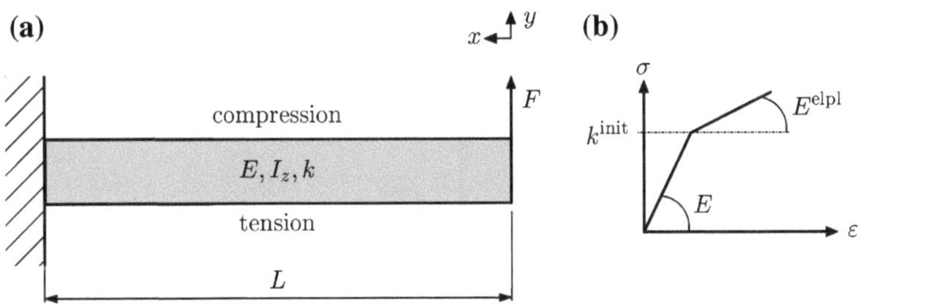


**Fig. 4.33** **a** Cantilevered BERNOULLI beam loaded by a single force $F$ at its end, **b** uniaxial stress-strain diagram with linear hardening

$$\alpha(x) = \left(-\frac{b}{2}+\sqrt{\frac{a(x)^3}{27}+\frac{b^2}{4}}\right)^{\frac{1}{3}} - \frac{a(x)}{3}\left(-\frac{b}{2}+\sqrt{\frac{a(x)^3}{27}+\frac{b^2}{4}}\right)^{-\frac{1}{3}}, \quad (4.134)$$

where

$$a(x) = \frac{2\frac{Fx}{M_{\text{lim}}^{\text{el}}} - 3 + 3\frac{E^{\text{elpl}}}{E}}{1 - \frac{E^{\text{elpl}}}{E}}, \quad (4.135)$$

$$b = -\frac{2\frac{E^{\text{elpl}}}{E}}{1 - \frac{E^{\text{elpl}}}{E}}. \quad (4.136)$$

To derive a closed-form solution for the bending line, the differential equation (cf. Eq. (4.130))

$$\kappa = \frac{\mathrm{d}^2 u_y(x)}{\mathrm{d}x^2} = \frac{2k^{\text{init}}}{Eh\alpha(x)} \quad (4.137)$$

has to be integrated twice. However, a closed form-solution cannot be obtained in this case and, for example, numerical approximate techniques could be applied. An alternative solution procedure can be derived based on the additional assumption that the elastic part of the material behavior can be neglected and a so-called rigid plastic material is obtained, see Fig. 4.34. This simplification might be reasonable for larger plastic strains.

For such a material, the stress is linearly distributed over the cross section as shown in Fig. 4.35a or constant as in the very first moment of any loading, see Fig. 4.35b. The strain is—as in all the previous derivations—linearly distributed as shown in

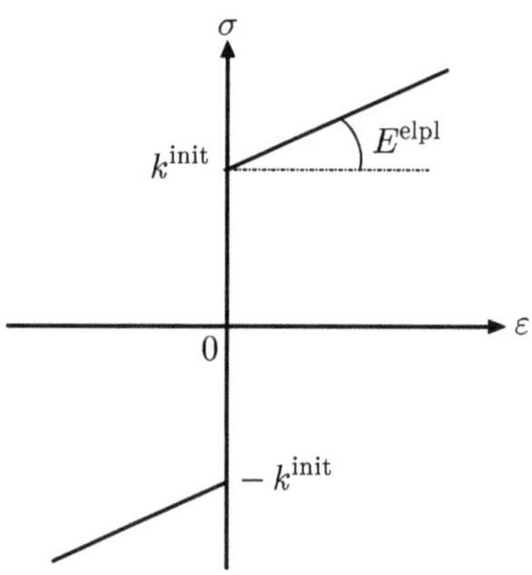


**Fig. 4.34** Uniaxial stress-strain diagram of a rigid plastic material with hardening

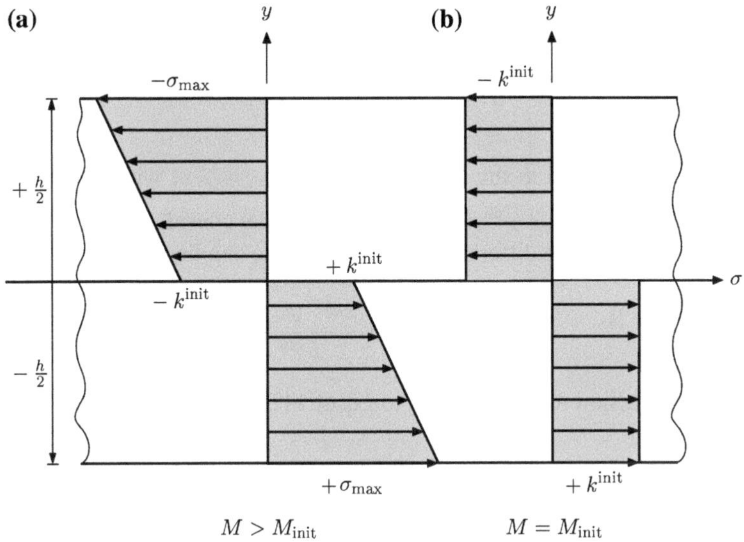


**Fig. 4.35** Stress distribution over the height of the cross section for a rigid plastic material with linear hardening: **a** for $M > M_{init}$ and **b** for $M = M_{init}$

Fig. 4.21a. Since there is no elastic strain, the total strain in this case is equal to the plastic strain: $\varepsilon = \varepsilon^{pl}$.

The slope of the line segments in Fig. 4.35a can be expressed as

$$\frac{d\sigma}{dy} = \frac{-\sigma_{max} + k^{init}}{\frac{h}{2} - 0}, \tag{4.138}$$

from which the stress course can be expressed as

$$\sigma(y, \sigma_{max}) = -k^{init} - \frac{\sigma_{max} - k^{init}}{\frac{h}{2}} \times y, \tag{4.139}$$

or by expressing the maximum stress as a fraction of the initial stress, i.e. $\sigma_{max} = \alpha k^{init}$:

$$\sigma(y, \alpha) = -k^{init} - \frac{\alpha k^{init} - k^{init}}{\frac{h}{2}} \times y. \tag{4.140}$$

Corresponding to Eq. (4.70), the acting bending moment can be calculated as:

$$M_z(\alpha) = -2b \int_0^{\frac{h}{2}} y \left( -k^{init} - \frac{(\alpha - 1)k^{init}}{\frac{h}{2}} \times y \right) dy = \frac{bk^{init}h^2}{12}(1 + 2\alpha), \tag{4.141}$$

from which the factor $\alpha$ is obtained as:

$$\alpha(M_z) = -\frac{1}{2} + \frac{6M_z}{bk^{\text{init}}h^2}. \tag{4.142}$$

Combining Eq. (4.71) at the outer fiber, i.e. $\varepsilon_{\text{max}} = \frac{h}{2}\kappa$, with the stress-strain relation in the form $\sigma_{\text{max}} = k^{\text{init}} + E^{\text{elpl}}\varepsilon_{\text{max}}$ gives under consideration of $\kappa = \frac{\mathrm{d}^2 u_y(x)}{\mathrm{d}x^2}$ the following form of the differential equation for the bending line:

$$\kappa = \frac{\mathrm{d}^2 u_y(x)}{\mathrm{d}x^2} = \frac{2k^{\text{init}}(\alpha - 1)}{E^{\text{elpl}}h}, \tag{4.143}$$

or with the expression for $\alpha$ according to Eq. (4.142) as:

$$\frac{\mathrm{d}^2 u_y(x)}{\mathrm{d}x^2} = \frac{2k^{\text{init}}(-\frac{3}{2} + \frac{6M_z}{bk^{\text{init}}h^2})}{E^{\text{elpl}}h}. \tag{4.144}$$

The last equation can be modified by introducing a so-called initial moment, $M_{\text{init}} = \frac{bh^2k^{\text{nit}}}{4} = \frac{3k^{\text{init}}I_z}{h}$, which corresponds to the very first moment of any loading, see Fig. 4.35b:

$$\frac{\mathrm{d}^2 u_y(x)}{\mathrm{d}x^2} = \frac{M_{\text{init}}}{E^{\text{elpl}}I_z}\left(-1 + \frac{M_z}{M_{\text{init}}}\right). \tag{4.145}$$

Integrating twice gives the following statement:

$$u_y(x) = \frac{M_{\text{init}}}{E^{\text{elpl}}I_z}\left(-1 + \frac{M_z}{M_{\text{init}}}\right)\frac{x^2}{2} + c_1 x + c_2. \tag{4.146}$$

If we consider, for example, in the following the case of a simply supported BERNOULLI beam (cf. Fig. 4.18) with a constant bending loading ($M_z = \text{const.}$) over the entire length, the consideration of the boundary conditions, i.e. $u_y(0) = 0$ and $u_y(L) = 0$, gives the following expression for the normalized bending line:

$$\frac{u_y(x)}{\frac{M_{\text{init}}L^2}{E^{\text{elpl}}I_z}} = \frac{1}{2}\left(-1 + \frac{M_z}{M_{\text{init}}}\right)\left[\left(\frac{x}{L}\right)^2 - \frac{x}{L}\right]. \tag{4.147}$$

It should be noted here that the important achievement of the above derivation is that an arbitrary moment distributions, i.e. $M_z = M_z(x)$, can be introduced into Eq. (4.145) and easily integrated.

At the end of this section, Fig. 4.36 compares the difference between the stress and strain state for an elasto-plastic material with linear hardening. As soon as the

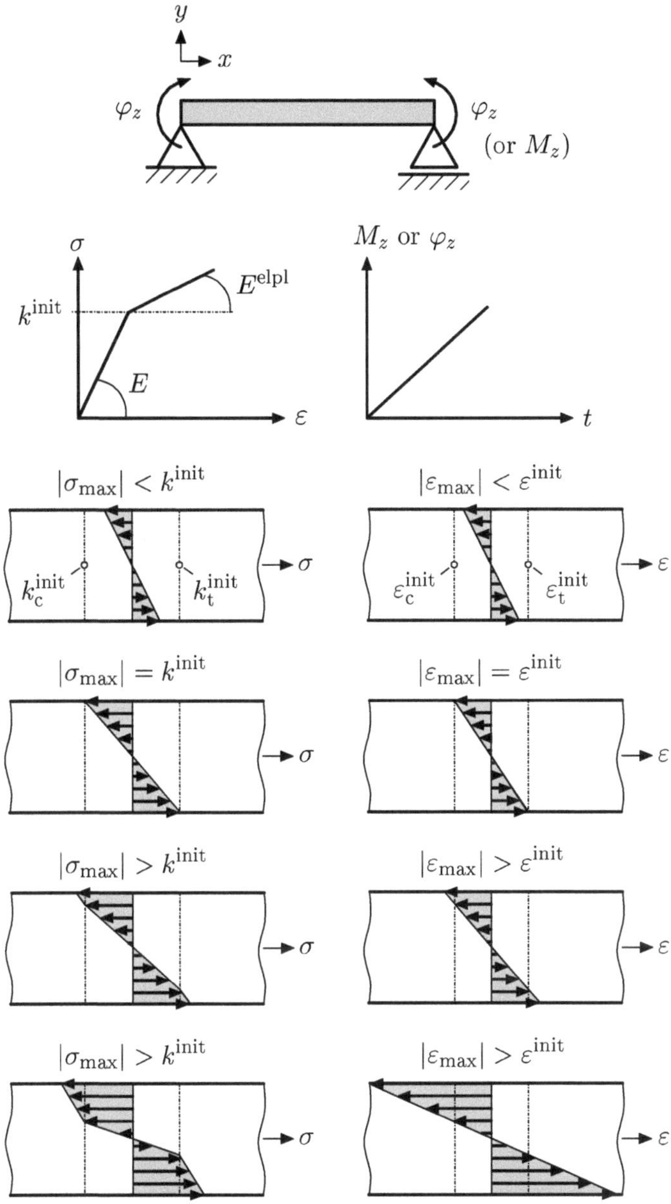


**Fig. 4.36** Comparison between stress and strain distribution over the height of a beam made of an elasto-plastic material with linear hardening

stress at the outer fiber is larger than the initial yield stress, the stress course becomes discontinuous while the strain remains linearly distributed.

## 4.3 Timoshenko Beam Theory

The general differences regarding the deformation and stress distribution of a beam with and without shear influence have already been discussed in Sect. 4.1. In this section, the shear influence is considered with the help of the TIMOSHENKO beam theory. Within the framework of the following remarks, first the definition of the shear strain and the relation between shear force and shear stress needs to be covered.

For the derivation of the equation for the shear strain in the $x$-$y$ plane, the infinitesimal rectangular beam element $ABCD$, shown in Fig. 4.37, is considered, which deforms under the influence of a pure shear stress. Here, a change of the angle of the original right angles as well as a change in the lengths of the edges occurs.

The deformation of the point $A$ can be described via the displacement fields $u_x(x, y)$ and $u_y(x, y)$. These two functions of *two* variables can be expanded in TAYLOR's series[22] of first order around point $A$ to calculate approximately the defor-

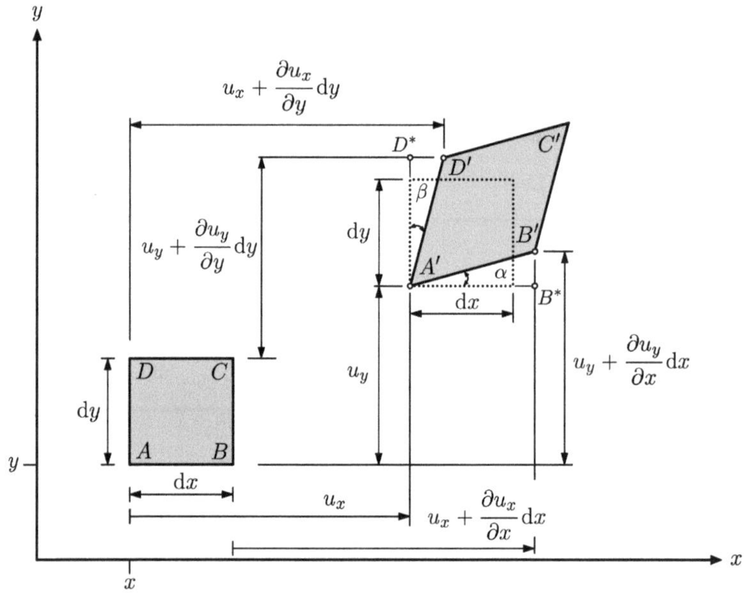


**Fig. 4.37** Definition of the shear strain $\gamma_{xy}$ in the $x$-$y$ plane at an infinitesimal beam element

[22] For a function $f(x, y)$ of two variables usually a TAYLOR's series expansion of first order is formulated around the point $(x_0, y_0)$ as follows: $f(x, y) = f(x_0 + dx, y_0 + dx) \approx f(x_0, y_0) + \left(\frac{\partial f}{\partial x}\right)_{x_0, y_0} \times (x - x_0) + \left(\frac{\partial f}{\partial y}\right)_{x_0, y_0} \times (y - y_0)$.

mations of the points $B$ and $D$ :

$$u_{x,B} = u_x(x + \mathrm{d}x, y) = u_x(x, y) + \frac{\partial u_x}{\partial x}\mathrm{d}x + \frac{\partial u_x}{\partial y}\mathrm{d}y, \tag{4.148}$$

$$u_{y,B} = u_y(x + \mathrm{d}x, y) = u_y(x, y) + \frac{\partial u_y}{\partial x}\mathrm{d}x + \frac{\partial u_y}{\partial y}\mathrm{d}y, \tag{4.149}$$

or alternatively

$$u_{x,D} = u_x(x, y + \mathrm{d}y) = u_x(x, y) + \frac{\partial u_x}{\partial x}\mathrm{d}x + \frac{\partial u_x}{\partial y}\mathrm{d}y, \tag{4.150}$$

$$u_{y,D} = u_y(x, y + \mathrm{d}y) = u_y(x, y) + \frac{\partial u_y}{\partial x}\mathrm{d}x + \frac{\partial u_y}{\partial y}\mathrm{d}y. \tag{4.151}$$

In Eqs. (4.148) up to (4.151) $u_x(x, y)$ and $u_y(x, y)$ represent the so-called rigid-body displacements, which do not cause a deformation. If one considers that point $B$ has the coordinates $(x + \mathrm{d}x, y)$ and $D$ the coordinates $(x, y + \mathrm{d}y)$, the following results:

$$u_{x,B} = u_x(x, y) + \frac{\partial u_x}{\partial x}\mathrm{d}x, \tag{4.152}$$

$$u_{y,B} = u_y(x, y) + \frac{\partial u_y}{\partial x}\mathrm{d}x, \tag{4.153}$$

or alternatively

$$u_{x,D} = u_x(x, y) + \frac{\partial u_x}{\partial y}\mathrm{d}y, \tag{4.154}$$

$$u_{y,D} = u_y(x, y) + \frac{\partial u_y}{\partial y}\mathrm{d}y. \tag{4.155}$$

The total shear strain $\gamma_{xy}$ of the deformed beam element $A'B'C'D'$ results, according to Fig. 4.37, from the sum of the angles $\alpha$ and $\beta$. The two angles can be identified in the rectangle, which is deformed to a rhombus. Under consideration of the two right-angled triangles $A'D^*D'$ and $A'B^*B'$, these two angles can be expressed as:

$$\tan\alpha = \frac{\frac{\partial u_y}{\partial x}\mathrm{d}x}{\mathrm{d}x + \frac{\partial u_x}{\partial x}\mathrm{d}x} \quad \text{and} \quad \tan\beta = \frac{\frac{\partial u_x}{\partial y}\mathrm{d}y}{\mathrm{d}y + \frac{\partial u_y}{\partial y}\mathrm{d}y}. \tag{4.156}$$

It holds approximately for small deformations that $\tan\alpha \approx \alpha$ and $\tan\beta \approx \beta$ or alternatively $\frac{\partial u_x}{\partial x} \ll 1$ and $\frac{\partial u_y}{\partial y} \ll 1$, so that the following expression results for the shear strain:

$$\gamma_{xy} = \alpha + \beta \approx \frac{\partial u_y}{\partial x} + \frac{\partial u_x}{\partial y}. \tag{4.157}$$

This total change of the angle is also called the engineering shear strain. In contrast to this, the expression $\varepsilon_{xy} = \frac{1}{2}\gamma_{xy} = \frac{1}{2}(\frac{\partial u_y}{\partial x} + \frac{\partial u_x}{\partial y})$ is known as the tensorial definition (tensor shear strain) in the literature [110]. Due to the symmetry of the strain tensor, the identity $\gamma_{ij} = \gamma_{ji}$ applies to the tensor elements outside the main diagonal.

The algebraic sign of the shear strain needs to be explained in the following with the help of Fig. 4.38 for the special case that only one shear force acts in parallel to the $y$-axis. If a shear force acts in the direction of the positive $y$-axis at the right-hand face—hence a positive shear force distribution is being assumed at this point —, according to Fig. 4.38a under consideration of Eq. (4.157) a positive shear strain results. In a similar way, a negative shear force distribution leads to a negative shear strain according to Fig. 4.38b.

It has already been mentioned in Sect. 4.1 that the shear stress distribution is variable over the cross-section. As an example, the parabolic shear stress distribution was illustrated over a rectangular cross section in Fig. 4.3. Based on HOOKE's law for a one-dimensional shear stress state, it can be derived that the shear stress has to exhibit a corresponding parabolic course. From the shear stress distribution in the cross-sectional area at location $x$ of the beam,[23] one receives the acting shear force through integration as:

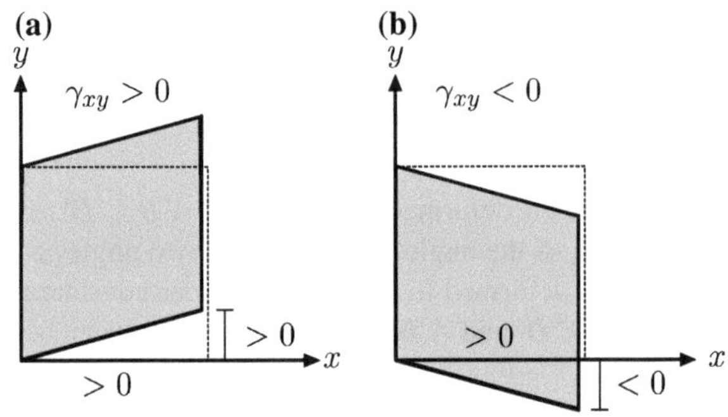


**Fig. 4.38** Definition of a **a** positive and **b** negative shear strain in the $x$-$y$ plane

[23] A closer analysis of the shear stress distribution in the cross-sectional area shows that the shear stress does not just alter over the height of the beam but also through the width of the beam. If the width of the beam is small compared to the height, only a small change along the width occurs and one can assume in the first approximation a constant shear stress throughout the width: $\tau_{xy}(y, z) \rightarrow \tau_{xy}(y)$. See for example [9, 106].

$$Q_y = \int_A \tau_{xy}(y, z)\,\mathrm{d}A. \tag{4.158}$$

To simplify the problem, it is however assumed for the TIMOSHENKO beam that an equivalent *constant* shear stress and strain act:

$$\tau_{xy}(y, z) \rightarrow \tau_{xy}. \tag{4.159}$$

This constant shear stress results from the shear force, which acts in an equivalent cross-sectional area, the so-called shear area $A_s$:

$$\tau_{xy} = \frac{Q_y}{A_s}, \tag{4.160}$$

whereupon the relation between the shear area $A_s$ and the actual cross-sectional area $A$ is referred to as the shear correction factor $k_s$:

$$k_s = \frac{A_s}{A}. \tag{4.161}$$

Different assumptions can be made for the calculation of the shear correction factor [28]. As an example, it can be demanded [7] that the elastic strain energy of the equivalent shear stress has to be identical with the energy, which results from the acting shear stress distribution in the actual cross-sectional-area. A comparison for a rectangular cross section is presented in Table 4.16.

Different geometric characteristics of simple geometric cross-sections—including the shear correction factor[24]—are collected in Table 4.17 [43, 114]. Further details

**Table 4.16** Comparison of shear correction factor values for a rectangular cross section based on different approaches

| $k_s$ | Comment | References |
|---|---|---|
| $\frac{2}{3}$ | – | [102,103] |
| $0.833\left(=\frac{5}{6}\right)$ | $\nu = 0.0$ | [28] |
| 0.850 | $\nu = 0.3$ | – |
| 0.870 | $\nu = 0.5$ | – |

[24] It should be noted that the so-called form factor for shear is also known in the literature. This results as the reciprocal of the shear correction factor.

**Table 4.17** Characteristics of different cross sections in the $y$-$z$ plane. $I_z$ and $I_y$: axial second moment of area; $A$: cross-sectional area; $k_s$: shear correction factor. Adapted from [114]

| cross-section | $I_z$ | $I_y$ | $A$ | $k_s$ |
|---|---|---|---|---|
| 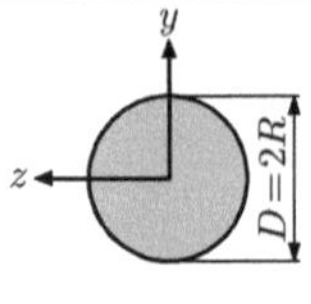 | $\frac{\pi R^4}{4}$ | $\frac{\pi R^4}{4}$ | $\pi R^2$ | $\frac{9}{10}$ |
| 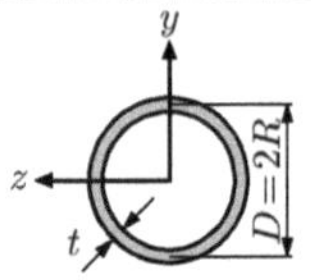 | $\pi R^3 t$ | $\pi R^3 t$ | $2\pi R t$ | 0.5 |
| 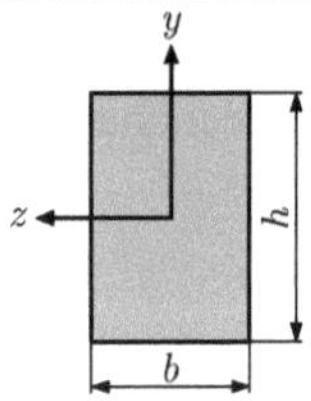 | $\frac{bh^3}{12}$ | $\frac{hb^3}{12}$ | $hb$ | $\frac{5}{6}$ |
| 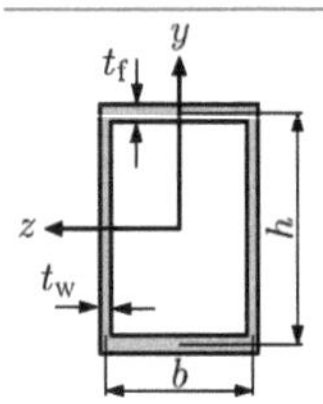 | $\frac{h^2}{6}(ht_w + 3bt_f)$ | $\frac{b^2}{6}(bt_f + 3ht_w)$ | $2(bt_f + ht_w)$ | $\frac{2ht_w}{A}$ |
| 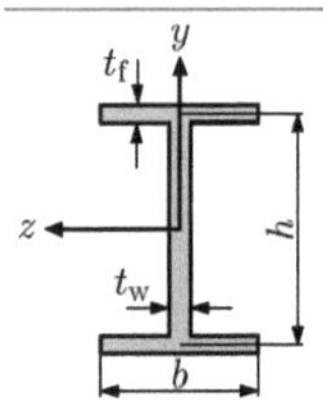 | $\frac{h^2}{12}(ht_w + 6bt_f)$ | $\frac{b^3 t_f}{6}$ | $ht_w + 2bt_f$ | $\frac{ht_w}{A}$ |

regarding the shear correction factor for arbitrary cross-sections can be taken from [47].

It is obvious that the equivalent constant shear stress can alter along the center line of the beam, in case the shear force along the center line of the beam changes. The attribute 'constant' thus just refers to the cross-sectional area at location $x$ and the equivalent constant shear stress is therefore in general a function of the coordinate of length for the TIMOSHENKO beam:

$$\tau_{xy} = \tau_{xy}(x). \tag{4.162}$$

The so-called TIMOSHENKO beam can be generated by superposing a shear deformation on a BERNOULLI beam according to Fig. 4.39.

One can see that the BERNOULLI hypothesis is partly no longer fulfilled for the TIMOSHENKO beam: Plane cross sections remain plane after the deformation. However, a cross section which stood at right angles on the beam axis before the deformation is not at right angles on the beam axis after the deformation. If the demand for planeness of the cross sections is also given up, one reaches theories of higher-order [65, 82, 84], at which, for example, a parabolic course of the shear strain and stress in the displacement field are considered, see Fig. 4.40. Therefore, a shear correction factor is not required for these theories of higher-order.

## 4.3.1 Kinematics

According to the alternative derivation in Sect. 4.2.1, the kinematics relation can also be derived for the beam with shear action, by considering the angle $\phi_z$ instead of the angle $\varphi_z$, see Fig. 4.39c. Following an equivalent procedure as in Sect. 4.2.1, corresponding relationships are obtained:

$$\sin\phi_z = \frac{-u_x}{y} \approx \phi_z \quad \text{or} \quad u_x = -y\phi_z, \tag{4.163}$$

wherefrom, via the general relation for the strain, meaning $\varepsilon_x = \mathrm{d}u_x/\mathrm{d}x$, the kinematics relation results through differentiation with respect to the $x$-coordinate:

$$\varepsilon_x = -y\frac{\mathrm{d}\phi_z}{\mathrm{d}x}. \tag{4.164}$$

Note that $\phi_z \rightarrow \varphi_z = \frac{\mathrm{d}u_y}{\mathrm{d}x}$ results from neglecting the shear deformation and a relation according to Eq. (4.15) results as a special case. Furthermore, the following relation between the angles can be derived from Fig. 4.39c

$$\phi_z = \frac{\mathrm{d}u_y}{\mathrm{d}x} - \gamma_{xy}, \tag{4.165}$$

which complements the set of the kinematics relations. It needs to be remarked that at this point the so-called bending line was considered. Therefore, the displacement field $u_y$ is only a function of *one* variable: $u_y = u_y(x)$.

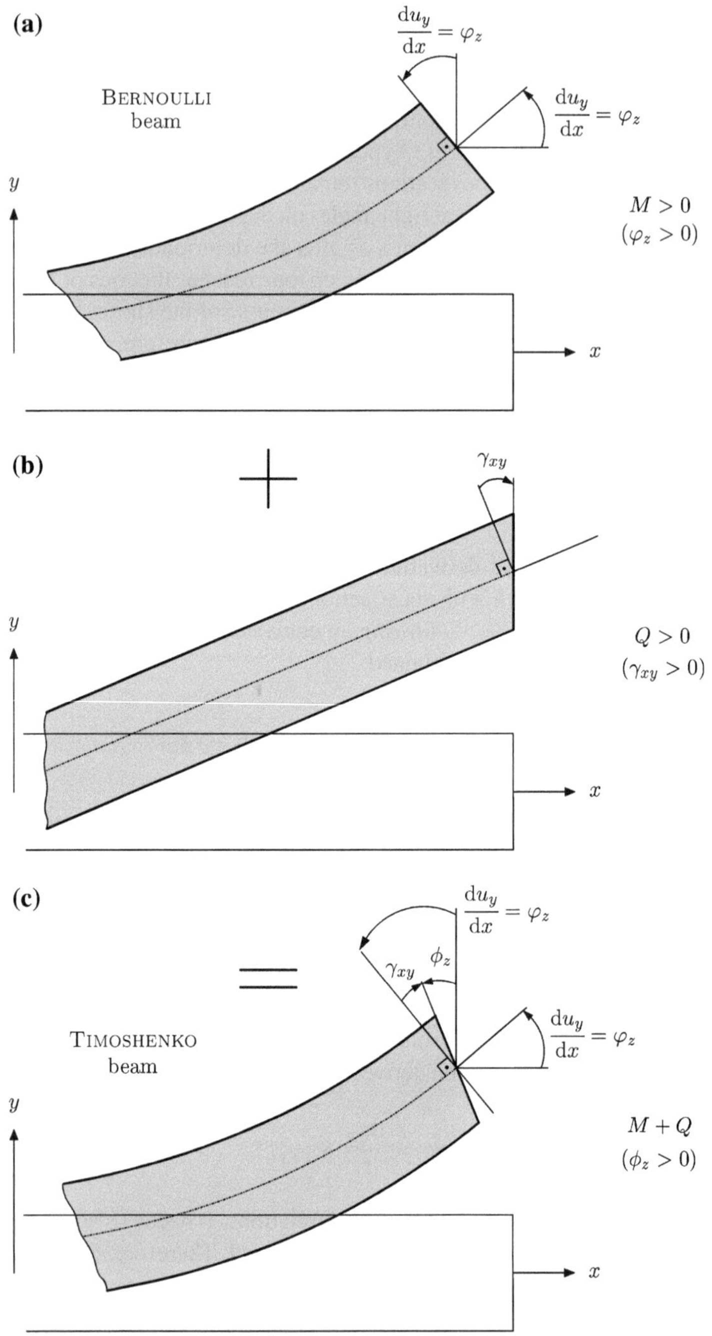


**Fig. 4.39** Superposition of the BERNOULLI beam **a** and the shear deformation, **b** to the TIMOSHENKO beam, **c** in the $x$-$y$ plane. The indicated orientations of the angles equal the positive definitions. Note that the deformation is overdrawn for better illustration

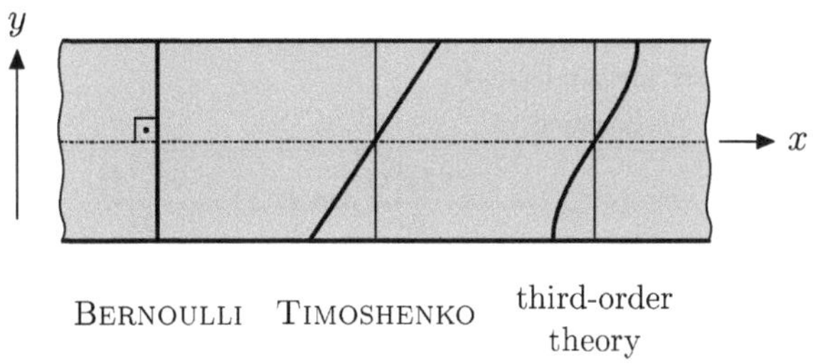


**Fig. 4.40** Deformation of originally plane cross-sections for the BERNOULLI beam (*left*), the TIMOSHENKO beam (*middle*) and a higher-order theory (*right*) [85]

### 4.3.2 Equilibrium

The derivation of the equilibrium condition for the TIMOSHENKO beam is identical with the derivation for the BERNOULLI beam according to Sect. 4.2.2:

$$\frac{\mathrm{d}Q_y(x)}{\mathrm{d}x} = -q_y(x), \tag{4.166}$$

$$\frac{\mathrm{d}M_z(x)}{\mathrm{d}x} = -Q_y(x). \tag{4.167}$$

### 4.3.3 Constitutive Equation

For the consideration of the constitutive relation, HOOKE's law for a one-dimensional normal stress state and for a one-dimensional shear stress state is used:

$$\sigma_x = E\varepsilon_x, \tag{4.168}$$

$$\tau_{xy} = G\gamma_{xy}, \tag{4.169}$$

whereupon the shear modulus $G$ can be calculated based on the YOUNG's modulus $E$ and the POISSON's ratio $\nu$ as:

$$G = \frac{E}{2(1+\nu)}. \tag{4.170}$$

According to the equilibrium configuration of Fig. 4.9 and Eq. (4.29), the relation between the internal moment and the bending stress can be used for the TIMOSHENKO beam as follows:

$$\mathrm{d}M_z = (+y)(-\sigma_x)\mathrm{d}A, \tag{4.171}$$

**Table 4.18** Elementary basic equations for the bending of a TIMOSHENKO beam in the $x$-$y$ plane ($x$-axis: right facing; $y$-axis: upward facing)

| Relation | Equation |
|---|---|
| Kinematics | $\varepsilon_x(x, y) = -y\dfrac{\mathrm{d}\phi_z(x)}{\mathrm{d}x}$ and $\phi_z(x) = \dfrac{\mathrm{d}u_y(x)}{\mathrm{d}x} - \gamma_{xy}(x)$ |
| Equilibrium | $\dfrac{\mathrm{d}Q_y(x)}{\mathrm{d}x} = -q_y(x)$; $\dfrac{\mathrm{d}M_z(x)}{\mathrm{d}x} = -Q_y(x)$ |
| Constitution | $\sigma_x(x, y) = E\varepsilon_x(x, y)$ and $\tau_{xy}(x) = G\gamma_{xy}(x)$ |

or alternatively after integration under the consideration of the constitutive equation (4.168) and the kinematics relation (4.164):

$$M_z(x) = EI_z\frac{\mathrm{d}\phi_z(x)}{\mathrm{d}x}. \tag{4.172}$$

The relation between shear force and cross-sectional rotation results from the equilibrium equation (4.167) as:

$$Q_y(x) = -\frac{\mathrm{d}M_z(x)}{\mathrm{d}x} = -EI_z\frac{\mathrm{d}^2\phi_z(x)}{\mathrm{d}x^2}. \tag{4.173}$$

Before looking in more detail at the differential equations of the bending line, let us summarize the basic equations for the TIMOSHENKO beam in Table 4.18. Note that the normal stress and normal strain are functions of both space coordinates, i.e. $x$ and $y$. However, the shear stress and shear strain are only dependent on the $x$-coordinate, since an equivalent *constant* shear stress has been introduced over the cross section as an approximation of the TIMOSHENKO beam theory.

### *4.3.4 Governing Differential Equation*

Within the previous section , the relation between the internal moment and the cross-sectional rotation was derived from the normal stress distribution with the help of HOOKE's law, see Eq. (4.172). Differentiation of this relation with respect to the $x$-coordinate leads to the following expression

$$\frac{\mathrm{d}M_z}{\mathrm{d}x} = \frac{\mathrm{d}}{\mathrm{d}x}\left(EI_z\frac{\mathrm{d}\phi_z}{\mathrm{d}x}\right), \tag{4.174}$$

which can be transformed with the help of the equilibrium relation (4.167) and the relation for the shear stress according to (4.160) and (4.161) to

$$\frac{\mathrm{d}}{\mathrm{d}x}\left(EI_z\frac{\mathrm{d}\phi_z}{\mathrm{d}x}\right) = -k_{\mathrm{s}}GA\gamma_{xy}. \tag{4.175}$$

If the kinematics relation (4.165) is considered in the last equation, the so-called bending differential equation results in:

$$\frac{\mathrm{d}}{\mathrm{d}x}\left(EI_z\frac{\mathrm{d}\phi_z}{\mathrm{d}x}\right) + k_{\mathrm{s}}GA\left(\frac{\mathrm{d}u_y}{\mathrm{d}x} - \phi_z\right) = 0. \tag{4.176}$$

Considering the shear stress according to Eqs. (4.160) and (4.161) in the expression of HOOKE's law according to (4.169), one obtains

$$Q_y = k_{\mathrm{s}}AG\gamma_{xy}. \tag{4.177}$$

Introducing the equilibrium relation (4.167) and the kinematics relation (4.165) in the last equation gives:

$$\frac{\mathrm{d}M_z}{\mathrm{d}x} = -k_{\mathrm{s}}AG\left(\frac{\mathrm{d}u_y}{\mathrm{d}x} - \phi_z\right). \tag{4.178}$$

After differentiation and the consideration of the equilibrium relations according to Eqs. (4.166) and (4.167), the so-called shear differential equation results finally in:

$$\frac{\mathrm{d}}{\mathrm{d}x}\left[k_{\mathrm{s}}AG\left(\frac{\mathrm{d}u_y}{\mathrm{d}x} - \phi_z\right)\right] = -q_y(x). \tag{4.179}$$

Therefore, the shear flexible TIMOSHENKO beam is described through the following two coupled differential equations of second order:

$$\frac{\mathrm{d}}{\mathrm{d}x}\left(EI_z\frac{\mathrm{d}\phi_z}{\mathrm{d}x}\right) + k_{\mathrm{s}}AG\left(\frac{\mathrm{d}u_y}{\mathrm{d}x} - \phi_z\right) = 0, \tag{4.180}$$

$$\frac{\mathrm{d}}{\mathrm{d}x}\left[k_{\mathrm{s}}AG\left(\frac{\mathrm{d}u_y}{\mathrm{d}x} - \phi_z\right)\right] = -q_y(x). \tag{4.181}$$

This system contains two unknown functions, namely the deflection $u_y(x)$ and the cross-sectional rotation $\phi_z(x)$. Boundary conditions must be formulated for both functions to be able to solve the system of differential equations for a specific problem.

Different formulations of these coupled differential equations are collected in Table 4.19 where different types of loading, geometry and bedding is differentiated. The last case in Table 4.19 refers again to the elastic or WINKLER foundation of a beam, [116]. The elastic foundation modulus $k$ has in the case of beams the units force per unit area.

A single-equation description for the TIMOSHENKO beam can be obtained under the assumption of constant material $(E, G)$ and geometrical $(I_z, A, k_s)$ properties: One- and three-times differentiation of Eq. (4.181) gives:

$$\frac{\mathrm{d}\phi_z}{\mathrm{d}x} = \frac{\mathrm{d}^2 u_y}{\mathrm{d}x^2} + \frac{q_y}{k_s G A}, \tag{4.182}$$

$$\frac{\mathrm{d}^3\phi_z}{\mathrm{d}x^3} = \frac{\mathrm{d}^4 u_y}{\mathrm{d}x^4} + \frac{\mathrm{d}^2 q_y}{k_s G A \mathrm{d}x^2}. \tag{4.183}$$

One-time differentiation of Eq. (4.180) gives:

$$E I_z \frac{\mathrm{d}^3\phi_z}{\mathrm{d}x^3} + k_s A G \left( \frac{\mathrm{d}^2 u_y}{\mathrm{d}x^2} - \frac{\mathrm{d}\phi_z}{\mathrm{d}x} \right) = 0. \tag{4.184}$$

Inserting Eq. (4.182) into (4.184) and consideration of (4.183) gives finally the following expression:

$$E I_z \frac{\mathrm{d}^4 u_y}{\mathrm{d}x^4} = q_y(x) - \frac{E I_z}{k_s A G} \frac{\mathrm{d}^2 q_y}{\mathrm{d}x^2}. \tag{4.185}$$

The last equation reduces for shear-rigid beams, i.e. $k_s A G \to \infty$, to the classical BERNOULLI formulation as given in Table 4.5.

### 4.3.5 Analytical Solutions in the Elastic Range

For the derivation of analytical solutions, the system of coupled differential equations as summarized in Table 4.19 has to be solved. Through the use of computer algebra systems (CAS) for the symbolic calculation of mathematical expressions,[25] the general solution of the system as given in Eqs. (4.180) and (4.181) results for constant $E I_z$ and $AG$ in:

[25] Maple®, Mathematica® and Matlab® can be mentioned as commercial examples of computer algebra systems.

**Table 4.19** Different formulations of the partial differential equation for a TIMOSHENKO beam in the $x$-$y$ plane ($x$-axis: right facing; $y$-axis: upward facing)

| Configuration | Partial Differential Equation |
|---|---|
| 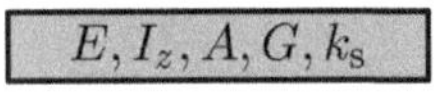  | $EI_z \frac{d^2\phi_z}{dx^2} + k_s GA \left( \frac{du_y}{dx} - \phi_z \right) = 0$ |
| | $k_s GA \left( \frac{d^2u_y}{dx^2} - \frac{d\phi_z}{dx} \right) = 0$ |
| 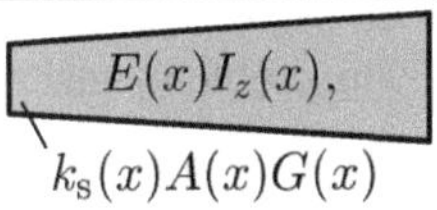  | $\frac{d}{dx}\left( E(x)I_z(x)\frac{d\phi_z}{dx} \right) + k_s(x)G(x)A(x)\left( \frac{du_y}{dx} - \phi_z \right) = 0$ |
| | $\frac{d}{dx}\left[ k_s(x)G(x)A(x)\left( \frac{du_y}{dx} - \phi_z \right) \right] = 0$ |
| 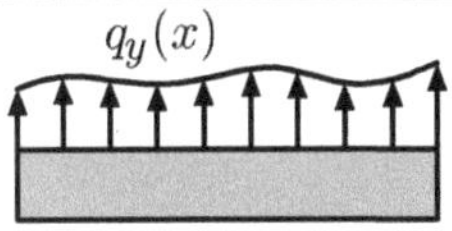  | $EI_z \frac{d^2\phi_z}{dx^2} + k_s GA \left( \frac{du_y}{dx} - \phi_z \right) = 0$ |
| | $k_s GA \left( \frac{d^2u_y}{dx^2} - \frac{d\phi_z}{dx} \right) = -q_y(x)$ |
| 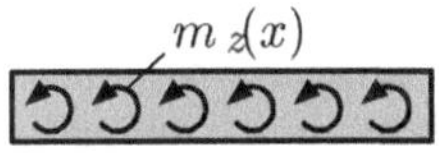  | $EI_z \frac{d^2\phi_z}{dx^2} + k_s GA \left( \frac{du_y}{dx} - \phi_z \right) = -m_z(x)$ |
| | $k_s GA \left( \frac{d^2u_y}{dx^2} - \frac{d\phi_z}{dx} \right) = 0$ |
| 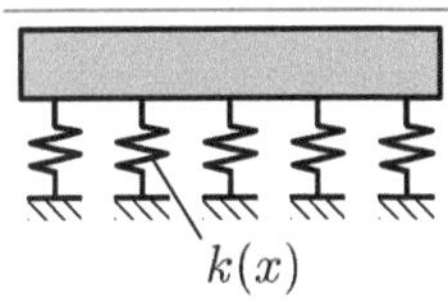  | $EI_z \frac{d^2\phi_z}{dx^2} + k_s GA \left( \frac{du_y}{dx} - \phi_z \right) = 0$ |
| | $k_s GA \left( \frac{d^2u_y}{dx^2} - \frac{d\phi_z}{dx} \right) = k(x)u_y$ |

$$u_y(x) = \frac{1}{EI_z}\left( \frac{q_y(x)x^4}{24} + c_1\frac{x^3}{6} + c_2\frac{x^2}{2} + c_3x + c_4 \right), \tag{4.186}$$

$$\phi_z(x) = \frac{1}{EI_z}\left( \frac{q_y(x)x^3}{6} + c_1\frac{x^2}{2} + c_2x + c_3 \right) + \frac{q_y(x)x}{k_sAG} + \frac{c_1}{k_sAG}. \tag{4.187}$$

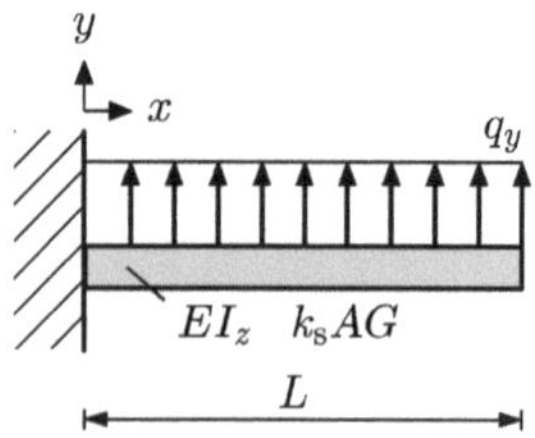


**Fig. 4.41** Analytical solution of a cantilevered TIMOSHENKO beam under constant distributed load

The constants of integration $c_1, \ldots, c_4$ must be defined through appropriate boundary conditions to calculate the specific solution of a given problem, meaning under consideration of the support and load conditions.

Consider the beam, which is illustrated in Fig. 4.41, as an example in the following. A constant distributed load $q_y$ acts in the positive $y$-direction and the boundary conditions are given as follows for this example:

$$u_y(x=0) = 0, \quad \phi_z(x=0) = 0, \tag{4.188}$$

$$M_z(x=0) = \frac{q_y L^2}{2}, \quad M_z(x=L) = 0. \tag{4.189}$$

The application of the boundary condition $(4.188)_1$ in the general analytical solution for the deflection according to Eq. (4.186) immediately yields $c_4 = 0$. With the second boundary condition in Eq. (4.188) and the general analytical solution for the rotation according to Eq. (4.187), the third constant of integration is obtained as $c_3 = -c_1 \frac{EI_z}{k_s AG}$. The further determination of the constants of integration demands that the bending moment is expressed with the help of the deformation. Application of Eq. (4.172) gives the moment distribution as

$$M_z(x) = EI_z \frac{\mathrm{d}\phi_z}{\mathrm{d}x} = \left( c_1 x + c_2 + \frac{3q_y x^2}{6} \right) + \frac{q_y EI_z}{k_s AG}, \tag{4.190}$$

and the consideration of boundary conditions $(4.189)_1$ yields $c_2 = \frac{q_y L^2}{2} - \frac{q_y EI_z}{k_s AG}$. In a similar way, consideration of the second boundary condition in Eq. (4.189) yields the first constant of integration to $c_1 = -q_y L$ and finally $c_3 = \frac{q_y LEI_z}{k_s AG}$. Therefore, the bending line results in

$$u_y(x) = \frac{1}{EI_z} \left( \frac{q_y x^4}{24} - q_y L \frac{x^3}{6} + \left[ \frac{q_y L^2}{2} - \frac{q_y EI_z}{k_s AG} \right] \frac{x^2}{2} + \frac{q_y LEI_z}{k_s AG} x \right), \tag{4.191}$$

or alternatively the maximal deflection on the right-hand boundary of the beam, meaning for $x = L$, to:

$$u_y(x = L) = \frac{q_y L^4}{8EI_z} + \frac{q_y L^2}{2k_s AG}. \tag{4.192}$$

Further analytical equations for the deflections and rotations and their maximal values of TIMOSHENKO beams are summarized in Tables 4.20 and 4.21. Through comparison with the analytical solutions in Sect. 4.2.5 it becomes obvious that the analytical solutions for the maximal deflection compose additively from the classical solution for the BERNOULLI beam and an additional shear part.

To highlight the influence of the shear contribution, the maximum deflection is presented in the following as a function of the fraction between beam height and beam length. As an example three different loading and boundary conditions for a rectangular cross section with the width $b$ and the height $h$ are presented in Fig. 4.42. It becomes obvious that the difference between the BERNOULLI and the TIMOSHENKO beam becomes smaller and smaller for a decreasing slenderness ratio, meaning for beams at which the length $L$ is significantly larger compared to the height $h$.

The relative difference between the BERNOULLI and the TIMOSHENKO solutions, for example for a POISSON's ratio of 0.3 and a slenderness ratio of 0.1—meaning for a beam, for which the length is ten times larger than the height—depending on the loading and boundary conditions is: 0.77 % for the cantilever beam with point load, 1.03 % for the cantilever beam with distributed load and 11.10 % for the simply supported beam. Further analytical solutions for the TIMOSHENKO beam can be obtained, for example, from [112].

Let us mention at the end of this section that for bending in the $x$-$z$ plane slightly modified equations occur compared to Table 4.18. The corresponding equations for bending in the $x$-$z$ plane with shear contribution are summarized in Table 4.22.

## 4.4 Higher-Order Beam Theories

### *4.4.1 Overview on Different Concepts*

The different beam theories can be introduced in different ways. One discrimination is the different consideration of the shear stress to derive the describing differential equations, see Fig. 4.43. The EULER- BERNOULLI beam theory (see Fig. 4.43a), which was introduced in Sect. 4.2, neglects the influence of the shear stress on the deformed shape. The TIMOSHENKO beam theory (see Fig. 4.43b) assumes an equivalent *constant* shear stress in the entire cross section as introduced in Sect 4.3. Higher-order beam theories consider a more realistic distribution of the shear stress (see Fig. 4.43c). A good introduction to this topic can be found in the monograph [113], the proceedings book [37] and the review article [44].

Systematics of the different theories can be given based on the assumed displacement fields $u_x(x, y)$ and $u_y(x, y)$ [44, 113]. This implies a different deformation of

**Table 4.20** Deflection and rotation of TIMOSHENKO beams subjected to simple load cases for bending in the $x$-$y$ plane. In the case of a point load at $x = \frac{L}{2}$, the equations are given for the domain $0 \leq x \leq \frac{L}{2}$

| Load case | Deflection | Rotation |
|---|---|---|
| 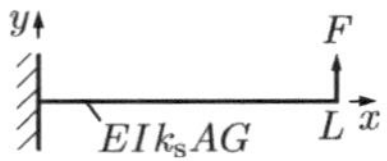 | $u_y(x) = \frac{1}{EI_z}\left(-\frac{Fx^3}{6} + \frac{FLx^2}{2} + \frac{FEI_z x}{k_s AG}\right)$ | $\phi_z(x) = \frac{1}{EI_z}\left(-\frac{Fx^2}{2} + FLx\right)$ |
|  | $u_y(x) = \frac{1}{EI_z}\left(\frac{q_y x^4}{24} - \frac{q_y L x^3}{6} + \left[\frac{q_y L^2}{2} - \frac{q_y EI_z}{k_s AG}\right]\frac{x^2}{2} + \frac{q_y L EI_z x}{k_s AG}\right)$ | $\phi_z(x) = \frac{1}{EI_z}\left(\frac{q_y x^3}{6} - \frac{q_y L x^2}{2} + \frac{q_y L^2 x}{2}\right)$ |
| 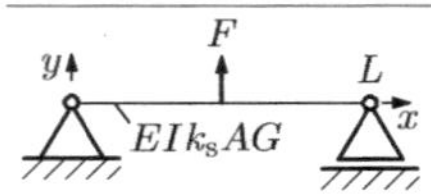 | $u_y(x) = \frac{1}{EI_z}\left(-\frac{Fx^3}{12} + \frac{FL^2 x}{16} + \frac{FEI_z x}{2k_s AG}\right)$ | $\phi_z(x) = \frac{1}{EI_z}\left(-\frac{Fx^2}{4} + \frac{FL^2}{16}\right)$ |
| 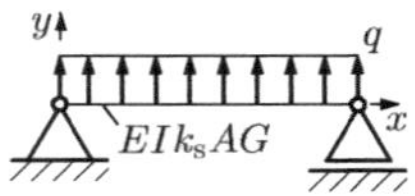 | $u_y(x) = \frac{1}{EI_z}\left(\frac{q_y x^4}{24} - \frac{q_y L x^3}{12} - \frac{EI_z q_y x^2}{2k_s AG} + \left[\frac{q_y L^3}{24} + \frac{EI_z q_y L}{2k_s AG}\right] x\right)$ | $\phi_z(x) = \frac{1}{EI_z}\left(\frac{q_y x^3}{6} - \frac{q_y L x^2}{4} + \frac{q_y L^3}{24}\right)$ |
| 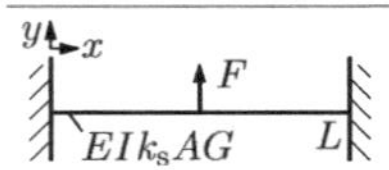 | $u_y(x) = \frac{1}{EI_z}\left(-\frac{Fx^3}{12} + \frac{FLx^2}{16} + \frac{FEI_z x}{2k_s AG}\right)$ | $\phi_z(x) = \frac{1}{EI_z}\left(-\frac{Fx^2}{4} + \frac{FLx}{8}\right)$ |
| 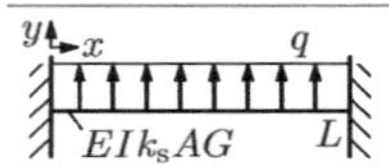 | $u_y(x) = \frac{1}{EI_z}\left(\frac{q_y x^4}{24} - \frac{q_y L x^3}{12} + \left[\frac{q_y L^2}{12} - \frac{q_y EI_z}{k_s AG}\right]\frac{x^2}{2} + \frac{q_y L EI_z x}{2k_s AG}\right)$ | $\phi_z(x) = \frac{1}{EI_z}\left(\frac{q_y x^3}{6} - \frac{q_y L x^2}{4} + \frac{q_y L^2 x}{12}\right)$ |

**Table 4.21** Maximum deflection and rotation of TIMOSHENKO beams subjected to simple load cases for bending in the $x$-$y$ plane

| Load Case | Maximum Deflection and Rotation |
|---|---|
| 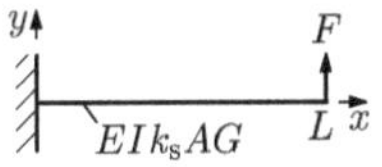 | $u_{y,\max} = u_y(L) = \dfrac{FL^3}{3EI_z} + \dfrac{FL}{k_s AG}$ |
| | $\phi_{z,\max} = \phi_z(L) = \dfrac{FL^2}{2EI_z}$ |
| 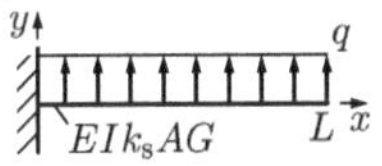 | $u_{y,\max} = u_y(L) = \dfrac{q_y L^4}{8EI_z} + \dfrac{q_y L^2}{2k_s AG}$ |
| | $\phi_{z,\max} = \phi_z(L) = \dfrac{q_y L^3}{6EI_z}$ |
| 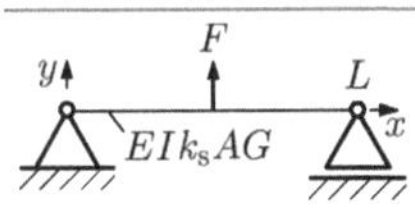 | $u_{y,\max} = u_y\left(\frac{L}{2}\right) = \dfrac{FL^3}{48EI_z} + \dfrac{FL}{4k_s AG}$ |
| | $\phi_{z,\max} = \phi_z(0) = \dfrac{FL^2}{16EI_z}, \quad (0 \le x \le \frac{L}{2})$ |
| 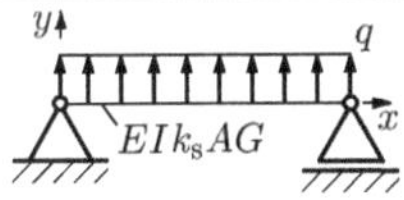 | $u_{y,\max} = u_y\left(\frac{L}{2}\right) = \dfrac{5q_y L^4}{384EI_z} + \dfrac{q_y L^2}{8k_s AG}$ |
| | $\phi_{z,\max} = \phi_z(0) = \dfrac{q_y L^3}{24EI_z}, \quad (0 \le x \le \frac{L}{2})$ |
| 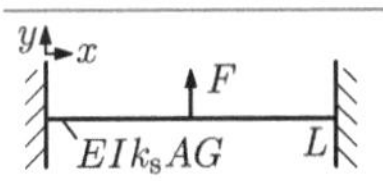 | $u_{y,\max} = u_y\left(\frac{L}{2}\right) = \dfrac{FL^3}{192EI_z} + \dfrac{FL}{4k_s AG}$ |
| | $\phi_{z,\max} = \phi_z(\frac{L}{4}) = \dfrac{FL^2}{64EI_z}, \quad (0 \le x \le \frac{L}{2})$ |
| 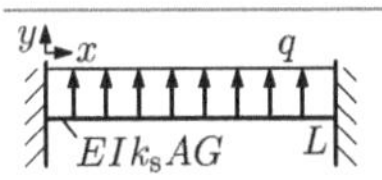 | $u_{y,\max} = u_y\left(\frac{L}{2}\right) = \dfrac{q_y L^4}{384EI_z}$ |
| | $\phi_{z,\max} = \phi_z\left(\frac{3-\sqrt{3}}{6}L\right) = \dfrac{\sqrt{3}q_y L^3}{216EI_z}, \quad (0 \le x \le \frac{L}{2})$ |

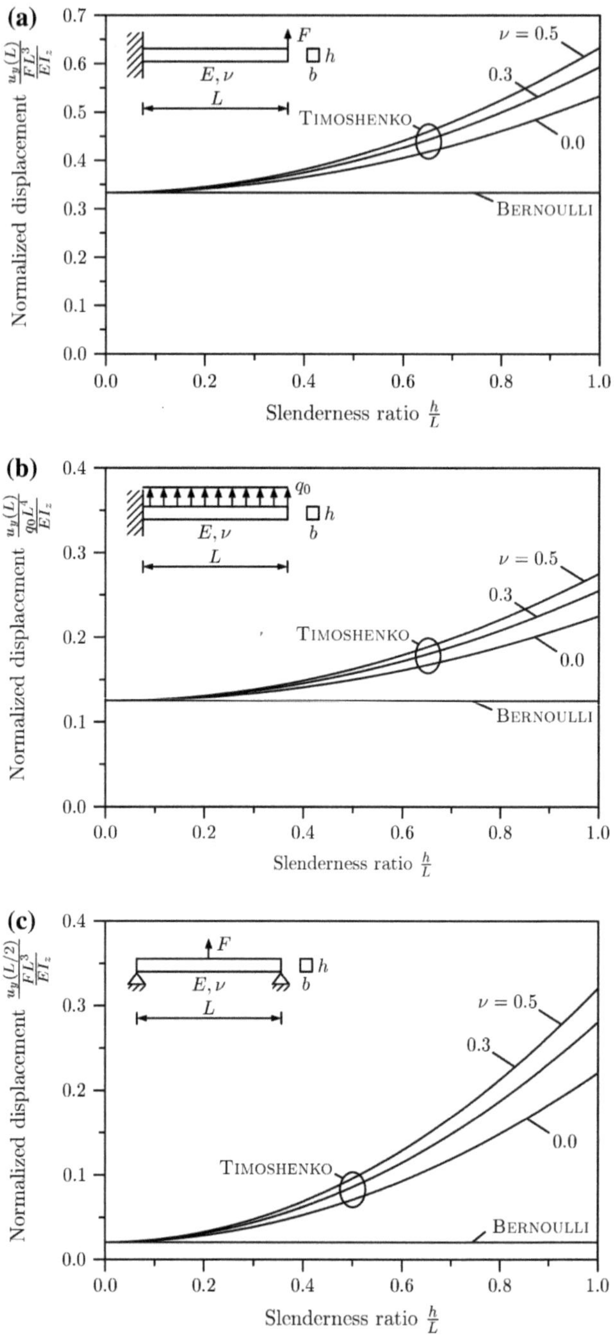


**Fig. 4.42** Comparison of the analytical solutions for the BERNOULLI and TIMOSHENKO beam for different loading and boundary conditions: **a** Cantilevered beam with end load, **b** cantilevered beam with distributed load, **c** simply supported beam with point load

**Table 4.22** Elementary basic equations for bending of a TIMOSHENKO beam in the $x$-$z$ plane ($x$-axis: right facing; $z$-axis: upward facing)

| Relation | Equation |
|---|---|
| Kinematics | $\varepsilon_x(x,z) = -z\frac{\mathrm{d}\phi_y(x)}{\mathrm{d}x}$ and $\phi_y(x) = -\frac{\mathrm{d}u_z(x)}{\mathrm{d}x} + \gamma_{xz}(x)$ |
| Equilibrium | $\frac{\mathrm{d}Q_z(x)}{\mathrm{d}x} = -q_z(x)\,;\ \frac{\mathrm{d}M_y(x)}{\mathrm{d}x} = +Q_z(x)$ |
| Constitution | $\sigma_x(x,z) = E\varepsilon_x(x,z)$ and $\tau_{xz}(x) = G\gamma_{xz}(x)$ |
| Diff. Equation | $-\frac{\mathrm{d}}{\mathrm{d}x}\left(EI_y\frac{\mathrm{d}\phi_y}{\mathrm{d}x}\right) + k_sAG\left(\frac{\mathrm{d}u_z}{\mathrm{d}x} + \phi_y\right) = 0$ |
| | $\frac{\mathrm{d}}{\mathrm{d}x}\left[k_sAG\left(\frac{\mathrm{d}u_z}{\mathrm{d}x} + \phi_y\right)\right] = -q_z(x)$ |

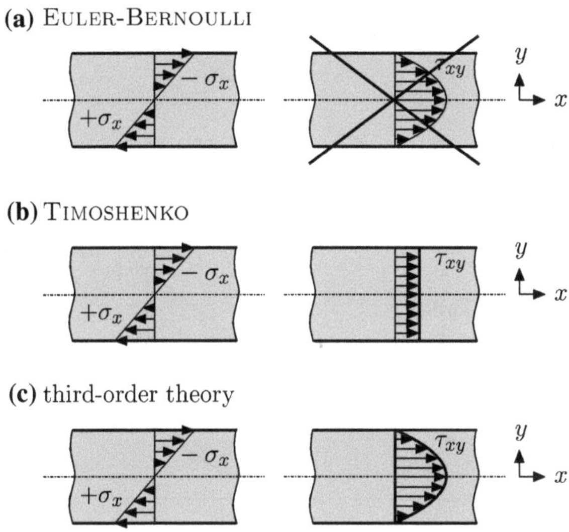


**Fig. 4.43** Comparison of different beam theories in regards to the contribution of the stress state on the deformation: **a** EULER- BERNOULLI, **b** TIMOSHENKO and **c** higher-order theory. Beam height $-\frac{h}{2} \leq y \leq +\frac{h}{2}$, beam width $b$

a typical transverse normal line which is in the undeformed state perpendicular to the center line (mid-plane), see Fig. 4.44.

- The EULER- BERNOULLI or elementary beam theory (EBT) assumes a displacement field of the form:

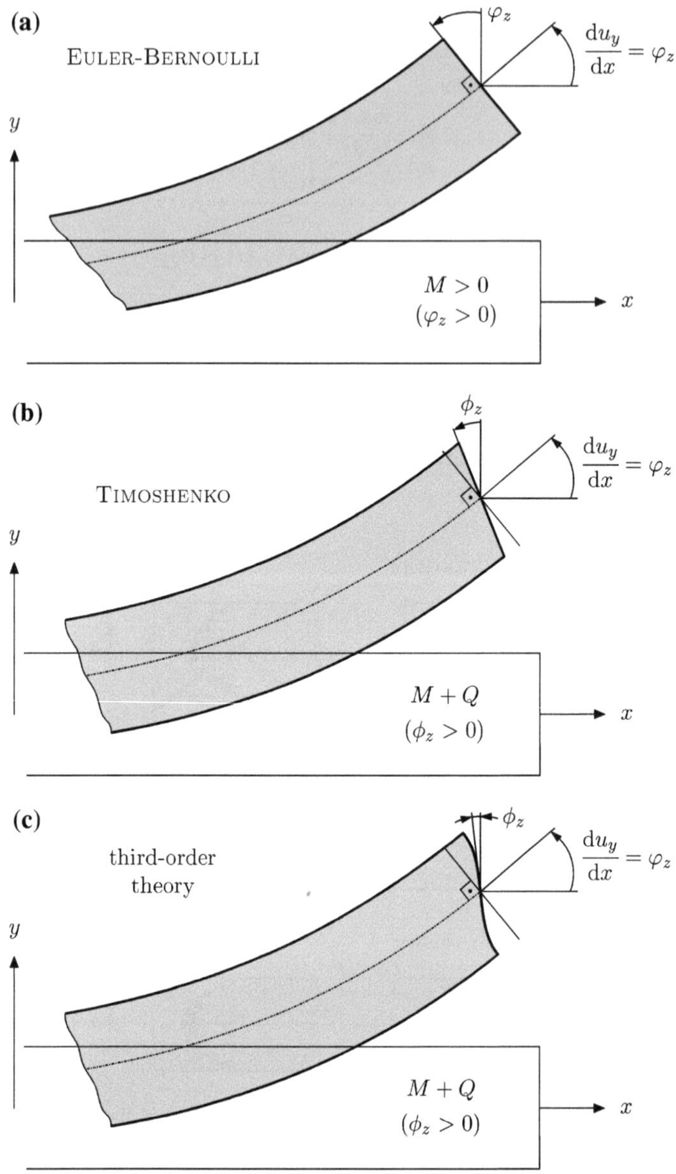


**Fig. 4.44** Comparison of different beam theories in regards to the deformation of a typical transverse normal line: **a** EULER- BERNOULLI, **b** TIMOSHENKO and **c** higher-order theory. Note that the deformation is overdrawn for better illustration

$$u_x(x, y) = -y\frac{du_y(x)}{dx} = -y\varphi_z, \quad (4.193)$$

$$u_y(x, y) = u_y(x). \quad (4.194)$$

This implies that the plane sections which are perpendicular to the neutral fiber before bending remain plane and perpendicular to the neutral fiber after bending, see Fig. 4.44a.

- The TIMOSHENKO beam or first-order shear deformation theory (FSDT) assumes a displacement field of the form:

$$u_x(x, y) = -y\phi_z, \tag{4.195}$$

$$u_y(x, y) = u_y(x). \tag{4.196}$$

This implies that the plane sections which are perpendicular to the neutral fiber before bending are no longer perpendicular to the neutral fiber after bending but they remain plane, see Fig. 4.44b.

- A second-order shear deformation theory (SSDT) assumes a displacement field of the form:

$$u_x(x, y) = -y\phi_z(x) + y^2\psi(x), \tag{4.197}$$

$$u_y(x, y) = u_y(x). \tag{4.198}$$

Now, an originally plane section is, after deformation, no longer perpendicular to the neutral fiber and no longer plane. The function $\phi_z(x)$ represents the rotation of the cross section of the beam at neutral fiber ($y = 0$) and $\phi_z(x)$ together with $\psi(x)$ defines the quadratic nature of the deformed line.

- A third-order shear deformation theory (TSDT) assumes a displacement field of the form[26]:

$$u_x(x, y) = -y\phi_z(x) + y^2\psi(x) + y^3\theta(x), \tag{4.199}$$

$$u_y(x, y) = u_y(x). \tag{4.200}$$

Here, $\phi_z(x)$ represents again the rotation of the cross section of the beam at neutral fiber ($y = 0$), see Fig. 4.44c.

It can be seen from the above that the second- and third-order theories are expressed as an extension of the TIMOSHENKO theory. In a slightly different form of systematics [83, 92], the higher-order theories may be expressed as extensions of the EULER-BERNOULLI beam theory as:

$$u_x(x, y) = -y\frac{\mathrm{d}u_y(x)}{\mathrm{d}x} + f(y)\phi_z(x), \tag{4.201}$$

$$u_y(x, y) = u_y(x). \tag{4.202}$$

[26] In a more general form, it could be stated that $u_y(x, y) = -y\phi_z(x) + y^2\psi(x) + y^3\theta(x)$, see [44].

Let us consider in the following a classical example of third-order theory which was proposed by LEVINSON in [65]. His proposed displacement field is generally expressed as:

$$u_x(x, y) = -y\phi_z(x) + y^3\theta(x), \tag{4.203}$$

together with the common simplification that $u_y(x, y) = u_y(x)$. The requirement that the shear stress must vanish at the outer surface layers (see Fig. 4.43c), i.e. at $y = \pm\frac{h}{2}$, yields:

$$\varepsilon_{xy}(x, \pm\tfrac{h}{2}) = \frac{1}{2}\left(\frac{\partial u_y(x)}{\partial x} + \frac{\partial u_x(x, y)}{\partial y}\right)\bigg|_{y=\pm\frac{h}{2}} \overset{!}{=} 0. \tag{4.204}$$

With

$$\frac{\partial u_x(x, y)}{\partial y}\bigg|_{y=\pm\frac{h}{2}} = -\phi_z(x) + 3y^2\theta(x)\Big|_{y=\pm\frac{h}{2}} = -\phi_z(x) + \frac{3h^2}{4}\theta(x), \tag{4.205}$$

the above stress condition (4.204) can be rearranged for $\theta(x)$ to obtain:

$$\theta(x) = -\frac{4}{3h^2}\left(-\phi_z(x) + \frac{\partial u_y(x)}{\partial x}\right). \tag{4.206}$$

The increment of the bending moment can be calculated based on (see Fig. 4.9b):

$$\mathrm{d}M_z = (+y)(-\sigma_x)\,\mathrm{d}A. \tag{4.207}$$

Integration and application of HOOKE's law (2.4) as well as the definition of the normal strain (3.2) gives the following expression:

$$M_z = -\int_A yE\frac{\partial u_x(x, y)}{\partial x}\mathrm{d}A. \tag{4.208}$$

Based on the general expression for the displacement field given in Eq. (4.203), the partial derivative of the displacement field with respect to the $x$-coordinate can be expressed as:

$$\frac{\partial u_x(x, y)}{\partial x} = -y\frac{\partial\phi_z(x)}{\partial x} + y^3\frac{\partial}{\partial x}\left[-\frac{4}{3h^2}\left(-\phi_z(x) + \frac{\partial u_y(x)}{\partial x}\right)\right] \tag{4.209}$$

$$= -y\frac{\partial\phi_z(x)}{\partial x} + y^3\frac{4}{3h^2}\frac{\partial\phi_x(x)}{\partial x} - y^3\frac{4}{3h^2}\frac{\partial^2 u_y(x)}{\partial x^2}. \tag{4.210}$$

Introducing this last expression in Eq. (4.213) and assuming a constant stiffness $E$, one obtains:

$$M_z = -E \int_{-\frac{b}{2}}^{+\frac{b}{2}} \int_{-\frac{h}{2}}^{+\frac{h}{2}} \left( -y^2 \frac{\partial \phi_z(x)}{\partial x} + y^4 \frac{4}{3h^2} \frac{\partial \phi_x(x)}{\partial x} - y^4 \frac{4}{3h^2} \frac{\partial^2 u_y(x)}{\partial x^2} \right) \mathrm{d}x\mathrm{d}y, \quad (4.211)$$

or

$$M_x = \frac{EI_z}{5} \left( 4 \frac{\partial \phi_z(x)}{\partial x} + \frac{\partial^2 u_y(x)}{\partial x^2} \right). \quad (4.212)$$

In a similar way, we can state the shear force as:

$$Q_y = \int_A \tau_{xy}(x, y) \mathrm{d}A = \int_A G\gamma_{x,y}(x, y) \mathrm{d}A = G \int_A \left( \frac{\partial u_y(x)}{\partial x} + \frac{\partial u_x(x, y)}{\partial y} \right) \mathrm{d}A. \quad (4.213)$$

Equation (4.203) allows us to express the partial derivative of the displacement field with respect to the $y$-coordinate:

$$\frac{\partial u_x(x, y)}{\partial y} = -\phi_z(x) - \frac{4y^2}{h^2} \left( -\phi_z(x) + \frac{\partial u_y(x)}{\partial x} \right). \quad (4.214)$$

Introducing this last expression in Eq. (4.213) and assuming a constant shear modulus $G$, one obtains:

$$Q_y = G \int_{-\frac{b}{2}}^{+\frac{b}{2}} \int_{-\frac{h}{2}}^{+\frac{h}{2}} \left( \frac{\partial u_y}{\partial x} - \phi_z + \frac{4y^2}{h^2} \phi_z - \frac{4y^2}{h^2} \frac{\partial u_y}{\partial x} \right) \mathrm{d}x\mathrm{d}y, \quad (4.215)$$

or

$$Q_y = \frac{2}{3} GA \left( -\phi_x(x) + \frac{\partial u_y(x)}{\partial x} \right). \quad (4.216)$$

The equilibrium equations are the same as in the case of the EULER- BERNOULLI or TIMOSHENKO beam, i.e. $\frac{\mathrm{d}Q_y(x)}{\mathrm{d}x} = -q_y(x)$ and $\frac{\mathrm{d}M_z(x)}{\mathrm{d}x} = -Q_y(x)$ in the case that no distributed moment is acting. Introducing the expressions for the bending moment and the shear force according to Eqs. (4.212) and (4.216) in these conditions gives finally the coupled systems of partial differential equations for the LEVINSON beam as:

$$\frac{2}{3}GA\left(\frac{\partial u_y}{\partial x} - \phi_z\right) + \frac{1}{5}\frac{\partial}{\partial x}\left[EI_z\left(\frac{\partial^2 u_y}{\partial x^2} + 4\frac{\partial \phi_z}{\partial x}\right)\right] = 0, \tag{4.217}$$

$$\frac{2}{3}\frac{\partial}{\partial x}\left[AG\left(\frac{\partial u_y}{\partial x} - \phi_z\right)\right] = -q_y(x). \tag{4.218}$$

This systems contains—as in the case of the TIMOSHENKO beam—two unknown functions, i.e. the deflection $u_y(x)$ and the cross-sectional rotation $\phi_z(x)$ at $y = 0$. The application of a computer algebra system gives the general solution of the system along the center line (mid-plane) $y = 0$ as:

$$u_y(x) = \frac{1}{EI_z}\left(\frac{q_y(x)x^4}{24} + c_1\frac{x^3}{6} + c_2\frac{x^2}{2} + c_3 x + c_4\right), \tag{4.219}$$

$$\phi_z(x) = \frac{1}{EI_z}\left(\frac{q_y(x)x^3}{6} + c_1\frac{x^2}{2} + c_2 x + c_3\right) + \frac{3q_y(x)x}{2AG} + \frac{3c_1}{2AG}. \tag{4.220}$$

The partial derivatives of these expressions can easily be obtained as

$$\frac{\partial u_y}{\partial x} = \frac{1}{EI_z}\left(\frac{q_y x^3}{6} + \frac{c_1 x^2}{2} + c_2 x + c_3\right), \tag{4.221}$$

$$\frac{\partial^2 u_y}{\partial x^2} = \frac{1}{EI_z}\left(\frac{q_y x^2}{2} + c_1 x + c_2\right), \tag{4.222}$$

$$\frac{\partial \phi_z}{\partial x} = \frac{1}{EI_z}\left(\frac{q_y x^2}{2} + c_1 x + c_2\right) + \frac{3}{2}\frac{q_y}{GA}, \tag{4.223}$$

and the expressions for the bending moment and shear force given in Eqs. (4.212) and (4.216) can be concretized to:

$$M_z(x) = \frac{q_y(x)x^2}{2} + c_1 x + c_2 + \frac{6EI_z}{5GA}q_y(x), \tag{4.224}$$

$$Q_y(x) = -GAq_y(x) - c_1. \tag{4.225}$$

**4.3 Example: Cantilevered Levinson beam loaded by a single force at its end**
Given is a cantilevered LEVINSON beam which is loaded by a single force $F$ as shown in Fig. 4.45. The cross section of the beam is rectangular with dimensions $b \times h$. Calculate the deflection $u_y(x)$ in the pure elastic range under the assumption that the material behavior, i.e. YOUNG's modulus $E$ and shear modulus $G$, is constant. Use for the derivations the coupled system of differential equations as given in Eqs. (4.219) and (4.220). Compare these results with the solution for a TIMOSHENKO beam.

**4.3 Solution**
The boundary conditions are given for this case as:

$$u_y(x=0)=0,\quad \phi_z(x=0)=0, \tag{4.226}$$
$$M_z(x=L)=0,\quad Q_y(x=L)=F. \tag{4.227}$$

Thus, the constants of integration are obtained as

$$c_1=-F,\quad c_2=FL, \tag{4.228}$$
$$c_3=\frac{3EI_z}{2GA}F,\quad c_4=0. \tag{4.229}$$

Thus, the deflection is obtained under consideration of $G=\frac{E}{2(1+\nu)}$ and $A=\frac{12I_z}{h^2}$ as:

$$u_y(x)=\frac{FL^3}{6EI_z}\left(-\frac{x^3}{L^3}+\frac{3x^2}{L^2}\right)+\frac{FL^3(1+\nu)}{4EI_z}\left(\frac{h^2}{L^2}\right)\left(\frac{x}{L}\right). \tag{4.230}$$

This is identical to the exact solution of the plane stress case according to the linear theory of elasticity [106]. The corresponding solution based on the TIMOSHENKO theory is (see Table 4.20):

$$u_y(x)=\frac{FL^3}{6EI_z}\left(-\frac{x^3}{L^3}+\frac{3x^2}{L^2}\right)+\frac{FL}{k_sAG}\left(\frac{x}{L}\right), \tag{4.231}$$

or

$$u_y(x)=\frac{FL^3}{6EI_z}\left(-\frac{x^3}{L^3}+\frac{3x^2}{L^2}\right)+\frac{FL^3(1+\nu)}{6k_sEI_z}\left(\frac{h^2}{L^2}\right)\left(\frac{x}{L}\right). \tag{4.232}$$

Assuming a shear correction factor of $k_s=\frac{5}{6}$ for a rectangular cross section as given in Table 4.17, the second expression on the right-hand side of the TIMOSHENKO solutions underestimates the correct value by 20 %.

At the end of this section, a comparison of the different beam theories is presented in Table 4.23.

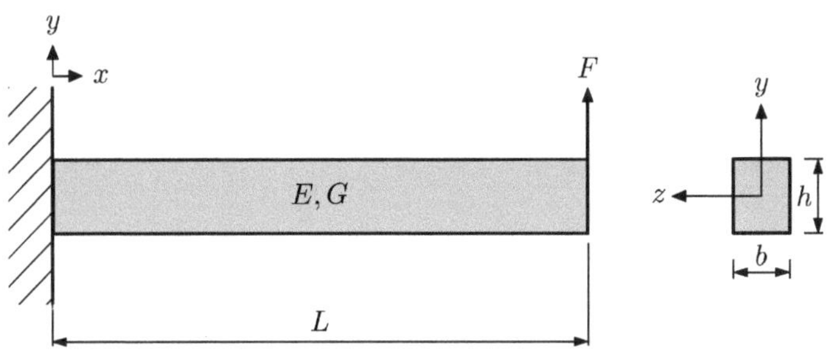


**Fig. 4.45** Cantilevered LEVINSON beam loaded by a single force $F$ at its end

**Table 4.23** Comparison of different beam theories

| | EULER- BERNOULLI | TIMOSHENKO | LEVINSON |
|---|---|---|---|
| Kinematics | $\varepsilon_x = -y \frac{\mathrm{d}^2 u_y}{\mathrm{d}x^2} = -y \frac{\mathrm{d}\varphi_z}{\mathrm{d}x}$ | $\varepsilon_x = -y \frac{\mathrm{d}\phi_z}{\mathrm{d}x}$ and $\phi_z = \frac{\mathrm{d}u_y}{\mathrm{d}x} - \gamma_{xy}$ | $\varepsilon_x = -y \frac{\mathrm{d}\phi_z}{\mathrm{d}x} + y^3 \frac{4}{3h^2}\left(\frac{\mathrm{d}\phi_z}{\mathrm{d}x} - \frac{\mathrm{d}^2 u_y}{\mathrm{d}x^2}\right)$ |
| Equilibrium | $\frac{\mathrm{d}Q_y}{\mathrm{d}x} = -q_y(x)\ ,\ \frac{\mathrm{d}M_z}{\mathrm{d}x} = -Q_y(x)$ | $\frac{\mathrm{d}Q_y}{\mathrm{d}x} = -q_y(x)\ ,\ \frac{\mathrm{d}M_z}{\mathrm{d}x} = -Q_y(x)$ | $\frac{\mathrm{d}Q_y}{\mathrm{d}x} = -q_y(x)\ ,\ \frac{\mathrm{d}M_z}{\mathrm{d}x} = -Q_y(x)$ |
| Constitution | $\sigma_x(x, y) = E\varepsilon_x(x, y)$ | $\sigma_x(x, y) = E\varepsilon_x(x, y)$ | $\sigma_x(x, y) = E\varepsilon_x(x, y)$ |
| | | and $\tau_{xy}(x, y) = G\gamma_{xy}(x, y)$ | and $\tau_{xy}(x, y) = G\gamma_{xy}(x, y)$ |
| Displ. field | $u_y(x) = -y\varphi_z(x)$ | $u_y(x) = -y\phi_z(x)$ | $u_y(x) = -y\phi_z(x) + y^3\theta(x)$ with $\theta(x) = \frac{4}{3h^2}\left(\phi_z(x) - \frac{\partial u_y(x)}{\partial x}\right)$ |
| Bending moment | $M_z(x) = EI_z \frac{\partial \varphi_z}{\partial x}$ | $M_z(x) = EI_z \frac{\partial \phi_z}{\partial x}$ | $M_z(x) = \frac{EI_z}{5}\left(4\frac{\partial \phi_z}{\partial x} + \frac{\partial^2 u_y}{\partial x^2}\right)$ |
| Shear force | – | $Q_y(x) = k_s AG\left(-\phi_z + \frac{\partial u_y}{\partial x}\right)$ | $Q_y(x) = \frac{2}{3}AG\left(-\phi_z + \frac{\partial u_y}{\partial x}\right)$ |
| Differential equation | $\frac{\mathrm{d}^2}{\mathrm{d}x^2}\left(EI_z \frac{\mathrm{d}\varphi_z}{\mathrm{d}x}\right) = q_y(x)$ | $\frac{\mathrm{d}}{\mathrm{d}x}\left(EI_z \frac{\mathrm{d}\phi_z}{\mathrm{d}x}\right) + k_s AG\left(\frac{\mathrm{d}u_y}{\mathrm{d}x} - \phi_z\right) = 0$ | $\frac{2}{3}GA\left(\frac{\partial u_y}{\partial x} - \phi_z\right) + \frac{1}{5}\frac{\partial}{\partial x}\left[EI_z\left(\frac{\partial^2 u_y}{\partial x^2} + 4\frac{\partial \phi_z}{\partial x}\right)\right] = 0$ |
| | | $\frac{\mathrm{d}}{\mathrm{d}x}\left[k_s AG\left(\frac{\mathrm{d}u_y}{\mathrm{d}x} - \phi_z\right)\right] = -q_y(x)$ | $\frac{2}{3}\frac{\partial}{\partial x}\left[AG\left(\frac{\partial u_y}{\partial x} - \phi_z\right)\right] = -q_y(x)$ |

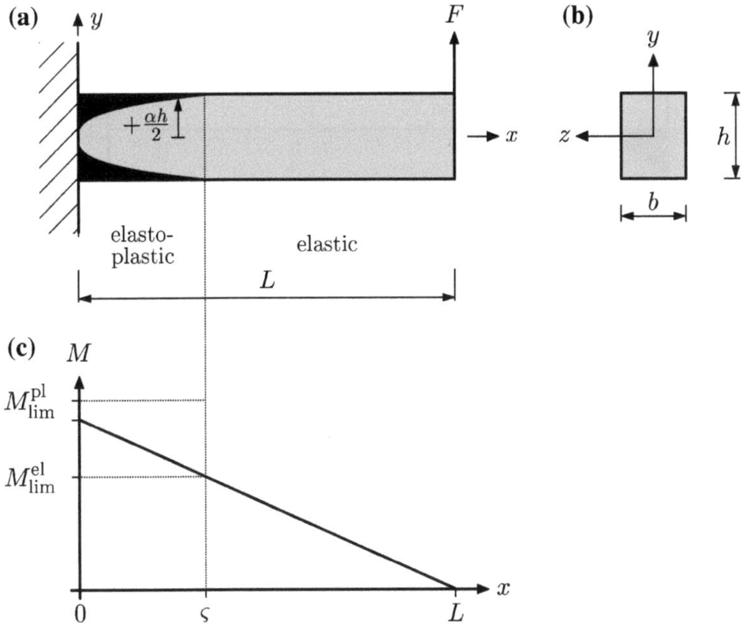


**Fig. 4.46** **a** Cantilevered beam loaded by a single force $F$ at its end, **b** rectangular cross-section and **c** bending moment distribution

## *4.4.2 Analytical Solutions in the Elasto-Plastic Range*

The following section[27] introduces into a common approach to consider the influence of the shear stress in the form of interaction relations between the shear force $Q_z$ and the plastic moment $M^{\text{pl}}_{\text{lim}}$. These interaction relations state how the plastic moment must be reduced under the influence of an acting shear force. The following derivations are obtained for the simple example of a cantilevered beam with rectangular cross section, cf. Fig. 4.46. This beam is loaded by a single force $F$ at its right-hand side and this force causes partial plasticity in the region $0 \leq x \leq \varsigma$.

The moment distribution shown in Fig. 4.46c indicates that the plastic moment is still not reached at the left-hand fixing point, i.e. $M_z(x = 0) = FL < M^{\text{pl}}_{\text{lim}}$. The objective is now to determine the plastic limit moment under the influence of a shear force, i.e. $M^{\text{pl}}_{\text{s, lim}}$, and to compare this value with the plastic limit moment $M^{\text{pl}}_{\text{lim}} = \frac{1}{4}bh^2k$ (cf. Eq. (4.78)) in the absence of any shear force. In the elasto-plastic zone, i.e. for $0 \leq x \leq \varsigma$, two regions can be distinguished. For $|y| < \frac{\alpha h}{2}$, the beam is till in the pure elastic range while the regions $|y| \geq \frac{\alpha h}{2}$ are characterized by plastic material behavior. Assuming ideal-plastic material behavior (cf. Fig. 4.19a), the stress state in this plastic part is characterized by

[27] Adapted from the classical textbook on plasticity [73].

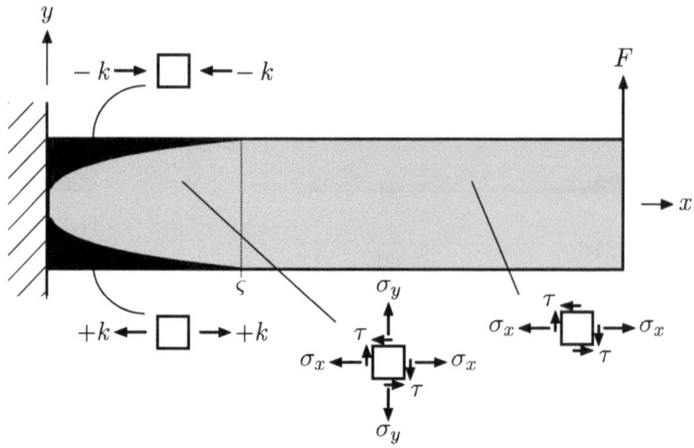


**Fig. 4.47** Different stress states in an elasto-plastic beam under consideration of the shear stress

$$\sigma_x = \pm k \ , \ \sigma_y = 0 \ , \ \tau_{xy} = 0. \tag{4.233}$$

The justification for the assumption that the shear stress is zero is given in [56, 79] or summarized in the Appendix B.4. As a consequence, the shear force must be carried entirely by the shear stress in the elastic region $|y| < \frac{\alpha h}{2}$. The different stress states in the elastic and elasto-plastic parts of a beam are shown in Fig. 4.47. As the entire height of this region, $\alpha h$, decreases for $x \to 0$, this shear stress must increase so that at some section yielding may start in the center of the elastic part due to the maximum of the shear stress distribution.

The VON MISES yield condition for a plane stress state will be used in the following, see [19]:

$$F = \sqrt{\sigma_x^2 - \sigma_x \sigma_y + \sigma_y^2 + 3\tau^2} = k. \tag{4.234}$$

The two equilibrium equations can be obtained from the general case given in Eqs. (8.1) and (8.2) as:

$$\frac{\partial \sigma_x}{\partial x} + \frac{\partial \tau}{\partial y} = 0, \tag{4.235}$$

$$\frac{\partial \sigma_y}{\partial y} + \frac{\partial \tau}{\partial x} = 0. \tag{4.236}$$

In the pure elastic zone of the beam, i.e. $\varsigma \leq x \leq L$, the normal stress component $\sigma_y$ is assumed to be zero, while the component $\sigma_x$ varies linearly across the section:

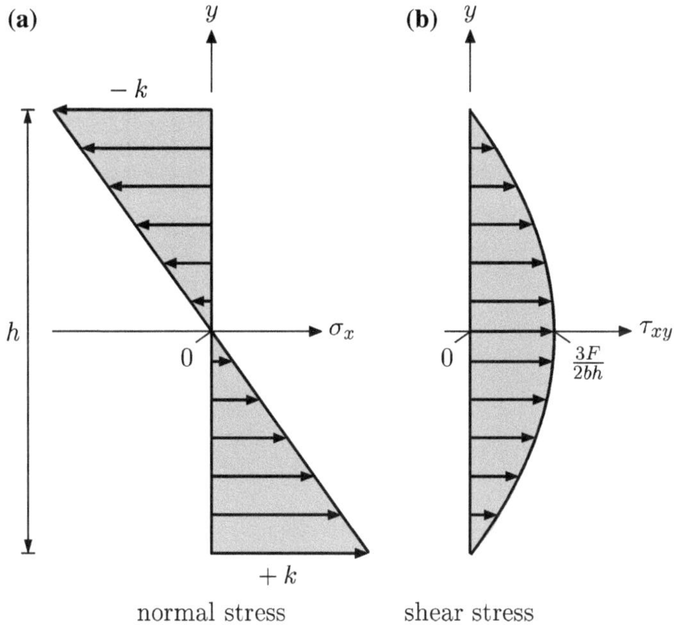


**Fig. 4.48** Stress distribution at $x = \varsigma$: **a** normal stress and **b** shear stress

$$\sigma_x(x, y) \overset{(4.33)}{=} -\frac{M_z(x)}{I_z} y = -\frac{F(L-x)}{\frac{1}{12}bh^3} y = -\frac{12F(L-x)y}{bh^3}, \tag{4.237}$$

$$\sigma_y(x, y) = 0. \tag{4.238}$$

The shear stress distribution in the pure elastic range (cf. suppl. problem 4.21) is given by

$$\tau = \frac{Q_y}{2I_z}\left[\left(\frac{h}{2}\right)^2 - y^2\right] = \frac{3F}{2bh}\left[1 - \left(\frac{2y}{h}\right)^2\right]. \tag{4.239}$$

The distribution of the stress components $\sigma_x$ and $\tau$ at $x = \varsigma$ is shown in Fig. 4.48 where it can be seen that the normal stress is linearly distributed and the shear stress follows a parabolic shape.

The normal stress reaches at $x = \varsigma$ at the outer fiber ($y = \frac{h}{2}$) the yield stress, i.e. $\sigma_x = -k$, and one can derive from Eq. (4.237):

$$-k = -\frac{12F(L-\varsigma)\frac{h}{2}}{bh^3} \quad \text{or} \quad F(L-\varsigma) = +\frac{1}{6}bh^2k. \tag{4.240}$$

Balancing the external and the internal moment (obtained from the stress distribution, cf. Eq. (4.240)) at $x = \varsigma$ gives:

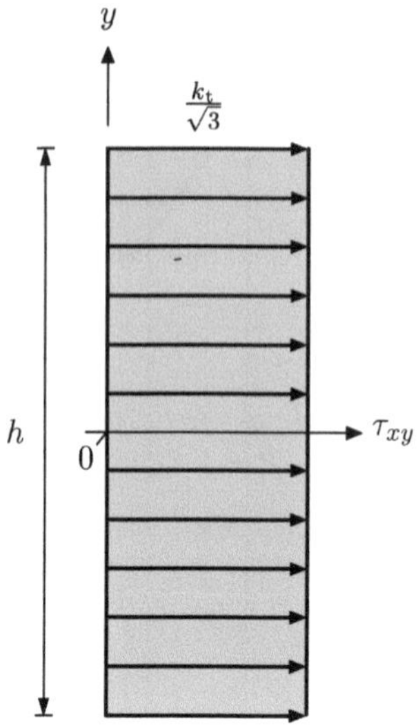


**Fig. 4.49** Entire cross section subjected to the shear yield stress

$$M_z(x = \varsigma) = F(L - \varsigma) = +\frac{1}{6}bh^2k. \tag{4.241}$$

Let us assume in the following derivation that the shear stress $\tau$ does not exceed the shear yield stress in the pure elastic range, i.e. $\tau(\varsigma \leq x \leq L) < k_s = \frac{k_t}{\sqrt{3}}$. Thus, it follows from Eq. (4.239) for $y = 0$ that:

$$\tau(y = 0) = \frac{3F}{2bh} < \frac{k_t}{\sqrt{3}} \quad \text{or} \quad F < \frac{2bhk}{3\sqrt{3}}. \tag{4.242}$$

The plastic shear force $Q^{pl}_{lim}$ is defined in the following[28] as the shear force which results in a cross section completely subjected to the shear yield stress, cf. Fig. 4.49, i.e.

$$Q^{pl}_{lim} = \underbrace{\frac{k}{\sqrt{3}}}_{\text{stress}} \times \underbrace{bh}_{\text{area}}. \tag{4.243}$$

Thus, the condition in Eq. (4.242) can be written as:

$$\frac{F}{F^{pl}_{lim}} < \frac{2}{3}, \tag{4.244}$$

which gives in the limiting case ('<' → '=') a distribution of the stress components at $x = \varsigma$ as:

[28] For the example under consideration, cf. Fig. 4.46a, it holds that $Q^{pl}_{lim} = F^{pl}_{lim}$.

$$\sigma_x(x=\varsigma,y) \overset{(4.237)}{=} -\frac{12F(L-\varsigma)y}{bh^3} \overset{(4.241)}{=} -k\left(\frac{2y}{h}\right), \tag{4.245}$$

$$\tau_x(x=\varsigma,y) \overset{(4.239)}{=} \frac{3F}{2bh}\left[1-\left(\frac{2y}{h}\right)^2\right] \overset{(4.242)}{=} \frac{k}{\sqrt{3}}\left[1-\left(\frac{2y}{h}\right)^2\right]. \tag{4.246}$$

Using these two expressions for the stress components and $\sigma_y(x=\varsigma,y)=0$ in the VON MISES yield condition (4.234) gives

$$\sqrt{k^2\left(\frac{2y}{h}\right)^2+\frac{3k^2}{3}\left[1-\left(\frac{2y}{h}\right)^2\right]^2}=k, \tag{4.247}$$

or

$$\sqrt{1-\left(\frac{2y}{h}\right)^2+\left(\frac{2y}{h}\right)^4}=1. \tag{4.248}$$

For $y$-coordinates in the range of $-\frac{h}{2}\leq y\leq +\frac{h}{2}$, the left-hand side of Eq. (4.248) is always smaller or equal than unity and only allowable stress states are obtained.

Let us focus in the following on the elasto-plastic part of the beam, i.e. the region $0\leq x\leq\varsigma$. The $x$-component of the normal stress is assumed to be still linearly distributed over the height of the cross section:

$$\sigma_x(0\leq x\leq\varsigma,-\tfrac{\alpha h}{2}\leq y\leq +\tfrac{\alpha h}{2})=-k\left(\frac{y}{\alpha\frac{h}{2}}\right). \tag{4.249}$$

The balance between the external and the internal moment can be written based on Sect. 4.2.6.1 as:

$$M_z=F(L-x)\overset{(4.82)}{=}\frac{1}{12}bh^2k\left(3-\alpha^2\right)=b\left[\frac{h^2}{4}-\frac{1}{3}\left(\frac{\alpha h}{2}\right)^2\right]k. \tag{4.250}$$

It follows from the last equation that

$$x=L-b\left[\frac{h^2}{4}-\frac{1}{3}\left(\frac{\alpha h}{2}\right)^2\right]\frac{k}{F}, \tag{4.251}$$

or

$$\frac{\mathrm{d}\left(\frac{\alpha h}{2}\right)}{\mathrm{d}x}=+\frac{3F}{2bk\frac{\alpha h}{2}}. \tag{4.252}$$

Calculation of the derivative of Eq. (4.249) with respect to the $x$-coordinate gives:

$$\frac{\partial \sigma_x}{\partial x} = \frac{\partial}{\partial x}\left(-ky\left(\frac{\alpha h}{2}\right)^{-1}\right) = ky\left(\frac{\alpha h}{2}\right)^{-2}\frac{\mathrm{d}\left(\frac{\alpha h}{2}\right)}{\mathrm{d}x} \overset{(4.252)}{=} \frac{3Fy}{2b\left(\frac{\alpha h}{2}\right)^3}. \tag{4.253}$$

The last equation can be introduced into the equilibrium equation (4.235) to give

$$\frac{3Fy}{2b\left(\frac{\alpha h}{2}\right)^3} + \frac{\partial \tau}{\partial y} = 0 \quad \text{or} \quad \int_0^\tau \mathrm{d}\tilde{\tau} = \int_{+\frac{\alpha h}{2}}^{y}\left(-\frac{3F\tilde{y}}{2b\left(\frac{\alpha h}{2}\right)^3}\right)\mathrm{d}\tilde{y}, \tag{4.254}$$

or finally:

$$\tau(0 \leq x \leq \varsigma, -\tfrac{\alpha h}{2} \leq y \leq +\tfrac{\alpha h}{2}) = \frac{3F}{4b\left(\frac{\alpha h}{2}\right)}\left[1 - \left(\frac{y}{\frac{\alpha h}{2}}\right)^2\right]. \tag{4.255}$$

It should be noted that in the last equation $\frac{\alpha h}{2} = \frac{\alpha h}{2}(x)$ holds. The last equation can be differentiated with respect to the $x$-coordinate to obtain:

$$\frac{\partial \tau}{\partial x} = \frac{\partial}{\partial x}\left\{\frac{3F}{4b}\left[\left(\frac{\alpha h}{2}\right)^{-1} - y^2\left(\frac{\alpha h}{2}\right)^{-3}\right]\right\} \tag{4.256}$$

$$\overset{(4.252)}{=} \frac{9F^2}{8b^2k\left(\frac{\alpha h}{2}\right)^3}\left[-1 + \frac{3y^2}{\left(\frac{\alpha h}{2}\right)^2}\right]. \tag{4.257}$$

The last equation can be introduced into the equilibrium equation (4.236) to give:

$$\frac{\partial \sigma_y}{\partial y} + \frac{9F^2}{8b^2k\left(\frac{\alpha h}{2}\right)^3}\left[-1 + \frac{3y^2}{\left(\frac{\alpha h}{2}\right)^2}\right] = 0, \tag{4.258}$$

or

$$\int_0^{\sigma_y} \mathrm{d}\tilde{\sigma_y} = -\frac{9F^2}{8b^2k\left(\frac{\alpha h}{2}\right)^3}\int_{+\frac{\alpha h}{2}}^{y}\left[-1 + \frac{3\tilde{y}^2}{\left(\frac{\alpha h}{2}\right)^2}\right]\mathrm{d}\tilde{y}, \tag{4.259}$$

or finally:

$$\sigma_y(0 \leq x \leq \varsigma, -\tfrac{\alpha h}{2} \leq y \leq +\tfrac{\alpha h}{2}) = \frac{9F^2y}{8b^2k\left(\frac{\alpha h}{2}\right)^3}\left[1 - \left(\frac{y}{\frac{\alpha h}{2}}\right)^2\right]. \tag{4.260}$$

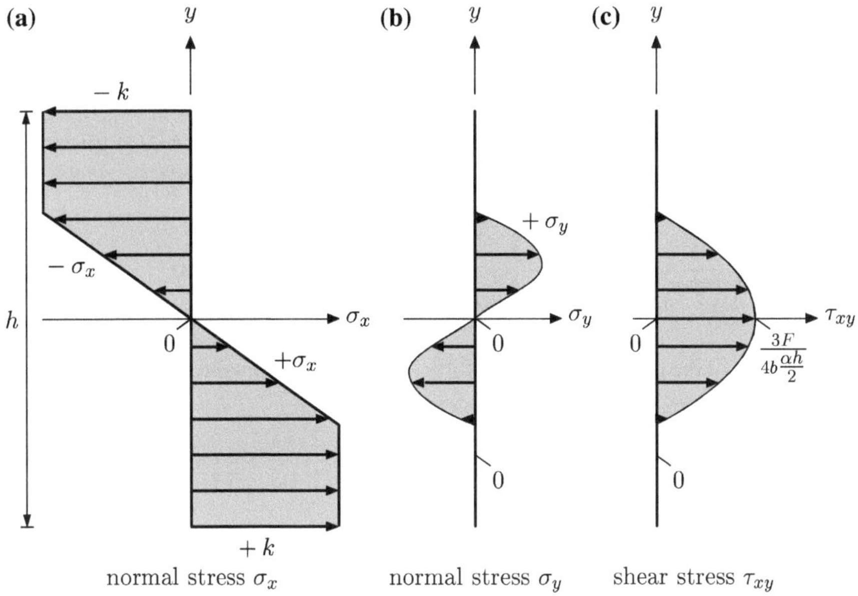


**Fig. 4.50** Stress distribution in the region $0 < x < \varsigma$: **a** normal $x$-stress, **b** normal $y$-stress and **c** shear stress

The graphical representations of Eqs. (4.249), (4.255) and (4.260) are given in Fig. 4.50.

It is now assumed in the following that the shear stress in the middle of the beam, i.e. $y = 0$, reaches the shear yield stress $k_s = \frac{k_t}{\sqrt{3}}$ at the fixed support, i.e. $x = 0$. The plastic zone extends in this case up to the vertical location $\frac{\alpha h}{2} = \left(\frac{\alpha h}{2}\right)_0$ which can be calculated from Eq. (4.255) as:

$$\frac{k}{\sqrt{3}} = \frac{3F}{4b\left(\frac{\alpha h}{2}\right)_0}[1-0] \quad \text{or} \quad \left(\frac{\alpha h}{2}\right)_0 = \frac{3\sqrt{3}F}{4bk}. \tag{4.261}$$

Thus, the distributions of the stress components can be obtained from Eqs. (4.249), (4.255) and (4.260) at $x = 0$ and for $\tau(y = 0) = \frac{k}{\sqrt{3}}$ as:

$$\sigma_x = -k\left(\frac{y}{\left(\frac{\alpha h}{2}\right)_0}\right), \tag{4.262}$$

$$\tau = \frac{k}{\sqrt{3}}\left[1-\left(\frac{y}{\left(\frac{\alpha h}{2}\right)_0}\right)^2\right], \tag{4.263}$$

$$\sigma_y = \frac{2k}{3}\left(\frac{y}{\left(\frac{\alpha h}{2}\right)_0}\right)\left[1-\left(\frac{y}{\left(\frac{\alpha h}{2}\right)_0}\right)^2\right]. \tag{4.264}$$

Substituting these stress components in the yield condition (4.234) gives after some transformation the following statement:

$$\frac{1}{9}\left(\frac{y}{\left(\frac{\alpha h}{2}\right)_0}\right)^2\left[1-\left(\frac{y}{\left(\frac{\alpha h}{2}\right)_0}\right)^2\right]\left[1-4\left(\frac{y}{\left(\frac{\alpha h}{2}\right)_0}\right)^2\right] \le 0. \tag{4.265}$$

The roots of the last equation in the form of $y/(\frac{\alpha h}{2})_0$ are obtained as $\{-1, -\frac{1}{2}, 0, 0, \frac{1}{2}, 1\}$ and the extreme points can be calculated as:

$$P_1\left(-\tfrac{1}{6}\sqrt{15+3\sqrt{13}}, \tfrac{1}{3888}(5+\sqrt{13})(\sqrt{13}-7)(2+\sqrt{13})\right), \tag{4.266}$$

$$P_2\left(-\tfrac{1}{6}\sqrt{15-3\sqrt{13}}, -\tfrac{1}{3888}(-5+\sqrt{13})(7+\sqrt{13})(-2+\sqrt{13})\right), \tag{4.267}$$

$$P_3\,(0,0), \tag{4.268}$$

$$P_4\left(+\tfrac{1}{6}\sqrt{15-3\sqrt{13}}, -\tfrac{1}{3888}(-5+\sqrt{13})(7+\sqrt{13})(-2+\sqrt{13})\right), \tag{4.269}$$

$$P_5\left(+\tfrac{1}{6}\sqrt{15+3\sqrt{13}}, \tfrac{1}{3888}(5+\sqrt{13})(\sqrt{13}-7)(2+\sqrt{13})\right). \tag{4.270}$$

The maximum value is 0.0061 for $y = \pm 0.34\left(\frac{\alpha h}{2}\right)_0$, i.e. points $P_2$ and $P_4$, and the yield condition is just not met. However, since the violation (>0) is quite small, the stress distribution can still be accepted. Thus, it could be shown that the section at $x = 0$ is the most critical section.

The value of the plastic limit moment under the influence of a shear force, i.e. $M^{\text{pl}}_{\text{s, lim}}$, can be obtained from Eq. (4.250) under consideration of $\left(\frac{\alpha h}{2}\right)_0$ as defined by Eq. (4.261)[29]:

$$M^{\text{pl}}_{\text{s, lim}} = b\left[\frac{h^2}{4} - \frac{1}{3}\times\underbrace{\frac{27F^2}{16b^2k^2}}_{\left(\frac{\alpha h}{2}\right)_0^2}\right]k. \tag{4.271}$$

[29] It should be mentioned here that the plastic limit moment $M^{\text{pl}}_{\text{lim}}$ is defined for the situation shown in Fig. 4.20d where the tensile yield stress $k_t$ is acting over the entire cross section. However, the plastic limit moment under the influence of a shear force $M^{\text{pl}}_{\text{s, lim}}$ considers a situation as shown in Fig. 4.50a together with a shear stress distribution where the maximum shear stress in the middle equals the shear yield stress $k_s$.

Introducing the plastic shear force (4.243) and the plastic limit moment (4.78) in the last equation gives after some transformations the following statement:

$$\frac{M_{\text{s, lim}}^{\text{pl}}}{M_{\text{lim}}^{\text{pl}}} = 1 - 0.75\left(\frac{F}{F_{\text{lim}}^{\text{pl}}}\right)^2. \tag{4.272}$$

The last relation is limited by Eq. (4.244), i.e. $\frac{F}{F_{\text{lim}}^{\text{pl}}} \leq \frac{2}{3}$. The shear stress $\tau$ equals the shear yield stress for $\frac{F}{F_{\text{lim}}^{\text{pl}}} = \frac{2}{3}$ in the middle of the beam and Eq. (4.272) gives $M_{\text{s, lim}}^{\text{pl}} = \frac{2}{3}M_{\text{lim}}^{\text{pl}}$. Under this condition, the value of $\left(\frac{\alpha h}{2}\right)_0$ as defined in Eq. (4.261) takes the following form:

$$\left(\frac{\alpha h}{2}\right)_0 = \frac{3\sqrt{3}F}{4bk} = \frac{3}{4}\frac{Fh}{\frac{1}{\sqrt{3}}bkh} = \frac{h}{2}. \tag{4.273}$$

This means that the locations $x = 0$ and $x = \varsigma$ coincide and that the length of the beam is $L - \varsigma$. It follows then from Eq. (4.240) that

$$\underbrace{\frac{2bhk}{3\sqrt{3}}}_{\frac{2F_{\text{lim}}^{\text{pl}}}{3}}(L - \varsigma) = \frac{bh^2k}{6} \quad \text{or} \quad \frac{L}{h} = \frac{\sqrt{3}}{4} = 0.433. \tag{4.274}$$

The condition $\frac{F}{F_{\text{lim}}^{\text{pl}}} \leq \frac{2}{3}$ is therefore identical to the dimensional requirement $\frac{L}{h} \geq 0.433$. However, the presented theory is hardly applicable to such short beams and this condition is not restrictive.

Typical interaction relations are collected in Table 4.24 and graphically represented in Fig. 4.51.

## 4.5 Supplementary Problems

### *4.5.1 Bernoulli Beam Problems*

**4.4 Equilibrium for a Differential Beam Element with Variable Distributed Load**
Derive the vertical force equilibrium and the moment equilibrium for the differential element of a BERNOULLI beam shown in Fig. 4.52.

**Table 4.24** Interactions relations for the reduction of the plastic moment due to the influence of a shear force

| Relation | Condition | Reference |
|---|---|---|
| $\dfrac{M^{pl}_{s,\,lim}}{M^{pl}_{lim}} = 1 - 0.75\left(\dfrac{Q}{Q^{pl}_{lim}}\right)^2$ | $\dfrac{Q}{Q^{pl}_{lim}} \leq 0.\bar{6}$ | Neal [73] |
| $\dfrac{M^{pl}_{s,\,lim}}{M^{pl}_{lim}} = 1 - 0.44\left(\dfrac{Q}{Q^{pl}_{lim}}\right)^2$ | $\dfrac{Q}{Q^{pl}_{lim}} \leq 0.79$ | Horne [56] |
| $\dfrac{M^{pl}_{s,\,lim}}{M^{pl}_{lim}} = 1 - 1.0\left(\dfrac{Q}{Q^{pl}_{lim}}\right)^2$ | $\dfrac{Q}{Q^{pl}_{lim}} \leq 1.0$ | Drucker [34] |

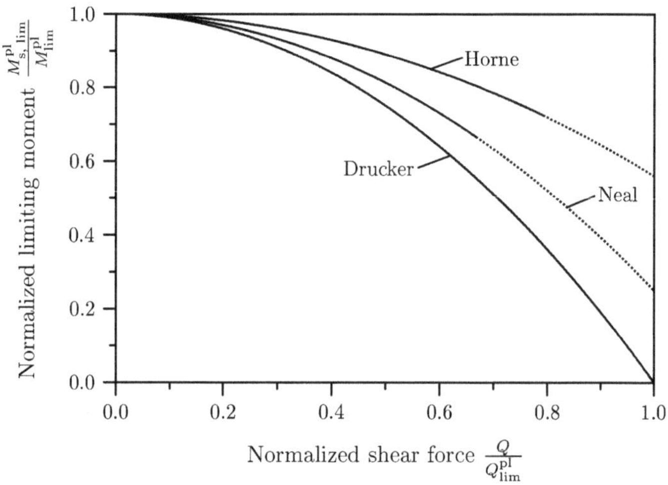


**Fig. 4.51** Interaction relations according to Horne [56], Neal [73], and Drucker [34]

**4.5 Bending Line Derived from Stress Distribution Over Cross Section**
Consider a linear-elastic BERNOULLI beam with a stress distribution over the height of the cross section as shown in Fig. 4.53. Calculate first the moment action resulting from this stress distribution and then use the kinematics relationship, i.e. $\varepsilon = -y\frac{d^2u_y(x)}{dx^2}$, to derive the expression for the bending line. Assume a rectangular cross section of $b \times h$ and a YOUNG's modulus of $E$.

**4.6 Simply Supported Beam with Centered Single Force**
Calculate the analytical solution for the deflection $u_y(x)$ and $u_y\left(\frac{L}{2}\right)$ of the simply supported BERNOULLI beam shown in Fig. 4.54 based on the fourth order differential

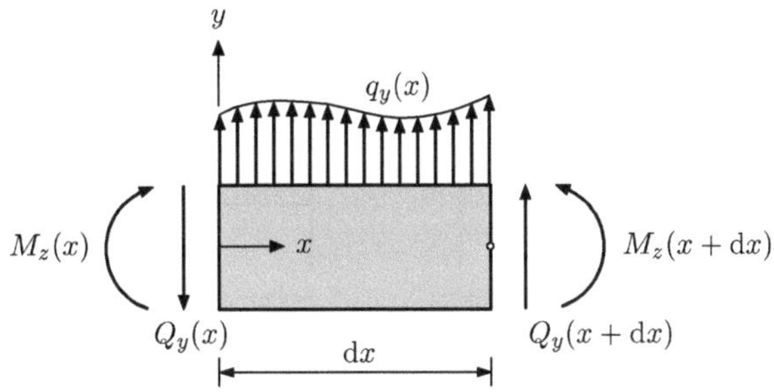


**Fig. 4.52** Differential element of a BERNOULLI beam with internal reactions and external variable load

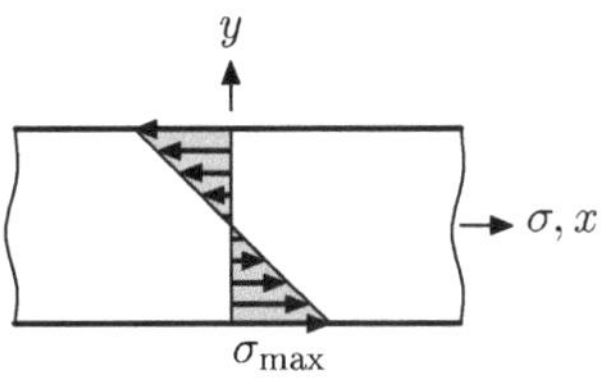


**Fig. 4.53** Stress distribution over the cross-section of a beam

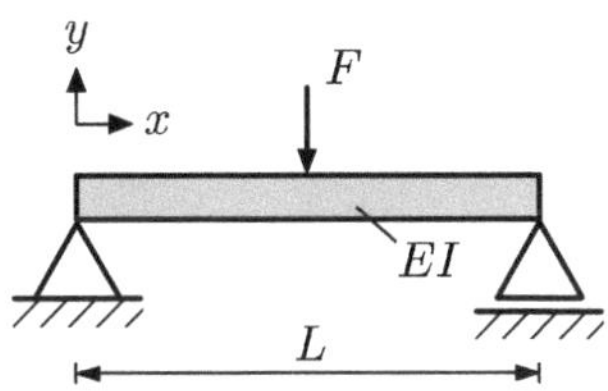


**Fig. 4.54** Simply supported BERNOULLI beam with centered single force

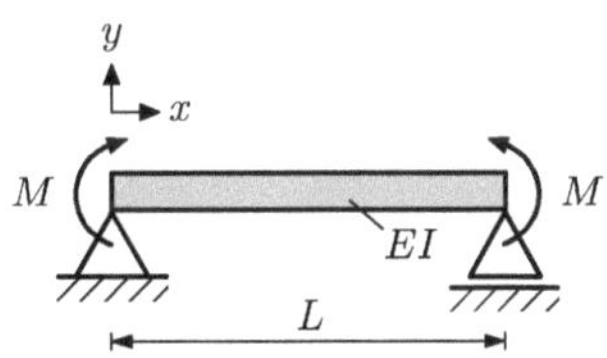


**Fig. 4.55** Simply supported BERNOULLI beam under pure bending load

equation given in Table 4.5. It can be assumed for this exercise that the bending stiffness $EI_z$ is constant.

**4.7 Simply Supported Beam Under Pure Bending Load**
Calculate the analytical solution for the deflection $u_y(x)$ and $u_y\left(\frac{L}{2}\right)$ of the simply supported BERNOULLI beam shown in Fig. 4.55 based on the second order differential equation for the bending moment distribution given in Eq. (4.37). It can be assumed for this exercise that the bending stiffness $EI_z$ is constant.

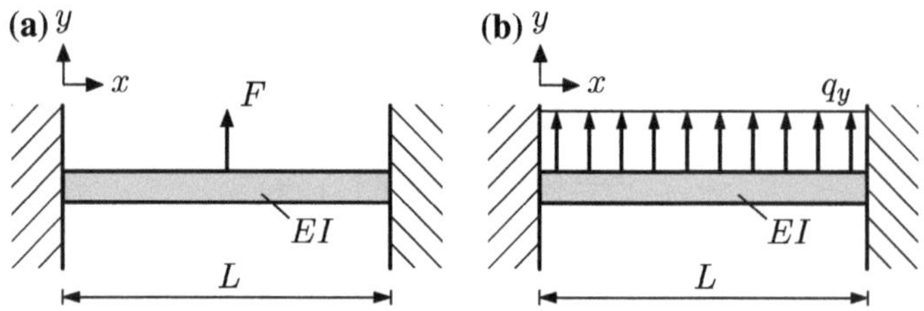


Fig. 4.56 BERNOULLI beam fixed at both ends: **a** single force case, **b** distributed load case

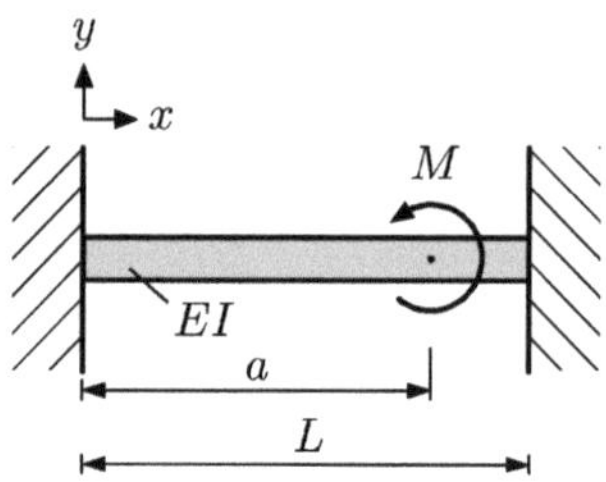


Fig. 4.57 BERNOULLI beam fixed at both ends and loaded by a single moment *M*

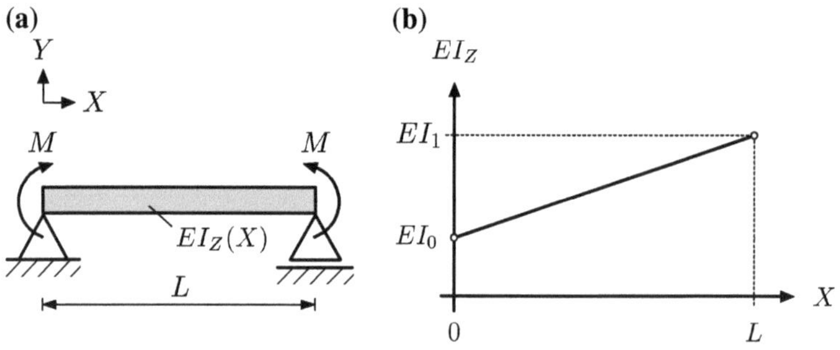


Fig. 4.58 BERNOULLI beam **(a)** with a functionally graded bending stiffness **(b)**

### 4.8 Bernoulli Beam Fixed at Both Ends

Calculate the analytical solution for the deflection $u_y(x)$ and slope $\varphi_z(x)$ of the BERNOULLI beams shown in Fig. 4.56 based on the fourth order differential equation given in Table 4.5. Determine in addition the maximum deflection and slope. It can be assumed for this exercise that the bending stiffness $EI_z$ is constant.

### 4.9 Bernoulli Beam Fixed at Both Ends and Loaded by a Single Moment

Calculate the analytical solution for the deflection $u_y(x)$ and slope $\varphi_z(x)$ of the BERNOULLI beam shown in Fig. 4.57 which is loaded by a single moment $M$ at a distance of $a$ from the left boundary. The beam is indeterminate to the second degree and the principle of superposition should be applied to solve the problem. Determine in addition the maximum deflection and slope for the special case $a = \frac{L}{2}$. It can be assumed for this exercise that the bending stiffness $EI_z$ is constant.

**4.10 Deflection of a Simply Supported Functionally Graded Beam**
Calculate the analytical solution for the deflection $u_y(x)$ for a simply supported BERNOULLI beam as shown in Fig. 4.58a. The bending stiffness of this functionally graded beam is described as, cf. Fig. 4.58b:

$$EI_z(x) = EI_0\left(1 + \frac{\frac{EI_1}{EI_0} - 1}{L}x\right) = EI_0(1 - \alpha x). \tag{4.275}$$

Sketch the normalized deflection $\frac{u_y(x)EI_0}{M}$ for the values $L = 10$ and $EI_1 = 2EI_0$. Add for comparison the normalized curve for a beam with constant bending stiffness $EI_1 = EI_0 = EI$, cf. suppl. problem 4.7.

**4.11 Equilibrium for a Differential Beam Element on Elastic Foundation with Distributed Load**
Derive the vertical force equilibrium and the moment equilibrium for the in Fig. 4.59 shown differential element of a BERNOULLI beam on an elastic foundation. Assume for the derivation that the elastic foundation modulus $k$ and the distributed load $q_0$ are constant.

**4.12 Cantilevered Beam on Elastic Foundation: Single Force Case**
Calculate the analytical solution for the deflection $u_y(x)$ and $u_y(L)$ of the cantilevered BERNOULLI beam shown in Fig. 4.60. It can be assumed for this exercise that the elastic foundation modulus $k$ and the bending stiffness $EI_z$ are constant. Calculate in addition the limiting case $u_y(L)$ for $\beta = \sqrt[4]{\frac{k}{4EI_z}} \rightarrow 0$.

**4.13 Simply Supported Beam on Elastic Foundation: Single Force Case**

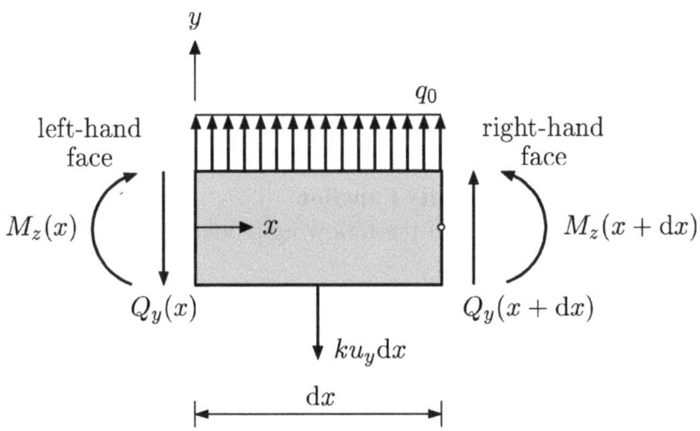


**Fig. 4.59** Differential element of a BERNOULLI beam on elastic foundation with internal reactions and external loads

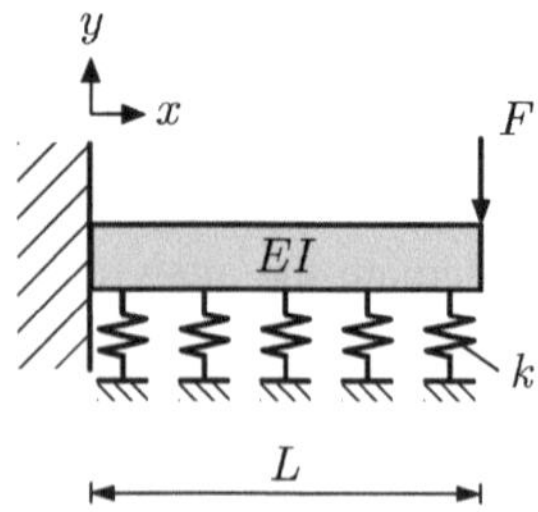


**Fig. 4.60** Cantilevered BERNOULLI beam on elastic foundation loaded by a single force

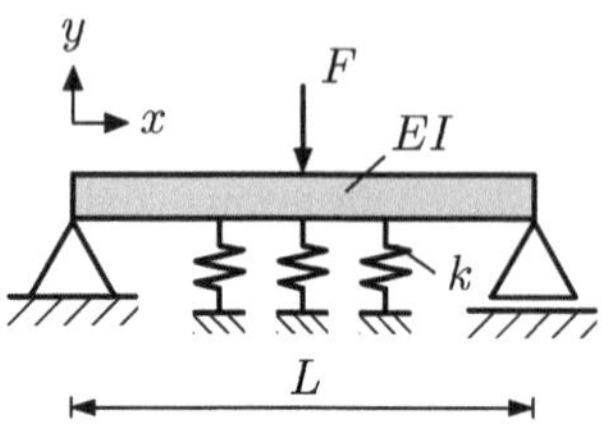


**Fig. 4.61** Simply supported BERNOULLI beam on elastic foundation loaded by a single force in the middle

Calculate the analytical solution for the deflection $u_y(x)$ and $u_y\left(\frac{L}{2}\right)$ of the simply supported BERNOULLI beam shown in Fig. 4.61. It can be assumed for this exercise that the elastic foundation modulus $k$ and the bending stiffness $EI_z$ are constant. Calculate in addition the limiting case $u_y\left(\frac{L}{2}\right)$ for $\beta = \sqrt[4]{\frac{k}{4EI_z}} \rightarrow 0$.

**4.14 Beam on Elastic Foundation: Distributed Load Case**
The partial differential equation which describes a BERNOULLI beam on an elastic foundation under the influence of a constant distributed load is given by:

$$EI_z \frac{\mathrm{d}^4 u_y(x)}{\mathrm{d}x^4} + k u_y(x) = q_y. \tag{4.276}$$

In this formulation it is assumed that the elastic foundation modulus $k$, the bending stiffness $EI_z$ and the distributed load $q_y = q_0$ are constant. In addition, it is assumed that the distributed load acts in the positive $y$-direction. Rewrite the PDE under consideration of the abbreviation $\beta = \sqrt[4]{\frac{k}{4EI_z}}$ and determine the general solution of the problem.

**4.15 Integration of a Discontinuity Function**
Calculate the indefinite integral of the following product of a polynomial of order two with a discontinuity function:

$$\int (x - a_1)^2 \langle x - a_2 \rangle^0 \mathrm{d}x. \tag{4.277}$$

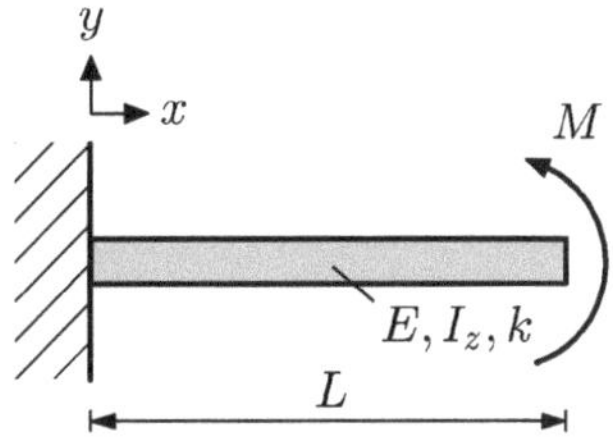


**Fig. 4.62** Cantilevered BERNOULLI beam in the elasto-plastic range loaded by a single moment

**4.16 Cantilevered Beam in the Elasto-Plastic Range Loaded by a Single Moment**
Given is a cantilevered BERNOULLI beam which is loaded at its right-hand end by a single moment $M$ as shown in Fig. 4.62. The cross section of the beam can be assumed to be rectangular ($b \times h$). Calculate the deflection $u_y(x)$ in the pure elastic and elasto-plastic range under the assumption that the material behavior is linear-elastic ideal-plastic with YOUNG's modulus $E$ and yield stress $k$.

**4.17 Cantilevered Beam in the Elasto-Plastic Range Loaded by a Distributed Load**
Given is a cantilevered BERNOULLI beam which is loaded by a constant distributed load of magnitude $q$ as shown in Fig. 4.63. The cross section of the beam can be assumed to be rectangular. Calculate the deflection $u_y(x)$ in the pure elastic and elasto-plastic range under the assumption that the material behavior is linear-elastic ideal-plastic with YOUNG's modulus $E$ and yield stress $k$.

**4.18 Cantilevered Beam in the Elasto-Plastic Range Loaded by a Distributed Load: Normalized Displacement Representation**
Problem 4.17 covered the displacement distribution of a cantilevered beam in the elasto-plastic range loaded by a distributed load. Convert the expression for the deflection $u_y(x)$ in a normalized form. Use the factor $\frac{M_{\text{lim}}^{\text{pl}} L^2}{E I_z}$ to normalize the deflection and give the expressions as a function of the normalized coordinate $\frac{x}{L}$ and the load ratio $\frac{M_z(L)}{M_{\text{lim}}^{\text{pl}}}$ with $M_z(L) = \frac{qL^2}{2}$.

**4.19 Cantilevered Beam in the Elasto-Plastic Range Loaded by a Distributed Load: Separation Between Pure Elastic and Elasto-Plastic Zone**

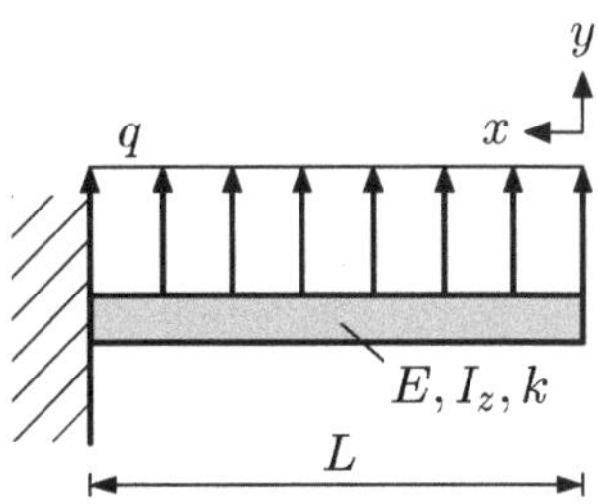


**Fig. 4.63** Cantilevered BERNOULLI beam in the elasto-plastic range loaded by a distributed load

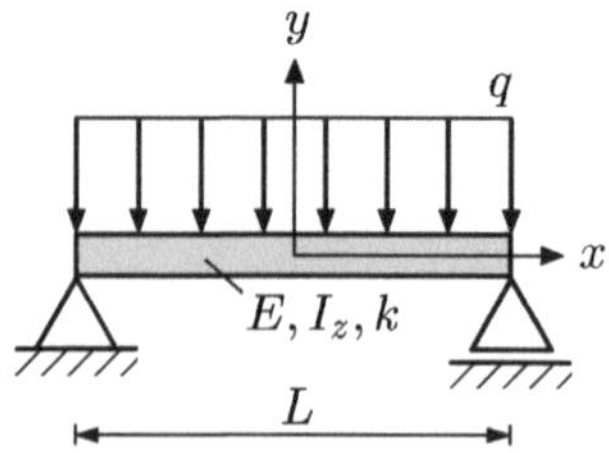


**Fig. 4.64** Simply supported BERNOULLI beam in the elasto-plastic range loaded by a distributed load

Reconsider again Problem 4.17 and derive the expression for the coordinate $x = \varsigma$ which separates the pure elastic and elasto-plastic zone. The expression should be given as a function of the load ratio $\frac{M_z(L)}{M_{\text{lim}}^{\text{pl}}}$. Discuss the limiting cases $M_z(L) = M_{\text{lim}}^{\text{el}}$ and $M_z(L) = M_{\text{lim}}^{\text{pl}}$.

**4.20 Simply Supported Beam in the Elasto-Plastic Range Loaded by a Distributed Load**
Given is a simply supported BERNOULLI beam which is loaded by a constant distributed load of magnitude $q$ as shown in Fig. 4.64. The cross section of the beam can be assumed to be rectangular (width $b$ and height $h$). Calculate the deflection $u_y(x)$ in the pure elastic and elasto-plastic range under the assumption that the material behavior is linear-elastic ideal-plastic with YOUNG's modulus $E$ and yield stress $k$.

### *4.5.2 Timoshenko Beam Problems*

**4.21 Calculation of the Shear Stress Distribution in a Rectangular Cross Section**
Given is a beam with rectangular cross section of width $b$ and height $h$. Calculate the distribution of the shear stress $\tau_{xy}$ over the cross section under the influence of a shear force $Q_y(x)$. Assume that the shear stress is constant along the width.

**4.22 Calculation of the Shear Correction Factor for Rectangular Cross Section**
For a rectangular cross section with width $b$ and height $h$, the shear stress distribution is given as follows [52]:

$$\tau_{xy}(y) = \frac{6Q_y}{bh^3}\left(\frac{h^2}{4} - y^2\right) \quad \text{with} \quad -\frac{h}{2} \leq y \leq \frac{h}{2}. \tag{4.278}$$

Compute the shear correction factor $k_s$ under the assumption that the constant—in the surface $A_s$ acting— equivalent shear stress $\tau_{xy} = Q_y/A_s$ yields the same shear strain energy as the actual shear stress distribution $\tau_{xy}(y)$, which acts in the actual cross-sectional area $A$ of the beam.

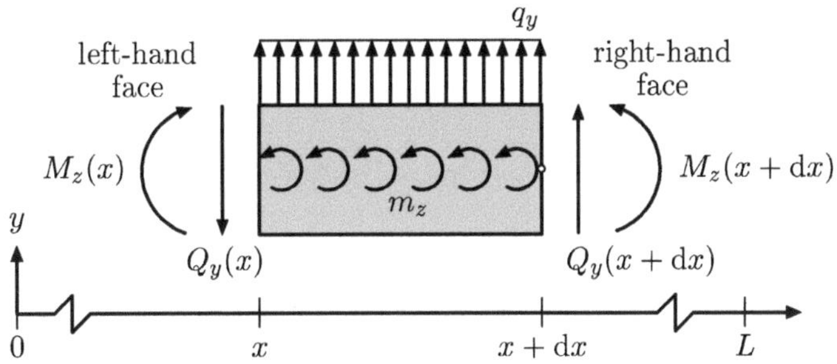


**Fig. 4.65** Infinitesimal beam element with internal reactions and distributed load

**Fig. 4.66** Cantilevered beam loaded by a single force

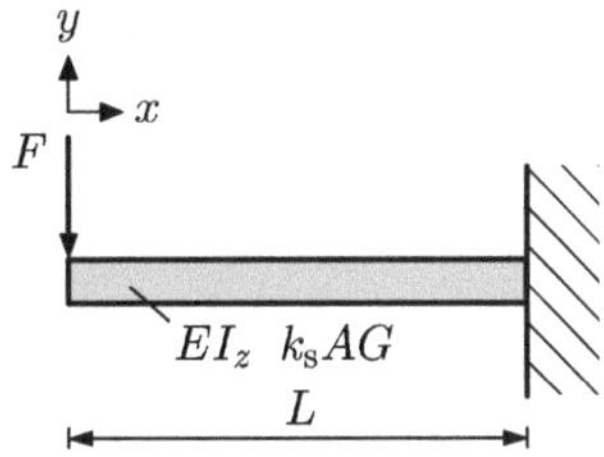


**4.23 Differential Equation Under Consideration of Distributed Moment**
For the derivation of the equilibrium condition, the infinitesimal beam element, illustrated in Fig. 4.65 needs to be considered, which is additionally loaded with a constant 'distributed moment' $m_z = \frac{\text{moment}}{\text{length}}$. Derive the differential equation for the TIMOSHENKO beam under consideration of a general moment distribution $m_z(x)$.

**4.24 Analytical Calculation of the Distribution of the Deflection and Rotation for a Cantilevered Beam Under Point Load**
For the illustrated cantilevered beam in Table 4.21, which is loaded with a point load $F$ at the right-hand end, calculate the distribution of the deflection $u_y(x)$ and the rotation $\phi_z(x)$ under consideration of the shear influence. Subsequently, the maximal deflection and the rotation at the loading point needs to be determined. Furthermore, the boundary value of the deflection at the loading point for slender ($h \ll L$) and compact ($h \gg L$) beams has to be determined.

**4.25 Analytical Calculation of the Normalized Deflection for Beams with Shear Contribution**
For the illustrated courses of the maximal normalized deflection $u_{y,\text{norm}}$ in Fig. 4.42 as a function of the slenderness ratio, derive the corresponding equations.

**4.26 Cantilevered beam loaded by a single force**
Calculate the analytical solution for the deflection $u_y(x)$ of a cantilevered TIMOSHENKO beam shown in Fig. 4.66 based on the general solution given in Eqs. (4.186) and (4.187). It can be assumed for this exercise that the bending stiffness $EI_z$ and the shear stiffness $GA$ are constant. The rectangular cross-sectional area is equal to $A = bh$.

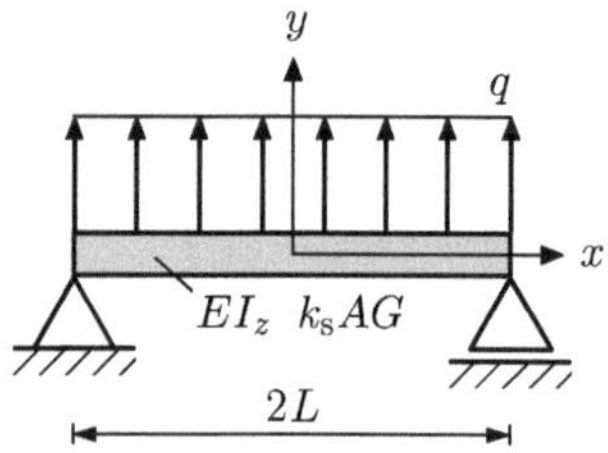


**Fig. 4.67** Simply supported TIMOSHENKO beam in the elastic range loaded by a distributed load

### 4.27 Simply Supported Beam in the Elastic Range Loaded by a Distributed Load

Given is a simply supported TIMOSHENKO beam which is loaded by a constant distributed load of magnitude $q$ as shown in Fig. 4.67. The cross section of the beam can be assumed rectangular (width $b$ and height $h$). Calculate the deflection $u_y(x)$ in the pure elastic range under the assumption that the bending stiffness $EI_z$ and the shear stiffness $GA$ are constant.

## *4.5.3 Third Order Beam Problems*

### 4.28 Cantilevered Beam Loaded by a Single Force

Calculate the analytical solution for the deflection $u_y(x)$ of a cantilevered LEVINSON beam shown in Fig. 4.68 based on the general solution given in Eqs. (4.219) and (4.220). It can be assumed for this exercise that the bending stiffness $EI_z$ and the shear stiffness $GA$ are constant. The rectangular cross-sectional area is equal to $A = bh$.

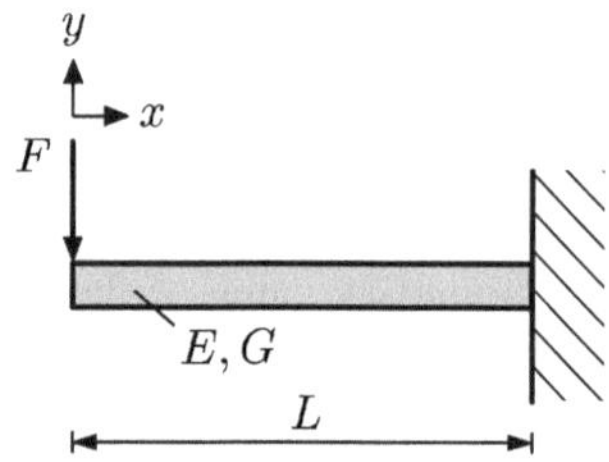


**Fig. 4.68** Cantilevered LEVINSON beam loaded by a single force

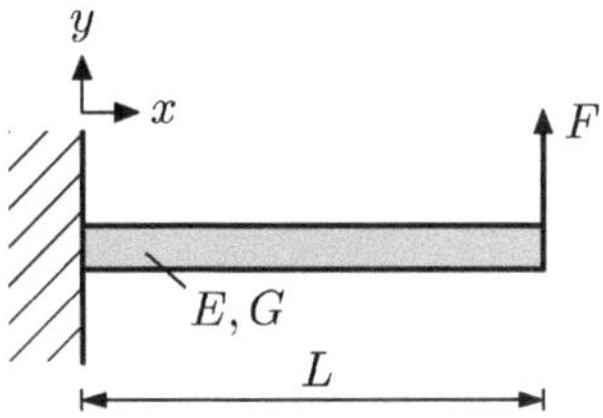


**Fig. 4.69** Cantilevered LEVINSON beam loaded by a single force

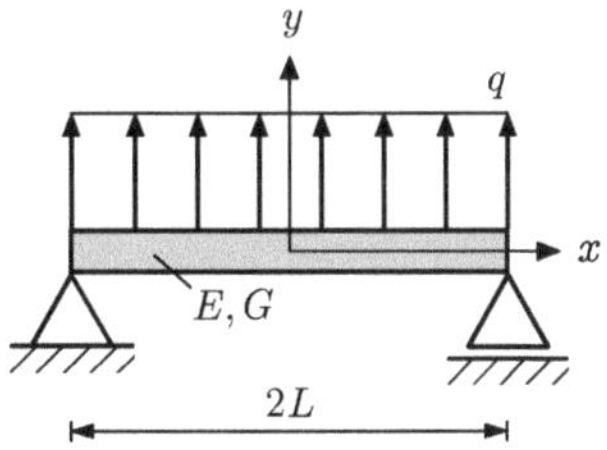


**Fig. 4.70** Simply supported LEVINSON beam in the elastic range loaded by a distributed load

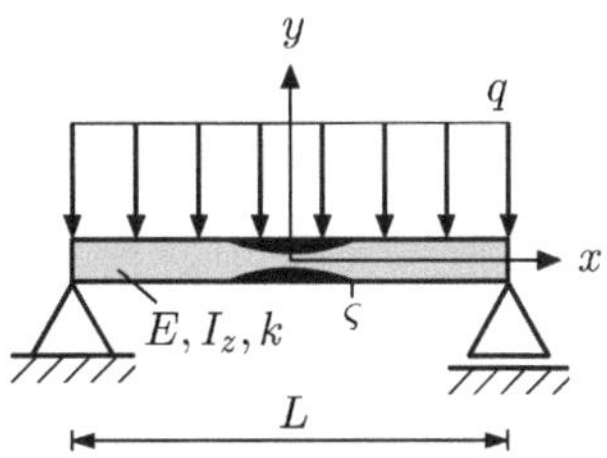


**Fig. 4.71** Simply supported beam in the elasto-plastic range loaded by a distributed load

**4.29 Shear Stress Distribution for a Cantilevered Beam Loaded by a Single Force**

Calculate the analytical solution for the shear stress distribution $\tau_{xy}(x, y)$ of a cantilevered LEVINSON beam ($A = bh$) shown in Fig. 4.69 based on Example 4.3. Compare the results with the analytical solution derived in supplementary problem 4.21.

**4.30 Simply Supported Beam in the Elastic Range Loaded by a Distributed Load**

Given is a simply supported LEVINSON beam which is loaded by a constant distributed load of magnitude $q$ as shown in Fig. 4.70. The cross section of the beam can be assumed rectangular (width $b$ and height $h$). Calculate the deflection $u_y(x)$ in the pure elastic range under the assumption that the bending stiffness $EI_z$ and the shear stiffness $GA$ are constant.

**4.31 Simply Supported Beam: Shear Stress Distribution in the Elasto-Plastic Range**

Given is a simply supported beam which is loaded by a constant distributed load of magnitude $q$ as shown in Fig. 4.71. The cross section of the beam can be assumed to be rectangular (width $b$ and height $h$). Calculate the shear stress distribution $\tau_{xy}(x, y)$ in the elasto-plastic range, i.e. $-\varsigma \leq x \leq +\varsigma$, for $-\frac{\alpha h}{2} \leq y \leq +\frac{\alpha h}{2}$. The material's behavior is linear-elastic ideal-plastic with YOUNG's modulus $E$ and yield stress $k$.

# Chapter 5
# Review of Linear-Elastic Finite Element Simulations

**Abstract** This chapter introduces the finite element theory for elastic problems based on rod, beam and generalized beam elements. First, the general concept of the weighted residual method, which allows to derive all classical approximation methods, is introduced. Then, each element type is considered separately in pure one-dimensional structures in the scope of the finite element method. Finally, the case of plane frame structures, which allows to arrange and combine different elements, is covered.

## 5.1 General Concept of the Weighted Residual Method

Starting point for the weighted residual method (WRM) is the differential equation which describes the physical problem. In the one-dimensional case, a physical problem can be generally described in a spatial domain $\Omega$ by the differential equation

$$\mathcal{L}\{u^0(x)\} = b \quad (x \in \Omega) \tag{5.1}$$

and by the conditions which are prescribed on the boundary $\Gamma$. The differential equation is also called the *strong form* or the *original statement* of the problem. The expression 'strong form' comes from the fact that the differential equation describes exact in each point $x$ of the domain the problem. The operator $\mathcal{L}\{\dots\}$ in Eq. (5.1) is an arbitrary differential operator which can take, for example, the following forms:

$$\mathcal{L}\{\dots\} = \frac{\mathrm{d}^2}{\mathrm{d}x^2}\{\dots\}, \tag{5.2}$$

$$\mathcal{L}\{\dots\} = \frac{\mathrm{d}^4}{\mathrm{d}x^4}\{\dots\}, \tag{5.3}$$

$$\mathcal{L}\{\dots\} = \frac{\mathrm{d}^4}{\mathrm{d}x^4}\{\dots\} + \frac{\mathrm{d}}{\mathrm{d}x}\{\dots\} + \{\dots\}. \tag{5.4}$$

Furthermore, variable $b$ in Eq. (5.1) is a given function and in the case of $b = 0$, the equation reduces to the *homogeneous differential equation*: $\mathcal{L}\{u^0(x)\} = 0$. The



A. Öchsner, *Elasto-Plasticity of Frame Structure Elements*,
DOI 10.1007/978-3-662-44225-8_5

exact or true solution, $u^0(x)$, fulfills in each point of the domain $x \in \Omega$ the differential equation and the prescribed geometrical and statical boundary conditions which are defined on the boundary $\Gamma$. Since the exact solution for most engineering problems is in general impossible to calculate, the aim of the following derivations is to find an as good as possible approximate solution

$$u(x) \approx u^0(x). \tag{5.5}$$

For the approximate solution in Eq. (5.5), the following shape can be chosen

$$u(x) = \alpha_0 + \sum_{k=1}^{n} \alpha_k \varphi_k(x), \tag{5.6}$$

where $\alpha_0$ should satisfy the nonhomogeneous part of the boundary conditions, $\varphi_k(x)$ are linearly independent basis functions and $\alpha_k$ are the undetermined parameters which must be identified by the respective approximation method to approximate at the best the exact solution $u^0(x)$ of the problem.

### 5.1.1 Methods Based on the Inner Product

Replacing the exact solution $u^0(x)$ in the differential equation (5.1) by an approximate solution $u(x)$ results in a local error, the so-called *residual*[1] $r$:

$$r = \mathcal{L}\{u(x)\} - b \neq 0. \tag{5.7}$$

This error is in the scope of the weighted residual method multiplied by a *weight function* $W(x)$ which distributes the error over the entire domain $\Omega$ in such a way that the error vanishes in average sense:

$$\int_{\Omega} rW\mathrm{d}\Omega = \int_{\Omega} (\mathcal{L}\{u(x)\} - b)W\mathrm{d}\Omega \stackrel{!}{=} 0. \tag{5.8}$$

This formulation is called the *inner product* and sometimes represented by a bracket expression $\langle \mathcal{L}\{u(x)\}, W(x)\rangle$ [16]. It should be mentioned here that the weight or test function $W(x)$ allows to distribute the error in the domain $\Omega$ in different ways. However, the total error must in average sense, i.e. integrated over the domain, become exactly zero. The shape of the weight functions is in the majority of cases in a similar way formulated as the approximate function $u(x)$, i.e.

[1] From Latin 'residuus' for left over or remaining.

$$W(x) = \sum_{k=1}^{n} \beta_k \psi_k(x), \tag{5.9}$$

where $\beta_k$ are arbitrary coefficients and $\psi_k(x)$ are linearly independent basis functions. The formulation (5.9) comprises—depending on the number of summands $k$ and the choice of functions $\psi_k(x)$—the class of methods which is based on the same basis functions for the approximate solution and the weight function ($\varphi_k(k) = \psi_k(x)$) and the class of methods which is based on different basis functions ($\varphi_k(k) \neq \psi_k(x)$). According to the choice of the weight function, the following classical methods can be distinguished [16, 120]:

• **Point Collocation Method:** $\psi_k(x) = \delta(x - x_k)$

The point collocation method uses the characteristic of the DIRAC delta function. The error must vanish at $n$ free points $x_1, x_2, \dots, x_n$, with $x_k \in \Omega$, the so-called collocation points. Thus, the approximate solution exactly fulfills the differential equation at these collocation points The weight function can be stated in this case as

$$W(x) = \beta_1 \underbrace{\delta(x - x_1)}_{\psi_1} + \dots + \beta_n \underbrace{\delta(x - x_n)}_{\psi_n} = \sum_{k=1}^{n} \beta_k \delta(x - x_k), \tag{5.10}$$

where the DIRAC[2] delta function is defined as:

$$\delta(x - x_k) = \begin{cases} 0 & \text{for}\, x \neq x_k \\ \infty & \text{for}\, x = x_k \end{cases}. \tag{5.11}$$

Substitution of this statement for the weight function in the inner product according to Eq. (5.8) and consideration of the characteristics of the DIRAC delta function, i.e.

$$\int_{-\infty}^{\infty} \delta(x - x_k)\,\mathrm{d}x = \int_{x_k-\varepsilon}^{x_k+\varepsilon} \delta(x - x_k)\,\mathrm{d}x = 1, \tag{5.12}$$

$$\int_{-\infty}^{\infty} f(x)\delta(x - x_k)\,\mathrm{d}x = \int_{x_k-\varepsilon}^{x_k+\varepsilon} f(x)\delta(x - x_k)\,\mathrm{d}x = f(x_k), \tag{5.13}$$

we obtain $n$ linearly independent equations for the determination of the parameters $\alpha_k$:

[2] Paul Adrien Maurice DIRAC (1902–1984), English theoretical physicist.

$$r(x_1) = \mathcal{L}\{u(x_1)\} - b = 0, \tag{5.14}$$
$$r(x_2) = \mathcal{L}\{u(x_2)\} - b = 0, \tag{5.15}$$
$$\vdots$$
$$r(x_n) = \mathcal{L}\{u(x_n)\} - b = 0. \tag{5.16}$$

It should be noted here that the approximate solution must satisfy all boundary conditions, i.e. the essential and natural conditions. Because of the characteristic of the DIRAC delta function, $\int_{\Omega} rW(\delta)\mathrm{d}\Omega = r = 0$, the point collocation method does not require the evaluation of any integral, i.e. no integration over the inner product. Therefore, one can save the integration—for example compared to the method of GALERKIN[3]—and the approximate solution is faster obtained. The disadvantage is that the collocation points can be freely selected and this selection can be unfavorable since it depends on the user.

- **Subdomain Collocation Method:** $\psi_k(x) = 1$ in $\Omega_k$ and otherwise zero

This method is also a collocation technique but instead of requiring that the error vanishes at certain points, the method requires that the integral of the error vanishes over a region, the subdomain:

$$\int_{\Omega_i} r\,\mathrm{d}\Omega_i = 0 \quad \text{for a subdomain } \Omega_i. \tag{5.17}$$

Based on this method, it is possible to derive, for example, the finite difference method, cf. Sect. 7.1.

- **Method of Least Squares:** $\psi_k(x) = \dfrac{\partial r}{\partial \alpha_k}$

The least square method optimizes the average squared error

$$\int_{\Omega} (\mathcal{L}\{u(x)\} - b)^2\,\mathrm{d}\Omega = \text{minimum}, \tag{5.18}$$

or

$$\frac{\mathrm{d}}{\mathrm{d}\alpha_k}\int_{\Omega} (\mathcal{L}\{u(x)\} - b)^2\,\mathrm{d}\Omega = 0, \tag{5.19}$$

$$\int_{\Omega} \frac{\mathrm{d}(\mathcal{L}\{u(x)\} - b)}{\mathrm{d}\alpha_k}(\mathcal{L}\{u(x)\} - b)\,\mathrm{d}\Omega = 0. \tag{5.20}$$

[3] Boris Grigoryevich GALERKIN (1871–1945), Russian mathematician.

- **Petrov-Galerkin Method:** $\psi_k(x) \neq \varphi_k(x)$

The PETROV[4]-GALERKIN approach collects all methods where the basis functions for the approximate solution and the weight function are different. Thus, the subdomain collocation method can be assigned for example to this method.

- **Galerkin Method:** $\psi_k(x) = \varphi_k(x)$

The basic idea of the GALERKIN or BUBNOV[5]-GALERKIN method is to choose the same basis functions for the approximate solution and the weight function. Thus, the weight function is obtained for this method as:

$$W(x) = \sum_{k=1}^{n} \beta_k \varphi_k(x). \tag{5.21}$$

Since for $u(x)$ and $W(x)$ the same basis functions $\varphi_k(x)$ were chosen and the coefficients $\beta_k$ are arbitrary, the function $W(x)$ can be written as a variation of $u(x)$ (with $\delta\alpha_0 = 0$):

$$W(x) = \delta u(x) = \delta\alpha_1\varphi_1(x) + \cdots + \delta\alpha_n\varphi_n(x) = \sum_{k=1}^{n} \delta\alpha_k \times \varphi_k(x). \tag{5.22}$$

These variations can be assigned to virtual quantities such as virtual displacements or velocities. Introduction of this formulation in the inner product according to Eq. (5.8) gives for a linear operator $\mathcal{L}\{\dots\}$ a set of $n$ linearly independent equations for the determination of the $n$ unknown coefficients $\alpha_k$:

$$\int_{\Omega} (\mathcal{L}\{u(x)\} - b) \times \varphi_1(x)\, \mathrm{d}\Omega = 0, \tag{5.23}$$

$$\int_{\Omega} (\mathcal{L}\{u(x)\} - b) \times \varphi_2(x)\, \mathrm{d}\Omega = 0, \tag{5.24}$$

$$\vdots$$

$$\int_{\Omega} (\mathcal{L}\{u(x)\} - b) \times \varphi_n(x)\, \mathrm{d}\Omega = 0. \tag{5.25}$$

Summary concerning methods based on the inner product:
These formulations require that the basis functions—which are here assumed to be defined over the entire domain $\Omega$—fulfil all, i.e. the essential and natural, boundary conditions. This requirement as well as the required differentiability from the $\mathcal{L}$-operator on the basis functions leads in practical applications to the difficulty

[4] Aleksei Zinovyevich PETROV (1910–1972), Russian mathematician.

[5] Ivan Grigoryevich BUBNOV (1872–1919), Russian engineer and scientist.

of finding appropriate functions. In addition, non-symmetric coefficient matrices are obtained (in the case of a symmetric $\mathcal{L}$-operator, the coefficient matrix of the GALERKIN method will be also symmetric).

### 5.1.2 Methods Based on the Weak Formulation

A further class of approximation methods can be derived by integration of parts of the inner product until the derivatives of $u(x)$ and $W(x)$ have the same order. As a result, the so-called *weak formulation* is obtained. The differentiability requirements for the approximate functions are reduced while the requirements for the weight functions are increased. If the idea of the GALERKIN method, i.e. the same basis functions for the approximate solution and the weight function, is used, then is the requirement on the differentiability of the basis functions overall reduced.
For a differential operator of order 2 or 4, i.e.

$$\int_{\Omega} \mathcal{L}_2\{u(x)\}W(x)\mathrm{d}\Omega, \tag{5.26}$$

$$\int_{\Omega} \mathcal{L}_4\{u(x)\}W(x)\mathrm{d}\Omega, \tag{5.27}$$

one-time integration by parts of Eq. (5.26) gives the weak form as

$$\int_{\Omega} \mathcal{L}_1\{u(x)\}\mathcal{L}_1\{W(x)\}\mathrm{d}\Omega = [\mathcal{L}_1\{u(x)\}W(x)]_{\Gamma}, \tag{5.28}$$

respectively two-time integration by parts of Eq. (5.27) the weak form as:

$$\int_{\Omega} \mathcal{L}_2\{u(x)\}\mathcal{L}_2\{W(x)\}\mathrm{d}\Omega = [\mathcal{L}_2\{u(x)\}\mathcal{L}_1\{W(x)\} - \mathcal{L}_3\{u(x)\}W(x)]_{\Gamma}. \tag{5.29}$$

Element-wise definitions of basis functions are used to derive the finite element method. For such a domain, i.e. a finite element with $\Omega^{\mathrm{e}} < \Omega$ and a local elemental coordinate $x^{\mathrm{e}}$, the weak form of Eq. (5.28) is, for example, obtained as:

$$\int_{\Omega^{\mathrm{e}}} \mathcal{L}_2\{u(x^{\mathrm{e}})\}\mathcal{L}_2\{W(x^{\mathrm{e}})\}\mathrm{d}\Omega^{\mathrm{e}} = \left[\mathcal{L}_2\{u(x^{\mathrm{e}})\}\mathcal{L}_1\{W(x^{\mathrm{e}})\} - \mathcal{L}_3\{u(x^{\mathrm{e}})\}W(x^{\mathrm{e}})\right]_{\Gamma^{\mathrm{e}}}. \tag{5.30}$$

Since the weak formulation comprises the natural boundary conditions, cf. the solved example 5.2, we can request in the following that the approximation[6] for $u(x)$ needs to satisfy only the essential boundary conditions. To derive the principal finite element equation, it is required according to the GALERKIN method that the same basis functions for the approximate solution and the weight function are chosen. In the framework of the finite element method, the coefficients $\alpha_k$ are chosen as the nodal values $u_k$ and the basis functions $\varphi_k(x)$ are named shape functions $N_k(x)$. Thus, the following representations for the approximate solution and the weight function are obtained:

$$u(x) = N_1(x)u_1 + N_2(x)u_2 + \cdots + N_n(x)u_n = \sum_{k=1}^{n} N_k(x)u_k, \tag{5.31}$$

$$W(x) = \delta u_1 N_1(x) + \delta u_2 N_2(x) + \cdots + \delta u_n N_n(x) = \sum_{k=1}^{n} \delta u_k N_k(x), \tag{5.32}$$

where $n$ is the number of nodes per element. It is important for this procedure that the error at the nodes—whose position is defined by the user—is minimized. This is a clear difference to the classical GALERKIN method based on the inner product where the points with $r = 0$ are found by the solution procedure. To further derive the finite element equation, it is necessary to write the representations (5.31) and (5.32) in matrix form and to introduce them in the weak form. For further details of the derivation, the reader should refer to the next sections concerning rod and beam elements in this chapter.

In the framework of the finite element method, the so-called RITZ[7] method is often mentioned. The classical method considers the total potential $\Pi$ of a structure. An approximate solution of the form (5.6), which is in the case of the RITZ method defined for the entire domain $\Omega$, is introduced in the formulation of the potential. The basis functions $\varphi_k$ must satisfy the geometrical but no the statical boundary conditions.[8] Derivation of the potential with respect to the unknown coefficients $\alpha_k$, i.e. determination of the extremum of $\Pi$, a system of equations is obtained to determine the $k$ coefficients, the so-called RITZ coefficients. However, it is difficult to find in the general case basis functions with unknown coefficients which satisfy *all* geometrical boundary conditions of the problem. Is the classical RITZ method modified in such a way that only the domain $\Omega^{\mathrm{e}}$ of a finite element is considered and is an approximate solution according to Eq. (5.31) utilized, then the classical finite element equation can be obtained.

---

[6] The superscript 'e' for the elemental coordinate will be omitted in the following where the understanding is not compromised.

[7] Walther RITZ (1878–1909), Swiss theoretical physicist.

[8] Since the total potential implicitly contains the statical boundary conditions, the basis functions do not need to satisfy them. Do the basis function in addition satisfy the statical boundary conditions, a higher accurate approximation can be expected.

### *5.1.3 Procedure on Basis of the Inverse Formulation*

Finally, it should be noted that the inner product can be partially further integrated to derive another class of approximate procedures. Shifting the derivative of $u(x)$ completely onto $W(x)$, one obtains the so-called *inverse formulation*. Depending on the choice of the weight function $W(x)$, one receives the following methods:

- Choice of $W$ so that $\mathcal{L}(W) = 0$ or $\mathcal{L}(u) \neq 0$.
  Procedure: *Boundary element method* (*Boundary integral equation of the first kind*).
- Use of a so-called fundamental solution $W = W^*$, i.e. a solution, which fulfills the equation $\mathcal{L}(W^*) = (-)\delta(\xi)$.
  Procedure: *Boundary element method* (*Boundary integral equation of the second kind*).
  The coefficient matrix of the corresponding system of equations is fully occupied and not symmetric. What is decisive for the application of the method is the knowledge about a fundamental solution for the $\mathcal{L}$ operator (in elasticity theory such an analytical solution is known through the KELVIN solution—concentrated load at a point of an infinite elastic medium).
- Equal shape functions for approximation approach and weight function approach.
  Procedure: TREFFTZ *method*.
- Equal shape functions for approximate approach and weight function and $\mathcal{L}(u) = \mathcal{L}(W) = 0$ is valid.
  Procedure: *Variation of the* TREFFTZ *method*.

The following Fig. 5.1 illustrates some common approximation methods in the context of the weighted residual method.

At the end of this section, a few comments on the different types of boundary conditions should be given. The boundary conditions can generally be divided into three groups:

- Boundary condition of the 1st kind or DIRICHLET[9] boundary condition (also referred to as the essential, fundamental, geometric or kinematic boundary condition):
  A boundary condition of the 1st kind exists, if the boundary condition is being expressed in parameters in which the differential equation is formulated.
- Boundary condition of the 2nd kind or NEUMANN[10] boundary condition (also referred to as the natural or static boundary condition):
  A boundary condition of the 2nd kind exists if the boundary condition is formulated in the derivative of the variable in which the differential equation is formulated.
- Boundary condition of the 3rd kind or CAUCHY[11] boundary condition (also referred to as the mixed or ROBIN boundary condition):
  Defines a weighted sum of DIRICHLET and NEUMANN boundary condition on the boundary.

---

[9] Johann Peter Gustav Lejeune DIRICHLET (1805–1859), German mathematician.

[10] Karl Gottfried NEUMANN (1832–1925), German mathematician.

[11] Augustin-Louis CAUCHY (1789–1857), French mathematician.

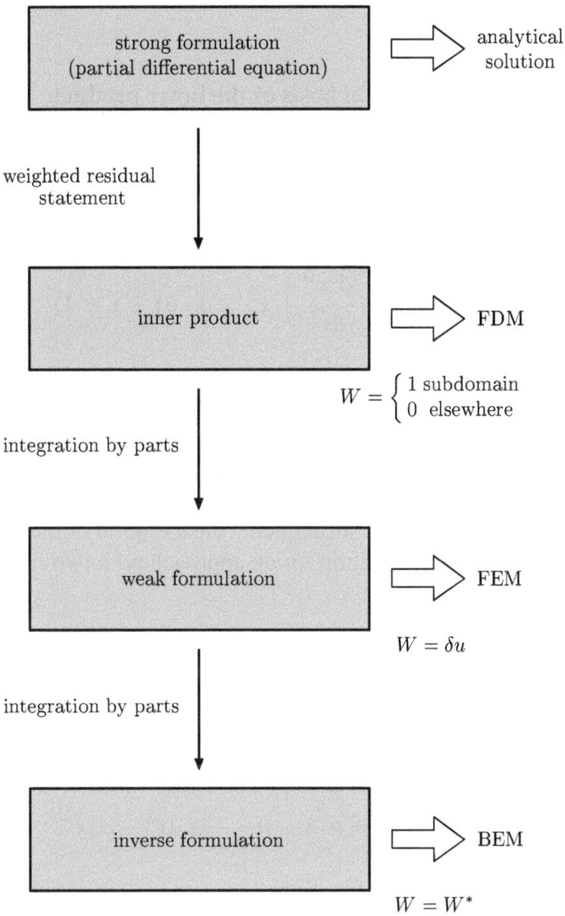


**Fig. 5.1** Some classical approximation methods in the context of the weighted residual method

**Table 5.1** Different boundary conditions of a differential equation

| Differential equation | DIRICHLET | NEUMANN | CAUCHY |
|---|---|---|---|
| $\mathcal{L}\{u(x)\} = b$ | $u$ | $\frac{du}{dx}$ | $\alpha u + \beta \frac{du}{dx}$ |

These three types of boundary conditions are summarized in Table 5.1.

One talks about *homogeneous boundary conditions* if the corresponding variables are zero on the boundary.

### 5.1.4 Solved Problems

**5.1 Example: Galerkin method on basis of the inner product**

Since the expression 'GALERKIN method' is an often used term in the finite element context, the original GALERKIN method needs to be explained in the following within the framework of this example. For this purpose, the following differential equation, which is defined in the domain $0 < x < 1$, is considered

$$\mathcal{L}\{u(x)\} - b = \frac{\mathrm{d}^2 u^0}{\mathrm{d}x^2} + x^2 = 0 \quad (0 < x < 1) \tag{5.33}$$

with the homogeneous boundary conditions $u^0(0) = u^0(1) = 0$. For this problem the exact solution

$$u^0(x) = \frac{x}{12}\left(-x^3 + 1\right) \tag{5.34}$$

can be defined via integration and subsequent consideration of the boundary conditions. Define the approximate solution for an approach with two free values.

**5.1 Solution**

For the construction of the approximate solution $u(x)$ according to the GALERKIN method, the following approach with two free parameters can be used:

$$u^0(x) \approx u(x) = \alpha_1 \varphi_1(x) + \alpha_2 \varphi_2(x), \tag{5.35}$$

$$= \alpha_1 x(1-x) + \alpha_2 x^2 (1-x), \tag{5.36}$$

$$= \alpha_1 x + (\alpha_2 - \alpha_1) x^2 - \alpha_2 x^3. \tag{5.37}$$

One needs to consider that the functions $\varphi_1(x)$ and $\varphi_2(x)$ are chosen in a way so that the boundary conditions, meaning $u(0) = u(1) = 0$, are fulfilled. Therefore, polynomials of first order are eliminated since a linear slope could only connect the two zero points as a horizontal line. Furthermore, both functions are chosen in a way so that they are linearly independent. The first two derivatives of the approximation approach result in

$$\frac{\mathrm{d}u(x)}{\mathrm{d}x} = \alpha_1 + 2(\alpha_2 - \alpha_1)x - 3\alpha_2 x^2, \tag{5.38}$$

$$\frac{\mathrm{d}^2 u(x)}{\mathrm{d}x^2} = 2(\alpha_2 - \alpha_1) - 6\alpha_2 x, \tag{5.39}$$

and the error function results in the following via the second derivative from Eq. (5.7):

$$r(x) = \frac{\mathrm{d}^2 u}{\mathrm{d}x^2} + x^2 = 2(\alpha_2 - \alpha_1) - 6\alpha_2 x + x^2. \tag{5.40}$$

The insertion of the weight function, meaning

$$W(x) = \delta u(x) = \delta\alpha_1 x(1-x) + \delta\alpha_2 x^2(1-x), \tag{5.41}$$

into the residual equation yields

$$\int_0^1 \underbrace{\left(2(\alpha_2-\alpha_1) - 6\alpha_2 x + x^2\right)}_{r(x)} \times \underbrace{\left(\delta\alpha_1 x(1-x) + \delta\alpha_2 x^2(1-x)\right)}_{W(x)} \mathrm{d}x = 0, \tag{5.42}$$

or generally split into two integrals:

$$\delta\alpha_1 \int_0^1 r(x)\varphi_1(x)\mathrm{d}x + \delta\alpha_2 \int_0^1 r(x)\varphi_2(x)\mathrm{d}x = 0. \tag{5.43}$$

Since the $\delta\alpha_i$ are arbitrary coefficients and the shape functions $\varphi_i(x)$ are linearly independent, the following system of equations results herefrom:

$$\delta\alpha_1 \int_0^1 \left(2(\alpha_2-\alpha_1) - 6\alpha_2 x + x^2\right) \times (x(1-x))\,\mathrm{d}x = 0, \tag{5.44}$$

$$\delta\alpha_2 \int_0^1 \left(2(\alpha_2-\alpha_1) - 6\alpha_2 x + x^2\right) \times \left(x^2(1-x)\right)\mathrm{d}x = 0. \tag{5.45}$$

After the integration, a system of equations results for the definition of the two unknown free parameters $\alpha_1$ and $\alpha_2$:

$$\frac{1}{20} - \frac{1}{6}\alpha_2 - \frac{1}{3}\alpha_1 = 0, \tag{5.46}$$

$$\frac{1}{30} - \frac{2}{15}\alpha_2 - \frac{1}{6}\alpha_1 = 0, \tag{5.47}$$

or alternatively in matrix notation:

$$\begin{bmatrix} \frac{1}{3} & \frac{1}{6} \\ \frac{1}{6} & \frac{2}{15} \end{bmatrix} \begin{bmatrix} \alpha_1 \\ \alpha_2 \end{bmatrix} = \begin{bmatrix} \frac{1}{20} \\ \frac{1}{30} \end{bmatrix}. \tag{5.48}$$

From this system of equations the free parameters result in $\alpha_1 = \frac{1}{15}$ and $\alpha_2 = \frac{1}{6}$. Therefore, the approximate solution and the error function finally result in:

$$u(x) = x\left(-\frac{1}{6}x^2 + \frac{1}{10}x + \frac{1}{15}\right), \tag{5.49}$$

$$r(x) = x^2 - x + \frac{1}{5}. \tag{5.50}$$

The comparison between the approximate solution and exact solution is illustrated in Fig. 5.2a. One can see that the two solutions coincide on the boundaries—one needs to consider that the approximate approach has to fulfill the boundary conditions—as well as on two other locations.

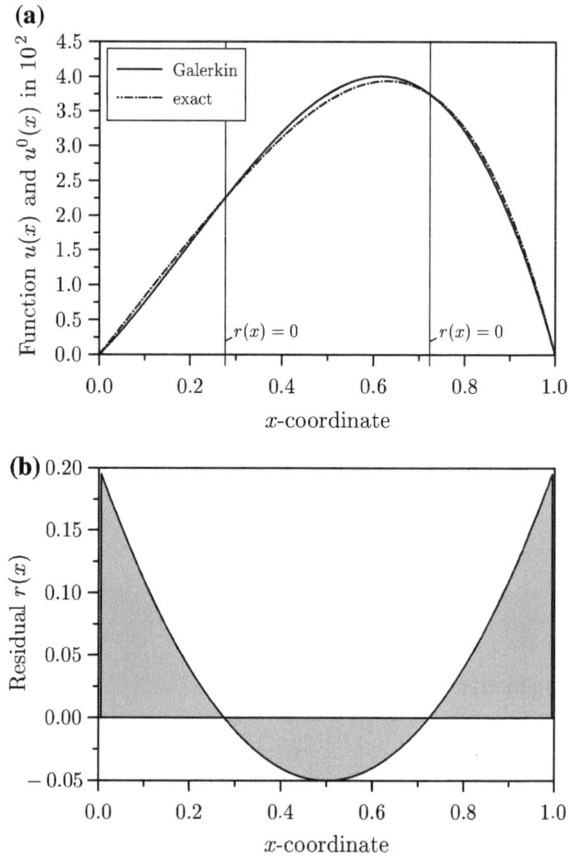


**Fig. 5.2** Approximate solution according to the GALERKIN method, **a** exact solution and **b** residual as a function of the coordinate

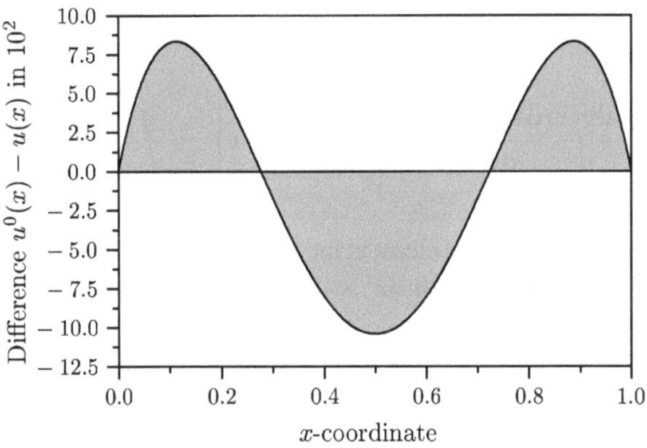


**Fig. 5.3** Absolute difference between exact solution and approximate solution as function of the coordinate

It needs to be remarked at this point that the error function—see Fig. 5.2b—does not illustrate the difference between the exact solution and the approximate solution. Rather it is about the error, which results from inserting the approximate solution into the differential equation. To illustrate this, Fig. 5.3 shows the absolute difference between exact solution and approximate solution.

Finally it can be summarized that the advantage of the GALERKIN method is that the procedure itself is in search of the points with $r = 0$. This is quite an advantage in comparison to the collocation method. However within the GALERKIN method the integration needs to be performed and therefore this method is in comparison to the collocation more complex and slower.

### 5.2 Example: Finite element method

For the differential equation (5.33) and the given boundary conditions one needs to calculate, based on the weak formulation, a finite element solution, based on two equidistant elements with linear shape functions.

### 5.2 Solution

The partial integration of the inner product yields the following formulation:

$$\int_0^1 \left( \frac{\mathrm{d}^2 u(x)}{\mathrm{d}x^2} + x^2 \right) W(x)\,\mathrm{d}x = 0, \tag{5.51}$$

$$\int_0^1 \frac{\mathrm{d}^2 u(x)}{\mathrm{d}x^2} W(x)\,\mathrm{d}x + \int_0^1 x^2 W(x)\,\mathrm{d}x = 0, \tag{5.52}$$

$$\left[ \frac{\mathrm{d}u(x)}{\mathrm{d}x} W(x) \right]_0^1 - \int_0^1 \frac{\mathrm{d}u(x)}{\mathrm{d}x}\frac{\mathrm{d}W(x)}{\mathrm{d}x}\,\mathrm{d}x + \int_0^1 x^2 W(x)\,\mathrm{d}x = 0, \tag{5.53}$$

or alternatively the weak form in its final form:

$$\int\limits_{0}^{1} \frac{du(x)}{dx}\frac{dW(x)}{dx}\,dx = \left[\frac{du(x)}{dx}W(x)\right]_{0}^{1} + \int\limits_{0}^{1} x^2 W(x)\,dx. \tag{5.54}$$

For the derivation of the finite element method, one considers domain-wise defined shape functions. For such a domain $\Omega^e < \Omega$, namely a finite element[12] of the length $L^e$, the weak formulation results in:

$$\int\limits_{0}^{L^e} \frac{du(x^e)}{dx^e}\frac{dW(x^e)}{dx^e}\,dx^e = \left[\frac{du(x^e)}{dx^e}W(x^e)\right]_{0}^{L^e} + \int\limits_{0}^{L^e} (x^e + c^e)^2 W(x^e)\,dx^e. \tag{5.55}$$

In the transition from Eq. (5.54) to (5.55), meaning from the global formulation to the consideration of the element level, in particular the quadratic expression on the right-hand side of Eq. (5.54) needs to be considered. To ensure that the in the global coordinate system defined expression $X^2$ is considered appropriately in the description on the element level, a coordinate transformation has to be performed for every element 'e' via a term $c^e$. From Fig. 5.4 it can be seen that the term $c$ turns zero for the first element (I) since global and local coordinate system coincide. For the second element (II) $c^{II} = X^{I,2} = \frac{1}{3}$ results with an equidistant division and $c^{III} = X^{II,2} = \frac{2}{3}$ results accordingly for the third element (III).

Since the weak formulation contains the natural boundary conditions—for this see the boundary expression in Eq. (5.55)—it can be demanded in the following that the approach for $u(x)$ has to fulfill the essential boundary conditions only. According to the GALERKIN method it is demanded for the derivation of the principal finite element equation that the same shape functions for the approximate and weight function are chosen. Within the framework of the finite element method the nodal values $u_k$ are

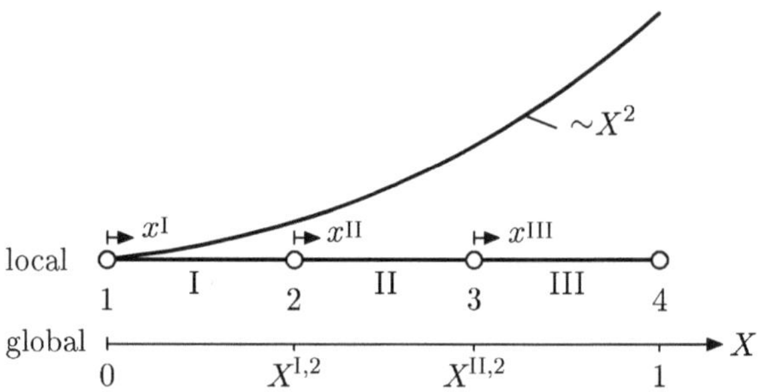


**Fig. 5.4** Global coordinate system $X$ and local coordinate system $x_i$ for every element

[12] Usually a separate *local* coordinate system $0 \leq x^e \leq L^e$ is introduced for each element 'e'. The coordinate in Eq. (5.54) is then referred to as the global coordinate and receives the symbol $X$.

chosen for the free parameters $\alpha_k$ and the shape functions $\varphi_k(x)$ are being referred to as the form and shape functions $N_k(x)$. For linear shape functions the following illustrations result for the approximate solution and the weight function:

$$u(x) = N_1(x)u_1 + N_2(x)u_2, \quad (5.56)$$

$$W(x) = \delta u_1 N_1(x) + \delta u_2 N_2(x). \quad (5.57)$$

For the chosen linear shape functions, an element-wise linear course of the approximate function and a difference between the exact solution and the approximate approach as shown in Fig. 5.5 are obtained. It is obvious that the error is minimal at the nodes, at the best identical with the exact solution.

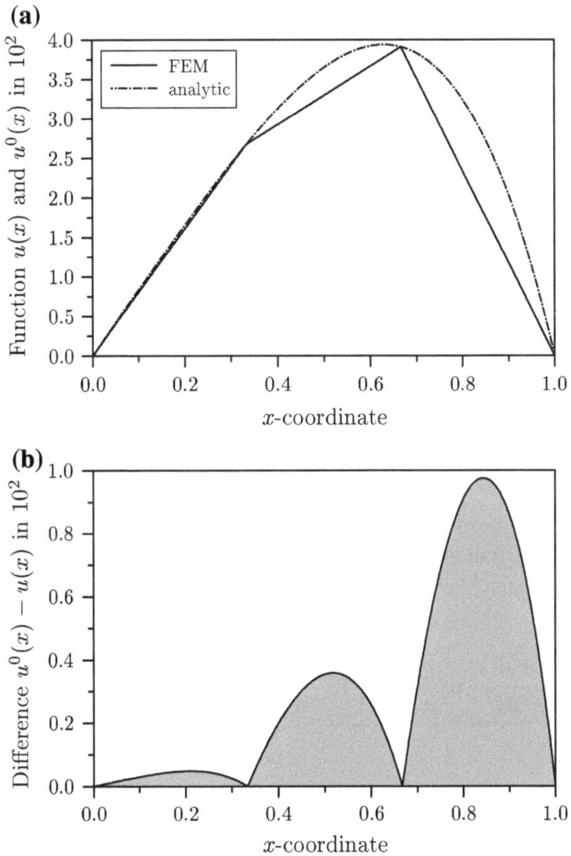


**Fig. 5.5** **a** Exact and finite element displacement distribution and **b** absolute difference between exact solution and finite element solution as a function of the coordinate

## 5.2 Rod Elements

### 5.2.1 Derivation of the Principal Finite Element Equation

Let us consider in the following the governing differential equation according to Eq. (3.11). This formulation assumes that the axial tensile stiffness *EA* is constant and we obtain

$$EA\frac{\mathrm{d}^2u^0(x)}{\mathrm{d}x^2}+p(x)=0, \tag{5.58}$$

where $u^0(x)$ represents the *exact* solution of the problem. The last equation which contains the exact solution of the problem is fulfilled at each location $x$ of the rod and is called the *strong formulation* of the problem. Replacing the exact solution in Eq. (5.58) by an approximate solution $u(x)$, a residual $r$ is obtained:

$$r(x)=EA\frac{\mathrm{d}^2u(x)}{\mathrm{d}x^2}+p(x)\neq 0. \tag{5.59}$$

As a consequence of the introduction of the approximate solution $u(x)$, it is in general no longer possible to satisfy the differential equation at each location $x$ of the rod. It is alternatively requested in the following that the differential equation is fulfilled over a certain length (an no longer at each location $x$) and the following integral statement is obtained

$$\int_0^L W(x)\left(EA\frac{\mathrm{d}^2u(x)}{\mathrm{d}x^2}+p(x)\right)\mathrm{d}x\overset{!}{=}0, \tag{5.60}$$

which is called the *inner product*. The function $W(x)$ in Eq. (5.60) is called the weight function which distributes the error or the residual in the considered domain.

Integrating by parts[13] of the first expression in the brackets of Eq. (5.60) gives

$$\int_0^L \underbrace{W}_{f}\, EA \underbrace{\frac{\mathrm{d}^2u(x)}{\mathrm{d}x^2}}_{g'}\mathrm{d}x = EA\left[W\frac{\mathrm{d}u(x)}{\mathrm{d}x}\right]_0^L - EA\int_0^L \frac{\mathrm{d}W(x)}{\mathrm{d}x}\frac{\mathrm{d}u(x)}{\mathrm{d}x}\mathrm{d}x. \tag{5.61}$$

Under consideration of Eq. (5.60), the so-called *weak formulation* of the problem is obtained as:

[13] A common representation of integration by parts of two functions $f(x)$ and $g(x)$ is: $\int fg'\mathrm{d}x = fg - \int f'g\,\mathrm{d}x$.

$$EA\int_0^L \frac{\mathrm{d}W(x)}{\mathrm{d}x}\frac{\mathrm{d}u(x)}{\mathrm{d}x}\,\mathrm{d}x = EA\left[W(x)\frac{\mathrm{d}u(x)}{\mathrm{d}x}\right]_0^L + \int_0^L W(x)p(x)\,\mathrm{d}x. \tag{5.62}$$

Looking at the weak formulation, it can be seen that the integration by parts shifted one derivative from the approximate solution to the weight function and a symmetrical formulation with respect to the derivatives is obtained. This symmetry with respect to the derivatives of the approximate solution and the weight function will guarantee in the following that a symmetric stiffness matrix is derived for the rod element.

In order to continue the derivation of the principal finite element equation, the displacement $u(x)$ and the weight function $W(x)$ must be expressed by some functions. The common way to express the unknown function $u(x)$ in the scope of the finite element method is the so-called nodal approach. This approach states that the unknown function within an element (superscript 'e') is given by

$$u^{\mathrm{e}}(x) = \boldsymbol{N}^{\mathrm{T}}(x)\,\boldsymbol{u}_{\mathrm{p}} = \begin{bmatrix} N_1 & N_2 & \cdots & N_n \end{bmatrix} \times \begin{bmatrix} u_1 \\ u_2 \\ \vdots \\ u_n \end{bmatrix}, \tag{5.63}$$

where $\boldsymbol{u}_{\mathrm{p}}$ is the column matrix of $n$ nodal unknowns and $\boldsymbol{N}(\boldsymbol{x})$ is the column matrix of the *shape functions*. Thus, the displacement at any point inside an element is approximated based on nodal values and shape functions which distribute these displacements between the nodes in a certain way. Equation (5.63) illustrates a basic idea of the finite element method where the unknown function is not approximated over the entire domain of the problem (in general $\Omega$) but in a sub-domain ($\Omega^{\mathrm{e}}$), the so-called finite element. In a similar way as the unknown function, the weight function is approximated as

$$W(x) = \left(\boldsymbol{N}^{\mathrm{T}}(x)\,\boldsymbol{u}_{\mathrm{p}}\right)^{\mathrm{T}} = \delta\boldsymbol{u}_{\mathrm{p}}^{\mathrm{T}}\boldsymbol{N}(\boldsymbol{x}) = \begin{bmatrix} \delta u_1 & \delta u_2 & \cdots & \delta u_n \end{bmatrix} \times \begin{bmatrix} N_1 \\ N_2 \\ \vdots \\ N_n \end{bmatrix}, \tag{5.64}$$

where the $\delta u_i$ represent the so-called arbitrary or virtual displacements. It will be shown in the following that the virtual displacements occur on both sides of Eq. (5.62) and can be eliminated. Thus, these virtual displacements do not need a deeper consideration at this point of the derivation. Equation (5.62) requires the derivatives of $u(x)$ and $W(x)$ which can be written on the element level as:

$$\frac{\mathrm{d}u^{\mathrm{e}}(x)}{\mathrm{d}x} = \frac{\mathrm{d}}{\mathrm{d}x}\left(\boldsymbol{N}^{\mathrm{T}}(x)\,\boldsymbol{u}_{\mathrm{p}}\right) = \frac{\mathrm{d}\boldsymbol{N}^{\mathrm{T}}(x)}{\mathrm{d}x}\boldsymbol{u}_{\mathrm{p}}, \tag{5.65}$$

$$\frac{\mathrm{d}W(x)}{\mathrm{d}x} = \frac{\mathrm{d}}{\mathrm{d}x}\left(\delta\boldsymbol{u}_{\mathrm{p}}^{\mathrm{T}}\boldsymbol{N}(\boldsymbol{x})\right) = \delta\boldsymbol{u}_{\mathrm{p}}^{\mathrm{T}}\frac{\mathrm{d}\boldsymbol{N}(\boldsymbol{x})}{\mathrm{d}x}. \tag{5.66}$$

It should be noted here that the nodal unknowns und their virtual counterparts are constant values, i.e. not a function of $x$, and are therefore not affected by the differential operator. It is common in some references (e.g. [120, 26]) to introduce the matrix which contains the derivatives of the shape functions as a matrix denoted by $\boldsymbol{B} = \frac{\mathrm{d}\boldsymbol{N}(x)}{\mathrm{d}x}$. Thus, the derivatives can be be written as:

$$\frac{\mathrm{d}u^{\mathrm{e}}(x)}{\mathrm{d}x} = \boldsymbol{B}^{\mathrm{T}}\,\boldsymbol{u}_{\mathrm{p}}, \tag{5.67}$$

$$\frac{\mathrm{d}W(x)}{\mathrm{d}x} = \delta\boldsymbol{u}_{\mathrm{p}}^{\mathrm{T}}\boldsymbol{B}. \tag{5.68}$$

#### 5.2.1.1 Linear Element Formulation

Let us consider in the following a rod element which is composed of two nodes as schematically shown in Fig. 5.6. Each node has only one degree of freedom, i.e. a displacement in the direction of the principal axis (cf. Fig. 5.6a) and each node can be only loaded by a single force acting in $x$-direction (cf. Fig. 5.6b).

Since there are only two nodes with two unknowns, the equation for the unknown displacement in the element and its virtual counterpart (cf. Eqs. (5.63) and (5.64)) are simplified to the following expressions:

$$u^{\mathrm{e}}(x) = \boldsymbol{N}^{\mathrm{T}}(x)\,\boldsymbol{u}_{\mathrm{p}} = \begin{bmatrix} N_1 \; N_2 \end{bmatrix} \times \begin{bmatrix} u_1 \\ u_2 \end{bmatrix}, \tag{5.69}$$

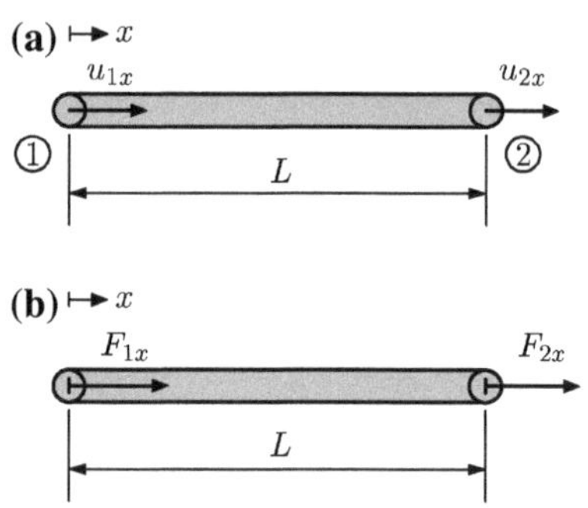


**Fig. 5.6** Definition of the one-dimensional linear rod element: **a** deformations; **b** external loads. The nodes are symbolized by the two *circles* at the ends (○)

and

$$W(x) = \delta \boldsymbol{u}_{\mathrm{p}}^{\mathrm{T}} \boldsymbol{N}(\boldsymbol{x}) = \begin{bmatrix} \delta u_1 \; \delta u_2 \end{bmatrix} \times \begin{bmatrix} N_1 \\ N_2 \end{bmatrix}. \tag{5.70}$$

Let us first consider in the following only the left-hand side of Eq. (5.62) in order to derive the expression for the elemental stiffness matrix $\boldsymbol{K}^{\mathrm{e}}$ of the linear rod element. Introduction of expressions (5.69) and (5.70) in the weak form gives

$$EA \int_0^L \left( \delta \boldsymbol{u}_{\mathrm{p}}^{\mathrm{T}} \frac{\mathrm{d}\boldsymbol{N}(\boldsymbol{x})}{\mathrm{d}x} \right) \left( \frac{\mathrm{d}\boldsymbol{N}^{\mathrm{T}}(x)}{\mathrm{d}x} \boldsymbol{u}_{\mathrm{p}} \right) \mathrm{d}x, \tag{5.71}$$

or under consideration that the column matrix of the nodal unknowns can be considered as constant as:

$$\delta \boldsymbol{u}_{\mathrm{p}}^{\mathrm{T}} \underbrace{EA \int_0^L \left( \frac{\mathrm{d}\boldsymbol{N}(\boldsymbol{x})}{\mathrm{d}x} \right) \left( \frac{\mathrm{d}\boldsymbol{N}^{\mathrm{T}}(x)}{\mathrm{d}x} \right) \mathrm{d}x}_{\boldsymbol{K}^{\mathrm{e}}} \, \boldsymbol{u}_{\mathrm{p}}. \tag{5.72}$$

It will be seen in the following that the expression $\delta \boldsymbol{u}_{\mathrm{p}}^{\mathrm{T}}$ can be canceled with an identical expression on the right-hand side of Eq. (5.62) and $\boldsymbol{u}_{\mathrm{p}}$ represents the column matrix of the unknown nodal displacements. Under consideration of the $\boldsymbol{B}$-matrix, the stiffness matrix can be expressed in a more general way for constant tensile stiffness $EA$ as:

$$\boldsymbol{K}^{\mathrm{e}} = EA \int_0^L \boldsymbol{B}\boldsymbol{B}^{\mathrm{T}} \mathrm{d}x. \tag{5.73}$$

In order to further evaluate Eq. (5.72), we can introduce the components of the derivatives to give:

$$EA \int_0^L \begin{bmatrix} \frac{\mathrm{d}N_1(x)}{\mathrm{d}x} \\ \frac{\mathrm{d}N_2(x)}{\mathrm{d}x} \end{bmatrix} \begin{bmatrix} \frac{\mathrm{d}N_1(x)}{\mathrm{d}x} & \frac{\mathrm{d}N_2(x)}{\mathrm{d}x} \end{bmatrix} \mathrm{d}x, \tag{5.74}$$

or after the matrix multiplication as:

$$EA \int_0^L \begin{bmatrix} \frac{\mathrm{d}N_1(x)}{\mathrm{d}x}\frac{\mathrm{d}N_1(x)}{\mathrm{d}x} & \frac{\mathrm{d}N_1(x)}{\mathrm{d}x}\frac{\mathrm{d}N_2(x)}{\mathrm{d}x} \\ \frac{\mathrm{d}N_2(x)}{\mathrm{d}x}\frac{\mathrm{d}N_1(x)}{\mathrm{d}x} & \frac{\mathrm{d}N_2(x)}{\mathrm{d}x}\frac{\mathrm{d}N_2(x)}{\mathrm{d}x} \end{bmatrix} \mathrm{d}x. \tag{5.75}$$

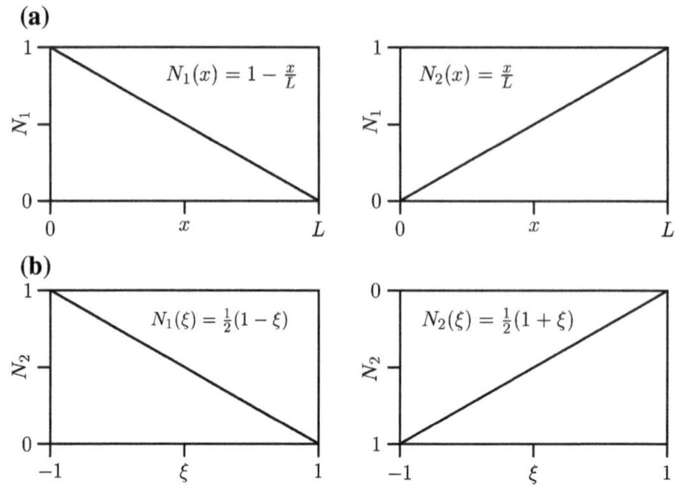


**Fig. 5.7** Shape functions for the linear rod element: **a** physical coordinate ($x$); **b** natural coordinate ($\xi$)

Any further evaluation of this equations requires now that the functional expressions $N_1(x)$ and $N_2(x)$ are known. The simplest assumption that can be done is that the nodal values are linearly distributed within the element, from its value at the node to zero at the opposite node. For such a linear superposition, the shape functions can be assumed as shown in Fig. 5.7a.

The derivatives of the shape functions can easily be calculated as $\frac{\mathrm{d}N_1(x)}{\mathrm{d}x} = -\frac{1}{L}$ and $\frac{\mathrm{d}N_2(x)}{\mathrm{d}x} = \frac{1}{L}$. Thus, the $\boldsymbol{B}$-matrix given in Eq. (5.67) takes the form:

$$\boldsymbol{B} = \frac{1}{L}\begin{bmatrix} -1 \\ 1 \end{bmatrix}. \tag{5.76}$$

The derivatives introduced into Eq. (5.77) give

$$EA \int_0^L \begin{bmatrix} \frac{1}{L^2} & -\frac{1}{L^2} \\ -\frac{1}{L^2} & \frac{1}{L^2} \end{bmatrix} \mathrm{d}x = \frac{EA}{L^2} \int_0^L \begin{bmatrix} 1 & -1 \\ -1 & 1 \end{bmatrix} \mathrm{d}x. \tag{5.77}$$

The integral in the last equation can be analytically integrated to obtain

$$\frac{EA}{L^2}\begin{bmatrix} x & -x \\ -x & x \end{bmatrix}\Bigg|_0^L = \frac{EA}{L^2}\begin{bmatrix} L & -L \\ -L & L \end{bmatrix} \tag{5.78}$$

and the stiffness matrix for a linear rod element is given by:

$$\boldsymbol{K}^{\mathrm{e}} = \frac{EA}{L}\begin{bmatrix} 1 & -1 \\ -1 & 1 \end{bmatrix}. \tag{5.79}$$

It must be noted here that an analytical integration as performed to obtain Eq. (5.78) cannot be performed in commercial finite element codes since they are written in traditional programming languages such as FORTRAN. Instead of the analytical integration, a numerical integration is performed (cf. Appendix A.11) where the integral is approximated by the evaluation and weighting of functional values at so-called integration or GAUSS[14] points. To this end, the Cartesian coordinate $x$ is transformed to the natural coordinate $\xi$ ranging from $-1$ to 1. Depending on the origin of the Cartesian coordinate system, the transformation can be performed based on the relations given in Table 5.2.

**Table 5.2** Transformation between Cartesian ($x$) and natural coordinates ($\xi$)

<table>
<tr><th>Configuration</th><th>Transformation</th></tr>
<tr><td>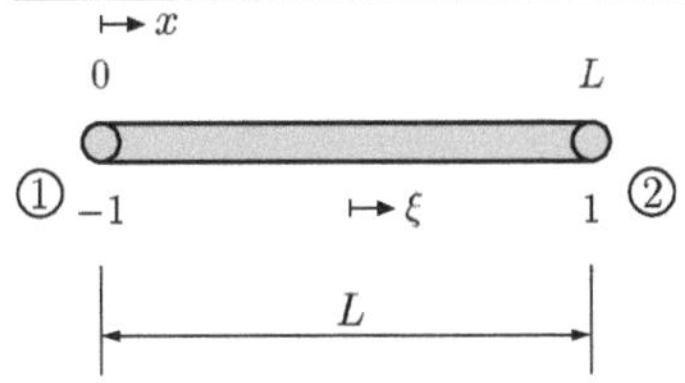
</td><td>$$\xi = \frac{2x}{L} - 1,$$ $$\frac{\mathrm{d}\xi}{\mathrm{d}x} = \frac{2}{L}.$$</td></tr>
<tr><td>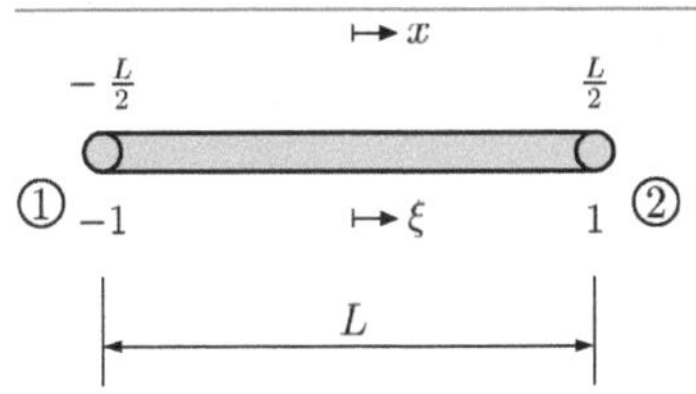
</td><td>$$\xi = \frac{2x}{L} - 1,$$ $$\frac{\mathrm{d}\xi}{\mathrm{d}x} = \frac{2}{L}.$$</td></tr>
<tr><td>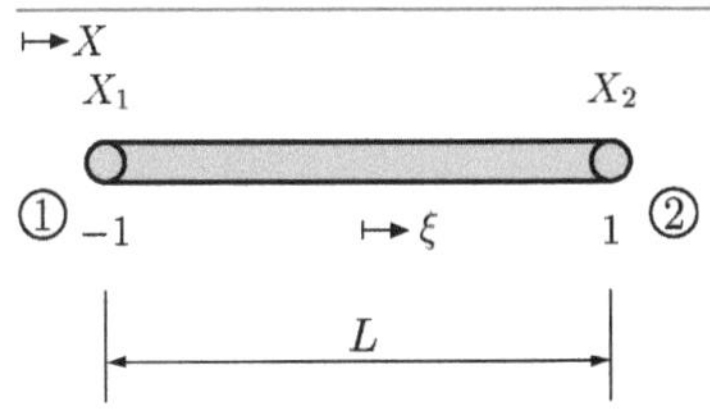
</td><td>$$\xi = \frac{2}{X_2 - X_1}(X - X_1) - 1,$$ $$\frac{\mathrm{d}\xi}{\mathrm{d}x} = \frac{2}{L}.$$</td></tr>
</table>

[14] Johann Carl Friedrich GAUSS (1777–1855), German mathematician and physical scientist.

The integral in Eq. (5.77) can be written in terms of the natural coordinate $\xi$ and approximated in terms of a GAUSS-LEGENDRE[15] quadrature as:

$$\boldsymbol{K}^{\mathrm{e}} = \frac{EA}{L^2}\int_{-1}^{1}\begin{bmatrix} 1 & -1 \\ -1 & 1 \end{bmatrix}\frac{L}{2}\,\mathrm{d}\xi \approx \sum_{i=1}^{n}\begin{bmatrix} \cdots\cdots \\ \cdots\cdots \end{bmatrix}(\xi_i)w(\xi_i), \tag{5.80}$$

where the matrix is to be evaluated at the $n$ integration points and multiplied by certain weights $w$, cf. Appendix A.11. Since the matrix is in this simple case only composed of constant values, it is sufficient to consider a one-point integration rule ($\xi = 0, w = 2$) to achieve the analytical result[16] as:

$$\boldsymbol{K}^{\mathrm{e}} = \frac{EA}{2L}\begin{bmatrix} 1 & -1 \\ -1 & 1 \end{bmatrix}\Bigg|_{\xi=0} \times 2 = \frac{EA}{L}\begin{bmatrix} 1 & -1 \\ -1 & 1 \end{bmatrix}. \tag{5.81}$$

Let us now consider the right-hand side of Eq. (5.62) in order to derive the expression for the elemental load column matrix $\boldsymbol{f}^{\mathrm{e}}$ of the linear rod element. The first part of the right-hand side, i.e.

$$EA\left[W(x)\frac{\mathrm{d}u(x)}{\mathrm{d}x}\right]_0^L \tag{5.82}$$

results with the definition of the weight function according to Eq. (5.70) in

$$EA\left[\delta\boldsymbol{u}_{\mathrm{p}}^{\mathrm{T}}\boldsymbol{N}(x)\frac{\mathrm{d}u(x)}{\mathrm{d}x}\right]_0^L, \tag{5.83}$$

or in components

$$\delta\boldsymbol{u}_{\mathrm{p}}^{\mathrm{T}}EA\left[\begin{bmatrix} N_1 \\ N_2 \end{bmatrix}\frac{\mathrm{d}u(x)}{\mathrm{d}x}\right]_0^L. \tag{5.84}$$

The virtual displacements $\delta\boldsymbol{u}_{\mathrm{p}}^{\mathrm{T}}$ in the last equation can be canceled with a corresponding expression in Eq. (5.72). Furthermore, the last equation constitutes a system of two equations which must be evaluated at the integration boundaries, i.e. at $x = 0$ and $x = L$. The first equation reads:

$$\left(N_1 EA\frac{\mathrm{d}u}{\mathrm{d}x}\right)_{x=L} - \left(N_1 EA\frac{\mathrm{d}u}{\mathrm{d}x}\right)_{x=0}. \tag{5.85}$$

[15] Adrien-Marie LEGENDRE (1752–1833), French mathematician.

[16] It must be noted here that in the general case only an *approximation* of the integral can be obtained and that the exact, i.e. analytical solution, is reserved for simple cases.

This gives under consideration of the boundary values of the shape functions, i.e. $N_1(L) = 0$ and $N_1(0) = 1$, the following statement:

$$-EA\left.\frac{\mathrm{d}u}{\mathrm{d}x}\right|_{x=0} \overset{(3.8)}{=} -N_x(x=0). \tag{5.86}$$

A corresponding expression can be derived for the second equation as:

$$EA\left.\frac{\mathrm{d}u}{\mathrm{d}x}\right|_{x=L} \overset{(3.8)}{=} N_x(x=L). \tag{5.87}$$

It must be noted here that the forces $N_x$ are the internal reactions according to Fig. 3.4. The external loads with their positive directions according to Fig. 5.6b can be obtained from the internal loads by inverting the sign at the left-hand boundary and by maintaining the positive direction of the internal reaction at the right-hand boundary. This can easily be shown by balancing the internal and external forces at each boundary node. Thus, the contribution to the load matrix due to single *external* forces $F_i$ at the nodes is expressed by:

$$\boldsymbol{f}_F^{\mathrm{e}} = \begin{bmatrix} F_{1x} \\ F_{2x} \end{bmatrix}. \tag{5.88}$$

The second part of Eq. (5.62), i.e. after eliminating of the virtual displacements $\delta\boldsymbol{u}^{\mathrm{T}}$

$$\int_0^L \boldsymbol{N}(\boldsymbol{x})\, p(x)\, \mathrm{d}x \tag{5.89}$$

presents the general rule to determine equivalent nodal loads in the case of arbitrarily distributed loads $p(x)$. As an example, the evaluation of Eq. (5.89) for a constant load $p$ results in the following load matrix:

$$\boldsymbol{f}_p^{\mathrm{e}} = p \int_0^L \begin{bmatrix} N_1 \\ N_2 \end{bmatrix} \mathrm{d}x = \frac{pL}{2} \begin{bmatrix} 1 \\ 1 \end{bmatrix}. \tag{5.90}$$

Further expressions for equivalent nodal loads can be taken from Table 5.3. Let us remind at this step that in the scope of the finite element method any type of load can be only introduced at nodes into the discretized structure.

Based on the derived results, the principal finite element equation for a single linear rod element with constant axial tensile stiffness $EA$ can be expressed in a general form as

$$\boldsymbol{K}^{\mathrm{e}}\boldsymbol{u}^{\mathrm{e}} = \boldsymbol{f}^{\mathrm{e}}, \tag{5.91}$$

**Table 5.3** Equivalent nodal loads for a linear rod element ($x$-axis: right facing)

| Loading | Axial force |
|---|---|
| 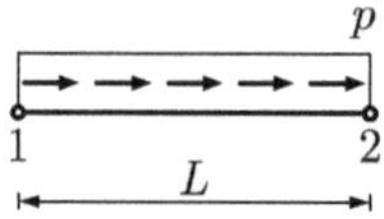 | $F_{1x} = \frac{pL}{2}$<br>$F_{2x} = \frac{pL}{2}$ |
| 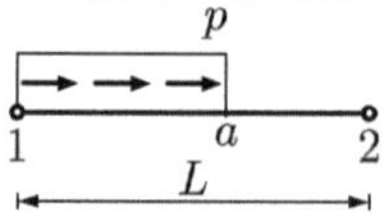 | $F_{1x} = -\frac{pa^2}{2L} + pa$<br>$F_{2x} = \frac{pa^2}{2L}$ |
| 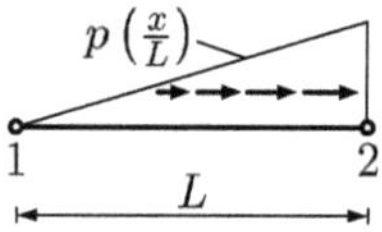 | $F_{1x} = \frac{pL}{6}$<br>$F_{2x} = \frac{pL}{3}$ |
| 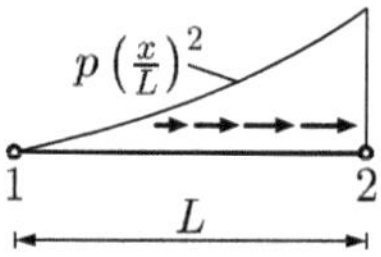 | $F_{1x} = \frac{pL}{12}$<br>$F_{2x} = \frac{pL}{4}$ |
| 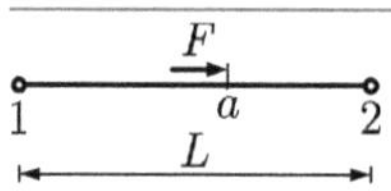 | $F_{1x} = \frac{F(L-a)}{L}$<br>$F_{2x} = \frac{Fa}{L}$ |

or in components as:

$$\frac{EA}{L}\begin{bmatrix} 1 & -1 \\ -1 & 1 \end{bmatrix}\begin{bmatrix} u_{1x} \\ u_{2x} \end{bmatrix} = \begin{bmatrix} F_{1x} \\ F_{2x} \end{bmatrix} + \int_0^L \begin{bmatrix} N_1 \\ N_2 \end{bmatrix} p_x(x)\,dx. \tag{5.92}$$

At the end of this section, a few comments on the accuracy of a linear rod element should be given, cf. Table 5.4. As can be seen, the linear rod element gives under

**Table 5.4** Comments on the accuracy of the finite element solution for a single cantilevered linear rod element

| Configuration | 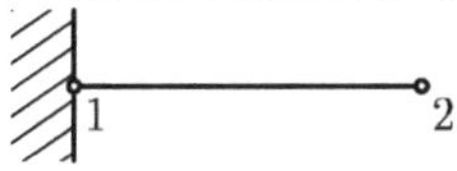  |
|---|---|
| Axial tensile stiffness and loading | Accuracy of $u(x)$ |
| $EA = \text{const.}$;<br>loaded by single force $F$ at node 2 | FE gives analytical solution at nodes<br>and between nodes |
| $EA = \text{const.}$;<br>displacement BC $u$ at node 2 | FE gives exact nodal values<br>and analytical solution between nodes |
| $EA = \text{const.}$;<br>distributed load $p$ | FE gives analytical solution at nodes<br>but only approximate solution between nodes |
| $EA \neq \text{const.}$;<br>loaded by single force $F$ at node 2 | FE gives approximate solution at nodes<br>and approximate solution between nodes |
| $EA \neq \text{const.}$;<br>displacement BC $u$ at node 2 | FE gives exact nodal values<br>but only approximate solution between nodes |

certain conditions the exact, i.e. the analytical solution. This is illustrated by several examples in the section 'Solved Problems' and 'Supplementary Problems'.

#### 5.2.1.2 Quadratic Element Formulation

Let us consider now a rod element which is composed of three nodes as schematically shown in Fig. 5.8. Each node has again only one degree of freedom, i.e. a displacement in $x$-direction and each node can be only loaded by a single force acting along the $x$-axis. It is assumed in the following that the second node is exactly located in the middle, i.e. at $x = \frac{L}{2}$, of the element.

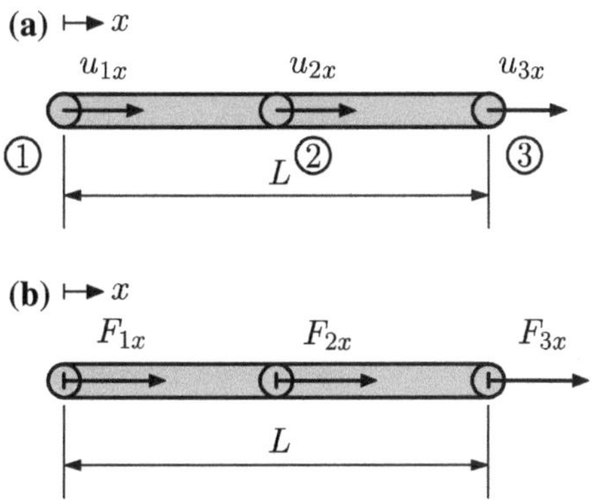


**Fig. 5.8** Definition of the one-dimensional quadratic rod element: **a** deformations; **b** external loads. The nodes are symbolized by *circles* at the ends and in the middle (○)

Since there are now three nodes with three unknowns, the equation for the unknown displacement in the element and its virtual counterpart (cf. Eqs. (5.63) and (5.64)) are now given by the expressions:

$$u^{\mathrm{e}}(x) = \boldsymbol{N}^{\mathrm{T}}(x)\,\boldsymbol{u}_{\mathrm{p}} = \begin{bmatrix} N_1 & N_2 & N_3 \end{bmatrix} \times \begin{bmatrix} u_1 \\ u_2 \\ u_3 \end{bmatrix}, \tag{5.93}$$

and

$$W(x) = \delta\boldsymbol{u}_{\mathrm{p}}^{\mathrm{T}}\boldsymbol{N}(x) = \begin{bmatrix} \delta u_1 & \delta u_2 & \delta u_3 \end{bmatrix} \times \begin{bmatrix} N_1 \\ N_2 \\ N_3 \end{bmatrix}. \tag{5.94}$$

Similar as in Eq. (5.77), the elemental stiffness matrix can be expressed before evaluating the integral as:

$$\boldsymbol{K}^{\mathrm{e}} = EA\int_0^L \begin{bmatrix} \frac{\mathrm{d}N_1(x)}{\mathrm{d}x}\frac{\mathrm{d}N_1(x)}{\mathrm{d}x} & \frac{\mathrm{d}N_1(x)}{\mathrm{d}x}\frac{\mathrm{d}N_2(x)}{\mathrm{d}x} & \frac{\mathrm{d}N_1(x)}{\mathrm{d}x}\frac{\mathrm{d}N_3(x)}{\mathrm{d}x} \\ \frac{\mathrm{d}N_2(x)}{\mathrm{d}x}\frac{\mathrm{d}N_1(x)}{\mathrm{d}x} & \frac{\mathrm{d}N_2(x)}{\mathrm{d}x}\frac{\mathrm{d}N_2(x)}{\mathrm{d}x} & \frac{\mathrm{d}N_2(x)}{\mathrm{d}x}\frac{\mathrm{d}N_3(x)}{\mathrm{d}x} \\ \frac{\mathrm{d}N_3(x)}{\mathrm{d}x}\frac{\mathrm{d}N_1(x)}{\mathrm{d}x} & \frac{\mathrm{d}N_3(x)}{\mathrm{d}x}\frac{\mathrm{d}N_2(x)}{\mathrm{d}x} & \frac{\mathrm{d}N_3(x)}{\mathrm{d}x}\frac{\mathrm{d}N_3(x)}{\mathrm{d}x} \end{bmatrix} \mathrm{d}x. \tag{5.95}$$

The shape functions $N_i$ in this case[17] are given by quadratic equations as shown in Fig. 5.9 in physical and natural coordinates.

From the functional expressions given in Fig. 5.9, the derivatives are obtained as $\frac{\mathrm{d}N_1}{\mathrm{d}x} = -\frac{3}{L} + \frac{4x}{L^2}$, $\frac{\mathrm{d}N_2}{\mathrm{d}x} = \frac{4}{L} - \frac{8x}{L^2}$ and $\frac{\mathrm{d}N_3}{\mathrm{d}x} = -\frac{1}{L} + \frac{4x}{L^2}$ and Eq. (5.95) can be evaluated by analytical or numerical integration to give the elemental stiffness matrix of the quadratic rod element as:

$$\boldsymbol{K}^{\mathrm{e}} = \frac{EA}{3L}\begin{bmatrix} 7 & -8 & 1 \\ -8 & 16 & -8 \\ 1 & -8 & 7 \end{bmatrix}. \tag{5.96}$$

Furthermore, it should be noted that the $\boldsymbol{B}$-matrix, cf. Eq. (5.67), takes the following form for the quadratic rod element:

$$\boldsymbol{B} = \frac{1}{L}\begin{bmatrix} -3 + \frac{4x}{L} \\ 4 - \frac{8x}{L} \\ -1 + \frac{4x}{L} \end{bmatrix} = \frac{1}{L}\begin{bmatrix} -1 + 2\xi \\ -4\xi \\ 1 + 2\xi \end{bmatrix}. \tag{5.97}$$

[17] A formal derivation of the functional expressions is presented in Sect. 5.2.2.

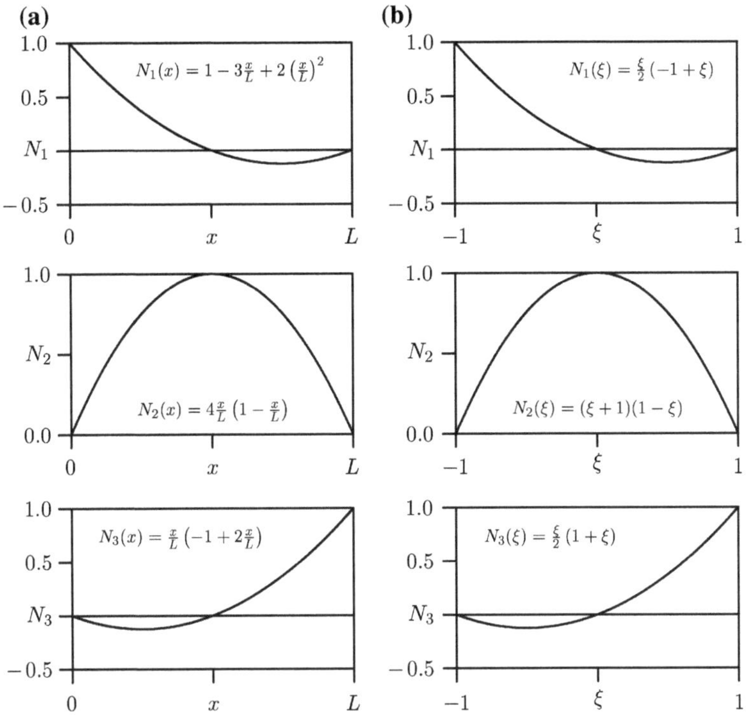


**Fig. 5.9** Shape functions for the quadratic rod element with equidistant nodes: **a** physical coordinate ($x$); **b** natural coordinate ($\xi$)

The right-hand side of Eq. (5.62) can be treated in a similar way as in Sect. 5.2.1.1 to obtain the elemental load vector in the form of:

$$\boldsymbol{f}^{\mathrm{e}} = \boldsymbol{f}_F^{\mathrm{e}} + \boldsymbol{f}_p^{\mathrm{e}} = \begin{bmatrix} F_{1x} \\ F_{2x} \\ F_{3x} \end{bmatrix} + \int_0^L \begin{bmatrix} N_1 \\ N_2 \\ N_3 \end{bmatrix} p(x)\,\mathrm{d}x. \tag{5.98}$$

Based on the derived results, the principal finite element equation for a single quadratic rod element with constant axial tensile stiffness $EA$ an be expressed in components as:

$$\frac{EA}{3L}\begin{bmatrix} 7 & -8 & 1 \\ -8 & 16 & -8 \\ 1 & -8 & 7 \end{bmatrix}\begin{bmatrix} u_{1x} \\ u_{2x} \\ u_{3x} \end{bmatrix} = \begin{bmatrix} F_{1x} \\ F_{2x} \\ F_{3x} \end{bmatrix} + \int_0^L \begin{bmatrix} N_1 \\ N_2 \\ N_3 \end{bmatrix} p(x)\,\mathrm{d}x. \tag{5.99}$$

This formulation of the principal finite element equation can be alternatively expressed by eliminating the expression for the second node. The system given in (5.99) can be written in the form of single equations as:

$$\frac{EA}{3L}(7u_{1x} - 8u_{2x} + 1u_{3x}) = F_{1x} + I_1, \tag{5.100}$$

$$\frac{EA}{3L}(-8u_{1x} + 16u_{2x} - 8u_{3x}) = F_{2x} + I_2, \tag{5.101}$$

$$\frac{EA}{3L}(1u_{1x} - 8u_{2x} + 7u_{3x}) = F_{3x} + I_3, \tag{5.102}$$

where $I_i$ is the abbreviation for the integral with the distributed load, e.g. $I_1 = \int N_1(x)p(x)\mathrm{d}x$.

The second equation can be rearranged for $u_{2x}$, i.e.

$$u_{2x} = \frac{1}{2}u_{1x} + \frac{1}{2}u_{3x} + \frac{1}{16}\frac{3L}{EA}(F_{2x} + I_2), \tag{5.103}$$

which can be introduced into Eqs. (5.100) and (5.102) to obtain:

$$\frac{EA}{3L}(3u_{1x} - 3u_{3x}) = F_{1x} + I_1 + \frac{1}{2}(F_{2x} + I_2), \tag{5.104}$$

$$\frac{EA}{3L}(-3u_{1x} + 3u_{3x}) = F_{3x} + I_3 + \frac{1}{2}(F_{2x} + I_2). \tag{5.105}$$

The last two equations can be written again in matrix form as:

$$\frac{EA}{L}\begin{bmatrix} 1 & -1 \\ -1 & 1 \end{bmatrix}\begin{bmatrix} u_{1x} \\ u_{3x} \end{bmatrix} = \begin{bmatrix} F_{1x} \\ F_{3x} \end{bmatrix} + \begin{bmatrix} I_1 \\ I_3 \end{bmatrix} + \begin{bmatrix} \frac{1}{2}F_{2x} \\ \frac{1}{2}F_{2x} \end{bmatrix} + \begin{bmatrix} \frac{1}{2}I_2 \\ \frac{1}{2}I_2 \end{bmatrix}. \tag{5.106}$$

This formulation looks similar to the expression for the linear rod element in Eq. (5.92). However, the right-hand side contains here in addition the contribution of the load from the middle load and it should be not forgotten that the distribution of the displacement $u^{\mathrm{e}}(x)$ inside the element is of quadratic shape. The values of equivalent nodal loads, i.e. the evaluation of the integral in Eq. (5.99), is given for some standard cases in Table 5.5. The reader should pay here attention to the fact that these equivalent nodal loads are different to the case of the linear rod element, cf. Table 5.3.

At the end of this section again a few words on the accuracy of the quadratic rod element. As can be seen in Table 5.6, the accuracy is for the investigated cases at least in the range of the linear element if we compare the general statements without investigating specific numerical values. For a constant distributed load, the quadratic element reproduces not only at the nodes but also between the nodes the analytical solution. However, it must be highlighted here that these results are element specific and that the finite element method calculates in the general case—even at nodes—only approximate solutions. Nevertheless, the comments presented in Table 5.6 can be helpful in special cases where a mesh refinement would not increase the accuracy but the computation time and the size of the results file. If the problem is such that

**Table 5.5** Equivalent nodal loads for a quadratic rod element ($x$-axis: right facing)

| Loading | Axial force |
|---|---|
| 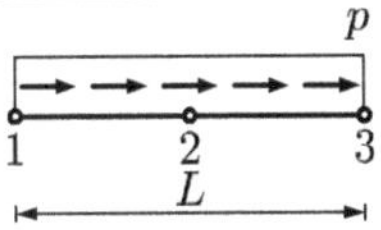 | $F_{1x} = \frac{pL}{6}, \quad F_{2x} = \frac{2pL}{3}, \quad F_{3x} = \frac{pL}{6}$ |
| 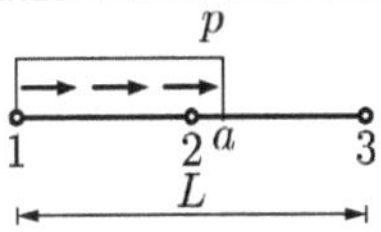 | $F_{1x} = \frac{2pa^3}{3L^2} - \frac{3pa^3}{2L} + pa, \quad F_{2x} = -\frac{4pa^3}{3L^2} + \frac{2pa^2}{L}, \quad F_{3x} = \frac{2pa^3}{3L^2} - \frac{pa^2}{2L}$ |
| 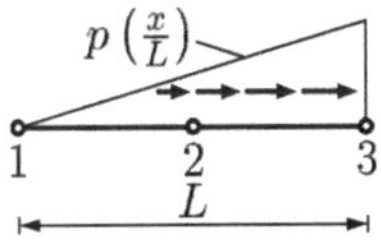 | $F_{1x} = 0, \quad F_{2x} = \frac{pL}{3}, \quad F_{3x} = \frac{pL}{6}$ |
| 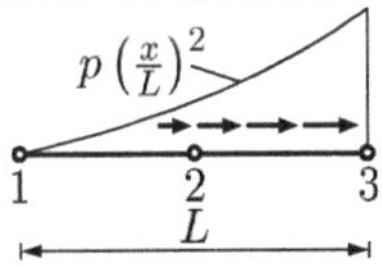 | $F_{1x} = -\frac{pL}{60}, \quad F_{2x} = \frac{pL}{5}, \quad F_{3x} = \frac{3pL}{20}$ |
| 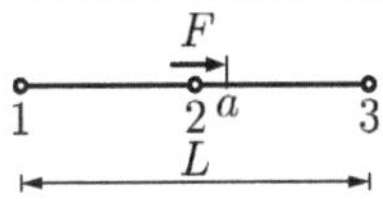 | $F_{1x} = FN_1(a), \quad F_{2x} = FN_2(a), \quad F_{3x} = FN_3(a)$ |

the exact solution is obtained at the nodes, a mesh refinement is in all likelihood not required in this case.

### 5.2.2 Derivation of Shape Functions

Let us introduce first a more natural way to derive linear shape functions for the displacement field in a linear rod element of length $L$ as shown in Fig. 5.6. If there are no distributed loads acting and if the axial tensile stiffness is constant, it follows that the stress and strain state in the element must be constant:

**Table 5.6** Comments on the accuracy of the finite element solution for a single cantilevered quadratic rod element

| Configuration | 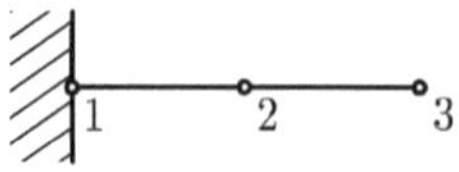  |
|---|---|
| Axial tensile stiffness and loading | Accuracy of $u(x)$ |
| $EA$ = const.; loaded by single force $F$ at node 3 | FE gives analytical solution at nodes and between nodes |
| $EA$ = const.; displacement BC $u$ at node 3 | FE gives exact nodal values and analytical solution between nodes |
| $EA$ = const.; distributed load $p$ = const. | FE gives analytical solution at nodes and between nodes |
| $EA$ = const.; distributed load $p(x)$ = linear | FE gives analytical solution at nodes but only approximate solution between nodes |
| $EA \neq$ const.; loaded by single force $F$ at node 3 | FE gives approximate solution at nodes and approximate solution between nodes |

$$\varepsilon(x) = \frac{\mathrm{d}u(x)}{\mathrm{d}x} = c, \tag{5.107}$$

or rearranged:

$$\mathrm{d}u(x) = c\,\mathrm{d}x. \tag{5.108}$$

The last equation can be integrated to obtain the solution for the displacement distribution as:

$$u(x) = c_1 \int_0^x \mathrm{d}\hat{x} + c_2 = c_1 x + c_2. \tag{5.109}$$

Since the displacement distribution $u(x)$ should be based on both nodal values, the integration can be performed from the other boundary to obtain a second solution for the displacement distribution as:

$$u(x) = c_3 \int_L^x \mathrm{d}\hat{x} + c_4 = c_3\,(x - L) + c_4. \tag{5.110}$$

If the distributions (5.109) and (5.110) are solutions of Eq. (5.108), then also their combination is a solution:

$$u(x) = c_2 + c_4 + c_1 x + c_3\,(x - L)\,. \tag{5.111}$$

Without loss of generality, the following expression can be used since the constants are arbitrary:

$$u(x) = c_1 x + c_3 (x - L)\,, \tag{5.112}$$

where the constants $c_1$ and $c_3$ can be determined by evaluating the nodal values:

$$u(0) = u_1 = -c_3 L \quad \rightarrow \quad c_3 = -\frac{u_1}{L}, \tag{5.113}$$

$$u(L) = u_2 = c_1 L \quad \rightarrow \quad c_1 = \frac{u_2}{L}. \tag{5.114}$$

Introducing these constants in the general displacement distribution according to Eq. (5.112) gives the following *linear* dependency of the displacement distribution along the $x$-axis of the element:

$$u(x) = \left(1 - \frac{x}{L}\right) u_1 + \frac{x}{L} u_2, \tag{5.115}$$

where the shape functions can be identified as given in Fig. 5.7.

At the end of this section, a more general concept based on basis functions will be introduced in order to derive the complete set of shape functions[18]. To this end, let us just assume the shape of the displacement distribution $u^{\mathrm{e}}(\xi)$ within an element without reference to the nodal values. It is obvious that this choice must be conform to the physical problem under consideration. We may assume that the distribution is given by a polynomial of the form

$$u^{\mathrm{e}}(\xi) = a_0 + a_1 \xi + a_2 \xi^2 + a_2 \xi^2 + \cdots + a_n \xi^n, \tag{5.116}$$

which can be expressed in matrix form as:

$$u^{\mathrm{e}}(\xi) = \boldsymbol{\chi}^{\mathrm{T}} \boldsymbol{a} = \begin{bmatrix} 1 & \xi & \xi^2 & \xi^3 & \cdots & \xi^n \end{bmatrix} \begin{bmatrix} a_1 \\ a_2 \\ a_3 \\ \vdots \\ a_n \end{bmatrix}. \tag{5.117}$$

The elements of $\boldsymbol{\chi}$ will be called *basis functions* and the elements of $\boldsymbol{a}$ will be called *basis coefficients*. If we assume that the number of basis functions equals the number of nodal variables associated with $u$, then the relationship between the basis coefficients $\boldsymbol{a}$ and the nodal values $\boldsymbol{u}_{\mathrm{p}}$ can be expressed as

[18] This approach is presented in Ref. [67] in a general way.

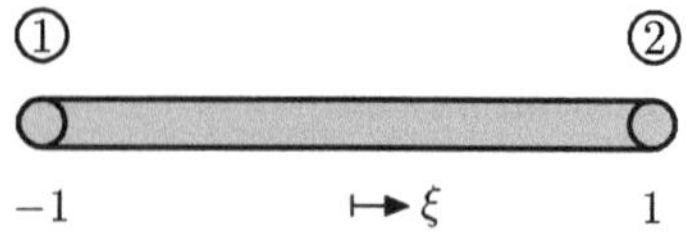


**Fig. 5.10** Linear rod element described based on the natural coordinate ($\xi$)

$$\boldsymbol{a} = \boldsymbol{A}\boldsymbol{u}_{\mathrm{p}}, \tag{5.118}$$

where $\boldsymbol{A}$ is a square matrix of constants. Equalizing the nodal approach given in Eq. (5.63) with the new expression in (5.117) and considering (5.118) results in:

$$\boldsymbol{N}^{\mathrm{T}}\boldsymbol{u}_{\mathrm{p}} = \boldsymbol{\chi}^{\mathrm{T}}\boldsymbol{a} \quad \text{or} \quad \boldsymbol{N}^{\mathrm{T}} = \boldsymbol{\chi}^{\mathrm{T}}\boldsymbol{A}. \tag{5.119}$$

Thus, the row matrix of the shape functions $\boldsymbol{N}^{\mathrm{T}}$ can be factored into a row vector of basis functions $\boldsymbol{\chi}^{\mathrm{T}}$ and a square matrix $\boldsymbol{A}$ of constant coefficients. To illustrate the procedure, let us have a look at a linear rod element as shown in Fig. 5.10 where the natural coordinate is used.

If the physical problem supports the assumption of a linear distribution of the displacement, the following linear description of the displacement field can be introduced:

$$u^{\mathrm{e}}(\xi) = a_0 + a_1\xi, \tag{5.120}$$

where the column matrix of the basis functions is given by $\boldsymbol{\chi} = \begin{bmatrix}1 & \xi\end{bmatrix}^{\mathrm{T}}$ and the column matrix of the basis coefficients by $\boldsymbol{a} = \begin{bmatrix}a_0 & a_1\end{bmatrix}^{\mathrm{T}}$. Evaluation of this function at both nodes gives:

$$\text{Node 1:}\;\; u_1 = u^{\mathrm{e}}(\xi = -1) = a_0 - a_1, \tag{5.121}$$

$$\text{Node 2:}\;\; u_2 = u^{\mathrm{e}}(\xi = +1) = a_0 + a_1. \tag{5.122}$$

The last two equations can be expressed in matrix form according to Eq. (5.118) as:

$$\begin{bmatrix} u_1 \\ u_2 \end{bmatrix} = \underbrace{\begin{bmatrix} 1 & -1 \\ 1 & 1 \end{bmatrix}}_{\boldsymbol{A}^{-1}} \begin{bmatrix} a_0 \\ a_1 \end{bmatrix}. \tag{5.123}$$

Solving this system of equations for the unknown basis functions $a_i$ gives

$$\underbrace{\begin{bmatrix} a_0 \\ a_1 \end{bmatrix}}_{\boldsymbol{a}} = \underbrace{\frac{1}{2}\begin{bmatrix} 1 & 1 \\ -1 & 1 \end{bmatrix}}_{\boldsymbol{A}} \underbrace{\begin{bmatrix} u_1 \\ u_2 \end{bmatrix}}_{\boldsymbol{u}_{\mathrm{p}}}, \tag{5.124}$$

and the matrix of the shape functions results according to Eq. (5.119) as:

$$\boldsymbol{N}^{\mathrm{T}} = \boldsymbol{\chi}^{\mathrm{T}}\boldsymbol{A} = \begin{bmatrix} 1 & \xi \end{bmatrix} \frac{1}{2}\begin{bmatrix} 1 & 1 \\ -1 & 1 \end{bmatrix} = \begin{bmatrix} \frac{1}{2}(1-\xi) & \frac{1}{2}(1+\xi) \end{bmatrix} = \begin{bmatrix} N_1 & N_2 \end{bmatrix}. \quad (5.125)$$

### 5.2.3 Assembly of Elements and Consideration of Boundary Conditions

Real structures of complex geometry (cf. Fig. 1.1) require the application of many finite elements in order to discretize the geometry. Thus, it is necessary to assemble the single elemental finite element equations $\boldsymbol{K}^{\mathrm{e}}\boldsymbol{u}_{\mathrm{p}}^{\mathrm{e}} = \boldsymbol{f}^{\mathrm{e}}$ to a global system of equations which can be symbolically written as $\boldsymbol{K}\boldsymbol{u}_{\mathrm{p}} = \boldsymbol{f}$, where $\boldsymbol{K}$ is the global stiffness matrix, $\boldsymbol{u}_{\mathrm{p}}$ the global column matrix of unknowns and $\boldsymbol{f}$ the global column matrix of loads. Let us illustrate the process to assemble the global system of equations due to a three-element axial structure as shown in Fig. 5.11. As can be seen in Fig. 5.11a, each element has its own coordinate system $x_i$ with $i = \mathrm{I}, \mathrm{II}, \mathrm{III}$ and its own nodal displacements $u_{1x}^{i}$ and $u_{2x}^{i}$. In order to assemble the single elements to a connected structure as shown in Fig. 5.11b, it is useful to introduce a global coordinate $X$ and global nodal displacements denoted by $u_{iX}$. Comparing the elemental and global nodal displacements shown in Fig. 5.11, the following mapping between the local and global displacements can be derived:

$$u_{1X} = u_{1x}^{\mathrm{I}}, \quad (5.126)$$

$$u_{2X} = u_{2x}^{\mathrm{I}} = u_{1x}^{\mathrm{II}}, \quad (5.127)$$

$$u_{3X} = u_{2x}^{\mathrm{II}} = u_{1x}^{\mathrm{III}}, \quad (5.128)$$

$$u_{4X} = u_{4x}^{\mathrm{I}}. \quad (5.129)$$

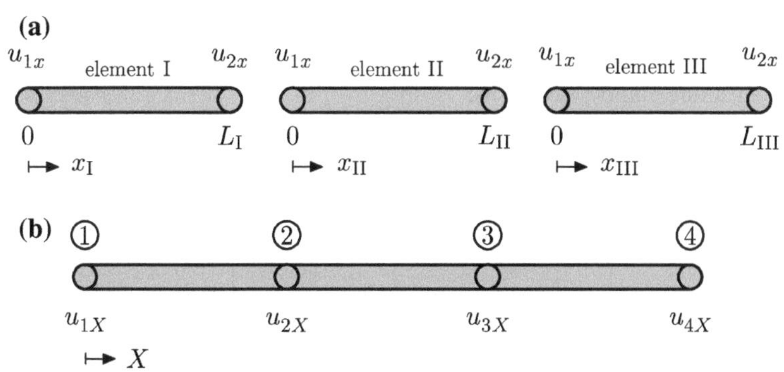


**Fig. 5.11** Relationship between **a** elemental and **b** global nodes and displacements in a horizontal rod structure

One possible way to assemble the elemental stiffness matrices to the global system will be illustrated in the following. In a first step, each single element is considered separately and its elemental stiffness matrix is written as, for example, given in Eq. (5.79). In addition, the corresponding *global* nodal displacements are written over the matrix and on the right-hand side which gives the following expressions:

$$\boldsymbol{K}^{\mathrm{e}}_{\mathrm{I}} = \frac{EA}{L}\begin{array}{c} \begin{array}{cc} u_{1X} & u_{2X} \end{array} \\ \begin{bmatrix} 1 & -1 \\ -1 & 1 \end{bmatrix} \end{array} \begin{array}{c} u_{1X} \\ u_{2X} \end{array}, \tag{5.130}$$

$$\boldsymbol{K}^{\mathrm{e}}_{\mathrm{II}} = \frac{EA}{L}\begin{array}{c} \begin{array}{cc} u_{2X} & u_{3X} \end{array} \\ \begin{bmatrix} 1 & -1 \\ -1 & 1 \end{bmatrix} \end{array} \begin{array}{c} u_{2X} \\ u_{3X} \end{array}, \tag{5.131}$$

$$\boldsymbol{K}^{\mathrm{e}}_{\mathrm{III}} = \frac{EA}{L}\begin{array}{c} \begin{array}{cc} u_{3X} & u_{4X} \end{array} \\ \begin{bmatrix} 1 & -1 \\ -1 & 1 \end{bmatrix} \end{array} \begin{array}{c} u_{3X} \\ u_{4X} \end{array}. \tag{5.132}$$

By indicating the global unknowns in the described manner at each elemental stiffness matrix, it is easy to assign to each element in a matrix a unique index. For example, the upper right element of the stiffness matrix $\boldsymbol{K}^{\mathrm{e}}_{\mathrm{I}}$ has the index[19] $(u_{1X}, u_{2X})$ and the value $-\frac{EA}{L}$. The next step consists of indicating the structure of the global stiffness matrix with its correct dimension. To this end, the total number of global unknowns[20] must be determined. In general, the global number of unknowns is given by number of nodes times degrees of freedom per node. Thus, the number of global unknowns for a structure of rod elements is simply the total number of nodes in the assembled structure. It should be noted here that the determination of the unknowns at this step of the procedure is without any consideration of boundary conditions. For the problem shown in Fig. 5.11a, the number of nodes is four which equals the number of unknowns. Thus, the dimension of the global stiffness matrix is given by (number global unknowns $\times$ number global unknowns) or for our example as $(4 \times 4)$ and the structure can be written as:

$$\boldsymbol{K} = \begin{array}{c} \begin{array}{cccc} u_{1X} & u_{2X} & u_{3X} & u_{4X} \end{array} \\ \left[\begin{array}{c|c|c|c} \quad & \quad & \quad & \quad \\ \hline \quad & \quad & \quad & \quad \\ \hline \quad & \quad & \quad & \quad \\ \hline \quad & \quad & \quad & \quad \end{array}\right] \end{array} \begin{array}{c} u_{1X} \\ u_{2X} \\ u_{3X} \\ u_{4X} \end{array}. \tag{5.133}$$

[19] We follow here the convention where the first expression specifies the row and the second one the column: (row, column).

[20] The total number of unknowns is alternatively named the total number of degrees of freedom (DOF).

It is now required to indicate the global unknowns over the empty global stiffness matrix and on its right-hand side. Any order can be chosen but it is common for the problem under consideration to start with $u_{1X}$ and simply move to the next node. The scheme for this consecutive use of the global unknowns from the lowest to the highest number is drawn on the matrix in Eq. (5.133). Each cell of the global stiffness matrix has now its unique index expressed by the global unknowns. Or in other words, each cell of each elemental stiffness matrix has a cell in the global stiffness matrix with the same index and each element of the elemental stiffness matrix must be placed in the global matrix based on this unique index scheme. As an example, the upper right element of the stiffness matrix $\boldsymbol{K}_{\mathrm{I}}^{\mathrm{e}}$ with the index $(u_{1X}, u_{2X})$ must be placed in the global stiffness matrix in the first row and the second column. If each entry of the elemental stiffness matrices is inserted into the global matrix based on the described index scheme, the assembly of the global stiffness matrix is completed. The process for the consecutive use of global unknowns is illustrated in Fig. 5.12a. As can be seen in this figure, there is an interaction at nodes where elements are connected and the corresponding entries of the elemental stiffness matrices are summed up. This interaction is illustrated in a different way in Fig. 5.12b where it can be seen that at each inner node two shape functions are acting, i.e. one from the left-hand element and one from the right-hand element.

A further important property of the global stiffness matrix can be seen in Fig. 5.12a. If an appropriate node numbering is chosen,[21] the global stiffness matrix reveals a strong band structure where all entries are grouped around the main diagonale and major parts grouped in the form of triangles contain only zeros. If there is such a clear boundary between the non-zero and the zero components, the border line is called the *skyline* of the matrix. As the elemental stiffness matrices, the global stiffness matrix is symmetric and commercial finite element codes store only half of the entries in order to reduce the requirements for data storage.

In order to complete the assembly of the global finite element equation, the global load vector $\boldsymbol{f}$ must be composed. Here, it is more advantageous to look from the beginning at the assembled structure and fill the external single loads $F_i$ which are acting at nodes in the proper order in the column matrix $\boldsymbol{f}$. A bit care must be taken if distributed loads were converted to equivalent nodal loads. For this case, components $f_i$ from both elements must be summed up at inner nodes:

$$\boldsymbol{f} = \begin{bmatrix} F_1 \\ F_2 \\ F_3 \\ F_4 \end{bmatrix} + \begin{bmatrix} f_{1,\mathrm{I}} \\ f_{2,\mathrm{I}} + f_{2,\mathrm{II}} \\ f_{3,\mathrm{II}} + f_{3,\mathrm{III}} \\ f_{4,\mathrm{III}} \end{bmatrix}. \tag{5.134}$$

The global system of equations for the problem shown in Fig. 5.11b is finally obtained as:

[21] Commercial finite element codes offer an option which is called 'bandwidth optimization' to achieve this structure. This is important if a direct solver is used in order to minimize the solution time and the amount of storage.

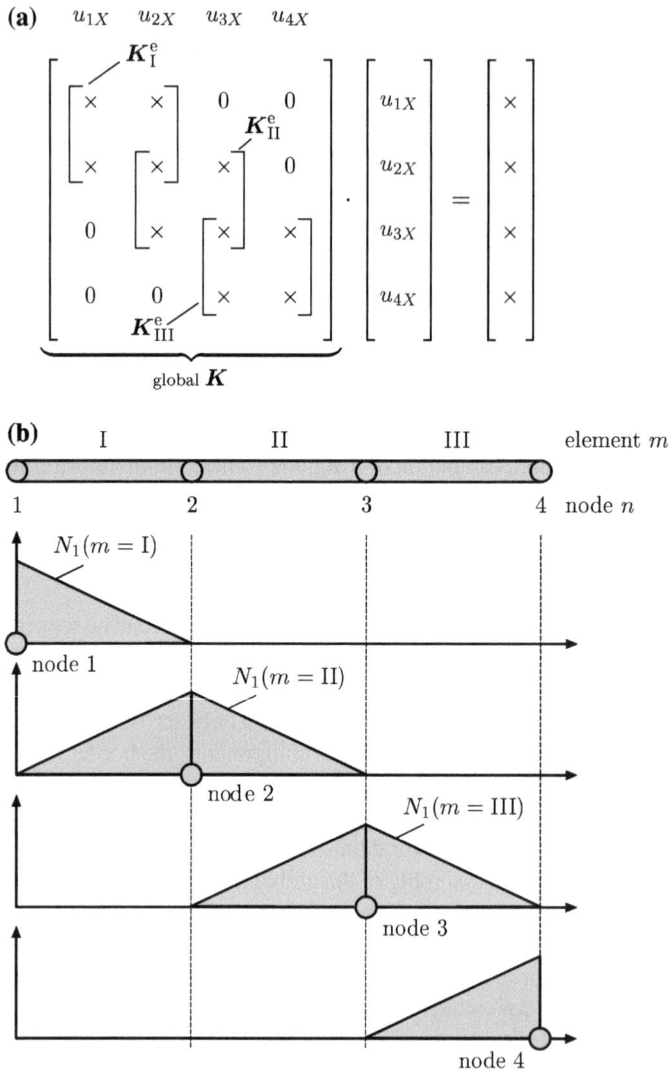


**Fig. 5.12** Assembly process to the global stiffness matrix: **a** composition of the elemental stiffness matrices to the global system; **b** interaction of shape functions at common nodes

$$\frac{EA}{L}\begin{bmatrix} 1 & -1 & 0 & 0 \\ -1 & 2 & -1 & 0 \\ 0 & -1 & 2 & -1 \\ 0 & 0 & -1 & 1 \end{bmatrix}\begin{bmatrix} u_{1X} \\ u_{2X} \\ u_{3X} \\ u_{4X} \end{bmatrix} = \begin{bmatrix} \cdots \\ \cdots \\ \cdots \\ \cdots \end{bmatrix}, \qquad (5.135)$$

where the right-hand side is not specified since nothing on the loading is indicated in Fig. 5.11. This system of equations without consideration of any boundary conditions

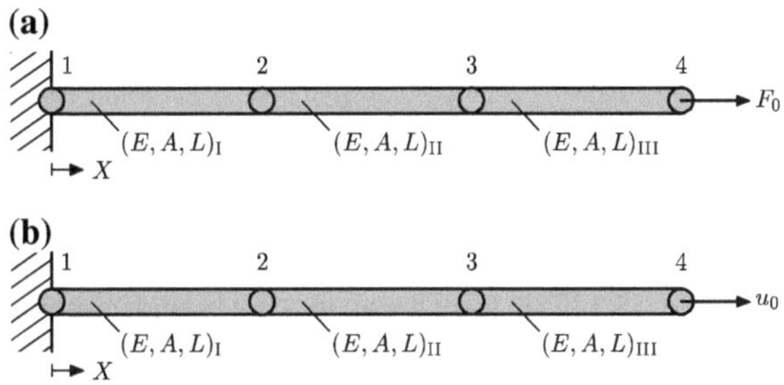


**Fig. 5.13** Consideration of boundary conditions for a cantilevered rod structure: **a** force boundary condition; **b** displacement boundary condition at the right-hand boundary node

is called the non-reduced system. For this system, the global stiffness matrix $\boldsymbol{K}$ is still singular and cannot be inverted in order to solve the global system of equations. Boundary conditions must be introduced in oder to make this matrix regular und thus invertible.

For the rod elements under consideration, two types of boundary conditions must be distinguished. The DIRICHLET boundary condition specifies the displacement $u$ at a node while the NEUMANN boundary condition $EA\frac{\mathrm{d}u}{\mathrm{d}x}$ assigns a force $F$ at a node. The different ways to handle these different types of boundary conditions will be explained in the following based on the problem shown in Fig. 5.13 where a cantilevered rod structure has different boundary conditions at its right-hand end node.

The consideration of the homogeneous DIRICHLET boundary condition, i.e. $u_1 = u(X = 0) = 0$, is the simplest case. To incorporate this boundary condition in the system (5.292), the first row and column can be canceled to obtain a reduced systems as:

$$\frac{EA}{L}\begin{bmatrix} 2 & -1 & 0 \\ -1 & 2 & -1 \\ 0 & -1 & 1 \end{bmatrix}\begin{bmatrix} u_{2X} \\ u_{3X} \\ u_{4X} \end{bmatrix} = \begin{bmatrix} \cdots \\ \cdots \\ \cdots \end{bmatrix}. \tag{5.136}$$

In general we can state that a homogenous DIRICHLET boundary condition at node $n$ ($u_{nX} = 0$) can be considered in the non-reduced system of equations by eliminating the $n$th row and $n$th column of the system. Let us consider next the case shown in Fig. 5.13a where the right-hand end node is subjected to a force $F_0$. This external force can simply be specified on the right-hand side and since no other external forces are acting, the reduced system of equations is finally obtained as:

$$\frac{EA}{L}\begin{bmatrix} 2 & -1 & 0 \\ -1 & 2 & -1 \\ 0 & -1 & 1 \end{bmatrix}\begin{bmatrix} u_{2X} \\ u_{3X} \\ u_{4X} \end{bmatrix} = \begin{bmatrix} 0 \\ 0 \\ F_0 \end{bmatrix}. \tag{5.137}$$

This system of equations can be solved, e.g. by inverting the reduced stiffness matrix and solving for the unknown nodal displacements in the form $\boldsymbol{u}_{\mathrm{p}} = \boldsymbol{K}^{-1}\boldsymbol{f}$:

$$\begin{bmatrix} u_{2X} \\ u_{3X} \\ u_{4X} \end{bmatrix} = \frac{LF_0}{EA} \begin{bmatrix} 1 \\ 2 \\ 3 \end{bmatrix}. \tag{5.138}$$

To incorporate a non-homogeneous DIRICHLET boundary condition ($u \neq 0$) as shown in Fig. 5.13b, three different strategies can be mentioned. The first one modifies the system shown in Eq. (5.137) in such a way that the boundary condition is directly introduced:

$$\frac{EA}{L} \begin{bmatrix} 2 & -1 & 0 \\ -1 & 2 & -1 \\ 0 & 0 & 1 \times \frac{L}{EA} \end{bmatrix} \begin{bmatrix} u_{2X} \\ u_{3X} \\ u_{4X} \end{bmatrix} = \begin{bmatrix} 0 \\ 0 \\ u_0 \end{bmatrix}, \tag{5.139}$$

where the last equation gives immediately the boundary condition as $u_{4X} = u_0$. The solution of the system of equations given in Eq. (5.152) can be obtained by inverting the coefficient matrix and multiplying it with the vector on the right-hand side as:

$$\begin{bmatrix} u_{2X} \\ u_{3X} \\ u_{4X} \end{bmatrix} = \begin{bmatrix} \frac{1}{3} u_0 \\ \frac{2}{3} u_0 \\ u_0 \end{bmatrix}. \tag{5.140}$$

In general we can state that a non-homogeneous DIRICHLET boundary condition at node $n$ can be introduced in the system of equations by modifying the $n$th line in such a way that at the position of the $n$th column a '1' is obtained while all other entries of the $n$th line are set to zero. On the right-hand side, the given value is introduced at the $n$th position of the column matrix.

The second way of considering a non-homogenous DIRICHLET boundary condition consists of the following step: The column of the stiffness matrix, which corresponds to the node where the boundary condition is given, is multiplied by the given displacement. In other words, if the boundary condition is specified at node $n$, the $n$th column of the stiffness matrix is multiplied by the given value $u_0$:

$$\frac{EA}{L} \begin{bmatrix} 2 & -1 & 0 \times u_0 \\ -1 & 2 & -1 \times u_0 \\ 0 & -1 & 1 \times u_0 \end{bmatrix} \begin{bmatrix} u_{2X} \\ u_{3X} \\ u_{4X} \end{bmatrix} = \begin{bmatrix} 0 \\ 0 \\ \cdots \end{bmatrix}. \tag{5.141}$$

Now we bring the $n$th column of the stiffness matrix to the right-hand side of the system

$$\frac{EA}{L} \begin{bmatrix} 2 & -1 \\ -1 & 2 \\ 0 & \end{bmatrix} \begin{bmatrix} u_{2X} \\ u_{3X} \\ u_{4X} \end{bmatrix} = \begin{bmatrix} 0 \\ 0 \\ \cdots \end{bmatrix} - \frac{EA}{L} \begin{bmatrix} 0 \times u_0 \\ -1 \times u_0 \\ 1 \times u_0 \end{bmatrix}, \tag{5.142}$$

and delete the $n$th row of the system:

$$\frac{EA}{L}\begin{bmatrix} 2 & -1 \\ -1 & 2 \end{bmatrix}\begin{bmatrix} u_{2X} \\ u_{3X} \end{bmatrix} = \begin{bmatrix} 0 \\ 0 \end{bmatrix} - \frac{EA}{L}\begin{bmatrix} 0 \times u_0 \\ -1 \times u_0 \end{bmatrix}. \tag{5.143}$$

As a results of this second approach, the dimension of the system of equations could be reduced compared to the first approach. However, this smaller matrix was not obtained for free since more steps have to be performed compared to the first possibility. The solution of Eq. (5.143) can be stated as:

$$\begin{bmatrix} u_{2X} \\ u_{3X} \end{bmatrix} = \begin{bmatrix} \frac{1}{3}\, u_0 \\ \frac{2}{3}\, u_0 \end{bmatrix}. \tag{5.144}$$

A third possible approach should be mentioned here since often the question arises by students why not simply replace in the column matrix of unknowns, i.e. on the left-hand side, the variable of the nodal value with the given value. This can be done but requires that the corresponding reaction force[22] is introduced on the right-hand side:

$$\frac{EA}{L}\begin{bmatrix} 2 & -1 & 0 \\ -1 & 2 & -1 \\ 0 & -1 & 1 \end{bmatrix}\underbrace{\begin{bmatrix} u_{2X} \\ u_{3X} \\ u_0 \end{bmatrix}}_{\boldsymbol{u}_\mathrm{p}} = \begin{bmatrix} 0 \\ 0 \\ -R_4 \end{bmatrix}. \tag{5.145}$$

However, the column matrix of the nodal displacements $\boldsymbol{u}_\mathrm{p}$ contains now unknown quantities ($u_{2X}$, $u_{3X}$) and the given nodal boundary condition ($u_0$). On the other hand, the right-hand side contains the unknown reaction force $R_4$. Thus, the structure of the linear system of equations is unfavorable for the solution. To rearrange the system to the classical structure where all unknowns are collected on the left and given quantities on the right-hand side, it is advised to write out the three single equations as:

$$\frac{EA}{L}(2u_{2X} - u_{3X}) = 0, \tag{5.146}$$

$$\frac{EA}{L}(-u_{2X} + 2u_{3X} - u_0) = 0, \tag{5.147}$$

$$\frac{EA}{L}(-u_{3X} + u_0) = -R_4. \tag{5.148}$$

After collecting unknown quantities on the left-hand side and known quantities on the right-hand side, one gets

$$\frac{EA}{L}(2u_{2X} - u_{3X}) = 0, \tag{5.149}$$

$$\frac{EA}{L}(-u_{2X} + 2u_{3X}) = \frac{EA}{L}u_0, \tag{5.150}$$

[22] Let us assume in the following that the reaction force $R_4$ is oriented in the negative $x$-direction.

$$\frac{EA}{L}\left(-u_{3X}+\frac{L}{EA}R_4\right)=-\frac{EA}{L}u_0, \tag{5.151}$$

or in matrix notation:

$$\frac{EA}{L}\begin{bmatrix} 2 & -1 & 0 \\ -1 & 2 & 0 \\ 0 & -1 & \frac{L}{EA} \end{bmatrix}\underbrace{\begin{bmatrix} u_{2X} \\ u_{3X} \\ R_4 \end{bmatrix}}_{\text{unknown}}=\underbrace{\frac{EA}{L}\begin{bmatrix} 0 \\ u_0 \\ -u_0 \end{bmatrix}}_{\text{given}}. \tag{5.152}$$

The solution of the last system of equations is obtained as:

$$\begin{bmatrix} u_{2X} \\ u_{3X} \\ R_4 \end{bmatrix}=\begin{bmatrix} \frac{1}{3}u_0 \\ \frac{2}{3}u_0 \\ -\frac{1}{3}\frac{EA}{L}u_0 \end{bmatrix}. \tag{5.153}$$

It should be noted here that this third approach is not the common way within the finite element method and is only shown for the sake of completeness.

A special type of 'boundary condition' can be realized by attaching a spring to a rod element a shown in Fig. 5.14. Let us have first a look at the configuration where the spring is attached to node 1 as shown in Fig. 5.14a. Assuming that node 2 is moved to the positive $x$-direction, the spring will cause a force on the rod element which can be expressed as $F_s = -ku_1$, where $k$ is the spring constant and $u_1$ the displacement of node 1, i.e. where the spring is attached to the rod element.[23] It should be mentioned here that the required force to elongate the spring by $u_1$ in the positive $x$-direction is equal to $ku_1$ but the force acting on the rod is oriented in the negative $x$-direction.

Thus, the principal finite element equation for the rod element can be written as:

$$\frac{EA}{L}\begin{bmatrix} 1 & -1 \\ -1 & 1 \end{bmatrix}\begin{bmatrix} u_1 \\ u_2 \end{bmatrix}=\begin{bmatrix} F_s \\ F_2 \end{bmatrix}=\begin{bmatrix} -ku_1 \\ F_2 \end{bmatrix}. \tag{5.154}$$

Looking at Eq. (5.154), it can be concluded that the expression $-ku_1$ on the right-hand side should be shifted to the left-hand side where the expressions with the nodal unknowns are collected. Thus, one can obtain the following expression:

$$\frac{EA}{L}\left[\begin{array}{c|c} 1+\frac{L}{EA}k & -1 \\ \hline -1 & 1 \end{array}\right]\begin{bmatrix} u_1 \\ u_2 \end{bmatrix}=\begin{bmatrix} 0 \\ F_2 \end{bmatrix}. \tag{5.155}$$

[23] It is assumed here that the spring is in its unstrained state in the sketched configuration, i.e. without the application of any force or displacement boundary conditions at the nodes of the rod.

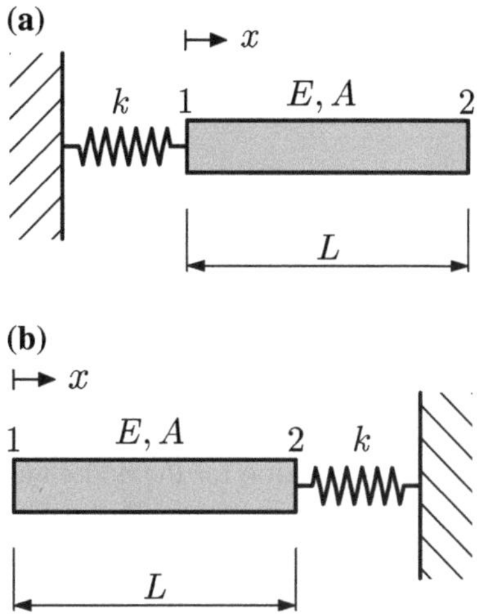


**Fig. 5.14** Consideration of a spring in a rod structure: **a** spring attached to node 1 or **b** to node 2

It can be seen from the last equation that a spring can simply be considered by adding the spring constant in the cell of the stiffness matrix with the index of the degree of freedom where the spring is attached, i.e. in our example the cell $(u_1, u_1)$. If the spring would be attached at the second node, cf. Fig. 5.14b, the spring constant should be added in the cell $(u_2, u_2)$ and the principal finite element equation for this case would finally read:

$$\frac{EA}{L}\left[\begin{array}{c|c} 1 & -1 \\ \hline -1 & 1+\frac{L}{EA}k \end{array}\right]\begin{bmatrix} u_1 \\ u_2 \end{bmatrix} = \begin{bmatrix} F_1 \\ 0 \end{bmatrix}. \tag{5.156}$$

If we like to consider that the springs shown in Fig. 5.14 are pre-strained,[24] i.e. elongated or compressed by a displacement of magnitude $u_s$, the force which acts on the rod element is given[25] by $F_s = -k(u_1 - u_s)$ or $F_s = -k(u_2 - u_s)$ and the principal finite element equations given in (5.155) and (5.156) are modified to:

$$\frac{EA}{L}\left[\begin{array}{c|c} 1+\frac{L}{EA}k & -1 \\ \hline -1 & 1 \end{array}\right]\begin{bmatrix} u_1 \\ u_2 \end{bmatrix} = \begin{bmatrix} ku_s \\ F_2 \end{bmatrix}, \tag{5.157}$$

[24] Such a pre-strained spring has its analogon in one-dimensional heat conduction in the form of a convective boundary condition: NEWTON's cooling law, i.e. $\dot{q} = h(T_\infty - T)$ where $\dot{q}$ is the heat flux in $\frac{\mathrm{W}}{\mathrm{m}^2}$, $h$ is the heat transfer coefficient in $\frac{\mathrm{W}}{\mathrm{m}^2\,\mathrm{K}}$, $T_\infty$ is the temperature of the environment and $T$ is the temperature of the object's surface, is in a similar manner treated as this type of spring. See also Table 5.7.

[25] Setting $u_s = 0$ results in an unstrained spring.

$$\frac{EA}{L}\left[\begin{array}{c|c} 1 & -1 \\ \hline -1 & 1+\frac{L}{EA}k \end{array}\right]\begin{bmatrix} u_1 \\ u_2 \end{bmatrix} = \begin{bmatrix} F_1 \\ ku_s \end{bmatrix}. \tag{5.158}$$

### 5.2.4 *Post-computation: Determination of Strain, Stress and Further Quantities*

The previous section explained how to compose the global system of equations from which the primary unknowns, i.e. the nodal displacements, can be obtained. After the solution for the nodal unknowns, further quantities can be calculated in a post-computational step. Based on the kinematics relationship for the continuum rod according to Eq. (3.2) together with the nodal approach (5.63) and the definition of the **B**-matrix (5.67), the following expression for the strain distribution inside a rod can be obtained:

$$\varepsilon_x^{\mathrm{e}}(x) = \frac{\mathrm{d}}{\mathrm{d}x} u^{\mathrm{e}}(x) = \frac{\mathrm{d}}{\mathrm{d}x} \boldsymbol{N}^{\mathrm{T}}(x)\, \boldsymbol{u}_{\mathrm{p}} = \boldsymbol{B}^{\mathrm{T}} \boldsymbol{u}_{\mathrm{p}}. \tag{5.159}$$

Considering the specific formulations of the **B**-matrices for a linear and a quadratic rod element according to Eqs. (5.76) and (5.97), the strain distribution can be expressed as

$$\varepsilon_x^{\mathrm{e}}(x) = \frac{1}{L}(-u_1 + u_2) \quad \text{(lin.)}, \tag{5.160}$$

$$\varepsilon_x^{\mathrm{e}}(x) = \frac{1}{L}\left(\left(-3 + \frac{4x}{L}\right)u_1 + \left(4 - \frac{8x}{L}\right)u_2 + \left(-1\frac{4x}{L}\right)u_3\right) \quad \text{(quad.)}, \tag{5.161}$$

or expressed in the natural coordinate $\xi$:

$$\varepsilon_x^{\mathrm{e}}(\xi) = \frac{1}{L}(-u_1 + u_2) \quad \text{(lin.)}, \tag{5.162}$$

$$\varepsilon_x^{\mathrm{e}}(\xi) = \frac{1}{L}((-1 + 2\xi)\,u_1 + (-4\xi)\,u_2 + (1 + 2\xi)\,u_3) \quad \text{(quad.)}. \tag{5.163}$$

Based on the obtained strain distribution, HOOKE's law (3.3) permits the calculation of the stress distribution inside a rod element as

$$\sigma_x^{\mathrm{e}}(x) = E\frac{\mathrm{d}}{\mathrm{d}x} u^{\mathrm{e}}(x) = E\frac{\mathrm{d}}{\mathrm{d}x} \boldsymbol{N}^{\mathrm{T}}(x)\, \boldsymbol{u}_{\mathrm{p}} = E\boldsymbol{B}^{\mathrm{T}} \boldsymbol{u}_{\mathrm{p}}, \tag{5.164}$$

or based on the nodal displacements for a linear and quadratic rod element as a function of the natural coordinate $\xi$:

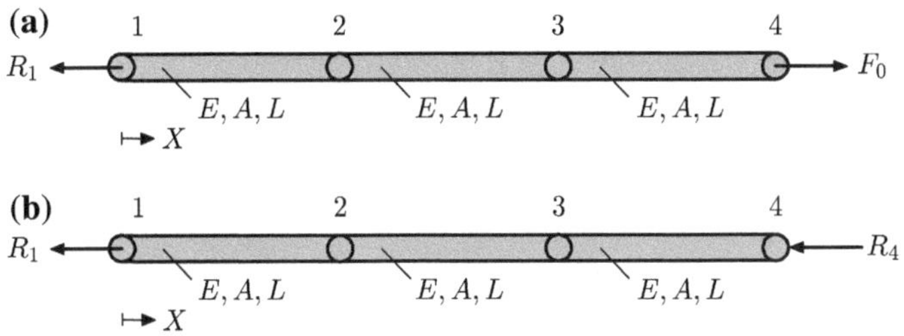


**Fig. 5.15** Free body diagram of the cantilevered rod structure shown in Fig. 5.13

$$\sigma_x^{\text{e}}(\xi) = \frac{E}{L}(-u_1 + u_2) \quad \text{(lin.)}, \tag{5.165}$$

$$\sigma_x^{\text{e}}(\xi) = \frac{E}{L}((-1 + 2\xi)\, u_1 + (-4\xi)\, u_2 + (1 + 2\xi)\, u_3)\,. \quad \text{(quad.)} \tag{5.166}$$

A final task is often to calculate the reaction forces at the supports or nodes of prescribed displacements. To explain the procedure, let us return to the example shown in Fig. 5.13. The free body diagram of the problem can be sketched as shown in Fig. 5.15.

Based on the indicated reaction forces, the global (non-reduced) system of equations can be stated for the configuration in Fig. 5.15a as

$$\frac{EA}{L}\begin{bmatrix} 1 & -1 & 0 & 0 \\ -1 & 2 & -1 & 0 \\ 0 & -1 & 2 & -1 \\ 0 & 0 & -1 & 1 \end{bmatrix}\begin{bmatrix} u_{1X} \\ u_{2X} \\ u_{3X} \\ u_{4X} \end{bmatrix} = \begin{bmatrix} -R_1 \\ 0 \\ 0 \\ F_0 \end{bmatrix}, \tag{5.167}$$

or for Fig. 5.15b as:

$$\frac{EA}{L}\begin{bmatrix} 1 & -1 & 0 & 0 \\ -1 & 2 & -1 & 0 \\ 0 & -1 & 2 & -1 \\ 0 & 0 & -1 & 1 \end{bmatrix}\begin{bmatrix} u_{1X} \\ u_{2X} \\ u_{3X} \\ u_{4X} \end{bmatrix} = \begin{bmatrix} -R_1 \\ 0 \\ 0 \\ -R_4 \end{bmatrix}. \tag{5.168}$$

Knowing all nodal displacements, the support reaction $R_1$ can be obtained for both cases by evaluating the first equation of the linear system as:

$$R_1 = -\frac{EA}{L}(u_{1X} - u_{2X})\,. \tag{5.169}$$

For the second case as shown in Fig. 5.15b, the reaction force $R_4$ is obtained by evaluating the fourth equation of the linear system (5.168) as:

$$R_4 = -\frac{EA}{L}(-u_{3X} + u_{4X})\,. \tag{5.170}$$

In general we can state that reactions forces are obtained from the non-reduced system of equations based on the prior to this calculated nodal displacements. Special attention must be given to the consideration of the reactions on the right-hand side of the system of equations since the pure calculation of the nodal displacements did not require an exact mentioning of these quantities.

### 5.2.5 Analogies to Other Field Problems

Further analogies to other field problems can be found, for example, in [88]. A comparison between solid mechanics and heat conduction is presented in Table 5.7.

**Table 5.7** Comparison of analogous properties in one-dimensional heat conduction and solid mechanics

| Solid mechanics | Heat conduction |
|---|---|
| Partial differential equation | |
| $EA\frac{\mathrm{d}^2 u_x}{\mathrm{d}x^2} = -p_0$ | $k\frac{\mathrm{d}^2 T}{\mathrm{d}x^2} = -\dot{\eta}_0$ |
| $\left(E\frac{\mathrm{d}^2 u_x}{\mathrm{d}x^2} = -\gamma_0\right)$ | $\left(kA\frac{\mathrm{d}^2 T}{\mathrm{d}x^2} = -\dot{\phi}_0\right)$ |
| Primary variable | |
| Displacement $u_x$ | Temperature $T$ |
| Derivative of primary variable | |
| Strain $\varepsilon_x = \frac{du_x}{dx}$ | Temperature gradient $\frac{\mathrm{d}T}{\mathrm{d}x}$ |
| Stress $\sigma_x = E\frac{du_x}{dx}$ | Heat flux $\dot{q}_x = -k\frac{\mathrm{d}T}{\mathrm{d}x}$ |
| Force $F_x = EA\frac{du_x}{dx}$ | Heat transfer rate $\dot{Q}_x = -kA\frac{\mathrm{d}T}{\mathrm{d}x}$ |
| Principal finite element equation | |
| $\frac{EA}{L}\begin{bmatrix} 1 & -1 \\ -1 & 1 \end{bmatrix}\begin{bmatrix} u_{1x} \\ u_{2x} \end{bmatrix} = \begin{bmatrix} F_{1x} \\ F_{2x} \end{bmatrix}$ | $\frac{k}{L}\begin{bmatrix} 1 & -1 \\ -1 & 1 \end{bmatrix}\begin{bmatrix} T_1 \\ T_2 \end{bmatrix} = \begin{bmatrix} \dot{q}_{1x} \\ \dot{q}_{2x} \end{bmatrix}$ |
| $\left(\frac{E}{L}\begin{bmatrix} 1 & -1 \\ -1 & 1 \end{bmatrix}\begin{bmatrix} u_{1x} \\ u_{2x} \end{bmatrix} = \begin{bmatrix} \sigma_{1x} \\ \sigma_{2x} \end{bmatrix}\right)$ | $\left(\frac{kA}{L}\begin{bmatrix} 1 & -1 \\ -1 & 1 \end{bmatrix}\begin{bmatrix} T_1 \\ T_2 \end{bmatrix} = \begin{bmatrix} \dot{Q}_{1x} \\ \dot{Q}_{2x} \end{bmatrix}\right)$ |

$p_0$ load per unit length in $\frac{\mathrm{N}}{\mathrm{m}}$, $\gamma_0$ load per unit volume in $\frac{\mathrm{N}}{\mathrm{m}^3}$, $\dot{\eta}_0$ rate of energy generation per unit volume in $\frac{\mathrm{W}}{\mathrm{m}^3}$, $\dot{\phi}_0$ rate of energy generation per unit length in $\frac{\mathrm{W}}{\mathrm{m}}$, $\dot{q}_x$ heat flux in $\frac{\mathrm{W}}{\mathrm{m}^2}$, $\dot{Q}_x$ heat transfer rate in W, $k$ thermal conductivity in $\frac{\mathrm{W}}{\mathrm{m\,K}}$

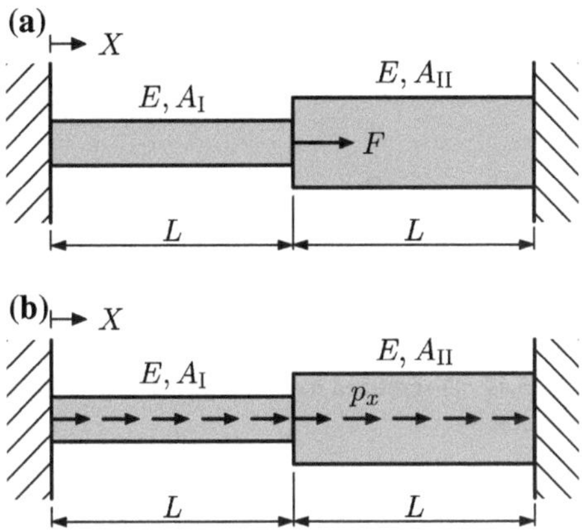


**Fig. 5.16** Rod structure fixed at both ends: **a** axial point load; **b** load per length

### *5.2.6 Solved Problems*

**5.3 Example:Rod structure fixed at both ends**
Given is a rod structure as sown in Fig. 5.16. The structure is composed of two rods of different cross-sectional areas $A_{\mathrm{I}}$ and $A_{\mathrm{II}}$. Length $L$ and YOUNG's modulus $E$ is the same for both rods. The structure is fixed at both ends and loaded by

(a) a point load $F$ in the middle and
(b) a uniform distributed load $p_x$, i.e. a force per unit length.

Model the rod structure with two linear finite elements and determine for both cases

- the displacement $u_2 = u(X = L)$ in the middle of the structure,
- the stresses and strains in both elements,
- the average stress and strain in the middle of the structure at $X = L$,
- the reaction forces at the supports and check the global force equilibrium.

Simplify all the results obtained for the special case of $A_{\mathrm{I}} = A_{\mathrm{II}} = A$.

**5.3 Solution**
The finite element discretization and all acting forces are shown in Fig. 5.17.

Case (a) point load:

• Displacement in the middle of the structure

The elemental stiffness matrix for each element is given by

$$\frac{EA_i}{L}\begin{bmatrix} 1 & -1 \\ -1 & 1 \end{bmatrix} \quad \text{with} \quad i = \mathrm{I}, \mathrm{II} \tag{5.171}$$

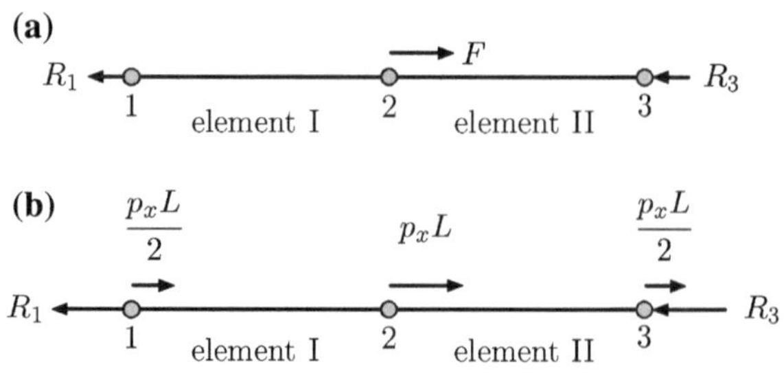


**Fig. 5.17** Discretized rod structure: **a** point load and reaction forces; **b** equivalent nodal loads and reaction forces

and can be composed to the global finite element equation to obtain:

$$\frac{E}{L}\begin{bmatrix} A_{\mathrm{I}} & -A_{\mathrm{I}} & 0 \\ -A_{\mathrm{I}} & A_{\mathrm{I}}+A_{\mathrm{II}} & -A_{\mathrm{II}} \\ 0 & -A_{\mathrm{II}} & A_{\mathrm{II}} \end{bmatrix}\begin{bmatrix} u_1 \\ u_2 \\ u_3 \end{bmatrix} = \begin{bmatrix} -R_1 \\ F \\ -R_3 \end{bmatrix}. \tag{5.172}$$

Consideration of the boundary conditions, i.e. $u_1 = u_3 = 0$, in the last system of equations allows to solve for the unknown displacement in the middle of the structure:

$$u_2 = \frac{FL}{E(A_{\mathrm{I}}+A_{\mathrm{II}})}. \tag{5.173}$$

- Stresses and strains in both elements

Based on the general definition of the strain in a rod element, i.e. $\varepsilon = \frac{1}{L}(u_{\mathrm{right}} - u_{\mathrm{left}})$, the constant strains in both elements can be derived under consideration of the boundary conditions as:

$$\varepsilon_{\mathrm{I}} = \frac{1}{L}(u_2 - 0) = \frac{F}{E(A_{\mathrm{I}}+A_{\mathrm{II}})}, \tag{5.174}$$

$$\varepsilon_{\mathrm{II}} = \frac{1}{L}(0 - u_2) = -\frac{F}{E(A_{\mathrm{I}}+A_{\mathrm{II}})}. \tag{5.175}$$

Application of HOOKE's law, i.e. $\sigma = E\varepsilon$, gives the constant stresses in each element:

$$\sigma_{\mathrm{I}} = E\varepsilon_{\mathrm{I}} = \frac{F}{(A_{\mathrm{I}}+A_{\mathrm{II}})}, \tag{5.176}$$

$$\sigma_{\mathrm{II}} = E\varepsilon_{\mathrm{II}} = -\frac{F}{(A_{\mathrm{I}}+A_{\mathrm{II}})}. \tag{5.177}$$

• Average stress and strain in the middle of the structure

As in the case of many finite element codes, the average stress and strain at the middle node can be calculated by the following averaging rule as:

$$\varepsilon_2 = \frac{\varepsilon_\mathrm{I} + \varepsilon_\mathrm{II}}{2} = 0, \tag{5.178}$$

$$\sigma_2 = \frac{\sigma_\mathrm{I} + \sigma_\mathrm{II}}{2} = 0. \tag{5.179}$$

As can be seen from this result, stress and strain values displayed at nodes should be taken with care.

• Reaction forces at the supports and check of the global force equilibrium

Evaluation of the first and third equation of the system (5.172) for known nodal displacements gives:

$$R_1 = \frac{EA_\mathrm{I}}{L} \times u_2 = \frac{A_\mathrm{I}}{A_\mathrm{I} + A_\mathrm{II}} \times F, \tag{5.180}$$

$$R_3 = \frac{EA_\mathrm{II}}{L} \times u_2 = \frac{A_\mathrm{II}}{A_\mathrm{I} + A_\mathrm{II}} \times F, \tag{5.181}$$

and the global force equilibrium

$$F - \frac{A_\mathrm{I}}{A_\mathrm{I} + A_\mathrm{II}} \times F - \frac{A_\mathrm{II}}{A_\mathrm{I} + A_\mathrm{II}} \times F = 0 \tag{5.182}$$

is fulfilled.

Case (b) distributed load:

• Displacement in the middle of the structure

The global finite element equation results under consideration of the equivalent nodal loads, cf. Fig. 5.17, as

$$\frac{E}{L}\begin{bmatrix} A_\mathrm{I} & -A_\mathrm{I} & 0 \\ -A_\mathrm{I} & A_\mathrm{I} + A_\mathrm{II} & -A_\mathrm{II} \\ 0 & -A_\mathrm{II} & A_\mathrm{II} \end{bmatrix} \begin{bmatrix} u_1 \\ u_2 \\ u_3 \end{bmatrix} = \begin{bmatrix} -R_1 + \frac{p_x L}{2} \\ p_x L \\ -R_3 + \frac{p_x L}{2} \end{bmatrix}, \tag{5.183}$$

from which the displacement at node 2 follows under consideration of the boundary conditions:

$$u_2 = \frac{(p_x L) L}{E(A_\mathrm{I} + A_\mathrm{II})}. \tag{5.184}$$

• Stresses and strains in both elements and at the middle node

Based on the procedure given in (a), the constant strains and stresses are given by:

$$\varepsilon_{\mathrm{I}} = \frac{p_x L}{E(A_{\mathrm{I}} + A_{\mathrm{II}})} \;,\; \sigma_I = \frac{p_x L}{A_{\mathrm{I}} + A_{\mathrm{II}}}, \tag{5.185}$$

$$\varepsilon_{\mathrm{II}} = -\frac{p_x L}{E(A_{\mathrm{I}} + A_{\mathrm{II}})} \;,\; \sigma_{II} = -\frac{p_x L}{A_{\mathrm{I}} + A_{\mathrm{II}}}, \tag{5.186}$$

$$\varepsilon_2 = 0 \;,\; \sigma_2 = 0. \tag{5.187}$$

- Reaction forces at the supports and check of the global force equilibrium

Evaluation of the first and third equation of the system (5.183) for known nodal displacements gives:

$$R_1 = \frac{p_x L}{2} + \frac{EA_{\mathrm{I}}}{L} \times u_2 = \left(\frac{1}{2} + \frac{A_{\mathrm{I}}}{A_{\mathrm{I}} + A_{\mathrm{II}}}\right) \times p_x L, \tag{5.188}$$

$$R_3 = \frac{p_x L}{2} + \frac{EA_{\mathrm{II}}}{L} \times u_2 = \left(\frac{1}{2} + \frac{A_{\mathrm{II}}}{A_{\mathrm{I}} + A_{\mathrm{II}}}\right) \times p_x L, \tag{5.189}$$

and the global force equilibrium is fulfilled. It can be concluded from this exercise that the equivalent loads applied at the supports do not influence the strains and stresses inside the rods but contribute to the reaction forces at the supports. Results for the special case $A_{\mathrm{I}} = A_{\mathrm{II}} = A$ are summarized in Table 5.8.

**Table 5.8** Results of the problem shown in Fig. 5.16 for the special case $A_{\mathrm{I}} = A_{\mathrm{II}} = A$

| Quantity | Point load $F$ | Distributed load $p_x$ |
|---|---|---|
| $u_2$ | $\frac{1}{2}\frac{FL}{EA}$ | $\frac{1}{2}\frac{(p_x L)L}{EA}$ |
| $\varepsilon_{\mathrm{I}}$ | $\frac{F}{2EA}$ | $\frac{p_x L}{2EA}$ |
| $\varepsilon_{\mathrm{II}}$ | $-\frac{F}{2EA}$ | $-\frac{p_x L}{2EA}$ |
| $\sigma_{\mathrm{I}}$ | $\frac{F}{2A}$ | $\frac{p_x L}{2A}$ |
| $\sigma_{\mathrm{II}}$ | $-\frac{F}{2A}$ | $-\frac{p_x L}{2A}$ |
| $\sigma_2$ | 0 | 0 |
| $\varepsilon_2$ | 0 | 0 |
| $R_1$ | $\frac{F}{2}$ | $p_x L$ |
| $R_3$ | $\frac{F}{2}$ | $p_x L$ |

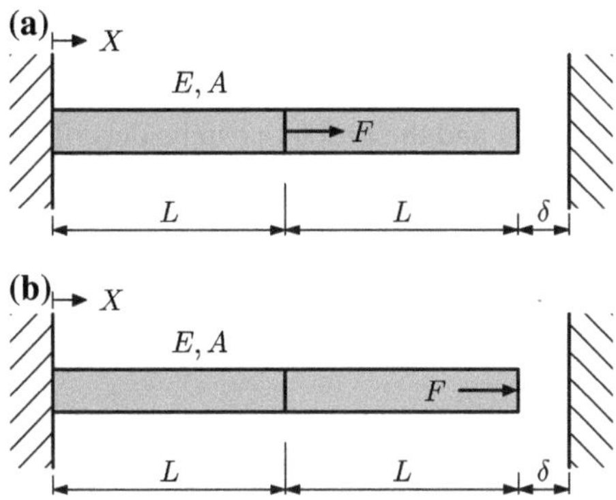


**Fig. 5.18** Rod structure with a gap at the right-hand end: **a** axial point load in the middle; **b** axial point load at the right-hand end

**5.4 Example: Rod structure with gap**

Given is a rod structure as sown in Fig. 5.18. The structure is composed of a rod with cross-sectional area $A$, length $L$ and YOUNG's modulus $E$. The structure is fixed at the left-hand end and a gap of distance $\delta$ is between the right-hand end and a rigid wall. The structure is loaded by

(a) a point load $F$ in the middle and
(b) a point load $F$ at the right end.

Model the rod structure with two linear finite elements and determine for both cases

- the displacement $u_2 = u(X = L)$ in the middle of the structure for the case of no contact and contact,
- the reaction forces at the supports and check the global force equilibrium,
- the stress distribution in the rod structure for increasing force $F$.

**5.4 Solution**

The finite element discretization and all acting forces are shown in Fig. 5.19.

Case (a) point load in the middle:

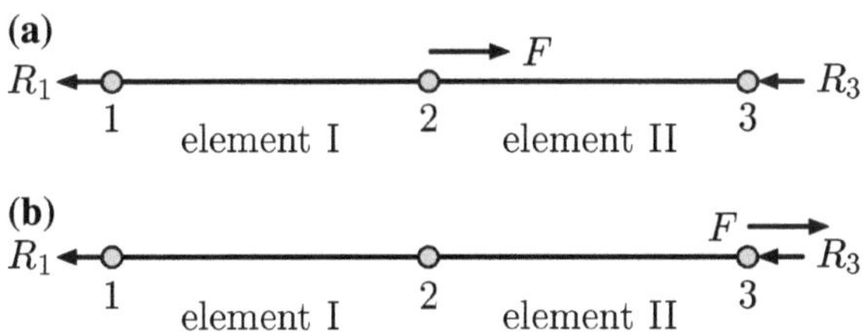


**Fig. 5.19** Discretized rod structure: **a** axial point load in the middle; **b** axial point load at the right-hand end. The reaction force $R_3$ is only acting in the case of contact

• Displacement in the middle of the structure

In the case that there is no contact, element II is not acting, i.e. $u_2 = u_3$, or contributing to the global stiffness matrix and the problem can be described by

$$\frac{EA}{L}\begin{bmatrix} 1 & -1 \\ -1 & 1 \end{bmatrix}\begin{bmatrix} u_1 \\ u_2 \end{bmatrix} = \begin{bmatrix} -R_1 \\ F \end{bmatrix}, \tag{5.190}$$

from which the displacement at node 2 can be obtained under consideration of the boundary condition ($u_1 = 0$) as:

$$u_2 = \frac{FL}{EA}. \tag{5.191}$$

If the force $F$ is further increased to a value of $F = \frac{EA\delta}{L}$, contact occurs, i.e. $u_2 = u_3 = \delta$, and the situation for the global system is different. Now, both elements contribute to the global system:

$$\frac{EA}{L}\begin{bmatrix} 1 & -1 & 0 \\ -1 & 2 & -1 \\ 0 & -1 & 1 \end{bmatrix}\begin{bmatrix} u_1 \\ u_2 \\ u_3 \end{bmatrix} = \begin{bmatrix} -R_1 \\ F \\ -R_3 \end{bmatrix}. \tag{5.192}$$

Consideration of the boundary conditions at the left- and right-hand end, i.e. $u_1 = 0$ and $u_3 = \delta$, gives

$$\frac{EA}{L}\begin{bmatrix} 2 & -1 \\ 0 & 1 \end{bmatrix}\begin{bmatrix} u_2 \\ u_3 \end{bmatrix} = \begin{bmatrix} F \\ \frac{EA\delta}{L} \end{bmatrix}, \tag{5.193}$$

and the displacement in the middle of the structure is obtained as:

$$u_2 = \frac{LF}{2EA} + \frac{\delta}{2}. \tag{5.194}$$

• Reaction forces

Based on the known values of the nodal displacements, Eq. (5.192) can be evaluated for the reaction forces:

$$R_1 = \left(\frac{F}{2} + \frac{EA\delta}{2L}\right), \tag{5.195}$$

$$R_3 = \left(\frac{F}{2} - \frac{EA\delta}{2L}\right). \tag{5.196}$$

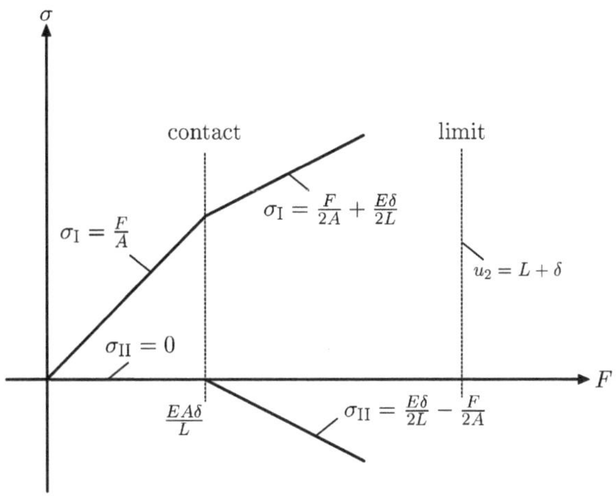


**Fig. 5.20** Stress distribution in the rod structure as a function of the external point load $F$ (point load in the middle)

It should be noted that both reaction forces are directed to the negative $X$-direction.

• Stress distribution in the rod structure for increasing force $F$

Since the nodal displacements are known, the strains can be obtained based on the general definition $\varepsilon = \frac{1}{L}(u_{\rm right} - u_{\rm left})$ and HOOKE's law gives the stresses. The results for the stress $\sigma$ in both elements as a function of the applied external force $F$ are shown in Fig. 5.20. As can be seen from this figure, the global stiffness changes as soon as the gap is closed.

Case (b) point load at the right-hand end:

• Displacement in the middle of the structure

In the case that there is no contact, the displacement of the middle node is simply half of the displacement obtained at the node of the right-hand end. Under the condition of contact, the global system of equations reads as:

$$\frac{EA}{L}\begin{bmatrix} 1 & -1 & 0 \\ -1 & 2 & -1 \\ 0 & -1 & 1 \end{bmatrix}\begin{bmatrix} u_1 \\ u_2 \\ u_3 \end{bmatrix} = \begin{bmatrix} -R_1 \\ 0 \\ F - R_3 \end{bmatrix}. \tag{5.197}$$

Consideration of the boundary conditions at the left- and right-hand end, i.e. $u_1 = 0$ and $u_3 = \delta$, gives

$$\frac{EA}{L}\begin{bmatrix} 2 & -1 \\ 0 & 1 \end{bmatrix}\begin{bmatrix} u_2 \\ u_3 \end{bmatrix} = \begin{bmatrix} 0 \\ \frac{EA\delta}{L} \end{bmatrix}, \tag{5.198}$$

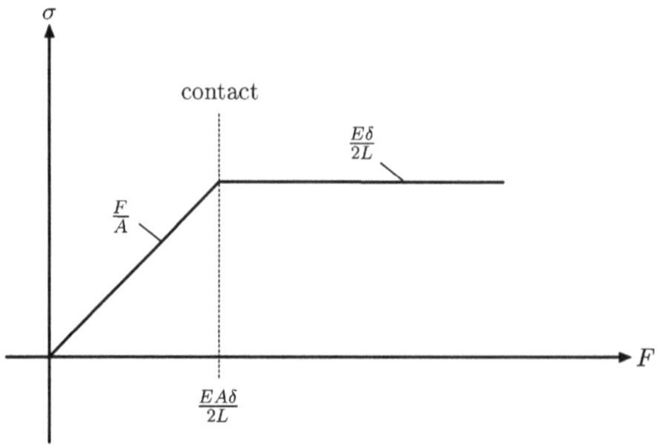


**Fig. 5.21** Stress distribution in the rod structure as a function of the external point load $F$ (point load at the right-hand end)

and the displacement in the middle of the structure is obtained as:

$$u_2 = \frac{\delta}{2}. \tag{5.199}$$

- Reaction forces

Based on the known values of the nodal displacements, Eq. (5.197) can be evaluated for the reaction forces:

$$R_1 = \frac{EA\delta}{2L}, \tag{5.200}$$

$$R_3 = F - \frac{EA\delta}{2L}. \tag{5.201}$$

- Stress distribution in the rod structure for increasing force $F$.

As mentioned in (a), strains and stresses can be calculated based on the known nodal displacements. The graphical representation of the stress in the rod is given Fig. 5.21. It can be concluded that any additional force after closing the gap is absorbed by the support and does not affect the stress state in the rod.

## 5.5 Example: Rod with changing cross-sectional area

Given is a rod structure as sown in Fig. 5.22. The structure reveals a linear changing cross-sectional area $A(x)$ while YOUNG's modulus $E$ is assumed to be constant. The structure is fixed at the left-hand end and loaded by a single force $F_0$ at the right-hand end. The ratio between the area $A_5$ and $A_1$ is given by the factor $a$.

Model the rod structure with four linear finite elements of constant cross-sectional area and determine for the stepped rod the nodal displacements. Each element should

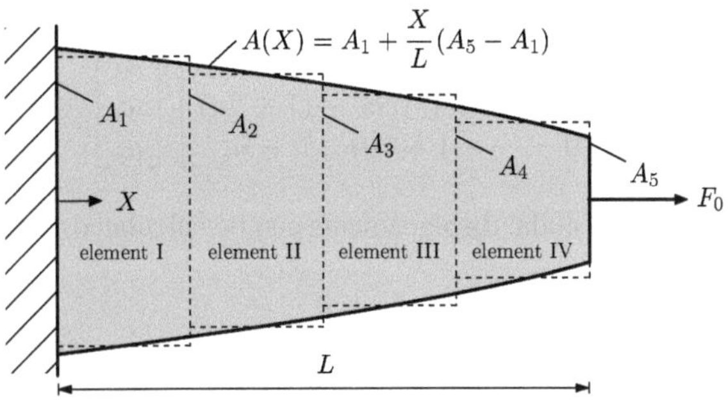


**Fig. 5.22** Rod with changing cross-sectional area $A = A(x)$

have the same length $\frac{L}{4}$ and the cross section should be the average of the cross-sectional area at the left- and right-hand end of each step.

### 5.5 Solution

Given the ratio between the cross section at the right- and left-hand end as $\frac{A_5}{A_1} = a$ and the functional dependency of the cross-sectional area as

$$A(X) = A_1 + \frac{X}{L}(A_5 - A_1)\,, \tag{5.202}$$

the areas $A_i$ $(i = 2, \ldots, 5)$ can be expressed as:

$$A_2 = \frac{3+a}{4}A_1\,,\; A_3 = \frac{1+a}{2}A_1\,,\; A_4 = \frac{1+3a}{4}A_1\,,\; A_5 = aA_1. \tag{5.203}$$

Based on these area relations, the averaged area for each element $A_{ij} = \frac{A_i+A_j}{2}$ is obtained as:

$$A_{12} = \frac{7+a}{8}A_1\,,\; A_{23} = \frac{5+3a}{8}A_1\,,\; A_{34} = \frac{3+5a}{8}A_1\,,\; A_{45} = \frac{1+7a}{8}A_1, \tag{5.204}$$

and the global stiffness matrix can be assembled to:

$$\frac{EA_1}{2L}\begin{bmatrix} 7+a & -(7+a) & 0 & 0 & 0 \\ -(7+a) & 12+4a & -(5+3a) & 0 & 0 \\ 0 & -(5+3a) & 8+8a & -(3+5a) & 0 \\ 0 & 0 & -(3+5a) & 4+12a & -(1+7a) \\ 0 & 0 & 0 & -(1+7a) & 1+7a \end{bmatrix}. \tag{5.205}$$

Considering the boundary condition at the left-hand end, i.e. $u_1 = 0$, the system of equations is obtained as

$$\frac{EA_1}{2L}\begin{bmatrix} 12+4a & -(5+3a) & 0 & 0 \\ -(5+3a) & 8+8a & -(3+5a) & 0 \\ 0 & -(3+5a) & 4+12a & -(1+7a) \\ 0 & 0 & -(1+7a) & 1+7a \end{bmatrix}\begin{bmatrix} u_2 \\ u_3 \\ u_4 \\ u_5 \end{bmatrix} = \begin{bmatrix} 0 \\ 0 \\ 0 \\ F_0 \end{bmatrix}, \quad (5.206)$$

from which the vector of nodal displacements can be calculated:

$$\begin{bmatrix} u_1 \\ u_2 \\ u_3 \\ u_4 \\ u_5 \end{bmatrix} = \frac{LF}{EA_1}\begin{bmatrix} 0 \\ \dfrac{2}{7+a} \\ \dfrac{8(3+a)}{3a^2+26a+35} \\ \dfrac{2(71+98a+23a^2)}{105+253a+139a^2+15a^3} \\ \dfrac{32(11+53a+53a^2+11a^3)}{(1+7a)(105+253a+139a^2+15a^3)} \end{bmatrix}. \quad (5.207)$$

## 5.3 Bernoulli Beam Elements

### 5.3.1 Derivation of the Principal Finite Element Equation

Let us consider in the following the governing differential equation according to Eq. (4.36). This formulation assumes that the bending stiffness $EI_z$ is constant and we obtain

$$EI_z \frac{d^4 u_y^0(x)}{dx^4} - q_y(x) = 0, \quad (5.208)$$

where $u_y^0(x)$ represents the exact solution of the problem. The last equation which contains the exact solution of the problem is fulfilled at each location $x$ of the beam and is called the *strong formulation* of the problem. Replacing the exact solution in Eq. (5.208) by an approximate solution $u_y(x)$, a residual $r$ is obtained:

$$r(x) = EI_z \frac{d^4 u_y(x)}{dx^4} - q_y(x) \neq 0. \quad (5.209)$$

As a consequence of the introduction of the approximate solution $u_y(x)$, it is in general no longer possible to satisfy the differential equation at each location $x$ of the beam. In the scope of the weighted residual method, it is alternatively requested that the differential equation is fulfilled over a certain length (and no longer at each location $x$) and the following integral statement is obtained

$$\int_0^L W(x)\left(EI_z\frac{\mathrm{d}^4u_y(x)}{\mathrm{d}x^4} - q_y(x)\right)\mathrm{d}x \stackrel{!}{=} 0, \tag{5.210}$$

which is called the *inner product*. The function $W(x)$ in Eq. (5.210) is called the weight function which distributes the error or the residual in the considered domain.

Integrating by parts[26] of the first expression in the parentheses of Eq. (5.210) gives:

$$\int_0^L \underbrace{W}_{f}\, EI_z \underbrace{\frac{\mathrm{d}^4u_y}{\mathrm{d}x^4}}_{g'}\mathrm{d}x = EI_z\left[W\frac{\mathrm{d}^3u_y}{\mathrm{d}x^3}\right]_0^L - EI_z\int_0^L \frac{\mathrm{d}W}{\mathrm{d}x}\frac{\mathrm{d}^3u_y}{\mathrm{d}x^3}\mathrm{d}x = 0. \tag{5.211}$$

Integrating by parts of the integral on the right-hand side of Eq. (5.211) results in:

$$EI_z\int_0^L \underbrace{\frac{\mathrm{d}W}{\mathrm{d}x}}_{f}\underbrace{\frac{\mathrm{d}^3u_y}{\mathrm{d}x^3}}_{g'}\mathrm{d}x = EI_z\left[\frac{\mathrm{d}W}{\mathrm{d}x}\frac{\mathrm{d}^2u_y}{\mathrm{d}x^2}\right]_0^L - EI_z\int_0^L \frac{\mathrm{d}^2W}{\mathrm{d}x^2}\frac{\mathrm{d}^2u_y}{\mathrm{d}x^2}\mathrm{d}x. \tag{5.212}$$

Combination of Eqs. (5.211) and (5.212) gives under consideration of Eq. (5.210) the so-called *weak formulation* of the problem as:

$$EI_z\int_0^L \frac{\mathrm{d}^2W}{\mathrm{d}x^2}\frac{\mathrm{d}^2u_y}{\mathrm{d}x^2}\mathrm{d}x = EI_z\left[-W\frac{\mathrm{d}^3u_y}{\mathrm{d}x^3} + \frac{\mathrm{d}W}{\mathrm{d}x}\frac{\mathrm{d}^2u_y}{\mathrm{d}x^2}\right]_0^L + \int_0^L Wq_y\,\mathrm{d}x. \tag{5.213}$$

Looking at the weak formulation, it can be seen that the integration by parts shifted two derivatives from the approximate solution to the weight function and a symmetrical formulation with respect to the derivatives is obtained. This symmetry with respect to the derivatives of the approximate solution and the weight function will again guarantee in the following—as in the case of the rod element—that a symmetric stiffness matrix is obtained for the beam element. In order to continue the derivation of the principal finite element equation, the displacement $u_y(x)$ and the weight function $W(x)$ must be expressed within an element (superscript 'e') in the form of the nodal approach. This nodal approach can be generally stated as

$$u_y^{\mathrm{e}}(x) = \boldsymbol{N}^{\mathrm{T}}(x)\,\boldsymbol{u}_{\mathrm{p}}, \tag{5.214}$$

$$W(x) = \delta\boldsymbol{u}_{\mathrm{p}}^{\mathrm{T}}\boldsymbol{N}(x), \tag{5.215}$$

[26] A common representation of integration by parts of two functions $f(x)$ and $g(x)$ is: $\int fg'\,\mathrm{d}x = fg - \int f'g\,\mathrm{d}x$.

where $\boldsymbol{N}(x)$ is the column matrix of the shape functions, $\boldsymbol{u}_{\mathrm{p}}$ is the column matrix of the nodal unknowns and $\delta\boldsymbol{u}_{\mathrm{p}}$ is the column matrix of the virtual displacements.

The kinematics relation according to Eq. (4.15) allows to express the strain distribution within an element based on the nodal approach as:

$$\varepsilon_x^{\mathrm{e}}(x,y) = -y\frac{\mathrm{d}^2 u_y^{\mathrm{e}}(x)}{\mathrm{d}x^2} = -y\frac{\mathrm{d}^2}{\mathrm{d}x^2}\left(\boldsymbol{N}^{\mathrm{T}}(x)\,\boldsymbol{u}_{\mathrm{p}}\right) = -y\frac{\mathrm{d}^2\boldsymbol{N}^{\mathrm{T}}(x)}{\mathrm{d}x^2}\boldsymbol{u}_{\mathrm{p}}. \tag{5.216}$$

Analogous to the procedure in Sect. 5.164, one can introduce for the beam element a generalized $\boldsymbol{B}$-matrix. Thus, an equivalent description as in Eq. (5.159), i.e. $\varepsilon_x^{\mathrm{e}}(x) = \boldsymbol{B}^{\mathrm{T}}\boldsymbol{u}_{\mathrm{p}}$, is obtained with:

$$\boldsymbol{B}^{\mathrm{T}} = -y\frac{\mathrm{d}^2\boldsymbol{N}^{\mathrm{T}}(x)}{\mathrm{d}x^2}. \tag{5.217}$$

This definition of the $\boldsymbol{B}$-matrix allows a formal derivation (cf. [75]) of the stiffness matrix based on the general formulation as presented in Eq. (8.34), i.e. $\boldsymbol{K}^{\mathrm{e}} = \int_V \boldsymbol{BCB}^{\mathrm{T}}\mathrm{d}V$. If a formulation corresponding to Eq. (5.73), i.e. $\boldsymbol{K}^{\mathrm{e}} = EI_z\int_x \boldsymbol{BB}^{\mathrm{T}}\mathrm{d}x$, is the goal of the derivation, the definition of the $\boldsymbol{B}$-matrix should be rather given as $\boldsymbol{B}^{\mathrm{T}} = \frac{\mathrm{d}^2\boldsymbol{N}^{\mathrm{T}}(x)}{\mathrm{d}x^2}$.

Let us introduce now the formulations for $u_y^{\mathrm{e}}(x)$ and $W(x)$ according to Eqs. (5.214) and (5.215) in the weak formulation (5.213):

$$\begin{aligned} EI_z\int_0^L \frac{\mathrm{d}^2}{\mathrm{d}x^2}\left(\delta\boldsymbol{u}_{\mathrm{p}}^{\mathrm{T}}\boldsymbol{N}(x)\right)\frac{\mathrm{d}^2}{\mathrm{d}x^2}\left(\boldsymbol{N}^{\mathrm{T}}(x)\,\boldsymbol{u}_{\mathrm{p}}\right)\mathrm{d}x &= EI_z\left[-\delta\boldsymbol{u}_{\mathrm{p}}^{\mathrm{T}}\boldsymbol{N}(x)\frac{\mathrm{d}^3u_y}{\mathrm{d}x^3}\right.\\ &\left.+\frac{\mathrm{d}}{\mathrm{d}x}\left(\delta\boldsymbol{u}_{\mathrm{p}}^{\mathrm{T}}\boldsymbol{N}(x)\right)\frac{\mathrm{d}^2u_y}{\mathrm{d}x^2}\right]_0^L + \int_0^L \delta\boldsymbol{u}_{\mathrm{p}}^{\mathrm{T}}\boldsymbol{N}(x)\,q_y(x)\,\mathrm{d}x. \end{aligned} \tag{5.218}$$

Since the virtual displacements $\delta\boldsymbol{u}_{\mathrm{p}}$ and the displacements $\boldsymbol{u}_{\mathrm{p}}$ are not a function of $x$, they can be considered as constants with respect to the integration and can be taken out of the integral on the left-hand side of Eq. (5.218). Furthermore, the virtual displacements $\delta\boldsymbol{u}_{\mathrm{p}}$ occur on both sides of Eq. (5.218) and can be canceled. Thus, the weak formulation takes the following form:

$$\underbrace{EI_z\int_0^L \frac{\mathrm{d}^2}{\mathrm{d}x^2}\left(\boldsymbol{N}(x)\right)\frac{\mathrm{d}^2}{\mathrm{d}x^2}\left(\boldsymbol{N}^{\mathrm{T}}(x)\right)\mathrm{d}x}_{\boldsymbol{K}^{\mathrm{e}}}\ \boldsymbol{u}_{\mathrm{p}}$$

$$= \underbrace{EI_z \left[ -N(x)\frac{\mathrm{d}^3 u_y}{\mathrm{d}x^3} + \frac{\mathrm{d}}{\mathrm{d}x}(N(x))\frac{\mathrm{d}^2 u_y}{\mathrm{d}x^2} \right]_0^L + \int_0^L N(x)\, q_y(x)\,\mathrm{d}x}_{f^e} . \qquad (5.219)$$

In order to continue the derivation of the principal finite element equation, it is at this point now required that the column matrix of the shape functions $\boldsymbol{N}$ and the column matrix of the nodal unknowns $\boldsymbol{u}_\mathrm{p}$ are further specified.

Let us consider in the following a BERNOULLI beam element which is composed of two nodes as schematically shown in Fig. 5.23. Each node has two degrees of freedom, i.e. a displacement $u_y$ and a rotation $\varphi_z = \frac{\mathrm{d}u_y}{\mathrm{d}x}$, and can be loaded by a single force $F_y$ and a bending moment $M_z$. Since the BERNOULLI beam element has two nodes with in total four nodal unknowns, the nodal approaches given in Eqs. (5.214) and (5.215) can be given as[27]

$$u_y^\mathrm{e}(x) = \boldsymbol{N}^\mathrm{T}(x)\boldsymbol{u}_\mathrm{p} = \begin{bmatrix} N_{1u} & N_{1\varphi} & N_{2u} & N_{2\varphi} \end{bmatrix} \times \begin{bmatrix} u_{1y} \\ \varphi_{1z} \\ u_{2y} \\ \varphi_{2z} \end{bmatrix} \qquad (5.220)$$

and

$$W(x) = \delta\boldsymbol{u}_\mathrm{p}^\mathrm{T}\boldsymbol{N}(x) = \begin{bmatrix} \delta u_{1y} & \delta\varphi_{1z} & \delta u_{2y} & \delta\varphi_{2z} \end{bmatrix} \times \begin{bmatrix} N_{1u} \\ N_{1\varphi} \\ N_{2u} \\ N_{2\varphi} \end{bmatrix}, \qquad (5.221)$$

where $N_u$ are the shape functions for the displacement field and $N_\varphi$ for the rotational field.

These four shape functions are commonly taken from the family of cubic HERMITE[28] interpolation functions as:

$$N_{1u}(x) = 1 - 3\left(\frac{x}{L}\right)^2 + 2\left(\frac{x}{L}\right)^3, \qquad (5.222)$$

$$N_{1\varphi}(x) = x - 2\frac{x^2}{L} + \frac{x^3}{L^2}, \qquad (5.223)$$

---

[27] A detailed description of the derivation of the nodal approach and the respective shape functions for the BERNOULLI beam element is given in Sect. 5.3.2.

[28] Charles HERMITE (1822–1901), French mathematician.

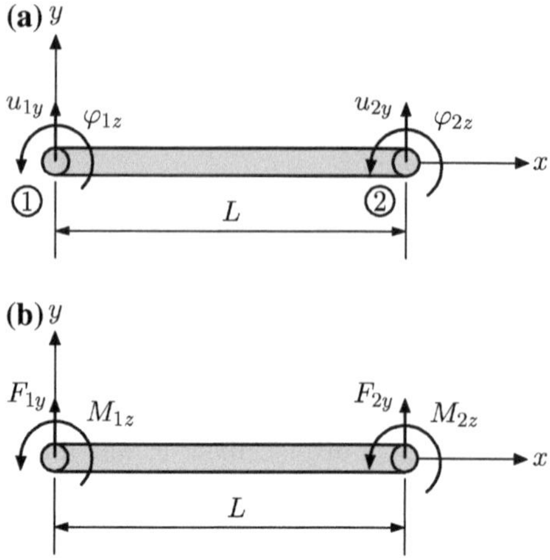


**Fig. 5.23** Definition of the BERNOULLI beam element for deformation in the $x$-$y$ plane: **a** deformations; **b** external loads. The nodes are symbolized by two *circles* at the ends (◯)

$$N_{2u}(x) = 3\left(\frac{x}{L}\right)^2 - 2\left(\frac{x}{L}\right)^3, \tag{5.224}$$

$$N_{2\varphi}(x) = -\frac{x^2}{L} + \frac{x^3}{L^2}. \tag{5.225}$$

The graphical representation of these cubic shape functions is given in Fig. 5.24.

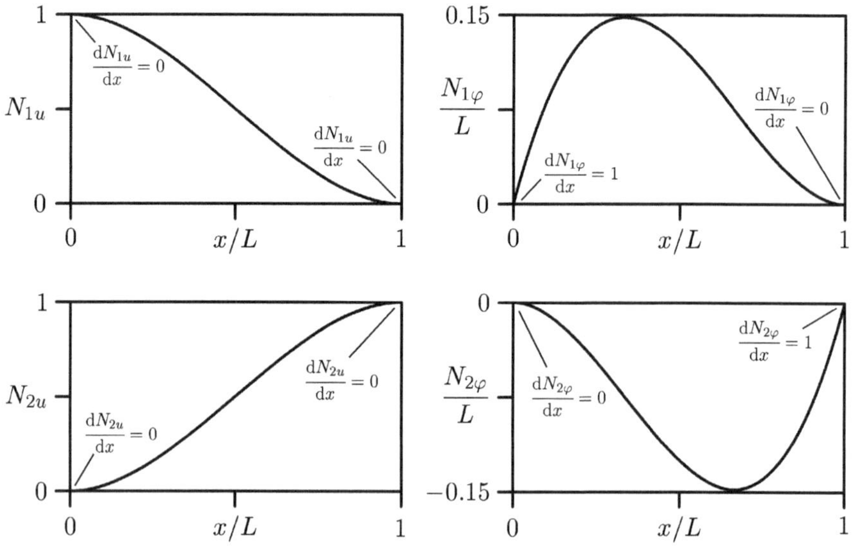


**Fig. 5.24** Shape functions for the BERNOULLI beam element given in the physical coordinate for bending in the $x$-$y$ plane

Let us first consider in the following only the left-hand side of Eq. (5.219) in order to derive the expression for the elemental stiffness matrix $\boldsymbol{K}^{\mathrm{e}}$ of the BERNOULLI beam element. Introducing the components of the shape function column matrix gives

$$\boldsymbol{K}^{\mathrm{e}} = EI_z \int_0^L \begin{bmatrix} \frac{\mathrm{d}N_{1u}^2}{\mathrm{d}x^2} \\ \frac{\mathrm{d}N_{1\varphi}^2}{\mathrm{d}x^2} \\ \frac{\mathrm{d}N_{2u}^2}{\mathrm{d}x^2} \\ \frac{\mathrm{d}N_{2\varphi}^2}{\mathrm{d}x^2} \end{bmatrix} \begin{bmatrix} \frac{\mathrm{d}N_{1u}^2}{\mathrm{d}x^2} & \frac{\mathrm{d}N_{1\varphi}^2}{\mathrm{d}x^2} & \frac{\mathrm{d}N_{2u}^2}{\mathrm{d}x^2} & \frac{\mathrm{d}N_{2\varphi}^2}{\mathrm{d}x^2} \end{bmatrix} \mathrm{d}x, \tag{5.226}$$

or after performing the multiplication as:

$$\boldsymbol{K}^{\mathrm{e}} = EI_z \int_0^L \left[\begin{array}{c|c|c|c} \frac{\mathrm{d}N_{1u}^2}{\mathrm{d}x^2}\frac{\mathrm{d}N_{1u}^2}{\mathrm{d}x^2} & \frac{\mathrm{d}N_{1u}^2}{\mathrm{d}x^2}\frac{\mathrm{d}N_{1\varphi}^2}{\mathrm{d}x^2} & \frac{\mathrm{d}N_{1u}^2}{\mathrm{d}x^2}\frac{\mathrm{d}N_{2u}^2}{\mathrm{d}x^2} & \frac{\mathrm{d}N_{1u}^2}{\mathrm{d}x^2}\frac{\mathrm{d}N_{2\varphi}^2}{\mathrm{d}x^2} \\ \hline \frac{\mathrm{d}N_{1\varphi}^2}{\mathrm{d}x^2}\frac{\mathrm{d}N_{1u}^2}{\mathrm{d}x^2} & \frac{\mathrm{d}N_{1\varphi}^2}{\mathrm{d}x^2}\frac{\mathrm{d}N_{1\varphi}^2}{\mathrm{d}x^2} & \frac{\mathrm{d}N_{1\varphi}^2}{\mathrm{d}x^2}\frac{\mathrm{d}N_{2u}^2}{\mathrm{d}x^2} & \frac{\mathrm{d}N_{1\varphi}^2}{\mathrm{d}x^2}\frac{\mathrm{d}N_{2\varphi}^2}{\mathrm{d}x^2} \\ \hline \frac{\mathrm{d}N_{2u}^2}{\mathrm{d}x^2}\frac{\mathrm{d}N_{1u}^2}{\mathrm{d}x^2} & \frac{\mathrm{d}N_{2u}^2}{\mathrm{d}x^2}\frac{\mathrm{d}N_{1\varphi}^2}{\mathrm{d}x^2} & \frac{\mathrm{d}N_{2u}^2}{\mathrm{d}x^2}\frac{\mathrm{d}N_{2u}^2}{\mathrm{d}x^2} & \frac{\mathrm{d}N_{2u}^2}{\mathrm{d}x^2}\frac{\mathrm{d}N_{2\varphi}^2}{\mathrm{d}x^2} \\ \hline \frac{\mathrm{d}N_{2\varphi}^2}{\mathrm{d}x^2}\frac{\mathrm{d}N_{1u}^2}{\mathrm{d}x^2} & \frac{\mathrm{d}N_{2\varphi}^2}{\mathrm{d}x^2}\frac{\mathrm{d}N_{1\varphi}^2}{\mathrm{d}x^2} & \frac{\mathrm{d}N_{2\varphi}^2}{\mathrm{d}x^2}\frac{\mathrm{d}N_{2u}^2}{\mathrm{d}x^2} & \frac{\mathrm{d}N_{2\varphi}^2}{\mathrm{d}x^2}\frac{\mathrm{d}N_{2\varphi}^2}{\mathrm{d}x^2} \end{array}\right] \mathrm{d}x. \tag{5.227}$$

The derivatives of the shape functions according to Eq. (5.222) till (5.225) can be calculated as

$$\frac{\mathrm{d}N_{1u}(x)}{\mathrm{d}x} = -\frac{6x}{L^2} + \frac{6x^2}{L^3}, \tag{5.228}$$

$$\frac{\mathrm{d}N_{1\varphi}(x)}{\mathrm{d}x} = 1 - \frac{4x}{L} + \frac{3x^2}{L^2}, \tag{5.229}$$

$$\frac{\mathrm{d}N_{2u}(x)}{\mathrm{d}x} = \frac{6x}{L^2} - \frac{6x^2}{L^3}, \tag{5.230}$$

$$\frac{\mathrm{d}N_{2\varphi}(x)}{\mathrm{d}x} = -\frac{2x}{L} + \frac{3x^2}{L^2}, \tag{5.231}$$

and accordingly the second-order derivatives as

$$\frac{\mathrm{d}^2N_{1u}(x)}{\mathrm{d}x^2} = -\frac{6}{L^2} + \frac{12x}{L^3}, \tag{5.232}$$

$$\frac{\mathrm{d}^2 N_{1\varphi}(x)}{\mathrm{d}x^2} = -\frac{4}{L} + \frac{6x}{L^2}, \tag{5.233}$$

$$\frac{\mathrm{d}^2 N_{2u}(x)}{\mathrm{d}x^2} = \frac{6}{L^2} - \frac{12x}{L^3}, \tag{5.234}$$

$$\frac{\mathrm{d}^2 N_{2\varphi}(x)}{\mathrm{d}x^2} = -\frac{2}{L} + \frac{6x}{L^2}. \tag{5.235}$$

These derivatives introduced into Eq. (5.227) give after analytical integration of the polynomials the elemental stiffness matrix of the BERNOULLI beam element as:

$$\boldsymbol{K}^{\mathrm{e}} = \frac{EI_z}{L^3} \begin{bmatrix} 12 & 6L & -12 & 6L \\ 6L & 4L^2 & -6L & 2L^2 \\ -12 & -6L & 12 & -6L \\ 6L & 2L^2 & -6L & 4L^2 \end{bmatrix}. \tag{5.236}$$

It must be noted here that the analytical integration as performed to obtain Eq. (5.236) cannot be performed in commercial finite element codes since they are written in traditional programming languages such as FORTRAN. As in the case of the rod element, a numerical integration is performed (cf. Appendix A.11) by GAUSS-LEGENDRE quadrature. Transforming the shape functions, cf. Eqs. (5.222) till (5.225), from the Cartesian coordinate $x$ to its natural coordinate $\xi$ based on the transformation given in Table 5.2 gives:

$$N_{1u}(\xi) = \frac{1}{4}\left[2 - 3\xi + \xi^3\right], \tag{5.237}$$

$$N_{1\varphi}(\xi) = \frac{1}{4}\left[1 - \xi - \xi^2 + \xi^3\right]\frac{L}{2}, \tag{5.238}$$

$$N_{2u}(\xi) = \frac{1}{4}\left[2 + 3\xi - \xi^3\right], \tag{5.239}$$

$$N_{1\varphi}(\xi) = \frac{1}{4}\left[-1 - \xi + \xi^2 + \xi^3\right]\frac{L}{2}. \tag{5.240}$$

The second-order derivatives which are required for Eq. (5.227) can easily be derived as:

$$\frac{\mathrm{d}^2 N_{1u}(x)}{\mathrm{d}x^2} = \frac{4}{L^2}\frac{\mathrm{d}^2 N_{1u}(\xi)}{\mathrm{d}\xi^2} = \frac{6}{L^2}\xi, \tag{5.241}$$

$$\frac{\mathrm{d}^2 N_{1\varphi}(x)}{\mathrm{d}x^2} = \frac{4}{L^2}\frac{\mathrm{d}^2 N_{1\varphi}(\xi)}{\mathrm{d}\xi^2} = \frac{1}{L}(-1 + 3\xi), \tag{5.242}$$

$$\frac{\mathrm{d}^2 N_{2u}(x)}{\mathrm{d}x^2} = \frac{4}{L^2}\frac{\mathrm{d}^2 N_{2u}(\xi)}{\mathrm{d}\xi^2} = -\frac{6}{L^2}\xi, \tag{5.243}$$

$$\frac{\mathrm{d}^2 N_{2\varphi}(x)}{\mathrm{d}x^2} = \frac{4}{L^2}\frac{\mathrm{d}^2 N_{2\varphi}(\xi)}{\mathrm{d}\xi^2} = \frac{1}{L}(1+3\xi)\,. \tag{5.244}$$

Introducing these relationships in Eq. (5.227), one receives

$$EI_z \int_{-1}^{1} \left[\begin{array}{c|c|c|c} \frac{36}{L^4}\xi^2 & \frac{6}{L^3}(-\xi+3\xi^2) & -\frac{36}{L^4}\xi^2 & \frac{6}{L^3}(\xi+3\xi^2) \\ \hline \frac{6}{L^3}(-\xi+3\xi^2) & \frac{1}{L^2}(-1+3\xi)^2 & -\frac{6}{L^3}(-\xi+3\xi^2) & \frac{1}{L^2}(9\xi^2-1) \\ \hline -\frac{36}{L^4}\xi^2 & -\frac{6}{L^3}(-\xi+3\xi^2) & \frac{36}{L^4}\xi^2 & -\frac{6}{L^3}(\xi+3\xi^2) \\ \hline \frac{6}{L^3}(\xi+3\xi^2) & \frac{1}{L^2}(9\xi^2-1) & -\frac{6}{L^3}(\xi+3\xi^2) & \frac{1}{L^2}(1+3\xi)^2 \end{array}\right] \frac{L\mathrm{d}\xi}{2}. \tag{5.245}$$

The polynomials included in Eq. (5.245) are of maximum order of three and thus, a two-point integration rule (cf. Table A.5) is sufficient to accurately integrate as:

$$\boldsymbol{K}^{\mathrm{e}} = EI_z \int_{-1}^{1} \left[\;\cdots\;\right] \frac{L}{2}\,\mathrm{d}\xi \approx \frac{EI_zL}{2}\left[\;\cdots\;\right]\bigg|_{\xi=\frac{1}{\sqrt{3}}} \times 1 + \frac{EI_zL}{2}\left[\;\cdots\;\right]\bigg|_{\xi=-\frac{1}{\sqrt{3}}} \times 1, \tag{5.246}$$

which gives the same result for the stiffness matrix as the analytical integration shown in Eq. (5.236).

Let us now consider the right-hand side of Eq. (5.219) in order to derive the expression for the load column matrix $\boldsymbol{f}^{\mathrm{e}}$ of the beam element. The first part of the right-hand side, i.e.

$$EI_z\left[-N(x)\frac{\mathrm{d}^3 u_y}{\mathrm{d}x^3} + \frac{\mathrm{d}}{\mathrm{d}x}(N(x))\frac{\mathrm{d}^2 u_y}{\mathrm{d}x^2}\right]_0^L \tag{5.247}$$

results with the definition of the column matrix of shape functions according to Eq. (5.220) in

$$EI_z \left[ - \begin{bmatrix} N_{1u} \\ N_{1\varphi} \\ N_{2u} \\ N_{2\varphi} \end{bmatrix} \frac{\mathrm{d}^3 u_y}{\mathrm{d}x^3} + \frac{\mathrm{d}}{\mathrm{d}x} \begin{bmatrix} N_{1u} \\ N_{1\varphi} \\ N_{2u} \\ N_{2\varphi} \end{bmatrix} \frac{\mathrm{d}^2 u_y}{\mathrm{d}x^2} \right]_0^L . \tag{5.248}$$

Equation (5.248) represents a system of four equations which are to be evaluated at the boundaries of the integration, i.e. $x = 0$ and $x = L$. The first row of Eq. (5.248) gives:

$$\left( -N_{1u} EI_z \frac{\mathrm{d}^3 u_y}{\mathrm{d}x^3} + \frac{\mathrm{d}N_{1u}}{\mathrm{d}x} \frac{\mathrm{d}^2 u_y}{\mathrm{d}x^2} \right)_{x=L} - \left( -N_{1u} EI_z \frac{\mathrm{d}^3 u_y}{\mathrm{d}x^3} + \frac{\mathrm{d}N_{1u}}{\mathrm{d}x} \frac{\mathrm{d}^2 u_y}{\mathrm{d}x^2} \right)_{x=0} . \tag{5.249}$$

Under consideration of the boundary values of the shape functions respectively their derivatives according to Fig. 5.24, i.e. $N_{1u}(L) = 0$, $\frac{\mathrm{d}N_{1u}}{\mathrm{d}x}(L) = \frac{\mathrm{d}N_{1u}}{\mathrm{d}x}(0) = 0$ and $N_{1u}(0) = 1$, one receives the following expression:

$$+ EI_z \left. \frac{\mathrm{d}^3 u_y}{\mathrm{d}x^3} \right|_{x=0} \overset{(4.38)}{=} -Q_y(0). \tag{5.250}$$

Corresponding expressions can be derived for the other rows of Eq. (5.248):

$$\text{Row 2:} \quad - EI_z \left. \frac{\mathrm{d}^2 u_y}{\mathrm{d}x^2} \right|_{x=0} \overset{(4.37)}{=} -M_z(0), \tag{5.251}$$

$$\text{Row 3:} \quad - EI_z \left. \frac{\mathrm{d}^3 u_y}{\mathrm{d}x^3} \right|_{x=L} \overset{(4.38)}{=} +Q_y(L), \tag{5.252}$$

$$\text{Row 4:} \quad + EI_z \left. \frac{\mathrm{d}^2 u_y}{\mathrm{d}x^2} \right|_{x=L} \overset{(4.37)}{=} +M_z(L). \tag{5.253}$$

It must be noted here that shear forces $Q_y$ and bending moments $M_z$ are the internal reactions according to Fig. (4.8). The external loads with their positive directions according to Fig. (5.23) can be obtained from the internal loads by inverting the sign at the left-hand boundary and by maintaining the positive direction of the internal reactions at the right-hand boundary. Thus, the contribution to the load matrix due to single forces $F_{iy}$ and moments $M_{iz}$ at the nodes is expressed by:

$$\boldsymbol{f}^{\mathrm{e}}_{FM} = \begin{bmatrix} F_{1y} \\ M_{1z} \\ F_{2y} \\ M_{2z} \end{bmatrix} . \tag{5.254}$$

The second part of Eq. (5.219), i.e.

$$\int_0^L \boldsymbol{N}(x)\, q_y(x)\, \mathrm{d}x \qquad (5.255)$$

represents the general rule to determine equivalent nodal loads in the case of arbitrarily distributed loads $q_y(x)$. As an example, the evaluation of Eq. (5.255) for a constant load $+q_y$ results in the following load matrix:

$$\boldsymbol{f}_q^{\mathrm{e}} = q_y \int_0^L \begin{bmatrix} N_{1u}(x) \\ N_{1\varphi}(x) \\ N_{2u}(x) \\ N_{2\varphi}(x) \end{bmatrix} \mathrm{d}x = \frac{q_y L}{2} \begin{bmatrix} 1 \\ \frac{L}{6} \\ 1 \\ -\frac{L}{6} \end{bmatrix}. \qquad (5.256)$$

Further expressions for equivalent nodal loads can be taken from Table 5.9. Let us remind at this step that in the scope of the finite element method any type of loading can be only introduced at nodes into the discretized structure.

Based on the derived results, the principal finite element equation for a single BERNOULLI beam element with constant bending stiffness $EI_z$ can be expressed in a general form as

$$\boldsymbol{K}^{\mathrm{e}} \boldsymbol{u}^{\mathrm{e}} = \boldsymbol{f}^{\mathrm{e}}, \qquad (5.257)$$

or in components as

$$\frac{EI_z}{L^3} \begin{bmatrix} 12 & 6L & -12 & 6L \\ 6L & 4L^2 & -6L & 2L^2 \\ -12 & -6L & 12 & -6L \\ 6L & 2L^2 & -6L & 4L^2 \end{bmatrix} \begin{bmatrix} u_{1y} \\ \varphi_{1z} \\ u_{2y} \\ \varphi_{2z} \end{bmatrix} = \begin{bmatrix} F_{1y} \\ M_{1z} \\ F_{2y} \\ M_{2z} \end{bmatrix} + \int_0^L q_y(x) \begin{bmatrix} N_{1u} \\ N_{1\varphi} \\ N_{2u} \\ N_{2\varphi} \end{bmatrix} \mathrm{d}x. \qquad (5.258)$$

## *5.3.2 Derivation of Shape Functions*

Looking at Fig. 5.23 which schematically shows the definition of the BERNOULLI beam element, one can see that four deformation quantities are the unknowns at the nodes. A nodal interpolation of the displacement and rotational field must therefore fulfill the following four conditions:

$$u_y(0) = u_{1y}\,,\ \varphi_z(0) = \varphi_{1z},\ u_y(L) = u_{2y}\,,\ \varphi_z(L) = \varphi_{2z}. \qquad (5.259)$$

Furthermore, the second-order derivative of the interpolation of the displacement field must be nonzero as can be concluded from the weak formulation given in Eq. (5.218). Thus, the following general third-order polynomial with four unknowns $(\alpha_0, \ldots, \alpha_3)$ can be introduced to describe the displacement field:

**Table 5.9** Equivalent joint loadings for a BERNOULLI beam element ($x$-axis: right facing; $y$-axis: upward facing), partially adapted from [17]

| Loading | Shear force | Bending moment |
|---|---|---|
| 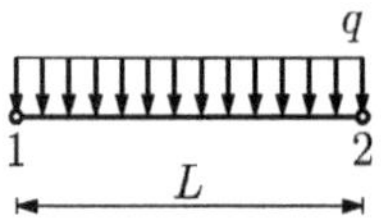 | $F_{1y} = -\frac{qL}{2}$ <br> $F_{2y} = -\frac{qL}{2}$ | $M_{1z} = -\frac{qL^2}{12}$ <br> $M_{2z} = +\frac{qL^2}{12}$ |
| 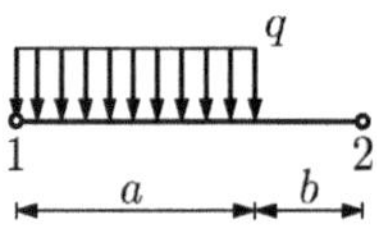 | $F_{1y} = -\frac{qa}{2L^3}(a^3 - 2a^2L + 2L^3)$ <br> $F_{2y} = -\frac{qa^3}{2L^3}(2L - a)$ | $M_{1z} = -\frac{qa^2}{12L^2}(3a^2 - 8aL + 6L^2)$ <br> $M_{2z} = +\frac{qa^3}{12L^2}(4L - 3a)$ |
| 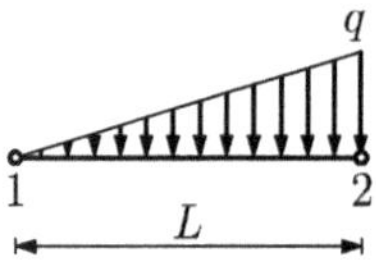 | $F_{1y} = -\frac{3}{20}qL$ <br> $F_{2y} = -\frac{7}{20}qL$ | $M_{1z} = -\frac{qL^2}{30}$ <br> $M_{2z} = +\frac{qL^2}{20}$ |
| 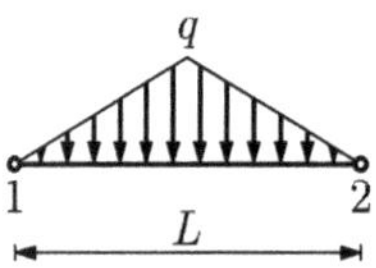 | $F_{1y} = -\frac{1}{4}qL$ <br> $F_{2y} = -\frac{1}{4}qL$ | $M_{1z} = -\frac{5qL^2}{96}$ <br> $M_{2z} = +\frac{5qL^2}{96}$ |
| 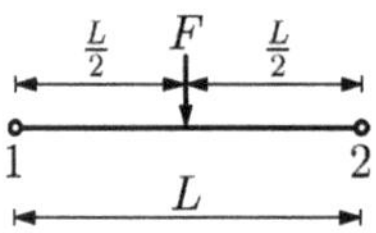 | $F_{1y} = -\frac{F}{2}$ <br> $F_{2y} = -\frac{F}{2}$ | $M_{1z} = -\frac{FL}{8}$ <br> $M_{2z} = +\frac{FL}{8}$ |
| 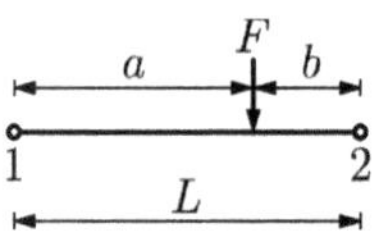 | $F_{1y} = -\frac{Fb^2(3a+b)}{L^3}$ <br> $F_{2y} = -\frac{Fa^2(a+3b)}{L^3}$ | $M_{1z} = -\frac{Fb^2a}{L^2}$ <br> $M_{2z} = +\frac{Fa^2b}{L^2}$ |
| 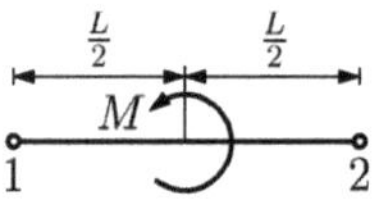 | $F_{1y} = -\frac{3}{2}\frac{M}{L}$ <br> $F_{2y} = +\frac{3}{2}\frac{M}{L}$ | $M_{1z} = -\frac{M}{4}$ <br> $M_{2z} = -\frac{M}{4}$ |
| 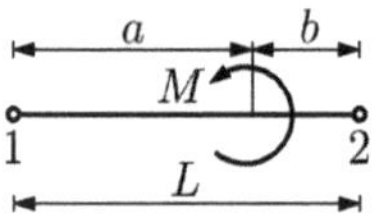 | $F_{1y} = -6M\frac{ab}{L^3}$ <br> $F_{2y} = +6M\frac{ab}{L^3}$ | $M_{1z} = -M\frac{b(2a-b)}{L^2}$ <br> $M_{2z} = -M\frac{a(2b-a)}{L^2}$ |

$$u_y(x) = a_0 + a_1 x + a_2 x^2 + a_3 x^3 = \begin{bmatrix} 1 \; x \; x^2 \; x^3 \end{bmatrix} \begin{bmatrix} a_0 \\ a_1 \\ a_2 \\ a_3 \end{bmatrix} = \boldsymbol{\chi}^{\mathrm{T}} \boldsymbol{a}. \tag{5.260}$$

Differentiation with respect to the $x$-coordinate gives the rotational field as:

$$\varphi_z(x) = \frac{\mathrm{d}u_y(x)}{\mathrm{d}x} = a_1 + 2a_2 x + 3a_3 x^2. \tag{5.261}$$

Evaluation of the functional expressions for the displacement, $u_y(x)$, and rotational, $\varphi_z(x)$, field at both nodes, i.e. for $x = 0$ and $x = L$, gives:

$$\text{Node 1:} \quad u_{1y}(0) = a_0, \tag{5.262}$$

$$\varphi_{1z}(0) = a_1, \tag{5.263}$$

$$\text{Node 2:} \quad u_{2y}(L) = a_0 + a_1 L + a_2 L^2 + a_3 L^3, \tag{5.264}$$

$$\varphi_{2z}(L) = a_1 + 2a_2 L + 3a_3 L^2. \tag{5.265}$$

The last four equations can be expressed in matrix form as:

$$\begin{bmatrix} u_{1y} \\ \varphi_{1z} \\ u_{2y} \\ \varphi_{2z} \end{bmatrix} = \underbrace{\begin{bmatrix} 1 & 0 & 0 & 0 \\ 0 & 1 & 0 & 0 \\ 1 & L & L^2 & L^3 \\ 0 & 1 & 2L & 3L^3 \end{bmatrix}}_{\boldsymbol{X}} \begin{bmatrix} a_0 \\ a_1 \\ a_2 \\ a_3 \end{bmatrix}. \tag{5.266}$$

Solving this system of equations for the unknown basis functions $a_i$ gives

$$\begin{bmatrix} a_0 \\ a_1 \\ a_2 \\ a_3 \end{bmatrix} = \underbrace{\begin{bmatrix} 1 & 0 & 0 & 0 \\ 0 & 1 & 0 & 0 \\ -\frac{3}{L^2} & -\frac{2}{L} & \frac{3}{L^2} & -\frac{1}{L} \\ \frac{2}{L^3} & \frac{1}{L^2} & -\frac{2}{L^3} & \frac{1}{L^2} \end{bmatrix}}_{\boldsymbol{A}} \begin{bmatrix} u_{1y} \\ \varphi_{1z} \\ u_{2y} \\ \varphi_{2z} \end{bmatrix}, \tag{5.267}$$

and the matrix of the shape functions results according to Eq. (5.119), i.e. $\boldsymbol{N}^{\mathrm{T}} = \boldsymbol{\chi}^{\mathrm{T}} \boldsymbol{A}$, as:

$$\begin{bmatrix} N_{1u} \; N_{1\varphi} \; N_{2u} \; N_{2\varphi} \end{bmatrix} = \begin{bmatrix} 1 \; x \; x^2 \; x^3 \end{bmatrix} \begin{bmatrix} 1 & 0 & 0 & 0 \\ 0 & 1 & 0 & 0 \\ -\frac{3}{L^2} & -\frac{2}{L} & \frac{3}{L^2} & -\frac{1}{L} \\ \frac{2}{L^3} & \frac{1}{L^2} & -\frac{2}{L^3} & \frac{1}{L^2} \end{bmatrix}. \tag{5.268}$$

The evaluation of the last equation gives the same results as in Eq. (5.222) till (5.225).

To conclude this section, a more descriptive approach to derive the shape functions is presented in the following. To this end, let us consider the general characteristic of a shape function, i.e. that a function $N_i$ takes the value 1 at node $i$ and is zero at all other nodes. Furthermore, it must be considered that the displacement and rotational fields are decoupled at the nodes in the case of a beam in bending. As a result, one obtains that a shape function for the displacement field takes at 'its' node a value of 1 whilst the slope must be equal to 0. At all other nodes $j$, the functional value and the slope are equal to zero:

$$N_{iu}(x_i) = 1, \tag{5.269}$$

$$N_{iu}(x_j) = 0, \tag{5.270}$$

$$\frac{dN_{iu}(x_i)}{dx} = 0, \tag{5.271}$$

$$\frac{dN_{iu}(x_j)}{dx} = 0. \tag{5.272}$$

In the same way, it is concluded that a shape function for the rotational field takes at 'its' node a value of 1 for the slope whilst the functional value must be equal to 0. At all other nodes, the functional values and slopes are identical zero. Thus, one obtains the in Fig. 5.25 represented boundary conditions for the four shape functions.

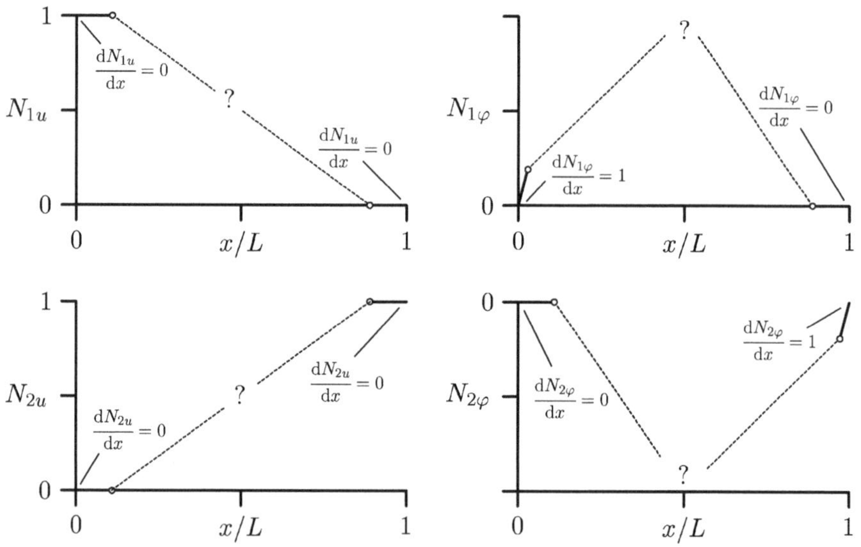


**Fig. 5.25** Boundary conditions for the shape functions of a BERNOULLI beam element for bending in the $x$-$y$ plane. Note that the regions for the given slopes are exaggerated for better illustration

Each shape function must change its curvature if there should be no geometrical discontinuities, i.e. kinks, in the course of the function. This can be achieved by a third-order polynomial whose curvature, i.e. the second-order derivative, is a linear function:

$$N(x) = a_0 + a_1 x + a_2 x^2 + a_3 x^3 . \tag{5.273}$$

Since a third-order polynomial contains in the general case four unknowns, $a_0, \ldots, a_3$, this approach allows via the four boundary conditions—two for the functional values and two for the slopes—to determine all unknowns. As an example, let us look at the first shape function for the displacement field. The boundary conditions for this case are as follows:

$$N_{1u}(0) = 1, \tag{5.274}$$

$$\frac{\mathrm{d}N_{1u}}{\mathrm{d}x}(0) = \frac{\mathrm{d}N_{1u}}{\mathrm{d}x}(L) = 0, \tag{5.275}$$

$$N_{1u}(L) = 0. \tag{5.276}$$

Evaluation of these boundary conditions based on the formulation (5.273) gives:

$$1 = a_0, \tag{5.277}$$

$$0 = a_1, \tag{5.278}$$

$$0 = a_0 + a_1 L + a_2 L^2 + a_3 L^3, \tag{5.279}$$

$$0 = a_1 + 2a_2 L + 3a_3 L^3, \tag{5.280}$$

or in matrix notation:

$$\begin{bmatrix} 1 \\ 0 \\ 0 \\ 0 \end{bmatrix} = \begin{bmatrix} 1 & 0 & 0 & 0 \\ 0 & 1 & 0 & 0 \\ 1 & L & L^2 & L^3 \\ 0 & 1 & 2L & 3L^3 \end{bmatrix} \begin{bmatrix} a_0 \\ a_1 \\ a_2 \\ a_3 \end{bmatrix} . \tag{5.281}$$

Solution of this system of equations for the unknowns gives $\boldsymbol{a} = \begin{bmatrix} 1 & 0 & -\frac{3}{L^2} & \frac{2}{L^3} \end{bmatrix}^{\mathrm{T}}$. Based on these constants, the formulation of the first shape function as given in Eq. (5.222) is obtained. Finally it should be noted here that the HERMITE interpolation considers in addition to the nodal value also the slope in the considered node. Thus, the displacements and the rotations are continuous at the nodes.

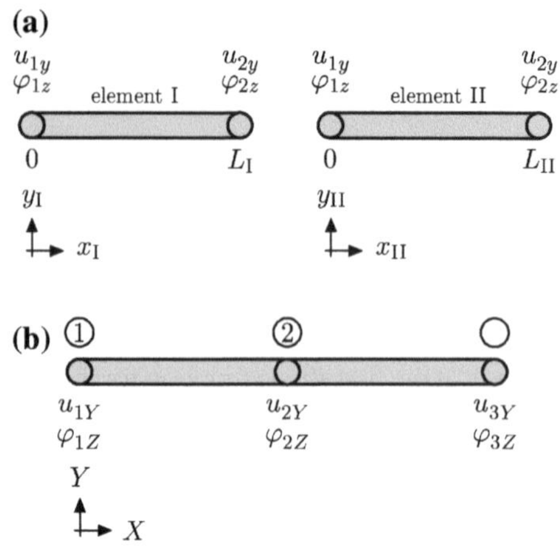


**Fig. 5.26** Relationship between **a** elemental and **b** global nodes and deformations in a horizontal beam structure. The nodes are symbolized by two *circles* at the ends (○)

## 5.3.3 *Assembly of Elements and Consideration of Boundary Conditions*

The assembly procedure of the single elemental finite element equations $\boldsymbol{K}^{\mathrm{e}}\boldsymbol{u}_{\mathrm{p}}^{\mathrm{e}} = \boldsymbol{f}^{\mathrm{e}}$ to a global system of equations, i.e. $\boldsymbol{K}\boldsymbol{u}_{\mathrm{p}} = \boldsymbol{f}$, where $\boldsymbol{K}$ is the global stiffness matrix, $\boldsymbol{u}_{\mathrm{p}}$ the global column matrix of unknowns and $\boldsymbol{f}$ the global column matrix of loads, can be treated as described in Sect. 5.2.3 for rod elements. Let us illustrate the process to assemble the global system of equations for BERNOULLI beam elements due to a two-element horizontal structure as shown in Fig. 5.26. As can be seen in Fig. 5.26a, each element has its own coordinate system $(x_i, y_i)$ with $i = \mathrm{I}, \mathrm{II}$ and its own nodal deformations $(u^i_{1y}, \varphi^i_{1z})$ and $(u^i_{2y}, \varphi^i_{2z})$. In order to assemble the single elements to a connected structure as shown in Fig. 5.26b, it is useful to introduce a global coordinate system $(X, Y)$ and global nodal deformations denoted by $(u_{iY}, \varphi_{iZ})$. Comparing the elemental and global nodal displacements shown in Fig. 5.26, the following mapping between the local and global displacements can be derived:

$$u_{1Y} = u^{\mathrm{I}}_{1y}, \tag{5.282}$$

$$u_{2Y} = u^{\mathrm{I}}_{2y} = u^{\mathrm{II}}_{1y}, \tag{5.283}$$

$$u_{3Y} = u^{\mathrm{II}}_{2y}, \tag{5.284}$$

and in the same way the relations between the rotations:

$$\varphi_{1Z} = \varphi^{\mathrm{I}}_{1z}, \tag{5.285}$$

$$\varphi_{2Z} = \varphi^{\mathrm{I}}_{2z} = \varphi^{\mathrm{II}}_{1z}, \tag{5.286}$$

$$\varphi_{3Z} = \varphi^{\mathrm{II}}_{2z}. \tag{5.287}$$

One possible way to assemble the elemental stiffness matrices to the global system will be illustrated in the following. In a first step, each single element is considered separately and its elemental stiffness matrix is written, for example, as given in Eq. (5.258). In addition, the corresponding *global* nodal deformations are written over the matrix and on the right-hand side which gives the following expressions:

$$\boldsymbol{K}_{\mathrm{I}}^{\mathrm{e}} = \frac{EI_Z}{L^3} \begin{array}{c} \begin{array}{cccc} u_{1Y} & \varphi_{1Z} & u_{2Y} & \varphi_{2Z} \end{array} \\ \begin{bmatrix} 12 & 6L & -12 & 6L \\ 6L & 4L^2 & -6L & 2L^2 \\ -12 & -6L & 12 & -6L \\ 6L & 2L^2 & -6L & 4L^2 \end{bmatrix} \end{array} \begin{array}{l} u_{1Y} \\ \varphi_{1Z}, \\ u_{2Y} \\ \varphi_{2Z} \end{array} \tag{5.288}$$

$$\boldsymbol{K}_{\mathrm{II}}^{\mathrm{e}} = \frac{EI_Z}{L^3} \begin{array}{c} \begin{array}{cccc} u_{2Y} & \varphi_{2Z} & u_{3Y} & \varphi_{3Z} \end{array} \\ \begin{bmatrix} 12 & 6L & -12 & 6L \\ 6L & 4L^2 & -6L & 2L^2 \\ -12 & -6L & 12 & -6L \\ 6L & 2L^2 & -6L & 4L^2 \end{bmatrix} \end{array} \begin{array}{l} u_{2Y} \\ \varphi_{2Z}. \\ u_{3Y} \\ \varphi_{3Z} \end{array} \tag{5.289}$$

By indicating the global unknowns in the described manner at each elemental stiffness matrix, it is easy to assign to each element in a matrix a unique index. For example, the upper right-hand element of the stiffness matrix $\boldsymbol{K}_{\mathrm{I}}^{\mathrm{e}}$ has the index[29] $(u_{1Y}, \varphi_{2Z})$ and the value $\frac{6EI_Z}{L^2}$. The next step consists of indicating the structure of the global stiffness matrix with its correct dimension. To this end, the total number of global unknowns must be determined. This is given here by the number of nodes times degrees of freedom per node. It should be noted here that the determination of the unknowns at this step of the procedure is without any consideration of boundary conditions. For the problem shown in Fig. 5.26a, the number of nodes is three and two unknowns per node. Thus, the dimension of the global stiffness matrix is given by (number global nodes × number global unknowns) or for our example as (3 × 2) and the structure can be written as a (6 × 6) matrix:

$$\boldsymbol{K} = \begin{array}{c} \begin{array}{cccccc} u_{1Y} & \varphi_{1Z} & u_{2Y} & \varphi_{2Z} & u_{3Y} & \varphi_{3Z} \end{array} \\ \left[\begin{array}{c|c|c|c|c|c} \quad & \quad & \quad & \quad & \quad & \quad \\ \hline & & & & & \\ \hline & & & & & \\ \hline & & & & & \\ \hline & & & & & \\ \hline & & & & & \end{array}\right] \end{array} \begin{array}{l} u_{1Y} \\ \varphi_{1Z} \\ u_{2Y}. \\ \varphi_{2Z} \\ u_{3Y} \\ \varphi_{3Z} \end{array} \tag{5.290}$$

It is now required to indicate the global unknowns over the empty global stiffness matrix and on its right-hand side. Any order can be chosen but it is common for the problem under consideration to start with $u_{1Y}$ and $\varphi_{1Z}$ and simply move to the next

[29] We follow here the convention where the first expression specifies the row and the second one the column: (row, column).

node. The scheme for this consecutive use of the global unknowns from the lowest to the highest number is drawn on the matrix in Eq. (5.290). Each cell of the global stiffness matrix has now its unique index expressed by the global unknowns and each element of the elemental stiffness matrix must be placed in the global matrix based on this unique index scheme. As an example, the upper right-hand element of the stiffness matrix $\boldsymbol{K}_{\mathrm{I}}^{\mathrm{e}}$ with the index $(u_{1Y}, \varphi_{2Z})$ must be placed in the global stiffness matrix in the first row and the fourth column. If each entry of the elemental stiffness matrices is inserted into the global matrix based on the described index scheme, the assembly of the global stiffness matrix is completed. As the elemental stiffness matrices, the global stiffness matrix is symmetric and commercial finite element codes store only half of the entries in order to reduce the requirements for data storage.

In order to complete the assembly of the global finite element equation, the global load vector $\boldsymbol{f}$ must be composed. Here, it is more advantageous to look from the beginning at the assembled structure and fill the external single loads $F_{iY}$ and moments $M_{iZ}$ which are acting at nodes in the proper order in the column matrix $\boldsymbol{f}$. A bit care must be taken if distributed loads were converted to equivalent nodal loads. For this case, components $f_i$ from both elements must be summed up at inner nodes:

$$\boldsymbol{f} = \begin{bmatrix} F_{1Y} \\ M_{1Z} \\ F_{2Y} \\ M_{2Z} \\ F_{3Y} \\ M_{3Z} \end{bmatrix} + \begin{bmatrix} f_{1\mathrm{Y,I}} \\ f_{1\mathrm{Z,I}} \\ f_{2\mathrm{Y,I}} + f_{2\mathrm{Y,II}} \\ f_{2\mathrm{Z,I}} + f_{2\mathrm{Z,II}} \\ f_{3\mathrm{Y,II}} \\ f_{3\mathrm{Z,II}} \end{bmatrix}. \tag{5.291}$$

The global system of equations for the problem shown in Fig. 5.26b is finally obtained as:

$$\frac{EI_Z}{L^3}\begin{bmatrix} 12 & 6L & -12 & 6L & 0 & 0 \\ 6L & 4L^2 & -6L & 2L^2 & 0 & 0 \\ -12 & -6L & 24 & 0 & -12 & 6L \\ 6L & 2L^2 & 0 & 8L^2 & -6L & 2L^2 \\ 0 & 0 & -12 & -6L & 12 & -6L \\ 0 & 0 & 6L & 2L^2 & -6L & 4L^2 \end{bmatrix}\begin{bmatrix} u_{1Y} \\ \varphi_{1Z} \\ u_{2Y} \\ \varphi_{2Z} \\ u_{3Y} \\ \varphi_{3Z} \end{bmatrix} = \begin{bmatrix} \cdots \\ \cdots \\ \cdots \\ \cdots \\ \cdots \\ \cdots \end{bmatrix}, \tag{5.292}$$

where the right-hand side is not specified since nothing on the loading is indicated in Fig. 5.26. This system of equations without consideration of any boundary conditions is called the non-reduced system. For this system, the global stiffness matrix $\boldsymbol{K}$ is still singular and cannot be inverted in order to solve the global system of equations. Boundary conditions must be introduced in oder to make this matrix regular und thus invertible.

For the beam elements under consideration, two types of boundary conditions must be distinguished. The DIRICHLET boundary condition specifies the displacement $u_Y$ or a rotation $\varphi_Z$ at a node while the NEUMANN boundary condition assigns a force

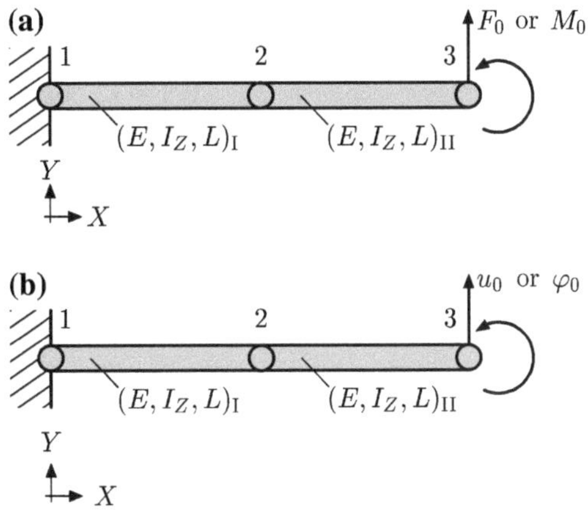


**Fig. 5.27** Consideration of boundary conditions for a cantilevered beam structure: **a** force or moment boundary condition; **b** displacement or rotation boundary condition at the right-hand boundary node. The nodes are symbolized by two *circles* at the ends (○)

$F_Y$ or a moment $M_Z$ at a node. The different ways to handle these different types of boundary conditions will be explained in the following based on the problem shown in Fig. 5.27 where a cantilevered beam structure has different boundary conditions at its right-hand end node.

The consideration of the homogeneous DIRICHLET boundary condition at the fixed support, i.e. $u_{1Y} = u(X = 0) = 0$ and $\varphi_{1Z} = \varphi(X = 0) = 0$, is the simplest case. To incorporate these boundary conditions in the system (5.135), the first two rows and columns can be canceled to obtain a reduced systems as:

$$\frac{EI_Z}{L^3}\begin{bmatrix} 24 & 0 & -12 & 6L \\ 0 & 8L^2 & -6L & 2L^2 \\ -12 & -6L & 12 & -6L \\ 6L & 2L^2 & -6L & 4L^2 \end{bmatrix}\begin{bmatrix} u_{2Y} \\ \varphi_{2Z} \\ u_{3Y} \\ \varphi_{3Z} \end{bmatrix} = \begin{bmatrix} \cdots \\ \cdots \\ \cdots \\ \cdots \end{bmatrix}. \tag{5.293}$$

In general we can state that a homogenous DIRICHLET boundary condition at node $n$ can be considered in the non-reduced system of equations by eliminating in the case of a displacement $u_{nY} = 0$ the $(2n - 1)$th row and $(2n - 1)$th column of the system. In the case of a rotation $\varphi_{nZ} = 0$, the $(2n)$th row and $(2n)$th column of the system can be eliminated. Let us consider next the case shown in Fig. 5.27a where the right-hand end node is subjected to an external load, i.e. a force $F_0$ or a moment $M_0$. This external force or moment can simply be specified on the right-hand side and since no other external forces are acting, the reduced system of equations is finally obtained as:

$$\frac{EI_Z}{L^3}\begin{bmatrix} 24 & 0 & -12 & 6L \\ 0 & 8L^2 & -6L & 2L^2 \\ -12 & -6L & 12 & -6L \\ 6L & 2L^2 & -6L & 4L^2 \end{bmatrix}\begin{bmatrix} u_{2Y} \\ \varphi_{2Z} \\ u_{3Y} \\ \varphi_{3Z} \end{bmatrix} = \begin{bmatrix} 0 \\ 0 \\ F_0 \\ 0 \end{bmatrix}, \tag{5.294}$$

or

$$\frac{EI_Z}{L^3}\begin{bmatrix} 24 & 0 & -12 & 6L \\ 0 & 8L^2 & -6L & 2L^2 \\ -12 & -6L & 12 & -6L \\ 6L & 2L^2 & -6L & 4L^2 \end{bmatrix}\begin{bmatrix} u_{2Y} \\ \varphi_{2Z} \\ u_{3Y} \\ \varphi_{3Z} \end{bmatrix} = \begin{bmatrix} 0 \\ 0 \\ 0 \\ M_0 \end{bmatrix}. \tag{5.295}$$

This system of equations can be solved, e.g. by inverting the reduced stiffness matrix and solving for the unknown nodal deformations in the form $\boldsymbol{u}_\mathrm{p} = \boldsymbol{K}^{-1}\boldsymbol{f}$:

$$\begin{bmatrix} u_{2Y} \\ \varphi_{2Z} \\ u_{3Y} \\ \varphi_{3Z} \end{bmatrix} = \frac{L^3 F_0}{EI_Z}\begin{bmatrix} \frac{5}{6} \\ \frac{3}{2L} \\ \frac{8}{3} \\ \frac{2}{L} \end{bmatrix}, \tag{5.296}$$

or

$$\begin{bmatrix} u_{2Y} \\ \varphi_{2Z} \\ u_{3Y} \\ \varphi_{3Z} \end{bmatrix} = \frac{L^2 M_0}{EI_Z}\begin{bmatrix} \frac{1}{2} \\ \frac{1}{L} \\ \frac{2}{1} \\ \frac{2}{L} \end{bmatrix}. \tag{5.297}$$

To incorporate a non-homogeneous DIRICHLET boundary condition $(u_Y \neq 0) \vee \varphi_Z \neq 0$ as shown in Fig. 5.27b, three different strategies can be mentioned, see Sect. 5.2.3. We will illustrate only the first approach and the interested reader may refer to the mentioned section for alternative strategies. The first one modifies the reduced system shown in Eqs. (5.294) or (5.295) in such a way that the boundary condition is directly introduced:

$$\frac{EI_Z}{L^3}\begin{bmatrix} 24 & 0 & -12 & 6L \\ 0 & 8L^2 & -6L & 2L^2 \\ 0 & 0 & 1 \times \frac{L^3}{EI_Z} & 0 \\ 6L & 2L^2 & -6L & 4L^2 \end{bmatrix}\begin{bmatrix} u_{2Y} \\ \varphi_{2Z} \\ u_{3Y} \\ \varphi_{3Z} \end{bmatrix} = \begin{bmatrix} 0 \\ 0 \\ u_0 \\ 0 \end{bmatrix}, \tag{5.298}$$

or

$$\frac{EI_Z}{L^3}\begin{bmatrix} 24 & 0 & -12 & 6L \\ 0 & 8L^2 & -6L & 2L^2 \\ -12 & -6L & 12 & -6L \\ 0 & 0 & 0 & 1 \times \frac{L^3}{EI_Z} \end{bmatrix}\begin{bmatrix} u_{2Y} \\ \varphi_{2Z} \\ u_{3Y} \\ \varphi_{3Z} \end{bmatrix} = \begin{bmatrix} 0 \\ 0 \\ 0 \\ \varphi_0 \end{bmatrix}, \tag{5.299}$$

where the third equation of (5.298) and the fourth equation of (5.299) gives immediately the boundary condition as $u_{3Y} = u_0$ or $\varphi_{3Z} = \varphi_0$. The solution of

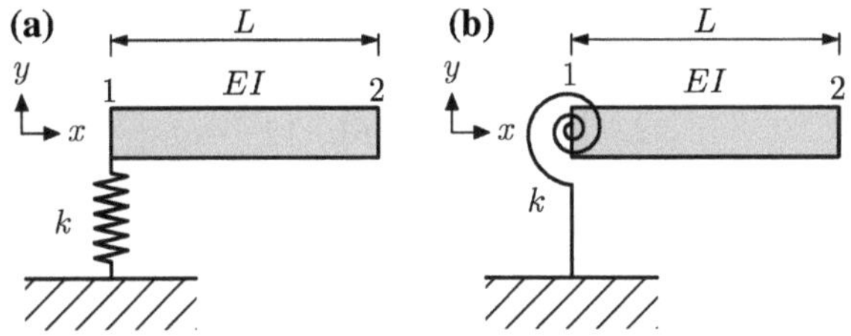


**Fig. 5.28** Consideration of springs in beam elements: **a** tension/compression spring; **b** torsion spring

the systems of equations given in Eqs. (5.298) and (5.299) can be obtained by inverting the coefficient matrices and multiplying them with the vectors on the right-hand sides as:

$$\begin{bmatrix} u_{2Y} \\ u_{3Z} \\ u_{2Y} \\ u_{3Z} \end{bmatrix} = u_0 \begin{bmatrix} \frac{5}{16} \\ \frac{9}{16L} \\ 1 \\ \frac{3}{4L} \end{bmatrix}, \quad \text{or} \quad \begin{bmatrix} u_{2Y} \\ u_{3Z} \\ u_{2Y} \\ u_{3Z} \end{bmatrix} = \varphi_0 \begin{bmatrix} \frac{L}{4} \\ \frac{1}{2} \\ \frac{L}{1} \\ 1 \end{bmatrix}. \tag{5.300}$$

In general we can state that a non-homogeneous DIRICHLET boundary condition at node $n$ can be introduced into the system of equations by the following modifications: In the case of a given displacement $u_0$: Modify the $(2n-1)$th line in such a way that at the position of the $(2n-1)$th column a '1' is obtained while all other entries of the $(2n-1)$th line are set to zero. On the right-hand side, the given displacement $u_0$ is introduced at the $(2n-1)$th position of the column matrix. In the case of a given rotation $\varphi_0$: Modify the $(2n)$th line in such a way that at the position of the $(2n)$th column a '1' is obtained while all other entries of the $(2n)$th line are set to zero. On the right-hand side, the given rotation $\varphi_0$ is introduced at the $(2n)$th position of the column matrix.

A special type of 'boundary condition' can be realized by attaching a spring to a beam element a shown in Fig. 5.28. The two different degrees of freedom at each node can be influenced by different types of springs: The translatory degree of freedom is addressed by a tension/compression spring (cf. Fig. 5.28a) while the rotatory degree of freedom is addressed by a torsion spring (cf. Fig. 5.28b).

Let us consider the configuration where the spring[30] is attached to node 1 as shown in Fig. 5.28. Then,[31] the elemental stiffness matrix in the case of the tension/compression spring is written as

[30] It is assumed here that the spring is in its unstrained state in the sketched configuration, i.e. without the application of any force or displacement boundary conditions at the nodes of the beam.

[31] See the comments in Sect. 5.2.3.

$$\boldsymbol{k}^{\mathrm{e}} = \frac{EI_z}{L^3}\begin{bmatrix} 12+\frac{L^3}{EI_z}k & 6L & -12 & 6L \\ 6L & 4L^2 & -6L & 2L^2 \\ -12 & -6L & 12 & -6L \\ 6L & 2L^2 & -6L & 4L^2 \end{bmatrix}, \tag{5.301}$$

or in the case of the torsion spring as:

$$\boldsymbol{k}^{\mathrm{e}} = \frac{EI_z}{L^3}\begin{bmatrix} 12 & 6L & -12 & 6L \\ 6L & 4L^2+\frac{L^3}{EI_z}k & -6L & 2L^2 \\ -12 & -6L & 12 & -6L \\ 6L & 2L^2 & -6L & 4L^2 \end{bmatrix}. \tag{5.302}$$

### *5.3.4 Post-computation: Determination of Stress, Strain and Further Quantities*

The previous section explained how to compose the global system of equations from which the primary unknowns, i.e. the nodal displacements and rotations, can be obtained. After the solution for the nodal unknowns, the distributions of these deformations and the curvature within an element can be obtained based on the equations provided in Table 5.10. These distributions allow then the calculation of strain and stress values at any location of the beam (cf. Table 4.4) as:

$$\varepsilon_x(x,y) = -y\frac{\mathrm{d}^2u_y(x)}{\mathrm{d}x^2} \quad \text{and} \quad \sigma_x(x,y) = E\varepsilon_x(x,y)\,. \tag{5.303}$$

### *5.3.5 Solved Problems*

**5.6 Example: Sample: Beam loaded by end force or moment—approximation through one single finite element**

Determine through one single finite element the displacement and the rotation of the right-hand end of the beam, which is illustrated in Fig. 5.29. Furthermore, determine the course of the bending line $u_y = u_y(x)$ and compare the finite element solution with the analytical solution.

**5.6 Solution**

(a) The finite element equation on element level according to Eq. (5.258) reduces for the illustrated load case to:

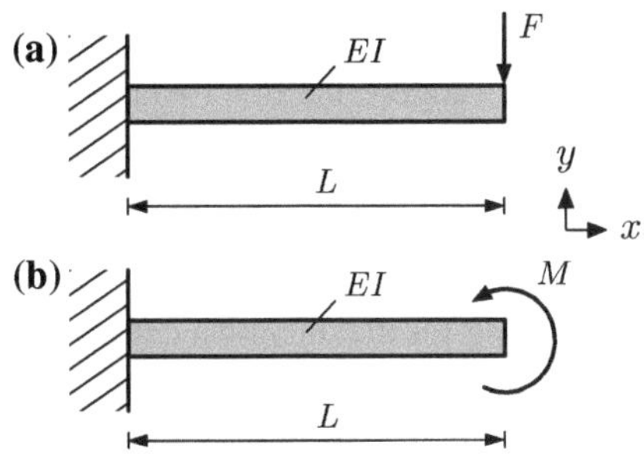


**Fig. 5.29** Sample problem beam with end load: **a** single force; **b** single moment

$$\frac{EI_z}{L^3}\begin{bmatrix} 12 & 6L & -12 & 6L \\ 6L & 4L^2 & -6L & 2L^2 \\ -12 & -6L & 12 & -6L \\ 6L & 2L^2 & -6L & 4L^2 \end{bmatrix}\begin{bmatrix} u_{1y} \\ \varphi_{1z} \\ u_{2y} \\ \varphi_{2z} \end{bmatrix} = \begin{bmatrix} 0 \\ 0 \\ -F \\ 0 \end{bmatrix}. \tag{5.304}$$

**Table 5.10** Displacement, rotation and curvature distribution for a BERNOULLI beam element given as being dependent on the nodal values as function of the physical coordinate $0 \leq x \leq L$ and natural coordinate $-1 \leq \xi \leq 1$

| Vertical displacement (Deflection) $u_y$ |
|---|
| $u_y(x) = \left[1 - 3\left(\frac{x}{L}\right)^2 + 2\left(\frac{x}{L}\right)^3\right] u_{1y} + \left[x - \frac{2x^2}{L} + \frac{x^3}{L^2}\right]\varphi_{1z} + \left[3\left(\frac{x}{L}\right)^2 - 2\left(\frac{x}{L}\right)^3\right] u_{2y} + \left[-\frac{x^2}{L} + \frac{x^3}{L^2}\right]\varphi_{2z}$ <br> $u_y(\xi) = \frac{1}{4}[2 - 3\xi + \xi^3]\, u_{1y} + \frac{1}{4}[1 - \xi - \xi^2 + \xi^3]\,\frac{L}{2}\varphi_{1z} + \frac{1}{4}[2 + 3\xi - \xi^3]\, u_{2y} + \frac{1}{4}[-1 - \xi + \xi^2 + \xi^3]\,\frac{L}{2}\varphi_{2z}$ |
| Rotation (Slope) $\varphi_z = \frac{\mathrm{d}u_y}{\mathrm{d}x} = \frac{2}{L}\frac{\mathrm{d}u_y}{\mathrm{d}\xi}$ |
| $\varphi_z(x) = \left[-\frac{6x}{L^2} + \frac{6x^2}{L^3}\right] u_{1y} + \left[1 - \frac{4x}{L} + \frac{3x^2}{L^2}\right]\varphi_{1z} + \left[\frac{6x}{L^2} - \frac{6x^2}{L^3}\right] u_{2y} + \left[-\frac{2x}{L} + \frac{3x^2}{L^2}\right]\varphi_{2z}$ <br> $\varphi_z(\xi) = \frac{1}{2L}[-3 + 3\xi^2]\, u_{1y} + \frac{1}{4}[-1 - 2\xi + 3\xi^2]\,\varphi_{1z} + \frac{1}{2L}[3 - 3\xi^2]\, u_{2y} + \frac{1}{4}[-1 + 2\xi + 3\xi^2]\,\varphi_{2z}$ |
| Curvature $\kappa_z = \frac{\mathrm{d}^2 u_y}{\mathrm{d}x^2} = \frac{4}{L^2}\frac{\mathrm{d}^2 u_y}{\mathrm{d}\xi^2}$ |
| $\kappa_z(x) = \left[-\frac{6}{L^2} + \frac{12x}{L^3}\right] u_{1y} + \left[-\frac{4}{L} + \frac{6x}{L^2}\right]\varphi_{1z} + \left[\frac{6}{L^2} - \frac{12x}{L^3}\right] u_{2y} + \left[-\frac{2}{L} + \frac{6x}{L^2}\right]\varphi_{2z}$ <br> $\kappa_z(\xi) = \frac{6}{L^2}[\,\xi\,]\, u_{1y} + \frac{1}{L}[-1 + 3\xi]\,\varphi_{1z} - \frac{6}{L^2}[\,\xi\,]\, u_{2y} + \frac{1}{L}[1 + 3\xi]\,\varphi_{2z}$ |

Since the displacement and the rotation are zero on the left-hand boundary due to the fixed support, the first two lines and columns of the system of equations can be eliminated:

$$\frac{EI_z}{L^3}\begin{bmatrix} 12 & -6L \\ -6L & 4L^2 \end{bmatrix}\begin{bmatrix} u_{2y} \\ \varphi_{2z} \end{bmatrix} = \begin{bmatrix} -F \\ 0 \end{bmatrix}. \tag{5.305}$$

Solving for the unknown deformations yields:

$$\begin{bmatrix} u_{2y} \\ \varphi_{2z} \end{bmatrix} = \frac{L^3}{EI_z}\begin{bmatrix} 12 & -6L \\ -6L & 4L^2 \end{bmatrix}^{-1}\begin{bmatrix} -F \\ 0 \end{bmatrix} \tag{5.306}$$

$$= \frac{L^3}{EI_z(48L^2 - 36L^2)}\begin{bmatrix} 4L^2 & 6L \\ 6L & 12 \end{bmatrix}\begin{bmatrix} -F \\ 0 \end{bmatrix} = \begin{bmatrix} -\frac{FL^3}{3EI_z} \\ -\frac{FL^2}{2EI_z} \end{bmatrix}. \tag{5.307}$$

According to Table 4.8, the analytical displacement results in:

$$u_y(x = L) = -\frac{F}{6EI_z}\left(3L^3 - L^3\right) = -\frac{FL^3}{3EI_z}. \tag{5.308}$$

The analytical solution for the rotation results from differentiation of the general displacement distribution according to Table 4.8 for $a = L$ to:

$$\varphi_z(x) = \frac{\mathrm{d}u_y(x)}{\mathrm{d}x} = -\frac{F}{6EI_z} \times \left[6Lx - 3x^2\right], \tag{5.309}$$

or alternatively on the right-hand boundary:

$$\varphi_z(x = L) = -\frac{F}{6EI_z} \times \left[6L^2 - 3L^2\right] = -\frac{FL^2}{2EI_z}. \tag{5.310}$$

The course of the bending line $u_y = u_y(x)$ results from Table 5.10 as:

$$u_y(x) = N_{2u}(x)u_{2y} + N_{2\varphi}(x)\varphi_{2z}$$

$$= \left[3\left(\frac{x}{L}\right)^2 - 2\left(\frac{x}{L}\right)^3\right]\left(-\frac{FL^3}{3EI_z}\right) + \left[-\frac{x^2}{L} + \frac{x^3}{L^2}\right]\left(-\frac{FL^2}{2EI_z}\right)$$

$$= \frac{F}{6EI_z}\left(x^3 - 3Lx^2\right). \tag{5.311}$$

According to Table 4.8, this course matches with the analytical solution.
Conclusion: Finite element solution and analytical solution are identical!

(b) The reduced system of equations in this case results in:

$$\frac{EI_z}{L^3}\begin{bmatrix} 12 & -6L \\ -6L & 4L^2 \end{bmatrix}\begin{bmatrix} u_{2y} \\ \varphi_{2z} \end{bmatrix} = \begin{bmatrix} 0 \\ M \end{bmatrix}. \tag{5.312}$$

Solving for the unknown deformations yields:

$$\begin{bmatrix} u_{2y} \\ \varphi_{2z} \end{bmatrix} = \frac{L^3}{12EI_zL^2}\begin{bmatrix} 4L^2 & 6L \\ 6L & 12 \end{bmatrix}\begin{bmatrix} 0 \\ M \end{bmatrix} = \begin{bmatrix} \frac{ML^2}{2EI_z} \\ \frac{ML}{EI_z} \end{bmatrix}. \tag{5.313}$$

The analytical solution according to Table 4.8 is

$$u_y(x = L) = -\frac{M}{2EI_z}\left(-L^2\right) = \frac{ML^2}{2EI_z}, \tag{5.314}$$

or alternatively the rotation in general for $a = L$:

$$\varphi_z(x) = \frac{\mathrm{d}u_y(x)}{\mathrm{d}x} = -\frac{M}{2EI_z}(-2x) \tag{5.315}$$

or only on the right-hand boundary:

$$\varphi_z(x = L) = -\frac{M}{2EI_z}(-L) = \frac{ML}{EI_z}. \tag{5.316}$$

The course of the bending line $u_y = u_y(x)$ results from Table 5.10 as:

$$\begin{aligned} u_y(x) &= N_{2u}(x)u_{2y} + N_{2\varphi}(x)\varphi_{2z} \\ &= \left[3\left(\frac{x}{L}\right)^2 - 2\left(\frac{x}{L}\right)^3\right]\left(\frac{ML^2}{2EI_z}\right) + \left[-\frac{x^2}{L} + \frac{x^3}{L^2}\right]\left(\frac{ML}{EI_z}\right) \\ &= \frac{Mx^2}{2EI_z}. \end{aligned} \tag{5.317}$$

According to Table 4.8, this course matches with the analytical solution. Conclusion: Finite element solution and analytical solution are identical!

**5.7 Example: Sample: Beam under constant distributed load—approximation through one single finite element**

Determine through one single finite element the displacement and the rotation (a) of the right-hand boundary and (b) in the middle for the beam under constant distributed load, which is illustrated in Fig. 5.30. Furthermore, determine the course of the bending line $u_y = u_y(x)$ and compare the finite element solution with the analytical solution.

## 5.7 Solution

To solve the problem, the constant distributed load has to be converted into equivalent nodal loads. These equivalent nodal loads can be extracted from Table 5.9 for the considered case, and the finite element equation results to:

$$\frac{EI_z}{L^3}\begin{bmatrix} 12 & 6L & -12 & 6L \\ 6L & 4L^2 & -6L & 2L^2 \\ -12 & -6L & 12 & -6L \\ 6L & 2L^2 & -6L & 4L^2 \end{bmatrix}\begin{bmatrix} u_{1y} \\ \varphi_{1z} \\ u_{2y} \\ \varphi_{2z} \end{bmatrix} = \begin{bmatrix} -\frac{qL}{2} \\ -\frac{qL^2}{12} \\ -\frac{qL}{2} \\ +\frac{qL^2}{12} \end{bmatrix}. \tag{5.318}$$

(a) Consideration of the boundary conditions shown in Fig. 5.30a, meaning the fixed support on the left-hand boundary, and solving for the unknowns yields:

$$\begin{bmatrix} u_{2y} \\ \varphi_{2z} \end{bmatrix} = \frac{L}{12EI_z}\begin{bmatrix} 4L^2 & 6L \\ 6L & 12 \end{bmatrix}\begin{bmatrix} -\frac{qL}{2} \\ +\frac{qL^2}{12} \end{bmatrix} = \begin{bmatrix} -\frac{qL^4}{8EI_z} \\ -\frac{qL^3}{6EI_z} \end{bmatrix}. \tag{5.319}$$

The analytical solution according to Table 4.8 yields

$$u_y(x = L) = -\frac{q}{24EI_z}\left(6L^4 - 4L^4 + L^4\right) = -\frac{qL^4}{8EI_z}, \tag{5.320}$$

or alternatively the rotation in general for $a_1 = 0$ and $a_2 = L$ to:

$$\varphi_z(x) = \frac{\mathrm{d}u_y(x)}{\mathrm{d}x} = -\frac{q}{24EI_z}\left(12L^2x - 12Lx^2 + 4x^3\right) \tag{5.321}$$

or only at the right-hand boundary:

$$\varphi_z(x = L) = -\frac{q}{24EI_z}\left(12L^3 - 12L^3 + 4L^3\right) = -\frac{qL^3}{6EI_z}. \tag{5.322}$$

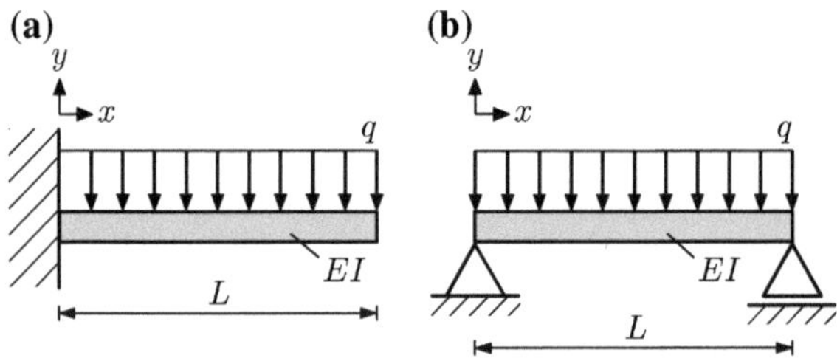


**Fig. 5.30** Sample problem beam under constant distributed load and different boundary conditions: **a** cantilevered and **b** simply supported beam

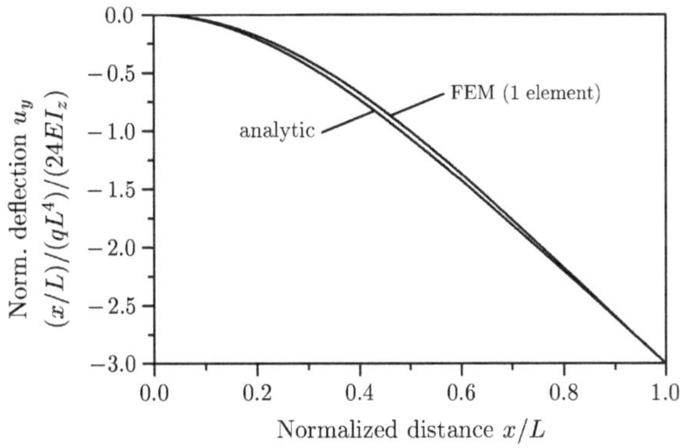


**Fig. 5.31** Comparison of the analytical and the finite element solution for the beam according to Fig. 5.30a

The course of the bending line $u_y = u_y(x)$ results from Table 5.10 as:

$$\begin{aligned} u_y(x) &= N_{2u}(x)u_{2y} + N_{2\varphi}(x)\varphi_{2z} \\ &= \left[3\left(\frac{x}{L}\right)^2 - 2\left(\frac{x}{L}\right)^3\right]\left(-\frac{qL^4}{8EI_z}\right) + \left[-\frac{x^2}{L} + \frac{x^3}{L^2}\right]\left(-\frac{qL^3}{6EI_z}\right) \\ &= -\frac{q}{24EI_z}\left(-2Lx^3 + 5L^2x^2\right), \end{aligned} \tag{5.323}$$

however the analytical course according to Table 4.8 results in $u_y(x) = -\frac{q}{24EI_z}$ $\left(x^4 - 4Lx^3 + 6L^2x^2\right)$, meaning the analytical and therefore the exact course is not identical with the numerical solution between the nodes $(0 < x < L)$, see Fig. 5.31. One can see that between the nodes a small difference between the two solutions arises. If a higher accuracy is demanded between those two nodes, the beam has to be divided into more elements.

Conclusion: Finite element solution and the analytical solution are only identical at the nodes!

(b) Consideration of the boundary conditions shown in Fig. 5.30b, meaning the simple support and the roller support, yields through the elimination of the first and third line and column of the system of equations (5.318):

$$\frac{EI_z}{L^3}\begin{bmatrix} 4L^2 & 2L^2 \\ 2L^2 & 4L^2 \end{bmatrix}\begin{bmatrix} \varphi_{1z} \\ \varphi_{2z} \end{bmatrix} = \begin{bmatrix} -\frac{qL^2}{12} \\ +\frac{qL^2}{12} \end{bmatrix}. \tag{5.324}$$

Solving for the unknowns yields:

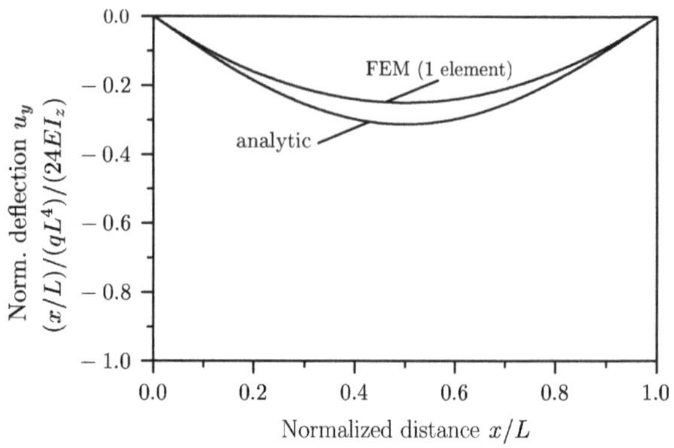


**Fig. 5.32** Comparison of the analytical and the finite element solution for the beam according to Fig. 5.30b

$$\begin{bmatrix} \varphi_{1z} \\ \varphi_{2z} \end{bmatrix} = \frac{1}{12EI_zL} \begin{bmatrix} 4L^2 & -2L^2 \\ -2L^2 & 4L^2 \end{bmatrix} \begin{bmatrix} -\frac{qL^2}{12} \\ +\frac{qL^2}{12} \end{bmatrix} = \begin{bmatrix} -\frac{qL^3}{24EI_z} \\ +\frac{qL^3}{24EI_z} \end{bmatrix}. \tag{5.325}$$

The course of the bending line $u_y = u_y(x)$ results from Table 5.10 as:

$$\begin{aligned} u_y(x) &= N_{1\varphi}(x)\varphi_{1z} + N_{2\varphi}(x)\varphi_{2z} \\ &= \left[x - 2\frac{x^2}{L} + \frac{x^3}{L^2}\right]\left(-\frac{qL^3}{24EI_z}\right) + \left[-\frac{x^2}{L} + \frac{x^3}{L^2}\right]\left(+\frac{qL^3}{24EI_z}\right) \\ &= -\frac{q}{24EI_z}\left(-L^2x^2 + L^3x\right), \end{aligned} \tag{5.326}$$

however the analytical course according to Table 4.8 results in $u_y(x) = -\frac{q}{24EI_z}$ $\left(x^4 - 2Lx^3 + L^3x\right)$, meaning the analytical and therefore exact course is also at this point not identical with the numerical solution between the nodes $(0 < x < L)$, see Fig. 5.32.
The numerical solution for the deflection in the middle of the beam yields $u_y(x = \frac{1}{2}L) = \frac{-4qL^4}{384EI_z}$, however the exact solution is $u_y(x = \frac{1}{2}L) = \frac{-5qL^4}{384EI_z}$.
Conclusion: Finite element solution and analytical solution are only identical at the nodes!

## 5.8 Example: Sample: Beam with variable cross section

The beam, which is shown in Fig. 5.33, has along the $x$-axis a variable cross section. Derive for

(a) a circular cross section,
(b) a square cross section

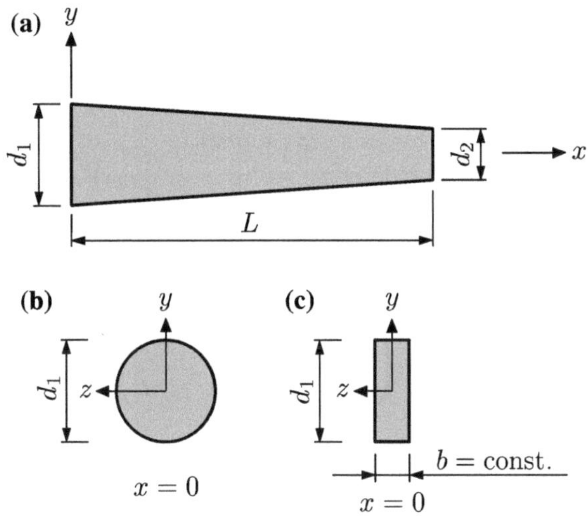


**Fig. 5.33** Sample problem beam with variable cross section: **a** change along the $x$-axis; **b** circular cross section; **c** square cross section

the element stiffness matrix for the case $d_1 = 2h$ and $d_2 = h$.

**5.8 Solution**

(a) Square cross section:

The following expression can be used as an initial point for the derivation of the stiffness matrix:

$$\boldsymbol{K}^{\mathrm{e}} = E \int_x I_z(x) \frac{\mathrm{d}^2 \boldsymbol{N}^{\mathrm{T}}(x)}{\mathrm{d}x^2} \frac{\mathrm{d}^2 \boldsymbol{N}(x)}{\mathrm{d}x^2} \mathrm{d}x. \tag{5.327}$$

Since the axial second moment of area changes along the $x$-axis, a corresponding function has to be derived at first. An elegant method would be to use the polar second moment of area of the circle, since in this case the function of the radius along the $x$-axis can be used. Hereby the relation, that the polar second moment of area composes additively of the two axial second moments of area $I_y$ and $I_z$, is being used:

$$I_{\mathrm{p}} = \int_A r^2 \mathrm{d}A = I_y + I_z. \tag{5.328}$$

Since the axial second moments of area of a circle are identical, the following expression can be derived for $I_z$:

$$I_z(x) = \frac{1}{2} I_{\mathrm{p}}(x) = \frac{1}{2} \int_A r^2 \mathrm{d}A = \frac{1}{2} \int_{\alpha=0}^{2\pi} \int_0^{r(x)} \hat{r}^2 \underbrace{\hat{r} \mathrm{d}\hat{r} \mathrm{d}\alpha}_{\mathrm{d}A} \tag{5.329}$$

$$= \pi \int_0^{r(x)} \hat{r}^3 \mathrm{d}\hat{r} = \pi \left[ \frac{1}{4} \hat{r}^4 \right]_0^{r(x)} = \frac{\pi}{4} r(x)^4. \tag{5.330}$$

The change of the radius along the $x$-axis can easily be derived from Fig. 5.33a:

$$r(x) = h \left( 1 - \frac{x}{2L} \right) = \frac{h}{2} \left( 2 - \frac{x}{L} \right). \tag{5.331}$$

Therefore, the axial second moment of area results in

$$I_z(x) = \frac{\pi h^4}{64} \left( 2 - \frac{x}{L} \right)^4 \tag{5.332}$$

and can be used in Eq. (5.327):

$$\boldsymbol{K}^{\mathrm{e}} = E \frac{\pi h^4}{64} \int_L \left( 2 - \frac{x}{L} \right)^4 \frac{\mathrm{d}^2 \boldsymbol{N}^{\mathrm{T}}(x)}{\mathrm{d}x^2} \frac{\mathrm{d}^2 \boldsymbol{N}(x)}{\mathrm{d}x^2} \mathrm{d}x. \tag{5.333}$$

The integration can be carried out through the second order derivatives of the shape function according to Eqs. (5.232) up to (5.235). As an example for the first component of the stiffness matrix

$$K_{11} = E \frac{\pi h^4}{64} \int_L \left( 2 - \frac{x}{L} \right)^4 \left( -\frac{6}{L^2} + \frac{12x}{L^3} \right)^2 \mathrm{d}x, \tag{5.334}$$

is being used and the entire stiffness matrix finally results after a short calculation:

$$\boldsymbol{K}^{\mathrm{e}}_{\text{circle}} = \frac{E}{L^3} \frac{\pi h^4}{64} \begin{bmatrix} \frac{2988}{35} & \frac{1998}{35} L & -\frac{2988}{35} & \frac{198}{7} L \\ \frac{1998}{35} L & \frac{1468}{35} L^2 & -\frac{1998}{35} L & \frac{106}{7} L^2 \\ -\frac{2988}{35} & -\frac{1998}{35} L & \frac{2988}{35} & -\frac{198}{7} L \\ \frac{198}{7} L & \frac{106}{7} L^2 & -\frac{198}{7} L & \frac{92}{7} L^2 \end{bmatrix}. \tag{5.335}$$

(b) Square cross section:

Regarding the square cross section, Eq. (5.327) serves as a basis as well. However, in this case it seems to be a good idea to go back to the definition of $I_z$:

$$I_z(x) = \int_A y^2 \mathrm{d}A = \int_{-y(x)}^{y(x)} \hat{y}^2 \underbrace{b\mathrm{d}\hat{y}}_{\mathrm{d}A} = b \left[ \frac{1}{3}\hat{y}^3 \right]_{-y(x)}^{y(x)} = \frac{2b}{3} y(x)^3 . \tag{5.336}$$

The course of the function $y(x)$ of the cross section is identical with the radius in part (a) meaning $y(x) = h(1 - \frac{x}{2L})$ and the second moment of area in this case results in:

$$I_z(x) = \frac{2bh^3}{3} \left( 1 - \frac{x}{2L} \right)^3 = \frac{bh^3}{12} \left( 2 - \frac{x}{L} \right)^3 . \tag{5.337}$$

Due to the special form of the second moment of area, the stiffness matrix therefore results in

$$\boldsymbol{K}^{\mathrm{e}} = E \frac{bh^3}{12} \int_L \left( 2 - \frac{x}{L} \right)^3 \frac{\mathrm{d}^2 \boldsymbol{N}^{\mathrm{T}}(x)}{\mathrm{d}x^2} \frac{\mathrm{d}^2 \boldsymbol{N}(x)}{\mathrm{d}x^2} \mathrm{d}x \tag{5.338}$$

or after the integration finally as:

$$\boldsymbol{K}^{\mathrm{e}}_{\text{square}} = \frac{E\,bh^3}{L^3\,12} \begin{bmatrix} \frac{243}{5} & \frac{156}{5}L & -\frac{243}{5} & \frac{87}{5}L \\ \frac{156}{5}L & \frac{114}{5}L^2 & -\frac{156}{5}L & \frac{42}{5}L^2 \\ -\frac{243}{5} & -\frac{156}{5}L & \frac{243}{5} & -\frac{87}{5}L \\ \frac{87}{5}L & \frac{42}{5}L^2 & -\frac{87}{5}L & 9L^2 \end{bmatrix} . \tag{5.339}$$

**5.9 Example: Advanced example: Beam element with nonlinear bending stiffness**

Derive the elemental stiffness matrix $\boldsymbol{K}^{\mathrm{e}}$ for a BERNOULLI beam element with nonlinear bending stiffness[32] $EI = EI(\kappa)$, cf. Fig. 5.34. Consider the case that the bending stiffness changes linearly with the curvature $\kappa$ as shown in Fig. 5.34b. The linear relationship of the bending stiffness should be defined by the two sampling points $EI(\kappa = 0) = I\!E_0$ and $EI(\kappa = \kappa_1) = I\!E_1$.

**5.9 Solution**

The relationship for the curvature-dependent bending stiffness can be derived based on the two given sampling points as:

$$EI(\kappa) = EI_0 - \frac{\kappa}{\kappa_1}(EI_0 - EI_1) = EI_0 \Bigg( 1 - \kappa \times \underbrace{\frac{1 - EI_1/EI_0}{\kappa_1}}_{\alpha_{01}} \Bigg)$$
$$= EI_0 \left( 1 - \kappa \times \alpha_{01} \right) . \tag{5.340}$$

[32] The product of YOUNG's modulus $E$ and moment of inertia $I$ is considered in the following as a single variable: $(EI) \rightarrow EI$.

It should be noted here that the constant factor $\alpha_{01}$ has the units of a length, e.g. meter, and the curvature $\kappa$ one divided by length.

The describing partial differential equation can be derived from Table 4.5 as

$$\frac{\mathrm{d}^2}{\mathrm{d}x^2}\left(EI(\kappa)\frac{\mathrm{d}^2 u_y}{\mathrm{d}x^2}\right) = 0. \tag{5.341}$$

Thus, the weighted residual statement (inner product) can be written as (cf. Eq. (5.210)):

$$\int_0^L W(x)\frac{\mathrm{d}^2}{\mathrm{d}x^2}\left(EI(\kappa)\frac{\mathrm{d}^2 u_y}{\mathrm{d}x^2}\right)\mathrm{d}x \stackrel{!}{=} 0. \tag{5.342}$$

Twice integrating by parts results in the following weak form of the problem:

$$\int_0^L \frac{\mathrm{d}^2 W}{\mathrm{d}x^2} EI(\kappa)\frac{\mathrm{d}^2 u_y}{\mathrm{d}x^2}\mathrm{d}x = \left[-W\frac{\mathrm{d}}{\mathrm{d}x}\left(EI(\kappa)\frac{\mathrm{d}^2 u_y}{\mathrm{d}x^2}\right) + \frac{\mathrm{d}W}{\mathrm{d}x} EI(\kappa)\frac{\mathrm{d}^2 u_y}{\mathrm{d}x^2}\right]_0^L. \tag{5.343}$$

The left-hand side of Eq. (5.343) represents the elemental stiffness matrix $\boldsymbol{K}^{\mathrm{e}}$. Introducing the nodal approach for the displacement, i.e. $u_y(x) = \boldsymbol{N}^{\mathrm{T}}(x)\boldsymbol{u}_{\mathrm{p}}$, and the weight function, i.e. $W(x) = \delta\boldsymbol{u}_{\mathrm{p}}^{\mathrm{T}}\boldsymbol{N}(x)$, the following statement for the stiffness matrix can be derived under consideration of Eq. (5.340):

$$\boldsymbol{K}^{\mathrm{e}} = \int_0^L \frac{\mathrm{d}^2\boldsymbol{N}}{\mathrm{d}x^2} \underbrace{EI_0\left(1 - \kappa \times \alpha_{01}\right)}_{\text{scalar}} \frac{\mathrm{d}^2\boldsymbol{N}^{\mathrm{T}}}{\mathrm{d}x^2}\mathrm{d}x. \tag{5.344}$$

Since the bending stiffness $EI(\kappa)$ is a scalar function, the last equation can be rearranged to obtain the following expression:

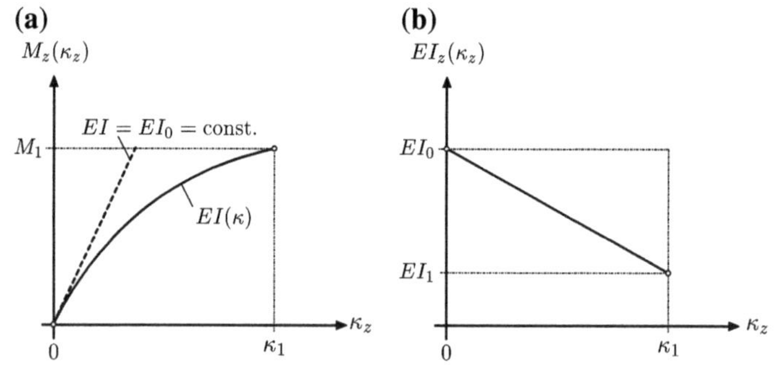


**Fig. 5.34** **a** Nonlinear moment-curvature diagram; **b** curvature-dependent bending stiffness

$$
\begin{aligned}
\boldsymbol{K}^{\mathrm{e}} &= \int_0^L EI_0\left(1-\kappa\times\alpha_{01}\right)\frac{\mathrm{d}^2\boldsymbol{N}}{\mathrm{d}x^2}\frac{\mathrm{d}^2\boldsymbol{N}^{\mathrm{T}}}{\mathrm{d}x^2}\mathrm{d}x \\
&= EI_0\int_0^L \frac{\mathrm{d}^2\boldsymbol{N}}{\mathrm{d}x^2}\frac{\mathrm{d}^2\boldsymbol{N}^{\mathrm{T}}}{\mathrm{d}x^2}\mathrm{d}x - EI_0\alpha_{01}\int_0^L \kappa\frac{\mathrm{d}^2\boldsymbol{N}}{\mathrm{d}x^2}\frac{\mathrm{d}^2\boldsymbol{N}^{\mathrm{T}}}{\mathrm{d}x^2}\mathrm{d}x,
\end{aligned}
\tag{5.345}
$$

where $\kappa$ is given by the expression (cf. Table 5.10):

$$
\kappa(x) = \frac{\mathrm{d}^2N_{1u}(x)}{\mathrm{d}x^2}u_{1y} + \frac{\mathrm{d}^2N_{1\varphi}(x)}{\mathrm{d}x^2}\varphi_{1z} + \frac{\mathrm{d}^2N_{2u}(x)}{\mathrm{d}x^2}u_{2y} + \frac{\mathrm{d}^2N_{2\varphi}(x)}{\mathrm{d}x^2}\varphi_{2z}. \tag{5.346}
$$

It should be noted here that the scalar function $\kappa(x)$ must be multiplied by each cell of the $4\times 4$ matrix $\frac{\mathrm{d}^2\boldsymbol{N}}{\mathrm{d}x^2}\frac{\mathrm{d}^2\boldsymbol{N}^{\mathrm{T}}}{\mathrm{d}x^2}$ in Eq. (5.345) and after performing the integration, the elemental stiffness matrix is obtained as shown in Eq. (5.347).

$$
\begin{aligned}
\boldsymbol{K}^{\mathrm{e}} &= \frac{(EI_z)_0}{L^3}\begin{bmatrix} 12 & 6L & -12 & 6L \\ 6L & 4L^2 & -6L & 2L^2 \\ -12 & -6L & 12 & -6L \\ 6L & 2L^2 & -6L & 4L^2 \end{bmatrix} - \frac{(EI_z)_0\times\alpha_{01}}{L^4} \\
&\times \left[\begin{matrix} -12(\varphi_{1z}-\varphi_{2z}) & -12(u_{1y}+\varphi_{1z}L-u_{2y}) \\ -12(u_{1y}+\varphi_{1z}L-u_{2y}) & -2(6u_{1y}+5\varphi_{1z}L+\varphi_{2z}L-6u_{2y})L \\ 12(\varphi_{1z}-\varphi_{2z}) & 12(u_{1y}+\varphi_{1z}L-u_{2y}) \\ 12(u_{1y}+\varphi_{2z}L-u_{2y}) & -2(\varphi_{1z}-\varphi_{2z})L^2 \end{matrix}\right. \\
&\qquad \left.\begin{matrix} 12(\varphi_{1z}-\varphi_{2z}) & 12(u_{1y}+\varphi_{2z}L-u_{2y}) \\ 12(u_{1y}+\varphi_{1z}L-u_{2y}) & -2(\varphi_{1z}-\varphi_{2z})L^2 \\ -12(\varphi_{1z}-\varphi_{2z}) & -12(u_{1y}+\varphi_{2z}L-u_{2y}) \\ -12(u_{1y}+\varphi_{2z}L-u_{2y}) & 2(6u_{1y}+\varphi_{1z}L+5\varphi_{2z}L-6u_{2y})L \end{matrix}\right]
\end{aligned}
\tag{5.347}
$$

It can be seen from Eq. (5.347) that the expression of the elemental stiffness matrix $\boldsymbol{K}^{\mathrm{e}}$ is composed of a part which is identical to the expression for a constant stiffness matrix and a second component which is dependent on the nodal unknowns. Because of this dependence on the nodal unknowns ($\boldsymbol{K}^{\mathrm{e}} = \boldsymbol{K}^{\mathrm{e}}(\boldsymbol{u}_{\mathrm{p}})$), the resulting system of equations is *nonlinear* and its solution requires the application of iteration techniques such as the NEWTON- RAPHSON[33,34] scheme.

[33] Isaac NEWTON (1642–1727), English physicist and mathematician.

[34] Joseph RAPHSON (ca. 1648–1715), English mathematician.

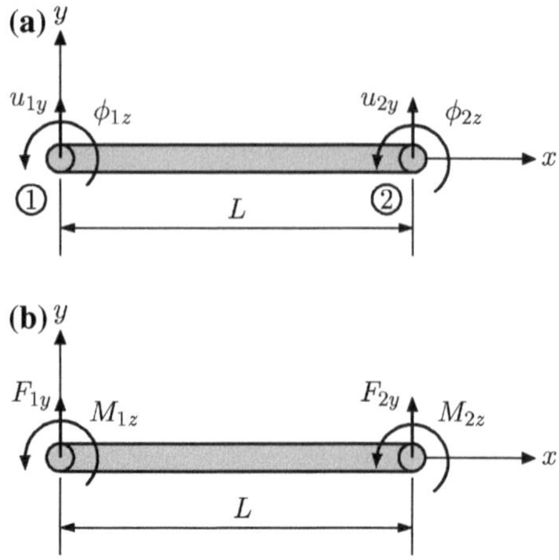


**Fig. 5.35** Definition of the TIMOSHENKO beam element for deformation in the $x$-$y$ plane: **a** deformations; **b** external loads. The nodes are symbolized by two *circles* at the end (○)

## 5.4 Timoshenko Beam Elements

### *5.4.1 Derivation of the Principal Finite Element Equation*

The TIMOSHENKO element is here defined as a prismatic body with the center line $x$ and the $y$-axis orthogonally to the center line. Nodes, at which displacements and rotations or alternatively forces and moments, as drafted in Fig. 5.35, are defined, will be introduced at both ends of the beam element. The deformation and load parameters are drafted in their positive orientation.

The two unknowns, meaning the deflection $u_y(x)$ and the herefrom independent cross-sectional rotation $\phi_z(x)$ are approximated with the help of the following nodal approaches:

$$u_y(x) = N_{1u}(x)u_{1y} + N_{2u}(x)u_{2y}, \tag{5.348}$$

$$\phi_z(x) = N_{1\phi}(x)\phi_{1z} + N_{2\phi}(x)\phi_{2z}, \tag{5.349}$$

or alternatively in matrix notation as

$$u_y(x) = \left[N_{1u}(x)\ 0\ N_{2u}(x)\ 0\right] \begin{bmatrix} u_{1y} \\ \phi_{1z} \\ u_{2y} \\ \phi_{2z} \end{bmatrix} = \boldsymbol{N}_u^{\mathrm{T}} \boldsymbol{u}_{\mathrm{p}}, \tag{5.350}$$

$$\phi_z(x) = \left[0\ N_{1\phi}\ 0\ N_{2\phi}(x)\right] \begin{bmatrix} u_{1y} \\ \phi_{1z} \\ u_{2y} \\ \phi_{2z} \end{bmatrix} = \boldsymbol{N}_\phi^{\mathrm{T}} \boldsymbol{u}_{\mathrm{p}}. \tag{5.351}$$

With these relations, the derivative of the cross-sectional rotation in the coupled differential equations (4.180) and (4.181) results in

$$\frac{\mathrm{d}\phi_z(x)}{\mathrm{d}x} = \frac{\mathrm{d}N_{1\phi}(x)}{\mathrm{d}x}\phi_{1z} + \frac{\mathrm{d}N_{2\phi}(x)}{\mathrm{d}x}\phi_{2z} = \frac{\mathrm{d}\boldsymbol{N}_\phi^{\mathrm{T}}}{\mathrm{d}x}\boldsymbol{u}_{\mathrm{p}}. \tag{5.352}$$

In the following, the shear differential equation (4.181) is first considered, which is multiplied by a deflection weight function $W_u(x)$ to obtain the following inner product:

$$\int_0^L \left\{ k_{\mathrm{s}}AG\left(\frac{\mathrm{d}^2 u_y}{\mathrm{d}x^2} - \frac{\mathrm{d}\phi_z}{\mathrm{d}x}\right) + q_y(x)\right\} W_u(x)\mathrm{d}x \stackrel{!}{=} 0. \tag{5.353}$$

Partial integrating of both expressions in the parentheses yields:

$$\int_0^L k_{\mathrm{s}}AG\frac{\mathrm{d}^2 u_y}{\mathrm{d}x^2}W_u\mathrm{d}x = \left[k_{\mathrm{s}}AG\frac{\mathrm{d}u_y}{\mathrm{d}x}W_u\right]_0^L - \int_0^L k_{\mathrm{s}}AG\frac{\mathrm{d}u_y}{\mathrm{d}x}\frac{\mathrm{d}W_u}{\mathrm{d}x}\mathrm{d}x, \tag{5.354}$$

$$-\int_0^L k_{\mathrm{s}}AG\frac{\mathrm{d}\phi_z}{\mathrm{d}x}W_u\mathrm{d}x = -\left[k_{\mathrm{s}}AG\phi_z W_u\right]_0^L + \int_0^L k_{\mathrm{s}}AG\phi_z\frac{\mathrm{d}W_u}{\mathrm{d}x}\mathrm{d}x. \tag{5.355}$$

Next, the bending differential equation (4.180) is multiplied by a rotation weight function $W_\phi(x)$ and is transformed in the inner product:

$$\int_0^L \left\{ \frac{\mathrm{d}}{\mathrm{d}x}\left(EI_z\frac{\mathrm{d}\phi_z}{\mathrm{d}x}\right) + k_{\mathrm{s}}AG\left(\frac{\mathrm{d}u_y}{\mathrm{d}x} - \phi_z\right)\right\} W_\phi(x)\mathrm{d}x \stackrel{!}{=} 0. \tag{5.356}$$

Partial integrating of the first expression yields

$$\int_0^L EI_z\frac{\mathrm{d}^2\phi_z}{\mathrm{d}x^2}W_\phi\mathrm{d}x = \left[EI_z\frac{\mathrm{d}\phi_z}{\mathrm{d}x}W_\phi\right]_0^L - \int_0^L EI_z\frac{\mathrm{d}\phi_z}{\mathrm{d}x}\frac{\mathrm{d}W_\phi}{\mathrm{d}x}\mathrm{d}x \tag{5.357}$$

and the bending differential equations results in:

$$\left[EI_z\frac{\mathrm{d}\phi_z}{\mathrm{d}x}W_\phi\right]_0^L - \int_0^L EI_z\frac{\mathrm{d}\phi_z}{\mathrm{d}x}\frac{\mathrm{d}W_\phi}{\mathrm{d}x}\mathrm{d}x + \int_0^L k_{\mathrm{s}}AG\left(\frac{\mathrm{d}u_y}{\mathrm{d}x} - \phi_z\right)W_\phi(x)\mathrm{d}x = 0. \tag{5.358}$$

Adding of the two converted differential equations yields

$$\left[k_s AG\frac{du_y}{dx}W_u\right]_0^L - \int_0^L k_s AG\frac{du_y}{dx}\frac{dW_u}{dx}dx - \left[k_s AG\phi_z W_u\right]_0^L$$
$$+\int_0^L k_s AG\phi_z\frac{dW_u}{dx}dx + \int_0^L q_y W_u dx + \int_0^L k_s AG\left(\frac{du_y}{dx}-\phi_z\right)W_\phi(x)dx$$
$$-\int_0^L EI_z\frac{d\phi_z}{dx}\frac{dW_\phi}{dx}dx + \left[EI_z\frac{d\phi_z}{dx}W_\phi\right]_0^L = 0, \tag{5.359}$$

or alternatively after a short conversion the weak form of the shear flexible bending beam:

$$\int_0^L EI_z\frac{d\phi_z}{dx}\frac{dW_\phi}{dx}dx + \int_0^L k_s AG\underbrace{\left(\frac{du_y}{dx}-\phi_z\right)}_{\gamma_{xy}}\underbrace{\left(\frac{dW_u}{dx}-W_\phi\right)}_{\delta\gamma_{xy}}dx$$
$$= \int_0^L q_y W_u dx + \left[k_s AG\left(\frac{du_y}{dx}-\phi_z\right)W_u\right]_0^L + \left[EI_z\frac{d\phi_z}{dx}W_\phi\right]_0^L. \tag{5.360}$$

One can see that the first part of the left-hand half represents the bending part and the second half the shear part. The right-hand side results from the external loads of the beam. In the following, the left-hand half of the weak form will be first considered to derive the stiffness matrix:

$$\int_0^L EI_z\frac{d\phi_z}{dx}\frac{dW_\phi}{dx}dx + \int_0^L k_s AG\left(\frac{dW_u}{dx}-W_\phi\right)\left(\frac{du_y}{dx}-\phi_z\right)dx. \tag{5.361}$$

In the next step, the approaches for the deflection and rotation of the nodes or alternatively their derivatives according to Eqs. (5.350) and (5.351), meaning

$$u_y(x) = \boldsymbol{N}_u^{\mathrm{T}}(x)\boldsymbol{u}_{\mathrm{p}}\ ,\quad \frac{du_y(x)}{dx} = \frac{d\boldsymbol{N}_u^{\mathrm{T}}(x)}{dx}\boldsymbol{u}_{\mathrm{p}}, \tag{5.362}$$
$$\phi_z(x) = \boldsymbol{N}_\phi^{\mathrm{T}}(x)\boldsymbol{u}_{\mathrm{p}}\ ,\quad \frac{d\phi_z(x)}{dx} = \frac{d\boldsymbol{N}_\phi^{\mathrm{T}}(x)}{dx}\boldsymbol{u}_{\mathrm{p}}, \tag{5.363}$$

have to be considered. The approaches for the weight functions are chosen analogous to the approaches for the unknowns:

$$W_u(x) = \delta \boldsymbol{u}_{\mathrm{p}}^{\mathrm{T}} \boldsymbol{N}_u(x), \tag{5.364}$$

$$W_\phi(x) = \delta \boldsymbol{u}_{\mathrm{p}}^{\mathrm{T}} \boldsymbol{N}_\phi(x), \tag{5.365}$$

or alternatively for the derivatives:

$$\frac{W_u(x)}{\mathrm{d}x} = \delta \boldsymbol{u}_{\mathrm{p}}^{\mathrm{T}} \frac{\boldsymbol{N}_u(x)}{\mathrm{d}x}, \tag{5.366}$$

$$\frac{W_\phi(x)}{\mathrm{d}x} = \delta \boldsymbol{u}_{\mathrm{p}}^{\mathrm{T}} \frac{\boldsymbol{N}_\phi(x)}{\mathrm{d}x}. \tag{5.367}$$

Therefore the left-hand half of Eq. (5.361)—under consideration that the rotation or alternatively the virtual rotation can be considered as constant respective to the integration—results in:

$$\delta \boldsymbol{u}_{\mathrm{p}}^{\mathrm{T}} \int_0^L EI_z \frac{\mathrm{d}\boldsymbol{N}_\phi}{\mathrm{d}x} \frac{\mathrm{d}\boldsymbol{N}_\phi^{\mathrm{T}}}{\mathrm{d}x} \mathrm{d}x \, \boldsymbol{u}_{\mathrm{p}} + \\ + \delta \boldsymbol{u}_{\mathrm{p}}^{\mathrm{T}} \int_0^L k_{\mathrm{s}} AG \left( \frac{\mathrm{d}\boldsymbol{N}_u}{\mathrm{d}x} - \boldsymbol{N}_\phi \right) \left( \frac{\mathrm{d}\boldsymbol{N}_u^{\mathrm{T}}}{\mathrm{d}x} - \boldsymbol{N}_\phi^{\mathrm{T}} \right) \mathrm{d}x \, \boldsymbol{u}_{\mathrm{p}}. \tag{5.368}$$

In the following, it remains to be seen that the virtual deformations $\delta \boldsymbol{u}^{\mathrm{T}}$ can be canceled with a corresponding expression on the right-hand side of Eq. (5.360). Therefore, on the left-hand side there remains

$$\underbrace{\int_0^L EI_z \frac{\mathrm{d}\boldsymbol{N}_\phi}{\mathrm{d}x} \frac{\mathrm{d}\boldsymbol{N}_\phi^{\mathrm{T}}}{\mathrm{d}x} \mathrm{d}x \, \boldsymbol{u}_{\mathrm{p}}}_{\boldsymbol{k}_{\mathrm{b}}^{\mathrm{e}}} + \underbrace{\int_0^L k_{\mathrm{s}} AG \left( \frac{\mathrm{d}\boldsymbol{N}_u}{\mathrm{d}x} - \boldsymbol{N}_\phi \right) \left( \frac{\mathrm{d}\boldsymbol{N}_u^{\mathrm{T}}}{\mathrm{d}x} - \boldsymbol{N}_\phi^{\mathrm{T}} \right) \mathrm{d}x \, \boldsymbol{u}_{\mathrm{p}}}_{\boldsymbol{k}_{\mathrm{s}}^{\mathrm{e}}} \tag{5.369}$$

and the bending or alternatively the shear stiffness matrix can be identified. The element stiffness matrices for constant bending stiffness $EI_z$ and constant shear stiffness $GA$ results herefrom in components to:

$$\boldsymbol{K}_{\mathrm{b}}^{\mathrm{e}} = EI_z \int_0^L \begin{bmatrix} 0 & 0 & 0 & 0 \\ 0 & \frac{\mathrm{d}N_{1\phi}}{\mathrm{d}x} \frac{\mathrm{d}N_{1\phi}}{\mathrm{d}x} & 0 & \frac{\mathrm{d}N_{1\phi}}{\mathrm{d}x} \frac{\mathrm{d}N_{2\phi}}{\mathrm{d}x} \\ 0 & 0 & 0 & 0 \\ 0 & \frac{\mathrm{d}N_{2\phi}}{\mathrm{d}x} \frac{\mathrm{d}N_{1\phi}}{\mathrm{d}x} & 0 & \frac{\mathrm{d}N_{2\phi}}{\mathrm{d}x} \frac{\mathrm{d}N_{2\phi}}{\mathrm{d}x} \end{bmatrix} \mathrm{d}x, \tag{5.370}$$

$$\boldsymbol{K}_{\mathrm{s}}^{\mathrm{e}} = k_{\mathrm{s}} GA \int_0^L \begin{bmatrix} \frac{\mathrm{d}N_{1u}}{\mathrm{d}x}\frac{\mathrm{d}N_{1u}}{\mathrm{d}x} & \frac{\mathrm{d}N_{1u}}{\mathrm{d}x}(-N_{1\phi}) & \frac{\mathrm{d}N_{1u}}{\mathrm{d}x}\frac{\mathrm{d}N_{2u}}{\mathrm{d}x} & \frac{\mathrm{d}N_{1u}}{\mathrm{d}x}(-N_{2\phi}) \\ (-N_{1\phi})\frac{\mathrm{d}N_{1u}}{\mathrm{d}x} & (-N_{1\phi})(-N_{1\phi}) & (-N_{1\phi})\frac{\mathrm{d}N_{2u}}{\mathrm{d}x} & (-N_{1\phi})(-N_{2\phi}) \\ \frac{\mathrm{d}N_{2u}}{\mathrm{d}x}\frac{\mathrm{d}N_{1u}}{\mathrm{d}x} & \frac{\mathrm{d}N_{2u}}{\mathrm{d}x}(-N_{1\phi}) & \frac{\mathrm{d}N_{2u}}{\mathrm{d}x}\frac{\mathrm{d}N_{2u}}{\mathrm{d}x} & \frac{\mathrm{d}N_{2u}}{\mathrm{d}x}(-N_{2\phi}) \\ (-N_{2\phi})\frac{\mathrm{d}N_{1u}}{\mathrm{d}x} & (-N_{2\phi})(-N_{1\phi}) & (-N_{2\phi})\frac{\mathrm{d}N_{2u}}{\mathrm{d}x} & (-N_{2\phi})(-N_{2\phi}) \end{bmatrix} \mathrm{d}x. \tag{5.371}$$

The two expressions for the bending and shear parts of the element stiffness matrix according to Eqs. (5.370) and (5.371) can be superposed for the principal finite element equation of the TIMOSHENKO beam on the element level

$$\boldsymbol{K}^{\mathrm{e}}\boldsymbol{u}_{\mathrm{p}} = \boldsymbol{f}^{\mathrm{e}}\,, \tag{5.372}$$

whereupon the total stiffness matrix according to Eq. (5.373) is given.

$$\underbrace{\begin{bmatrix} k_{\mathrm{s}}GA\int_0^L \frac{\mathrm{d}N_{1u}}{\mathrm{d}x}\frac{\mathrm{d}N_{1u}}{\mathrm{d}x}\mathrm{d}x & k_{\mathrm{s}}GA\int_0^L \frac{\mathrm{d}N_{1u}}{\mathrm{d}x}(-N_{1\phi})\,\mathrm{d}x & k_{\mathrm{s}}GA\int_0^L \frac{\mathrm{d}N_{1u}}{\mathrm{d}x}\frac{\mathrm{d}N_{2u}}{\mathrm{d}x}\mathrm{d}x & k_{\mathrm{s}}GA\int_0^L \frac{\mathrm{d}N_{1u}}{\mathrm{d}x}(-N_{2\phi})\,\mathrm{d}x \\ k_{\mathrm{s}}GA\int_0^L(-N_{1\phi})\frac{\mathrm{d}N_{1u}}{\mathrm{d}x}\mathrm{d}x & \begin{matrix} k_{\mathrm{s}}GA\int_0^L(-N_{1\phi})(-N_{1\phi})\,\mathrm{d}x \\ +EI_z\int_0^L \frac{\mathrm{d}N_{1\phi}}{\mathrm{d}x}\frac{\mathrm{d}N_{1\phi}}{\mathrm{d}x}\mathrm{d}x \end{matrix} & k_{\mathrm{s}}GA\int_0^L(-N_{1\phi})\frac{\mathrm{d}N_{2u}}{\mathrm{d}x}\mathrm{d}x & \begin{matrix} k_{\mathrm{s}}GA\int_0^L(-N_{1\phi})(-N_{2\phi})\,\mathrm{d}x \\ +EI_z\int_0^L \frac{\mathrm{d}N_{2\phi}}{\mathrm{d}x}\frac{\mathrm{d}N_{1\phi}}{\mathrm{d}x}\mathrm{d}x \end{matrix} \\ k_{\mathrm{s}}GA\int_0^L \frac{\mathrm{d}N_{2u}}{\mathrm{d}x}\frac{\mathrm{d}N_{1u}}{\mathrm{d}x}\mathrm{d}x & k_{\mathrm{s}}GA\int_0^L \frac{\mathrm{d}N_{2u}}{\mathrm{d}x}(-N_{1\phi})\,\mathrm{d}x & k_{\mathrm{s}}GA\int_0^L \frac{\mathrm{d}N_{2u}}{\mathrm{d}x}\frac{\mathrm{d}N_{2u}}{\mathrm{d}x}\mathrm{d}x & k_{\mathrm{s}}GA\int_0^L \frac{\mathrm{d}N_{2u}}{\mathrm{d}x}(-N_{2\phi})\,\mathrm{d}x \\ k_{\mathrm{s}}GA\int_0^L(-N_{2\phi})\frac{\mathrm{d}N_{1u}}{\mathrm{d}x}\mathrm{d}x & \begin{matrix} k_{\mathrm{s}}GA\int_0^L(-N_{2\phi})(-N_{1\phi})\,\mathrm{d}x \\ +EI_z\int_0^L \frac{\mathrm{d}N_{1\phi}}{\mathrm{d}x}\frac{\mathrm{d}N_{2\phi}}{\mathrm{d}x}\mathrm{d}x \end{matrix} & k_{\mathrm{s}}GA\int_0^L(-N_{2\phi})\frac{\mathrm{d}N_{2u}}{\mathrm{d}x}\mathrm{d}x & \begin{matrix} k_{\mathrm{s}}GA\int_0^L(-N_{2\phi})(-N_{2\phi})\,\mathrm{d}x \\ +EI_z\int_0^L \frac{\mathrm{d}N_{2\phi}}{\mathrm{d}x}\frac{\mathrm{d}N_{2\phi}}{\mathrm{d}x}\mathrm{d}x \end{matrix} \end{bmatrix}}_{\boldsymbol{K}^{\mathrm{e}}} \tag{5.373}$$

Any further evaluation of these stiffness matrices requires the introduction of the shape functions $N_i$.

Finally, the right-hand side of the weak form according to Eq. (5.360) is considered:

$$\int_0^L q_y W_u \mathrm{d}x + \left[k_{\mathrm{s}}AG\left(\frac{\mathrm{d}u_y}{\mathrm{d}x} - \phi_z\right)W_u\right]_0^L + \left[EI_z\frac{\mathrm{d}\phi_z}{\mathrm{d}x}W_\phi\right]_0^L. \tag{5.374}$$

Consideration of the relations for the shear force and the internal moment according to Eqs. (4.177) and (4.172) in the right-hand side of the weak form yields

$$\int_0^L q_y W_u \mathrm{d}x + \left[Q_y(x)W_u(x)\right]_0^L + \left[M_z(x)W_\phi(x)\right]_0^L, \tag{5.375}$$

or alternatively after the introduction of the approaches for the deflection and rotation of the nodes according to Eqs. (5.350) and (5.351):

$$\delta \boldsymbol{u}_{\mathrm{p}}^{\mathrm{T}} \int_{0}^{L} q_y \boldsymbol{N}_u \mathrm{d}x + \delta \boldsymbol{u}_{\mathrm{p}}^{\mathrm{T}} \left[Q_y(x) \boldsymbol{N}_u(x)\right]_0^L + \delta \boldsymbol{u}_{\mathrm{p}}^{\mathrm{T}} \left[M_z(x) \boldsymbol{N}_\phi(x)\right]_0^L . \tag{5.376}$$

$\delta \boldsymbol{u}_{\mathrm{p}}^{\mathrm{T}}$ can be eliminated with a corresponding expression in Eq. (5.368) and the following remains

$$\int_{0}^{L} q_y \boldsymbol{N}_u \mathrm{d}x + \left[Q_y(x) \boldsymbol{N}_u(x)\right]_0^L + \left[M_z(x) \boldsymbol{N}_\phi(x)\right]_0^L , \tag{5.377}$$

or alternatively in components:

$$\int_{0}^{L} q_y(x) \begin{bmatrix} N_{1u} \\ 0 \\ N_{2u} \\ 0 \end{bmatrix} \mathrm{d}x + \begin{bmatrix} -Q_y(0) \\ 0 \\ +Q_y(L) \\ 0 \end{bmatrix} + \begin{bmatrix} 0 \\ -M_z(0) \\ 0 \\ +M_z(L) \end{bmatrix} . \tag{5.378}$$

One notes that the general characteristics of the shape function have been used during the evaluation of the boundary integrals:

$$\text{1st row:} \quad Q_y(L) \underbrace{N_{1u}(L)}_{0} - Q_y(0) \underbrace{N_{1u}(0)}_{1} , \tag{5.379}$$

$$\text{2nd row:} \quad M_z(L) \underbrace{N_{1\phi}(L)}_{0} - M_z(0) \underbrace{N_{1\phi}(0)}_{1} , \tag{5.380}$$

$$\text{3rd row:} \quad Q_y(L) \underbrace{N_{2u}(L)}_{1} - Q_y(0) \underbrace{N_{2u}(0)}_{0} , \tag{5.381}$$

$$\text{4th row:} \quad M_z(L) \underbrace{N_{2\phi}(L)}_{1} - M_z(0) \underbrace{N_{2\phi}(0)}_{0} . \tag{5.382}$$

### 5.4.2 Linear Shape Functions for the Deflection and Displacement Field

Only the first derivatives of the shape functions appear in the element stiffness matrices $\boldsymbol{K}_{\mathrm{b}}^{\mathrm{e}}$ and $\boldsymbol{K}_{\mathrm{s}}^{\mathrm{e}}$ according to Eqs. (5.370) and (5.371). This demand on the differentiability of the shape functions leads to polynomials of minimum first order (linear functions) for the deflection and displacement field, so that in the approaches according to Eqs. (5.348) and (5.349) the following linear shape functions can be used:

$$N_{1u}(x) = N_{1\phi}(x) = 1 - \frac{x}{L}, \tag{5.383}$$

$$N_{2u}(x) = N_{2\phi}(x) = \frac{x}{L}. \tag{5.384}$$

The necessary derivatives result in:

$$\frac{\mathrm{d}N_{1u}}{\mathrm{d}x} = \frac{\mathrm{d}N_{1\phi}}{\mathrm{d}x} = -\frac{1}{L}, \tag{5.385}$$

$$\frac{\mathrm{d}N_{2u}}{\mathrm{d}x} = \frac{\mathrm{d}N_{2\phi}}{\mathrm{d}x} = \frac{1}{L}. \tag{5.386}$$

A graphical illustration of the shape function is given in Fig. 5.36. Additionally, the shape functions in the natural coordinate $\xi \in [-1, 1]$ are given. This formulation is more beneficial for the numerical integration of the stiffness matrices.

The integrals of the element stiffness matrices $\boldsymbol{K}_{\mathrm{b}}^{\mathrm{e}}$ and $\boldsymbol{K}_{\mathrm{s}}^{\mathrm{e}}$ according to Eqs. (5.370) and (5.371) are analytically calculated in the following. Using the linear approaches for the shape functions, the following results for the bending stiffness matrix:

$$\boldsymbol{K}_{\mathrm{b}}^{\mathrm{e}} = EI_z \int\limits_0^L \begin{bmatrix} 0 & 0 & 0 & 0 \\ 0 & \frac{1}{L^2} & 0 & -\frac{1}{L^2} \\ 0 & 0 & 0 & 0 \\ 0 & -\frac{1}{L^2} & 0 & \frac{1}{L^2} \end{bmatrix} \mathrm{d}x = EI_z \begin{bmatrix} 0 & 0 & 0 & 0 \\ 0 & \frac{x}{L^2} & 0 & -\frac{x}{L^2} \\ 0 & 0 & 0 & 0 \\ 0 & -\frac{x}{L^2} & 0 & \frac{x}{L^2} \end{bmatrix}_0^L , \tag{5.387}$$

or alternatively under consideration of the integration boundaries:

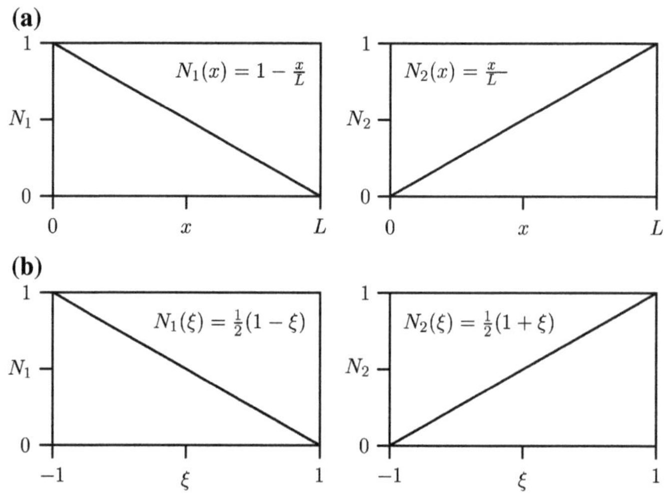


**Fig. 5.36** Linear shape functions $N_1 = N_{1u} = N_{1\phi}$ and $N_2 = N_{2u} = N_{2\phi}$ for the TIMOSHENKO element: **a** physical coordinate ($x$); **b** natural coordinate ($\xi$)

$$\boldsymbol{K}_{\mathrm{b}}^{\mathrm{e}} = EI_z \begin{bmatrix} 0 & 0 & 0 & 0 \\ 0 & \frac{1}{L} & 0 & -\frac{1}{L} \\ 0 & 0 & 0 & 0 \\ 0 & -\frac{1}{L} & 0 & \frac{1}{L} \end{bmatrix}. \tag{5.388}$$

Using the linear approaches for the shape functions, the following results for the shear stiffness matrix:

$$\boldsymbol{K}_{\mathrm{s}}^{\mathrm{e}} = k_{\mathrm{s}}AG \int\limits_0^L \begin{bmatrix} \frac{1}{L^2} & \left(1-\frac{x}{L}\right)\frac{1}{L} & -\frac{1}{L^2} & \frac{x}{L^2} \\ \left(1-\frac{x}{L}\right)\frac{1}{L} & \left(1-\frac{x}{L}\right)^2 & -\left(1-\frac{x}{L}\right)\frac{1}{L} & \left(1-\frac{x}{L}\right)\frac{x}{L} \\ -\frac{1}{L^2} & -\left(1-\frac{x}{L}\right)\frac{1}{L} & \frac{1}{L^2} & -\frac{x}{L^2} \\ \frac{x}{L^2} & \left(1-\frac{x}{L}\right)\frac{x}{L} & -\frac{x}{L^2} & \frac{x^2}{L^2} \end{bmatrix} \mathrm{d}x \tag{5.389}$$

$$= k_{\mathrm{s}}AG \begin{bmatrix} \frac{x}{L^2} & \frac{x(-2L+x)}{2L^2} & -\frac{x}{L^2} & \frac{x^2}{2L^2} \\ \frac{x(-2L+x)}{2L^2} & \frac{(-L+x)^3}{3L^2} & \frac{x(-2L+x)}{2L^2} & \frac{x^2(2x-3L)}{6L^2} \\ -\frac{x}{L^2} & \frac{x(-2L+x)}{2L^2} & \frac{x}{L^2} & -\frac{x^2}{2L^2} \\ \frac{x^2}{2L^2} & \frac{x^2(2x-3L)}{6L^2} & -\frac{x^2}{2L^2} & \frac{x^3}{3L^2} \end{bmatrix}_0^L \tag{5.390}$$

and finally after considering the constants of integration:

$$\boldsymbol{K}_{\mathrm{s}}^{\mathrm{e}} = k_{\mathrm{s}}AG \begin{bmatrix} +\frac{1}{L} & +\frac{1}{2} & -\frac{1}{L} & +\frac{1}{2} \\ +\frac{1}{2} & +\frac{L}{3} & -\frac{1}{2} & +\frac{L}{6} \\ -\frac{1}{L} & -\frac{1}{2} & +\frac{1}{L} & -\frac{1}{2} \\ +\frac{1}{2} & +\frac{L}{6} & -\frac{1}{2} & +\frac{L}{3} \end{bmatrix}. \tag{5.391}$$

The two stiffness matrices according to Eqs. (5.388) and (5.391) can be summarized additively to the total stiffness matrix of the TIMOSHENKO beam:

$$\boldsymbol{K}^{\mathrm{e}} = \begin{bmatrix} \frac{k_{\mathrm{s}}AG}{L} & \frac{k_{\mathrm{s}}AG}{2} & -\frac{k_{\mathrm{s}}AG}{L} & \frac{k_{\mathrm{s}}AG}{2} \\ \frac{k_{\mathrm{s}}AG}{2} & \frac{k_{\mathrm{s}}AGL}{3} + \frac{EI_z}{L} & -\frac{k_{\mathrm{s}}AG}{2} & \frac{k_{\mathrm{s}}AGL}{6} - \frac{EI_z}{L} \\ -\frac{k_{\mathrm{s}}AG}{L} & -\frac{k_{\mathrm{s}}AG}{2} & \frac{k_{\mathrm{s}}AG}{L} & -\frac{k_{\mathrm{s}}AG}{2} \\ \frac{k_{\mathrm{s}}AG}{2} & \frac{k_{\mathrm{s}}AGL}{6} - \frac{EI_z}{L} & -\frac{k_{\mathrm{s}}AG}{2} & \frac{k_{\mathrm{s}}AGL}{3} + \frac{EI_z}{L} \end{bmatrix}, \tag{5.392}$$

or alternatively via the abbreviation $\alpha = \frac{4EI_z}{k_{\mathrm{s}}AG}$

$$\boldsymbol{K}^{\mathrm{e}} = \frac{k_{\mathrm{s}}AG}{4L} \begin{bmatrix} 4 & 2L & -4 & 2L \\ 2L & \frac{4}{3}L^2 + \alpha & -2L & \frac{4}{6}L^2 - \alpha \\ -4 & -2L & 4 & -2L \\ 2L & \frac{4}{6}L^2 - \alpha & -2L & \frac{4}{3}L^2 + \alpha \end{bmatrix}, \tag{5.393}$$

or alternatively via the abbreviation $\Lambda = \frac{EI_z}{k_{\mathrm{s}}AGL^2}$

$$\boldsymbol{K}^{\mathrm{e}} = \frac{EI_z}{6\Lambda L^3} \begin{bmatrix} 6 & 3L & -6 & 3L \\ 3L & L^2(2+6\Lambda) & -3L & L^2(1-6\Lambda) \\ -6 & -3L & 6 & -3L \\ 3L & L^2(1-6\Lambda) & -3L & L^2(2+6\Lambda) \end{bmatrix}. \tag{5.394}$$

In the following, the deformation behavior of this analytically integrated[35] TIMOSHENKO element needs to be analyzed. For this, the configuration in Fig. 5.37 needs to be considered for which the beam has a fixed support on the left-hand side and a point load on the right-hand side. The displacement of the loading point has to be analyzed.

Through the stiffness matrix according to Eq. (5.393), the principal finite element equation for a single element results in

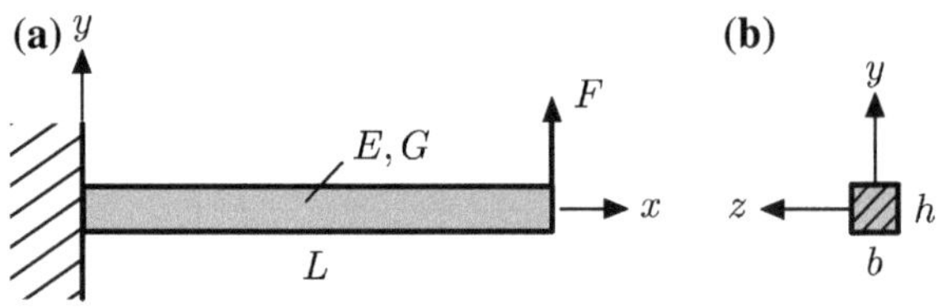


**Fig. 5.37** Analysis of a TIMOSHENKO element under point load

[35] A numerical GAUSS integration with two integration points yields the same results as the exact analytical integration.

$$\frac{k_s AG}{4L}\begin{bmatrix} 4 & 2L & -4 & 2L \\ 2L & \frac{4}{3}L^2+\alpha & -2L & \frac{4}{6}L^2-\alpha \\ -4 & -2L & 4 & -2L \\ 2L & \frac{4}{6}L^2-\alpha & -2L & \frac{4}{3}L^2+\alpha \end{bmatrix}\begin{bmatrix} u_{1y} \\ \phi_{1z} \\ u_{2y} \\ \phi_{2z} \end{bmatrix} = \begin{bmatrix} \dots \\ \dots \\ F \\ 0 \end{bmatrix}, \tag{5.395}$$

or alternatively after considering the fixed support ($u_{1y} = 0$, $\phi_{1z} = 0$) of the left-hand side:

$$\frac{k_s AG}{4L}\begin{bmatrix} 4 & -2L \\ -2L & \frac{4}{3}L^2+\alpha \end{bmatrix}\begin{bmatrix} u_{2y} \\ \phi_{2z} \end{bmatrix} = \begin{bmatrix} F \\ 0 \end{bmatrix}. \tag{5.396}$$

Solving this $2 \times 2$ system of equations for the unknown parameters on the right-hand boundary yields:

$$\begin{bmatrix} u_{2y} \\ \phi_{2z} \end{bmatrix} = \frac{4L}{k_s AG} \times \frac{1}{4(\frac{4}{3}L^2+\alpha) - (-2L)(-2L)}\begin{bmatrix} \frac{4}{3}L^2+\alpha & 2L \\ 2L & 4 \end{bmatrix}\begin{bmatrix} F \\ 0 \end{bmatrix}, \tag{5.397}$$

or alternatively solved for the unknown displacement at the right-hand boundary:

$$u_{2y}(L) = \frac{12EI_z + 4k_s AGL^2}{12EI_z + k_s AGL^2} \times \left(\frac{FL}{k_s AG}\right). \tag{5.398}$$

Considering the rectangular cross section, illustrated in Fig. 5.37, meaning $A = hb$ and $k_s = \frac{5}{6}$ and furthermore the relation for the shear modulus according to Eq. (4.170), after a short calculation the displacement on the right-hand end results:

$$u_{2y}(L) = \frac{12(1+\nu)\left(\frac{h}{L}\right)^2 + 20}{60 + 25\left(\frac{L}{h}\right)^2 \frac{1}{1+\nu}} \times \left(\frac{FL^3}{EI_z}\right). \tag{5.399}$$

For very compact beams, meaning $h \gg L$, $\frac{L}{h} \to 0$ results and Eq. (5.399) converges against the analytical solution.[36] For very slender beams, however, meaning $h \ll L$, a boundary value[37] of $\frac{4FL}{k_s AG}$ results from Eq. (5.398).

This boundary value only contains the shear part without bending and runs against a wrong solution. This phenomenon is called *shear locking*. A graphical illustration of this behavior is given in Fig. 5.38 via the normalized deflection with the BERNOULLI solution. One can clearly see the different convergence behaviors for different domains of the slenderness ratio, meaning for slender and compact beams.

For the improvement of the convergence behavior, the literature suggests [26, 88] to conduct the integration via numerical GAUSS integration with only one integration

---

[36] For this see Fig. 4.42 and the supplementary problem 4.25.

[37] One considers the definition of $I_z$ and $A$ in Eq. (5.398) and divides the fraction by $h^3$.

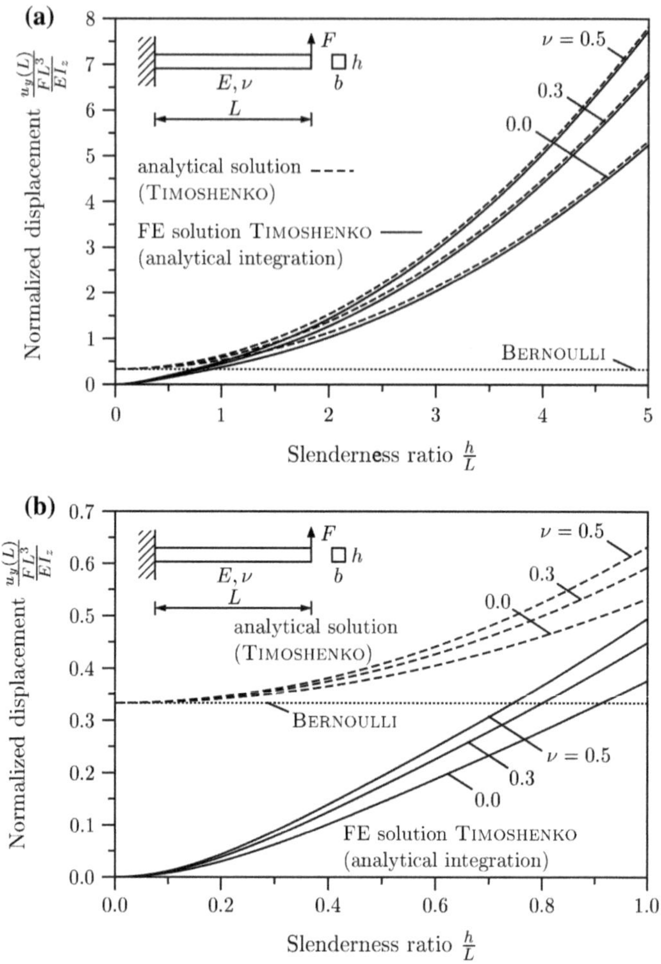


**Fig. 5.38** Comparison of the analytical solution for a TIMOSHENKO beam and the corresponding discretization via one single finite element with analytical integration of the stiffness matrix: **a** general view and **b** magnification for small slenderness ratios

point. Therefore, the arguments and the integration boundaries in the formulations of the element stiffness matrices for $\boldsymbol{K}_{\mathrm{b}}^{\mathrm{e}}$ and $\boldsymbol{K}_{\mathrm{s}}^{\mathrm{e}}$ according to Eqs. (5.370) and (5.371) have to be transformed into the natural coordinate $-1 \leq \xi \leq 1$. Furthermore, the shape functions need to be used according to Fig. 5.36. Via the transformation of the derivative to the new coordinate, meaning $\frac{\mathrm{d}N}{\mathrm{d}x} = \frac{\mathrm{d}N}{\mathrm{d}\xi}\frac{\mathrm{d}\xi}{\mathrm{d}x}$ and the transformation of the coordinate $\xi = -1 + 2\frac{x}{L}$ or alternatively $\mathrm{d}\xi = \frac{2}{L}\mathrm{d}x$, the bending stiffness matrix results in:

$$\boldsymbol{K}_{\text{b}}^{\text{e}} = EI_z \int_0^L \frac{4}{L^2} \begin{bmatrix} 0 & 0 & 0 & 0 \\ 0 & \frac{\mathrm{d}N_{1\phi}}{\mathrm{d}\xi}\frac{\mathrm{d}N_{1\phi}}{\mathrm{d}\xi} & 0 & \frac{\mathrm{d}N_{1\phi}}{\mathrm{d}\xi}\frac{\mathrm{d}N_{2\phi}}{\mathrm{d}\xi} \\ 0 & 0 & 0 & 0 \\ 0 & \frac{\mathrm{d}N_{2\phi}}{\mathrm{d}\xi}\frac{\mathrm{d}N_{1\phi}}{\mathrm{d}\xi} & 0 & \frac{\mathrm{d}N_{2\phi}}{\mathrm{d}\xi}\frac{\mathrm{d}N_{2\phi}}{\mathrm{d}\xi} \end{bmatrix} \frac{L}{2}\mathrm{d}\xi, \tag{5.400}$$

$$\boldsymbol{K}_{\text{b}}^{\text{e}} = \frac{2EI_z}{L} \int_0^L \frac{4}{L^2} \begin{bmatrix} 0 & 0 & 0 & 0 \\ 0 & \frac{1}{4} & 0 & -\frac{1}{4} \\ 0 & 0 & 0 & 0 \\ 0 & -\frac{1}{4} & 0 & \frac{1}{4} \end{bmatrix} \mathrm{d}\xi = \frac{EI_z}{2L} \sum_{i=1}^{1} \begin{bmatrix} 0 & 0 & 0 & 0 \\ 0 & 1 & 0 & -1 \\ 0 & 0 & 0 & 0 \\ 0 & -1 & 0 & 1 \end{bmatrix} \times 2 \tag{5.401}$$

and after all in the final formulation in:

$$\boldsymbol{K}_{\text{b}}^{\text{e}} = EI_z \begin{bmatrix} 0 & 0 & 0 & 0 \\ 0 & \frac{1}{L} & 0 & -\frac{1}{L} \\ 0 & 0 & 0 & 0 \\ 0 & -\frac{1}{L} & 0 & \frac{1}{L} \end{bmatrix}. \tag{5.402}$$

One can see that the same result for the bending stiffness matrix results as for the analytical integration. In the case of the bending stiffness matrix therefore the GAUSS integration with just one integration point is accurate.

The following expression results for the shear stiffness matrix under the use of the natural coordinate:

$$\frac{2k_{\text{s}}GA}{L} \int_0^L \begin{bmatrix} \frac{\mathrm{d}N_{1u}}{\mathrm{d}\xi}\frac{\mathrm{d}N_{1u}}{\mathrm{d}\xi} & \frac{L}{2}\frac{\mathrm{d}N_{1u}}{\mathrm{d}\xi}(-N_{1\phi}) & \frac{\mathrm{d}N_{1u}}{\mathrm{d}\xi}\frac{\mathrm{d}N_{2u}}{\mathrm{d}\xi} & \frac{L}{2}\frac{\mathrm{d}N_{1u}}{\mathrm{d}\xi}(-N_{2\phi}) \\ \frac{L}{2}(-N_{1\phi})\frac{\mathrm{d}N_{1u}}{\mathrm{d}\xi} & \frac{L^2}{4}(N_{1\phi})(N_{1\phi}) & \frac{L}{2}(-N_{1\phi})\frac{\mathrm{d}N_{2u}}{\mathrm{d}\xi} & \frac{L^2}{4}(N_{1\phi})(N_{2\phi}) \\ \frac{\mathrm{d}N_{2u}}{\mathrm{d}\xi}\frac{\mathrm{d}N_{1u}}{\mathrm{d}\xi} & \frac{L}{2}\frac{\mathrm{d}N_{2u}}{\mathrm{d}\xi}(-N_{1\phi}) & \frac{\mathrm{d}N_{2u}}{\mathrm{d}\xi}\frac{\mathrm{d}N_{2u}}{\mathrm{d}\xi} & \frac{L}{2}\frac{\mathrm{d}N_{2u}}{\mathrm{d}\xi}(-N_{2\phi}) \\ \frac{L}{2}(-N_{2\phi})\frac{\mathrm{d}N_{1u}}{\mathrm{d}\xi} & \frac{L^2}{4}(N_{2\phi})(N_{1\phi}) & \frac{L}{2}(-N_{2\phi})\frac{\mathrm{d}N_{2u}}{\mathrm{d}\xi} & \frac{L^2}{4}(N_{2\phi})(N_{2\phi}) \end{bmatrix} \mathrm{d}\xi, \tag{5.403}$$

or alternatively after the introduction of the shape functions

$$\frac{2k_{\text{s}}GA}{L} \int_0^L \begin{bmatrix} \frac{1}{4} & \frac{L}{2}\left(\frac{1}{4}-\frac{x}{4}\right) & -\frac{1}{4} & \frac{L}{2}\left(\frac{1}{4}+\frac{x}{4}\right) \\ \frac{L}{2}\left(\frac{1}{4}-\frac{x}{4}\right) & \frac{L^2}{4}\left(\frac{(-1+x)^2}{4}\right) & \frac{L}{2}\left(-\frac{1}{4}+\frac{x}{4}\right) & \frac{L^2}{4}\left(\frac{1}{4}-\frac{x^2}{4}\right) \\ -\frac{1}{4} & \frac{L}{2}\left(-\frac{1}{4}+\frac{x}{4}\right) & \frac{1}{4} & \frac{L}{2}\left(-\frac{1}{4}-\frac{x}{4}\right) \\ \frac{L}{2}\left(\frac{1}{4}+\frac{x}{4}\right) & \frac{L^2}{4}\left(\frac{1}{4}-\frac{x^2}{4}\right) & \frac{L}{2}\left(-\frac{1}{4}-\frac{x}{4}\right) & \frac{L^2}{4}\left(\frac{(1+x)^2}{4}\right) \end{bmatrix} \mathrm{d}\xi \tag{5.404}$$

or after the transition to the numerical integration

$$\frac{2k_sGA}{L}\begin{bmatrix} \frac{1}{4} & \frac{L}{2}\frac{1}{4} & -\frac{1}{4} & \frac{L}{2}\frac{1}{4} \\ \frac{L}{2}\frac{1}{4} & \frac{L^2}{4}\frac{1}{4} & \frac{L}{2}\left(-\frac{1}{4}\right) & \frac{L^2}{4}\frac{1}{4} \\ -\frac{1}{4} & \frac{L}{2}\left(-\frac{1}{4}\right) & \frac{1}{4} & \frac{L}{2}\left(-\frac{1}{4}\right) \\ \frac{L}{2}\frac{1}{4} & \frac{L^2}{4}\frac{1}{4} & \frac{L}{2}\left(-\frac{1}{4}\right) & \frac{L^2}{4}\frac{1}{4} \end{bmatrix}_{\xi_i=0} \times 2 \tag{5.405}$$

and after all in the final formulation as:

$$\boldsymbol{K}_s^e = k_sAG\begin{bmatrix} \frac{1}{L} & \frac{1}{2} & -\frac{1}{L} & \frac{1}{2} \\ \frac{1}{2} & \frac{L}{4} & -\frac{1}{2} & \frac{L}{4} \\ -\frac{1}{L} & -\frac{1}{2} & \frac{1}{L} & -\frac{1}{2} \\ \frac{1}{2} & \frac{L}{4} & -\frac{1}{2} & \frac{L}{4} \end{bmatrix}. \tag{5.406}$$

The two stiffness matrices according to Eqs. (5.402) and (5.406) can be summarized additively to the total stiffness matrix of the TIMOSHENKO beam and with the abbreviation $\alpha = \frac{4EI_z}{k_sAG}$ the following results:

$$\boldsymbol{K}^e = \frac{k_sAG}{4L}\begin{bmatrix} 4 & 2L & -4 & 2L \\ 2L & L^2+\alpha & -2L & L^2-\alpha \\ -4 & -2L & 4 & -2L \\ 2L & L^2-\alpha & -2L & L^2+\alpha \end{bmatrix}, \tag{5.407}$$

or alternatively via the abbreviation $\Lambda = \dfrac{EI_z}{k_sAGL^2}$:

$$\boldsymbol{K}^e = \frac{EI_z}{6\Lambda L^3}\begin{bmatrix} 6 & 3L & -6 & 3L \\ 3L & L^2(1,5+6\Lambda) & -3L & L^2(1,5-6\Lambda) \\ -6 & -3L & 6 & -3L \\ 3L & L^2(1,5-6\Lambda) & -3L & L^2(2+6\Lambda) \end{bmatrix}. \tag{5.408}$$

With the help of this formulation for the stiffness matrix the example according to Fig. 5.37 needs to be analyzed once again in the following to investigate the differences to the analytical integration. Via the stiffness matrix according to Eq. (5.407) the principal finite element equation for a single element under consideration of the fixed support ($u_{1y} = 0$, $\phi_{1z} = 0$) on the left-hand side results in:

$$\frac{k_sAG}{4L}\begin{bmatrix} 4 & -2L \\ -2L & L^2+\alpha \end{bmatrix}\begin{bmatrix} u_{2y} \\ \phi_{2z} \end{bmatrix} = \begin{bmatrix} F \\ 0 \end{bmatrix}. \tag{5.409}$$

Solving of this $2 \times 2$ system of equations for the unknown displacement on the right-hand side yields:

$$u_{2y}(L) = \left(1 + \frac{4EI_z}{k_s AGL^2}\right) \times \frac{FL^3}{4EI_z}. \tag{5.410}$$

If the illustrated rectangular cross section in Fig. 5.37 is being considered at this point as well, after a short calculation the displacement on the right-hand side, via $A = hb$, $k_s = \frac{5}{6}$ and the relation for the shear modulus according to Eq. (4.170) results in:

$$u_{2y}(L) = \left(\frac{1}{4} + \frac{1}{5}(1+\nu)\left(\frac{h}{L}\right)^2\right) \times \left(\frac{FL^3}{EI_z}\right). \tag{5.411}$$

For very compact beams, meaning $h \gg L$, the solution converges against the analytical solution.[38] For very slender beams, however, meaning $h \ll L$, a boundary value of $\frac{FL^3}{4EI_z}$ results from Eq. (5.411), whereupon the analytical solution yields a value of $\frac{FL^3}{3EI_z}$. However, the phenomenon of shear locking does not occur and therefore, compared to the stiffness matrix based on the analytical integration, an improvement of the element formulation has been achieved.

A graphical illustration of this behavior via the normalized deflection is given in Fig. 5.39. One can clearly see the improved convergence behavior for small slenderness ratios. For larger slenderness ratios the behavior remains according to the result of the analytical integration, since both approaches converge against the analytical solution.

When the differential equations according to (4.180) and (4.181) are considered, it becomes obvious that the derivative $\frac{du_y}{dx}$ and the function $\phi_z$ itself are contained there. If linear shape functions are being used for $u_y$ and $\phi_z$, the degree for polynomials for $\frac{du_y}{dx}$ and $\phi_z$ is different. In the limiting case of slender beams, however, the relation $\phi \approx \frac{du_y}{dx}$ has to be fulfilled and the consistency of the polynomials for $\frac{du_y}{dx}$ and $\phi_z$ is of importance. The linear approach for $u_y$ yields for $\frac{du_y}{dx}$ a constant function and therefore also for $\phi_z$ a constant would be desirable. However, it needs to be considered at this point that the demand for the differentiability of $\phi_z$ at least results in a linear function. The one-point integration[39] in the case of the shear stiffness matrix with the expressions $N_{i\phi}N_{j\phi}$ causes however that the linear approach for $\phi_z$ is treated as a constant term, since two integration points would have to be used for an exact integration. A one-point integration can at most integrate a polynomial of first order exactly, meaning proportional to $x^1$, and therefore the following point of view results $(N_{i\phi}N_{j\phi}) \sim x^1$. However, this means that at most $N_{i\phi} \sim x^{0.5}$ or alternatively $N_{j\phi} \sim x^{0.5}$ holds. Since the polynomial approach solely allows integer values for the exponent of $x$, $N_{i\phi} \sim x^0$ or alternatively $N_{j\phi} \sim x^0$ results and the rotation needs to be seen as a constant term. This is consistent with the demand that the shear strain

[38] For this see Fig. 4.42 and the supplementary problem 4.25.

[39] The numerical integration according to the GAUSS-LEGENDRE method with $n$ integration points integrates a polynomial, which degree is at most $2n-1$, exactly.

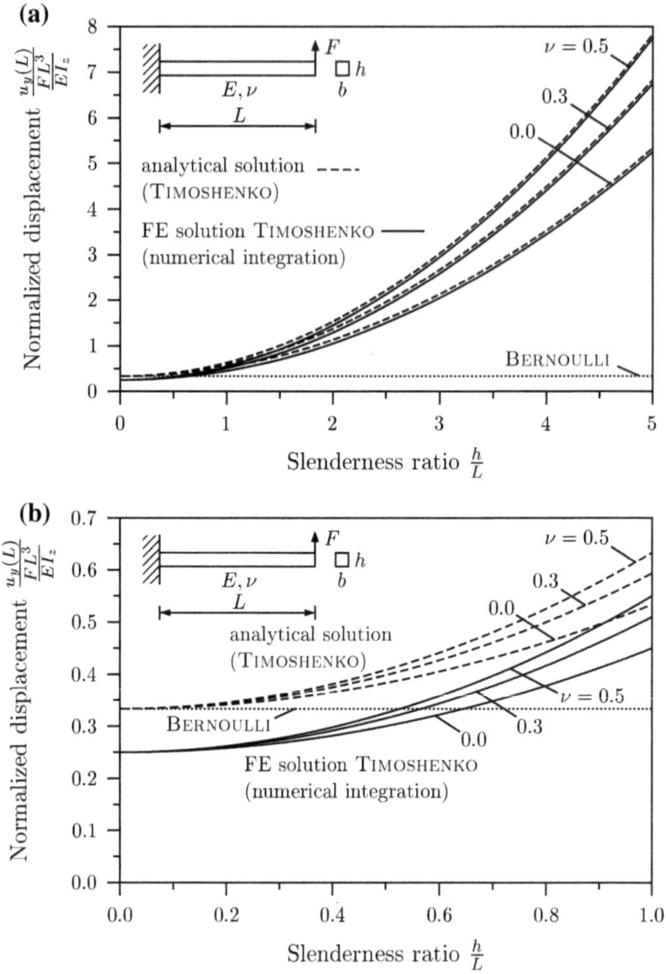


**Fig. 5.39** Comparison of the analytical solution for a TIMOSHENKO beam and the appropriate discretization via one single finite element with one-point numerical integration of the corresponding matrix: **a** general view and **b** magnification for small slenderness ratios

$\gamma_{xy} = \frac{\mathrm{d}u_y}{\mathrm{d}x} - \phi_z$ has to be constant in an element for constant bending stiffness $EI_z$. Therefore, *shear locking* does not occur in this case.

As another option for the improvement of the convergence behavior of linear TIMOSHENKO elements with numerical one-point integrations, [26, 67] suggests to correct the shear stiffness $k_sAG$ according to the analytical correct solution.[40] To this end, the elastic strain energy is being regarded (see [75]), which results as follows:

[40] MACNEAL herefore uses the expression 'residual bending flexibility' [90, 66].

$$\Pi_{\text{int}} = \frac{1}{2}\int_0^L EI_z \left(\frac{\mathrm{d}\phi_z(x)}{\mathrm{d}x}\right)^2 \mathrm{d}x + \frac{1}{2}\int_0^L k_s AG \left(\frac{\mathrm{d}u_y(x)}{\mathrm{d}x} - \phi_z(x)\right)^2 \mathrm{d}x. \tag{5.412}$$

It is now demanded that the strain energy for the analytical solution and the finite element solution under the use of the corrected shear stiffness $(k_s AG)^*$ are identical. The analytical solution[41] for the problem in Fig. 5.37 results in

$$u_y(x) = \frac{1}{EI_z}\left(-F\frac{x^3}{6} + FL\frac{x^2}{2} + \frac{EI_z F}{k_s AG}x\right), \tag{5.413}$$

$$\phi_z(x) = \frac{1}{EI_z}\left(-F\frac{x^2}{2} + FLx\right), \tag{5.414}$$

and the elastic strain energy for the analytical solution therefore results in:

$$\Pi_{\text{int}} = \frac{F^2}{2EI_z}\int_0^L (L-x)^2 \mathrm{d}x + \frac{F^2 (EI_z)^2}{2k_s AG}\int_0^L \mathrm{d}x = \frac{F^2 L^3}{6EI_z} + \frac{F^2 L}{2k_s AG}. \tag{5.415}$$

Via Eq. (5.409) the finite element solution of the elastic strain energy results in:

$$\begin{aligned}\Pi_{\text{int}} =& \frac{EI_z}{2}\int_0^L \left(\frac{FL}{2EI_z}\right)^2 \mathrm{d}x \\ &+ \frac{(k_s AG)^*}{2}\int_0^L \left(\left(1 + \frac{4EI_z}{(k_s AG)^* L^2}\right)\frac{FL^2}{4EI_z} - \frac{FLx}{2EI_z}\right)^2 \mathrm{d}x.\end{aligned} \tag{5.416}$$

This integral has to be evaluated numerically with a one-point integration rule and it is therefore necessary to introduce the natural coordinate via the transformation $x = \frac{L}{2}(\xi + 1)$:

$$\begin{aligned}\Pi_{\text{int}} =& \frac{F^2 L^3}{8EI_z} \\ &+ \frac{(k_s AG)^*}{2}\int_0^L \left(\left(1 + \frac{4EI_z}{(k_s AG)^* L^2}\right)\frac{FL^2}{4EI_z} - \frac{FL}{2EI_z}(\xi+1)\frac{L}{2}\right)^2 \frac{L}{2}\mathrm{d}\xi \\ =& \frac{F^2 L^3}{8EI_z} + \frac{(k_s AG)^*}{2}\left(\frac{4EI_z}{(k_s AG)^* L^2} \times \frac{FL^2}{4EI_z}\right)^2 \frac{L}{2} 2\end{aligned} \tag{5.417}$$

[41] For this see the supplementary problem 4.24.

and finally

$$\Pi_{\text{int}} = \frac{F^2 L^3}{8EI_z} + \frac{F^2 L}{2(k_s AG)^*}. \tag{5.418}$$

Equalizing of the two energy expressions according to Eqs. (5.415) and (5.418) finally yields the corrected shear stiffness:

$$(k_s AG)^* = \left( \frac{L^2}{12EI_z} + \frac{1}{k_s AG} \right)^{-1}. \tag{5.419}$$

By inserting this—with the 'residual bending flexibility' $\frac{L^2}{12EI_z}$ corrected—shear stiffness into the finite element solution according to Eq. (5.410), the analytically exact solution results. The same result is derived in [67], starting from the general—meaning without considering a certain support of the beam—solution for the beam deflection, and in [26] the derivation for the equality of the deflection on the loading point according to the analytical and the corrected finite element solution takes place. It is to be considered that the derived corrected shear stiffness is not just valid for the cantilevered beam under point load, but yields the same value for arbitrary supports and loads on the ends of the beam. However, the derivation of the corrected shear stiffness for nonhomogeneous, anisotropic and non-linear materials appears problematic [26].

### *5.4.3 Higher-Order Shape Functions for the Beam with Shear Contribution*

This subsection follows the derivations presented in [88] and derives first a general approach for a TIMOSHENKO element with an arbitrary number of nodes. In generalization of Eqs. (5.348) and (5.349), the following approach results for the unknowns at the nodes:

$$u_y(x) = \sum_{i=1}^{m} N_{iu}(x) u_{iy}, \tag{5.420}$$

$$\phi_z(x) = \sum_{i=1}^{n} N_{i\phi}(x) \phi_{iz}, \tag{5.421}$$

or alternatively in matrix notation as

$$u_y(x) = \begin{bmatrix} N_{1u} \ldots N_{mu} \; 0 \ldots 0 \end{bmatrix} \begin{bmatrix} u_{1y} \\ \vdots \\ u_{my} \\ \phi_{1z} \\ \vdots \\ \phi_{nz} \end{bmatrix} = \boldsymbol{N}_u^{\mathrm{T}} \boldsymbol{u}_{\mathrm{p}}, \tag{5.422}$$

$$\phi_z(x) = \begin{bmatrix} 0 \ldots 0 \; N_{1\phi} \ldots N_{n\phi} \end{bmatrix} \begin{bmatrix} u_{1y} \\ \vdots \\ u_{my} \\ \phi_{1z} \\ \vdots \\ \phi_{nz} \end{bmatrix} = \boldsymbol{N}_\phi^{\mathrm{T}} \boldsymbol{u}_{\mathrm{p}}. \tag{5.423}$$

With this generalized approach, the deflection can be evaluated at $m$ nodes and the rotation at $n$ nodes. For the shape functions $N_i$ usually LAGRANGE polynomials[42] are used, which in general are calculated in the case of the deflection as follows:

$$\begin{aligned} N_i &= \prod_{j=0 \wedge j \neq i}^{m} \frac{x_j - x}{x_j - x_i} \\ &= \frac{(x_1 - x)(x_2 - x) \cdots [x_i - x] \cdots (x_m - x)}{(x_1 - x_i)(x_2 - x_i) \cdots [x_i - x_i] \cdots (x_m - x_i)}, \end{aligned} \tag{5.424}$$

whereupon the expressions in the square brackets for the $i$-th shape function remains unconsidered. The abscissa values $x_1, \ldots, x_m$ represent the $x$-coordinates of the $m$ nodes. In the case of the rotation, the variable $m$ has to be replaced by $n$ in Eq. (5.424).

For the derivation of the general stiffness matrix, the weighted residual method is considered. One can use the new approaches (5.422) and (5.423) in Eq. (5.369). Execution of the multiplication for the bending stiffness matrix yields

[42] In the case of the so-called LAGRANGE interpolation, $m$ points are approximated via the ordinate values with the help of a polynomial of the order $m - 1$. In the case of the HERMITE interpolation, the slope of the regarded points is considered in addition to the ordinate value.

$$
\boldsymbol{K}_{\mathrm{b}}^{\mathrm{e}} = \int_{0}^{L} EI_z \left[ \begin{array}{ccc|ccc}
0 & \cdots & 0 & 0 & \cdots & 0 \\
\vdots & (m \times m) & \vdots & \vdots & (m \times n) & \vdots \\
0 & \cdots & 0 & 0 & \cdots & 0 \\
\hline
0 & \cdots & 0 & \frac{\mathrm{d}N_{1\phi}}{\mathrm{d}x}\frac{\mathrm{d}N_{1\phi}}{\mathrm{d}x} & \cdots & \frac{\mathrm{d}N_{1\phi}}{\mathrm{d}x}\frac{\mathrm{d}N_{n\phi}}{\mathrm{d}x} \\
\vdots & (n \times m) & \vdots & \vdots & (n \times n) & \vdots \\
0 & \cdots & 0 & \frac{\mathrm{d}N_{n\phi}}{\mathrm{d}x}\frac{\mathrm{d}N_{1\phi}}{\mathrm{d}x} & \cdots & \frac{\mathrm{d}N_{n\phi}}{\mathrm{d}x}\frac{\mathrm{d}N_{n\phi}}{\mathrm{d}x}
\end{array} \right] \mathrm{d}x \tag{5.425}
$$

and a corresponding execution of the multiplication for the shear stiffness matrix $\boldsymbol{K}_{\mathrm{s}}^{\mathrm{e}}$ yields

$$
\int_{0}^{L} k_{\mathrm{s}}AG \left[ \begin{array}{ccc|ccc}
\frac{\mathrm{d}N_{1u}}{\mathrm{d}x}\frac{\mathrm{d}N_{1u}}{\mathrm{d}x} & \cdots & \frac{\mathrm{d}N_{1u}}{\mathrm{d}x}\frac{\mathrm{d}N_{mu}}{\mathrm{d}x} & \frac{\mathrm{d}N_{1u}}{\mathrm{d}x}(-N_{1\phi}) & \cdots & \frac{\mathrm{d}N_{1u}}{\mathrm{d}x}(-N_{n\phi}) \\
\vdots & (m\times m) & \vdots & \vdots & (m\times n) & \vdots \\
\frac{\mathrm{d}N_{mu}}{\mathrm{d}x}\frac{\mathrm{d}N_{1u}}{\mathrm{d}x} & \cdots & \frac{\mathrm{d}N_{mu}}{\mathrm{d}x}\frac{\mathrm{d}N_{mu}}{\mathrm{d}x} & \frac{\mathrm{d}N_{mu}}{\mathrm{d}x}(-N_{1\phi}) & \cdots & \frac{\mathrm{d}N_{mu}}{\mathrm{d}x}(-N_{m\phi}) \\
\hline
-N_{1\phi}\frac{\mathrm{d}N_{1u}}{\mathrm{d}x} & \cdots & -N_{1\phi}\frac{\mathrm{d}N_{mu}}{\mathrm{d}x} & N_{1\phi}N_{1\phi} & \cdots & N_{1\phi}N_{n\phi} \\
\vdots & (n\times m) & \vdots & \vdots & (n\times n) & \vdots \\
-N_{n\phi}\frac{\mathrm{d}N_{1u}}{\mathrm{d}x} & \cdots & -N_{n\phi}\frac{\mathrm{d}N_{mu}}{\mathrm{d}x} & N_{n\phi}N_{1\phi} & \cdots & N_{n\phi}N_{n\phi}
\end{array} \right] \mathrm{d}x. \tag{5.426}
$$

These two stiffness matrices can be be superposed additively at this point and the following general structure for the total stiffness matrix is obtained:

$$
\boldsymbol{K}^{\mathrm{e}} = \begin{bmatrix} \boldsymbol{K}^{11} & \boldsymbol{K}^{12} \\ \boldsymbol{K}^{21} & \boldsymbol{K}^{22} \end{bmatrix}, \tag{5.427}
$$

with

$$
\boldsymbol{K}_{kl}^{11} = \int_{0}^{L} k_{\mathrm{s}}AG \frac{\mathrm{d}N_{ku}}{\mathrm{d}x}\frac{\mathrm{d}N_{lu}}{\mathrm{d}x}\mathrm{d}x, \tag{5.428}
$$

$$
\boldsymbol{K}_{kl}^{12} = \int_{0}^{L} k_{\mathrm{s}}AG \frac{\mathrm{d}N_{ku}}{\mathrm{d}x}(-N_{l\phi})\mathrm{d}x, \tag{5.429}
$$

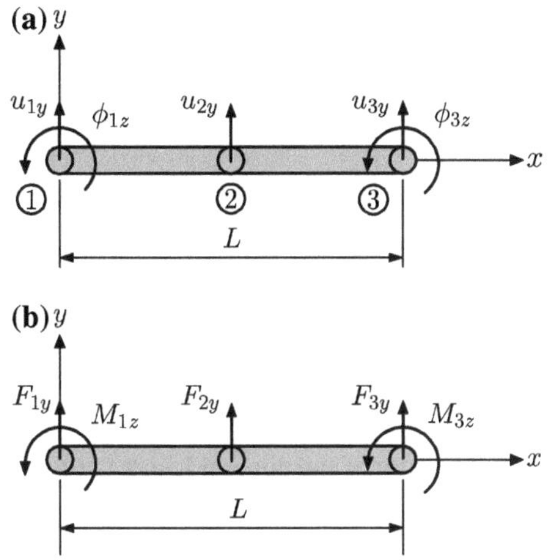


**Fig. 5.40** TIMOSHENKO beam element with quadratic shape functions for the deflection and linear shape functions for the rotation: **a** deformation parameters; **b** load parameters

$$\boldsymbol{K}_{kl}^{21} = \boldsymbol{k}_{kl}^{12,\mathrm{T}} = \int_0^L k_{\mathrm{s}}AG(-N_{k\phi})\frac{\mathrm{d}N_{lu}}{\mathrm{d}x}\,\mathrm{d}x, \tag{5.430}$$

$$\boldsymbol{K}_{kl}^{22} = \int_0^L \left( k_{\mathrm{s}}AGN_{k\phi}N_{l\phi} + EI_z\frac{\mathrm{d}N_{k\phi}}{\mathrm{d}x}\frac{\mathrm{d}N_{l\phi}}{\mathrm{d}x} \right)\mathrm{d}x. \tag{5.431}$$

The derivation of the right-hand side can be performed according to Eq. (5.378) and the following column matrix of the loads results:

$$\boldsymbol{f}^{\mathrm{e}} = \int_0^L q_y(x) \begin{bmatrix} N_{1u} \\ \vdots \\ N_{mu} \\ 0 \\ \vdots \\ 0 \end{bmatrix} \mathrm{d}x + \begin{bmatrix} F_{1y} \\ \vdots \\ F_{my} \\ M_{1z} \\ \vdots \\ M_{nz} \end{bmatrix}. \tag{5.432}$$

In the following, a quadratic interpolation for $u_y(x)$ as well as a linear interpolation for $\phi_z(x)$ are chosen [88]. Therefore, for $\frac{\mathrm{d}u_y(x)}{\mathrm{d}x}$ and $\phi_z(x)$ functions of the same order result and the phenomenon of *shear locking* can be avoided. Quadratic interpolation for the deflection means that the deflection will be evaluated at three nodes. The linear approach for the rotation means that the unknowns will be evaluated at only two nodes. Therefore, the illustrated configuration in Fig. 5.40 for this TIMOSHENKO element results.

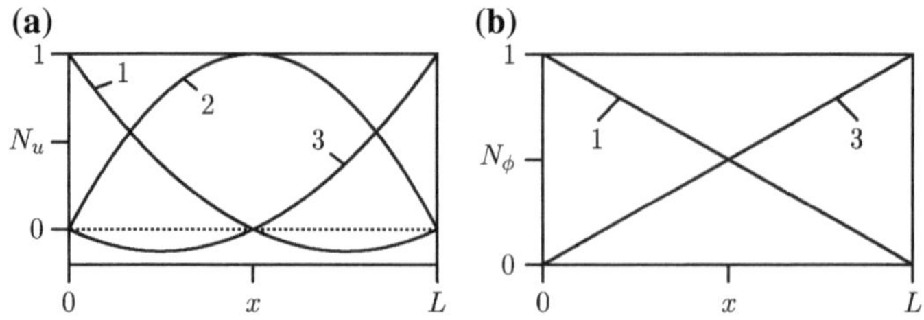


**Fig. 5.41** Shape functions for a TIMOSHENKO beam element with **a** quadratic approach for the deflection and **b** linear approach for the rotation

Evaluation of the general LAGRANGE polynomial according to Eq. (5.424) for the deflection, meaning under consideration of three nodes, yields

$$N_{1u} = \frac{(x_2 - x)(x_3 - x)}{(x_2 - x_1)(x_3 - x_1)} = 1 - 3\frac{x}{L} + 2\left(\frac{x}{L}\right)^2, \tag{5.433}$$

$$N_{2u} = \frac{(x_1 - x)(x_3 - x)}{(x_1 - x_2)(x_3 - x_2)} = 4\frac{x}{L} - 4\left(\frac{x}{L}\right)^2, \tag{5.434}$$

$$N_{3u} = \frac{(x_1 - x)(x_2 - x)}{(x_1 - x_3)(x_2 - x_3)} = -\frac{x}{L} + 2\left(\frac{x}{L}\right)^2, \tag{5.435}$$

or alternatively for both nodes for the rotation:

$$N_{1\phi} = \frac{(x_2 - x)}{(x_2 - x_1)} = 1 - \frac{x}{L}, \tag{5.436}$$

$$N_{2\phi} = \frac{(x_1 - x)}{(x_1 - x_2)} = \frac{x}{L}. \tag{5.437}$$

A graphical illustration of the shape function is given in Fig. 5.41. One can see that the typical characteristics for shape functions, meaning $N_i(x_i) = 1 \wedge N_i(x_j) = 0$ and $\sum_i N_i = 1$ are fulfilled.

With these shape functions the submatrices $\boldsymbol{K}^{11}, \ldots, \boldsymbol{K}^{22}$ in Eq. (5.427) result in the following via analytical integration as:

$$\boldsymbol{K}^{11} = \frac{k_s AG}{3L} \begin{bmatrix} 7 & -8 & 1 \\ -8 & 16 & -8 \\ 1 & -8 & 7 \end{bmatrix}, \tag{5.438}$$

$$\boldsymbol{K}^{12} = \frac{k_s AG}{6} \begin{bmatrix} 5 & 1 \\ -4 & 4 \\ -1 & -5 \end{bmatrix} = (\boldsymbol{K}^{21})^{\mathrm{T}}, \tag{5.439}$$

$$\boldsymbol{K}^{22}=\frac{k_sAGL}{6}\begin{bmatrix}2&1\\1&2\end{bmatrix}+\frac{EI_z}{L}\begin{bmatrix}1&-1\\-1&1\end{bmatrix}, \tag{5.440}$$

which can be assembled to the principal finite element equation by making use of the abbreviation $\Lambda=\frac{EI_z}{k_sAGL^2}$:

$$\frac{k_sAG}{6L}\begin{bmatrix}14&-16&2&5L&1L\\-16&32&-16&-4L&4L\\2&-16&14&-1L&-5L\\5L&-4L&-1L&2L^2(1+3\Lambda)&L^2(1-6\Lambda)\\1L&4L&-5L&L^2(1-6\Lambda)&2L^2(1+3\Lambda)\end{bmatrix}\begin{bmatrix}u_{1y}\\u_{2y}\\u_{3y}\\\phi_{1z}\\\phi_{3z}\end{bmatrix}=\begin{bmatrix}F_{1y}\\F_{2y}\\F_{3y}\\M_{1z}\\M_{3z}\end{bmatrix}. \tag{5.441}$$

Since only a displacement is evaluated at the middle node, the number of unknowns is not the same at each node. This circumstance complicates the creation of the global system of equations for several of these elements. However, the degree of freedom $u_{2y}$ can be expressed via the remaining unknowns and therefore the possibility exists to eliminate this node from the system of equations. For this, the second Eq. (5.441)[43] has to be evaluated:

$$\frac{k_sAG}{6L}\left(-16u_{1y}+32u_{2y}-16u_{3y}-4L\phi_{1z}+4L\phi_{3z}\right)=F_{2y}, \tag{5.442}$$

$$u_{2y}=\frac{6L}{32k_sAG}F_{2y}+\frac{u_{1y}+u_{3y}}{2}+\frac{\phi_{1z}-\phi_{3z}}{8}L. \tag{5.443}$$

Furthermore, it can be demanded that no external force should act at the middle node, so that the relation between the deflection at the middle node and the other unknowns yields as follows:

$$u_{2y}=\frac{u_{1y}+u_{3y}}{2}+\frac{\phi_{1z}-\phi_{3z}}{8}L. \tag{5.444}$$

This relation can be introduced into the system of equations (5.441) to eliminate the degree of freedom $u_{2u}$. Finally, after a new arrangement of the unknowns, the following principal finite element equation results, which is reduced by one column and one row:

$$\frac{EI_z}{6\Lambda L^3}\begin{bmatrix}6&3L&-6&3L\\3L&L^2(1,5+6\Lambda)&-3L&L^2(1,5-6\Lambda)\\-6&-3L&6&-3L\\3L&L^2(1,5-6\Lambda)&-3L&L^2(1,5+6\Lambda)\end{bmatrix}=\begin{bmatrix}u_{1y}\\\phi_{1z}\\u_{3y}\\\phi_{3z}\end{bmatrix}=\begin{bmatrix}F_{1y}\\M_{1z}\\F_{3y}\\M_{3z}\end{bmatrix}. \tag{5.445}$$

[43] It needs to be remarked that the influence of distributed loads is disregarded in this derivation. If distributed loads occur, the equivalent nodal loads have to be distributed on the remaining nodes.

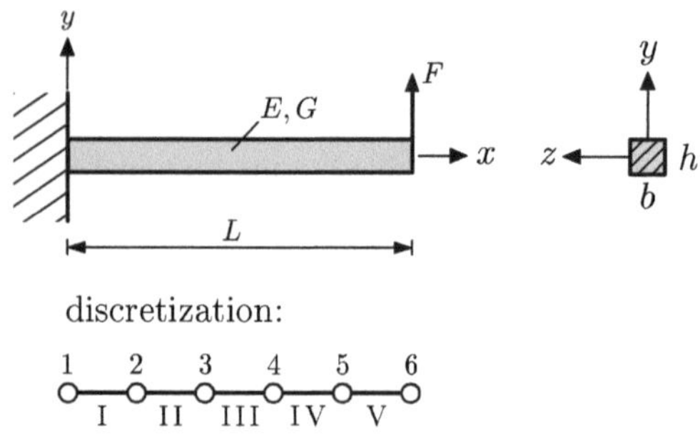


**Fig. 5.42** Discretization of a beam structure with elements under consideration of the shear contribution

This element formulation is identical with Eq. (5.408), which was derived for linear shape functions and numerical one-point integration. However, it should be considered that the interpolation between the nodes during the use of (5.445) takes place with quadratic functions.

Further details and formulations regarding the TIMOSHENKO beam element can be found in the scientific papers [85, 86].

## *5.4.4 Solved Problems*

**5.10 Example: Discretization of a beam with five linear elements with shear contribution**

Discretize equally with five linear TIMOSHENKO elements the beam[44] which is illustrated in Fig. 5.42. Discuss the displacement of the loading point as a function of the slenderness ratio and the POISSON's ratio. Consider the case of (a) an analytically and (b) a numerically (one integration point) integrated stiffness matrix.

**5.10 Solution**

(a) Stiffness matrix via analytical integration

The element stiffness matrix according to Eq. (5.393) can be used for each of the five elements, whereupon it has to be considered that the single element length is $\frac{L}{5}$. The resulting total stiffness matrix has the dimension $12 \times 12$, which reduces to a $10 \times 10$ matrix due to the consideration of the fixed support on the left-hand boundary ($u_{1y} = 0, \phi_{1z} = 0$). Through inversion of the stiffness matrix, the reduced system of equations can be solved via $\boldsymbol{u} = \boldsymbol{K}^{-1}\boldsymbol{F}$. The following extract shows the most important entries in this system of equations:

[44] A similar example is presented in [98].

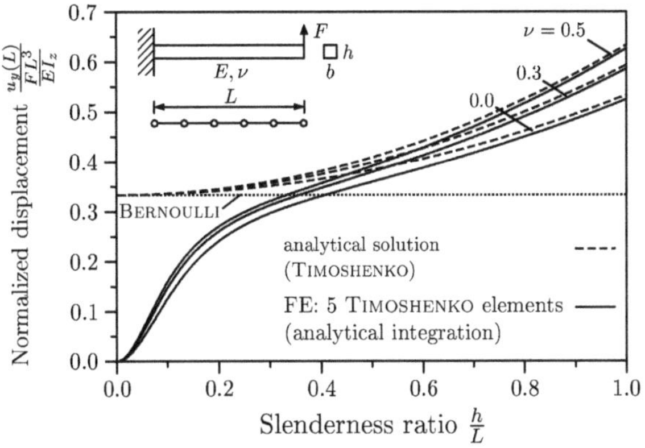


**Fig. 5.43** Discretization of a beam with five linear TIMOSHENKO elements and analytical integration of the stiffness matrix

$$\begin{bmatrix} u_{2y} \\ \\ \vdots \\ \\ u_{6y} \\ \phi_{6z} \end{bmatrix} = \frac{\frac{4}{5}L}{k_sAG} \underbrace{\begin{bmatrix} \mathsf{x} & \cdots & \mathsf{x} & \mathsf{x} \\ \vdots & & \vdots & \vdots \\ \mathsf{x} & \cdots & \frac{125(3\alpha+4L^2)}{4(75\alpha+L^2)} & \mathsf{x} \\ \mathsf{x} & \cdots & \mathsf{x} & \mathsf{x} \end{bmatrix}}_{10\times 10 \text{ matrix}} \begin{bmatrix} 0 \\ \vdots \\ F \\ 0 \end{bmatrix}. \tag{5.446}$$

Multiplication of the 9th row of the matrix with the load vector yields the displacement of the loading point to:

$$u_{6y} = \frac{25(3\alpha + 4L^2)}{75\alpha + L^2} \times \frac{FL}{k_sAG}, \tag{5.447}$$

or alternatively via $A = hb$, $k_s = \frac{5}{6}$ and the relation for the shear modulus according to Eq. (4.170) after a short calculation:

$$u_{6y} = \frac{12(1+\nu)\left(\frac{h}{L}\right)^2 + 20}{60 + \left(\frac{L}{h}\right)^2 \frac{1}{1+\nu}} \times \frac{FL^3}{EI_z}. \tag{5.448}$$

A graphical illustration of the displacement dependent on the slenderness ratio can be seen in Fig. 5.43. A comparison with Fig. 5.38 shows that the convergence behavior in the lower domain of the slenderness ratio for $0.2 < \frac{h}{L} < 1.0$ has significantly improved through the fine discretization. However, the phenomenon of *shear locking* for $\frac{h}{L} \to 0$ still occurs.

(b) Stiffness matrix via numerical integration with one integration point

According to the procedure in part (a) of this problem, the following $10 \times 10$ system of equations results at this point via the stiffness matrix according to Eq. (5.407)

$$\begin{bmatrix} u_{2y} \\ \\ \vdots \\ \\ u_{6y} \\ \phi_{6z} \end{bmatrix} = \frac{\frac{4}{5}L}{k_s AG} \underbrace{\begin{bmatrix} \mathsf{X} & \cdots & \mathsf{X} & \mathsf{X} \\ \vdots & & \vdots & \vdots \\ \mathsf{X} & \cdots & \frac{25\alpha+33L^2}{20\alpha} & \mathsf{X} \\ \mathsf{X} & \cdots & \mathsf{X} & \mathsf{X} \end{bmatrix}}_{10\times 10 \text{ matrix}} \begin{bmatrix} 0 \\ \\ \vdots \\ \\ F \\ 0 \end{bmatrix}, \tag{5.449}$$

from which the displacement on the right boundary can be defined as the following

$$u_{6y} = \frac{4}{5}\left(\frac{5}{4} + \frac{33L^2}{20\alpha}\right) \times \frac{FL}{k_s AG}. \tag{5.450}$$

With the use of $A = hb$, $k_s = \frac{5}{6}$ and the relation for the shear modulus according to Eq. (4.170), the following results after a short calculation:

$$u_{6y} = \left(\frac{33}{100} + \frac{1}{5}(1+\nu)\left(\frac{h}{L}\right)^2\right) \times \frac{FL^3}{EI_z}. \tag{5.451}$$

The graphical illustration of the displacement in Fig. 5.44 shows that an excellent conformity with the analytical solution throughout the entire domain of the slenderness ratio results through the mesh refinement. Therefore, the accuracy of a TIMOSHENKO element with linear shape functions and reduced numerical integration can be increased considerably through mesh refinement.

## 5.11 Example: Timoshenko bending element with quadratic shape functions for the deflection and the rotation

Derive the stiffness matrix and the principal finite element equation $\boldsymbol{K}^e \boldsymbol{u}_p = \boldsymbol{f}^e$ for the illustrated TIMOSHENKO beam element in Fig. 5.45 with quadratic shape functions. Distinguish in the derivation between the analytical and numerical integration. Analyze the convergence behavior of an element for the illustrated configuration in Fig. 5.37.

## 5.11 Solution

Evaluation of the general LAGRANGE polynomial according to Eq. (5.424) under consideration of three nodes yields the following shape functions for the deflection and the rotation:

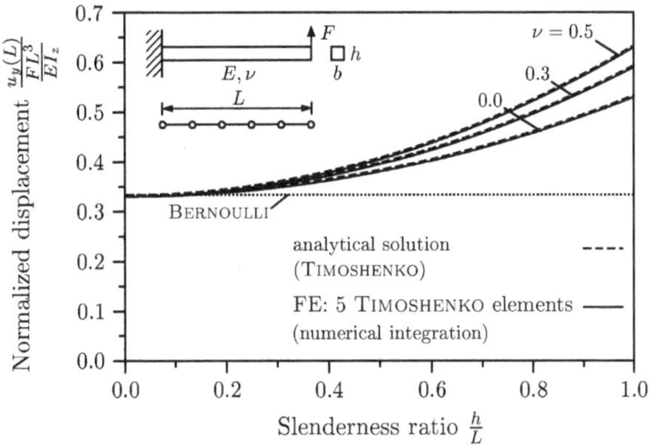


**Fig. 5.44** Discretization of a beam via five linear TIMOSHENKO elements and numerical integration of the stiffness matrix with one integration point

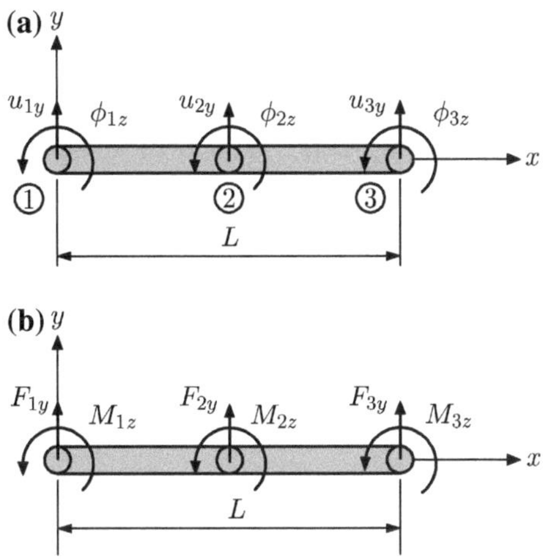


**Fig. 5.45** TIMOSHENKO beam element with quadratic shape functions for the deflection and the rotation: **a** deformation parameters; **b** load parameters

$$N_{1u} = N_{1\phi} = \frac{(x_2 - x)(x_3 - x)}{(x_2 - x_1)(x_3 - x_1)} = 1 - 3\frac{x}{L} + 2\left(\frac{x}{L}\right)^2, \tag{5.452}$$

$$N_{2u} = N_{2\phi} = \frac{(x_1 - x)(x_3 - x)}{(x_1 - x_2)(x_3 - x_2)} = 4\frac{x}{L} - 4\left(\frac{x}{L}\right)^2, \tag{5.453}$$

$$N_{3u} = N_{3\phi} = \frac{(x_1 - x)(x_2 - x)}{(x_1 - x_3)(x_2 - x_3)} = -\frac{x}{L} + 2\left(\frac{x}{L}\right)^2. \tag{5.454}$$

With these shape functions the submatrices $\boldsymbol{K}^{11}, \ldots, \boldsymbol{K}^{22}$ in Eq. (5.427) result as follows through *analytical* integration:

$$\boldsymbol{K}^{11} = \frac{k_s AG}{6L} \begin{bmatrix} 14 & -16 & 2 \\ -16 & 32 & -16 \\ 2 & -16 & 14 \end{bmatrix}, \tag{5.455}$$

$$\boldsymbol{K}^{12} = \frac{k_s AG}{6L} \begin{bmatrix} 3L & 4L & -1L \\ -4L & 0 & 4L \\ 1L & -4L & -3L \end{bmatrix} = (\boldsymbol{K}^{21})^{\mathrm{T}}, \tag{5.456}$$

$$\boldsymbol{K}^{22} = \frac{k_s AGL}{30} \begin{bmatrix} 4 & 2 & -1 \\ 2 & 16 & 2 \\ -1 & 2 & 4 \end{bmatrix} + \frac{EI_z}{3L} \begin{bmatrix} 7 & -8 & 1 \\ -8 & 16 & -8 \\ 1 & -8 & 7 \end{bmatrix}, \tag{5.457}$$

which can be composed to the stiffness matrix $\boldsymbol{K}^{\mathrm{e}}$ via the use of the abbreviation $\Lambda = \frac{EI_z}{k_s AGL^2}$:

$$\frac{k_s AG}{6L} \left[ \begin{array}{ccc|ccc} 14 & -16 & 2 & 3L & 4L & -1L \\ -16 & 32 & -16 & -4L & 0 & 4L \\ 2 & -16 & 14 & 1L & -4L & -3L \\ \hline 3L & -4L & 1L & L^2(\frac{4}{5}+14\Lambda) & L^2(\frac{2}{5}-16\Lambda) & L^2(-\frac{1}{5}+2\Lambda) \\ 4L & 0 & -4L & L^2(\frac{2}{5}-16\Lambda) & L^2(\frac{16}{5}+32\Lambda) & L^2(\frac{2}{5}-16\Lambda) \\ -1L & 4L & -3L & L^2(-\frac{1}{5}+2\Lambda) & L^2(\frac{2}{5}-16\Lambda) & L^2(\frac{4}{5}+14\Lambda) \end{array} \right]. \tag{5.458}$$

With this stiffness matrix the principal finite element equation results in $\boldsymbol{K}^{\mathrm{e}}\boldsymbol{u}_{\mathrm{p}} = \boldsymbol{f}^{\mathrm{e}}$, at which the deformation and load vector contains the following components:

$$\boldsymbol{u}_{\mathrm{p}} = \left[u_{1y} \; u_{2y} \; u_{3y} \; \phi_{1z} \; \phi_{2z} \; \phi_{3z}\right]^{\mathrm{T}}, \tag{5.459}$$

$$\boldsymbol{f}^{\mathrm{e}} = \left[F_{1y} \; F_{2y} \; F_{3y} \; M_{1z} \; M_{2z} \; M_{3z}\right]^{\mathrm{T}}. \tag{5.460}$$

For the analysis of the convergence behavior of an element for the illustrated beam in Fig. 5.37 with point load, the columns and rows for the entries $u_{1y}$ and $\phi_{1z}$ in Eq. (5.458) can be canceled due to the fixed support on this node. This reduced $4 \times 4$ stiffness matrix can be inverted and the following system of equations for the definition of the unknown degrees of freedom results:

$$\begin{bmatrix} u_{2y} \\ u_{3y} \\ \vdots \\ \phi_{3z} \end{bmatrix} = \frac{6L}{k_s AG} \underbrace{\begin{bmatrix} \mathsf{x} & \cdots & \cdots \mathsf{x} \\ \mathsf{x} & \frac{-3+340\Lambda+1200\Lambda^2}{8(-1-45\Lambda+900\Lambda^2)} & \cdots \mathsf{x} \\ \vdots & & \vdots \\ \mathsf{x} & & \cdots \mathsf{x} \end{bmatrix}}_{4\times 4 \text{ matrix}} \begin{bmatrix} 0 \\ F \\ \vdots \\ 0 \end{bmatrix}, \tag{5.461}$$

from which, through evaluation of the second row, the displacement at the right-hand boundary can be defined as:

$$u_{3y} = \underbrace{\frac{6L}{k_s AG}}_{\frac{6\Lambda L^3}{EI_z}} \times \frac{-3 + 340\Lambda + 1200\Lambda^2}{8(-1 - 45\Lambda + 900\Lambda^2)} \times F. \tag{5.462}$$

For a rectangular cross section $\Lambda = \frac{1}{5}(1+\nu)\left(\frac{h}{L}\right)^2$ results, and one can see that *shear locking* occurs also at this point for slender beams with $L \gg h$, since in the limit case $u_{3y} \to 0$ occurs.

In the following, the reduced numerical integration of the stiffness matrix needs to be analyzed. For the definition of a reasonable amount of integration points one takes into account the following consideration:

If quadratic shape functions are used for $u_y$ and $\phi_z$, the degree of the polynomials for $\frac{du_y}{dx}$ and $\phi_z$ differs. The quadratic approach for $u_y$ yields for $\frac{du_y}{dx}$ a linear function and thus a linear function would also be desirable for $\phi_z$. The two-point integration, however, determines that the quadratic approach for $\phi_z$ is treated as a linear function. A two-point integration can exactly integrate a polynomial of third order, meaning proportional to $x^3$, at most and therefore the following view results: $(N_{i\phi}N_{j\phi}) \sim x^3$. This however means that $N_{i\phi} \sim x^{1.5}$ or alternatively $N_{j\phi} \sim x^{1.5}$ applies at most. Since the polynomial approach only allows integer values for the exponent, $N_{i\phi} \sim x^1$ or alternatively $N_{j\phi} \sim x^1$ results and the rotation needs to be considered as a linear function.

The integration via numerical GAUSS integration with two integration points demands that the arguments and the integration boundaries in the formulations of the submatrices $\boldsymbol{K}^{11}, \ldots, \boldsymbol{K}^{22}$ in Eq. (5.427) have to be transformed to the natural coordinate $-1 \leq \xi \leq 1$. Via the transformation of the derivative to the new coordinate, meaning $\frac{dN}{dx} = \frac{dN}{d\xi}\frac{d\xi}{dx}$ and the transformation of the coordinate $\xi = -1 + 2\frac{x}{L}$ or alternatively $d\xi = \frac{2}{L}dx$, the numerical approximation of the submatrices for two integration points $\xi_{1,2} = \pm\frac{1}{\sqrt{3}}$ results in:

$$\boldsymbol{K}^{11} = \sum_{i=1}^{2} \frac{2k_s AG}{L} \begin{bmatrix} \frac{dN_{1u}}{d\xi}\frac{dN_{1u}}{d\xi} & \frac{dN_{1u}}{d\xi}\frac{dN_{2u}}{d\xi} & \frac{dN_{1u}}{d\xi}\frac{dN_{3u}}{d\xi} \\ \frac{dN_{2u}}{d\xi}\frac{dN_{1u}}{d\xi} & \frac{dN_{2u}}{d\xi}\frac{dN_{2u}}{d\xi} & \frac{dN_{2u}}{d\xi}\frac{dN_{3u}}{d\xi} \\ \frac{dN_{3u}}{d\xi}\frac{dN_{1u}}{d\xi} & \frac{dN_{3u}}{d\xi}\frac{dN_{2u}}{d\xi} & \frac{dN_{3u}}{d\xi}\frac{dN_{3u}}{d\xi} \end{bmatrix} \times 1, \tag{5.463}$$

$$\boldsymbol{K}^{12} = \sum_{i=1}^{2} k_s AG \begin{bmatrix} \frac{dN_{1u}}{d\xi}(-N_{1\phi}) & \frac{dN_{1u}}{d\xi}(-N_{2\phi}) & \frac{dN_{1u}}{d\xi}(-N_{3\phi}) \\ \frac{dN_{2u}}{d\xi}(-N_{1\phi}) & \frac{dN_{2u}}{d\xi}(-N_{2\phi}) & \frac{dN_{2u}}{d\xi}(-N_{3\phi}) \\ \frac{dN_{3u}}{d\xi}(-N_{1\phi}) & \frac{dN_{3u}}{d\xi}(-N_{2\phi}) & \frac{dN_{3u}}{d\xi}(-N_{3\phi}) \end{bmatrix} \times 1, \tag{5.464}$$

$$\boldsymbol{K}^{22} = \sum_{i=1}^{2} \frac{k_s AGL}{2} \begin{bmatrix} N_{1\phi}N_{1\phi} & N_{1\phi}N_{2\phi} & N_{1\phi}N_{3\phi} \\ N_{2\phi}N_{1\phi} & N_{2\phi}N_{2\phi} & N_{2\phi}N_{3\phi} \\ N_{3\phi}N_{1\phi} & N_{3\phi}N_{2\phi} & N_{3\phi}N_{3\phi} \end{bmatrix} \times 1 \tag{5.465}$$

$$+ \sum_{i=1}^{2} \frac{2EI_z}{L} \begin{bmatrix} \frac{dN_{1\phi}}{d\xi}\frac{dN_{1\phi}}{d\xi} & \frac{dN_{1\phi}}{d\xi}\frac{dN_{2\phi}}{d\xi} & \frac{dN_{1\phi}}{d\xi}\frac{dN_{3\phi}}{d\xi} \\ \frac{dN_{2\phi}}{d\xi}\frac{dN_{1\phi}}{d\xi} & \frac{dN_{2\phi}}{d\xi}\frac{dN_{2\phi}}{d\xi} & \frac{dN_{2\phi}}{d\xi}\frac{dN_{3\phi}}{d\xi} \\ \frac{dN_{3\phi}}{d\xi}\frac{dN_{1\phi}}{d\xi} & \frac{dN_{3\phi}}{d\xi}\frac{dN_{2\phi}}{d\xi} & \frac{dN_{3\phi}}{d\xi}\frac{dN_{3\phi}}{d\xi} \end{bmatrix} \times 1. \tag{5.466}$$

The quadratic shape functions, which have already been introduced into Eqs. (5.433) up to (5.435), still have to be transformed to the new coordinates via the transformation $x = (\xi + 1)\frac{L}{2}$. Therefore for the shape functions or alternatively their derivatives the following results:

$$N_1(\xi) = -\frac{1}{2}(\xi - \xi^2), \qquad \frac{dN_1}{d\xi} = -\frac{1}{2}(1 - 2\xi), \tag{5.467}$$

$$N_2(\xi) = 1 - \xi^2, \qquad \frac{dN_2}{d\xi} = -2\xi, \tag{5.468}$$

$$N_3(\xi) = \frac{1}{2}(\xi + \xi^2), \qquad \frac{dN_3}{d\xi} = \frac{1}{2}(1 + 2\xi). \tag{5.469}$$

The use of these shape functions or alternatively their derivatives finally leads to the following submatrices

$$\boldsymbol{K}^{11} = \frac{k_s AG}{6L} \begin{bmatrix} 14 & -16 & 2 \\ -16 & 32 & -16 \\ 2 & -16 & 14 \end{bmatrix}, \tag{5.470}$$

$$\boldsymbol{K}^{12} = \frac{k_s AG}{6L} \begin{bmatrix} 3L & 4L & -L \\ -4L & 0 & 4L \\ +L & -4L & -3L \end{bmatrix}, \tag{5.471}$$

$$\boldsymbol{K}^{22} = \frac{k_{\mathrm{s}}AG}{6L}\begin{bmatrix} \frac{2}{3}L^2 & \frac{2}{3}L^2 & -\frac{1}{3}L^2 \\ \frac{2}{3}L^2 & \frac{8}{3}L^2 & \frac{2}{3}L^2 \\ -\frac{1}{3}L^2 & \frac{2}{3}L^2 & \frac{2}{3}L^2 \end{bmatrix} + \frac{EI_z}{L^3}\begin{bmatrix} \frac{7}{3}L^2 & -\frac{8}{3}L^2 & \frac{1}{3}L^2 \\ -\frac{8}{3}L^2 & \frac{16}{3}L^2 & -\frac{8}{3}L^2 \\ \frac{1}{3}L^2 & -\frac{8}{3}L^2 & \frac{7}{3}L^2 \end{bmatrix}, \tag{5.472}$$

which can be put together to the stiffness matrix $\boldsymbol{K}^{\mathrm{e}}$ under the use of the abbreviation $\Lambda = \frac{EI_z}{k_{\mathrm{s}}AGL^2}$:

$$\frac{k_{\mathrm{s}}AG}{6L}\left[\begin{array}{ccc|ccc} 14 & -16 & 2 & 3L & 4L & -1L \\ -16 & 32 & -16 & -4L & 0 & 4L \\ 2 & -16 & 14 & 1L & -4L & -3L \\ \hline 3L & -4L & 1L & L^2(\frac{2}{3}+14\Lambda) & L^2(\frac{2}{3}-16\Lambda) & L^2(-\frac{1}{3}+2\Lambda) \\ 4L & 0 & -4L & L^2(\frac{2}{3}-16\Lambda) & L^2(\frac{8}{3}+32\Lambda) & L^2(\frac{2}{3}-16\Lambda) \\ -1L & 4L & -3L & L^2(-\frac{1}{3}+2\Lambda) & L^2(\frac{2}{3}-16\Lambda) & L^2(\frac{2}{3}+14\Lambda) \end{array}\right], \tag{5.473}$$

whereupon the deformation and load vector also contains the following components at this point:

$$\boldsymbol{u}_{\mathrm{p}} = \left[u_{1y}\ u_{2y}\ u_{3y}\ \phi_{1z}\ \phi_{2z}\ \phi_{3z}\right]^{\mathrm{T}}, \tag{5.474}$$

$$\boldsymbol{f}^{\mathrm{e}} = \left[F_{1y}\ F_{2y}\ F_{3y}\ M_{1z}\ M_{2z}\ M_{3z}\right]^{\mathrm{T}}. \tag{5.475}$$

For the analysis of the convergence behavior for the beam according to Fig. 5.37 the columns and rows for the entries $u_{1y}$ and $\phi_{1z}$ in the present system of equations can be canceled. The inverted $4 \times 4$ stiffness matrix can be used for the definition of the unknown degrees of freedom:

$$\begin{bmatrix} u_{2y} \\ u_{3y} \\ \vdots \\ \phi_{3z} \end{bmatrix} = \frac{6L}{k_{\mathrm{s}}AG}\underbrace{\begin{bmatrix} \mathsf{x} & \cdots & \cdots & \mathsf{x} \\ \mathsf{x} & \frac{1+3\Lambda}{18\Lambda} & \cdots & \mathsf{x} \\ \vdots & & & \vdots \\ \mathsf{x} & & \cdots & \mathsf{x} \end{bmatrix}}_{4\times4 \text{ matrix}}\begin{bmatrix} 0 \\ F \\ \vdots \\ 0 \end{bmatrix}, \tag{5.476}$$

from which, through evaluation of the second row, the deformation on the right-hand boundary can be defined as:

$$u_{3y} = \underbrace{\frac{6L}{k_{\mathrm{s}}AG}}_{\frac{6\Lambda L^3}{EI_z}} \times \frac{1+3\Lambda}{18\Lambda} \times F = \left(\frac{1}{3}+\Lambda\right)\frac{FL^3}{EI_z}. \tag{5.477}$$

For a rectangular cross section $\Lambda = \frac{1}{5}(1+\nu)\left(\frac{h}{L}\right)^2$ results and one receives the exact solution[45] of the problem as:

$$u_{3y} = \left(\frac{1}{3} + \frac{1+\nu}{5}\left(\frac{h}{L}\right)^2\right) \times \frac{FL^3}{EI_z}. \tag{5.478}$$

According to the procedure for the TIMOSHENKO element with quadratic-linear shape functions in Sect. 5.4.3, the middle node can be eliminated. Under the assumption that no forces or moments should have an effect on the middle node, the 2nd and 5th row of Eq. (5.473) yields the following relation for the unknowns at the middle node:

$$u_{2y} = \frac{1}{2}u_{1y} + \frac{1}{2}u_{3y} + \frac{1}{8}L\phi_{1z} - \frac{1}{8}L\phi_{3z}, \tag{5.479}$$

$$\phi_{2z} = \frac{-4u_{1y}}{L\left(\frac{8}{3}+32\Lambda\right)} + \frac{4u_{3y}}{L\left(\frac{8}{3}+32\Lambda\right)} - \frac{\left(\frac{2}{3}-16\lambda\right)\phi_{1z}}{\left(\frac{8}{3}+32\Lambda\right)} - \frac{\left(\frac{2}{3}-16\lambda\right)\phi_{3z}}{\left(\frac{8}{3}+32\Lambda\right)}. \tag{5.480}$$

These two relations can be considered in Eq. (5.473) so that the following principal finite element equation results after a short conversion:

$$\frac{2EI_z}{L^3(1+12\Lambda)}\begin{bmatrix} 6 & 3L & -6 & 3L \\ 3L & 2L^2(1+3\Lambda) & -3L & L^2(1-6\Lambda) \\ -6 & -3L & 6 & -3L \\ 3L & L^2(1-6\Lambda) & -3L & 2L^2(1+3\Lambda) \end{bmatrix}\begin{bmatrix} u_{1y} \\ \phi_{1z} \\ u_{3y} \\ \phi_{3z} \end{bmatrix} = \begin{bmatrix} F_{1y} \\ M_{1z} \\ F_{3y} \\ M_{3z} \end{bmatrix}. \tag{5.481}$$

With this formulation the one-beam problem according to Fig. 5.37 can be solved a little bit faster since after the consideration of the boundary conditions only a $2 \times 2$ matrix needs to be inverted. In this case, for the definition of the unknown the following results:

$$\frac{L^3(1+12\Lambda)}{2EI_z}\begin{bmatrix} \frac{2(1+3\Lambda)}{3(1+12\Lambda)} & \frac{1}{L(1+12\Lambda)} \\ \frac{1}{L(1+12\Lambda)} & \frac{2}{L^2(1+12\Lambda)} \end{bmatrix}\begin{bmatrix} F \\ 0 \end{bmatrix} = \begin{bmatrix} u_{3y} \\ \phi_{3z} \end{bmatrix}, \tag{5.482}$$

which results from the exact solution for the deflection according to Eq. (5.478).

[45] For this see the supplementary problem 4.25.

# 5.5 Assembly of Elements to Plane Truss Structures

## 5.5.1 Rotational Transformation in a Plane

Let us consider in the following a rod element which can deform in the global $X$-$Y$ plane. The local $x$-coordinate is rotated by an angle $\alpha$ against the global coordinate system $(X, Y)$, cf. Fig. 5.46.

Each node has now in the global coordinate system two degrees of freedom, i.e. a displacement in the $X$- and a displacement in the $Y$-direction. These two global displacements at each node can be used to calculate the displacement in the direction of the rod axis, i.e. in the direction of the local $x$-axis. Based on the right-angled triangles shown in Fig. 5.46, the displacements in the local coordinate system are given based on the global displacements as:

$$u_{1x} = \cos\alpha \, u_{1X} + \sin\alpha \, u_{1Y}, \tag{5.483}$$

$$u_{2x} = \cos\alpha \, u_{2X} + \sin\alpha \, u_{2Y}. \tag{5.484}$$

It is possible to derive in a similar way the global displacements based on the local displacements as:

$$u_{1X} = \cos\alpha \, u_{1x}, \qquad u_{2X} = \cos\alpha \, u_{2x}, \tag{5.485}$$

$$u_{1Y} = \sin\alpha \, u_{1x}, \qquad u_{2Y} = \sin\alpha \, u_{2x}. \tag{5.486}$$

The last relationship between the global and local displacements can be written in matrix form as

$$\begin{bmatrix} u_{1X} \\ u_{1Y} \\ u_{2X} \\ u_{2Y} \end{bmatrix} = \begin{bmatrix} \cos\alpha & 0 \\ \sin\alpha & 0 \\ 0 & \cos\alpha \\ 0 & \sin\alpha \end{bmatrix} \begin{bmatrix} u_{1x} \\ u_{2x} \end{bmatrix}, \tag{5.487}$$

or in abbreviated matrix notation as:

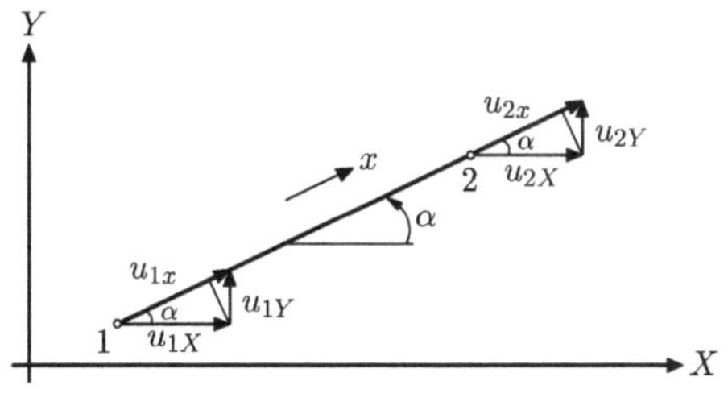


**Fig. 5.46** Rotational transformation of a rod element in the $X$-$Y$ plane

$$\boldsymbol{u}_{XY} = \boldsymbol{T}^{\mathrm{T}}\boldsymbol{u}_{xy}, \tag{5.488}$$

where $\boldsymbol{u}_{XY}$ is the displacement column matrix in the global coordinate system and $\boldsymbol{u}_{xy}$ the local displacement column matrix. The last equation can be solved for the displacements in the local coordinate system and inverting[46] the transformation matrix gives

$$\boldsymbol{u}_{xy} = \boldsymbol{T}\boldsymbol{u}_{XY}, \tag{5.489}$$

or in components:

$$\begin{bmatrix} u_{1x} \\ u_{2x} \end{bmatrix} = \underbrace{\begin{bmatrix} \cos\alpha & \sin\alpha & 0 & 0 \\ 0 & 0 & \cos\alpha & \sin\alpha \end{bmatrix}}_{\boldsymbol{T}} \begin{bmatrix} u_{1X} \\ u_{1Y} \\ u_{2X} \\ u_{2Y} \end{bmatrix}. \tag{5.490}$$

It is possible to transform in a similar way the vector of the external loads as:

$$\boldsymbol{f}_{XY} = \boldsymbol{T}^{\mathrm{T}}\boldsymbol{f}_{xy}, \tag{5.491}$$

$$\boldsymbol{f}_{xy} = \boldsymbol{T}\boldsymbol{f}_{XY}. \tag{5.492}$$

Considering the transformation of the local displacements into the global coordinate system in the weak form according to Eq. (5.62), the transformation of the stiffness matrix into the global coordinate system is given as:

$$\boldsymbol{K}^{\mathrm{e}}_{XY} = \boldsymbol{T}^{\mathrm{T}}\boldsymbol{K}^{\mathrm{e}}_{xy}\boldsymbol{T}, \tag{5.493}$$

or in components

$$\begin{bmatrix} \cos\alpha & 0 \\ \sin\alpha & 0 \\ 0 & \cos\alpha \\ 0 & \sin\alpha \end{bmatrix} \frac{EA}{L} \begin{bmatrix} 1 & -1 \\ -1 & 1 \end{bmatrix} \begin{bmatrix} \cos\alpha & \sin\alpha & 0 & 0 \\ 0 & 0 & \cos\alpha & \sin\alpha \end{bmatrix}. \tag{5.494}$$

The evaluation of this triple product results finally in the stiffness matrix in the global coordinate system as:

[46] Since the transformation matrix is orthogonal, it follows that $\boldsymbol{T}^{\mathrm{T}} = \boldsymbol{T}^{-1}$.

**Table 5.11** Elemental stiffness matrices for truss elements given for different rotation angles $\alpha$, cf. Eq. (5.495)

| 0° | 180° |
|---|---|
| $\frac{EA}{L}\begin{bmatrix} 1 & 0 & -1 & 0 \\ 0 & 0 & 0 & 0 \\ -1 & 0 & 1 & 0 \\ 0 & 0 & 0 & 0 \end{bmatrix}$ | $\frac{EA}{L}\begin{bmatrix} 1 & 0 & -1 & 0 \\ 0 & 0 & 0 & 0 \\ -1 & 0 & 1 & 0 \\ 0 & 0 & 0 & 0 \end{bmatrix}$ |
| −30° | 30° |
| $\frac{EA}{L}\begin{bmatrix} \frac{3}{4} & -\frac{1}{4}\sqrt{3} & -\frac{3}{4} & \frac{1}{4}\sqrt{3} \\ -\frac{1}{4}\sqrt{3} & \frac{1}{4} & \frac{1}{4}\sqrt{3} & -\frac{1}{4} \\ -\frac{3}{4} & \frac{1}{4}\sqrt{3} & \frac{3}{4} & -\frac{1}{4}\sqrt{3} \\ \frac{1}{4}\sqrt{3} & -\frac{1}{4} & -\frac{1}{4}\sqrt{3} & \frac{1}{4} \end{bmatrix}$ | $\frac{EA}{L}\begin{bmatrix} \frac{3}{4} & \frac{1}{4}\sqrt{3} & -\frac{3}{4} & -\frac{1}{4}\sqrt{3} \\ \frac{1}{4}\sqrt{3} & \frac{1}{4} & -\frac{1}{4}\sqrt{3} & -\frac{1}{4} \\ -\frac{3}{4} & -\frac{1}{4}\sqrt{3} & \frac{3}{4} & \frac{1}{4}\sqrt{3} \\ -\frac{1}{4}\sqrt{3} & -\frac{1}{4} & \frac{1}{4}\sqrt{3} & \frac{1}{4} \end{bmatrix}$ |
| −45° | 45° |
| $\frac{EA}{L}\begin{bmatrix} \frac{1}{2} & -\frac{1}{2} & -\frac{1}{2} & \frac{1}{2} \\ -\frac{1}{2} & \frac{1}{2} & \frac{1}{2} & -\frac{1}{2} \\ -\frac{1}{2} & \frac{1}{2} & \frac{1}{2} & -\frac{1}{2} \\ \frac{1}{2} & -\frac{1}{2} & -\frac{1}{2} & \frac{1}{2} \end{bmatrix}$ | $\frac{EA}{L}\begin{bmatrix} \frac{1}{2} & \frac{1}{2} & -\frac{1}{2} & -\frac{1}{2} \\ \frac{1}{2} & \frac{1}{2} & -\frac{1}{2} & -\frac{1}{2} \\ -\frac{1}{2} & -\frac{1}{2} & \frac{1}{2} & \frac{1}{2} \\ -\frac{1}{2} & -\frac{1}{2} & \frac{1}{2} & \frac{1}{2} \end{bmatrix}$ |
| −90° | 90° |
| $\frac{EA}{L}\begin{bmatrix} 0 & 0 & 0 & 0 \\ 0 & 1 & 0 & -1 \\ 0 & 0 & 0 & 0 \\ 0 & -1 & 0 & 1 \end{bmatrix}$ | $\frac{EA}{L}\begin{bmatrix} 0 & 0 & 0 & 0 \\ 0 & 1 & 0 & -1 \\ 0 & 0 & 0 & 0 \\ 0 & -1 & 0 & 1 \end{bmatrix}$ |

$$\frac{AE}{L}\begin{bmatrix} \cos^2\alpha & \cos\alpha\,\sin\alpha & -\cos^2\alpha & -\cos\alpha\,\sin\alpha \\ \cos\alpha\,\sin\alpha & \sin^2\alpha & -\cos\alpha\,\sin\alpha & -\sin^2\alpha \\ -\cos^2\alpha & -\cos\alpha\,\sin\alpha & \cos^2\alpha & \cos\alpha\,\sin\alpha \\ -\cos\alpha\,\sin\alpha & -\sin^2\alpha & \cos\alpha\,\sin\alpha & \sin^2\alpha \end{bmatrix}\begin{bmatrix} u_{1X} \\ u_{1Y} \\ u_{2X} \\ u_{2Y} \end{bmatrix} = \begin{bmatrix} F_{1X} \\ F_{1Y} \\ F_{2X} \\ F_{2Y} \end{bmatrix}. \tag{5.495}$$

To simplify the solution of simple truss structures, Table 5.11 collects expressions for the global stiffness matrix for some common angles $\alpha$.

## 5.5.2 Solved Problems

**5.12 Example: Truss structure arranged as an equilateral triangle**

Given is the two-dimensional truss structure as shown in Fig. 5.47 where the trusses are arranged in the form of an equilateral triangle (internal angles $\beta = 60°$). The three trusses have the same length $L$, the same YOUNG's modulus $E$ and the same cross-sectional area $A$. The structure is loaded by

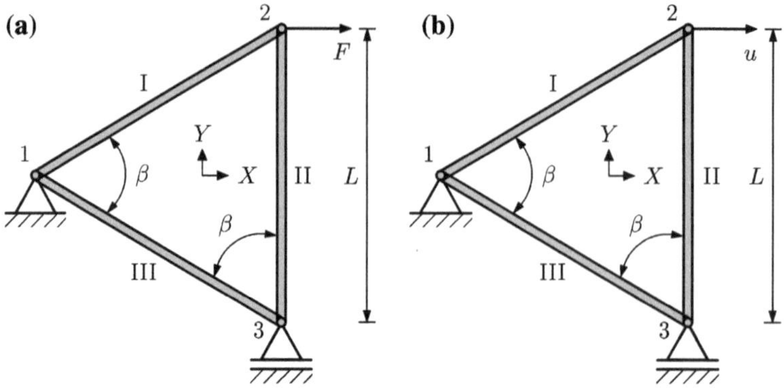


**Fig. 5.47** Truss structure in the form of an equilateral triangle: **a** force boundary condition; **b** displacement boundary condition

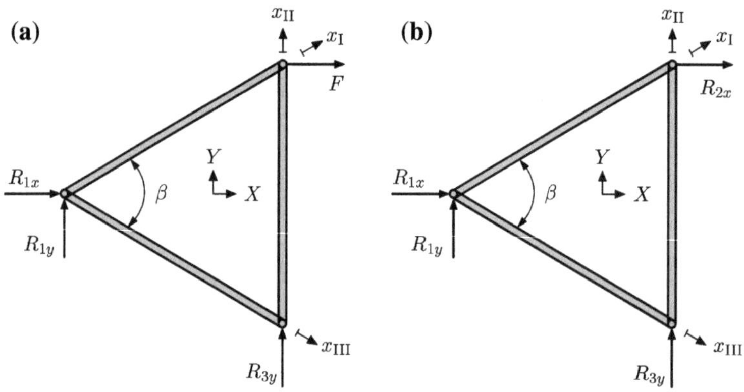


**Fig. 5.48** Free body diagram of the truss structure: **a** force boundary condition; **b** displacement boundary condition

(a) a horizontal force $F$ at node 2,
(b) a prescribed displacement $u$ at node 2.

Determine for both cases

- the global system of equations,
- the reduced system of equations,
- all nodal displacements,
- all reaction forces,
- the force in each rod.

**5.12 Solution**

The free body diagram and the local coordinate axes of each element are shown in Fig. 5.48. From this figure, the rotational angles from the global to the local

**Table 5.12** Angle of rotation $\alpha_i$ and sine and cosine values for the problem shown in Fig. 5.48

| Element | Angle of rotation | sine | cosine |
|---|---|---|---|
| I | 30° | $\frac{1}{2}$ | $\frac{\sqrt{3}}{2}$ |
| II | 90° | 1 | 0 |
| III | 330° | $-\frac{1}{2}$ | $\frac{\sqrt{3}}{2}$ |

coordinate system can be determined and the sine and cosine values calculated as given in Table 5.12.

(a) Force boundary condition

Based on Eq. (5.495) and the values given in Table 5.12, the elemental stiffness matrices can be calculated as:

$$\boldsymbol{K}_{\mathrm{I}}^{\mathrm{e}} = \underbrace{\frac{EA}{L}}_{k_{\mathrm{I}}} \begin{bmatrix} \frac{3}{4} & \frac{\sqrt{3}}{4} & -\frac{3}{4} & -\frac{\sqrt{3}}{4} \\ \frac{\sqrt{3}}{4} & \frac{1}{4} & -\frac{\sqrt{3}}{4} & -\frac{1}{4} \\ -\frac{3}{4} & -\frac{\sqrt{3}}{4} & \frac{3}{4} & \frac{\sqrt{3}}{4} \\ -\frac{\sqrt{3}}{4} & -\frac{1}{4} & \frac{\sqrt{3}}{4} & \frac{1}{4} \end{bmatrix}, \tag{5.496}$$

$$\boldsymbol{K}_{\mathrm{II}}^{\mathrm{e}} = \underbrace{\frac{EA}{L}}_{k_{\mathrm{II}}} \begin{bmatrix} 0 & 0 & 0 & 0 \\ 0 & 1 & 0 & -1 \\ 0 & 0 & 0 & 0 \\ 0 & -1 & 0 & 1 \end{bmatrix}, \tag{5.497}$$

$$\boldsymbol{K}_{\mathrm{III}}^{\mathrm{e}} = \underbrace{\frac{EA}{L}}_{k_{\mathrm{III}}} \begin{bmatrix} \frac{3}{4} & -\frac{\sqrt{3}}{4} & -\frac{3}{4} & \frac{\sqrt{3}}{4} \\ -\frac{\sqrt{3}}{4} & \frac{1}{4} & \frac{\sqrt{3}}{4} & -\frac{1}{4} \\ -\frac{3}{4} & \frac{\sqrt{3}}{4} & \frac{3}{4} & -\frac{\sqrt{3}}{4} \\ \frac{\sqrt{3}}{4} & -\frac{1}{4} & -\frac{\sqrt{3}}{4} & \frac{1}{4} \end{bmatrix}, \tag{5.498}$$

which can be assembled to the global stiffness matrix as:

$$\frac{EA}{L} \begin{bmatrix} \frac{3}{4}+\frac{3}{4} & \frac{\sqrt{3}}{4}-\frac{\sqrt{3}}{4} & -\frac{3}{4} & -\frac{\sqrt{3}}{4} & -\frac{3}{4} & \frac{\sqrt{3}}{4} \\ \frac{\sqrt{3}}{4}-\frac{\sqrt{3}}{4} & \frac{1}{4}+\frac{1}{4} & -\frac{\sqrt{3}}{4} & -\frac{1}{4} & \frac{\sqrt{3}}{4} & -\frac{1}{4} \\ -\frac{3}{4} & -\frac{\sqrt{3}}{4} & \frac{3}{4} & \frac{\sqrt{3}}{4} & 0 & 0 \\ -\frac{\sqrt{3}}{4} & -\frac{1}{4} & \frac{\sqrt{3}}{4} & \frac{1}{4}+1 & 0 & -1 \\ -\frac{3}{4} & \frac{\sqrt{3}}{4} & 0 & 0 & \frac{3}{4} & -\frac{\sqrt{3}}{4} \\ \frac{\sqrt{3}}{4} & -\frac{1}{4} & 0 & -1 & -\frac{\sqrt{3}}{4} & 1+\frac{1}{4} \end{bmatrix} \begin{matrix} u_{1X} \\ u_{1Y} \\ u_{2X} \\ u_{2Y} \\ u_{3X} \\ u_{3Y} \end{matrix} . \tag{5.499}$$

Introducing the boundary conditions, i.e. $u_{1X} = u_{1Y} = u_{3Y} = 0$, gives the reduced system of equations as:

$$\frac{EA}{L}\begin{bmatrix} \frac{3}{4} & \frac{\sqrt{3}}{4} & 0 \\ \frac{\sqrt{3}}{4} & \frac{5}{4} & 0 \\ 0 & 0 & \frac{3}{4} \end{bmatrix}\begin{bmatrix} u_{2X} \\ u_{2Y} \\ u_{3X} \end{bmatrix} = \begin{bmatrix} F \\ 0 \\ 0 \end{bmatrix}. \tag{5.500}$$

The solution of this system can be obtained, for example, by inverting the reduced stiffness matrix to give the reduced result vector as:

$$\begin{bmatrix} u_{2X} \\ u_{2Y} \\ u_{3X} \end{bmatrix} = \frac{L}{EA}\begin{bmatrix} \frac{5}{3} & -\frac{\sqrt{3}}{3} & 0 \\ -\frac{\sqrt{3}}{3} & 1 & 0 \\ 0 & 0 & \frac{4}{3} \end{bmatrix}\begin{bmatrix} F \\ 0 \\ 0 \end{bmatrix} = \frac{LF}{EA}\begin{bmatrix} \frac{5}{3} \\ -\frac{\sqrt{3}}{3} \\ 0 \end{bmatrix}. \tag{5.501}$$

The reaction forces can be obtained by multiplying the stiffness matrix according to Eq. (5.499) with the total displacement vector, i.e.

$$\boldsymbol{u}^{\mathrm{T}} = \begin{bmatrix} 0 \; 0 \; u_{2X} \; u_{2Y} \; u_{3X} \; 0 \end{bmatrix}, \tag{5.502}$$

to give:

$$R_{1X} = -F,\ R_{1Y} = -\frac{\sqrt{3}F}{3},\ R_{3X} = 0,\ R_{3Y} = \frac{\sqrt{3}F}{3}. \tag{5.503}$$

The rod forces can be obtained from the global coordinates as:

$$F_{\mathrm{I}} = k_{\mathrm{I}}\left(-\cos\alpha_{\mathrm{I}}u_{1X} - \sin\alpha_{\mathrm{I}}u_{1Y} + \cos\alpha_{\mathrm{I}}u_{2X} + \sin\alpha_{\mathrm{I}}u_{2Y}\right) = \frac{2\sqrt{3}F}{3}, \tag{5.504}$$

$$F_{\mathrm{II}} = k_{\mathrm{II}}\left(-\cos\alpha_{\mathrm{II}}u_{3X} - \sin\alpha_{\mathrm{II}}u_{3Y} + \cos\alpha_{\mathrm{II}}u_{2X} + \sin\alpha_{\mathrm{II}}u_{2Y}\right) = \frac{\sqrt{3}F}{3}, \tag{5.505}$$

$$F_{\mathrm{III}} = k_{\mathrm{III}}\left(-\cos\alpha_{\mathrm{III}}u_{1X} - \sin\alpha_{\mathrm{III}}u_{1Y} + \cos\alpha_{\mathrm{III}}u_{3X} + \sin\alpha_{\mathrm{III}}u_{3Y}\right) = 0. \tag{5.506}$$

(b) Displacement boundary condition

Considering the displacement boundary condition $u$ at node 2, the reduced system of equations reads:

$$\frac{EA}{L}\begin{bmatrix} \frac{L}{EA} & 0 & 0 \\ \frac{\sqrt{3}}{4} & \frac{5}{4} & 0 \\ 0 & 0 & \frac{3}{4} \end{bmatrix}\begin{bmatrix} u_{2X} \\ u_{2Y} \\ u_{3X} \end{bmatrix} = \begin{bmatrix} u \\ 0 \\ 0 \end{bmatrix}. \tag{5.507}$$

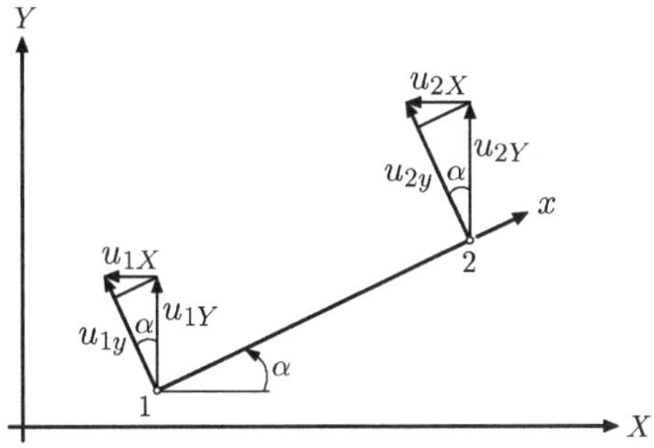


**Fig. 5.49** Rotatory transformation of a beam element in the $X$-$Y$ plane

Inverting the reduced stiffness matrix can be used to calculate the unknown displacements as:

$$\begin{bmatrix} u_{2X} \\ u_{2Y} \\ u_{3X} \end{bmatrix} = \frac{L}{EA} \begin{bmatrix} \frac{EA}{L} & 0 & 0 \\ -\frac{\sqrt{3}EA}{5L} & \frac{4}{5} & 0 \\ 0 & 0 & \frac{4}{3} \end{bmatrix} \begin{bmatrix} u \\ 0 \\ 0 \end{bmatrix} = u \begin{bmatrix} 1 \\ -\frac{\sqrt{3}}{5} \\ 0 \end{bmatrix}. \tag{5.508}$$

Reaction and rod forces can be obtained as described in part (a) as:

$$R_{1X} = -\frac{3}{5} \times \frac{EAu}{L}, \quad R_{1Y} = -\frac{\sqrt{3}}{5} \times \frac{EAu}{L}, \quad R_{2X} = \frac{3}{5} \times \frac{EAu}{L}, \tag{5.509}$$

$$R_{3X} = 0, \quad R_{3Y} = \frac{\sqrt{3}}{5} \times \frac{EAu}{L}. \tag{5.510}$$

$$F_{\mathrm{I}} = \frac{2\sqrt{3}}{5} \times \frac{EAu}{L}, \quad F_{\mathrm{II}} = \frac{\sqrt{3}}{5} \times \frac{EAu}{L}, \quad F_{\mathrm{III}} = 0. \tag{5.511}$$

## 5.6 Assembly of Elements to Plane Frame Structures

### *5.6.1 Rotation of a Beam Element*

We consider in the following the planar rotation of a beam element. As a result, an angle $\alpha$ is obtained between the global $(X, Y)$ and the local $(x, y)$ coordinate system, see Fig. 5.49.

Every node in the global coordinate system now has two degrees of freedom, i.e. a displacement in the $X$- and a displacement in the $Y$-direction. These two displacements at a node can be used to determine the displacement perpendicular to the beam axis, meaning in the direction of the local $y$-axis. By means of the right-angled triangles illustrated in Fig. 5.49, the displacements in the local coordinate system based on the global displacement values result in

$$u_{1y} = -\underbrace{\sin\alpha}_{>0}\,\underbrace{u_{1X}}_{<0} + \underbrace{\cos\alpha}_{>0}\,\underbrace{u_{1Y}}_{>0}\,, \tag{5.512}$$

$$u_{2y} = -\sin\alpha\,\underbrace{u_{2X}}_{<0} + \cos\alpha\,\underbrace{u_{2Y}}_{>0}\,. \tag{5.513}$$

Corresponding to the above, the global displacements can be calculated from the local displacements:

$$u_{1X} = -u_{1y}\sin\alpha, \qquad u_{2X} = -u_{2y}\sin\alpha, \tag{5.514}$$

$$u_{1Y} = u_{1y}\cos\alpha, \qquad u_{2Y} = u_{2y}\cos\alpha. \tag{5.515}$$

The last relations between global and local displacements can be written in matrix form as:

$$\begin{bmatrix} u_{1X} \\ u_{1Y} \\ u_{2X} \\ u_{2Y} \end{bmatrix} = \begin{bmatrix} -\sin\alpha & 0 \\ \cos\alpha & 0 \\ 0 & -\sin\alpha \\ 0 & \cos\alpha \end{bmatrix} \begin{bmatrix} u_{1y} \\ u_{2y} \end{bmatrix}. \tag{5.516}$$

The nodal rotations do not need a transformation and the general transformation rule for the calculation of the global parameters from the local deformations results in abbreviated notation in

$$\boldsymbol{u}_{XY} = \boldsymbol{T}^{\mathrm{T}}\boldsymbol{u}_{xy}, \tag{5.517}$$

or alternatively in components:

$$\begin{bmatrix} u_{1X} \\ u_{1Y} \\ \varphi_{1Z} \\ u_{2X} \\ u_{2Y} \\ \varphi_{1Z} \end{bmatrix} = \begin{bmatrix} -\sin\alpha & 0 & 0 & 0 \\ \cos\alpha & 0 & 0 & 0 \\ 0 & 1 & 0 & 0 \\ 0 & 0 & -\sin\alpha & 0 \\ 0 & 0 & \cos\alpha & 0 \\ 0 & 0 & 0 & 1 \end{bmatrix} \begin{bmatrix} u_{1y} \\ \varphi_{1z} \\ u_{2y} \\ \varphi_{2z} \end{bmatrix}. \tag{5.518}$$

The last equation can also be solved for the deformations in the local coordinate system and through inversion[47] the transformation matrix results

$$\boldsymbol{u}_{xy} = \boldsymbol{T}\boldsymbol{u}_{XY} \tag{5.519}$$

or alternatively in components:

[47] Since the transformation matrix $\boldsymbol{T}$ is an orthogonal matrix, the following applies: $\boldsymbol{T}^{\mathrm{T}} = \boldsymbol{T}^{-1}$.

$$\begin{bmatrix} u_{1y} \\ \varphi_{1z} \\ u_{2y} \\ \varphi_{2z} \end{bmatrix} = \begin{bmatrix} -\sin\alpha & \cos\alpha & 0 & 0 & 0 & 0 \\ 0 & 0 & 1 & 0 & 0 & 0 \\ 0 & 0 & 0 & -\sin\alpha & \cos\alpha & 0 \\ 0 & 0 & 0 & 0 & 0 & 1 \end{bmatrix} \begin{bmatrix} u_{1X} \\ u_{1Y} \\ \varphi_{1Z} \\ u_{2X} \\ u_{2Y} \\ \varphi_{1Z} \end{bmatrix}. \tag{5.520}$$

The vector of the external load can be transformed in the same way:

$$\boldsymbol{f}_{XY} = \boldsymbol{T}^{\mathrm{T}} \boldsymbol{f}_{xy}, \tag{5.521}$$

$$\boldsymbol{f}_{xy} = \boldsymbol{T} \boldsymbol{f}_{XY}. \tag{5.522}$$

If the transformation of the local deformation into the global coordinate system is considered in the expression for the weak formulation according to Eq. (5.219), the transformation of the stiffness matrix into the global coordinate system results in

$$\boldsymbol{K}^{\mathrm{e}}_{XY} = \boldsymbol{T}^{\mathrm{T}} \boldsymbol{K}^{\mathrm{e}}_{xy} \boldsymbol{T}, \tag{5.523}$$

or alternatively in components:

$$\boldsymbol{K}^{\mathrm{e}}_{XYZ} = \frac{EI_z}{L^3} \begin{bmatrix} 12\mathrm{s}^2\alpha & -12\mathrm{s}\alpha\mathrm{c}\alpha & -6L\mathrm{s}\alpha & -12\mathrm{s}^2\alpha & 12\mathrm{s}\alpha\mathrm{c}\alpha & -6L\mathrm{s}\alpha \\ -12\mathrm{s}\alpha\mathrm{c}\alpha & 12\mathrm{c}^2\alpha & 6L\mathrm{c}\alpha & 12\mathrm{s}\alpha\mathrm{c}\alpha & -12\mathrm{c}^2\alpha & 6L\mathrm{c}\alpha \\ -6L\mathrm{s}\alpha & 6L\mathrm{c}\alpha & 4L^2 & 6L\mathrm{s}\alpha & -6L\mathrm{c}\alpha & 2L^2 \\ -12\mathrm{s}^2\alpha & 12\mathrm{s}\alpha\mathrm{c}\alpha & 6L\mathrm{s}\alpha & -12\mathrm{s}^2\alpha & -12\mathrm{s}\alpha\mathrm{c}\alpha & 6L\mathrm{s}\alpha \\ 12\mathrm{s}\alpha\mathrm{c}\alpha & -12\mathrm{c}^2\alpha & -6L\mathrm{c}\alpha & -12\mathrm{s}\alpha\mathrm{c}\alpha & 12\mathrm{c}^2\alpha & -6L\mathrm{c}\alpha \\ -6L\mathrm{s}\alpha & 6L\mathrm{c}\alpha & 2L^2 & 6L\mathrm{s}\alpha & -6L\mathrm{c}\alpha & 4L^2 \end{bmatrix}. \tag{5.524}$$

The sines and cosines values of the rotation angle $\alpha$ can be calculated through the global node coordinates via

$$\mathrm{s}\alpha \mathrel{\hat{=}} \sin\alpha = \frac{Y_2 - Y_1}{L} \quad \text{or} \quad \mathrm{c}\alpha \mathrel{\hat{=}} \cos\alpha = \frac{X_2 - X_1}{L} \tag{5.525}$$

and

$$L = \sqrt{(X_2 - X_1)^2 + (Y_2 - Y_1)^2}. \tag{5.526}$$

It needs to be remarked at this point, that in a mathematical positive sense the angle $\alpha$ always should be plotted from the global to the local coordinate system. The mathematical positive direction of rotation and therefore the algebraic sign of $\alpha$ is illustrated in Fig. 5.50. However, independent from the algebraic sign of $\alpha$ the calculation can always occur according to Eq. (5.525).

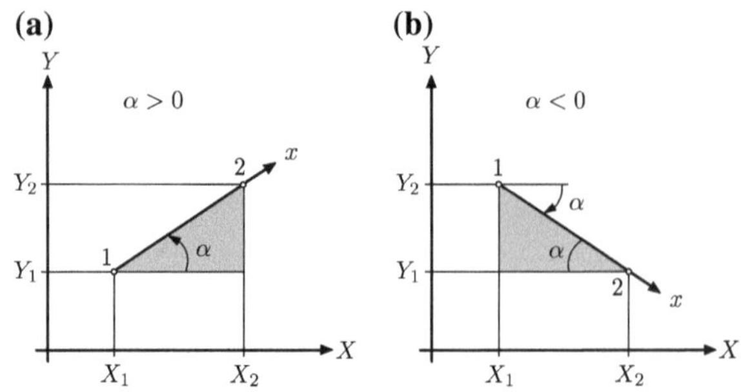


**Fig. 5.50** Rotation angle: **a** α positive; **b** α negative

## 5.6.2 Generalized Beam Element

A generalized beam element, i.e. an element which can deform in direction and perpendicular to the principal beam axis, can be obtained by superposition of the stiffness matrix of a rod and a beam element as given in Eqs. (5.79) and (5.236) as:

$$\begin{bmatrix} \frac{EA}{L} & 0 & 0 & -\frac{EA}{L} & 0 & 0 \\ 0 & \frac{12EI}{L^3} & \frac{6EI}{L^2} & 0 & -\frac{12EI}{L^3} & \frac{6EI}{L^2} \\ 0 & \frac{6EI}{L^2} & \frac{4EI}{L} & 0 & -\frac{6EI}{L^2} & \frac{2EI}{L} \\ -\frac{EA}{L} & 0 & 0 & \frac{EA}{L} & 0 & 0 \\ 0 & -\frac{12EI}{L^3} & -\frac{6EI}{L^2} & 0 & \frac{12EI}{L^3} & -\frac{6EI}{L^2} \\ 0 & \frac{6EI}{L^2} & \frac{2EI}{L} & 0 & -\frac{6EI}{L^2} & \frac{4EI}{L} \end{bmatrix} \begin{bmatrix} u_{1x} \\ u_{1y} \\ \varphi_{1z} \\ u_{2x} \\ u_{2y} \\ \varphi_{2z} \end{bmatrix} = \begin{bmatrix} F_{1x} \\ F_{1y} \\ M_{1z} \\ F_{2x} \\ F_{2y} \\ M_{2z} \end{bmatrix}. \tag{5.527}$$

The equation of the stiffness matrix in a global coordinate system, i.e. after a rotation by an angle $\alpha$, can be obtained by superposition of the deformations of a truss element (cf. Eqs. (5.485) and (5.486)) and a beam element (cf. Eqs. (5.514) and (5.515)) for the first node as

$$u_{1X} = \cos\alpha\, u_{1x} - \sin\alpha\, u_{1y}, \tag{5.528}$$

$$u_{1Y} = \sin\alpha\, u_{1x} + \cos\alpha\, u_{1y}, \tag{5.529}$$

and for the second node in a similar way as:

$$u_{2X} = \cos\alpha\, u_{12x} - \sin\alpha\, u_{2y}, \tag{5.530}$$

$$u_{2Y} = \sin\alpha\, u_{2x} + \cos\alpha\, u_{2y}. \tag{5.531}$$

These four transformation relationships can be arranged in matrix form as:

$$\begin{bmatrix} u_{1X} \\ u_{1Y} \\ u_{2X} \\ u_{2Y} \end{bmatrix} = \begin{bmatrix} \cos\alpha & -\sin\alpha & 0 & 0 \\ \sin\alpha & \cos\alpha & 0 & 0 \\ 0 & 0 & \cos\alpha & -\sin\alpha \\ 0 & 0 & \sin\alpha & \cos\alpha \end{bmatrix} \begin{bmatrix} u_{1x} \\ u_{1y} \\ u_{2x} \\ u_{2y} \end{bmatrix}, \tag{5.532}$$

or in abbreviated form as:

$$\boldsymbol{u}_{XY} = \boldsymbol{T}^{\mathrm{T}} \boldsymbol{u}_{xy}. \tag{5.533}$$

Since the rotational degrees of freedom do not need to be transformed, the final transformation for all six degrees of freedom can be obtained as:

$$\begin{bmatrix} u_{1X} \\ u_{1Y} \\ \varphi_{1Z} \\ \hline u_{2X} \\ u_{2Y} \\ \varphi_{2Z} \end{bmatrix} = \underbrace{\left[\begin{array}{ccc|ccc} \cos\alpha & -\sin\alpha & 0 & 0 & 0 & 0 \\ \sin\alpha & \cos\alpha & 0 & 0 & 0 & 0 \\ 0 & 0 & 1 & 0 & 0 & 0 \\ \hline 0 & 0 & 0 & \cos\alpha & -\sin\alpha & 0 \\ 0 & 0 & 0 & \sin\alpha & \cos\alpha & 0 \\ 0 & 0 & 0 & 0 & 0 & 1 \end{array}\right]}_{\boldsymbol{T}^{\mathrm{T}}} \begin{bmatrix} u_{1x} \\ u_{1y} \\ \varphi_{1z} \\ \hline u_{2x} \\ u_{2y} \\ \varphi_{2z} \end{bmatrix}. \tag{5.534}$$

The last equation can be rearranged for the unknowns in the local coordinate system as:

$$\begin{bmatrix} u_{1x} \\ u_{1y} \\ \varphi_{1z} \\ \hline u_{2x} \\ u_{2y} \\ \varphi_{2z} \end{bmatrix} = \underbrace{\left[\begin{array}{ccc|ccc} \cos\alpha & \sin\alpha & 0 & 0 & 0 & 0 \\ -\sin\alpha & \cos\alpha & 0 & 0 & 0 & 0 \\ 0 & 0 & 1 & 0 & 0 & 0 \\ \hline 0 & 0 & 0 & \cos\alpha & \sin\alpha & 0 \\ 0 & 0 & 0 & -\sin\alpha & \cos\alpha & 0 \\ 0 & 0 & 0 & 0 & 0 & 1 \end{array}\right]}_{\boldsymbol{T}} \begin{bmatrix} u_{1X} \\ u_{1Y} \\ \varphi_{1Z} \\ \hline u_{2X} \\ u_{2Y} \\ \varphi_{2Z} \end{bmatrix}, \tag{5.535}$$

or

$$\boldsymbol{u}_{xy} = \boldsymbol{T} \boldsymbol{u}_{XY}, \tag{5.536}$$

where the transformation matrix $\boldsymbol{T}$ can be expressed as:

$$\boldsymbol{T} = \begin{bmatrix} \boldsymbol{T}_1 & \boldsymbol{0} \\ \boldsymbol{0} & \boldsymbol{T}_2 \end{bmatrix}. \tag{5.537}$$

It should be noted here that the submatrix $\boldsymbol{T}_1$ in Eq. (5.537) performs the transformation at node 1 while submatrix $\boldsymbol{T}_2$ handles everything at node 2. The global stiffness matrix, i.e. the stiffness matrix in the global coordinate system $(X, Y)$, can be obtained by multiplication of the transformation matrix (5.537) with the beam stiffness matrix (5.236) according to relation (5.493), i.e. $\boldsymbol{K}^{\mathrm{e}}_{XY} = \boldsymbol{T}^{\mathrm{T}} \boldsymbol{K}^{\mathrm{e}}_{xy} \boldsymbol{T}$. The result of

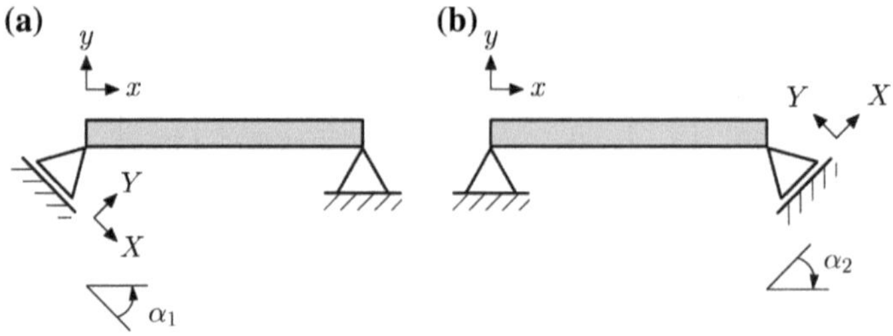


**Fig. 5.51** Beam element with rotated support: **a** rotation at node 1; **b** rotation at node 2

this multiplication is shown in Eq. (5.538) where the principal finite element equation in the global coordinate system is presented.

To simplify the solution of simple frame structures, Tables 5.13 and 5.14 collect expressions for the global stiffness matrix for some common angles $\alpha$.

$$E\begin{bmatrix} \frac{12I}{L^3}\sin^2\alpha+\frac{A}{L}\cos^2\alpha & \left(-\frac{12I}{L^3}+\frac{A}{L}\right)\sin\alpha\cos\alpha & -\frac{6I}{L^2}\sin\alpha & -\frac{12I}{L^3}\sin^2\alpha-\frac{A}{L}\cos^2\alpha & \left(\frac{12I}{L^3}-\frac{A}{L}\right)\sin\alpha\cos\alpha & -\frac{6I}{L^2}\sin\alpha \\ \left(-\frac{12I}{L^3}+\frac{A}{L}\right)\sin\alpha\cos\alpha & \frac{12I}{L^3}\cos^2\alpha+\frac{A}{L}\sin^2\alpha & \frac{6I}{L^2}\cos\alpha & \left(\frac{12I}{L^3}-\frac{A}{L}\right)\sin\alpha\cos\alpha & -\frac{12I}{L^3}\cos^2\alpha-\frac{A}{L}\sin^2\alpha & \frac{6I}{L^2}\cos\alpha \\ -\frac{6I}{L^2}\sin\alpha & \frac{6I}{L^2}\cos\alpha & \frac{4I}{L} & \frac{6I}{L^2}\sin\alpha & -\frac{6I}{L^2}\cos\alpha & \frac{2I}{L} \\ -\frac{12I}{L^3}\sin^2\alpha-\frac{A}{L}\cos^2\alpha & \left(\frac{12I}{L^3}-\frac{A}{L}\right)\sin\alpha\cos\alpha & \frac{6I}{L^2}\sin\alpha & \frac{12I}{L^3}\sin^2\alpha+\frac{A}{L}\cos^2\alpha & \left(-\frac{12I}{L^3}+\frac{A}{L}\right)\sin\alpha\cos\alpha & \frac{6I}{L^2}\sin\alpha \\ \left(\frac{12I}{L^3}-\frac{A}{L}\right)\sin\alpha\cos\alpha & -\frac{12I}{L^3}\cos^2\alpha-\frac{A}{L}\sin^2\alpha & -\frac{6I}{L^2}\cos\alpha & \left(-\frac{12I}{L^3}+\frac{A}{L}\right)\sin\alpha\cos\alpha & \frac{12I}{L^3}\cos^2\alpha+\frac{A}{L}\sin^2\alpha & -\frac{6I}{L^2}\cos\alpha \\ -\frac{6I}{L^2}\sin\alpha & \frac{6I}{L^2}\cos\alpha & \frac{2I}{L} & \frac{6I}{L^2}\sin\alpha & -\frac{6I}{L^2}\cos\alpha & \frac{4I}{L} \end{bmatrix}\begin{bmatrix} u_{1X} \\ u_{1Y} \\ \varphi_{1Z} \\ u_{2X} \\ u_{2Y} \\ \varphi_{2Z} \end{bmatrix}=\begin{bmatrix} F_{1X} \\ F_{1Y} \\ M_{1Z} \\ F_{2X} \\ F_{2Y} \\ M_{2Z} \end{bmatrix} \tag{5.538}$$

It might be required for certain problems to apply the transformation given in Eq. (5.537) only at one node of the element. This can be the case if a support is rotated only at one node as shown in Fig. 5.51. Let us have a look, for example, on the case shown in Fig. 5.51a where it would be quite difficult to describe the boundary condition in the local $(x, y)$ system. However, the global $(X, Y)$ system easily allows to specify the boundary conditions at the first node as: $u_{1Y} = 0 \wedge u_{1X} \neq 0$.

Thus, the transformation (5.537) can be individually applied at each node with a different transformation angle $\alpha_1$ at node 1 and $\alpha_2$ at node 2 as:

$$\boldsymbol{T} = \begin{bmatrix} \boldsymbol{T}_1(\alpha_1) & \boldsymbol{0} \\ \boldsymbol{0} & \boldsymbol{T}_2(\alpha_2) \end{bmatrix}. \tag{5.539}$$

**Table 5.13** Elemental stiffness matrices for plane frame elements given for different rotation angles $\alpha$, cf. Eq. (5.538)

| 0° | 180° |
|---|---|
| $E\begin{bmatrix} \frac{A}{L} & 0 & 0 & -\frac{A}{L} & 0 & 0 \\ 0 & \frac{12I}{L^3} & \frac{6I}{L^2} & 0 & -\frac{12I}{L^3} & \frac{6I}{L^2} \\ 0 & \frac{6I}{L^2} & \frac{4I}{L} & 0 & -\frac{6I}{L^2} & \frac{2I}{L} \\ -\frac{A}{L} & 0 & 0 & \frac{A}{L} & 0 & 0 \\ 0 & -\frac{12I}{L^3} & -\frac{6I}{L^2} & 0 & \frac{12I}{L^3} & -\frac{6I}{L^2} \\ 0 & \frac{6I}{L^2} & \frac{2I}{L} & 0 & -\frac{6I}{L^2} & \frac{4I}{L} \end{bmatrix}$ | $E\begin{bmatrix} \frac{A}{L} & 0 & 0 & -\frac{A}{L} & 0 & 0 \\ 0 & \frac{12I}{L^3} & -\frac{6I}{L^2} & 0 & -\frac{12I}{L^3} & -\frac{6I}{L^2} \\ 0 & -\frac{6I}{L^2} & \frac{4I}{L} & 0 & \frac{6I}{L^2} & \frac{2I}{L} \\ -\frac{A}{L} & 0 & 0 & \frac{A}{L} & 0 & 0 \\ 0 & -\frac{12I}{L^3} & \frac{6J}{L^2} & 0 & \frac{12I}{L^3} & \frac{6I}{L^2} \\ 0 & -\frac{6I}{L^2} & \frac{2I}{L} & 0 & \frac{6I}{L^2} & \frac{4I}{L} \end{bmatrix}$ |
| −90° | 90° |
| $E\begin{bmatrix} \frac{12I}{L^3} & 0 & \frac{6I}{L^2} & -\frac{12I}{L^3} & 0 & \frac{6I}{L^2} \\ 0 & \frac{A}{L} & 0 & 0 & -\frac{A}{L} & 0 \\ \frac{6I}{L^2} & 0 & \frac{4I}{L} & -\frac{6I}{L^2} & 0 & \frac{2I}{L} \\ -\frac{12I}{L^3} & 0 & -\frac{6I}{L^2} & \frac{12I}{L^3} & 0 & -\frac{6I}{L^2} \\ 0 & -\frac{A}{L} & 0 & 0 & \frac{A}{L} & 0 \\ \frac{6I}{L^2} & 0 & \frac{2I}{L} & -\frac{6I}{L^2} & 0 & \frac{4I}{L} \end{bmatrix}$ | $E\begin{bmatrix} \frac{12I}{L^3} & 0 & -\frac{6I}{L^2} & -\frac{12I}{L^3} & 0 & -\frac{6I}{L^2} \\ 0 & \frac{A}{L} & 0 & 0 & -\frac{A}{L} & 0 \\ -\frac{6I}{L^2} & 0 & \frac{4I}{L} & \frac{6I}{L^2} & 0 & \frac{2I}{L} \\ -\frac{12I}{L^3} & 0 & \frac{6I}{L^2} & \frac{12I}{L^3} & 0 & \frac{6I}{L^2} \\ 0 & -\frac{A}{L} & 0 & 0 & \frac{A}{L} & 0 \\ -\frac{6I}{L^2} & 0 & \frac{2I}{L} & \frac{6I}{L^2} & 0 & \frac{4I}{L} \end{bmatrix}$ |
| −45° | |
| $E\begin{bmatrix} \frac{6I}{L^3}+\frac{1}{2}\frac{A}{L} & \frac{6I}{L^3}-\frac{1}{2}\frac{A}{L} & \frac{3I\sqrt{2}}{L^2} & -\frac{6I}{L^3}-\frac{1}{2}\frac{A}{L} & -\frac{6I}{L^3}+\frac{1}{2}\frac{A}{L} & \frac{3I\sqrt{2}}{L^2} \\ \frac{6I}{L^3}-\frac{1}{2}\frac{A}{L} & \frac{6I}{L^3}+\frac{1}{2}\frac{A}{L} & \frac{3I\sqrt{2}}{L^2} & -\frac{6I}{L^3}+\frac{1}{2}\frac{A}{L} & -\frac{6I}{L^3}-\frac{1}{2}\frac{A}{L} & \frac{3I\sqrt{2}}{L^2} \\ \frac{3I\sqrt{2}}{L^2} & \frac{3I\sqrt{2}}{L^2} & \frac{4I}{L} & -\frac{3I\sqrt{2}}{L^2} & -\frac{3I\sqrt{2}}{L^2} & \frac{2I}{L} \\ -\frac{6I}{L^3}-\frac{1}{2}\frac{A}{L} & -\frac{6I}{L^3}+\frac{1}{2}\frac{A}{L} & -\frac{3I\sqrt{2}}{L^2} & \frac{6I}{L^3}+\frac{1}{2}\frac{A}{L} & \frac{6I}{L^3}-\frac{1}{2}\frac{A}{L} & -\frac{3I\sqrt{2}}{L^2} \\ -\frac{6I}{L^3}+\frac{1}{2}\frac{A}{L} & -\frac{6I}{L^3}-\frac{1}{2}\frac{A}{L} & -\frac{3I\sqrt{2}}{L^2} & \frac{6I}{L^3}-\frac{1}{2}\frac{A}{L} & \frac{6I}{L^3}+\frac{1}{2}\frac{A}{L} & -\frac{3I\sqrt{2}}{L^2} \\ \frac{3I\sqrt{2}}{L^2} & \frac{3I\sqrt{2}}{L^2} & \frac{2I}{L} & -\frac{3I\sqrt{2}}{L^2} & -\frac{3I\sqrt{2}}{L^2} & \frac{4I}{L} \end{bmatrix}$ | |
| 45° | |
| $E\begin{bmatrix} \frac{6I}{L^3}+\frac{1}{2}\frac{A}{L} & -\frac{6I}{L^3}+\frac{1}{2}\frac{A}{L} & -\frac{3I\sqrt{2}}{L^2} & -\frac{6I}{L^3}-\frac{1}{2}\frac{A}{L} & \frac{6I}{L^3}-\frac{1}{2}\frac{A}{L} & -\frac{3I\sqrt{2}}{L^2} \\ -\frac{6I}{L^3}+\frac{1}{2}\frac{A}{L} & \frac{6I}{L^3}+\frac{1}{2}\frac{A}{L} & \frac{3I\sqrt{2}}{L^2} & \frac{6I}{L^3}-\frac{1}{2}\frac{A}{L} & -\frac{6I}{L^3}-\frac{1}{2}\frac{A}{L} & \frac{3I\sqrt{2}}{L^2} \\ -\frac{3I\sqrt{2}}{L^2} & \frac{3I\sqrt{2}}{L^2} & \frac{4I}{L} & \frac{3I\sqrt{2}}{L^2} & -\frac{3I\sqrt{2}}{L^2} & \frac{2I}{L} \\ -\frac{6I}{L^3}-\frac{1}{2}\frac{A}{L} & \frac{6I}{L^3}-\frac{1}{2}\frac{A}{L} & \frac{3I\sqrt{2}}{L^2} & \frac{6I}{L^3}+\frac{1}{2}\frac{A}{L} & -\frac{6I}{L^3}+\frac{1}{2}\frac{A}{L} & \frac{3I\sqrt{2}}{L^2} \\ \frac{6I}{L^3}-\frac{1}{2}\frac{A}{L} & -\frac{6I}{L^3}-\frac{1}{2}\frac{A}{L} & -\frac{3I\sqrt{2}}{L^2} & -\frac{6I}{L^3}+\frac{1}{2}\frac{A}{L} & \frac{6I}{L^3}+\frac{1}{2}\frac{A}{L} & -\frac{3I\sqrt{2}}{L^2} \\ -\frac{3I\sqrt{2}}{L^2} & \frac{3I\sqrt{2}}{L^2} & \frac{2I}{L} & \frac{3I\sqrt{2}}{L^2} & -\frac{3I\sqrt{2}}{L^2} & \frac{4I}{L} \end{bmatrix}$ | |

**Table 5.14** Elemental stiffness matrices for plane frame elements given for different rotation angles $\alpha$, cf. Eq. (5.538)

| | |
|---|---|
| $-30°$ | $E\begin{bmatrix} \frac{3I}{L^3}+\frac{3}{4}\frac{A}{L} & -\frac{\sqrt{3}}{4}\left(-\frac{12I}{L^3}+\frac{A}{L}\right) & \frac{3I}{L^2} & -\frac{3I}{L^3}-\frac{3}{4}\frac{A}{L} & -\frac{\sqrt{3}}{4}\left(\frac{12I}{L^3}-\frac{A}{L}\right) & \frac{3I}{L^2} \\ -\frac{\sqrt{3}}{4}\left(-\frac{12I}{L^3}+\frac{A}{L}\right) & \frac{9I}{L^3}+\frac{1}{4}\frac{A}{L} & \frac{3I\sqrt{3}}{L^2} & -\frac{\sqrt{3}}{4}\left(\frac{12I}{L^3}-\frac{A}{L}\right) & -\frac{9I}{L^3}-\frac{1}{4}\frac{A}{L} & \frac{3I\sqrt{3}}{L^2} \\ \frac{3I}{L^2} & \frac{3I\sqrt{3}}{L^2} & \frac{4I}{L} & -\frac{3I}{L^2} & -\frac{3I\sqrt{3}}{L^2} & \frac{2I}{L} \\ -\frac{3I}{L^3}-\frac{3}{4}\frac{A}{L} & -\frac{\sqrt{3}}{4}\left(\frac{12I}{L^3}-\frac{A}{L}\right) & -\frac{3I}{L^2} & \frac{3I}{L^3}+\frac{3}{4}\frac{A}{L} & -\frac{\sqrt{3}}{4}\left(-\frac{12I}{L^3}+\frac{A}{L}\right) & -\frac{3I}{L^2} \\ -\frac{\sqrt{3}}{4}\left(\frac{12I}{L^3}-\frac{A}{L}\right) & -\frac{9I}{L^3}-\frac{1}{4}\frac{A}{L} & -\frac{3I\sqrt{3}}{L^2} & -\frac{\sqrt{3}}{4}\left(-\frac{12I}{L^3}+\frac{A}{L}\right) & \frac{9I}{L^3}+\frac{1}{4}\frac{A}{L} & -\frac{3I\sqrt{3}}{L^2} \\ \frac{3I}{L^2} & \frac{3I\sqrt{3}}{L^2} & \frac{2I}{L} & -\frac{3I}{L^2} & -\frac{3I\sqrt{3}}{L^2} & \frac{4I}{L} \end{bmatrix}$ |
| $30°$ | $E\begin{bmatrix} \frac{3I}{L^3}+\frac{3}{4}\frac{A}{L} & \frac{\sqrt{3}}{4}\left(-\frac{12I}{L^3}+\frac{A}{L}\right) & -\frac{3I}{L^2} & -\frac{3I}{L^3}-\frac{3}{4}\frac{A}{L} & \frac{\sqrt{3}}{4}\left(\frac{12I}{L^3}-\frac{A}{L}\right) & -\frac{3I}{L^2} \\ \frac{\sqrt{3}}{4}\left(-\frac{12I}{L^3}+\frac{A}{L}\right) & \frac{9I}{L^3}+\frac{1}{4}\frac{A}{L} & \frac{3I\sqrt{3}}{L^2} & \frac{\sqrt{3}}{4}\left(\frac{12I}{L^3}-\frac{A}{L}\right) & -\frac{9I}{L^3}-\frac{1}{4}\frac{A}{L} & \frac{3I\sqrt{3}}{L^2} \\ -\frac{3I}{L^2} & \frac{3I\sqrt{3}}{L^2} & \frac{4I}{L} & \frac{3I}{L^2} & -\frac{3I\sqrt{3}}{L^2} & \frac{2I}{L} \\ -\frac{3I}{L^3}-\frac{3}{4}\frac{A}{L} & \frac{\sqrt{3}}{4}\left(\frac{12I}{L^3}-\frac{A}{L}\right) & \frac{3I}{L^2} & \frac{3I}{L^3}+\frac{3}{4}\frac{A}{L} & \frac{\sqrt{3}}{4}\left(-\frac{12I}{L^3}+\frac{A}{L}\right) & \frac{3I}{L^2} \\ \frac{\sqrt{3}}{4}\left(\frac{12I}{L^3}-\frac{A}{L}\right) & -\frac{9I}{L^3}-\frac{1}{4}\frac{A}{L} & -\frac{3I\sqrt{3}}{L^2} & \frac{\sqrt{3}}{4}\left(-\frac{12I}{L^3}+\frac{A}{L}\right) & \frac{9I}{L^3}+\frac{1}{4}\frac{A}{L} & -\frac{3I\sqrt{3}}{L^2} \\ -\frac{3I}{L^2} & \frac{3I\sqrt{3}}{L^2} & \frac{2I}{L} & \frac{3I}{L^2} & -\frac{3I\sqrt{3}}{L^2} & \frac{4I}{L} \end{bmatrix}$ |

The last equation implies that the global coordinate system can be differently chosen at each node. In the case that the rotation is only required at the first node as shown in Fig. 5.51a, Eq. (5.539) can be simplified to

$$\boldsymbol{T} = \begin{bmatrix} \boldsymbol{T}_1(\alpha_1) & \mathbf{0} \\ \mathbf{0} & \mathbf{I} \end{bmatrix}, \tag{5.540}$$

where $\mathbf{I}$ is the identity matrix. Thus, the elemental stiffness matrix in the global coordinate system can be obtained for this special case based on the following relationship:

$$\boldsymbol{K}^{\mathrm{e}}_{XY} = \begin{bmatrix} \boldsymbol{T}_1(\alpha_1) & \mathbf{0} \\ \mathbf{0} & \mathbf{I} \end{bmatrix}^{\mathrm{T}} \boldsymbol{K}_{xy} \begin{bmatrix} \boldsymbol{T}_1(\alpha_1) & \mathbf{0} \\ \mathbf{0} & \mathbf{I} \end{bmatrix}. \tag{5.541}$$

In a similar way, the transformation can be only performed at node 2. The elemental stiffness matrices for these two special cases are summarized in Eqs. (5.542) and (5.543).

$$\boldsymbol{K}^{\mathrm{e}}_{XY}(\alpha_1, \alpha_2 = 0) = E \times \begin{bmatrix} \frac{12I - 12I\cos^2\alpha_1 + AL^2\cos^2\alpha_1}{L^3} & \frac{(-12I + AL^2)\cos\alpha_1\sin\alpha_1}{L^3} & -\frac{6I\sin\alpha_1}{L^2} & -\frac{A\cos\alpha_1}{L} & \frac{12I\sin\alpha_1}{L^3} & -\frac{6I\sin\alpha_1}{L^2} \\ \frac{(-12I + AL^2)\cos\alpha_1\sin\alpha_1}{L^3} & \frac{12I\cos^2\alpha_1 + AL^2 - AL^2\cos^2\alpha_1}{L^3} & \frac{6I\cos\alpha_1}{L^2} & -\frac{A\sin\alpha_1}{L} & -\frac{12I\cos\alpha_1}{L^3} & \frac{6I\cos\alpha_1}{L^2} \\ -\frac{6I\sin\alpha_1}{L^2} & \frac{6I\cos\alpha_1}{L^2} & \frac{4I}{L} & 0 & -\frac{6I}{L^2} & \frac{2I}{L} \\ -\frac{A\cos\alpha_1}{L} & -\frac{A\sin\alpha_1}{L} & 0 & \frac{A}{L} & 0 & 0 \\ \frac{12I\sin\alpha_1}{L^3} & -\frac{12I\cos\alpha_1}{L^3} & -\frac{6I}{L^2} & 0 & \frac{12I}{L^3} & -\frac{6I}{L^2} \\ -\frac{6I\sin\alpha_1}{L^2} & \frac{6I\cos\alpha_1}{L^2} & \frac{2I}{L} & 0 & -\frac{6I}{L^2} & \frac{4I}{L} \end{bmatrix} \tag{5.542}$$

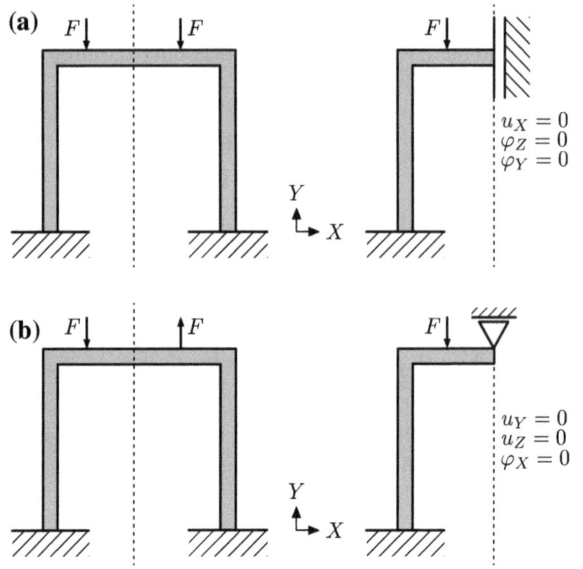


**Fig. 5.52** Portal frame structure with **a** symmetry and **b** anti-symmetry

$$\boldsymbol{K}^{\mathrm{e}}_{XY}(\alpha_1 = 0, \alpha_2) = E \times \begin{bmatrix} \frac{A}{L} & 0 & 0 & -\frac{A\cos\alpha_2}{L} & -\frac{A\sin\alpha_2}{L} & 0 \\ 0 & \frac{12I}{L^3} & \frac{6I}{L^2} & \frac{12I\sin\alpha_2}{L^3} & -\frac{6I\cos\alpha_2}{L^2} & \frac{2I}{L} \\ 0 & \frac{6I}{L^2} & \frac{4I}{L} & \frac{6I\sin\alpha_2}{L^2} & -\frac{12I\cos\alpha_2}{L^3} & \frac{6I}{L^2} \\ -\frac{A\cos\alpha_2}{L} & \frac{12I\sin\alpha_2}{L^3} & \frac{6I\sin\alpha_2}{L^2} & \frac{12I-12I\cos^2\alpha_2+AL^2\cos^2\alpha_2}{L^3} & \frac{(-12I+AL^2)\cos\alpha_2\sin\alpha_2}{L^3} & \frac{6I\sin\alpha_2}{L^2} \\ -\frac{A\sin\alpha_2}{L} & -\frac{12I\cos\alpha_2}{L^3} & -\frac{6I\cos\alpha_2}{L^2} & \frac{(-12I+AL^2)\cos\alpha_2\sin\alpha_2}{L^3} & \frac{12I\cos^2\alpha_2+AL^2-AL^2\cos^2\alpha_2}{L^3} & -\frac{6I\cos\alpha_2}{L^2} \\ 0 & \frac{6I}{L^2} & \frac{2I}{L} & \frac{6I\sin\alpha_2}{L^2} & -\frac{6I\cos\alpha_2}{L^2} & \frac{4I}{L} \end{bmatrix} \tag{5.543}$$

The size of some structures can be reduced by consideration of symmetry and anti-symmetry boundary conditions as shown in Fig. 5.52. The case of symmetry requires that the geometry and the load case is symmetric with respect to a certain plane while

**Table 5.15** Conditions for symmetry and anti-symmetry

| Case | $u_X$ | $u_Y$ | $u_Z$ | $\varphi_X$ | $\varphi_Y$ | $\varphi_Z$ |
|---|---|---|---|---|---|---|
| $X$-symmetry (resp. $Y$-$Z$ plane) | 0 | free | free | free | 0 | 0 |
| $Y$-symmetry (resp. $X$-$Y$ plane) | free | 0 | free | 0 | free | 0 |
| $Z$-symmetry (resp. $X$-$Y$ plane) | free | free | 0 | 0 | 0 | free |
| $X$-anti-symmetry (resp. $Y$-$Z$ plane) | free | 0 | 0 | 0 | free | free |
| $Y$-anti-symmetry (resp. $X$-$Y$ plane) | 0 | free | 0 | free | 0 | free |
| $Z$-anti-symmetry (resp. $X$-$Y$ plane) | 0 | 0 | free | free | free | 0 |

the case of anti-symmetry requires a symmetric geometry and anti-symmetric loading and results with respect to the same plane. A systematic summary of symmetric and anti-symmetric boundary conditions is given in Table 5.15.

### 5.6.3 Solved Problems

**5.13 Example: Portal frame structure**
The portal frame structure shown in Fig. 5.53 is loaded by a constant distributed load $q$ and a single horizontal force $F$. The three parts of the frame have the same length $L$, the same YOUNG's modulus $E$, the same cross-sectional area $A$, and the same second moment of area $I$.
Determine
(a)
- the elemental stiffness matrices in the global $X$-$Y$ system,
- the reduced system of equations,
- all nodal displacements and rotations,
- all reaction forces.

(b) Consider now the case $F = 0$, i.e. only the distributed load is acting on the frame. Develop a simpler model under consideration of the symmetry and determine the quantities as requested in part (a).

**5.13 Solution** (a) The free body diagram and the local coordinate axes of each element are shown in Fig. 5.54. In addition, the equivalent nodal loads resulting from the distributed load $q$ (cf. Table 5.9) are introduced at nodes 2 and 3. It should be noted that the unknown reactions at the supports are introduced in the positive direction of the respective coordinate axes. From this figure, the rotational angles

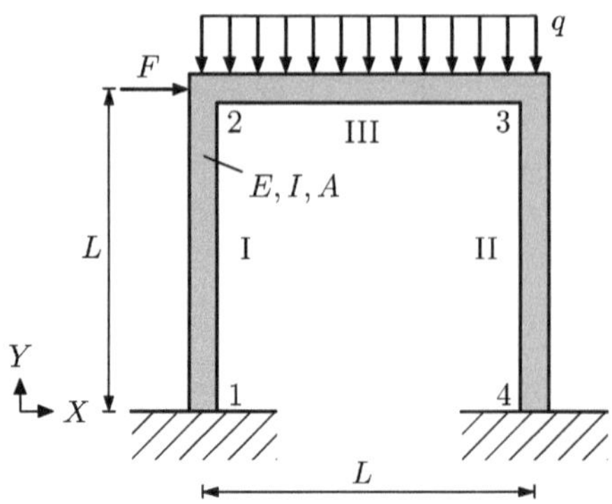


**Fig. 5.53** Portal frame structure

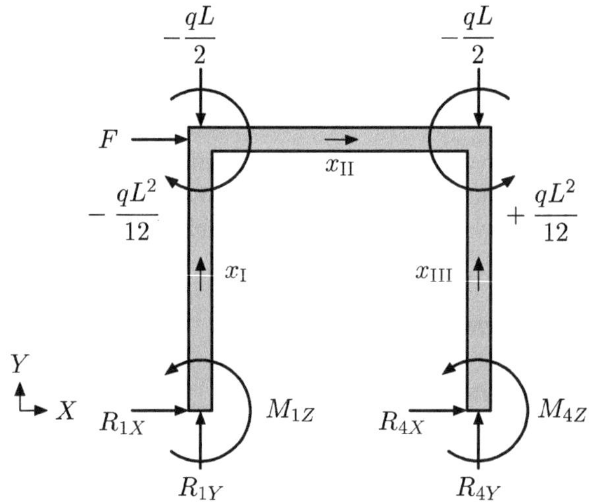


**Fig. 5.54** Free body diagram of the portal frame structure, see Fig. 5.53

from the global to the local coordinate system can be determined as $\alpha_{\mathrm{I}} = \alpha_{\mathrm{III}} = 90°$ and $\alpha_{\mathrm{II}} = 0°$.

The elemental stiffness matrices in the global $X$-$Y$ system can be obtained from Eq. (5.538) as:

$$\boldsymbol{K}_{\mathrm{I}}^{\mathrm{e}} = E \begin{array}{c} \begin{array}{cccccc} u_{1X} & u_{1Y} & \varphi_{1Z} & u_{2X} & u_{2Y} & \varphi_{2Z} \end{array} \\ \begin{bmatrix} \frac{12I}{L^3} & 0 & -\frac{6I}{L^2} & -\frac{12I}{L^3} & 0 & -\frac{6I}{L^2} \\ 0 & \frac{A}{L} & 0 & 0 & -\frac{A}{L} & 0 \\ -\frac{6I}{L^2} & 0 & \frac{4I}{L} & \frac{6I}{L^2} & 0 & \frac{2I}{L} \\ -\frac{12I}{L^3} & 0 & \frac{6I}{L^2} & \frac{12I}{L^3} & 0 & \frac{6I}{L^2} \\ 0 & -\frac{A}{L} & 0 & 0 & \frac{A}{L} & 0 \\ -\frac{6I}{L^2} & 0 & \frac{2I}{L} & \frac{6I}{L^2} & 0 & \frac{4I}{L} \end{bmatrix} \end{array} \begin{array}{c} u_{1X} \\ u_{1Y} \\ \varphi_{1Z} \\ u_{2X} \\ u_{2Y} \\ \varphi_{2Z} \end{array} \tag{5.544}$$

$$\boldsymbol{K}_{\mathrm{II}}^{\mathrm{e}} = E \begin{array}{c} \begin{array}{cccccc} u_{2X} & u_{2Y} & \varphi_{2Z} & u_{3X} & u_{3Y} & \varphi_{3Z} \end{array} \\ \begin{bmatrix} \frac{A}{L} & 0 & 0 & -\frac{A}{L} & 0 & 0 \\ 0 & \frac{12I}{L^3} & \frac{6I}{L^2} & 0 & -\frac{12I}{L^3} & \frac{6I}{L^2} \\ 0 & \frac{6I}{L^2} & \frac{4I}{L} & 0 & -\frac{6I}{L^2} & \frac{2I}{L} \\ -\frac{A}{L} & 0 & 0 & \frac{A}{L} & 0 & 0 \\ 0 & -\frac{12I}{L^3} & -\frac{6I}{L^2} & 0 & \frac{12I}{L^3} & -\frac{6I}{L^2} \\ 0 & \frac{6I}{L^2} & \frac{2I}{L} & 0 & -\frac{6I}{L^2} & \frac{4I}{L} \end{bmatrix} \end{array} \begin{array}{c} u_{2X} \\ u_{2Y} \\ \varphi_{2Z} \\ u_{3X} \\ u_{3Y} \\ \varphi_{3Z} \end{array} . \tag{5.545}$$

The third stiffness matrix is the same as given in Eq. (5.544), only the degrees of freedom are different:

$$\boldsymbol{K}_{\mathrm{III}}^{\mathrm{e}} = E \begin{array}{c} \begin{array}{cccccc} u_{4X} & u_{4Y} & \varphi_{4Z} & u_{3X} & u_{3Y} & \varphi_{3Z} \end{array} \\ \begin{bmatrix} \frac{12I}{L^3} & 0 & -\frac{6I}{L^2} & -\frac{12I}{L^3} & 0 & -\frac{6I}{L^2} \\ 0 & \frac{A}{L} & 0 & 0 & -\frac{A}{L} & 0 \\ -\frac{6I}{L^2} & 0 & \frac{4I}{L} & \frac{6I}{L^2} & 0 & \frac{2I}{L} \\ -\frac{12I}{L^3} & 0 & \frac{6I}{L^2} & \frac{12I}{L^3} & 0 & \frac{6I}{L^2} \\ 0 & -\frac{A}{L} & 0 & 0 & \frac{A}{L} & 0 \\ -\frac{6I}{L^2} & 0 & \frac{2I}{L} & \frac{6I}{L^2} & 0 & \frac{4I}{L} \end{bmatrix} \end{array} \begin{array}{c} u_{4X} \\ u_{4Y} \\ \varphi_{4Z} \\ u_{3X} \\ u_{3Y} \\ \varphi_{3Z} \end{array} . \tag{5.546}$$

The global system of equations, i.e. $\boldsymbol{K}\boldsymbol{u}_{\mathrm{p}} = \boldsymbol{f}$, can be assembled based on the scheme of unknowns which is indicated at each elemental stiffness matrix. Introducing the boundary conditions, i.e. $u_{1X} = u_{1Y} = u_{4X} = u_{4Y} = 0$ and $\varphi_{1Z} = \varphi_{4Z} = 0$, gives the following reduced system of equations:

$$\begin{bmatrix} \frac{12I}{L^3}+\frac{A}{L} & 0+0 & \frac{6I}{L^2}+0 & -\frac{A}{L} & 0 & 0 \\ 0+0 & \frac{A}{L}+\frac{12I}{L^3} & 0+\frac{6I}{L^2} & 0 & -\frac{12I}{L^3} & \frac{6I}{L^2} \\ \frac{6I}{L^2}+0 & 0+\frac{6I}{L^2} & \frac{4I}{L}+\frac{4I}{L} & 0 & -\frac{6I}{L^2} & \frac{2I}{L} \\ -\frac{A}{L} & 0 & 0 & \frac{A}{L}+\frac{12I}{L^3} & 0+0 & 0+\frac{6I}{L^2} \\ 0 & -\frac{12I}{L^3} & -\frac{6I}{L^2} & 0+0 & \frac{12I}{L^3}+\frac{A}{L} & -\frac{6I}{L^2}+0 \\ 0 & \frac{6I}{L^2} & \frac{2I}{L} & 0+\frac{6I}{L^2} & -\frac{6I}{L^2}+0 & \frac{4I}{L}+\frac{4I}{L} \end{bmatrix} \begin{bmatrix} u_{2X} \\ u_{2Y} \\ \varphi_{2Z} \\ u_{3X} \\ u_{3Y} \\ \varphi_{3Z} \end{bmatrix} = \begin{bmatrix} F \\ -\frac{qL}{2} \\ -\frac{qL^2}{12} \\ 0 \\ -\frac{qL}{2} \\ \frac{qL^2}{12} \end{bmatrix}. \tag{5.547}$$

The solution of this linear (6×6) system can be obtained, for example, by inverting the reduced stiffness matrix to give the reduced result vector $\left[u_{2X}\ u_{2Y}\ \varphi_{2Z}\ u_{3X}\ u_{3Y}\ \varphi_{3Z}\right]^{\mathrm{T}}$ as:

$$\begin{bmatrix} \frac{L^3\left(10FL^4A^2+7qAJL^3+168FAJL^2+24qJ^2L+432FJ^2\right)}{24\left(7A^2L^4+45AL^2J+72J^2\right)EJ} \\ \frac{L^2\left(6LFA-7\,qAL^2-24qJ\right)}{2\left(7AL^2+24J\right)EA} \\ -\frac{L^2\left(18FL^4A^2+612FAJL^2+1728FJ^2+7qL^5A^2+66qAJL^3+144qJ^2L\right)}{72\left(AL^2+3J\right)\left(7AL^2+24J\right)EJ} \\ \frac{L^3\left(10FL^4A^2+84FAJL^2+144FJ^2-7qAJL^3-24qJ^2L\right)}{24\left(7A^2L^4+45AL^2J+72J^2\right)EJ} \\ -\frac{L^2\left(6FLA+24qJ+7qAL^2\right)}{2\left(7AL^2+24J\right)EA} \\ -\frac{L^2\left(18FL^4A^2+360FAJL^2+864FJ^2-7qL^5A^2-66qAJL^3-144qJ^2L\right)}{72\left(7A^2L^4+45AL^2J+72J^2\right)EJ} \end{bmatrix}. \tag{5.548}$$

The reaction forces can be obtained by multiplying the global (non-reduced) stiffness matrix with the total displacement vector, i.e.

$$\boldsymbol{u}^{\mathrm{T}} = [0\ \ 0\ \ 0\ \ u_{2X}\ \ u_{2Y}\ \ \varphi_{2Z}\ \ u_{3X}\ \ u_{3Y}\ \ \varphi_{3Z}\ \ 0\ \ 0\ \ 0]\,, \tag{5.549}$$

to give $\left[R_{1X}\ R_{1Y}\ M_{1Z}\ R_{4X}\ R_{4Y}\ M_{4Z}\ \right]^{\mathrm{T}}$ as:

$$\begin{bmatrix} -\dfrac{-qL^3A+6FL^2A+36FJ}{12\left(AL^2+3J\right)} \\ -\dfrac{L\left(6LFA-7qAL^2-24qJ\right)}{2\left(7AL^2+24J\right)} \\ \dfrac{\left(-7qL^5A^2+72FL^4A^2-3qAJL^3+900FAJL^2+72qJ^2L+2160FJ^2\right)L}{36\left(AL^2+3J\right)\left(7AL^2+24J\right)} \\ -\dfrac{\left(qL+6F\right)AL^2}{12\left(AL^2+3J\right)} \\ \dfrac{L\left(6FLA+24qJ+7qAL^2\right)}{2\left(7AL^2+24J\right)} \\ \dfrac{L\left(72FL^4A^2+396FAJL^2+432FJ^2+3qAJL^3-72qJ^2L+7qL^5A^2\right)}{36\left(7A^2L^4+45AL^2J+72J^2\right)} \end{bmatrix}. \quad (5.550)$$

(b) In the case of $F=0$, the symmetry in regards to the geometry *and* the load case can be used to create a simplified model under consideration of appropriate symmetry conditions. As can be seen in Fig. 5.55a, only half of the structure needs to be modeled if at the symmetry line ($X=\frac{L}{2}$) the condition $u_{3X}=\varphi_{3Z}=0$ is imposed. The free body diagram of this structure is shown in Fig. 5.55b where now only two finite elements are required to simulate the structure.

The elemental stiffness matrices can be taken from Eqs. (5.544) and (5.545) in which the transformation $L_{\mathrm{II}}=\frac{L}{2}$ must be applied. Assembling to the global system of equations and consideration of the boundary condition, i.e. $u_{1X}=u_{1Y}=u_{3X}=0$ and $\varphi_{1Z}=\varphi_{3Z}=0$, gives the following reduced system of equations:

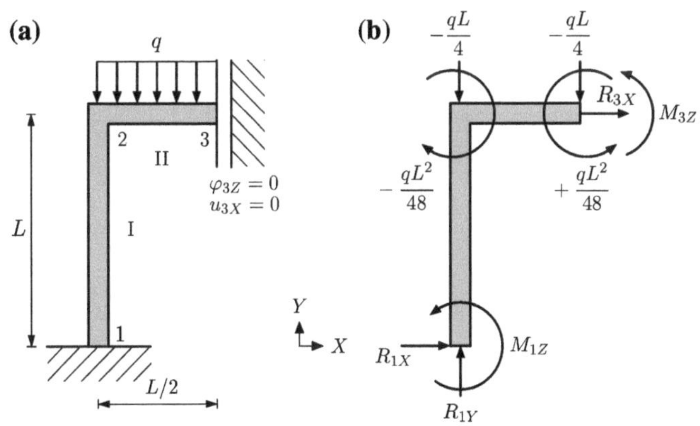


**Fig. 5.55** Symmetrical portal frame structure: **a** half model; **b** free body diagram

$$\begin{bmatrix} \frac{12I}{L^3}+\frac{2A}{L} & 0+0 & \frac{6I}{L^2}+0 & 0 \\ 0+0 & \frac{A}{L}+\frac{96I}{L^3} & 0+\frac{24I}{L^2} & -\frac{96I}{L^3} \\ \frac{6I}{L^2}+0 & 0+\frac{24I}{L^2} & \frac{4I}{L}+\frac{8I}{L} & -\frac{24I}{L^2} \\ 0 & -\frac{96I}{L^3} & -\frac{24I}{L^2} & \frac{96I}{L^3} \end{bmatrix} \begin{bmatrix} u_{2X} \\ u_{2Y} \\ \varphi_{2Z} \\ u_{3Y} \end{bmatrix} = \begin{bmatrix} 0 \\ -\frac{qL}{4} \\ -\frac{qL^2}{48} \\ -\frac{qL}{4} \end{bmatrix}. \tag{5.551}$$

The solution of this systems of equations is obtained, for example, by inverting the reduced stiffness matrix as:

$$\begin{bmatrix} u_{2X} \\ u_{2Y} \\ \varphi_{2Z} \\ u_{3Y} \end{bmatrix} = \begin{bmatrix} \frac{qL^4}{24\left(AL^2+3I\right)E} \\ -\frac{qL^2}{2EA} \\ -\frac{qL^3\left(6I+AL^2\right)}{72\left(AL^2+3I\right)EI} \\ -\frac{qL^2\left(609IAL^2+1728I^2+7A^2L^4\right)}{1152\left(AL^2+3I\right)AEI} \end{bmatrix}. \tag{5.552}$$

Multiplying the global (non-reduced) stiffness matrix with the known vector of nodal displacements and rotations serves to calculate the reactions as:

$$\begin{bmatrix} R_{1X} \\ R_{1Y} \\ M_{1Z} \\ R_{3x} \\ M_{3Z} \end{bmatrix} = \begin{bmatrix} \frac{qAL^3}{12\left(AL^2+3I\right)} \\ \frac{qL}{2} \\ -\frac{qL^2\left(-3I+AL^2\right)}{36\left(AL^2+3I\right)} \\ -\frac{qAL^3}{12\left(AL^2+3I\right)} \\ \frac{qL^2\left(21I+5AL^2\right)}{72\left(AL^2+3I\right)} \end{bmatrix}. \tag{5.553}$$

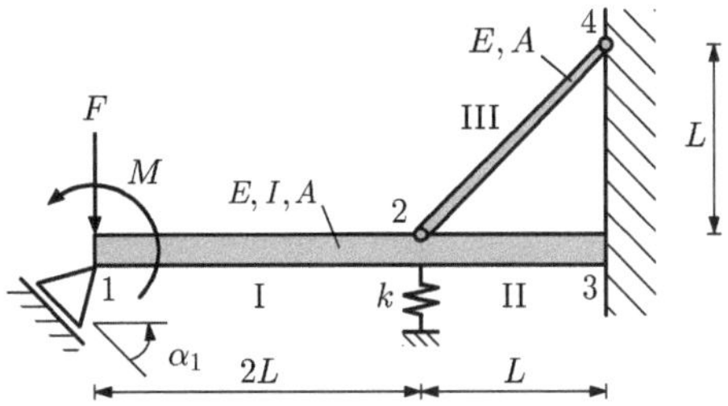


**Fig. 5.56** Plane frame structure with rotated support and spring

**5.14 Example: Plane frame structure with rotated support and spring**

The plane frame structure shown in Fig. 5.56 is loaded by a single vertical force $F$ and a single moment $M = FL$ at point 1. The structure is composed of a generalized beam of length $3L$ between point 1 and 3 and a rod element between point 2 and 4. The members of the frame have the same YOUNG's modulus $E$ and the same cross-sectional area $A$. The generalized beam is in addition characterized by the second moment of area $I$. The structure is supported at point 2 by a spring with a spring constant of $k = \frac{EA}{L}$. The simple support at point 1 is rotated by $\alpha_1 = 45°$. Use three elements as indicated in the figure to model the problem. Determine

- the elemental stiffness matrices of the three elements,
- the reduced system of equations,
- all nodal displacements and rotations,
- all reaction forces and reaction moments at the supports.
- Validate your results by checking the global equilibrium of forces and moments.

**5.14 Solution**

The free body diagram and the local coordinate systems are shown in Fig. 5.57. The unknown reactions at the supports are introduced in the positive direction of the respective global coordinate system. Let us have first a look at element I. At node 1, it is required to introduce a rotated global coordinate system in order to be able to impose the boundary condition resulting from the rotated support. Thus, the global coordinate system is rotated by $\alpha_1 = 45°$ as indicated in Fig. 5.57. At the second node, there is no transformation between the local an global coordinate

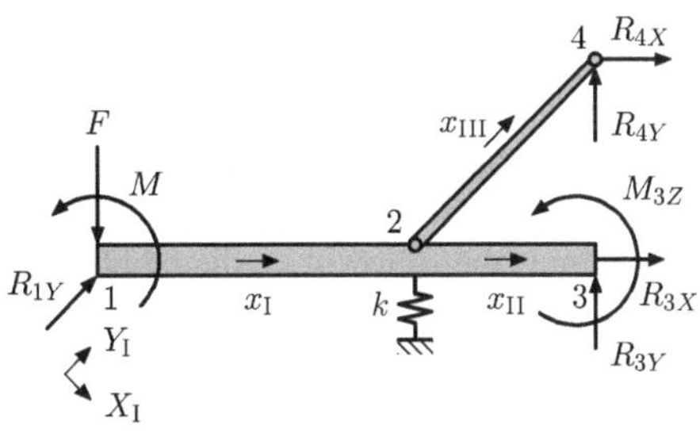


**Fig. 5.57** Free body diagram of the plane frame structure with rotated support and spring, see Fig. 5.56

system required ($\alpha_2 = 0$). Thus, the global elemental stiffness matrix of element I can be obtained from Eq. (5.542) with $\alpha_{\mathrm{I}} = 45°$ and $L_{\mathrm{I}} = 2L$ as:

$$\boldsymbol{K}_{\mathrm{I}}^{\mathrm{e}} = E \times \begin{bmatrix} \frac{AL^2+12I}{16L^3} & \frac{AL^2-12I}{16L^3} & -\frac{3\sqrt{2}I}{4L^2} & -\frac{\sqrt{2}A}{4L} & \frac{3\sqrt{2}I}{4L^3} & -\frac{3\sqrt{2}I}{4L^2} \\ \frac{AL^2-12I}{16L^3} & \frac{AL^2+12I}{16L^3} & \frac{3\sqrt{2}I}{4L^2} & -\frac{\sqrt{2}A}{4L} & -\frac{3\sqrt{2}I}{4L^3} & \frac{3\sqrt{2}I}{4L^2} \\ -\frac{3\sqrt{2}I}{4L^2} & \frac{3\sqrt{2}I}{4L^2} & \frac{2I}{L} & 0 & -\frac{3I}{2L^2} & \frac{I}{L} \\ -\frac{\sqrt{2}A}{4L} & -\frac{\sqrt{2}A}{4L} & 0 & \frac{A}{2L} & 0 & 0 \\ \frac{3\sqrt{2}I}{4L^3} & -\frac{3\sqrt{2}I}{4L^3} & -\frac{3I}{2L^2} & 0 & \frac{3I}{2L^3} & -\frac{3I}{2L^2} \\ -\frac{3\sqrt{2}I}{4L^2} & \frac{3\sqrt{2}I}{4L^2} & \frac{I}{L} & 0 & -\frac{3I}{2L^2} & \frac{2I}{L} \end{bmatrix}. \tag{5.554}$$

Considering the boundary condition at node 1, i.e. $u_{1Y} = 0$, the second column and second row can be canceled to form the global stiffness matrix. The second element does not require transformation and the stiffness matrix in the global system is given by Eq. (5.527), i.e.

$$\boldsymbol{K}_{\mathrm{II}}^{\mathrm{e}} = \begin{bmatrix} \frac{EA}{L} & 0 & 0 & -\frac{EA}{L} & 0 & 0 \\ 0 & \frac{12EI}{L^3} & \frac{6EI}{L^2} & 0 & -\frac{12EI}{L^3} & \frac{6EI}{L^2} \\ 0 & \frac{6EI}{L^2} & \frac{4EI}{L} & 0 & -\frac{6EI}{L^2} & \frac{2EI}{L} \\ -\frac{EA}{L} & 0 & 0 & \frac{EA}{L} & 0 & 0 \\ 0 & -\frac{12EI}{L^3} & -\frac{6EI}{L^2} & 0 & \frac{12EI}{L^3} & -\frac{6EI}{L^2} \\ 0 & \frac{6EI}{L^2} & \frac{2EI}{L} & 0 & -\frac{6EI}{L^2} & \frac{4EI}{L} \end{bmatrix}. \tag{5.555}$$

The fixed support at node 3 allows to cancel the last three columns and rows to form the global stiffness matrix. The rod element is rotated by 45° and the stiffness matrix in the global coordinate system can be taken from Table 5.11 as:

$$\boldsymbol{K}_{\mathrm{III}}^{\mathrm{e}} = \frac{EA}{L} \begin{bmatrix} \frac{1}{2} & \frac{1}{2} & -\frac{1}{2} & -\frac{1}{2} \\ \frac{1}{2} & \frac{1}{2} & -\frac{1}{2} & -\frac{1}{2} \\ -\frac{1}{2} & -\frac{1}{2} & \frac{1}{2} & \frac{1}{2} \\ -\frac{1}{2} & -\frac{1}{2} & \frac{1}{2} & \frac{1}{2} \end{bmatrix}. \tag{5.556}$$

Considering the support at node 4, the last two columns and rows can be canceled to form the global matrix. Assembling the three matrices and considering the boundary conditions, the following reduced global system of equations can be obtained:

$$\begin{bmatrix} \frac{(6I+2AL^2)E}{8L^3} & -\frac{3IE\sqrt{2}}{4L^2} & -\frac{AE\sqrt{2}}{4L} & \frac{3IE\sqrt{2}}{4L^3} & -\frac{3IE\sqrt{2}}{4L^2} \\ -\frac{3IE\sqrt{2}}{4L^2} & \frac{2IE}{L} & 0 & -\frac{3IE}{2L^2} & \frac{IE}{L} \\ -\frac{AE\sqrt{2}}{4L} & 0 & \frac{3AE}{2L}+\frac{AE\sqrt{2}}{4L} & \frac{AE\sqrt{2}}{4L} & 0 \\ \frac{3IE\sqrt{2}}{4L^3} & -\frac{3IE}{2L^2} & \frac{AE\sqrt{2}}{4L} & \frac{27IE}{2L^3}+\frac{AE\sqrt{2}}{4L}+\frac{AE}{L} & \frac{9IE}{2L^2} \\ -\frac{3IE\sqrt{2}}{4L^2} & \frac{IE}{L} & 0 & \frac{9IE}{2L^2} & \frac{6IE}{L} \end{bmatrix} \begin{bmatrix} u_{1X} \\ \varphi_{1Z} \\ u_{2X} \\ u_{2Y} \\ \varphi_{2Z} \end{bmatrix}$$
$$= \left[\tfrac{1}{2}\sqrt{2}F \;\; M \;\; 0 \;\; 0 \;\; 0\right]^{\mathrm{T}}. \tag{5.557}$$

It should be noted here that the spring constant $k = \frac{EA}{L}$ was added in the cell $(u_{2Y}, u_{2Y})$. The solution of this system of equations is obtained, for example, by inverting the reduced stiffness matrix as:

$$\begin{bmatrix} u_{1X} \\ \varphi_{1Z} \\ u_{2X} \\ u_{2Y} \\ \varphi_{2Z} \end{bmatrix} = \begin{bmatrix} \frac{L^3\sqrt{2}\left(1458\,I+243\,I\sqrt{2}+79\,A\sqrt{2}L^2+222\,AL^2\right)F}{2E\left(9\sqrt{2}I^2+54\,I^2+36\sqrt{2}AL^2I+180\,AL^2I+11\sqrt{2}A^2L^4+22\,A^2L^4\right)} \\ \frac{3L^2\left(45\sqrt{2}I^2+270\,I^2+29\sqrt{2}AL^2I+138\,AL^2I+8\,A^2L^4+4\sqrt{2}A^2L^4\right)F}{2EI\left(9\sqrt{2}I^2+54\,I^2+36\sqrt{2}AL^2I+180\,AL^2I+11\sqrt{2}A^2L^4+22\,A^2L^4\right)} \\ \frac{L^3\left(33\,I\sqrt{2}+486\,I+5\,A\sqrt{2}L^2+74\,AL^2\right)F}{2E\left(9\sqrt{2}I^2+54\,I^2+36\sqrt{2}AL^2I+180\,AL^2I+11\sqrt{2}A^2L^4+22\,A^2L^4\right)} \\ -\frac{L^3\left(33\,I\sqrt{2}+198\,I+5\,A\sqrt{2}L^2-54\,AL^2\right)F}{2E\left(9\sqrt{2}I^2+54\,I^2+36\sqrt{2}AL^2I+180\,AL^2I+11\sqrt{2}A^2L^4+22\,A^2L^4\right)} \\ \frac{L^2\left(63\sqrt{2}I^2+378\,I^2+9\sqrt{2}AL^2I-54\,AL^2I-2\sqrt{2}A^2L^4-4\,A^2L^4\right)F}{2EI\left(9\sqrt{2}I^2+54\,I^2+36\sqrt{2}AL^2I+180\,AL^2I+11\sqrt{2}A^2L^4+22\,A^2L^4\right)} \end{bmatrix}. \tag{5.558}$$

The reaction forces can be obtained by multiplying the global (non-reduced) stiffness matrix with the total displacement vector, i.e.

$$\boldsymbol{u}^{\mathrm{T}} = [u_{1X} \;\; 0 \;\; \varphi_{1Z} \;\; u_{2X} \;\; u_{2Y} \;\; \varphi_{2Z} \;\; 0 \;\; 0 \;\; 0 \;\; 0 \;\; 0]\,, \tag{5.559}$$

to give

$$\begin{bmatrix} R_{1Y} \\ R_{3X} \\ R_{3Y} \\ M_{3Z} \\ R_{4X} \\ R_{4Y} \end{bmatrix} = \begin{bmatrix} \frac{\sqrt{2}AL^2\left(105I\sqrt{2}+486I+37A\sqrt{2}L^2+74AL^2\right)F}{2\left(9\sqrt{2}I^2+54I^2+36\sqrt{2}AL^2I+180AL^2I+11\sqrt{2}A^2L^4+22A^2L^4\right)} \\ -\frac{AL^2\left(33I\sqrt{2}+486I+5A\sqrt{2}L^2+74AL^2\right)F}{2\left(9\sqrt{2}I^2+54I^2+36\sqrt{2}AL^2I+180AL^2I+11\sqrt{2}A^2L^4+22A^2L^4\right)} \\ \frac{3\left(3\sqrt{2}I^2+18I^2+\sqrt{2}AL^2I-54AL^2I+2\sqrt{2}A^2L^4+4A^2L^4\right)F}{9\sqrt{2}I^2+54I^2+36\sqrt{2}AL^2I+180AL^2I+11\sqrt{2}A^2L^4+22A^2L^4} \\ -\frac{4A\sqrt{2}L^2(9I+4AL^2)F}{9\sqrt{2}I^2+54I^2+36\sqrt{2}AL^2I+180AL^2I+11\sqrt{2}A^2L^4+22A^2L^4} \\ -\frac{4A\sqrt{2}L^2(9I+4AL^2)F}{9\sqrt{2}I^2+54I^2+36\sqrt{2}AL^2I+180AL^2I+11\sqrt{2}A^2L^4+22A^2L^4} \end{bmatrix} . \tag{5.560}$$

with $R_{1X} = R_{2X} = R_{2Y} = 0$ and $M_{1Z} = M_{2Z} = 0$. The global equilibrium can be stated in the following way:

$$\sum_i F_{X,i} = \frac{\sqrt{2}}{2}R_{1X} + \frac{\sqrt{2}}{2}R_{1Y} + R_{2X} + R_{3X} + R_{4X} = 0 \quad \checkmark . \tag{5.561}$$

Here, it is important that the reaction forces at node 1 are transformed to a common coordinate system, i.e. with a horizontal $X$-axis and a vertical $Y$-axis. To state the force equilibrium in the vertical direction, it is required to consider the force which acts from the spring to the beam element. If a negative displacement at node is obtained, an upward vertical force of magnitude $u_{2Y}k$ should be considered. Thus, the force equilibrium in the vertical direction reads as:

$$\sum_i F_{Y,i} = -F + \frac{\sqrt{2}}{2}R_{1Y} - \frac{\sqrt{2}}{2}R_{1X} \\ + R_{2Y} + R_{3Y} + R_{4Y} - u_{2Y}k = 0 \quad \checkmark . \tag{5.562}$$

In a similar way, it is required to consider the force from the spring to the beam when formulating the global moment balance. Taking node 1 as the reference point, the global moment equilibrium reads as:

$$\sum_i M_{Z,i} = M + M_{3Z} + R_{3Y}(L_{\mathrm{I}} + L_{\mathrm{II}}) + R_{4Y}(L_{\mathrm{I}} + L_{\mathrm{II}}) \\ - R_{4X}L_{\mathrm{II}} - u_{2Y}kL_{\mathrm{I}} = 0 \quad \checkmark . \tag{5.563}$$

## 5.7 Supplementary Problems

### 5.7.1 Weighted Residual Method Problems

**5.15 Example: Method of least squares**
The method of least squares minimizes the integral of the square of the residuals, i.e. $\int_\Omega r^2 \mathrm{d}\Omega$ = minimum. Why is an approach which would just aim to minimize the integral of the residuals, i.e.

$$\int_\Omega r \mathrm{d}\Omega = \text{minimum}, \tag{5.564}$$

not the best approach?

### 5.7.2 Rod Element Problems

**5.16 Example: Weighted residual method with arbitrary distributed load for a rod**
Derive the principal finite element equation for a rod element based on the weighted residual method. Starting point should be the partial differential equation with an arbitrary distributed load $p_x(x)$. In addition, it can be assumed that the axial tensile stiffness $EA$ is constant.

**5.17 Example: Finite element approximation with a single linear rod element**
Given is a rod with different load cases as shown in Fig. 5.58. The axial tensile stiffness $EA$ is constant and the length is equal to $L$. Derive the finite element solution based on a single linear element and compare the elongation $u_x(x)$ and $u_x(L)$ with the analytical solution.

**5.18 Example: Finite element approximation with a single quadratic rod element**
Solve problem 5.17 with a single quadratic rod element.

**5.19 Example: Equivalent nodal loads for a quadratic distribution (linear rod element)**
Given are the following two formulations for a distributed quadratic load. Calculate the equivalent nodal loads for a linear rod element, cf. Fig. 5.59.

(a) $p_x(x) = p_0^* x^2$,

(b) $p_x(x) = p_0 \left(\frac{x}{L}\right)^2$.

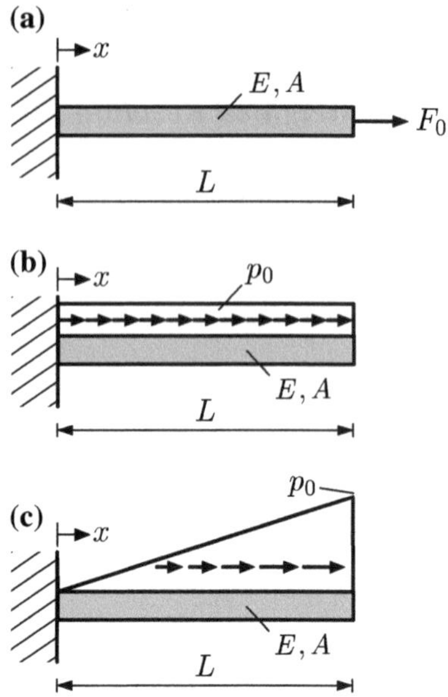


**Fig. 5.58** Finite element approximation with a single element for different load cases: **a** single force, **b** constant distributed load and **c** linear distributed load

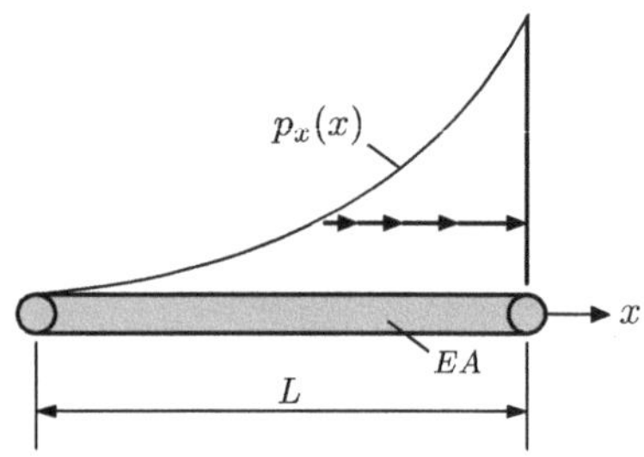


**Fig. 5.59** Distributed load with quadratic function

The dimension of the constant $p_0^*$ is equal to force per length to power 3 while the dimension of $p_0$ is force per length.

**5.20 Example: Derivation of shape functions for a quadratic rod element**
Derive the three shape functions for a quadratic rod element under the following assumption for the displacement field:

$$u^{\mathrm{e}}(\xi) = a_0 + a_1\xi + a_2\xi^2. \tag{5.565}$$

Assume for the derivation that the third node is exactly in the middle of the element ($\xi = 0$). Plot the four shape functions in dependence of the natural coordinate $\xi$.

**5.21 Example: Derivation of shape functions and stiffness matrix for a quadratic rod element with unevenly distributed nodes**
Derive the three shape functions for a quadratic rod element with unevenly distributed nodes (cf. Fig. 5.60) under the following assumption for the displacement field:

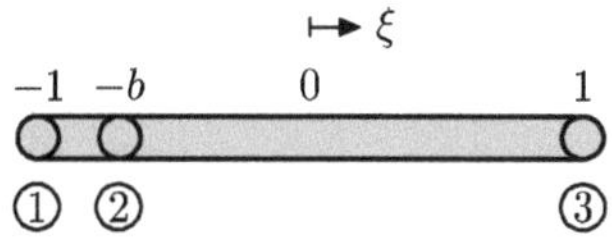


**Fig. 5.60** Quadratic rod element with unevenly distributed nodes

$$u^{\mathrm{e}}(\xi) = a_0 + a_1\xi + a_2\xi^2. \tag{5.566}$$

Derive as a function of position $b$ the stiffness matrix for an axial stiffness $EA$ and length $L$ of the element. Which problems can occur if the second node is close to the boundary, e.g. $\xi = -0.9$?

**5.22 Example: Derivation of shape functions for a cubic rod element**
Derive the four shape functions for a cubic rod element under the following assumption for the displacement field:

$$u^{\mathrm{e}}(\xi) = a_0 + a_1\xi + a_2\xi^2 + a_3\xi^3. \tag{5.567}$$

Assume for the derivation that the nodes are equally spaced.

**5.23 Example: Structure composed of three linear rod elements**
Calculate for the two structures shown in Fig. 5.61 the unknown displacement vector and the reaction forces for $L_{\mathrm{I}} = L_{\mathrm{II}} = L_{\mathrm{III}} = L$ and

$$(EA)_{\mathrm{I}} = 3EA, \tag{5.568}$$

$$(EA)_{\mathrm{II}} = 2EA, \tag{5.569}$$

$$(EA)_{\mathrm{III}} = 1EA. \tag{5.570}$$

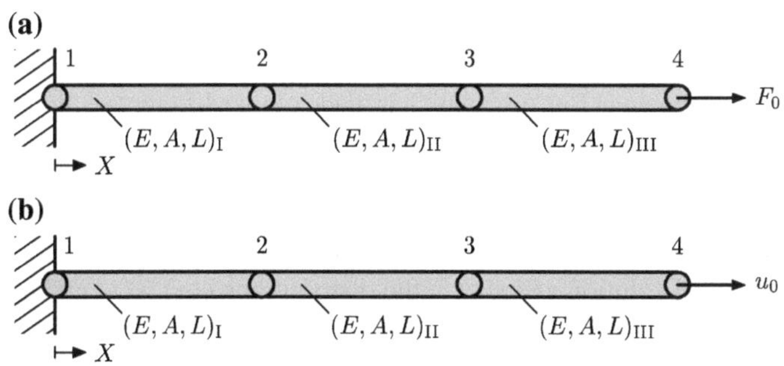


**Fig. 5.61** Structure composed of three rod elements: **a** force boundary condition; **b** displacement boundary condition

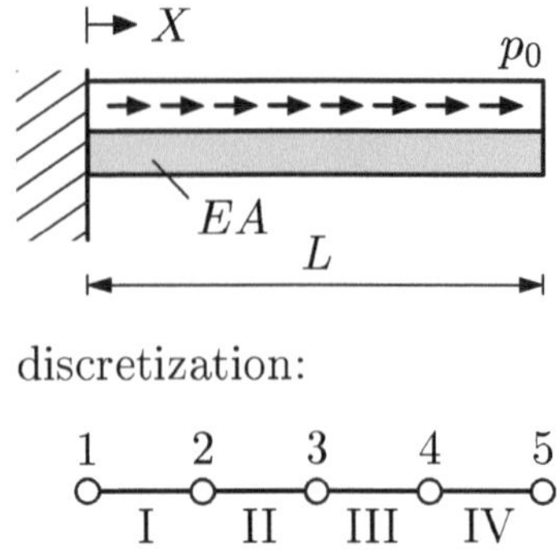


**Fig. 5.62** Rod element discretized by four elements

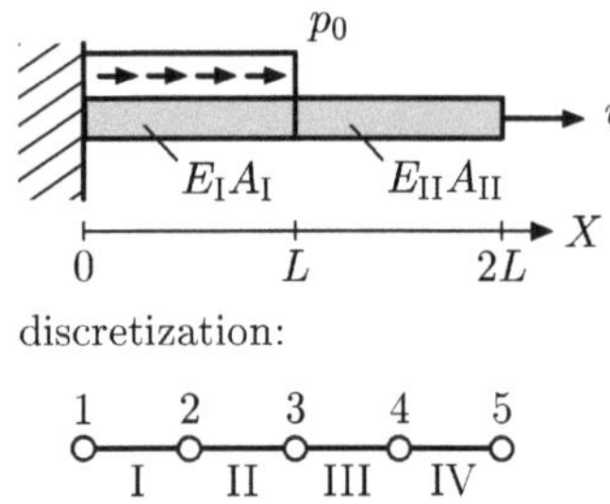


**Fig. 5.63** Bi-material rod discretized by four elements

**5.24 Example: Finite element approximation of a rod with four elements: comparison of displacement, strain and stress distribution with analytical solution**
Given is a rod of length $L$ and tensile stiffness $EA$ which is loaded by a constant distributed load $p_0$ as shown in Fig. 5.62. Use four linear rod elements of length $\frac{L}{4}$ to discretize the rod and calculate the nodal displacements, strains and stresses. Compare the results with the analytical solution and sketch the normalized finite element and analytical solutions $u(X)/\frac{p_0L^2}{EA}$, $\varepsilon(X)/\frac{p_0L}{EA}$ and $\sigma(X)/\frac{p_0L}{A}$ over the normalized coordinate $X/L$.

**5.25 Example: Elongation of a bi-material rod: finite element solution and comparison with analytical solution**
Given is a rod as shown in Fig. 5.63 which is made of two different sections with axial stiffness $k_I = E_I A_I$ and $k_{II} = E_{II} A_{II}$. Each section is of length $L$ and in the left-hand section, i.e. $0 \leq x \leq L$, is a constant distributed load $p_0$ acting while the right-hand end is elongated by $u_0$. Use four linear rod elements of length $\frac{L}{4}$ to discretize the rod and calculate the nodal displacements, strains and stresses. Compare the results with the analytical solution and sketch the distributions $u(x)$, $\varepsilon(x)$ and $\sigma(x)$ for for the case $k_I = 2k_{II} = 1$, $L_I = L_{II} = 1$, $p_0 = 1$, $u_0 = 1$ and $E_I = 2E_{II} = 1$.

**5.26 Example: Stress distribution for a fixed-fixed rod structure**
Given is a rod structure as sown in Fig. 5.64. The structure is of length $2L$, cross-sectional area $A$ and Youngs modulus $E$. The structure is fixed at both ends and loaded

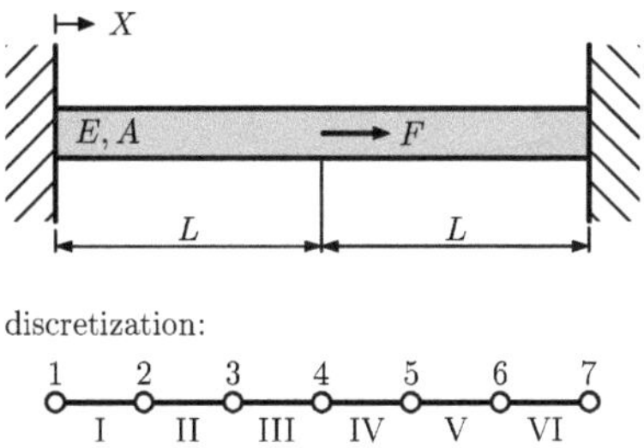


**Fig. 5.64** Rod structure fixed at both ends

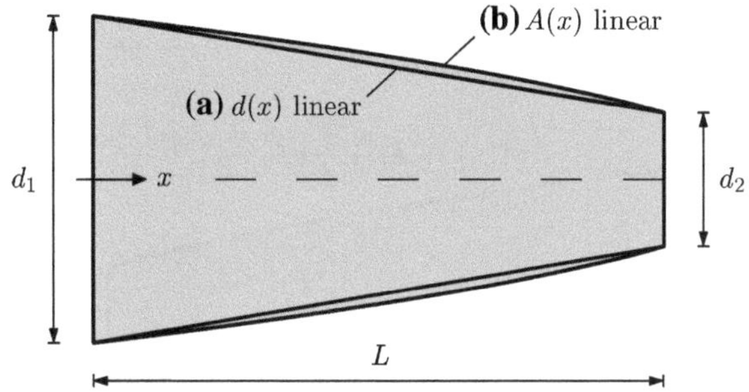


**Fig. 5.65** Rod element with variable cross section: **a** linear changing diameter; **b** linear changing cross-sectional area

by a point load $F$ in the middle, i.e. $X = L$. Calculate based on six finite elements of length $\frac{L}{3}$ the stress distribution. Show the difference between the elemental stress values and the averaged nodal values.

**5.27 Example: Linear rod element with variable cross section: derivation of stiffness matrix**
Determine the elemental stiffness matrix for a linear rod element with changing cross-sectional area as shown in Fig. 5.65. Consider the following two relationships for a linear changing diameter and a linear changing area:

$$\text{a)}\ d(x) = d_1 + \frac{x}{L}(d_2 - d_1), \tag{5.571}$$

$$\text{b)}\ A(x) = A_1 + \frac{x}{L}(A_2 - A_1). \tag{5.572}$$

Use analytical integration to obtain the stiffness matrix and compare the results with a two-point GAUSS integration rule. A circular cross section can be assumed in case (a).

**5.28 Example: Quadratic rod element with variable cross section: derivation of stiffness matrix**
Solve problem 5.27 with a single quadratic rod element.

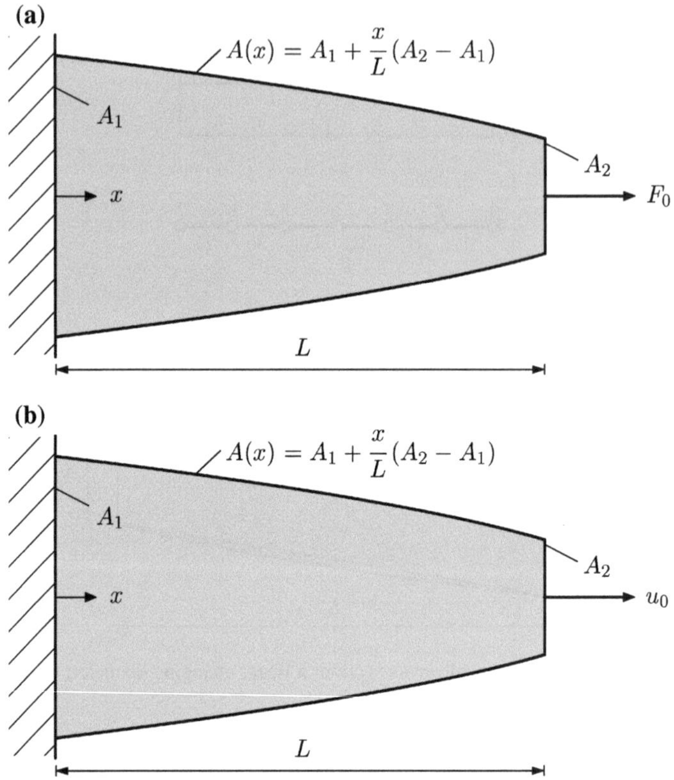


**Fig. 5.66** Rod element with variable cross-section and different boundary conditions at the right end: **a** external force $F_0$; **b** displacement $u_0$

**5.29 Example: Linear rod element with variable cross-section: comparison of displacements between FE and analytical solution for a single element**
Determine for the rod element shown in Fig. 5.66 the end displacement based on a single finite element. The elemental stiffness matrix from problem 5.27 can be used. Distinguish two different cases of boundary conditions, i.e. a single force $F_0$ or a prescribed displacement $u_0$ at the right-hand end. Compare the results obtained with the analytical solution.

**5.30 Example: Quadratic rod element with variable cross section: comparison of end displacement between FE and analytical solution for single element**
Recalculate problem 5.29 (a) for a single quadratic rod element.

**5.31 Example: Subdivided structure with variable cross section: comparison of displacements and stresses between FE and analytical solution for four elements**
Calculate for the rod shown in Fig. 5.67 the distribution of the elongation $u(X)$ and the stress $\sigma(X)$. To this end, subdivide the structure in four elements of length $\frac{L}{4}$ and

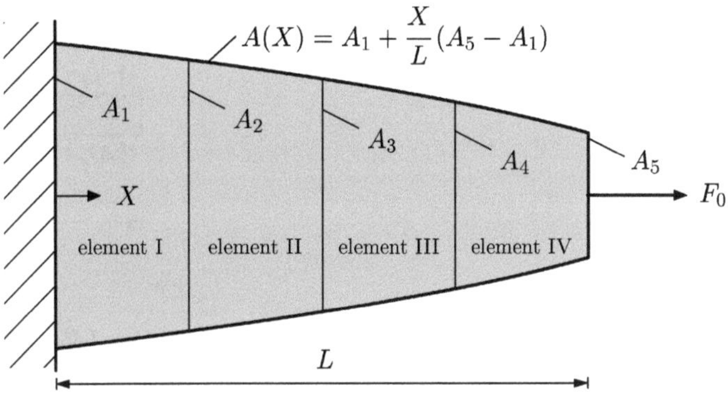


**Fig. 5.67** Rod element with variable cross-section discretized by four finite elements

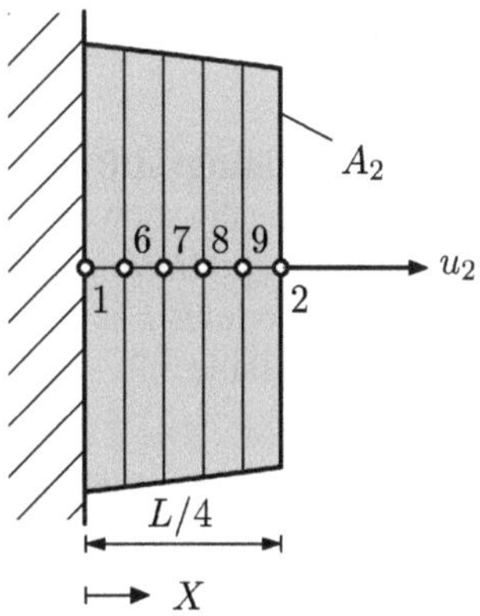


**Fig. 5.68** Submodel of the structure with variable cross section, cf. Fig. 5.67

use the expression for the elemental stiffness matrix which was derived in problem 5.27. The ratio between the end and initial cross section area is equal to $\frac{A_5}{A_1} = 0.2$. Compare the results obtained with the analytical solution.

**5.32 Example: Submodel of a structure with variable cross section**

To increase the accuracy of a stress approximation, the submodeling technique can be applied. Let us come back to Problem 5.31 and assume that the area of interest is near the left support, i.e. $X \rightarrow 0$. In the first step of a submodeling analysis, the structure is simulated with a coarse mesh as indicated in Fig. 5.67. In the next step, the area of element I is separately considered and discretized with a finer mesh. The nodal displacement $u_2$ from Problem 5.31 is applied as boundary condition at the right-hand end of the submodel. Calculate the stress distribution based on the submodel and compare with the analytical solution and the result from the coarse mesh (Fig. 5.68).

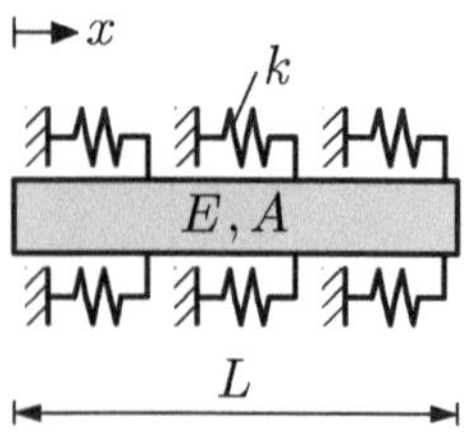


**Fig. 5.69** Rod with elastic embedding

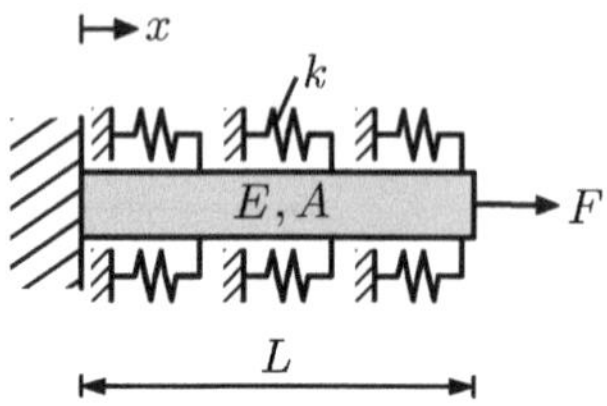


**Fig. 5.70** Rod with elastic embedding loaded by a single force

**5.33 Example: Rod with elastic embedding: stiffness matrix**
A rod with elastic embedding is schematically shown in Fig. 5.69. Derive the elemental stiffness matrix for a rod element with (a) linear and (b) quadratic shape functions under the assumption that the elastic modulus $k$ is constant. The describing partial differential equation can be taken from Table 3.2.

**5.34 Example: Rod with elastic embedding: single force case**
A cantilevered rod with elastic embedding is loaded by a single force $F$ as shown in Fig. 5.70. Assume that the elastic modulus $k$ and the axial tensile stiffness $EA$ is constant. Use a single (a) linear and (b) quadratic rod element to determine

- the reduced system of equations,
- the elongation of the rod at $x = L$,
- simplify your result for the special case $k = 0$,
- simplify your result for the special case $EA = 0$.
- Compare the finite element solution with the analytical solution for the case $k = 3$, $EA = 1$ and $L = 1$.

## 5.7.3 *Bernoulli Beam Element Problems*

**5.35 Example: Weighted residual method with arbitrary distributed load for a beam**
Derive the principal finite element equation for a BERNOULLI beam element based on the weighted residual method. Starting point should be the partial differential equation with an arbitrary distributed load $q_y(x)$. In addition, it can be assumed that the bending stiffness $EI_z$ is constant.

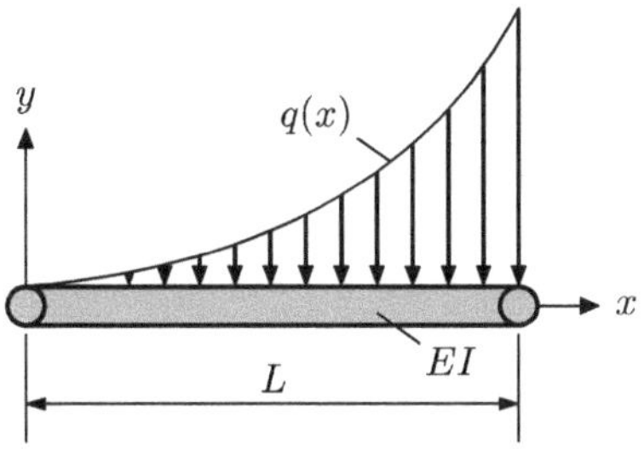


**Fig. 5.71** Quadratic distributed load

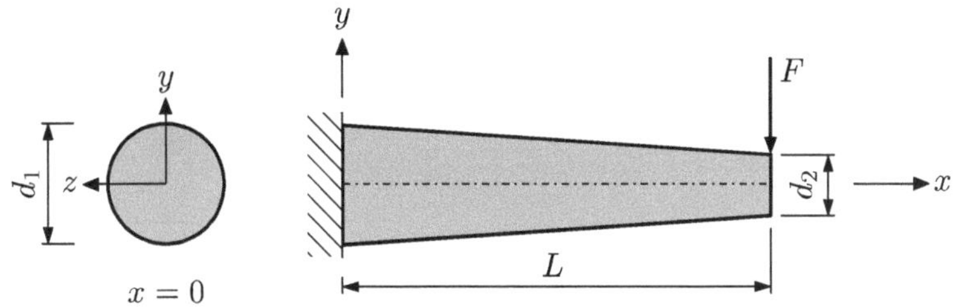


**Fig. 5.72** Beam with variable cross-sectional area loaded by a single force

**5.36 Example: Stiffness matrix for bending in the $x - z$ plane**
Derive the stiffness matrix for a BERNOULLI beam element for bending in the $x$-$z$ plane.

**5.37 Example: Beam with variable cross-sectional area**
Solve example 5.8 for arbitrary values of $d_1$ and $d_2$!

**5.38 Example: Equivalent nodal loads for quadratic distributed load**
Determine the equivalent nodal loads for a BERNOULLI beam, cf. Fig. 5.71, for the cases:

(a) $q(x) = q_0 x^2$,

(b) $q(x) = q_0 \left(\frac{x}{L}\right)^2$.

**5.39 Example: Beam with variable cross section loaded by a single force**
Calculate for the BERNOULLI beam shown in Fig. 5.72 the vertical displacement of the right-hand boundary for $d_1 = 2h$ and $d_2 = h$. Use one single finite element and compare the numerical solution with the analytical solution. Advice: The stiffness matrix can be taken from problem 5.8.

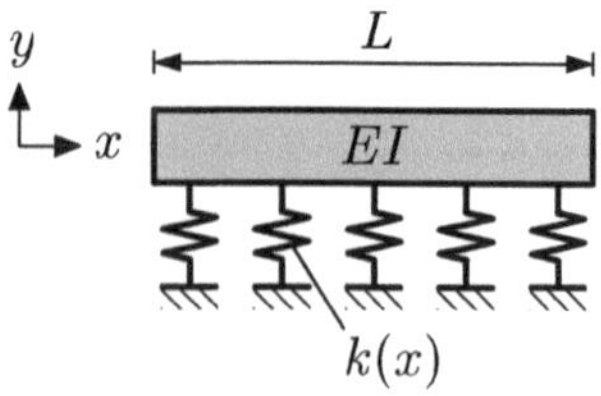


**Fig. 5.73** Beam on elastic foundation

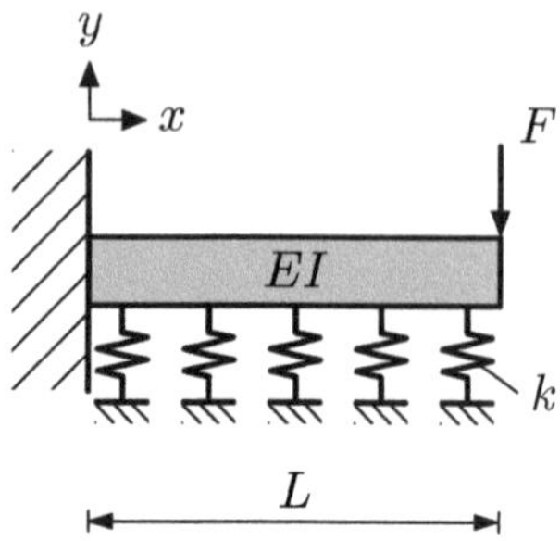


**Fig. 5.74** Beam on elastic foundation loaded by a single force

**5.40 Example: Beam on elastic foundation: stiffness matrix**

A beam on an elastic foundation is schematically shown in Fig. 5.73. Derive the elemental stiffness matrix for a BERNOULLI beam element under the assumption that the elastic foundation modulus $k$ is constant. The describing partial differential equation can be taken from Table 4.5.

**5.41 Example: Beam on elastic foundation: single force case**

A cantilevered beam on an elastic foundation is loaded by a single force $F$ as shown in Fig. 5.74. Assume that the elastic foundation modulus $k$ and the bending stiffness $EI_z$ is constant. Use one single BERNOULLI beam element to determine

- the reduced system of equations,
- the deflection and rotation of the beam at $x = L$,
- simplify your result for the special case $k = 0$,
- simplify your result for the special case $EI_z = 0$.
- Compare the finite element solution with the analytical solution for the case $k = 4$, $EI_x = 1$ and $L = 1$.

**5.42 Example: Beam on nonlinear elastic foundation: stiffness matrix**

Derive the elemental stiffness matrix $\boldsymbol{K}^{\mathrm{e}}$ for a BERNOULLI beam element with constant bending stiffness $EI_z$ and nonlinear elastic foundation modulus $k = k(u_y)$, cf. Fig. 5.73. Consider the case that the elastic foundation modulus changes linearly with the vertical displacement $u_y$ as shown in Fig. 5.75b. The linear relationship of the elastic foundation modulus should be defined by two sampling points $k(u_y = 0) = k_0$ and $k(u_y = u_1) = k_1$.

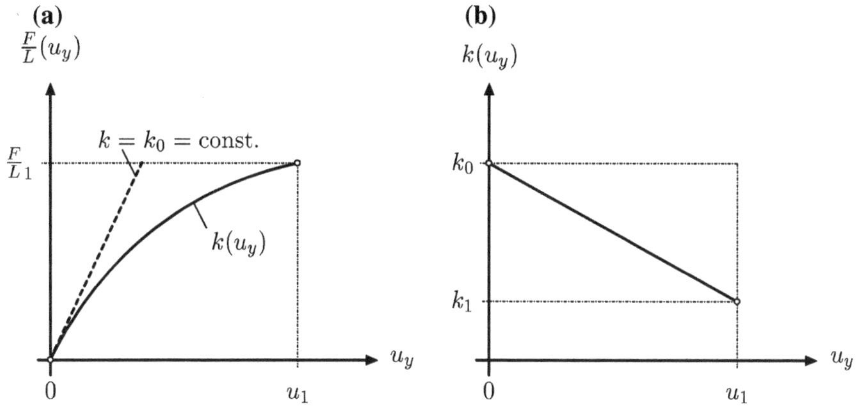


**Fig. 5.75** **a** Nonlinear load-displacement diagram; **b** displacement-dependent elastic foundation modulus

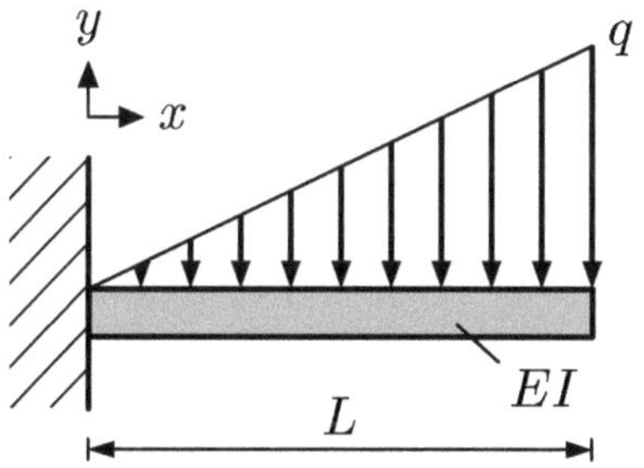


**Fig. 5.76** Cantilevered beam with triangular shaped distributed load

**5.43 Example: Cantilevered beam with triangular shaped distributed load**
Given is a cantilevered beam of length $L$ and constant bending stiffness $EI$, cf. Fig. 5.76. The beam is loaded with a triangular shaped distributed load with a maximum value of $q$. Use one single finite element to calculate

- the maximum deflection of the beam,
- the distribution of the deflection between the nodes: $u_y = u_y(x)$.
- Calculate based on the appropriate partial differential equation the analytical solution for the deflection $u_y = u_y(x)$ and compare this result with the finite element solution at $x = L$.
- Sketch the analytical and finite element solution for $u_y = u_y(x)$ in the range $0 \leq x \leq L$.

**5.44 Example: Finite element approximation with a single beam element**
Given is a beam with different supports as shown in Fig. 5.77. The bending stiffness $EI$ is constant and the length is equal to $L = a + b$. The beam is loaded by a single force at location $x = a$. Derive the finite element solution based on one single beam

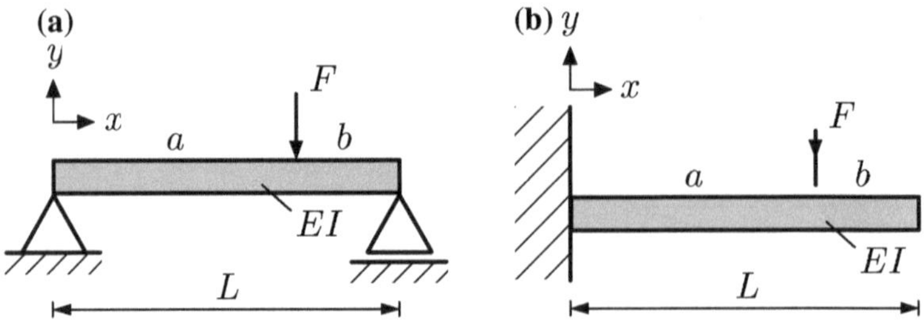


**Fig. 5.77** Finite element approximation with a single beam element: **a** simply supported beam; **b** cantilevered beam

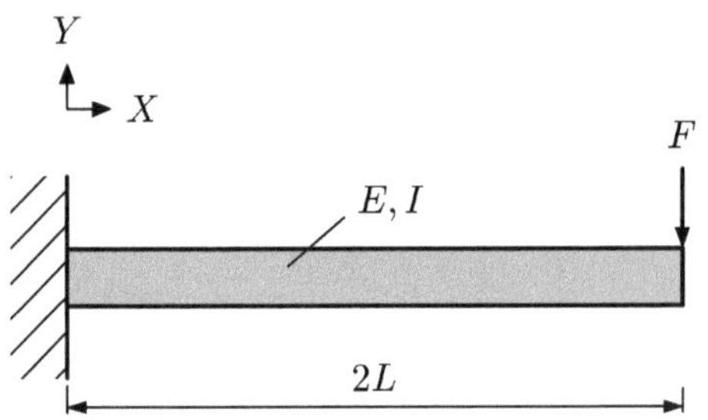


**Fig. 5.78** Cantilevered beam: moment curvature relationship

element and compare the displacements $u_y(0)$, $u_y(L)$ and $u_y(a)$ with the analytical solution.

**5.45 Example: Cantilevered beam: moment curvature relationship**
The cantilevered beam shown in Fig. 5.78 is loaded by a single force $F$ in the negative $Y$-direction at its right-hand end. The total length of the beam is $2L$ and the bending stiffness is $EI$. Use two elements of length $L$ to determine

- the displacements and rotations at each node,
- the curvature at each node,
- the bending moment at each node,
- check at each node the moment curvature relationship $\frac{M_z}{\kappa_z} = EI_z$.
- Compare the results with the analytical solution.

**5.46 Example: Cantilevered stepped beam**
The cantilevered beam shown in Fig. 5.79 is loaded by a single force $F$ in negative $Y$-direction at its right-hand end. The beam is divided in two sections of length $L_{\text{I}}$ and $L_{\text{II}}$. The geometrical and material properties of the structure are given as

(a) $L_{\text{I}} = L_{\text{II}} = L$, $E_{\text{I}} = E_{\text{II}} = E$, $I_{\text{I}} = 2I$ and $I_{\text{II}} = I$,
(b) $L_{\text{I}} = L_{\text{II}} = L$, $E_{\text{I}} = E_{\text{II}} = E$, $I_{\text{I}} = I_{\text{II}} = I$.

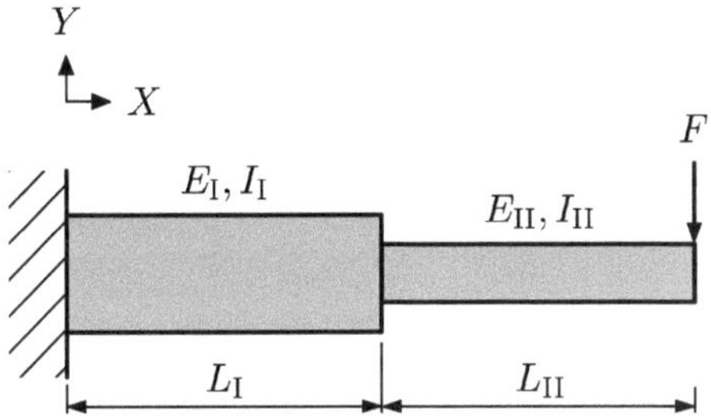


**Fig. 5.79** Cantilevered stepped beam

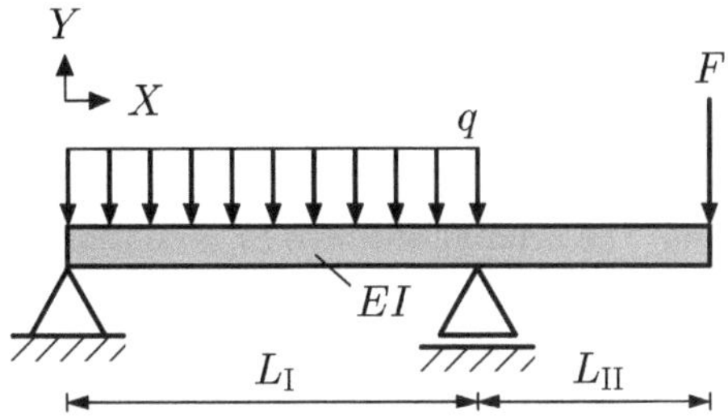


**Fig. 5.80** Overhang beam with different types of vertical loads

Determine for both cases

- the global system of equations,
- the reduced system of equations,
- all nodal unknowns,
- all reactions at the support,
- the bending moment at the nodes and its distribution in each element,
- the shear force in each element,
- the stress distribution in $Y$-direction at locations $X = 0, \frac{L}{2}, L, \frac{3L}{2}, L$ which results from the bending moment.

**5.47 Example: Overhang beam with distributed load and single force**

The beam shown in Fig. 5.80 is loaded by a constant distributed load $q$ and a single force $F$. The bending stiffness $EI$ is constant and the total length of the beam is equal to $(L_{\mathrm{I}} + L_{\mathrm{II}})$. Model the beam with two elements to determine:

- the unknown rotations and displacements at the nodes,
- the reaction forces at the supports,
- the vertical deflection in the middle of the section with the distributed load, i.e. $X = \frac{1}{2}L_{\mathrm{I}}$,
- the bending moment and shear force at the midpoint of the section with the distributed load, i.e. $X = \frac{1}{2}L_{\mathrm{I}}$.
- Improve the approximate solution for $u_Y(X = \frac{1}{2}L_{\mathrm{I}})$ by subdividing the section with the distributed load in two elements of equal length.

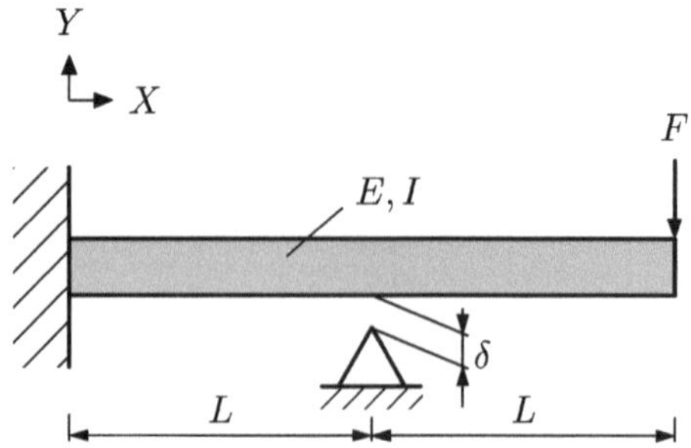


**Fig. 5.81** Beam structure with a gap in the middle and vertical point load at the right end

**5.48 Example: Beam structure with a gap**

The beam shown in Fig. 5.81 is loaded by a single vertical force $F$ at its right-hand end. The bending stiffness $EI$ is constant and the total length of the beam is equal to $2L$. In the middle of the entire structure, there is a pap of length $\delta$ between the beam and a simple support. Model the beam with two elements to determine:

- the force to close the gap,
- the deflection and rotation at the free end, i.e. $X = 2L$, as a function of the increasing force $F$,
- the maximum normalized stress $\frac{\sigma_x(y_{max})}{y_{max}}$ at the nodes for the situation $u_{2y}(X = L) < \delta$ and $u_{2y}(X = L) \geq \delta$.

**5.49 Example: Advanced example: beam element with nonlinear bending stiffness**

Derive the elemental stiffness matrix $\boldsymbol{K}^e$ for a BERNOULLI beam element with nonlinear bending stiffness $EI = EI(\kappa)$. Consider the case that the bending stiffness changes quadratically with the curvature $\kappa$ as shown in Fig. 5.82. The quadratic relationship of the bending stiffness should be defined by the three sampling points $EI(\kappa = 0) = IE_0$, $EI(\kappa = \frac{1}{2}\kappa_1) = \beta_{05} IE_0$ and $EI(\kappa = \kappa_1) = \beta_1 IE_0$. To simplify the

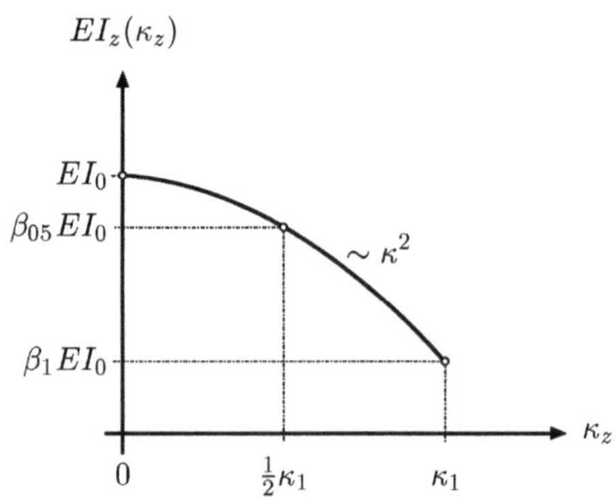


**Fig. 5.82** Curvature-dependent bending stiffness

notation, the product of YOUNG's modulus $E$ and moment of inertia $I$ is considered as a single variable: $(EI) \rightarrow EI$.

### 5.7.4 Timoshenko Beam Element Problems

**5.50 Example: Timoshenko beam element with quadratic shape functions for the deflection and linear shape functions for the rotation**
For a TIMOSHENKO beam element with quadratic shape functions for the deflection and linear shape functions for the rotation, the stiffness matrix, after elimination of the middle node according to Eq. (5.445), is given. Derive the additional load vector on the right-hand side of the principal finite element equation which results from a distributed load $q_y(x)$. Subsequently simplify the result for a constant load.

**5.51 Example: Timoshenko beam element with cubic shape functions for the deflection and quadratic shape functions for the rotation**
Derive the stiffness matrix and the principal finite element equation $\boldsymbol{K}^{\text{e}}\boldsymbol{u}_{\text{p}} = \boldsymbol{f}^{\text{e}}$ for a TIMOSHENKO beam element with cubic shape functions for the deflection and quadratic shape functions for the rotation. Use the exact solution for the integration. Subsequently analyze the convergence behavior of an element configuration, which is illustrated in Fig. 5.37. The element deforms in the $x$-$y$ plane. How does the principal finite element equation change, when the deformation occurs in the $x$-$z$ plane?

### 5.7.5 Plane Truss Structure Problems

**5.52 Example: Plane truss structure arranged in a square**

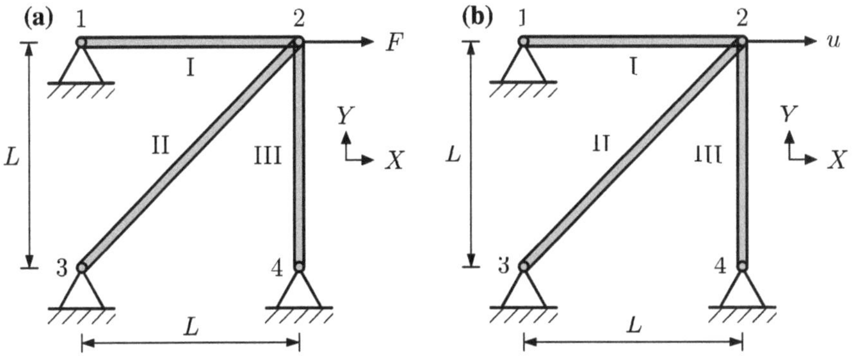


**Fig. 5.83** Three-element truss structure with different external loading: **a** force boundary condition; **b** displacement boundary condition

Given is the two-dimensional truss structure as shown in Fig. 5.83. The three truss elements have the same cross-sectional area $A$ and YOUNG's modulus $E$. The length of each element can be taken from the dimensions given in the figure. The structure is loaded by

(a) a horizontal force $F$ at node 2,
(b) a prescribed horizontal displacement $u$ at node 2.

Determine for both cases

- the global system of equations,
- the reduced system of equations,
- all nodal displacements,
- all reaction forces,
- the force in each rod.

**5.53 Example: Plane truss structure arranged in a triangle**
Given is the two-dimensional truss structure as shown in Fig. 5.84. The three truss elements have the same cross-sectional area $A$ and YOUNG's modulus $E$. The length of each element can be taken from the dimensions given in the figure. The structure is loaded by

(a) single forces $F_X$ and $F_Y$ at node 4,
(b) prescribed displacements $u_X$ and $u_Y$ at node 4.

Determine for both cases

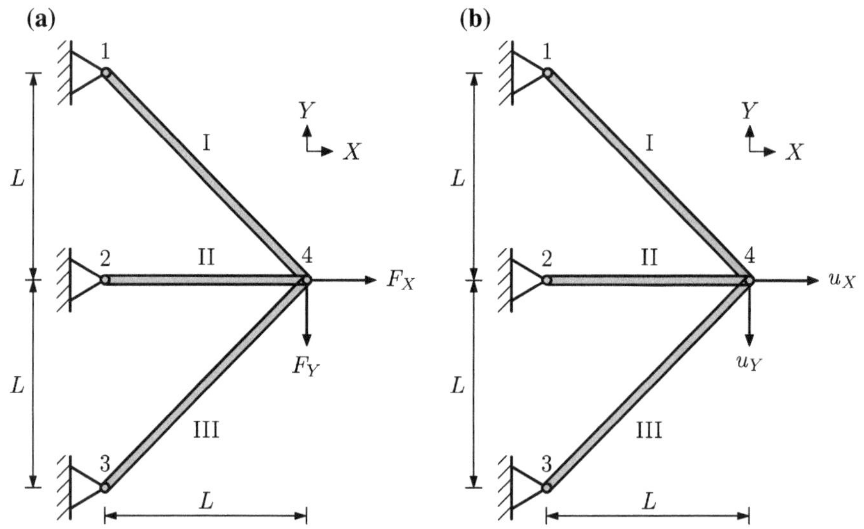


**Fig. 5.84** Three-element truss structure with different external loading: **a** force boundary conditions; **b** displacement boundary conditions

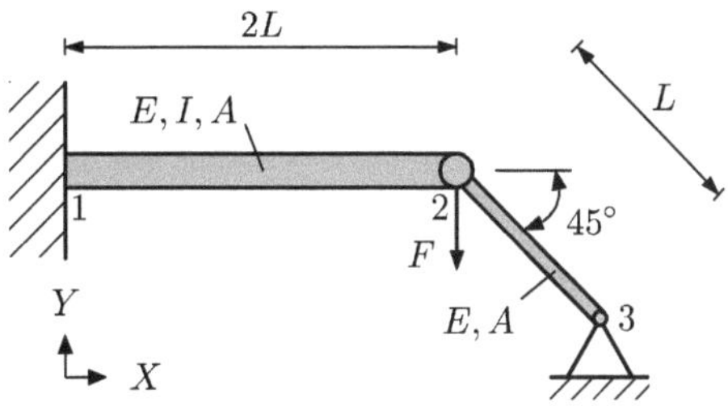


**Fig. 5.85** Plane beam rod structure

- the global system of equations,
- the reduced system of equations,
- all nodal displacements,
- all reaction forces,
- the force in each rod.

## *5.7.6 Plane Frame Structure Problems*

**5.54 Example: Plane beam-rod structure**
Given is the two-dimensional beam-rod structure as shown in Fig. 5.85. The generalized beam element of length $2L$ has the bending stiffness $EI$ and the tensile stiffness $EA$. The rod of length $L$ has the tensile stiffness $EA$. The structure is loaded by a vertical force $F$ at node 2.

Determine

- the global system of equations,
- the reduced system of equations,
- all nodal displacements and rotations,
- all reaction forces and reaction moments.
- Validate your results by checking the global equilibrium of forces and moments.

**5.55 Example: Stiffness matrix for a generalized beam element for different rotation angles at the nodes**
Derive the stiffness matrix $\boldsymbol{K}^{\mathrm{e}}_{XY}(\alpha_1, \alpha_1)$ for a generalized beam (BERNOULLI) element which can deform in the global $X$-$Y$ plane. Consider that the rotation angle between the local and global coordinate system is different at both nodes.

**5.56 Example: Mechanical properties of squared frame structure**
Given is a squared frame structure made of generalized beam elements with side length $L$ as shown in Fig. 5.86. Two different orientations, i.e. flat or angled orientation, should be considered in the following. Calculate the displacement (BC: $F$) or reaction force (BC: $u$) of the point of load application and estimate the macroscopic stiffness $E_{\text{struct}}$ of the frame structure. Simplify your results for the macroscopic

stiffness for the special case $A = \frac{\pi d^2}{4}$ and $I = \frac{\pi d^4}{64}$, i.e. a circular cross section of the beam elements. The derivation should be performed first for the full model, then for the half model and finally for the quarter model.

**5.57 Example: Squared frame structure: different ways of load application**
Given is a squared frame structure made of generalized beam elements with side length $L$ as shown in Fig. 5.87. Two different load cases, i.e. central and on the sides, should be compared. Calculate the displacement of the point of load application and estimate the macroscopic stiffness $E_{\text{struct}}$ of the frame structure. Simplify your results for the macroscopic stiffness for the special case $A = \frac{\pi d^2}{4}$ and $I = \frac{\pi d^4}{64}$, i.e. a circular cross section of the beam elements. The derivation should be performed for the quarter model under consideration of the double symmetry.

**5.58 Example: Mechanical properties of idealized honeycomb structure**
A honeycomb structure should be idealized by a regular hexagon as shown in Fig. 5.88. Such a regular hexagon has all sides of the same length $L$, and all internal angles are 120°. Assume for this simplified approach that the honeycomb structure is represented by a single cell, either in flat or angled orientation. Furthermore, a two-dimensional approach based on a frame structure made of generalized beam elements is to consider. Calculate the displacement of the point of load application and estimate the macroscopic stiffness $E_{\text{struct}}$ of the idealized honeycomb structure. Simplify your results for the macroscopic stiffness for the special case $A = \frac{\pi d^2}{4}$ and $I = \frac{\pi d^4}{64}$, i.e. a circular cross section of the beam elements. To facilitate the finite element approach, exploit the symmetry of the problem.

**5.59 Example: Bridge structure (computational problem)**
Given is a simplified plane bridge structure over a valley as shown in Fig. 5.89. The bridge structure is idealized in the $X$-$Y$ plane based on thirteen line elements (I, ..., XIII) which are connected at eight nodes (1, ..., 8). Consider the following numerical values for the geometrical and material parameters: $L = 4000\,\text{mm}$; $E = 200000\,\text{MPa}$; $I = 80\,\text{mm}^4$ and $A = 10\,\text{mm}^2$.
Consider the following cases that

(a) all elements are generalized beams (i.e. with $EA$ and $EI$),
(b) only elements I, II, III, IV are generalized beams, all other elements are rod elements (only $EA$),
(c) all elements are rod elements,
(d) the structure is only composed of the beam elements I, II, III, IV,

to calculate the unknowns at the nodes. Sketch in addition the vertical deformation of elements I, II, III, IV and use the shape functions to interpolate between the nodes. Consider the following load cases:

(1) a vertical force of magnitude $-F$ at node 3 (the force $F$ is given in general, i.e. not as a numerical value),

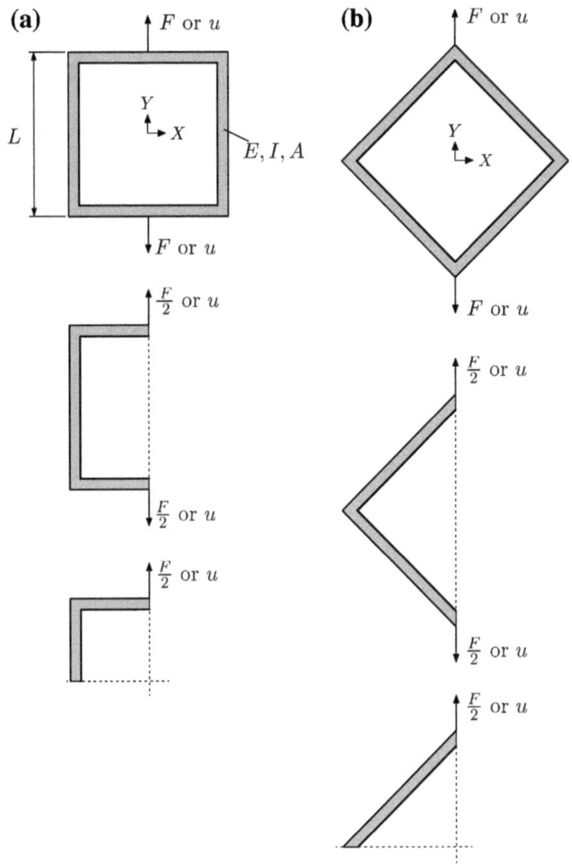


**Fig. 5.86** Squared frame structure in **a** flat and **b** angled orientation. Top: full model; middle: half model and bottom: quarter model

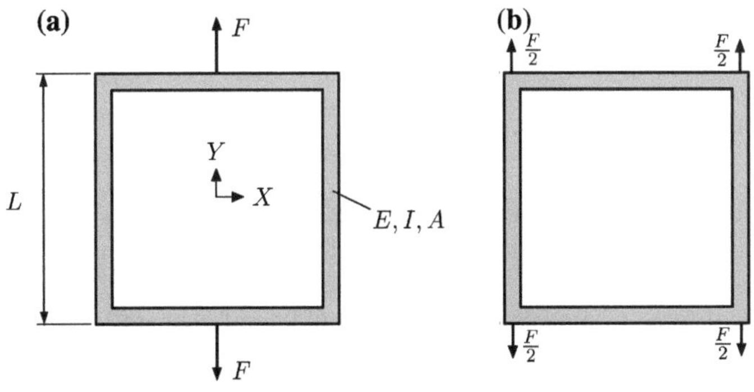


**Fig. 5.87** Squared frame structure: **a** central load application and **b** loads on the sides

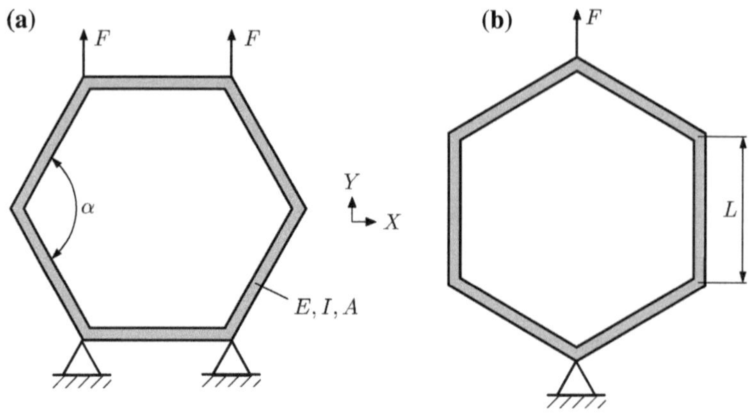


**Fig. 5.88** Honeycomb structure approximated by a regular hexagon: **a** flat orientation and **b** pointy or angled orientation

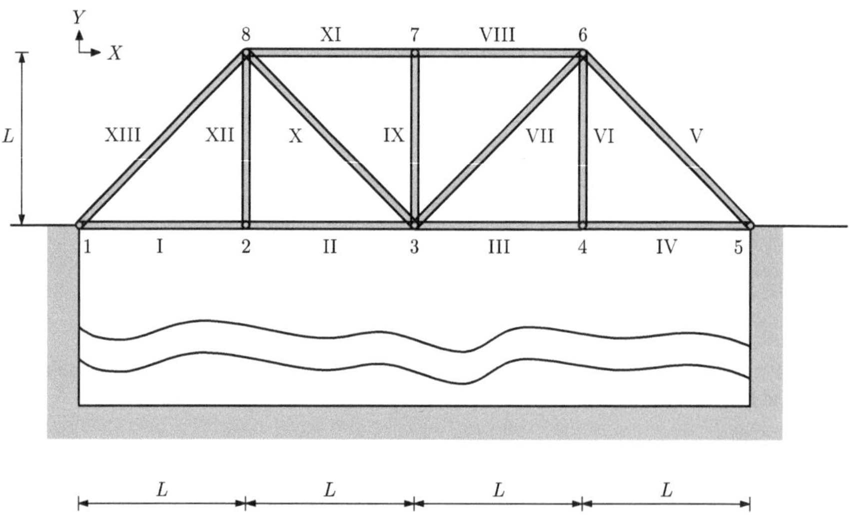


**Fig. 5.89** Plane bridge structure over a valley

(2) a vertical force of magnitude $-F$ at node 2,
(3) a vertical force of magnitude $-F$ in the middle of element I, i.e. for $X = \frac{L}{2}$ (only case (a), (b) and (d)),
(4) a vertical force of magnitude $-F$ in the middle of element II, i.e. for $X = \frac{3L}{2}$ (only case (a), (b) and (d)).

At the supports, i.e. at node 1 and 5, the horizontal and vertical displacements are zero but the beam element can rotate ($\varphi_Z \neq 0$). In the cases (c) and (d), it is possible to derive general expressions for the unknowns, i.e. as a general function of $L$, $E$, ($A$), ($I$) and $F$, which are not too complicated.

# Chapter 6
# Elasto-Plastic Finite Element Simulations

**Abstract** The continuum mechanics basics for plasticity, i.e. the yield condition, the flow rule, the hardening law, and the elasto-plastic modulus, for the one-dimensional rod will be used to describe the elasto-plastic behavior for monotonic and cyclic loading. Isotropic kinematic and combined hardening is considered within the scope of the hardening law. Perfect elasto-plastic materials and different types of damage formulations are considered. For the integration of the elasto-plastic constitutive equation in the scope of the finite element approach, the incremental predictor-corrector method is generally introduced and derived for the fully implicit backward-Euler algorithm. On crucial points the difference between one- and two-dimensional descriptions is pointed out, to guarantee a simple transfer of the derived methods to general problems. Calculated examples and supplementary problems with short solutions serve as an introduction for the theoretical description.

## 6.1 Integration of the Constitutive Equations

The treatment of elasto-plastic finite element simulation is covered in many articles and book publications. The following Table 6.1 summarizes a few recommended textbooks on this topic.

In comparison to a finite element calculation with pure linear-elastic material behavior, a calculation with plastic material behavior cannot be conducted by applying the loads in a single step. Since there is no longer a unique relationship between stress and strain[1], the load must be applied in incremental steps and a non-linear system of equations must be solved for each increment (for example with the NEWTON- RAPHSON algorithm). Therefore, the principal finite element equation has to be written in the following incremental form:

$$\boldsymbol{K}\Delta\boldsymbol{u} = \Delta\boldsymbol{f}\,. \tag{6.1}$$

[1] In the general case with six stress and strain components (under consideration of the symmetry of the stress and strain tensor) an obvious relation only exists between the effective stress and effective plastic strain. In the one-dimensional case, however, these parameters reduce to: $\sigma_{\text{eff}} = |\sigma|$ and $\varepsilon^{\text{pl}}_{\text{eff}} = |\varepsilon^{\text{pl}}|$.



A. Öchsner, *Elasto-Plasticity of Frame Structure Elements*,
DOI 10.1007/978-3-662-44225-8_6

**Table 6.1** Recommended textbooks for the treatment of elasto-plastic finite element simulations

| Year (1st ed.) | Author | Title | Refs. |
|---|---|---|---|
| 1980 | D.R.J. Owen, E. Hinton | Finite Elements in Plasticity: Theory and Practice | [77] |
| 1991 | M.A. Crisfield | Non-Linear Finite Element Analysis of Solids and Structures. Volume 1: Essentials | [30] |
| 1997 | M.A. Crisfield | Non-Linear Finite Element Analysis of Solids and Structures. Volume 2: Advanced Topics | [29] |
| 1998 | J.C. Simo, T.J.R. Hughes | Computational Inelasticity | [96] |
| 2000 | T. Belytschko, W.K.L. Liu, B. Moran | Nonlinear Finite Elements for Continua and Structures | [10] |
| | I. Doltsinis | Elements of Plasticity: Theory and Computation | [32] |
| 2004 | J.N. Reddy | An Introduction to Nonlinear Finite Element Analysis | [87] |
| 2005 | F. Dunne, N. Petrinic | Introduction to Computational Plasticity | [36] |
| 2008 | P. Wriggers | Nonlinear Finite Element Methods | [115] |
| 2008 | E.A. de Souza Neto, D. Perić, D.R.J. Owen | Computational Methods for Plasticity | [74] |
| 2010 | A. Anandarajah | Computational Methods in Elasticity and Plasticity: Solids and Porous Media | [5] |
| 2013 | R.I. Borja | Plasticity: Modeling and Computation | [15] |

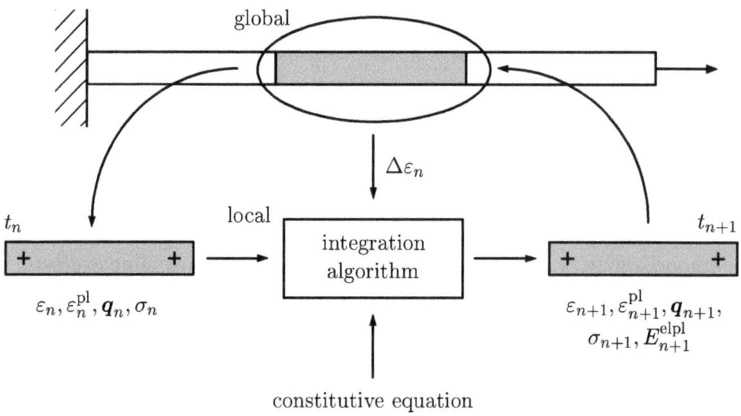


**Fig. 6.1** Schematic illustration of the integration algorithm for plastic material behavior in the FEM; adopted from [115]. Integration points are marked schematically by the symbol '+'

Additionally, the state variables—for example the stress $\sigma_{n+1}$—have to be calculated for each increment $(n+1)$ in each integration point (GAUSS point), based on the state variables at the end of the previous increment $(n)$ and the given strain increment $(\Delta\varepsilon_n)$ (see Fig. 6.1).

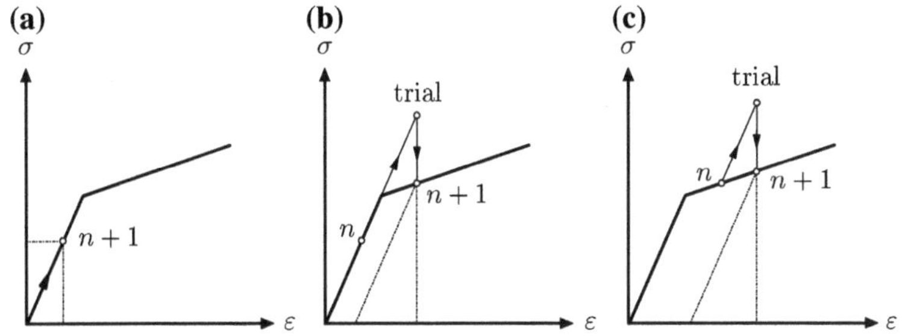


**Fig. 6.2** Schematic illustration of the predictor-corrector method in the stress-strain diagram: **a** elastic predictor in the elastic area; **b** and **c** elastic predictor outside the yield surface boundary

For this purpose, the explicit material law in infinitesimal form has to be integrated numerically according to Eqs. (2.6) and (2.68). Explicit integration methods, as for example the EULER procedure,[2] however, are inaccurate and possibly unstable, since a global error could accumulate [30]. Within the FEM one uses so-called predictor-corrector methods (see Fig. 6.2), in which, first a so-called predictor is explicitly determined and afterwards implicitly corrected instead of explicit integration procedures. In a first step, a test stress state (the so-called trial stress condition) is calculated under the assumption of pure linear-elastic material behavior via an elastic predictor[3]:

$$\sigma_{n+1}^{\text{trial}} = \sigma_n + \underbrace{E\Delta\varepsilon_n}_{\text{predictor } \Delta\sigma_n^{\text{el}}} \,. \tag{6.2}$$

The given hardening condition[4] in this test stress state equals the condition at the end of the previous increment. Therefore, it is assumed that the load step occurs pure elastically, meaning without plastic deformation and therefore without hardening:

$$\kappa_{n+1}^{\text{trial}} = \kappa_n \,. \tag{6.3}$$

Based on the location of the test stress state in the stress space, two elementary conditions can be distinguished based on the yield condition:

(a) The stress state is in the elastic area (see Fig. 6.2a) or on the yield surface boundary (valid stress state):

$$F(\sigma_{n+1}^{\text{trial}}, \kappa_{n+1}^{\text{trial}}) \le 0 \,. \tag{6.4}$$

---

[2] The explicit EULER procedure or polygon method (also EULER- CAUCHY method) is the most simple procedure for the numerical solution of an initial value problem. The new stress state results according to this procedure as $\sigma_{n+1} = \sigma_n + E_n^{\text{elpl}}\Delta\varepsilon$, whereupon the initial value problem can be stated as $\frac{d\sigma}{d\varepsilon} = E^{\text{elpl}}(\sigma, \varepsilon)$ with $\sigma(\varepsilon_0) = \sigma_0$.

[3] In the general three-dimensional case the relation is applied on the stress column matrix and the increment of the strain column matrix: $\boldsymbol{\sigma}_{n+1}^{\text{trial}} = \boldsymbol{\sigma}_n + \boldsymbol{C}\Delta\boldsymbol{\varepsilon}_n$, see Sect. 8.7.

[4] Let us consider for simplicity only isotropic hardening at this stage of the derivation.

In this case, the trial state can be taken as the new stress/hardening state, since it equals the real state:

$$\sigma_{n+1} = \sigma_{n+1}^{\text{trial}} \,, \tag{6.5}$$

$$\kappa_{n+1} = \kappa_{n+1}^{\text{trial}} \,. \tag{6.6}$$

In conclusion, one moves to the next increment.

(b) The stress state is outside the yield surface boundary (invalid stress state), see Fig. 6.2b, c:

$$F(\sigma_{n+1}^{\text{trial}}, \kappa_{n+1}^{\text{trial}}) > 0 \,. \tag{6.7}$$

If this case occurs, a valid state ($F(\sigma_{n+1}, \kappa_{n+1}) = 0$) on the yield surface boundary is calculated in the second part of the procedure from the invalid test state. Therefore, the necessary stress difference

$$\Delta\sigma^{\text{pl}} = \sigma_{n+1}^{\text{trial}} - \sigma_{n+1} \tag{6.8}$$

is referred to as the plastic corrector.

For the calculation of the plastic corrector, the terms back projection, return mapping or catching up are used. Figure 6.3 compares the predictor-corrector method schematically in the one-dimensional and two-component stress space.

The back projection is considered in the following. Detailed descriptions can be found in [10, 29, 30, 36, 74, 96].

The stress difference between the initial and final state (stress increment)

$$\Delta\sigma_n = \sigma_{n+1} - \sigma_n \tag{6.9}$$

results, according to HOOKE's law from the elastic part of the strain increment, which results as the difference from the total strain increment and its plastic part:

$$\Delta\sigma_n = E\Delta\varepsilon_n^{\text{el}} = E(\Delta\varepsilon_n - \Delta\varepsilon_n^{\text{pl}}) \,. \tag{6.10}$$

Figure 6.4 shows that the total strain increment in dependence on the test stress state can be expressed as follows:

$$\Delta\varepsilon_n = \varepsilon_{n+1} - \varepsilon_n = \frac{1}{E}(\sigma_{n+1}^{\text{trial}} - \sigma_n). \tag{6.11}$$

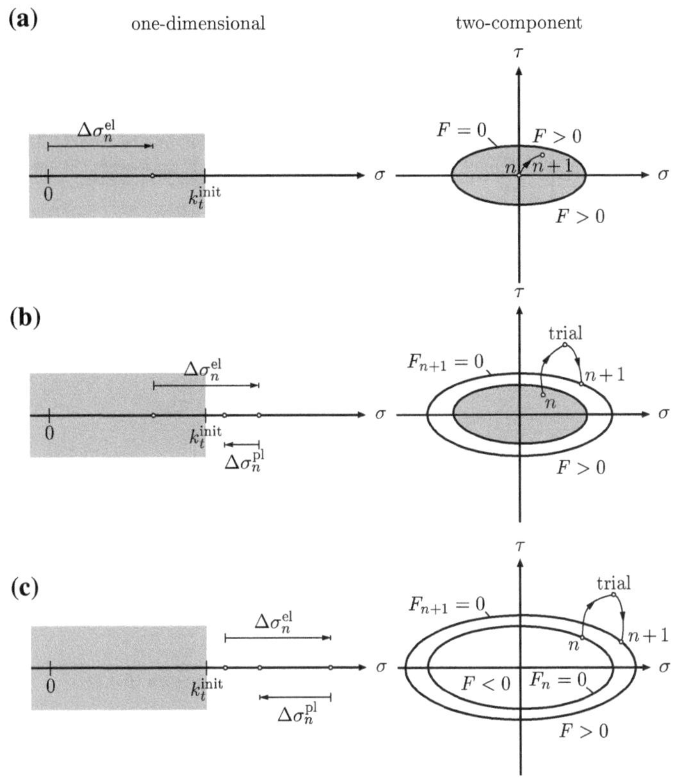


**Fig. 6.3** Schematic illustration of the predictor-corrector method in the one-dimensional and two-component stress space. **a** elastic predictor in the elastic area; **b** and **c** elastic predictor outside of the yield surface

If the last equation as well as the flow rule[5] according to (2.21) are inserted into Eq. (6.10), the final stress state $\sigma_{n+1}$ in dependence on the test stress state $\sigma_{n+1}^{\text{trial}}$ results in:

$$\sigma_{n+1} = \sigma_{n+1}^{\text{trial}} - \Delta\lambda_{n+1} E \,\text{sgn}(\sigma)\,. \tag{6.12}$$

Dependent on the location of the evaluation of the function sgn($\sigma$) different methods result (see Table 6.2) to calculate the initial value for the plastic corrector or alternatively to iteratively define the final stress state. To obtain an initial value for the plastic corrector, sgn($\sigma$) can either be evaluated in the test stress state (backward-EULER) or on the yield surface (forward-EULER; at the transition from the elastic to the plastic area this is the initial yield surface, see Fig. 6.2b). If the function sgn($\sigma$) is evaluated in the final state at the iterative calculation, the normal rule (see Sect. 2.3)

[5] At this point within the notation it is formally switched from d$\lambda$ to $\Delta\lambda$. Therefore, the transition from the differential to the incremental notation is performed.

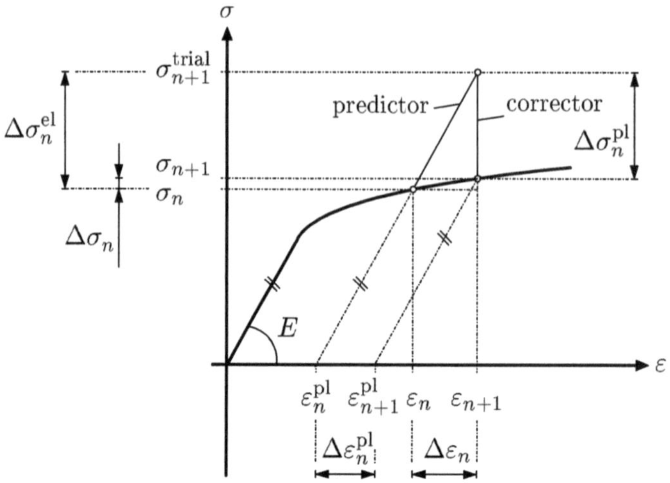


**Fig. 6.4** Schematic representation of the integration algorithm in the stress-strain diagram, adapted from [96]

**Table 6.2** Summary of different predictor-corrector methods

| Location of the evaluation of sgn($n$) | Equation (6.12) |
|---|---|
| *Initial value for corrector* | |
| Trial condition | $\sigma_{n+1} = \sigma_{n+1}^{\text{trial}} - \Delta\lambda_{n+1} E\ \text{sgn}(\sigma_{n+1}^{\text{trial}})$ |
| On the flow curve | $\sigma_{n+1} = \sigma_{n+1}^{\text{trial}} - \Delta\lambda_{n+1} E\ \text{sgn}(\sigma_n)$ |
| *During the iteration* | |
| In the final stress state<br>(fully impl. backward-EULER alg.)<br>(closest point projection) | $\sigma_{n+1} = \sigma_{n+1}^{\text{trial}} - \Delta\lambda_{n+1} E\ \text{sgn}(\sigma_{n+1})$ |
| On the flow curve<br>(semi-impl. backward-EULER alg.) | $\sigma_{n+1} = \sigma_{n+1}^{\text{trial}} - \Delta\lambda_{n+1} E\ \text{sgn}(\sigma_n)$ |
| In the final state and<br>on the flow curve<br>(mid-point rule) | $\sigma_{n+1} = \sigma_{n+1}^{\text{trial}} - \Delta\lambda_{n+1} E\ \frac{1}{2}\times$<br>$(\text{sgn}(\sigma_{n+1}) + \text{sgn}(\sigma_n))$ |
| Stress state of $i$-th<br>iteration step<br>(cutting-plane algorithm) | $\sigma_{n+1} = \sigma_{n+1}^{\text{trial}} - \Delta\lambda_{n+1} E\ \text{sgn}(\sigma^{(i)})$ |

is fulfilled in the final state. For this fully implicit backward-EULER algorithm (also referred to as the closest point projection (CPP)) [96], however, in the general three-dimensional case derivatives of higher order must be calculated. In the case of the so-called cutting-plane algorithm [95], the function $\mathrm{sgn}(\sigma)$ is calculated in the stress state of the $i$-th iteration step. The normal rule is not exactly fulfilled in the final state. However, no calculations of derivatives of higher order are necessary. In the case of the so-called mid-point rule [76], the function $\mathrm{sgn}(\sigma)$ is evaluated in the final state and on the yield surface as well as weighted in equal parts. If the function $\mathrm{sgn}(\sigma)$ is only evaluated on the yield surface, this leads to the semi-implicit backward-EULER algorithm [71], for which only derivatives of first order are necessary.

If one considers the dependency of the yield condition from the hardening variable, one needs another equation, which describes the hardening. From the evolution equation of the hardening variables (2.8) the following incremental relation results

$$\kappa_{n+1} = \kappa_n + \Delta\lambda_{n+1} \tag{6.13}$$

for the definition of the hardening variable.

Finally it can be remarked that three of the listed integration rules in Table 6.2 can be summarized via the following equation:

$$\sigma_{n+1} = \sigma_{n+1}^{\text{trial}} - \Delta\lambda_{n+1} E\left([1-\eta]\mathrm{sgn}(\sigma_n) + \eta\, \mathrm{sgn}(\sigma_{n+1})\right). \tag{6.14}$$

The parameter $\eta$ then becomes 1, 0 or $\frac{1}{2}$.

## 6.2 Derivation of the Fully Implicit Backward-Euler Algorithm for Isotropic Hardening

In the case of the back projection method, the stress location on the yield surface, which is energetically closest to the test state (see [75]) is calculated. Therefore, this does not involve, as the name suggests, a calculation of the closest geometric point. The assumption that the plastic work takes on a maximum at a given strain serves as a basis for the method. Together with the elementary demand that the calculated stress state has to lie on the flow curve (yield surface), the CPP method can be interpreted in the mathematical sense as a solution of an extremal value problem (maximum of the plastic work) with side-conditions (the unknown stress condition has to be on the yield surface) [96]. The method is hereby implicit in the calculation of the function $\mathrm{sgn}(\sigma)$ since the evaluation occurs in the final state $n+1$. Because of this, the CPP algorithm is also referred to as the fully implicit backward-EULER algorithm. Therefore, the following equations are fulfilled in the final state:

$$\sigma_{n+1} = \sigma_{n+1}^{\text{trial}} - \Delta\lambda_{n+1} E\, \mathrm{sgn}(\sigma_{n+1})\,, \tag{6.15}$$

$$\kappa_{n+1} = \kappa_n + \Delta\lambda_{n+1}\,, \tag{6.16}$$

$$F_{n+1} = |\sigma_{n+1}| - k(\kappa_{n+1}) = 0\,. \tag{6.17}$$

Outside of the final state, however, at each of these equations a residual $r$ remains:

$$r_\sigma(\sigma, \Delta\lambda) = \sigma - \sigma_{n+1}^{\text{trial}} + \Delta\lambda E \,\text{sgn}(\sigma) \neq 0 \quad \text{or}$$
$$= E^{-1}\sigma - E^{-1}\sigma_{n+1}^{\text{trial}} + \Delta\lambda \,\text{sgn}(\sigma) \neq 0\,, \tag{6.18}$$
$$r_\kappa(\kappa, \Delta\lambda) = \kappa - \kappa_n - \Delta\lambda \neq 0 \quad \text{or}$$
$$= -\kappa + \kappa_n + \Delta\lambda \neq 0\,, \tag{6.19}$$
$$r_F(\sigma, \kappa) = F(\sigma, \kappa) = |\sigma| - k(\kappa) \neq 0\,. \tag{6.20}$$

The unknown stress/hardening state therefore represents the root of a vector function $\boldsymbol{m}$, which consists of the single residual functions. Furthermore, it seems to make sense to also summarize the arguments for a single vector argument $\boldsymbol{v}$:

$$\boldsymbol{m}(\boldsymbol{v}) \in (\mathbb{R}^3 \to \mathbb{R}^3) = \begin{bmatrix} r_\sigma(\boldsymbol{v}) \\ r_\kappa(\boldsymbol{v}) \\ r_F(\boldsymbol{v}) \end{bmatrix}, \quad \boldsymbol{v} = \begin{bmatrix} \sigma \\ \kappa \\ \Delta\lambda \end{bmatrix}. \tag{6.21}$$

The NEWTON method (iteration index: $i$) is used for the definition of the root[6]:

$$\boldsymbol{v}^{(i+1)} = \boldsymbol{v}^{(i)} - \left(\frac{\mathrm{d}\boldsymbol{m}}{\mathrm{d}\boldsymbol{v}}(\boldsymbol{v}^{(i)})\right)^{-1} \times \boldsymbol{m}(\boldsymbol{v}^{(i)})\,, \tag{6.22}$$

whereupon the following

$$\boldsymbol{v}^{(0)} = \begin{bmatrix} \sigma^{(0)} \\ \kappa^{(0)} \\ \Delta\lambda^{(0)} \end{bmatrix} = \begin{bmatrix} \sigma_{n+1}^{\text{trial}} \\ \kappa_n \\ 0 \end{bmatrix} \tag{6.23}$$

has to be used as the initial value. The Jacobian matrix $\frac{\partial \boldsymbol{m}}{\partial \boldsymbol{v}}$ of the residual functions results from the partial derivatives of Eqs. (6.56)–(6.59) to:

$$\frac{\partial \boldsymbol{m}}{\partial \boldsymbol{v}}(\sigma, \kappa, \Delta\lambda) = \begin{bmatrix} \frac{\partial r_\sigma}{\partial \sigma} & \frac{\partial r_\sigma}{\partial \kappa} & \frac{\partial r_\sigma}{\partial \Delta\lambda} \\ \frac{\partial r_q}{\partial \sigma} & \frac{\partial r_\kappa}{\partial \kappa} & \frac{\partial r_\kappa}{\partial \Delta\lambda} \\ \frac{\partial r_F}{\partial \sigma} & \frac{\partial r_F}{\partial \kappa} & \frac{\partial r_F}{\partial \Delta\lambda} \end{bmatrix}$$
$$= \begin{bmatrix} E^{-1} & 0 & \text{sgn}(\sigma) \\ 0 & -1 & 1 \\ \text{sgn}(\sigma) & -\frac{\mathrm{d}k}{\mathrm{d}\kappa} & 0 \end{bmatrix}. \tag{6.24}$$

[6] The NEWTON method is usually used as follows for a one-dimensional function: $x^{(i+1)} = x^{(i)} - \left(\frac{\mathrm{d}f}{\mathrm{d}x}(x^{(i)})\right)^{-1} \times f(x^{(i)})$.

Next to the fulfillment of Eqs. (6.52)–(6.55), which are given due to plasticity, in each integration point also the global force equilibrium hast to be fulfilled. In order to make use of the NEWTON method, it is even necessary at small strains in the general three-dimensional case to define the elasto-plastic stiffness matrix,[7] which is consistent to the integration algorithm [115]. The consistent elasto-plastic modulus results from the following in the one-dimensional case:

$$E_{n+1}^{\text{elpl}} = \frac{\partial \sigma_{n+1}}{\partial \varepsilon_{n+1}} = \frac{\partial \Delta \sigma_n}{\partial \varepsilon_{n+1}}. \tag{6.25}$$

With the inversion of the Jacobian matrix $\frac{\partial \boldsymbol{m}}{\partial \boldsymbol{v}}$, which has to be evaluated in the converged condition of the above listed NEWTON iteration,

$$\left[\left(\frac{\partial \boldsymbol{m}}{\partial \boldsymbol{v}}\right)_{n+1}\right]^{-1} = \begin{bmatrix} \tilde{m}_{11} & \tilde{m}_{12} & \tilde{m}_{13} \\ \tilde{m}_{21} & \tilde{m}_{22} & \tilde{m}_{23} \\ \tilde{m}_{31} & \tilde{m}_{32} & \tilde{m}_{33} \end{bmatrix}_{n+1} = \frac{E}{E + \frac{\mathrm{d}k}{\mathrm{d}\kappa}}$$
$$\times \begin{bmatrix} \frac{\mathrm{d}k}{\mathrm{d}\kappa} & -\text{sgn}(\sigma)\frac{\mathrm{d}k}{\mathrm{d}\kappa} & \text{sgn}(\sigma) \\ \text{sgn}(\sigma) & -1 & -E^{-1} \\ \text{sgn}(\sigma) & E^{-1}\frac{\mathrm{d}k}{\mathrm{d}\kappa} & -E^{-1} \end{bmatrix}_{n+1} \tag{6.26}$$

the elasto-plastic modulus can be defined from

$$E_{n+1}^{\text{elpl}} = \tilde{m}_{11}. \tag{6.27}$$

For this consider the definition of the elasto-plastic modulus for isotropic hardening and one-dimensional state (see suppl. Problem 2.2) and note that under the assumption of Eq. (2.11) the relation $\frac{\partial F}{\partial \kappa} = -\frac{\partial k}{\partial \kappa}$ results. As can be seen from Eq. (6.26), the consistent elasto-plastic modulus in the *one-dimensional* case does not depend on the chosen integration algorithm and equals the continuum form given in Eq. (2.73). However, it needs be considered at this point that this identity does not have to exist at higher dimensions any longer.

For the special case of linear isotropic hardening, i.e. $\frac{\partial k}{\partial \kappa} = E^{\text{pl}} = \text{const.}$, Eq. (6.22) is not solved iteratively and the unknown solution vector $\boldsymbol{v}_{n+1}$ results directly with the help of the initial value (6.23) in:

$$\boldsymbol{v}_{n+1} = \boldsymbol{v}^{(0)} - \left(\frac{\mathrm{d}\boldsymbol{m}}{\mathrm{d}\boldsymbol{v}}(\boldsymbol{v}^{(0)})\right)^{-1} \boldsymbol{m}(\boldsymbol{v}^{(0)}), \tag{6.28}$$

[7] Also referred to as the consistent elasto-plastic tangent modulus matrix, consistent tangent stiffness matrix or algorithmic stiffness matrix.

or in components as:

$$\begin{bmatrix} \sigma_{n+1} \\ \kappa_{n+1} \\ \Delta\lambda_{n+1} \end{bmatrix} = \begin{bmatrix} \sigma_{n+1}^{\text{trial}} \\ \kappa_n \\ 0 \end{bmatrix} - \frac{E}{E+E^{\text{pl}}}$$

$$\times \begin{bmatrix} E^{\text{pl}} & -\text{sgn}(\sigma_{n+1}^{\text{trial}})E^{\text{pl}} & \text{sgn}(\sigma_{n+1}^{\text{trial}}) \\ \text{sgn}(\sigma_{n+1}^{\text{trial}}) & -1 & -E^{-1} \\ \text{sgn}(\sigma_{n+1}^{\text{trial}}) & E^{-1}E^{\text{pl}} & -E^{-1} \end{bmatrix} \begin{bmatrix} 0 \\ 0 \\ F_{n+1}^{\text{trial}} \end{bmatrix}. \tag{6.29}$$

The third equation of (6.29) yields the consistency parameter in the case of linear hardening to:

$$\Delta\lambda_{n+1} = \frac{F_{n+1}^{\text{trial}}}{E + E^{\text{pl}}}. \tag{6.30}$$

The insertion of the consistency parameter into the first equation of (6.29) yields the stress in the final stress state to:

$$\sigma_{n+1} = \left(1 - \frac{F_{n+1}^{\text{trial}}}{E + E^{\text{pl}}} \times \frac{E}{|\sigma_{n+1}^{\text{trial}}|}\right)\sigma_{n+1}^{\text{trial}}. \tag{6.31}$$

From the second equation of (6.29), the isotropic hardening variable in the final stress state is given in the following with the last two results:

$$\kappa_{n+1} = \kappa_n + \frac{F_{n+1}^{\text{trial}}}{E + E^{\text{pl}}} = \kappa_n + \Delta\lambda_{n+1}. \tag{6.32}$$

Finally, the plastic strain results into the following via the flow rule

$$\varepsilon_{n+1}^{\text{pl}} = \varepsilon_n^{\text{pl}} + \Delta\lambda_{n+1}\,\text{sgn}(\sigma_{n+1}), \tag{6.33}$$

and the consistent elasto-plastic modulus can be defined according to $E^{\text{elpl}} = \frac{E \times E^{\text{pl}}}{E+E^{\text{pl}}}$.

The difference in the iteration scheme for linear and nonlinear hardening is illustrated in Fig. 6.5 where it can be seen that a single step is required for linear hardening while nonlinear hardening requires $i$ iteration steps to obtain a converged solution.

To conclude, the calculation steps of the CPP algorithm are presented in compact form:

**I. Calculation of the Test State**

$$\sigma_{n+1}^{\text{trial}} = \sigma_n + E\Delta\varepsilon_n$$
$$\kappa_{n+1}^{\text{trial}} = \kappa_n$$

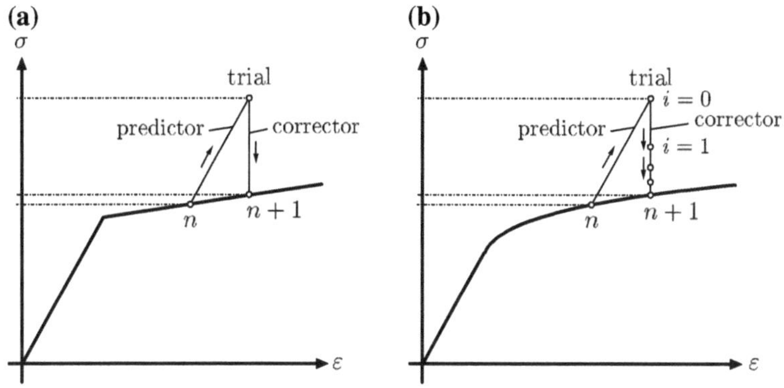


**Fig. 6.5** Corrector step in the case of **a** linear hardening and **b** nonlinear hardening

## II. Test of Validity of the Stress State

$F(\sigma_{n+1}^{\text{trial}}, \kappa_{n+1}^{\text{trial}}) \leq 0$

Y:

$\sigma_{n+1} = \sigma_{n+1}^{\text{trial}}$
$\kappa_{n+1} = \kappa_{n+1}^{\text{trial}}$
End CPP

N:

Back Projection necessary ($\Rightarrow$ Step III)

## III. Back Projection

Initial Values:

$$\boldsymbol{v}^{(0)} = \begin{bmatrix} \sigma^{(0)} \\ \kappa^{(0)} \\ \Delta\lambda^{(0)} \end{bmatrix} = \begin{bmatrix} \sigma_{n+1}^{\text{trial}} \\ \kappa_n \\ 0 \end{bmatrix}$$

Root Finding with the NEWTON Method:

$$\boldsymbol{v}^{(i+1)} = \boldsymbol{v}^{(i)} - \left( \frac{\partial \boldsymbol{m}}{\partial \boldsymbol{v}} (\boldsymbol{v}^{(i)}) \right)^{-1} \boldsymbol{m}(\boldsymbol{v}^{(i)})$$

As long $\|\boldsymbol{v}^{(i+1)} - \boldsymbol{v}^{(i)}\| < t_{\text{end}}^{\boldsymbol{v}}$

In the termination criterion the vector norm (or length) of a vector has been used, which results in the following $\|\boldsymbol{x}\| = \left( \sum_{i=1}^{n} x_i^2 \right)^{0.5}$ for an arbitrary vector $\boldsymbol{x}$.

### IV. Actualization of the Parameters

$$\sigma_{n+1} = \sigma^{(i+1)}$$
$$\kappa_{n+1} = \kappa^{(n+1)}$$
$$E^{\text{elpl}}_{n+1} = \tilde{m}_{11}$$

The internally used calculation precision in the FE system appears perfect as termination precision $t^{v}_{\text{end}}$ in the NEWTON method.

Figure 6.4 shows that the entire strain increment in dependence on the test stress state can be expressed as

$$\Delta\varepsilon_n = \varepsilon_{n+1} - \varepsilon_n = E^{-1}(\sigma^{\text{trial}}_{n+1} - \sigma_n). \tag{6.34}$$

If one introduces the last equation and the flow rule[8] according to (2.21) in Eq. (6.10), the final stress state $\sigma_{n+1}$ in dependence on the test stress state $\sigma^{\text{trial}}_{n+1}$ results in:

$$\sigma_{n+1} = \sigma^{\text{trial}}_{n+1} - \Delta\lambda_{n+1} E \,\text{sgn}(\sigma^{\text{trial}}_{n+1}). \tag{6.35}$$

The update of the state variables alternatively based on $\Delta\varepsilon_n$ and $\Delta\lambda_{n+1}$ is summarized in Table 6.3.

In an alternative solution approach, the constitutive relations can be combined to obtain a single nonlinear equation to solve this problem. To this end, let us introduce the equation for the stress update (6.15) and the update of the internal variable according to Eq. (6.16) in the yield condition (6.17) to obtain:

$$F_{n+1}(\Delta\lambda_{n+1}) = \left|\sigma^{\text{trial}}_{n+1} - \Delta\lambda_{n+1} E \text{sgn}(\sigma^{\text{trial}}_{n+1})\right| - k(\kappa_n + \Delta\lambda_{n+1}) = 0. \tag{6.36}$$

In should be noted here that in the derivation of the last equation the fact was used that the updated and trial stress are co-linear: $\text{sgn}(\sigma_{n+1}) = \text{sgn}(\sigma^{\text{trial}}_{n+1})$. Equation (6.36) can be expressed for arbitrary $\Delta\lambda$ in the form of a residual $r_F$ as:

$$r_F(\Delta\lambda) = F(\Delta\lambda) = \left|\sigma^{\text{trial}}_{n+1} - \Delta\lambda E \text{sgn}(\sigma^{\text{trial}}_{n+1})\right| - k(\kappa_n + \Delta\lambda) \neq 0. \tag{6.37}$$

**Table 6.3** Update of state variables based on $\Delta\varepsilon_n$ and $\Delta\lambda_{n+1}$ for the case of a material with linear isotropic hardening

| Update | cf. Eq. Nr. |
|---|---|
| $\varepsilon_{n+1} = \varepsilon_n + \Delta\varepsilon_n$ | (6.10) |
| $\text{d}\lambda_{n+1} = \text{d}\lambda_n + \Delta\lambda_{n+1}$ | – |
| $\varepsilon^{\text{pl}}_{n+1} = \varepsilon^{\text{pl}}_n + \Delta\lambda_{n+1}\text{sgn}(\sigma_{n+1})$ | (2.21) |
| $\varepsilon^{\text{pl}}_{\text{eff},n+1} = \varepsilon^{\text{pl}}_{\text{eff},n} + \Delta\lambda_{n+1}$ | (2.61) |

[8] At this point it is switched from $\text{d}\lambda$ to $\Delta\lambda$.

A TAYLOR's series expansion of first order, i.e.

$$r_F\left(\Delta\lambda^{(i+1)}\right) = r_F\left(\Delta\lambda^{(i)}\right) + \left(\frac{\mathrm{d}r_F}{\mathrm{d}\Delta\lambda}\right)^{(i)} \times \delta(\Delta\lambda) + \cdots \overset{!}{=} 0\,, \tag{6.38}$$

with $\delta(\Delta\lambda) = \Delta\lambda^{(i+1)} - \Delta\lambda^{(i)}$ gives finally the numerical scheme to solve this equation:

$$\Delta\lambda^{(i+1)} = \Delta\lambda^{(i)} + \left[\left(\frac{\mathrm{d}r_F}{\mathrm{d}\Delta\lambda}\right)^{(i)}\right]^{-1} \times \left[-r_F(\Delta\lambda^{(i)})\right], \tag{6.39}$$

with

$$r_F(\Delta\lambda^{(i)}) = \left|\sigma_{n+1}^{\text{trial}} - \Delta\lambda^{(i)} E \operatorname{sgn}(\sigma_{n+1}^{\text{trial}})\right| - k(\kappa_n + \Delta\lambda^{(i)})\,, \tag{6.40}$$

$$\left(\frac{\mathrm{d}r_F}{\mathrm{d}\Delta\lambda}\right)^{(i)} = \operatorname{sgn}\left(\sigma_{n+1}^{\text{trial}} - \Delta\lambda^{(i)} E \operatorname{sgn}(\sigma_{n+1}^{\text{trial}})\right) \times \left(-E \operatorname{sgn}(\sigma_{n+1}^{\text{trial}})\right) - \left(\frac{\mathrm{d}k}{\mathrm{d}\Delta\lambda}\right)^{(i)}, \tag{6.41}$$

where in the last equation the following simplification can be applied:

$$\operatorname{sgn}\left(\sigma_{n+1}^{\text{trial}} - \Delta\lambda^{(i)} E \operatorname{sgn}(\sigma_{n+1}^{\text{trial}})\right) \times \left(-E \operatorname{sgn}(\sigma_{n+1}^{\text{trial}})\right) \overset{!}{=} -E\,. \tag{6.42}$$

## 6.3 Derivation of the Fully Implicit Backward-Euler Algorithm for Kinematic Hardening

The derivation for pure kinematic hardening is quite similar to the approach for isotropic hardening, see Sect. 6.2. Replacing the equation for the isotropic hardening variable by the expression for the kinematic hardening variable according to PRAGER gives the following set of three equations in the final state:

$$\sigma_{n+1} = \sigma_{n+1}^{\text{trial}} - \Delta\lambda_{n+1} E \operatorname{sgn}(\sigma_{n+1} - \alpha_{n+1})\,, \tag{6.43}$$

$$\alpha_{n+1} = \alpha_n + \Delta\lambda_{n+1} H \operatorname{sgn}(\sigma_{n+1} - \alpha_{n+1})\,, \tag{6.44}$$

$$F_{n+1} = |\sigma_{n+1} - \alpha_{n+1}| - k = 0\,. \tag{6.45}$$

Outside of the final state, at each of these equations a residual $r$ remains again:

$$\begin{aligned} r_\sigma(\sigma, \Delta\lambda, \alpha) &= \sigma - \sigma_{n+1}^{\text{trial}} + \Delta\lambda E \,\text{sgn}(\sigma - \alpha) \neq 0 \quad \text{or} \\ &= E^{-1}\sigma - E^{-1}\sigma_{n+1}^{\text{trial}} + \Delta\lambda \,\text{sgn}(\sigma - \alpha) \neq 0 \,, \qquad (6.46) \\ r_\alpha(\alpha, \Delta\lambda, \sigma) &= \alpha - \alpha_n - \Delta\lambda H \,\text{sgn}(\sigma - \alpha) \neq 0 \quad \text{or} \\ &= -\alpha + \alpha_n + \Delta\lambda H \,\text{sgn}(\sigma - \alpha) \neq 0 \,, \qquad (6.47) \\ r_F(\sigma, \alpha) &= F(\sigma, \alpha) = |\sigma - \alpha| - k \neq 0 \,. \qquad (6.48) \end{aligned}$$

We define again a vector function $\boldsymbol{m}$, a vector argument $\boldsymbol{v}$ and initial values $\boldsymbol{v}^{(0)}$:

$$\boldsymbol{m}(\boldsymbol{v}) \in (\mathbb{R}^3 \to \mathbb{R}^3) = \begin{bmatrix} r_\sigma(\boldsymbol{v}) \\ r_\alpha(\boldsymbol{v}) \\ r_F(\boldsymbol{v}) \end{bmatrix}, \quad \boldsymbol{v} = \begin{bmatrix} \sigma \\ \alpha \\ \Delta\lambda \end{bmatrix}, \boldsymbol{v}^{(0)} = \begin{bmatrix} \sigma_{n+1}^{\text{trial}} \\ \alpha_n \\ 0 \end{bmatrix}, \qquad (6.49)$$

to perform a NEWTON iteration as given in Eq. (6.22). The Jacobian matrix $\frac{\partial \boldsymbol{m}}{\partial \boldsymbol{v}}$ of the residual functions reads for this case:

$$\begin{aligned} \frac{\partial \boldsymbol{m}}{\partial \boldsymbol{v}}(\sigma, \alpha, \Delta\lambda) &= \begin{bmatrix} \frac{\partial r_\sigma}{\partial \sigma} & \frac{\partial r_\sigma}{\partial \alpha} & \frac{\partial r_\sigma}{\partial \Delta\lambda} \\ \frac{\partial r_\alpha}{\partial \sigma} & \frac{\partial r_\alpha}{\partial \alpha} & \frac{\partial r_\alpha}{\partial \Delta\lambda} \\ \frac{\partial r_F}{\partial \sigma} & \frac{\partial r_F}{\partial \alpha} & \frac{\partial r_F}{\partial \Delta\lambda} \end{bmatrix} \\ &= \begin{bmatrix} E^{-1} & 0 & \text{sgn}(\sigma - \alpha) \\ 0 & -1 & H\text{sgn}(\sigma - \alpha) \\ \text{sgn}(\sigma - \alpha) & -\text{sgn}(\sigma - \alpha) & 0 \end{bmatrix}. \qquad (6.50) \end{aligned}$$

Thus, the inversion of the Jacobian matrix $\frac{\partial \boldsymbol{m}}{\partial \boldsymbol{v}}$, which has to be evaluated in the converged condition of the NEWTON iteration (see Eq. 6.22), takes the following form:

$$\begin{aligned} \left[\left(\frac{\partial \boldsymbol{m}}{\partial \boldsymbol{v}}\right)_{n+1}\right]^{-1} &= \begin{bmatrix} \tilde{m}_{11} & \tilde{m}_{12} & \tilde{m}_{13} \\ \tilde{m}_{21} & \tilde{m}_{22} & \tilde{m}_{23} \\ \tilde{m}_{31} & \tilde{m}_{32} & \tilde{m}_{33} \end{bmatrix}_{n+1} = \frac{1}{E + H} \\ &\times \begin{bmatrix} HE & -E & E\text{sgn}(\sigma - \alpha) \\ HE & -E & -H\text{sgn}(\sigma - \alpha) \\ E\text{sgn}(\sigma - \alpha) & \text{sgn}(\sigma - \alpha) & -1 \end{bmatrix}_{n+1}. \end{aligned} \qquad (6.51)$$

## 6.4 Derivation of the Fully Implicit Backward-Euler Algorithm for Combined Hardening

The derivation of the algorithm for combined hardening requires the combination of the approaches from the two previous sections. Thus, the following set of four equations is obtained in the final state:

$$\sigma_{n+1} = \sigma_{n+1}^{\text{trial}} - \Delta\lambda_{n+1} E \,\text{sgn}(\sigma_{n+1} - \alpha_{n+1})\,, \tag{6.52}$$

$$\kappa_{n+1} = \kappa_n + \Delta\lambda_{n+1}\,, \tag{6.53}$$

$$\alpha_{n+1} = \alpha_n + \Delta\lambda_{n+1} H \,\text{sgn}(\sigma_{n+1} - \alpha_{n+1})\,, \tag{6.54}$$

$$F_{n+1} = F(\sigma_{n+1}, \kappa_{n+1}, \alpha_{n+1}) = 0\,. \tag{6.55}$$

The corresponding four residual equations read now:

$$\begin{aligned} r_\sigma(\sigma, \Delta\lambda, \alpha) &= \sigma - \sigma_{n+1}^{\text{trial}} + \Delta\lambda E \,\text{sgn}(\sigma - \alpha) \neq 0 \quad \text{or} \\ &= E^{-1}\sigma - E^{-1}\sigma_{n+1}^{\text{trial}} + \Delta\lambda \,\text{sgn}(\sigma - \alpha) \neq 0\,, \end{aligned} \tag{6.56}$$

$$\begin{aligned} r_\kappa(\kappa, \Delta\lambda) &= \kappa - \kappa_n - \Delta\lambda \neq 0 \quad \text{or} \\ &= -\kappa + \kappa_n + \Delta\lambda \neq 0\,, \end{aligned} \tag{6.57}$$

$$\begin{aligned} r_\alpha(\alpha, \Delta\lambda, \sigma) &= \alpha - \alpha_n - \Delta\lambda H \,\text{sgn}(\sigma - \alpha) \neq 0 \quad \text{or} \\ &= -\alpha + \alpha_n + \Delta\lambda H \,\text{sgn}(\sigma - \alpha) \neq 0\,, \end{aligned} \tag{6.58}$$

$$r_F(\sigma, \alpha, \kappa) = F(\sigma, \alpha, \kappa) = |\sigma - \alpha| - (k^{\text{init}} + E^{\text{pl}}\kappa) \neq 0\,. \tag{6.59}$$

We define again a vector function $\boldsymbol{m}$, a vector argument $\boldsymbol{v}$ and initial values $\boldsymbol{v}^{(0)}$

$$\boldsymbol{m}(\boldsymbol{v}) \in (\mathbb{R}^4 \to \mathbb{R}^4) = \begin{bmatrix} r_\sigma(\boldsymbol{v}) \\ r_\kappa(\boldsymbol{v}) \\ r_\alpha(\boldsymbol{v}) \\ r_F(\boldsymbol{v}) \end{bmatrix}, \quad \boldsymbol{v} = \begin{bmatrix} \sigma \\ \kappa \\ \alpha \\ \Delta\lambda \end{bmatrix}, \boldsymbol{v}^{(0)} = \begin{bmatrix} \sigma_{n+1}^{\text{trial}} \\ \kappa_n \\ \alpha_n \\ 0 \end{bmatrix}, \tag{6.60}$$

to perform the NEWTON iteration as given in Eq. (6.22). The Jacobian matrix $\frac{\partial \boldsymbol{m}}{\partial \boldsymbol{v}}$ of the residual functions reads for this case:

$$\frac{\partial \boldsymbol{m}}{\partial \boldsymbol{v}}(\sigma, \kappa, \alpha, \Delta\lambda) = \begin{bmatrix} \frac{\partial r_\sigma}{\partial \sigma} & \frac{\partial r_\sigma}{\partial \kappa} & \frac{\partial r_\sigma}{\partial \alpha} & \frac{\partial r_\sigma}{\partial \Delta\lambda} \\ \frac{\partial r_\kappa}{\partial \sigma} & \frac{\partial r_\kappa}{\partial \kappa} & \frac{\partial r_\kappa}{\partial \alpha} & \frac{\partial r_\kappa}{\partial \Delta\lambda} \\ \frac{\partial r_\alpha}{\partial \sigma} & \frac{\partial r_\alpha}{\partial \kappa} & \frac{\partial r_\alpha}{\partial \alpha} & \frac{\partial r_\alpha}{\partial \Delta\lambda} \\ \frac{\partial r_F}{\partial \sigma} & \frac{\partial r_F}{\partial \kappa} & \frac{\partial r_F}{\partial \alpha} & \frac{\partial r_F}{\partial \Delta\lambda} \end{bmatrix}$$

$$= \begin{bmatrix} E^{-1} & 0 & 0 & \mathrm{sgn}(\sigma-\alpha) \\ 0 & -1 & 0 & 1 \\ 0 & 0 & -1 & H\mathrm{sgn}(\sigma-\alpha) \\ \mathrm{sgn}(\sigma-\alpha) & -E^{\mathrm{pl}} & -\mathrm{sgn}(\sigma-\alpha) & 0 \end{bmatrix}, \tag{6.61}$$

and its inverse can be stated as:

$$\left[\left(\frac{\partial \boldsymbol{m}}{\partial \boldsymbol{v}}\right)_{n+1}\right]^{-1} = \begin{bmatrix} \tilde{m}_{11} & \tilde{m}_{12} & \tilde{m}_{13} & \tilde{m}_{14} \\ \tilde{m}_{21} & \tilde{m}_{22} & \tilde{m}_{23} & \tilde{m}_{24} \\ \tilde{m}_{31} & \tilde{m}_{32} & \tilde{m}_{33} & \tilde{m}_{34} \\ \tilde{m}_{41} & \tilde{m}_{42} & \tilde{m}_{43} & \tilde{m}_{44} \end{bmatrix}_{n+1} = \frac{1}{E+E^{\mathrm{pl}}+H}$$

$$\times \begin{bmatrix} (E^{\mathrm{pl}}+H)E & -E^{\mathrm{pl}}E\mathrm{sgn}(\sigma-\alpha) & -E & E\mathrm{sgn}(\sigma-\alpha) \\ E\mathrm{sgn}(\sigma-\alpha) & -(E+H) & \mathrm{sgn}(\sigma-\alpha) & -1 \\ HE & E^{\mathrm{pl}}H\mathrm{sgn}(\sigma-\alpha) & -(E+E^{\mathrm{pl}}) & -H\mathrm{sgn}(\sigma-\alpha) \\ E\mathrm{sgn}(\sigma-\alpha) & E^{\mathrm{pl}} & \mathrm{sgn}(\sigma-\alpha) & -1 \end{bmatrix}_{n+1}. \tag{6.62}$$

It should be noted here that some authors introduce an auxiliary variable, the so-called *relative stress* $\xi = \sigma - \alpha$. Then, the partial derivatives in Eq. (6.61) can be obtained by respecting the derivative of the transformed coordinate, for example, as:

$$\frac{\partial r_F}{\partial \alpha} = \frac{\partial r_F}{\partial \xi}\frac{\partial \xi}{\partial \alpha} = \mathrm{sgn}(\xi) \times (-1) = -\mathrm{sgn}(\sigma-\alpha)\,. \tag{6.63}$$

The complete iteration scheme for the case of combined hardening reads finally:

$$\begin{bmatrix} \sigma_{n+1} \\ \kappa_{n+1} \\ \alpha_{n+1} \\ \mathrm{d}\lambda_{n+1} \end{bmatrix} = \begin{bmatrix} \sigma_{n+1}^{\mathrm{trial}} \\ \kappa_n \\ \alpha_n \\ 0 \end{bmatrix} - \frac{1}{E+E^{\mathrm{pl}}+H}$$

$$\times \begin{bmatrix} (E^{\mathrm{pl}}+H)E & -E^{\mathrm{pl}}E\mathrm{sgn}(\sigma-\alpha) & -E & E\mathrm{sgn}(\sigma-\alpha) \\ E\mathrm{sgn}(\sigma-\alpha) & -(E+H) & \mathrm{sgn}(\sigma-\alpha) & -1 \\ HE & E^{\mathrm{pl}}H\mathrm{sgn}(\sigma-\alpha) & -(E+E^{\mathrm{pl}}) & -H\mathrm{sgn}(\sigma-\alpha) \\ E\mathrm{sgn}(\sigma-\alpha) & E^{\mathrm{pl}} & \mathrm{sgn}(\sigma-\alpha) & -1 \end{bmatrix}_{n+1}$$

$$\begin{bmatrix} 0 \\ 0 \\ 0 \\ F_{n+1}^{\mathrm{trial}} \end{bmatrix}. \tag{6.64}$$

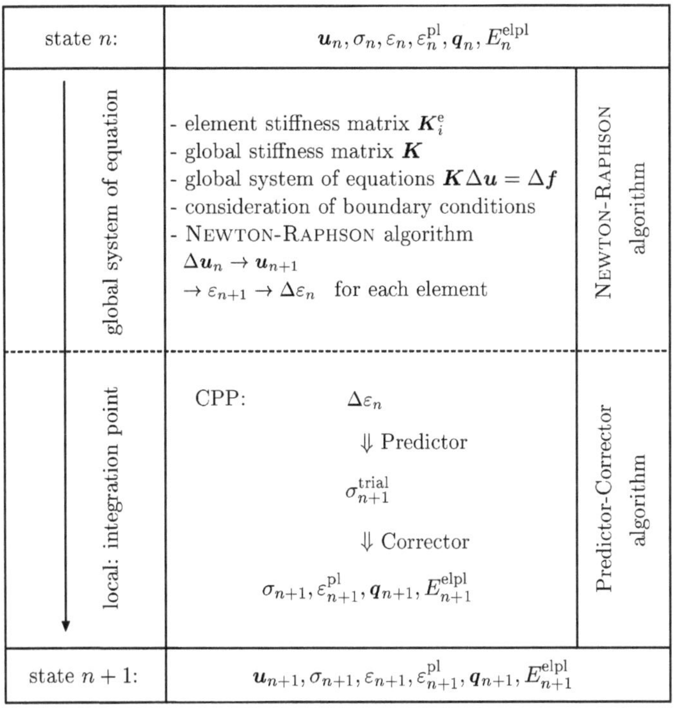


**Fig. 6.6** General procedure of an elasto-plastic finite element calculation

The general procedure of an elasto-plastic finite element calculation is illustrated at the end of this section in Fig. 6.6.

One can see that the solution of an elasto-plastic problem occurs on two levels. On the global level, meaning for the global system of equations under consideration of the boundary conditions, the NEWTON- RAPHSON iteration scheme is made use of to define the incremental global displacement vector $\Delta\boldsymbol{u}_n$. By summing the displacement increments, the total global displacement vector $\boldsymbol{u}_{n+1}$ of the unknown nodal displacement of a structure consisting of finite elements results. Via the strain-displacement relation, the strain $\varepsilon_{n+1}$ or, alternatively the strain increment $\Delta\varepsilon_n$ per element, can be determined from the nodal displacement.[9] The strain increment of an element is now used on the level of the integration points, to define the remaining state variables iteratively with the help of a predictor-corrector method.

[9] At this point, under consideration of linear bar elements, a constant strain distribution per element results. In general, the strain results as a function of the element coordinates which is usually evaluated at the integration points. Therefore, one would in the general case normally define a strain *vector* $\boldsymbol{\varepsilon}$ per element, which combines the different strain values at the integration points. However, this is unnecessary for a linear bar element. A scalar strain or alternative stress value is enough for the description.

## 6.5 Consideration of Damage

### 6.5.1 Lemaitre Damage Model

The numerical scheme to solve the problem is again derived from the general finite element approach which is based on the predictor-corrector scheme. Let us have a look at the schematic representation of the predictor-corrector integration algorithm in the stress-strain diagram as shown in Fig. 6.7 where damage effects are considered.

It can be concluded from Fig. 6.7 that the trial stress can be expressed as:

$$\sigma_{n+1}^{\text{trial}} = \sigma_n + \Delta\sigma_n^{\text{el}}, \tag{6.65}$$

$$= \sigma_n + \bar{E}_n \Delta\varepsilon_n . \tag{6.66}$$

Under consideration of the definition of $\bar{E}$ according to Eq. (2.93), the last equation can be written as:

$$\sigma_{n+1}^{\text{trial}} = \sigma_n + (1 - D_n) E \varepsilon_n . \tag{6.67}$$

Looking at the strains indicated in Fig. 6.7, it can be concluded that the elastic trial strain can be expressed as

$$\varepsilon_{n+1}^{\text{el,trial}} = \underbrace{\varepsilon_n - \varepsilon_n^{\text{pl}}}_{\varepsilon_n^{\text{el}}} + \Delta\varepsilon_n = \varepsilon_{n+1} - \varepsilon_n^{\text{pl}} \tag{6.68}$$

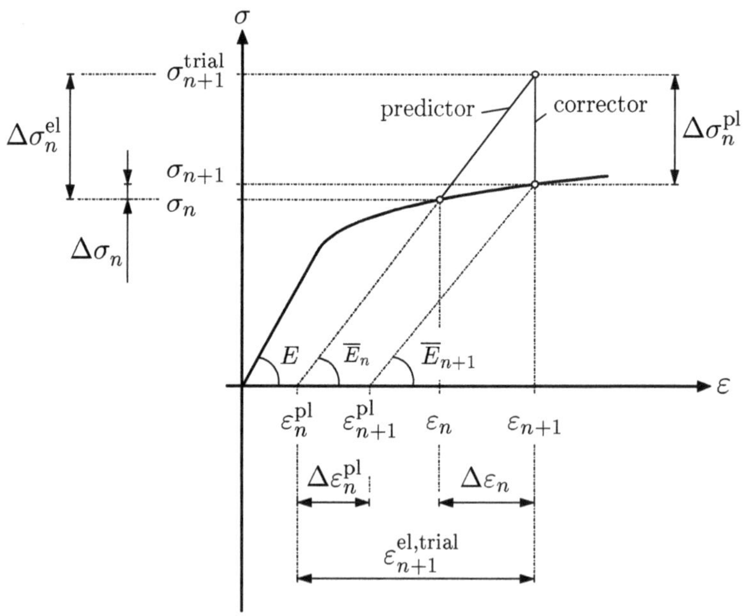


**Fig. 6.7** Schematic representation of the integration algorithm in the stress-strain diagram under consideration of damage effects

or as

$$\varepsilon_{n+1}^{\text{el,trial}} = \varepsilon_{n+1} - \varepsilon_{n+1}^{\text{pl}} + \Delta\varepsilon_{n}^{\text{pl}} = \varepsilon_{n+1}^{\text{el}} + \Delta\varepsilon_{n}^{\text{pl}} . \tag{6.69}$$

The elastic strain in the state $n+1$ is given according to Fig. 6.7 and HOOKE's law as $\varepsilon_{n+1}^{\text{el}} = \frac{\sigma_{n+1}}{(1-D_{n+1})E}$ and Eq. (6.69) can be rearranged to obtain the following expression for the stress in the final state:

$$\sigma_{n+1} = (1 - D_{n+1})E\varepsilon_{n+1}^{\text{el,trial}} - (1 - D_{n+1})E\,\Delta\varepsilon_{n}^{\text{pl}} . \tag{6.70}$$

In the last equation, the elastic trial strain can be calculated according to Eq. (6.68) and the plastic strain increment can be replaced by the flow rule for a fully implicit scheme, i.e. $\Delta\varepsilon_{n}^{\text{pl}} = \Delta\lambda_{n+1}\frac{\text{sgn}(\sigma_{n+1})}{1-D_{n+1}} = \Delta\lambda_{n+1}\frac{\text{sgn}(\sigma_{n+1}^{\text{trial}})}{1-D_{n+1}}$, to obtain:

$$\sigma_{n+1} = (1 - D_{n+1})E\varepsilon_{n+1}^{\text{el,trial}} - \Delta\lambda_{n+1}E\,\text{sgn}(\sigma_{n+1}^{\text{trial}}). \tag{6.71}$$

It should be highlighted here that the first expression on the right-hand side of Eq. (6.71) is unequal to the trial stress which is given according to Fig. 6.7 by:

$$(1 - D_{n})E\varepsilon_{n+1}^{\text{el,trial}} = \sigma_{n+1}^{\text{trial}} . \tag{6.72}$$

Only if there is no damage, Eq. (6.71) reduces to the statement given by Eq. (6.12) in Sect. 6.1.

Dividing Eq. (6.71) by $(1 - D_{n+1})$ and introducing relationship (6.72) gives the updated stress expressed by the trial stress as:

$$\frac{\sigma_{n+1}}{1 - D_{n+1}} = \frac{\sigma_{n+1}^{\text{trial}}}{1 - D_{n}} - \frac{\Delta\lambda_{n+1}E\,\text{sgn}(\sigma_{n+1}^{\text{trial}})}{1 - D_{n+1}} . \tag{6.73}$$

Summarizing, it can be stated that in the final state $(n+1)$, the following four equations must be fulfilled:

$$\sigma_{n+1} = (1 - D_{n+1})E\varepsilon_{n+1}^{\text{el,trial}} - \Delta\lambda_{n+1}E\,\text{sgn}(\sigma_{n+1}^{\text{trial}}) , \tag{6.74}$$

$$\kappa_{n+1} = \kappa_{n} + \Delta\lambda_{n+1} , \tag{6.75}$$

$$F_{n+1} = \frac{|\sigma_{n+1}|}{1 - D_{n+1}} - k(\kappa_{n+1}) = 0 , \tag{6.76}$$

$$D_{n+1} = D_{n} + \frac{\Delta\lambda_{n+1}}{1 - D_{n+1}}\left(\frac{-\,Y(\sigma_{n+1}, D_{n+1})}{r}\right)^{s} . \tag{6.77}$$

Given the set $(\sigma_{n}, \varepsilon_{n}^{\text{pl}}, \kappa_{n}, D_{n})$ at time $n$ and the global strain increment $\Delta\varepsilon$, the last four expressions are a set of nonlinear algebraic equations for $(\sigma_{n+1}, \varepsilon_{n+1}^{\text{pl}}, \kappa_{n+1}, D_{n+1})$.

#### 6.5.1.1 Matrix Solution Scheme

The set of equations given in (6.74)–(6.77) are only fulfilled in the final state $(n+1)$. Out of this final state, a residual $r_i$ is obtained for each of these equations:

$$r_\sigma = E^{-1}\sigma - (1-D)\varepsilon^{\text{el,trial}}_{n+1} + \Delta\lambda \text{sgn}(\sigma) \neq 0\,, \tag{6.78}$$

$$r_\kappa = -\kappa + \kappa_n + \Delta\lambda \neq 0\,, \tag{6.79}$$

$$r_F = \frac{|\sigma|}{1-D} - k(\kappa) \neq 0\,, \tag{6.80}$$

$$r_D = -D + D_n + \frac{\Delta\lambda}{1-D}\left(\frac{-Y(\sigma,D)}{r}\right)^s \neq 0\,, \tag{6.81}$$

with

$$Y(\sigma,D) = -\frac{\sigma^2}{2E(1-D)^2}\,. \tag{6.82}$$

Thus, the final state is the root of a vector function $\boldsymbol{m}$ which is composed of the residual functions. Furthermore, it is useful to collect all variables in a single variable vector $\boldsymbol{v}$:

$$\boldsymbol{m}(\boldsymbol{v}) \in (\mathbb{R}^4 \to \mathbb{R}^4) = \begin{bmatrix} r_\sigma(\boldsymbol{v}) \\ r_\kappa(\boldsymbol{v}) \\ r_F(\boldsymbol{v}) \\ r_D(\boldsymbol{v}) \end{bmatrix}, \quad \boldsymbol{v} = \begin{bmatrix} \sigma \\ \kappa \\ \Delta\lambda \\ D \end{bmatrix}. \tag{6.83}$$

The root is found by a NEWTON iteration (iteration index: $i$):

$$\boldsymbol{v}^{(i+1)} = \boldsymbol{v}^{(i)} - \left(\frac{\partial \boldsymbol{m}}{\partial \boldsymbol{v}}\boldsymbol{v}^{(i)}\right)^{-1} \boldsymbol{m}\left(\boldsymbol{v}^{(i)}\right), \tag{6.84}$$

where

$$\boldsymbol{v}^{(0)} = \begin{bmatrix} \sigma^{(0)} \\ \kappa^{(0)} \\ \Delta\lambda^{(0)} \\ D^{(0)} \end{bmatrix} = \begin{bmatrix} \sigma^{\text{trial}}_{n+1} \\ \kappa_n \\ 0 \\ D_n \end{bmatrix} \tag{6.85}$$

can be used as initial value of the argument. The Jacobian matrix of the residual functions is obtained from partial derivatives of (6.78)–(6.81) as

$$\frac{\partial \boldsymbol{m}}{\partial \boldsymbol{v}}(\sigma, \kappa, \Delta\lambda, D)$$
$$= \begin{bmatrix} \frac{\partial r_\sigma}{\partial \sigma} & \frac{\partial r_\sigma}{\partial \kappa} & \frac{\partial r_\sigma}{\partial \Delta\lambda} & \frac{\partial r_\sigma}{\partial D} \\ \frac{\partial r_\kappa}{\partial \sigma} & \frac{\partial r_\kappa}{\partial \kappa} & \frac{\partial r_\kappa}{\partial \Delta\lambda} & \frac{\partial r_\kappa}{\partial D} \\ \frac{\partial r_F}{\partial \sigma} & \frac{\partial r_F}{\partial \kappa} & \frac{\partial r_F}{\partial \Delta\lambda} & \frac{\partial r_F}{\partial D} \\ \frac{\partial r_D}{\partial \sigma} & \frac{\partial r_D}{\partial \kappa} & \frac{\partial r_D}{\partial \Delta\lambda} & \frac{\partial r_D}{\partial D} \end{bmatrix}$$
$$= \begin{bmatrix} E^{-1} & 0 & \operatorname{sgn}(\sigma) & \varepsilon_{n+1}^{\text{el,trial}} \\ 0 & -1 & 1 & 0 \\ \frac{\operatorname{sgn}(\sigma)}{1-D} & -\frac{\mathrm{d}k}{\mathrm{d}\kappa} & 0 & \frac{|\sigma|}{(1-D)^2} \\ \frac{2\Delta\lambda s}{(1-D)^{2s+1}} \frac{\sigma^{2s-1}}{(2rE)^s} & 0 & \frac{1}{(1-D)^{2s+1}}\left(\frac{\sigma^2}{2Er}\right)^s & -1+\frac{(2s+1)\Delta\lambda}{(1-D)^{2(s+1)}}\left(\frac{\sigma^2}{2rE}\right)^s \end{bmatrix}. \tag{6.86}$$

The inversion of the Jacobian matrix which must be evaluated in the converged state of the above mentioned NEWTON iteration and is given in Eq. (6.87). It should be noted here that neglecting of damage in Eq. (6.86), i.e. assigning $D = 0$, $s = 1$ and $r \to \infty$, reduces this formulation to the expression given in Eq. (6.24) for isotropic hardening.

$$\begin{bmatrix} a_{11} & 0 & a_{13} & a_{14} \\ 0 & -1 & 1 & 0 \\ a_{31} & a_{32} & 0 & a_{34} \\ a_{41} & 0 & a_{43} & a_{44} \end{bmatrix}^{-1}$$
$$= \frac{1}{det}\left[\begin{array}{c|c|c|c} -a_{44}a_{32}+a_{34}a_{43} & a_{32}\left(a_{44}a_{13}-a_{14}a_{43}\right) & a_{44}a_{13}-a_{14}a_{43} & a_{14}a_{32}-a_{34}a_{13} \\ a_{44}a_{31}-a_{34}a_{41} & -a_{44}a_{13}a_{31}-a_{34}a_{11}a_{43} +a_{34}a_{13}a_{41}+a_{14}a_{31}a_{43} & -a_{44}a_{11}+a_{14}a_{41} & a_{34}a_{11}-a_{14}a_{31} \\ a_{44}a_{31}-a_{34}a_{41} & -\left(a_{44}a_{11}-a_{14}a_{41}\right)a_{32} & -a_{44}a_{11}+a_{14}a_{41} & a_{34}a_{11}-a_{14}a_{31} \\ a_{41}a_{32}-a_{31}a_{43} & \left(a_{43}a_{11}-a_{13}a_{41}\right)a_{32} & a_{43}a_{11}-a_{13}a_{41} & a_{13}a_{31}-a_{32}a_{11} \end{array}\right],$$

where

$$det = a_{44}a_{13}a_{31} - a_{44}a_{11}a_{32} + a_{14}a_{41}a_{32} + a_{34}a_{11}a_{43} - a_{34}a_{13}a_{41} - a_{14}a_{31}a_{43}\,. \tag{6.87}$$

#### 6.5.1.2 Single Equation Return Mapping

It is possible to combine the constitutive relations in a different way to obtain a single nonlinear equation to solve the problem. This is even for the general three-dimensional case possible as long as the contribution of kinematic hardening is disregarded. To start this derivation, let us introduce Eqs. (6.73) and (6.75) in (6.76) to obtain the following statement for the yield condition at state $(n+1)$:

$$F_{n+1}(\Delta\lambda_{n+1}, D_{n+1}) = \left| \frac{\sigma_{n+1}^{\text{trial}}}{1-D_n} - \frac{\Delta\lambda_{n+1} E \operatorname{sgn}(\sigma_{n+1}^{\text{trial}})}{1-D_{n+1}} \right| - k(\kappa_n + \Delta\lambda_{n+1}) = 0\,. \tag{6.88}$$

Combining the damage energy release rate according to Eq. (6.82) with the expression for the updated stress according to Eq. (6.73), one can obtain the following relationship:

$$Y(\Delta\lambda_{n+1}, D_{n+1}) = -\frac{\left( \dfrac{\sigma_{n+1}^{\text{trial}}}{1-D_n} - \dfrac{\Delta\lambda_{n+1} E \operatorname{sgn}(\sigma_{n+1}^{\text{trial}})}{1-D_{n+1}} \right)^2}{2E}\,. \tag{6.89}$$

Furthermore, Eq. (6.88) can be rearranged to obtain the squared value of the yield stress $k$ which can be introduced into Eq. (6.89) to obtain the damage energy release rate only as a function of a single variable:

$$Y(\Delta\lambda_{n+1}) = -\frac{k^2(\kappa_n + \Delta\lambda_{n+1})}{2E}\,. \tag{6.90}$$

In the following derivation, it is advantageous to introduce a new variable, the so-called material integrity

$$\omega = 1 - D\,, \tag{6.91}$$

which is for the undamaged state equal to 1 and in the case of rupture equal to 0. Furthermore, Eq. (6.88) can be solved for the damage variable in the form $(1-D_{n+1})$ to obtain together with the new definition the expression:

$$\omega_{n+1}(\Delta\lambda_{n+1}) = 1 - D_{n+1} = \frac{\Delta\lambda_{n+1} E \operatorname{sgn}(\sigma_{n+1}^{\text{trial}})}{\frac{\sigma_{n+1}^{\text{trial}}}{1-D_n} - \operatorname{sgn}(\sigma_{n+1}^{\text{trial}}) k(\kappa_n + \Delta\lambda_{n+1})}\,. \tag{6.92}$$

It must be noted here that the last equation was derived under the assumption of plastic yield since Eq. (6.88) was written with an equal sign. In the case $F_{n+1} < 0$, i.e. $\Delta\lambda_{n+1} = 0$, we must consider that $\omega_{n+1}(\Delta\lambda_{n+1} = 0) := 1$. Thus, the relationship for the material integrity should be written as:

$$\omega_{n+1}(\Delta\lambda_{n+1}) = \begin{cases} \dfrac{\Delta\lambda_{n+1} E \operatorname{sgn}(\sigma_{n+1}^{\text{trial}})}{\frac{\sigma_{n+1}^{\text{trial}}}{1-D_n} - \operatorname{sgn}(\sigma_{n+1}^{\text{trial}}) k(\kappa_n + \Delta\lambda_{n+1})} & \text{for} \quad \Delta\lambda_{n+1} > 0 \\[2ex] 1 & \text{for} \quad \Delta\lambda_{n+1} = 0 \end{cases}\,. \tag{6.93}$$

Let us write now the evolution equation for the damage variable transformed to the material integrity, i.e. multiply Eq. (6.77) by $-1$ and add 1 on both sides to obtain:

$$\omega_{n+1} = \omega_n - \frac{\Delta\lambda_{n+1}}{1 - D_{n+1}} \left( \frac{-Y(\Delta\lambda_{n+1})}{r} \right)^s , \tag{6.94}$$

which can be expressed for arbitrary $\Delta\lambda$ in the form of a residual $r_\omega$ as:

$$r_\omega(\Delta\lambda) = \omega(\Delta\lambda) - \omega_n + \frac{\Delta\lambda}{\omega(\Delta\lambda)} \left( \frac{-Y(\Delta\lambda)}{r} \right)^s = 0 . \tag{6.95}$$

A TAYLOR's series expansion of first order, i.e.

$$r_\omega \left( \Delta\lambda^{(i+1)} \right) = r_\omega \left( \Delta\lambda^{(i)} \right) + \left( \frac{\mathrm{d} r_\omega}{\mathrm{d}\Delta\lambda} \right)^{(i)} \times \delta(\Delta\lambda) + \cdots = 0 , \tag{6.96}$$

with $\delta(\Delta\lambda) = \Delta\lambda^{(i+1)} - \Delta\lambda^{(i)}$, gives finally the numerical scheme to solve this equation:

$$\Delta\lambda^{(i+1)} = \Delta\lambda^{(i)} + \left[ \left( \frac{\mathrm{d} r_\omega}{\mathrm{d}\Delta\lambda} \right)^{(i)} \right]^{-1} \times \left[ -r_\omega(\Delta\lambda^{(i)}) \right] . \tag{6.97}$$

This iterative scheme results in the value of $\Delta\lambda_{n+1}$ which can be used to update the state variable $\sigma$, $D$, $\kappa$, $\varepsilon^{\mathrm{pl}}$ and $\varepsilon^{\mathrm{pl}}_{\mathrm{eff}}$, cf. Table 6.4.

It is useful to have a closer look at the iteration scheme in order to obtain a stable iteration. Due to the discontinuous definition of the material integrity $\omega$ according to Eq. (6.93), the residual equation (6.95) exhibits a discontinuity for $\Delta\lambda \to 0^+$, cf. Fig. 6.8. Calculating $\omega$ based on $(6.93)_1$ for $\Delta\lambda \to 0$, expression (6.95) gives

**Table 6.4** Update of state variables based on $\Delta\lambda_{n+1}$ for the LEMAITRE damage model

| Update | Reference |
|---|---|
| $D_{n+1} = 1 - \dfrac{\Delta\lambda_{n+1} E \operatorname{sgn}(\sigma^{\mathrm{trial}}_{n+1})}{\frac{\sigma^{\mathrm{trial}}_{n+1}}{1-D_n} - \operatorname{sgn}(\sigma^{\mathrm{trial}}_{n+1}) k(\kappa_n + \Delta\lambda_{n+1})}$ | Eq. (6.93) |
| $\sigma_{n+1} = \dfrac{1 - D_{n+1}}{1 - D_n} \sigma^{\mathrm{trial}}_{n+1} - \Delta\lambda_{n+1} E \operatorname{sgn}(\sigma^{\mathrm{trial}}_{n+1})$ | Eq. (6.73) |
| $\kappa_{n+1} = \kappa_n + \Delta\lambda_{n+1}$ | Eq. (6.75) |
| $\varepsilon^{\mathrm{pl}}_{n+1} = \varepsilon^{\mathrm{pl}}_n + \dfrac{\Delta\lambda_{n+1}}{1 - D_{n+1}} \operatorname{sgn}(\sigma_{n+1})$ | Table 2.4 |
| $\varepsilon^{\mathrm{pl}}_{\mathrm{eff},n+1} = \varepsilon^{\mathrm{pl}}_{\mathrm{eff},n} + \dfrac{\Delta\lambda_{n+1}}{1 - D_{n+1}}$ | Table 2.4 |

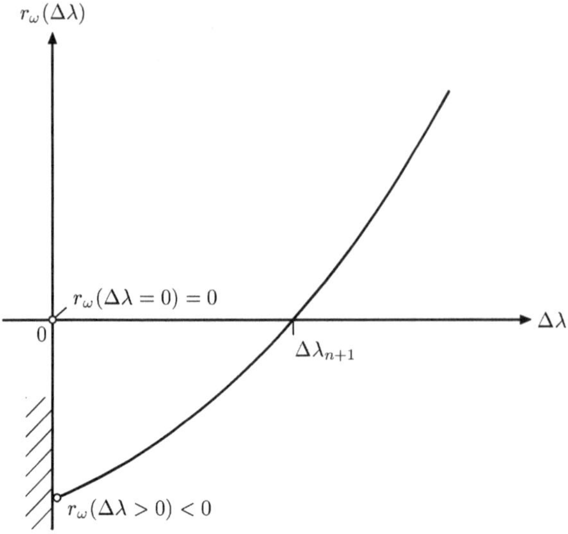


**Fig. 6.8** Schematic representation of the residual equation (6.95) as a function of $\Delta\lambda$. There is a discontinuity indicated by (∘) for $\Delta\lambda \to 0^+$

approximately $-\omega_n$ and not zero as the application of Eq. $(6.93)_2$ would give. Thus, the iteration scheme (6.97) should not start with an initial value of $\Delta\lambda^{(i=0)} = 0$. Any value $\Delta\lambda^{(i=0)} > 0$ may be suitable and a good guess can be obtained from Eq. $(6.93)_1$ under the assumption of a perfectly plastic material for the step $n \to n+1$ with frozen (damaged) yield surface at step $n$:

$$\Delta\lambda^{(i=0)} = \frac{\left(\frac{\sigma_{n+1}^{\text{trial}}}{1-D_n} - \text{sgn}(\sigma_{n+1}^{\text{trial}})k(\kappa_n)\right)\omega_n}{E\,\text{sgn}(\sigma_{n+1}^{\text{trial}})}. \tag{6.98}$$

The iterative scheme given in Eq. (6.97) requires the calculation of the derivative $\frac{\mathrm{d}\omega(\Delta\lambda)}{\mathrm{d}\Delta\lambda}^{(i)}$. To this end, let us introduce the expression for the damage energy release rate according to Eq. (6.90) in the expression for the residual (6.95) to obtain

$$r_\omega(\Delta\lambda) = \omega(\Delta\lambda) - \omega_n + \frac{\Delta\lambda k^{2s}(\kappa_n + \Delta\lambda)}{\omega(\Delta\lambda)} \times \left(\frac{1}{2Er}\right)^s, \tag{6.99}$$

from which the derivative can be obtained as:

$$\frac{\mathrm{d}r_\omega}{\mathrm{d}\Delta\lambda} = \underbrace{\frac{\mathrm{d}\omega(\Delta\lambda)}{\mathrm{d}\Delta\lambda}}_{A(\Delta\lambda)} + \underbrace{\left(\frac{1}{2Er}\right)^{s} \frac{\mathrm{d}}{\mathrm{d}\Delta\lambda}\left(\frac{\Delta\lambda k^{2s}(\kappa_n + \Delta\lambda)}{\omega(\Delta\lambda)}\right)}_{B(\Delta\lambda)}. \tag{6.100}$$

Application of the quotient rule to calculate the derivative of $\omega$ according to Eq. $(6.93)_1$ gives the first expression in the last equation as:

$$A(\Delta\lambda) = \frac{\mathrm{d}\omega(\Delta\lambda)}{\mathrm{d}\Delta\lambda} = \frac{\left(\frac{|\sigma_{n+1}^{\mathrm{trial}}|}{1-D_n} E - Ek(\kappa_n + \Delta\lambda)\right) + \left(\Delta\lambda E \frac{\mathrm{d}k}{\mathrm{d}\Delta\lambda}\right)}{\left(\frac{\sigma_{n+1}^{\mathrm{trial}}}{1-D_n} - \mathrm{sgn}(\sigma_{n+1}^{\mathrm{trial}})k(\kappa_n + \Delta\lambda)\right)^2}. \tag{6.101}$$

In a similar way, the derivative of the second expression in Eq. (6.100) can be calculated as:

$$\begin{aligned} B(\Delta\lambda) &= \left(\frac{1}{2Er}\right)^{s} \times \frac{\frac{\mathrm{d}}{\mathrm{d}\Delta\lambda}\left(\Delta\lambda k^{2s}\right) \times \omega(\Delta\lambda) - \Delta\lambda k^{2s} \times \frac{\mathrm{d}\omega(\Delta\lambda)}{\mathrm{d}\Delta\lambda}}{(\omega(\Delta\lambda))^2} \\ &= \left(\frac{1}{2Er}\right)^{s} \times \frac{\left[1k^{2s} + \Delta\lambda 2sk^{2s-1}\frac{\mathrm{d}k}{\mathrm{d}\Delta\lambda}\right] \times \omega(\Delta\lambda) - \Delta\lambda k^{2s} \times \frac{\mathrm{d}\omega(\Delta\lambda)}{\mathrm{d}\Delta\lambda}}{(\omega(\Delta\lambda))^2}. \end{aligned} \tag{6.102}$$

## *6.5.2 Gurson Damage Model*

The initial execution of the derivation can be performed as in the case of Sect. 6.2, i.e. the fully implicit backward-EULER algorithm for isotropic hardening without any damage. Similar to Eq. (6.15), one can obtain in a more general form the following expression

$$\sigma_{n+1} = \sigma_{n+1}^{\mathrm{trial}} - E\Delta\varepsilon_n^{\mathrm{pl}}, \tag{6.103}$$

where $\Delta\varepsilon_n^{\mathrm{pl}}$ can be expressed based on the flow rule (cf. Table 2.6) for a fully implicit scheme as:

$$\Delta\varepsilon_n^{\mathrm{pl}} = \Delta\lambda_{n+1}\left(\frac{2\sigma_{n+1}}{k_{\mathrm{t}}^2} + \frac{D_{n+1}}{k_{\mathrm{t}}}\sinh\left(\frac{\sigma_{n+1}}{2k_{\mathrm{t}}}\right)\right). \tag{6.104}$$

Thus, it can be stated that in the final state $(n+1)$, the following four equations (cf. Table 2.6) must be fulfilled:

$$\sigma_{n+1} = \sigma_{n+1}^{\text{trial}} - \Delta\lambda_{n+1} E \left( \frac{2\sigma_{n+1}}{k_{\text{t}}^2} + \frac{D_{n+1}}{k_{\text{t}}} \sinh\left( \frac{\sigma_{n+1}}{2k_{\text{t}}} \right) \right), \tag{6.105}$$

$$\kappa_{n+1} = \kappa_n + \frac{\Delta\lambda_{n+1}}{1 - D_{n+1}} \left( \frac{2\sigma_{n+1}^2}{k_{\text{t}}^3} + \frac{D_{n+1}\sigma_{n+1}}{k_{\text{t}}^2} \sinh\left( \frac{\sigma_{n+1}}{2k_{\text{t}}} \right) \right), \tag{6.106}$$

$$F_{n+1} = \left( \frac{|\sigma_{n+1}|}{k_{\text{t}}} \right)^2 + 2D_{n+1} \cosh\left( \frac{\sigma_{n+1}}{2k_{\text{t}}} \right) - (1 + D_{n+1}^2) = 0, \tag{6.107}$$

$$D_{n+1} = D_n + \Delta\lambda_{n+1} \frac{3(D_{n+1} - D_{n+1}^2)}{k_{\text{t}}} \sinh\left( \frac{\sigma_{n+1}}{2k_{\text{t}}} \right). \tag{6.108}$$

Again, given the set $(\sigma_n, \varepsilon_n^{\text{pl}}, \kappa_n, D_n)$ at time $n$ and the global strain increment $\Delta\varepsilon$, the last four expressions are a set of nonlinear algebraic equations for $(\sigma_{n+1}, \varepsilon_{n+1}^{\text{pl}}, \kappa_{n+1}, D_{n+1})$. However, they are only fulfilled in the final state $(n+1)$ and outside this final state, a residual $r_i$ is obtained for each of these equations:

$$r_\sigma = \sigma - \sigma_{n+1}^{\text{trial}} + \Delta\lambda E \left( \frac{2\sigma}{k_{\text{t}}^2} + \frac{D}{k_{\text{t}}} \sinh\left( \frac{\sigma}{2k_{\text{t}}} \right) \right) \neq 0, \tag{6.109}$$

$$r_\kappa = \kappa - \kappa_n - \frac{\Delta\lambda}{1 - D} \left( \frac{2\sigma^2}{k_{\text{t}}^3} + \frac{D\sigma}{k_{\text{t}}^2} \sinh\left( \frac{\sigma}{2k_{\text{t}}} \right) \right) \neq 0, \tag{6.110}$$

$$r_F = \left( \frac{|\sigma|}{k_{\text{t}}} \right)^2 + 2D \cosh\left( \frac{\sigma}{2k_{\text{t}}} \right) - (1 + D^2) \neq 0, \tag{6.111}$$

$$r_D = D - D_n - \Delta\lambda \frac{3(D - D^2)}{k_{\text{t}}} \sinh\left( \frac{\sigma}{2k_{\text{t}}} \right) \neq 0. \tag{6.112}$$

Thus, the final state is the root of a vector function $\boldsymbol{m}$ which is composed of the residual functions. Furthermore, it is useful to collect all variables in a single variable vector $\boldsymbol{v}$:

$$\boldsymbol{m}(\boldsymbol{v}) \in (\mathbb{R}^4 \to \mathbb{R}^4) = \begin{bmatrix} r_\sigma(\boldsymbol{v}) \\ r_\kappa(\boldsymbol{v}) \\ r_F(\boldsymbol{v}) \\ r_D(\boldsymbol{v}) \end{bmatrix}, \quad \boldsymbol{v} = \begin{bmatrix} \sigma \\ \kappa \\ \Delta\lambda \\ D \end{bmatrix}. \tag{6.113}$$

The root is found by a NEWTON iteration (iteration index: $i$):

$$\boldsymbol{v}^{(i+1)} = \boldsymbol{v}^{(i)} - \left( \frac{\partial \boldsymbol{m}}{\partial \boldsymbol{v}} \boldsymbol{v}^{(i)} \right)^{-1} \boldsymbol{m}\left( \boldsymbol{v}^{(i)} \right), \tag{6.114}$$

where

$$\boldsymbol{v}^{(0)} = \begin{bmatrix} \sigma^{(0)} \\ \kappa^{(0)} \\ \Delta\lambda^{(0)} \\ D^{(0)} \end{bmatrix} = \begin{bmatrix} \sigma_{n+1}^{\text{trial}} \\ \kappa_n \\ 0 \\ D_n \end{bmatrix} \tag{6.115}$$

can be used as initial value of the argument. The Jacobian matrix of the residual functions is obtained from partial derivatives of (6.109)–(6.112) as:

$$\frac{\partial \boldsymbol{m}}{\partial \boldsymbol{v}}(\sigma, \kappa, \Delta\lambda, D) = \begin{bmatrix} \frac{\partial r_\sigma}{\partial \sigma} & \frac{\partial r_\sigma}{\partial \kappa} & \frac{\partial r_\sigma}{\partial \Delta\lambda} & \frac{\partial r_\sigma}{\partial D} \\ \frac{\partial r_\kappa}{\partial \sigma} & \frac{\partial r_\kappa}{\partial \kappa} & \frac{\partial r_\kappa}{\partial \Delta\lambda} & \frac{\partial r_\kappa}{\partial D} \\ \frac{\partial r_F}{\partial \sigma} & \frac{\partial r_F}{\partial \kappa} & \frac{\partial r_F}{\partial \Delta\lambda} & \frac{\partial r_F}{\partial D} \\ \frac{\partial r_D}{\partial \sigma} & \frac{\partial r_D}{\partial \kappa} & \frac{\partial r_D}{\partial \Delta\lambda} & \frac{\partial r_D}{\partial D} \end{bmatrix}. \tag{6.116}$$

The partial derivatives in the matrix of Eq. (6.116) are explicitly written in Eq. (6.117) for the special case[10] of ideal plasticity, i.e. $k_\text{t} = \text{const}$.

$$\left[\begin{array}{c|c|c|c} 1 + \Delta\lambda E\left(\frac{2}{k_\text{t}^2} + \frac{D\cosh\left(\frac{\sigma}{2k_\text{t}}\right)}{2k_\text{t}^2}\right) & 0 & E\left(\frac{2\sigma}{k_\text{t}^2} + \frac{D\sinh\left(\frac{\sigma}{2k_\text{t}}\right)}{k_\text{t}}\right) & \frac{\Delta\lambda E\sinh\left(\frac{\sigma}{2k_\text{t}}\right)}{k_\text{t}} \\ \hline -\frac{\Delta\lambda}{1-D}\left(\frac{4\sigma}{k_\text{t}^3} + \frac{D\sinh\left(\frac{\sigma}{2k_\text{t}}\right)}{k_\text{t}^2} + \frac{\sigma D\cosh\left(\frac{\sigma}{2k_\text{t}}\right)}{2k_\text{t}^3}\right) & 1 & -\frac{1}{1-D}\left(\frac{2\sigma^2}{k_\text{t}^3} + \frac{\sigma D\sinh\left(\frac{\sigma}{2k_\text{t}}\right)}{k_\text{t}^2}\right) & -\frac{\Delta\lambda}{(1-D)}\left(\frac{1}{1-D}\left(\frac{2\sigma^2}{k_\text{t}^3} + \frac{\sigma D\sinh\left(\frac{\sigma}{2k_\text{t}}\right)}{k_\text{t}^2}\right) + \frac{\sigma\sinh\left(\frac{\sigma}{2k_\text{t}}\right)}{k_\text{t}^2}\right) \\ \hline \frac{2\sigma}{k_\text{t}^2} + \frac{D}{k_\text{t}}\sinh\left(\frac{\sigma}{2k_\text{t}}\right) & 0 & 0 & 2\cosh\left(\frac{\sigma}{2k_\text{t}}\right) - 2D \\ \hline -\frac{3\Delta\lambda(D-D^2)}{2k_\text{t}^2}\cosh\left(\frac{\sigma}{2k_\text{t}}\right) & 0 & -\frac{3(D-D^2)}{k_\text{t}}\sinh\left(\frac{\sigma}{2k_\text{t}}\right) & 1 - \frac{3\Delta\lambda(1-2D)}{k_\text{t}}\sinh\left(\frac{\sigma}{2k_\text{t}}\right) \end{array}\right] \tag{6.117}$$

[10] This corresponds to the original assumption of GURSON, see Sect. 2.8.2.

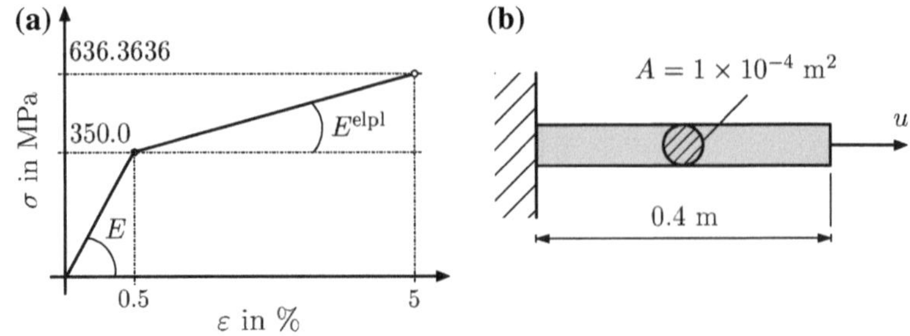


**Fig. 6.9** Sample problem back projection at linear hardening: **a** stress-strain distribution; **b** geometry and boundary conditions

## 6.6 Solved Examples

**6.1 Example: Back projection at linear hardening—continuum rod**
Figure 6.9a shows an idealized stress-strain diagram as it is derived, for example, experimentally from an uniaxial tensile test. With the help of this material behavior the deformation of a tension rod (see Fig. 6.9b) can be simulated. Thereby the right-hand end is displaced by $u = 8 \times 10^{-3}$ m in total, whereupon the deformation is applied in 10 equal increments. The rod needs to be considered as a continuum hereby and should not be discretized with finite elements.

(a) Calculate the stress state with the help of the CPP algorithm in each increment and mark all values, which have to be updated.
(b) Illustrate graphically the stress distribution.

**6.1 Solution**
(a) Various material properties of the elastic and plastic regime are needed for the back projection. The modulus of elasticity $E$ results as the quotient from the stress and strain increment in the elastic regime in:

$$E = \frac{\Delta\sigma}{\Delta\varepsilon} = \frac{350 \text{ MPa}}{0.005} = 70000 \text{ MPa}. \tag{6.118}$$

The plastic modulus $E^{\text{pl}}$ results as the quotient from the yield stress and the plastic strain increment in:

$$E^{\text{pl}} = \frac{\Delta k}{\Delta\varepsilon^{\text{pl}}} = \frac{636.3636 \text{ MPa} - 350 \text{ MPa}}{(0.05 - 636.3636 \text{ MPa}/E) - (0.005 - 350 \text{ MPa}/E)}$$
$$= 7000 \text{ MPa}. \tag{6.119}$$

Therefore, the elasto-plastic material modulus can be calculated from Eq. (2.73) as

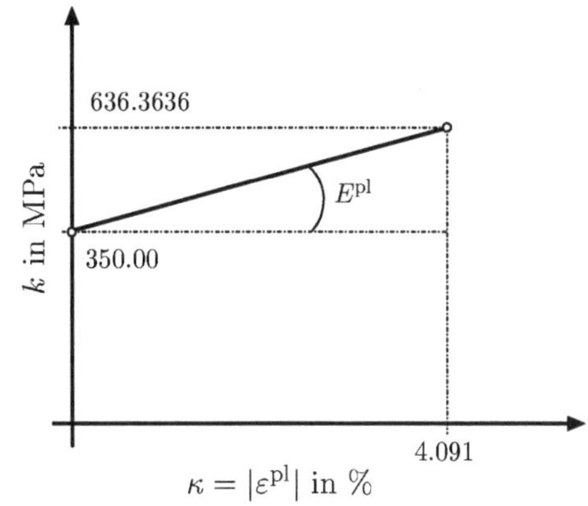


**Fig. 6.10** Flow curve for continuum rod

$$E^{elpl} = \frac{E \times E^{pl}}{E + E^{pl}} = \frac{70000 \text{ MPa} \times 7000 \text{ MPa}}{70000 \text{ MPa} + 7000 \text{ MPa}} = 6363.636 \text{ MPa}. \tag{6.120}$$

Finally, the equation of the flow curve results in the following via the initial yield stress:

$$k(\kappa) = 350 \text{ MPa} + 7000 \text{ MPa} \times \kappa. \tag{6.121}$$

A graphical illustration of the flow curve is given in Fig. 6.10. For the integration algorithm the strain increment is additionally necessary. At a total displacement of $8 \times 10^{-3}$ m and 10 equidistant steps, the strain increment can be determined via:

$$\Delta\varepsilon = \frac{1}{10} \times \frac{0.008 \text{ m}}{0.4 \text{ m}} = 0.002. \tag{6.122}$$

For the first two increments, trial stress states in the elastic regime ($F < 0$) result, and the resulting stress can be calculated from Eq. (6.2), i.e. HOOKE's law. From the third increment on, an invalid trial stress state ($F > 0$) results for the first time and the stress has to be calculated via Eq. (6.29), whereupon the constant matrix expression results in:

$$\begin{bmatrix} 7000 & -7000 & 1 \\ 1 & -1 & -(70000)^{-1} \\ 1 & 0,1 & -(70000)^{-1} \end{bmatrix}. \tag{6.123}$$

Table 6.5 summarizes the numerical results for the 10 increments.
(b) A graphical illustration of the stress distribution is given in Fig. 6.11. Due to the linear hardening, the back projection for each increment occurs in one step.

Finally it needs to be remarked at this point that for the special case of *linear* hardening at *uniaxial* stress states, the stress in the plastic area (inc $\geq 3$) can directly be calculated via (see Fig. 2.2b)

**Table 6.5** Numerical values of the back projection for a continuum rod with linear hardening (10 increments, $\Delta\varepsilon = 0.002$)

| Inc | $\varepsilon$ | $\sigma^{\text{trial}}$ | $\sigma$ | $\kappa$ | $\Delta\lambda$ | $E^{\text{elpl}}$ |
|---|---|---|---|---|---|---|
| – | – | MPa | MPa | $10^{-3}$ | $10^{-3}$ | MPa |
| 1 | 0.002 | 140.0 | 140.0 | 0.0 | 0.0 | 0.0 |
| 2 | 0.004 | 280.0 | 280.0 | 0.0 | 0.0 | 0.0 |
| 3 | 0.006 | 420.0 | 356.364 | 0.909091 | 0.909091 | 6363.636 |
| 4 | 0.008 | 496.364 | 369.091 | 2.727273 | 1.818182 | 6363.636 |
| 5 | 0.010 | 509.091 | 381.818 | 4.545455 | 1.818182 | 6363.636 |
| 6 | 0.012 | 521.818 | 394.545 | 6.363636 | 1.818182 | 6363.636 |
| 7 | 0.014 | 534.545 | 407.273 | 8.181818 | 1.818182 | 6363.636 |
| 8 | 0.016 | 547.273 | 420.000 | 10.000000 | 1.818182 | 6363.636 |
| 9 | 0.018 | 560.000 | 432.727 | 11.818182 | 1.818182 | 6363.636 |
| 10 | 0.020 | 572.727 | 445.455 | 13.636364 | 1.818182 | 6363.636 |

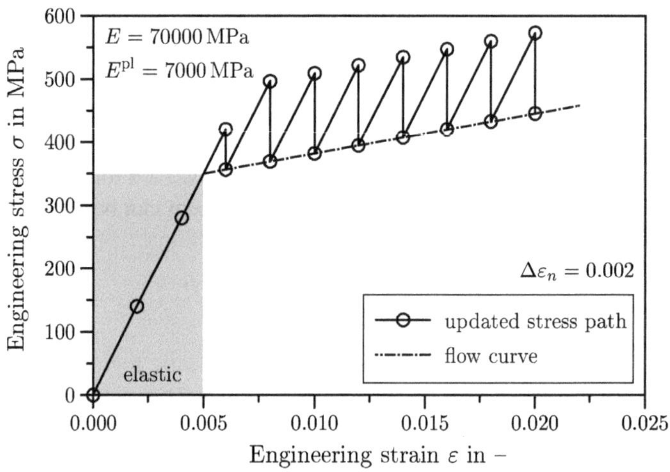


**Fig. 6.11** Stress distribution in the case of the back projection for a continuum rod with linear hardening (10 increments, $\Delta\varepsilon = 0.002$)

$$
\begin{aligned}
\sigma(\varepsilon) &= k_{\text{t}}^{\text{init}} + E^{\text{elpl}} \times (\varepsilon - \varepsilon_{\text{t}}^{\text{init}}) \\
&= E\varepsilon_{\text{t}}^{\text{init}} + E^{\text{elpl}}\varepsilon - E^{\text{elpl}}\varepsilon_{\text{t}}^{\text{init}} \\
\sigma(\varepsilon) &= (E - E^{\text{elpl}}) \times \varepsilon_{\text{t}}^{\text{init}} + E^{\text{elpl}} \times \varepsilon .
\end{aligned}
\tag{6.124}
$$

However, the intention of this example is to illustrate the concept of the back projection and not to define the stress according to the simplest method.

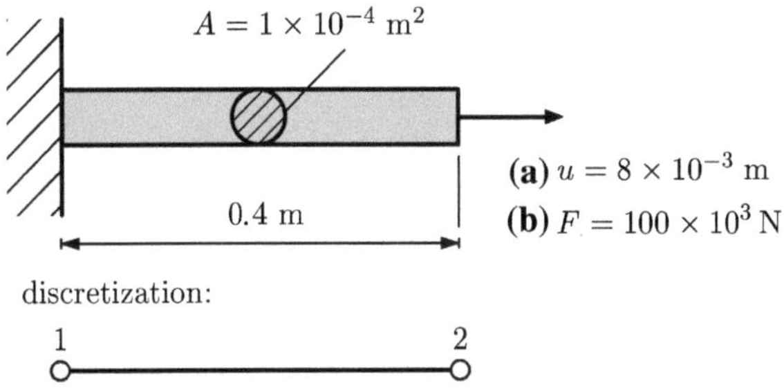


**Fig. 6.12** Sample problem back projection in the case of linear hardening: **a** displacement boundary condition; **b** force boundary condition

**6.2 Example: Back projection at linear hardening—discretization via one finite element, displacement and force boundary condition**

The continuum rod from Example 6.1 is discretized within this example via one single finite element. The material behavior is assumed as shown in Fig. 6.9a. The load on the right-hand end of the rod is applied in 10 equal increments, whereupon

(a) $u = 8 \times 10^{-3}$ m,
(b) $F = 100$ kN

can be applied, see Fig. 6.12. Calculate the stress state in each increment with the help of the CPP algorithm and mark all values, which have to be updated. As convergence criteria, an absolute displacement difference of $1 \times 10^{-5}$ mm can be assigned.

**6.2 Solution**

When using solely one element, the global system of equations results in the following — without the consideration of the boundary conditions:

$$\frac{A\tilde{E}}{L}\begin{bmatrix} 1 & -1 \\ -1 & 1 \end{bmatrix}\begin{bmatrix} \Delta u_1 \\ \Delta u_2 \end{bmatrix} = \begin{bmatrix} \Delta F_1 \\ \Delta F_2 \end{bmatrix}. \tag{6.125}$$

Since this is in general a nonlinear system of equations, an incremental form has been assigned. The modulus $\tilde{E}$ equals the elasticity modulus $E$ in the elastic range and the elasto-plastic material modulus $E^{\text{elpl}}$ in the plastic range. Since a fixed support is given on the left-hand node ($\Delta u_1 = 0$), Eq. (6.125) can be simplified to

$$\frac{A\tilde{E}}{L} \times \Delta u_2 = \Delta F_2. \tag{6.126}$$

Case (a) Displacement boundary condition $u = 8 \times 10^{-3}$ m on the right-hand node:

In the case of the displacement boundary condition Eq. (6.126) must not be solved, since for each increment $\Delta u_2 = 8 \times 10^{-3}/10 = 8 \times 10^{-4}$ m is known. Via the

**Table 6.6** Numerical values of the displacement and strain for displacement boundary condition (10 increments, $\Delta\varepsilon = 0.002$)

| Inc | $\Delta u_2$ | $\Delta\varepsilon$ | $u_2$ | $\varepsilon$ |
|---|---|---|---|---|
| – | $10^{-4}$ m | – | $10^{-4}$ m | – |
| 1 | 8.0 | 0.002 | 8.0 | 0.002 |
| 2 | 8.0 | 0.002 | 16.0 | 0.004 |
| 3 | 8.0 | 0.002 | 24.0 | 0.006 |
| 4 | 8.0 | 0.002 | 32.0 | 0.008 |
| 5 | 8.0 | 0.002 | 40.0 | 0.010 |
| 6 | 8.0 | 0.002 | 48.0 | 0.012 |
| 7 | 8.0 | 0.002 | 56.0 | 0.014 |
| 8 | 8.0 | 0.002 | 64.0 | 0.016 |
| 9 | 8.0 | 0.002 | 72.0 | 0.018 |
| 10 | 8.0 | 0.002 | 80.0 | 0.020 |

equation for the strain in the element, meaning $\varepsilon = \frac{1}{L}(u_2 - u_1)$, the strain increment results in the following in the case of the fixed support at the left-hand node:

$$\Delta\varepsilon = \frac{1}{L} \times \Delta u_2 = \frac{8 \times 10^{-4}\ \text{m}}{0.4\ \text{m}} = 0.002. \tag{6.127}$$

The entire displacement or alternatively strain can be calculated through the summation of the incremental displacement or alternatively strain values, see Table 6.6. It can be remarked at this point that for this case of displacement boundary condition at one element, the calculation of the displacement or alternatively the strain for all increments can be done without a stress calculation.

To calculate the stress and plastic strain in each increment, the calculation via the CPP algorithm for each increment has to be conducted via the strain increment $\Delta\varepsilon$ from Table 6.6. This is exactly what is calculated in Example 6.1 and the numerical results can be taken from Table 6.5.

Case (b) Force boundary condition $F = 100$ kN on the right-hand node:

In the case of the force boundary condition, Eq. (6.126) can be solved via the NEWTON- RAPHSON method. To do so, this equation has to be written in the form of a residual $r$ as

$$r = \frac{A\tilde{E}}{L} \times \Delta u_2 - \Delta F_2 = \tilde{E}(u_2) \times \frac{A}{L} \times \Delta u_2 - \Delta F_2 = 0. \tag{6.128}$$

If one develops the last equation into a TAYLOR's series and neglects the terms of higher order, the following form results

**Table 6.7** Numerical values for one element in the case of linear hardening (10 increments; $\Delta F = 1 \times 10^4$ N)

| Inc | ex. Force | $\Delta u_2$ | $u_2$ | $\varepsilon$ | $\Delta\varepsilon$ | $\sigma$ | $\varepsilon^{\mathrm{pl}}$ |
|---|---|---|---|---|---|---|---|
| – | $10^4$ N | mm | mm | $10^{-2}$ | $10^{-2}$ | MPa | $10^{-2}$ |
| 1 | 1.0 | 0.5714 | 0.5714 | 0.1429 | 0.1429 | 100.0 | 0.0 |
| 2 | 2.0 | 0.5714 | 1.1427 | 0.2857 | 0.1429 | 200.0 | 0.0 |
| 3 | 3.0 | 0.5714 | 1.7143 | 0.4286 | 0.1429 | 300.0 | 0.0 |
| 4 | 4.0 | 4.4286 | 5.1429 | 1.2857 | 0.8571 | 400.0 | 0.7143 |
| 5 | 5.0 | 6.2857 | 11.4286 | 2.8571 | 1.5714 | 500.0 | 2.1429 |
| 6 | 6.0 | 6.2857 | 17.7143 | 4.4286 | 1.5714 | 600.0 | 3.5714 |
| 7 | 7.0 | 6.2857 | 24.0000 | 6.0000 | 1.5714 | 700.0 | 5.0000 |
| 8 | 8.0 | 6.2857 | 30.2857 | 7.5714 | 1.5714 | 800.0 | 6.4286 |
| 9 | 9.0 | 6.2857 | 36.5714 | 9.1429 | 1.5714 | 900.0 | 7.8571 |
| 10 | 10.0 | 6.2857 | 42.8571 | 10.7143 | 1.5714 | 1000.0 | 9.2857 |

$$r(\Delta u_2^{(i+1)}) = r(\Delta u_2^{(i)}) + \left(\frac{\partial r}{\partial \Delta u_2}\right)^{(i)} \times \delta(\Delta u_2) + \cdots, \tag{6.129}$$

whereupon

$$\delta(\Delta u_2) = \Delta u_2^{(i+1)} - \Delta u_2^{(i)} \tag{6.130}$$

is valid and

$$\left(\frac{\partial r}{\partial \Delta u_2}\right)^{(i)} = K_{\mathrm{T}}^{(i)} \tag{6.131}$$

represents the tangent stiffness matrix[11] in the $i$-th iteration step. Then Eq. (6.129) can also be written as

$$\delta(\Delta u_2) K_{\mathrm{T}}^{(i)} = -r(\Delta u_2^{(i)}) = \Delta F^{(i)} - \frac{\tilde{E}A}{L}\Delta u_2^{(i)}. \tag{6.132}$$

Multiplication via $(K_{\mathrm{T}}^{(i)})^{-1}$ and the use of Eq. (6.130) finally leads to

$$\Delta u_2^{(i+1)} = \Delta F^{(i)} \times \frac{L}{\tilde{E}A}, \tag{6.133}$$

[11] In the considered example with linear hardening, $\tilde{E}$ is constant in the elastic range (increment 1–3) and in the plastic range (increment 4–10) and therefore not a function of $u_2$. However, $\tilde{E}$ has to be differentiated as well in the general case.

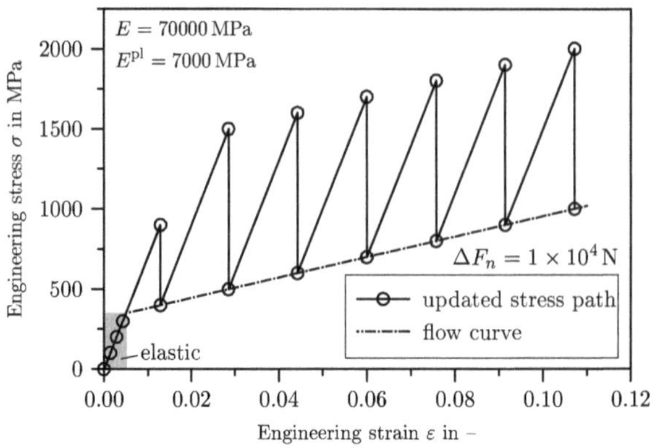


**Fig. 6.13** Stress distribution in the case of the back projection for a continuum rod with linear hardening (10 increments, $\Delta F = 1 \times 10^4$)

whereupon $\tilde{E} = E$ is valid in the elastic range (increment 1 to 3) and $\tilde{E} = E^{elpl}$ in the plastic range (increment 4 to 10).

Application of Eq. (6.133) leads to a value of $\Delta u_2 = 0.571429$ mm in the elastic range (increment 1 to 3) and in the plastic range (increment 4 to 10) a displacement increment of $\Delta u_2 = 6.285715$ mm occurs. It can be remarked at this point that the calculation of the displacement increments (increments 1 to 3 and 4 to 10) does not need an iteration and the application of Eq. (6.133) directly yields the desired result. As soon as the displacement increments ($\Delta u_2$) are calculated, the entire displacement on node 2 results via summation of the incremental values. Subsequently the strain in the element can be calculated via the relation $\varepsilon = \frac{1}{L}(u_2 - u_1)$, and the strain increments result through subtraction of two consecutive strain values (see Table 6.8).

The calculation of the stress and the plastic strain now requires that the CPP algorithm is used in each increment, based on the strain increment $\Delta\varepsilon$. The graphical illustration of the back projection is given in Fig. 6.13. One can see clearly that the strain increments at a force boundary condition differ in the elastic and plastic range. As a consequence, quite high values result for the test stress state in the plastic range. Thus for the increments 5 to 10 $\sigma_{n+1}^{trial} = \sigma_{n+1} + 1000$ MPa is valid.

Particular attention is required at the transition from the elastic to the plastic regime, meaning from increment 3 to 4. Here the modulus $\tilde{E}$ is not defined clearly and Eq. (6.133) has to be solved iteratively. For the first cycle of the calculation (cycle $j = 0$) the arithmetic mean between the modulus of elasticity $E$ and the elasto-plastic material modulus $E^{elpl}$ can be applied. For the further cycles (cycle $j \geq 1$) $\tilde{E}$ can be approximated via an intermediate modulus (secant modulus, see Fig. 6.14). Therefore the following relation results as a calculation requirement for the intermediate modulus $\tilde{E}$ in the elasto-plastic transition regime (at this point in the considered example at the transition from increment 3 to 4):

$$\tilde{E} = \begin{cases} \dfrac{E + E^{\text{elpl}}}{2} & \text{for } j = 0 \\ \dfrac{\sigma_{n+1}^{(j)} - \sigma_n}{\varepsilon_{n+1}^{(j)} - \varepsilon_n} & \text{for } j > 0 \end{cases} . \tag{6.134}$$

The numerical values of the intermediate modulus $\tilde{E}$ at the transition from increment 3 to 4 and the differences of the displacement, which result herefrom on node 2 are summarized in Table 6.8. Since an absolute displacement difference of $1 \times 10^{-5}$ mm was required as convergence criteria, 18 cycles are necessary to iterate the difference between the new and the old displacement on node 2 under these values. It can be considered at this point that the difference between the displacements at node 2 in the first cycle (cycle 0) result in $u_2^{\text{new}} - u_2^{\text{old}} = u_2^{(j=0)} - u_2|_{\text{inc } 3}$ and for the following cycles of the iteration via $u_2^{(j+1)} - u_2^{(j)}$.

The convergence behavior of the iteration rule is illustrated graphically in Fig. 6.15. An equidistant division was chosen in Fig. 6.15a and a logarithmic division (to be base 10) in Fig. 6.15b. One can see that quite a high convergence rate occurs at the beginning of the iteration, which flattens throughout the different cycles. At the chosen convergence criteria here of $10^{-5}$ the 18 iteration steps are therefore necessary, to finally reach the required absolute displacement difference. If one would require an absolute difference of $10^{-6}$ as convergence criteria, 21 iteration steps would be necessary.

### 6.3 Example: Back projection for bimaterial rod

Two different material behaviors (see Fig. 6.16a) can be considered in the following to model a bimaterial rod (see Fig. 6.16b) via the FE method. The right-hand end is displaced by $u = 8 \times 10^{-3}$ m hereby, whereupon the deformation is applied in

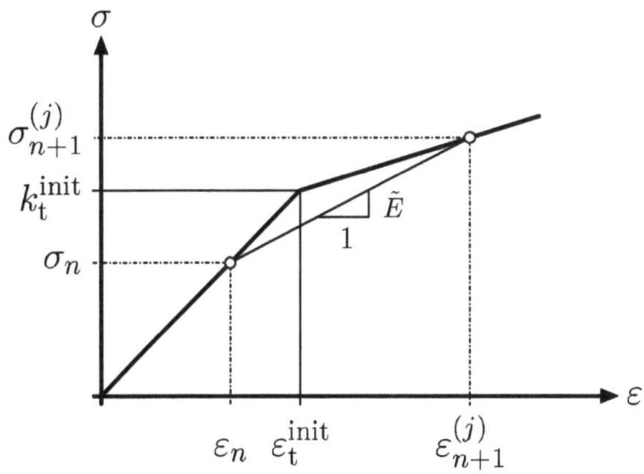


**Fig. 6.14** Definition of the intermediate modulus $\tilde{E}$ in the case of the transition from the elastic to the plastic range

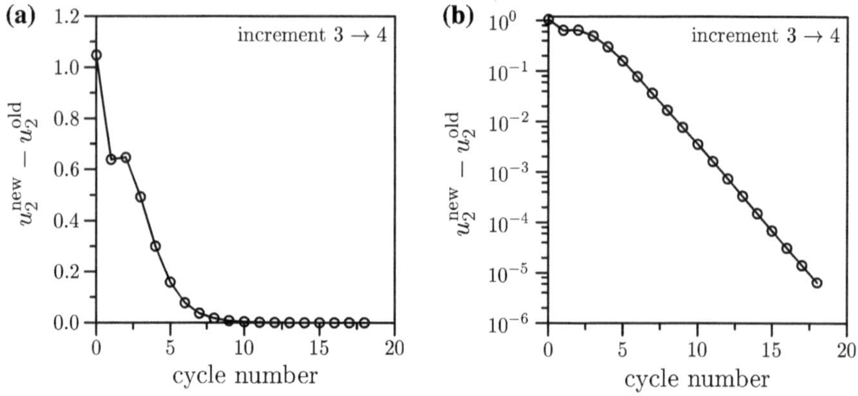


**Fig. 6.15** Convergence behavior in the case of the transition from increment 3 to 4: **a** equidistant division; **b** logarithmic division of the absolute displacement difference

10 equal increments. The rod can be discretized with two finite elements hereby. Examine the following material combinations:

**Table 6.8** Numerical values for the transition from increment 3 to 4

| Cycle | $\tilde{E}^{(i)}$ | $u_2^{new} - u_2^{old}$ |
|---|---|---|
| – | MPa | mm |
| 0 | 38181.84 | $1.048 \times 10^{-0}$ |
| 1 | 23719.00 | $6.388 \times 10^{-1}$ |
| 2 | 17145.00 | $6.466 \times 10^{-1}$ |
| 3 | 14157.16 | $4.925 \times 10^{-1}$ |
| 4 | 12798.56 | $2.999 \times 10^{-1}$ |
| 5 | 12181.16 | $1.584 \times 10^{-1}$ |
| 6 | 11900.52 | $7.744 \times 10^{-2}$ |
| 7 | 11772.96 | $3.642 \times 10^{-2}$ |
| 8 | 11715.00 | $1.682 \times 10^{-2}$ |
| 9 | 11688.64 | $7.699 \times 10^{-3}$ |
| 10 | 11676.64 | $3.511 \times 10^{-3}$ |
| 11 | 11671.20 | $1.598 \times 10^{-3}$ |
| 12 | 11668.72 | $7.270 \times 10^{-4}$ |
| 13 | 11667.60 | $3.305 \times 10^{-4}$ |
| 14 | 11667.08 | $1.503 \times 10^{-4}$ |
| 15 | 11666.88 | $6.831 \times 10^{-5}$ |
| 16 | 11666.76 | $3.105 \times 10^{-5}$ |
| 17 | 11666.72 | $1.411 \times 10^{-5}$ |
| 18 | 11666.68 | $6.416 \times 10^{-6}$ |

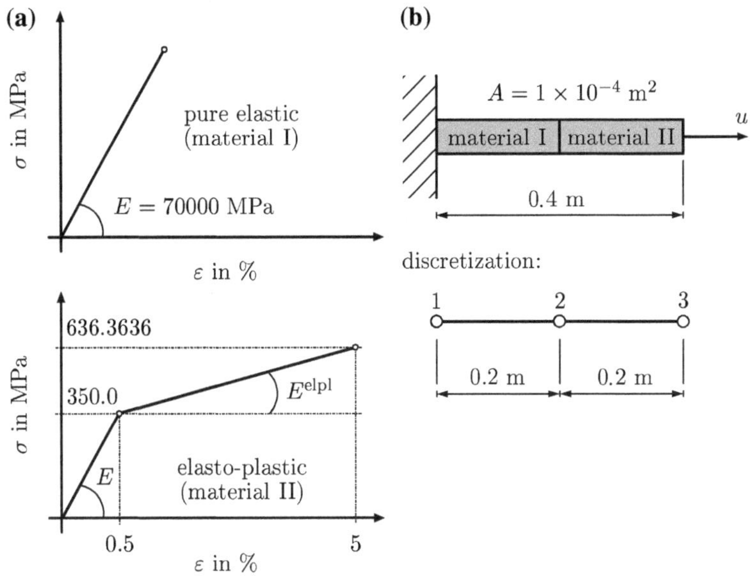


**Fig. 6.16** Sample problem back projection in the case of a rod with different materials: **a** stress-strain distributions, **b** geometry and boundary conditions

(a) Material I: pure elastic; Material II: pure elastic,
(b) Material I: elasto-plastic; Material II: elasto-plastic,
(c) Material I: pure elastic; Material II: elasto-plastic,

to calculate the displacement of the middle node. Furthermore, calculate the stress state in each element with the help of the CPP algorithm and mark all values, which have to be updated.

**6.3 Solution**
When using two finite elements, the global system of equations, without consideration of the boundary conditions, results in the following incremental form for this example:

$$\frac{A}{L} \times \begin{bmatrix} \tilde{E}^{\text{I}} & -\tilde{E}^{\text{I}} & 0 \\ -\tilde{E}^{\text{I}} & \tilde{E}^{\text{I}} + \tilde{E}^{\text{II}} & -\tilde{E}^{\text{II}} \\ 0 & -\tilde{E}^{\text{II}} & \tilde{E}^{\text{II}} \end{bmatrix} \begin{bmatrix} \Delta u_1 \\ \Delta u_2 \\ \Delta u_3 \end{bmatrix} = \begin{bmatrix} \Delta F_1 \\ \Delta F_2 \\ \Delta F_3 \end{bmatrix}. \tag{6.135}$$

The consideration of the boundary condition on the left-hand side, meaning $u_1 = 0$, yields the following reduced global system of equations:

$$\frac{A}{L} \times \begin{bmatrix} \tilde{E}^{\text{I}} + \tilde{E}^{\text{II}} & -\tilde{E}^{\text{II}} \\ -\tilde{E}^{\text{II}} & \tilde{E}^{\text{II}} \end{bmatrix} \begin{bmatrix} \Delta u_2 \\ \Delta u_3 \end{bmatrix} = \begin{bmatrix} \Delta F_2 \\ \Delta F_3 \end{bmatrix}. \tag{6.136}$$

The consideration of the displacement boundary condition on the right-hand side, meaning $u_3 = u(t)$, and that $\Delta F_2 = \Delta F_3 = 0$ is valid, yields:

$$\frac{A}{L} \times (\tilde{E}^{\mathrm{I}} + \tilde{E}^{\mathrm{II}})\Delta u_2 = \frac{\tilde{E}_2 A}{L} \Delta u_3 \,, \tag{6.137}$$

or

$$\left( \frac{\tilde{E}^{\mathrm{I}}}{\tilde{E}^{\mathrm{II}}} + 1 \right) \Delta u_2 = \Delta u_3 \,. \tag{6.138}$$

For the application of the NEWTON- RAPHSON method, Eq. (6.138) is written as a residual equation:

$$r = \left( \frac{\tilde{E}^{\mathrm{I}}}{\tilde{E}^{\mathrm{II}}} + 1 \right) \Delta u_2 - \Delta u_3 = 0 \,. \tag{6.139}$$

If one develops the last equation into a TAYLOR's series and neglects the terms of higher order according to the procedure in Example 6.2, finally the following iteration rule results for the definition of the displacement of the middle node:

$$\Delta u_2^{(i+1)} = \left( \frac{\tilde{E}^{\mathrm{I}}}{\tilde{E}^{\mathrm{II}}} + 1 \right)^{-1} \times \Delta u_3^{(i)} \,. \tag{6.140}$$

In Eq. (6.140) in the elastic regime the elasticity modulus $E$ has to be used for $\tilde{E}$ and in the plastic regime the elasto-plastic material modulus $E^{\mathrm{elpl}}$.
Case (a) Material I: pure elastic; Material II: pure elastic:

For the case that both sections exhibit pure elastic material behavior with $E^{\mathrm{I}} = E^{\mathrm{II}}$, Eq. (6.140) simplifies in:

$$\Delta u_2^{(i+1)} = (1+1)^{-1} \times \Delta u_3^{(i)} = \frac{1}{2} \times \Delta u_3^{(i)} = 4\,\mathrm{mm} \,. \tag{6.141}$$

The strain in the left-hand element—which is identical with the strain in the right-hand element—can easily be defined via $\Delta\varepsilon = \frac{1}{200\,\mathrm{mm}} \times \Delta u_2$, and the stress results from the strain through multiplication with the modulus of elasticity. The results of this pure elastic calculation are summarized in Table 6.9.

In addition to the displacement, strain and stress values,[12] Table 6.9 also contains the reaction forces at node 3. These reaction forces result due to multiplication of the stiffness matrix with the resultant vector of the displacements and have to be in equilibrium with the resulting forces from the stress: $F_{\mathrm{r},3} = \sigma A$.

Case (b) Material I: elasto-plastic; Material II: elasto-plastic:

[12] One considers that in both sections or alternatively elements, the stress and strain are identical.

**Table 6.9** Numerical values for a bimaterial rod in the case of pure linear-elastic behavior (10 increments; $\Delta u_3 = 0.8\,\text{mm}$; $A = 100\ \text{mm}^2$)

| Inc | $\Delta u_2$ | $u_2$ | $\Delta\varepsilon$ | $\varepsilon$ | $\sigma$ | $F_{r,3}$ |
|---|---|---|---|---|---|---|
| – | mm | mm | – | – | MPa | $10^4$ N |
| 1 | 0.4 | 0.4 | 0.002 | 0.002 | 140.0 | 1.4 |
| 2 | 0.4 | 0.8 | 0.002 | 0.004 | 280.0 | 2.8 |
| 3 | 0.4 | 1.2 | 0.002 | 0.006 | 420.0 | 4.2 |
| 4 | 0.4 | 1.6 | 0.002 | 0.008 | 560.0 | 5.6 |
| 5 | 0.4 | 2.0 | 0.002 | 0.010 | 700.0 | 7.0 |
| 6 | 0.4 | 2.4 | 0.002 | 0.012 | 840.0 | 8.4 |
| 7 | 0.4 | 2.8 | 0.002 | 0.014 | 980.0 | 9.8 |
| 8 | 0.4 | 3.2 | 0.002 | 0.016 | 1120.0 | 11.2 |
| 9 | 0.4 | 3.6 | 0.002 | 0.018 | 1260.0 | 12.6 |
| 10 | 0.4 | 4.0 | 0.002 | 0.020 | 1400.0 | 14.0 |

**Table 6.10** Numerical values for a bimaterial rod in the case of elasto-plastic behavior (10 Increments; $\Delta u_3 = 0.8\,\text{mm}$; $A = 100\ \text{mm}^2$)

| Inc | $\Delta u_2$ | $u_2$ | $\Delta\varepsilon$ | $\varepsilon$ | $\sigma$ | $\varepsilon^{\text{pl}}$ | $F_{r,3}$ |
|---|---|---|---|---|---|---|---|
| – | mm | mm | – | – | MPa | $10^{-3}$ | $10^4$ N |
| 1 | 0.4 | 0.4 | 0.002 | 0.002 | 140.0 | 0.0 | 1.4 |
| 2 | 0.4 | 0.8 | 0.002 | 0.004 | 280.0 | 0.0 | 2.8 |
| 3 | 0.4 | 1.2 | 0.002 | 0.006 | 356.364 | 0.909091 | 3.56364 |
| 4 | 0.4 | 1.6 | 0.002 | 0.008 | 369.091 | 2.727273 | 3.69091 |
| 5 | 0.4 | 2.0 | 0.002 | 0.010 | 381.818 | 4.545455 | 3.81818 |
| 6 | 0.4 | 2.4 | 0.002 | 0.012 | 394.545 | 6.363636 | 3.94545 |
| 7 | 0.4 | 2.8 | 0.002 | 0.014 | 407.273 | 8.181818 | 4.07273 |
| 8 | 0.4 | 3.2 | 0.002 | 0.016 | 420.000 | 10.000000 | 4.20000 |
| 9 | 0.4 | 3.6 | 0.002 | 0.018 | 432.727 | 11.818182 | 4.32727 |
| 10 | 0.4 | 4.0 | 0.002 | 0.020 | 445.455 | 13.636364 | 4.45455 |

For the case that both sections exhibit the same elasto-plastic material behavior, $\tilde{E}^{\text{I}} = \tilde{E}^{\text{II}}$ is always valid, and Eq. (6.140) also at this point yields a displacement increment on the middle node from $\Delta u_2^{(i+1)} = \frac{1}{2} \times \Delta u_3^{(i)} = 4\,\text{mm}$ or alternatively a strain increment of $\Delta\varepsilon = 0.002$. For the calculation of the stress and the plastic strain for element II in the nonlinear regime, the CPP algorithm has to be made use of, as in Example 6.1. The corresponding values are summarized in Table 6.10.

Case (c) Material I: pure elastic; Material II: elasto-plastic:

At this point in the elastic regime of both elements, it occurs that the displacement at the middle node amounts to half of the displacement, which occurred on the right-hand node. As soon as the plastic flow occurs in the right-hand half of the bar, $\tilde{E}^{\mathrm{I}} \neq \tilde{E}^{\mathrm{II}}$ occurs and for the right-hand half of the rod, the elasto-plastic material modulus has to be made use of. Therefore, the calculation procedure for the displacement increment can be summarized as follows:

$$\Delta u_2^{(i+1)} = \begin{cases} \dfrac{1}{2} \times \Delta u_3^{(i)} & \text{in the elastic region} \\ \left( \dfrac{E^{\mathrm{I}}}{E^{\mathrm{elpl,II}}} + 1 \right)^{-1} \times \Delta u_3^{(i)} & \text{in the plastic region} \end{cases} . \tag{6.142}$$

The total displacement at the middle node results from the displacement increments through summation and the strain for each element can be defined via $\varepsilon = \frac{1}{L}(-u_{\mathrm{l}} + u_{\mathrm{r}})$ (Index 'l' for left-hand and index 'r' for right-hand node of the rod element). As soon as the right-hand part of the rod enters the plastic region, the predictor-corrector method has to be made use of to be able to calculate the state variables. The numerical values of the incremental solution method are summarized in Table 6.11. At this point it can be remarked that in the plastic region a similar relation as in the elastic region can be set to calculate the stress increment from the strain increment (see Eq. (6.143)). However, it has to be considered that the modulus of elasticity has to be substituted by the elasto-plastic material modulus:

$$\Delta\sigma = \begin{cases} \Delta\varepsilon \times E & \text{in the elastic region} \\ \Delta\varepsilon \times E^{\mathrm{elpl}} & \text{in the plastic region} \end{cases} . \tag{6.143}$$

The transition from the elastic to the plastic region, meaning from increment 2 to 3, demands a special consideration at this point. The intermediate modulus $\tilde{E}^{\mathrm{II}}$ hereby has to be calculated according to Eq. (6.134), to be able to define the displacement increment subsequently according to Eq. $(6.142)_2$. The absolute displacement on the middle node results from the summation, meaning $u_2^{(i)} = \Delta u_2^{(i)} + u_2|_{\mathrm{inc}\,2}$. The difference of the displacement on the middle node has to be determined in the first cycle (cycle 0) via $|u_2^{(i=0)} - u_2|_{\mathrm{inc}\,2}$ and for each further cycle $(i)$ via $|u_2^{(i)} - u_2^{(i-1)}|$. The calculation of the strain in the right-hand half of the rod can occur via the given boundary condition $u_3$ via $\varepsilon^{\mathrm{II},(i)} = \frac{1}{L}(-u_2^{(i)} + u_3)$. Finally the stress results via the CPP algorithm, based on the strain increment $\Delta\varepsilon^{\mathrm{II},(i)} = \varepsilon^{\mathrm{II},(i)} - \varepsilon^{\mathrm{II}}|_{\mathrm{inc}\,2}$. If the stress and strains are known, the new intermediate modulus can be determined for the next cycle via Eq. (6.134). As convergence criteria for the absolute displacement difference a value of $10^{-5}$ was given within this example (Table 6.12).

## 6.4 Example: Cyclic loading of an elastic-plastic continuum rod

Consider a pure one-dimensional tensile test as schematically shown in Fig. 6.17. Use the matrix solution scheme as given in Sect. 6.2 to simulate a cycling load in the form of an end displacement boundary condition (cf. Fig. 6.17b) which is applied in

**Table 6.11** Numerical values for a bimaterial rod in the case of different material behavior (10 Increments; $\Delta u_3 = 0.8$ mm; $A = 100$ mm$^2$)

| Inc | $\Delta u_2$ | $u_2$ | $\varepsilon^{\text{I}}$ | $\varepsilon^{\text{II}}$ | $\sigma^{\text{I}}$ | $\sigma^{\text{II}}$ | $\varepsilon^{\text{pl,I}}$ | $\varepsilon^{\text{pl,II}}$ | $F_{\text{r},3}$ |
|---|---|---|---|---|---|---|---|---|---|
| – | mm | mm | $10^{-3}$ | $10^{-3}$ | MPa | MPa | $10^{-3}$ | $10^{-3}$ | $10^4$ N |
| 1 | 0.4 | 0.4 | 2.0 | 2.0 | 140.0 | 140.0 | 0.0 | 0.0 | 1.4 |
| 2 | 0.4 | 0.8 | 4.0 | 4.0 | 280.0 | 280.0 | 0.0 | 0.0 | 2.8 |
| 3 | 0.23333 | 1.03333 | 5.16667 | 6.83333 | 361.667 | 361.667 | 0.0 | 1.66667 | 3.61667 |
| 4 | 0.06667 | 1.10000 | 5.50000 | 10.50000 | 385.000 | 385.000 | 0.0 | 5.00000 | 3.85000 |
| 5 | 0.06667 | 1.16667 | 5.83333 | 14.16667 | 408.333 | 408.333 | 0.0 | 8.33333 | 4.08333 |
| 6 | 0.06667 | 1.23333 | 6.16667 | 17.83333 | 431.667 | 431.667 | 0.0 | 11.66667 | 4.31667 |
| 7 | 0.06667 | 1.30000 | 6.50000 | 21.50000 | 455.000 | 455.000 | 0.0 | 15.00000 | 4.55000 |
| 8 | 0.06667 | 1.36667 | 6.83333 | 25.16667 | 478.333 | 478.333 | 0.0 | 18.33333 | 4.78333 |
| 9 | 0.06667 | 1.43333 | 7.16667 | 28.83333 | 501.667 | 501.667 | 0.0 | 21.66667 | 5.01667 |
| 10 | 0.06667 | 1.50000 | 7.50000 | 32.50000 | 525.000 | 525.000 | 0.0 | 25.00000 | 5.25000 |

**Table 6.12** Numerical values for transition from increment 2 to 3

| Cycle | $\tilde{E}^{\text{II},(i)}$ | $\Delta_2^{(i)}$ | $u_2^{(i)}$ | $\lvert u_2^{\text{new}} - u_2^{\text{old}}\rvert$ | $\varepsilon^{\text{II},(i)}$ | $\Delta\varepsilon^{\text{II},(i)}$ | $\sigma^{\text{II},(i)}$ |
|---|---|---|---|---|---|---|---|
| – | MPa | mm | mm | $10^{-2}$mm | $10^{-3}$ | $10^{-3}$ | MPa |
| 0 | 38181.82 | 0.282353 | 1.082353 | 28.235294 | 6.588235 | 2.588235 | 360.1070 |
| 1 | 30950.41 | 0.245272 | 1.045272 | 3.708073 | 6.773639 | 2.773639 | 361.2868 |
| 2 | 29306.91 | 0.236092 | 1.036092 | 0.918059 | 6.819542 | 2.819542 | 361.5789 |
| 3 | 28933.39 | 0.233962 | 1.033963 | 0.212904 | 6.830187 | 2.830187 | 361.6466 |
| 4 | 28848.50 | 0.233476 | 1.033476 | 0.0486115 | 6.832618 | 2.832618 | 361.6621 |
| 5 | 28829.20 | 0.233366 | 1.033366 | 0.0110598 | 6.833171 | 2.833171 | 361.6656 |
| 6 | 28824.82 | 0.233341 | 1.033341 | 0.0025142 | 6.833296 | 2.833296 | 361.6664 |
| 7 | 28823.82 | 0.233335 | 1.033335 | 0.0005714 | 6.833325 | 2.833325 | 361.6666 |

50 equidistant increments ($u_{\text{max}} = 8$ mm). The cross-sectional area is $A = 100$ mm$^2$ and the length equals $L = 400$ mm. The elastic range is characterized by YOUNG's modulus $E = 70000$ MPa and the plastic range should be modeled as

(a) an ideal plastic material with a yield stress of $k = 350$ MPa or
(b) a linear hardening material with $k^{\text{init}} = 350$ MPa and $E^{\text{pl}} = 7000$ MPa.

Assume that the material behavior is the same in the tensile and compressive regime of the isotropic hardening material to determine all the state variables ($\sigma, \varepsilon, \varepsilon^{\text{pl}}, \varepsilon^{\text{pl}}_{\text{eff}}, E^{\text{elpl}}$) in each increment.

**Table 6.13** Numerical results for cyclic loading with $u_{max} \leq u \leq -u_{max}$ for ideal plastic material

| Inc | $u$ | $\varepsilon$ | $\sigma^{trial}$ | $\sigma$ | $\varepsilon^{pl}$ | $\varepsilon^{pl}_{eff}$ | $\Delta\lambda$ | $E^{elpl}$ |
|---|---|---|---|---|---|---|---|---|
| – | mm | $10^{-3}$ | MPa | MPa | $10^{-3}$ | $10^{-3}$ | $10^{-3}$ | MPa |
| 1 | 0.8 | 2.0 | 140.0 | 140.0 | 0.0 | 0.0 | 0.0 | 0.0 |
| 2 | 1.6 | 4.0 | 280.0 | 280.0 | 0.0 | 0.0 | 0.0 | 0.0 |
| 3 | 2.4 | 6.0 | 420.0 | 350.0 | 1.0 | 1.0 | 1.0 | 0.0 |
| 4 | 3.2 | 8.0 | 490.0 | 350.0 | 3.0 | 3.0 | 2.0 | 0.0 |
| 5 | 4.0 | 10.0 | 490.0 | 350.0 | 5.0 | 5.0 | 2.0 | 0.0 |
| 6 | 4.8 | 12.0 | 490.0 | 350.0 | 7.0 | 7.0 | 2.0 | 0.0 |
| 7 | 5.6 | 14.0 | 490.0 | 350.0 | 9.0 | 9.0 | 2.0 | 0.0 |
| 8 | 6.4 | 16.0 | 490.0 | 350.0 | 11.0 | 11.0 | 2.0 | 0.0 |
| 9 | 7.2 | 18.0 | 490.0 | 350.0 | 13.0 | 13.0 | 2.0 | 0.0 |
| 10 | 8.0 | 20.0 | 490.0 | 350.0 | 15.0 | 15.0 | 2.0 | 0.0 |
| 11 | 7.2 | 18.0 | 210.0 | 210.0 | 15.0 | 15.0 | 0.0 | 0.0 |
| 12 | 6.4 | 16.0 | 70.0 | 70.0 | 15.0 | 15.0 | 0.0 | 0.0 |
| 13 | 5.6 | 14.0 | −70.0 | −70.0 | 15.0 | 15.0 | 0.0 | 0.0 |
| 14 | 4.8 | 12.0 | −210.0 | −210.0 | 15.0 | 15.0 | 0.0 | 0.0 |
| 15 | 4.0 | 10.0 | −350.0 | −350.0 | 15.0 | 15.0 | 0.0 | 0.0 |
| 16 | 3.2 | 8.0 | −490.0 | −350.0 | 13.0 | 17.0 | 2.0 | 0.0 |
| 17 | 2.4 | 6.0 | −490.0 | −350.0 | 11.0 | 19.0 | 2.0 | 0.0 |
| 18 | 1.6 | 4.0 | −490.0 | −350.0 | 9.0 | 21.0 | 2.0 | 0.0 |
| 19 | 0.8 | 2.0 | −490.0 | −350.0 | 7.0 | 23.0 | 2.0 | 0.0 |
| 20 | 0.0 | 0.0 | −490.0 | −350.0 | 5.0 | 25.0 | 2.0 | 0.0 |
| 21 | −0.8 | −2.0 | −490.0 | −350.0 | 3.0 | 27.0 | 2.0 | 0.0 |
| 22 | −1.6 | −4.0 | −490.0 | −350.0 | 1.0 | 29.0 | 2.0 | 0.0 |
| 23 | −2.4 | −6.0 | −490.0 | −350.0 | −1.0 | 31.0 | 2.0 | 0.0 |
| 24 | −3.2 | −8.0 | −490.0 | −350.0 | −3.0 | 33.0 | 2.0 | 0.0 |
| 25 | −4.0 | −10.0 | −490.0 | −350.0 | −5.0 | 35.0 | 2.0 | 0.0 |
| 26 | −4.8 | −12.0 | −490.0 | −350.0 | −7.0 | 37.0 | 2.0 | 0.0 |
| 27 | −5.6 | −14.0 | −490.0 | −350.0 | −9.0 | 39.0 | 2.0 | 0.0 |
| 28 | −6.4 | −16.0 | −490.0 | −350.0 | −11.0 | 41.0 | 2.0 | 0.0 |
| 29 | −7.2 | −18.0 | −490.0 | −350.0 | −13.0 | 43.0 | 2.0 | 0.0 |
| 30 | −8.0 | −20.0 | −490.0 | −350.0 | −15.0 | 45.0 | 2.0 | 0.0 |
| 31 | −7.2 | −18.0 | −210.0 | −210.0 | −15.0 | 45.0 | 0.0 | 0.0 |
| 32 | −6.4 | −16.0 | −70.0 | −70.0 | −15.0 | 45.0 | 0.0 | 0.0 |
| 33 | −5.6 | −14.0 | 70.0 | 70.0 | −15.0 | 45.0 | 0.0 | 0.0 |
| 34 | −4.8 | −12.0 | 210.0 | 210.0 | −15.0 | 45.0 | 0.0 | 0.0 |
| 35 | −4.0 | −10.0 | 350.0 | 350.0 | −15.0 | 45.0 | 0.0 | 0.0 |
| 36 | −3.2 | −8.0 | 490.0 | 350.0 | −13.0 | 47.0 | 2.0 | 0.0 |
| 37 | −2.4 | −6.0 | 490.0 | 350.0 | −11.0 | 49.0 | 2.0 | 0.0 |
| 38 | −1.6 | −4.0 | 490.0 | 350.0 | −9.0 | 51.0 | 2.0 | 0.0 |

(continued)

**Table 6.13** (continued)

| Inc | $u$ | $\varepsilon$ | $\sigma^{\text{trial}}$ | $\sigma$ | $\varepsilon^{\text{pl}}$ | $\varepsilon^{\text{pl}}_{\text{eff}}$ | $\Delta\lambda$ | $E^{\text{elpl}}$ |
|---|---|---|---|---|---|---|---|---|
| – | mm | $10^{-3}$ | MPa | MPa | $10^{-3}$ | $10^{-3}$ | $10^{-3}$ | MPa |
| 39 | −0.8 | −2.0 | 490.0 | 350.0 | −7.0 | 53.0 | 2.0 | 0.0 |
| 40 | 0.0 | 0.0 | 490.0 | 350.0 | −5.0 | 55.0 | 2.0 | 0.0 |
| 41 | 0.8 | 2.0 | 490.0 | 350.0 | −3.0 | 57.0 | 2.0 | 0.0 |
| 42 | 1.6 | 4.0 | 490.0 | 350.0 | −1.0 | 59.0 | 2.0 | 0.0 |
| 43 | 2.4 | 6.0 | 490.0 | 350.0 | 1.0 | 61.0 | 2.0 | 0.0 |
| 44 | 3.2 | 8.0 | 490.0 | 350.0 | 3.0 | 63.0 | 2.0 | 0.0 |
| 45 | 4.0 | 10.0 | 490.0 | 350.0 | 5.0 | 65.0 | 2.0 | 0.0 |
| 46 | 4.8 | 12.0 | 490.0 | 350.0 | 7.0 | 67.0 | 2.0 | 0.0 |
| 47 | 5.6 | 14.0 | 490.0 | 350.0 | 9.0 | 69.0 | 2.0 | 0.0 |
| 48 | 6.4 | 16.0 | 490.0 | 350.0 | 11.0 | 71.0 | 2.0 | 0.0 |
| 49 | 7.2 | 18.0 | 490.0 | 350.0 | 13.0 | 73.0 | 2.0 | 0.0 |
| 50 | 8.0 | 20.0 | 490.0 | 350.0 | 15.0 | 75.0 | 2.0 | 0.0 |

**Table 6.14** Numerical results for cyclic loading with $u_{\text{max}} \leq u \leq -u_{\text{max}}$ for linear hardening material

| Inc | $u$ | $\varepsilon$ | $\sigma^{\text{trial}}$ | $\sigma$ | $\varepsilon^{\text{pl}}$ | $\varepsilon^{\text{pl}}_{\text{eff}}$ | $\Delta\lambda$ | $E^{\text{elpl}}$ |
|---|---|---|---|---|---|---|---|---|
| – | mm | $10^{-3}$ | MPa | MPa | $10^{-3}$ | $10^{-3}$ | $10^{-3}$ | MPa |
| 1 | 0.8 | 2.0 | 140.0 | 140.0 | 0.0 | 0.0 | 0.0 | 0.0 |
| 2 | 1.6 | 4.0 | 280.0 | 280.0 | 0.0 | 0.0 | 0.0 | 0.0 |
| 3 | 2.4 | 6.0 | 420.000 | 356.364 | 0.909 | 0.909 | 0.909 | 6363.636 |
| 4 | 3.2 | 8.0 | 496.364 | 369.091 | 2.727 | 2.727 | 1.818 | 6363.636 |
| 5 | 4.0 | 10.0 | 509.091 | 381.818 | 4.545 | 4.545 | 1.818 | 6363.636 |
| 6 | 4.8 | 12.0 | 521.818 | 394.545 | 6.364 | 6.364 | 1.818 | 6363.636 |
| 7 | 5.6 | 14.0 | 534.545 | 407.273 | 8.182 | 8.182 | 1.818 | 6363.636 |
| 8 | 6.4 | 16.0 | 547.273 | 420.000 | 10.000 | 10.000 | 1.818 | 6363.636 |
| 9 | 7.2 | 18.0 | 560.000 | 432.727 | 11.818 | 11.818 | 1.818 | 6363.636 |
| 10 | 8.0 | 20.0 | 572.727 | 445.455 | 13.636 | 13.636 | 1.818 | 6363.636 |
| 11 | 7.2 | 18.0 | 305.455 | 305.455 | 13.636 | 13.636 | 0.0 | 0.0 |
| 12 | 6.4 | 16.0 | 165.455 | 165.455 | 13.636 | 13.636 | 0.0 | 0.0 |
| 13 | 5.6 | 14.0 | 25.455 | 25.455 | 13.636 | 13.636 | 0.0 | 0.0 |
| 14 | 4.8 | 12.0 | −114.545 | −114.545 | 13.636 | 13.636 | 0.0 | 0.0 |
| 15 | 4.0 | 10.0 | −254.545 | −254.545 | 13.636 | 13.636 | 0.0 | 0.0 |
| 16 | 3.2 | 8.0 | −394.545 | −394.545 | 13.636 | 13.636 | 0.0 | 0.0 |
| 17 | 2.4 | 6.0 | −534.554 | −453.554 | 12.479 | 14.793 | 1.157 | 6363.636 |
| 18 | 1.6 | 4.0 | −593.545 | −466.281 | 10.661 | 16.612 | 1.818 | 6363.636 |
| 19 | 0.8 | 2.0 | −606.281 | −479.008 | 8.843 | 18.430 | 1.818 | 6363.636 |
| 20 | 0.0 | 0.0 | −619.008 | −491.736 | 7.025 | 20.248 | 1.818 | 6363.636 |

(continued)

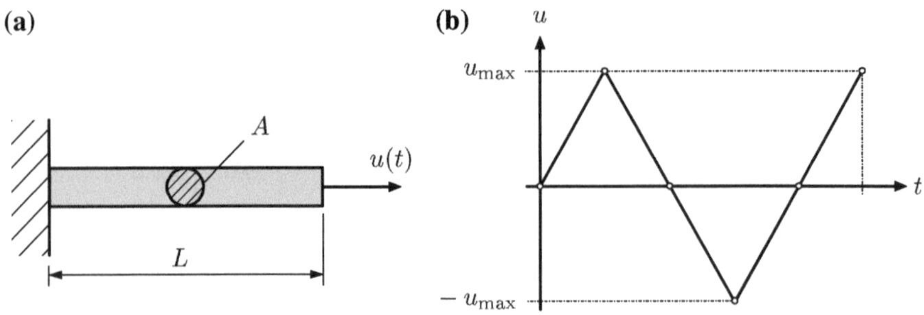


**Fig. 6.17** Cyclic loading of an elastic-plastic continuum rod: **a** configuration and **b** applied displacement boundary condition

**6.4 Solution**

The strain increment for each load step can be calculated as

$$|\Delta\varepsilon_n| = \frac{1}{10} \times \frac{8\text{ mm}}{400\text{ mm}} = 0.002\,. \tag{6.144}$$

Application of the fully implicit backward-EULER algorithm for linear hardening as given in Eq. (6.29) and Table 6.3 results in the numerical values as given in Table 6.13 and 6.14.

## 6.5 Example: Cyclic loading of an elastic-plastic continuum rod

Consider a pure one-dimensional tensile test as schematically shown in Fig. 6.18. Use the matrix solution scheme as given in Sect. 6.2 to simulate a cycling load in the form of an end displacement boundary condition (cf. Fig. 6.17b) which is applied in 40 equidistant increments. The cross-sectional area is $A = 100\text{ mm}^2$ and the length equals $L = 400$ mm. The elastic range is characterized by YOUNG's modulus $E = 70000$ MPa and the plastic range and the displacement boundary condition should be modeled as

(a) ideal plastic with a yield stress of $k = 350$ MPa, and $u_{\text{max}} = 8$ mm and $u_{\text{min}} = -2$ mm,
(b) linear hardening $k^{\text{init}} = 350$ MPa and $E^{\text{pl}} = 7000$ MPa, and $u_{\text{max}} = 8$ mm and $u_{\text{min}} = -3.0909$ mm.

Assume that the material behavior is the same in the tensile and compressive regime of the isotropic hardening material to determine all the state variables ($\sigma$, $\varepsilon$, $\varepsilon^{\text{pl}}$, $\varepsilon^{\text{pl}}_{\text{eff}}$, $E^{\text{elpl}}$) in each increment.

**6.5 Solution**

The strain increment in the range $0 \leq inc \leq 20$ can be calculated as

**Table 6.14** (continued)

| Inc | $u$ | $\varepsilon$ | $\sigma^{\text{trial}}$ | $\sigma$ | $\varepsilon^{\text{pl}}$ | $\varepsilon^{\text{pl}}_{\text{eff}}$ | $\Delta\lambda$ | $E^{\text{elpl}}$ |
|---|---|---|---|---|---|---|---|---|
| – | mm | $10^{-3}$ | MPa | MPa | $10^{-3}$ | $10^{-3}$ | $10^{-3}$ | MPa |
| 21 | −0.8 | −2.0 | −631.736 | −504.463 | 5.207 | 22.066 | 1.818 | 6363.636 |
| 22 | −1.6 | −4.0 | −644.463 | −517.190 | 3.388 | 23.884 | 1.818 | 6363.636 |
| 23 | −2.4 | −6.0 | −657.190 | −529.917 | 1.570 | 25.703 | 1.818 | 6363.636 |
| 24 | −3.2 | −8.0 | −669.917 | −542.645 | −0.248 | 27.521 | 1.818 | 6363.636 |
| 25 | −4.0 | −10.0 | −682.645 | −555.372 | −2.066 | 29.339 | 1.818 | 6363.636 |
| 26 | −4.8 | −12.0 | −695.372 | −568.099 | −3.884 | 31.157 | 1.818 | 6363.636 |
| 27 | −5.6 | −14.0 | −708.099 | −580.826 | −5.702 | 32.975 | 1.818 | 6363.636 |
| 28 | −6.4 | −16.0 | −720.826 | −593.554 | −7.521 | 34.793 | 1.818 | 6363.636 |
| 29 | −7.2 | −18.0 | −733.554 | −606.281 | −9.339 | 36.612 | 1.818 | 6363.636 |
| 30 | −8.0 | −20.0 | −746.281 | −619.008 | −11.157 | 38.430 | 0.0 | 0.0 |
| 31 | −7.2 | −18.0 | −479.008 | −479.008 | −11.157 | 38.430 | 0.0 | 0.0 |
| 32 | −6.4 | −16.0 | −339.008 | −339.008 | −11.157 | 38.430 | 0.0 | 0.0 |
| 33 | −5.6 | −14.0 | −199.008 | −199.008 | −11.157 | 38.430 | 0.0 | 0.0 |
| 34 | −4.8 | −12.0 | −59.008 | −59.008 | −11.157 | 38.430 | 0.0 | 0.0 |
| 35 | −4.0 | −10.0 | 80.992 | 80.992 | −11.157 | 38.430 | 0.0 | 0.0 |
| 36 | −3.2 | −8.0 | 220.992 | 220.992 | −11.157 | 38.430 | 0.0 | 0.0 |
| 37 | −2.4 | −6.0 | 360.992 | 360.992 | −11.157 | 38.430 | 0.0 | 0.0 |
| 38 | −1.6 | −4.0 | 500.992 | 500.992 | −11.157 | 38.430 | 0.0 | 0.0 |
| 39 | −0.8 | −2.0 | 640.992 | 621.007 | −10.872 | 38.715 | 0.285 | 6363.636 |
| 40 | 0.0 | 0.0 | 761.007 | 633.734 | −9.053 | 40.533 | 1.818 | 6363.636 |
| 41 | 0.8 | 2.0 | 773.734 | 646.461 | −7.235 | 42.352 | 1.818 | 6363.636 |
| 42 | 1.6 | 4.0 | 786.461 | 659.189 | −5.417 | 44.170 | 1.818 | 6363.636 |
| 43 | 2.4 | 6.0 | 799.189 | 671.916 | −3.599 | 45.988 | 1.818 | 6363.636 |
| 44 | 3.2 | 8.0 | 811.916 | 684.643 | −1.781 | 47.806 | 1.818 | 6363.636 |
| 45 | 4.0 | 10.0 | 824.643 | 697.37 | 0.375 | 49.624 | 1.818 | 6363.636 |
| 46 | 4.8 | 12.0 | 837.370 | 710.098 | 1.856 | 51.443 | 1.818 | 6363.636 |
| 47 | 5.6 | 14.0 | 850.098 | 722.825 | 3.674 | 53.261 | 1.818 | 6363.636 |
| 48 | 6.4 | 16.0 | 862.825 | 735.552 | 5.492 | 55.079 | 1.818 | 6363.636 |
| 49 | 7.2 | 18.0 | 875.552 | 748.279 | 7.310 | 56.897 | 1.818 | 6363.636 |
| 50 | 8.0 | 20.0 | 888.279 | 761.007 | 9.124 | 58.715 | 1.818 | 6363.636 |

$$|\Delta\varepsilon_n| = \frac{1}{10} \times \frac{8\ \text{mm}}{400\ \text{mm}} = 0.002\,. \tag{6.145}$$

while the range $21 \leq inc \leq 40$ requires

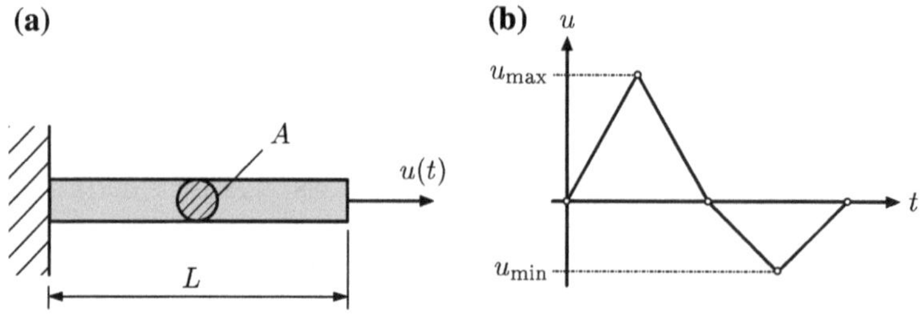


**Fig. 6.18** Cyclic loading of an elastic-plastic continuum rod: **a** configuration and **b** applied displacement boundary condition

$$|\Delta\varepsilon_n| = \frac{1}{10} \times \frac{2 \text{ mm}}{400 \text{ mm}} = 0.0005 \quad \text{(ideal)}, \tag{6.146}$$

$$|\Delta\varepsilon_n| = \frac{1}{10} \times \frac{3.0909 \text{ mm}}{400 \text{ mm}} = 0.000773 \quad \text{(linear)}. \tag{6.147}$$

Application of the fully implicit backward-EULER algorithm for linear hardening as outlined in Eq. (6.29) and Table 6.3 results in the numerical values as given in Tables 6.15 and 6.16.

## 6.6 Example: Monotonic loading of an elastic-plastic continuum rod with non-linear material behavior

Consider a pure one-dimensional tensile test as schematically shown in Fig. 6.19. Use the matrix solution scheme as given in Sect. 6.2 to simulate a monotonic load in the form of a displacement boundary condition (cf. Fig. 6.19b) which is applied in equidistant increments up to $u_{max} = 8$ mm. The cross-sectional area is $A = 100$ mm$^2$ and the length equals $L = 400$ mm. The elastic range is characterized by a YOUNG's modulus of $E = 70000$ MPa and the plastic range by a quadratic flow curve in the form of $k(\kappa) = \left(350 + 12900\kappa - 1.25 \times 10^5 \kappa^2\right)$ MPa where the equivalent plastic strain $|\varepsilon^{pl}|$ should be used as the hardening variable. Determine all the state variables ($\sigma$, $\varepsilon$, $\varepsilon^{pl}$, $\varepsilon^{pl}_{eff}$, $E^{elpl}$) in each increment for the case of

(a) 10 and 20 equidistant increments for the maximum displacement $u_{max}$,
(b) a termination criterion $t^{v}_{end}$ of 0.1. 0.001, 0.00001 and 0.0000001 for the vector norm $\|\boldsymbol{v}^{(i+1)}$-$\boldsymbol{v}^{(i)}\|$ during the iteration.

Derive first the analytical expression for the stress-strain distribution $\sigma = \sigma(\varepsilon)$ and sketch the stress-strain diagram and the flow curve in a strain range of up to 0.07. Furthermore, determine the strain at maximum load in both diagrams. Compare and analyze how the increment number and termination criterion influences the accuracy of the iteration.

### 6.6 Solution

**Table 6.15** Numerical results for cyclic loading with $u_{max} \leq u \leq -u_{min}$ for ideal plastic material

| Inc | $u$ | $\varepsilon$ | $\sigma^{trial}$ | $\sigma$ | $\varepsilon^{pl}$ | $\varepsilon^{pl}_{eff}$ | $\Delta\lambda$ | $E^{elpl}$ |
|---|---|---|---|---|---|---|---|---|
| – | mm | $10^{-3}$ | MPa | MPa | $10^{-3}$ | $10^{-3}$ | $10^{-3}$ | MPa |
| 1 | 0.8 | 2.0 | 140.0 | 140.0 | 0.0 | 0.0 | 0.0 | 0.0 |
| ⋮ | | | | | | | | ⋮ |
| 20 | 0.0 | 0.0 | −385.0 | −350.0 | 5.0 | 25.0 | 2.0 | 0.0 |
| 21 | −0.2 | −0.5 | −385.0 | −350.0 | 4.5 | 25.5 | 0.5 | 0.0 |
| 22 | −0.4 | −1.0 | −385.0 | −350.0 | 4.0 | 26.0 | 0.5 | 0.0 |
| 23 | −0.6 | −1.5 | −385.0 | −350.0 | 3.5 | 26.5 | 0.5 | 0.0 |
| 24 | −0.8 | −2.0 | −385.0 | −350.0 | 3.0 | 27.0 | 0.5 | 0.0 |
| 25 | −1.0 | −2.5 | −385.0 | −350.0 | 2.5 | 27.5 | 0.5 | 0.0 |
| 26 | −1.2 | −3.0 | −385.0 | −350.0 | 2.0 | 28.0 | 0.5 | 0.0 |
| 27 | −1.4 | −3.5 | −385.0 | −350.0 | 1.5 | 28.5 | 0.5 | 0.0 |
| 28 | −1.6 | −4.0 | −385.0 | −350.0 | 1.0 | 29.0 | 0.5 | 0.0 |
| 29 | −1.8 | −4.5 | −385.0 | −350.0 | 0.5 | 29.5 | 0.5 | 0.0 |
| 30 | −2.0 | −5.0 | −385.0 | −350.0 | 0.0 | 30.0 | 0.5 | 0.0 |
| 31 | −1.8 | −4.5 | −315.0 | −315.0 | 0.0 | 30.0 | 0.0 | 0.0 |
| 32 | −1.6 | −4.0 | −280.0 | −280.0 | 0.0 | 30.0 | 0.0 | 0.0 |
| 33 | −1.4 | −3.5 | −245.0 | −245.0 | 0.0 | 30.0 | 0.0 | 0.0 |
| 34 | −1.2 | −3.0 | −210.0 | −210.0 | 0.0 | 30.0 | 0.0 | 0.0 |
| 35 | −1.0 | −2.5 | −175.0 | −175.0 | 0.0 | 30.0 | 0.0 | 0.0 |
| 36 | −0.8 | −2.0 | −140.0 | −140.0 | 0.0 | 30.0 | 0.0 | 0.0 |
| 37 | −0.6 | −1.5 | −105.0 | −105.0 | 0.0 | 30.0 | 0.0 | 0.0 |
| 38 | −0.4 | −1.0 | −70.0 | −70.0 | 0.0 | 30.0 | 0.0 | 0.0 |
| 39 | −0.2 | −0.5 | −35.0 | −35.0 | 0.0 | 30.0 | 0.0 | 0.0 |
| 40 | 0.0 | 0.0 | 0.0 | 0.0 | 0.0 | 30.0 | 0.0 | 0.0 |

The strain increment for each load step can be calculated based on the maximum displacement, the original rod length and the increment number as

$$|\Delta\varepsilon_n| = \frac{1}{10} \times \frac{8\ \text{mm}}{400\ \text{mm}} = 0.002\,, \tag{6.148}$$

or as

$$|\Delta\varepsilon_n| = \frac{1}{20} \times \frac{8\ \text{mm}}{400\ \text{mm}} = 0.001\,. \tag{6.149}$$

The elastic limit strain $\varepsilon_t$ results from HOOKE's law and the initial yield stress $k_t^{init}$ as

$$\varepsilon_t = \frac{k_t^{init}}{E} = \frac{350\ \text{MPa}}{70000\ \text{MPa}} = 0.005\,. \tag{6.150}$$

**Table 6.16** Numerical results for cyclic loading with $u_{\text{max}} \leq u \leq -u_{\text{min}}$ for linear hardening material

| Inc | $u$ | $\varepsilon$ | $\sigma^{\text{trial}}$ | $\sigma$ | $\varepsilon^{\text{pl}}$ | $\varepsilon^{\text{pl}}_{\text{eff}}$ | $\Delta\lambda$ | $E^{\text{elpl}}$ |
|---|---|---|---|---|---|---|---|---|
| -- | mm | $10^{-3}$ | MPa | MPa | $10^{-3}$ | $10^{-3}$ | $10^{-3}$ | MPa |
| 1 | 0.8 | 2.0 | 140.0 | 140.0 | 0.0 | 0.0 | 0.0 | 0.0 |
| ⋮ | | | | | | | | ⋮ |
| 20 | 0.0 | 0.0 | −619.008 | −491.736 | 7.025 | 20.248 | 1.818 | 6363.636 |
| 21 | −0.309 | −0.773 | −545.827 | −496.653 | 6.322 | 20.950 | 0.702 | 6363.636 |
| 22 | −0.618 | −1.545 | −550.744 | −501.570 | 5.620 | 21.653 | 0.702 | 6363.636 |
| 23 | −0.927 | −2.318 | −555.661 | −506.488 | 4.917 | 22.355 | 0.702 | 6363.636 |
| 24 | −1.236 | −3.091 | −560.579 | −511.405 | 4.215 | 23.058 | 0.702 | 6363.636 |
| 25 | −1.545 | −3.864 | −565.496 | −516.322 | 3.512 | 23.760 | 0.702 | 6363.636 |
| 26 | −1.855 | −4.636 | −570.413 | −521.240 | 2.810 | 24.463 | 0.702 | 6363.636 |
| 27 | −1.164 | −5.409 | −575.331 | −526.157 | 2.107 | 25.165 | 0.702 | 6363.636 |
| 28 | −2.473 | −6.182 | −580.248 | −531.074 | 1.405 | 25.868 | 0.702 | 6363.636 |
| 29 | −2.782 | −6.955 | −585.165 | −535.992 | 0.702 | 26.570 | 0.702 | 6363.636 |
| 30 | −3.091 | −7.727 | −590.083 | −540.909 | 0.0 | 27.273 | 0.702 | 6363.636 |
| 31 | −2.782 | −6.955 | −486.818 | −486.818 | 0.0 | 27.273 | 0.0 | 0.0 |
| 32 | −2.473 | −6.182 | −432.727 | −432.727 | 0.0 | 27.273 | 0.0 | 0.0 |
| 33 | −2.164 | −5.409 | −378.637 | −378.637 | 0.0 | 27.273 | 0.0 | 0.0 |
| 34 | −1.855 | −4.636 | −324.546 | −324.546 | 0.0 | 27.273 | 0.0 | 0.0 |
| 35 | −1.545 | −3.864 | −270.455 | −270.455 | 0.0 | 27.273 | 0.0 | 0.0 |
| 36 | −1.236 | −3.091 | −216.364 | −216.364 | 0.0 | 27.273 | 0.0 | 0.0 |
| 37 | −0.927 | −2.318 | −162.274 | −162.274 | 0.0 | 27.273 | 0.0 | 0.0 |
| 38 | −0.618 | −1.545 | −108.182 | −108.182 | 0.0 | 27.273 | 0.0 | 0.0 |
| 39 | −0.309 | −0.773 | −54.092 | −54.092 | 0.0 | 27.273 | 0.0 | 0.0 |
| 40 | 0.0 | 0.0 | −14.465 | −14.465 | 0.0 | 27.273 | 0.0 | 0.0 |

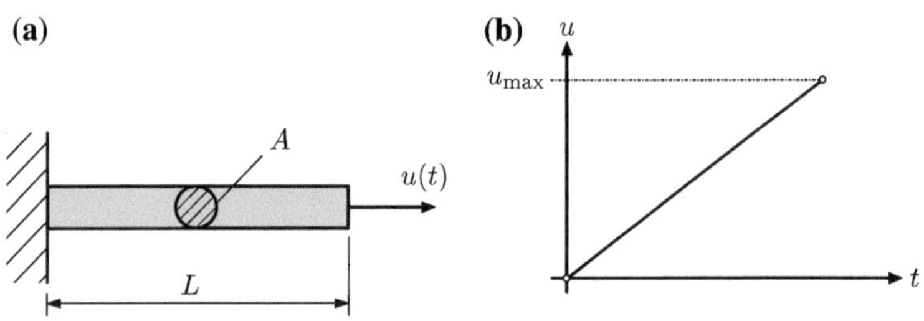


**Fig. 6.19** Monotonic loading of an elastic−plastic continuum rod: **a** configuration and **b** applied displacement boundary condition

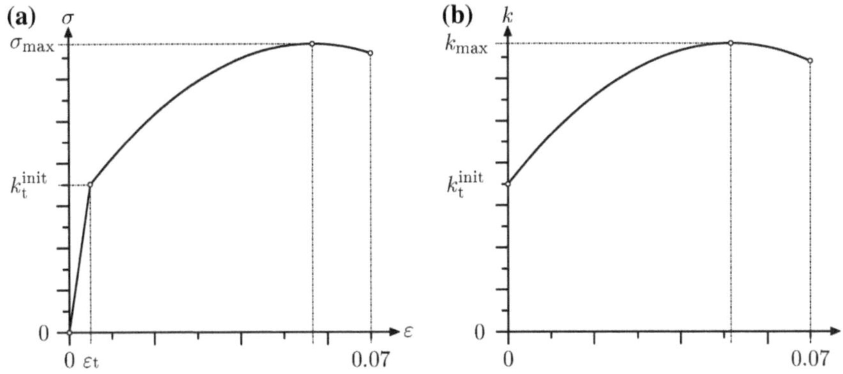


**Fig. 6.20** **a** Stress-strain diagram and **b** flow *curve*

This value can be used to horizontally shift the flow curve in a stress-strain diagram to obtain the following relationship:

$$\sigma(\varepsilon) = \begin{cases} E\varepsilon & \text{for } \varepsilon \leq \varepsilon_t \\ 350 + 12900(\varepsilon\text{-}\varepsilon_t)\text{-}1.25 \times 10^5(\varepsilon\text{-}\varepsilon_t)^2 & \text{for } \varepsilon \geq \varepsilon_t \end{cases}. \tag{6.151}$$

The graphical representation of the stress-strain diagram and the flow curve is given in Fig. 6.20. The strain at maximum stress results from the conditions $\frac{d\sigma}{d\varepsilon} = 0$ and $\frac{dk}{d\kappa} = 0$ as $\varepsilon(\sigma_{max}) = 0.0566$ and $|\varepsilon^{pl}|(k_{max}) = 0.0516$.

The vector norm $\|\boldsymbol{v}^{(i+1)}\text{-}\boldsymbol{v}^{(i)}\|$ for each iteration step $i$ is given according to the equations in Sect. 6.2 as:

$$\|\boldsymbol{v}^{(i+1)} - \boldsymbol{v}^{(i)}\| = \left\| \left( \frac{\partial \boldsymbol{m}}{\partial \boldsymbol{v}} \left( \boldsymbol{v}^{(i)} \right) \right)^{-1} \boldsymbol{m} \left( \boldsymbol{v}^{(i)} \right) \right\| = \tag{6.152}$$

$$= \left\| \frac{E}{E + \frac{dk}{d\kappa}} \times \begin{bmatrix} \frac{dk}{d\kappa} & -\text{sgn}(\sigma)\frac{dk}{d\kappa} & \text{sgn}(\sigma) \\ \text{sgn}(\sigma) & -1 & -E^{-1} \\ \text{sgn}(\sigma) & E^{-1}\frac{dk}{d\kappa} & -E^{-1} \end{bmatrix}_{(i)} \begin{bmatrix} r_\sigma(\boldsymbol{v}^{(i)}) \\ r_\kappa(\boldsymbol{v}^{(i)}) \\ r_F(\boldsymbol{v}^{(i)}) \end{bmatrix} \right\|. \tag{6.153}$$

The results for 10 increments are presented in Table 6.17 ($t^{v}_{end} = 0.1$) and Table 6.18 ($t^{v}_{end} = 0.001$ and $0.00001$). Looking, for example, at the results for the last increment, it can bee seen that there is no difference in the results. Only for the 3rd increment, there is a very small difference for both approaches. For the stricter criteria, two cycles are required to achieve the requested termination value. However, the stress changes only from 360.893 to 360.817 MPa, i.e a change of only 0.021 % and negligible from a practical point of view. Furthermore, it should be highlighted

**Table 6.17** Numerical results for monotonic loading with $0 \leq u \leq u_{max}$ for nonlinear hardening material

| inc $n$ | cycle $i$ | $\varepsilon$ | $\sigma^{trial}$ | $\sigma$ | $\varepsilon^{pl}$ | $\varepsilon^{pl}_{eff}$ | $\Delta\lambda$ | $E^{elpl}$ | $\lVert v^{(i+1)}$-$v^{(i)} \rVert$ |
|---|---|---|---|---|---|---|---|---|---|
| – | – | $10^{-3}$ | MPa | MPa | $10^{-3}$ | $10^{-3}$ | $10^{-3}$ | MPa | – |
| 1 | 1 | 2.0 | 140.0 | 140.0 | 0.0 | 0.0 | 0.0 | 0.0 | 0.0 |
| 2 | 1 | 4.0 | 280.0 | 280.0 | 0.0 | 0.0 | 0.0 | 0.0 | 0.0 |
| 3 | 1 | 6.0 | 420.0 | 360.893 | 0.844 | 0.844 | 0.844 | 10741.746 | 0.0754480 |
| 4 | 1 | 8.0 | 500.893 | 382.301 | 2.539 | 2.539 | 1.694 | 10436.654 | 0.3052850 |
| | 2 | 8.0 | 500.893 | 381.995 | 2.543 | 2.543 | 1.699 | 10435.866 | 0.00000202307 |
| 5 | 1 | 10.0 | 521.995 | 402.867 | 4.245 | 4.245 | 1.702 | 10126.207 | 0.309658 |
| | 2 | 10.0 | 521.995 | 402.557 | 4.249 | 4.249 | 1.706 | 10125.398 | 0.00000209230 |
| 6 | 1 | 12.0 | 542.557 | 422.808 | 5.960 | 5.960 | 1.711 | 9810.856 | 0.314543 |
| | 2 | 12.0 | 542.557 | 422.494 | 5.964 | 5.964 | 1.715 | 9810.025 | 0.00000217019 |
| 7 | 1 | 14.0 | 562.494 | 442.114 | 7.684 | 7.684 | 1.720 | 9490.469 | 0.319557 |
| | 2 | 14.0 | 562.494 | 441.794 | 7.689 | 7.689 | 1.724 | 9489.616 | 0.00000225186 |
| 8 | 1 | 16.0 | 581.794 | 460.773 | 9.418 | 9.418 | 1.729 | 9164.910 | 0.324706 |
| | 2 | 16.0 | 581.794 | 460.449 | 9.422 | 9.422 | 1.734 | 9164.034 | 0.00000233752 |
| 9 | 1 | 18.0 | 600.449 | 478.777 | 11.160 | 11.160 | 1.738 | 8834.040 | 0.329994 |
| | 2 | 18.0 | 600.449 | 478.447 | 11.165 | 11.165 | 1.743 | 8833.140 | 0.00000242742 |
| 10 | 1 | 20.0 | 618.447 | 496.113 | 12.913 | 12.913 | 1.748 | 8497.712 | 0.335428 |
| | 2 | 20.0 | 618.447 | 495.778 | 12.917 | 12.917 | 1.752 | 8496.787 | 0.00000252181 |

Case: 10 increments, $t^{v}_{end} = 0.1$

here that an increase of the termination value from 0.001 to 0.00001, i.e. by a factor of 100, does not change any of the results obtained. Further increasing the termination value by a factor of 100 ($t^{v}_{end} = 0.0000001$) results in the fact that in the elasto-plastic regime 3 cycles are required to meet the criterion, cf. Table 6.19. However, the final values of the state variables do not change from a practical point of view.

The results for 20 increments are presented in Table 6.20 ($t^{v}_{end} = 0.1$), Table 6.21 ($t^{v}_{end} = 0.001$ and 0.00001), and Table 6.22 ($t^{v}_{end} = 0.0000001$). Comparing the results for 10 and 20 increments, it can be concluded that the same values are obtained from a practical point of view and that a finer incrementation of the load (20 increments) does not provide any advantage for the chosen example and parameters.

## 6.7 Example: Reversed loading of an elastic-plastic rod element under damage effects (Lemaitre)

Consider a pure one-dimensional tensile test as schematically shown in Fig. 2.11 or Fig. 6.21a where no effects from contraction or clamping of the specimen are considered [39]. Damage can develop according to the model by LEMAITRE. Use the matrix solution scheme as given in Sect. 6.5.1.1 to consider a reversed load in the form of an end displacement boundary condition (cf. Fig. 6.21b) to simulate experiments with cyclic loading and unloading in the tensile regime, i.e. $\sigma \geq 0$.

**Table 6.18** Numerical results for monotonic loading with $0 \leq u \leq u_{max}$ for nonlinear hardening material

| inc $n$ | cycle $i$ | $\varepsilon$ | $\sigma^{trial}$ | $\sigma$ | $\varepsilon^{pl}$ | $\varepsilon^{pl}_{eff}$ | $\Delta\lambda$ | $E^{elpl}$ | $\Vert \boldsymbol{v}^{(i+1)}$-$\boldsymbol{v}^{(i)} \Vert$ |
|---|---|---|---|---|---|---|---|---|---|
| – | – | $10^{-3}$ | MPa | MPa | $10^{-3}$ | $10^{-3}$ | $10^{-3}$ | MPa | – |
| 1 | 1 | 2.0 | 140.0 | 140.0 | 0.0 | 0.0 | 0.0 | 0.0 | 0.0 |
| 2 | 1 | 4.0 | 280.0 | 280.0 | 0.0 | 0.0 | 0.0 | 0.0 | 0.0 |
| 3 | 1 | 6.0 | 420.0 | 360.893 | 0.844 | 0.844 | 0.844 | 10741.746 | 0.0754480 |
| | 2 | 6.0 | 420.0 | 360.817 | 0.845 | 0.845 | 0.845 | 10741.553 | 0.000000122931 |
| 4 | 1 | 8.0 | 500.817 | 382.300 | 2.539 | 2.539 | 1.693 | 10436.654 | 0.304899 |
| | 2 | 8.0 | 500.817 | 381.995 | 2.543 | 2.543 | 1.697 | 10435.866 | 0.00000201796 |
| 5 | 1 | 10.0 | 521.995 | 402.867 | 4.245 | 4.245 | 1.702 | 10126.207 | 0.309658 |
| | 2 | 10.0 | 521.995 | 402.557 | 4.249 | 4.249 | 1.706 | 10125.398 | 0.00000209230 |
| 6 | 1 | 12.0 | 542.557 | 422.808 | 5.960 | 5.960 | 1.711 | 9810.856 | 0.314543 |
| | 2 | 12.0 | 542.557 | 422.494 | 5.964 | 5.964 | 1.715 | 9810.025 | 0.00000217019 |
| 7 | 1 | 14.0 | 562.494 | 442.114 | 7.684 | 7.684 | 1.720 | 9490.469 | 0.319557 |
| | 2 | 14.0 | 562.494 | 441.794 | 7.689 | 7.689 | 1.724 | 9489.616 | 0.00000225186 |
| 8 | 1 | 16.0 | 581.794 | 460.773 | 9.418 | 9.418 | 1.729 | 9164.910 | 0.324706 |
| | 2 | 16.0 | 581.794 | 460.449 | 9.422 | 9.422 | 1.734 | 9164.034 | 0.00000233752 |
| 9 | 1 | 18.0 | 600.449 | 478.777 | 11.160 | 11.160 | 1.738 | 8834.040 | 0.329994 |
| | 2 | 18.0 | 600.449 | 478.447 | 11.165 | 11.165 | 1.743 | 8833.140 | 0.00000242742 |
| 10 | 1 | 20.0 | 618.447 | 496.113 | 12.913 | 12.913 | 1.748 | 8497.712 | 0.335428 |
| | 2 | 20.0 | 618.447 | 495.778 | 12.917 | 12.917 | 1.752 | 8496.787 | 0.00000252181 |

Case: 10 increments, $t^{v}_{end} = 0.001$ and 0.00001

Staying in the tensile regime implies that any effects of kinematic hardening can be disregarded for this material.

The considered material behavior for AISI low carbon steel as given in Table 6.23 is taken from [11, 74]. In addition to these values, the case of ideal plasticity with $k = 620$ MPa is to be considered to investigate a different hardening behavior. A reference solution is to be calculated first where all damage effects are disregarded, i.e. $D = 0$ holds during the entire simulation.

## 6.7 Solution

Due to the assumption of an idealized one-dimensional test, all the state variables will have the same value in the specimen without any spatial variation. The reference solution without damage can easily be introduced into the numerical scheme given in Sect. 6.5.1.1 by setting $D = 0$, $s = 1$ and $r \to \infty$.

The results for the 'perfect' material in the case of hardening and ideal plasticity in the form of stress-strain diagrams are presented in Figs. 6.22a and 6.23a and opposed to diagrams considering damage effects shown in Figs. 6.22b and 6.23b. In the scale of these two diagrams, it is clear that the consideration of the damage clearly reduces the value of the stress which is obtained for a given strain. However, the estimation

**Table 6.19** Numerical results for monotonic loading with $0 \leq u \leq u_{\max}$ for nonlinear hardening material

| inc $n$ | cycle $i$ | $\varepsilon$ | $\sigma^{\text{trial}}$ | $\sigma$ | $\varepsilon^{\text{pl}}$ | $\varepsilon^{\text{pl}}_{\text{eff}}$ | $\Delta\lambda$ | $E^{\text{elpl}}$ | $\Vert \boldsymbol{v}^{(i+1)} - \boldsymbol{v}^{(i)} \Vert$ |
|---|---|---|---|---|---|---|---|---|---|
| – | – | $10^{-3}$ | MPa | MPa | $10^{-3}$ | $10^{-3}$ | $10^{-3}$ | MPa | – |
| 1 | 1 | 2.0 | 140.0 | 140.0 | 0.0 | 0.0 | 0.0 | 0.0 | 0.0 |
| 2 | 1 | 4.0 | 280.0 | 280.0 | 0.0 | 0.0 | 0.0 | 0.0 | 0.0 |
| 3 | 1 | 6.0 | 420.0 | 360.893 | 0.844 | 0.844 | 0.844 | 10741.746 | 0.0754480 |
| | 2 | 6.0 | 420.0 | 360.817 | 0.845 | 0.845 | 0.845 | 10741.553 | 0.000000122931 |
| | 3 | 6.0 | 420.0 | 360.817 | 0.845 | 0.845 | 0.845 | 10741.553 | $0.537078 \times 10^{-18}$ |
| 4 | 1 | 8.0 | 500.817 | 382.300 | 2.539 | 2.539 | 1.693 | 10436.654 | 0.304899 |
| | 2 | 8.0 | 500.817 | 381.995 | 2.543 | 2.543 | 1.697 | 10435.866 | 0.00000201796 |
| | 3 | 8.0 | 500.817 | 381.995 | 2.543 | 2.543 | 1.697 | 10435.865 | $0.956880 \times 10^{-16}$ |
| 5 | 1 | 10.0 | 521.995 | 402.867 | 4.245 | 4.245 | 1.702 | 10126.207 | 0.309658 |
| | 2 | 10.0 | 521.995 | 402.557 | 4.249 | 4.249 | 1.706 | 10125.398 | 0.00000209230 |
| | 3 | 10.0 | 521.995 | 402.557 | 4.249 | 4.249 | 1.706 | 10125.398 | $0.930761 \times 10^{-16}$ |
| 6 | 1 | 12.0 | 542.557 | 422.808 | 5.960 | 5.960 | 1.711 | 9810.856 | 0.314543 |
| | 2 | 12.0 | 542.557 | 422.494 | 5.964 | 5.964 | 1.715 | 9810.025 | 0.00000217019 |
| | 3 | 12.0 | 542.557 | 422.494 | 5.964 | 5.964 | 1.715 | 9810.025 | $0.111781 \times 10^{-15}$ |
| 7 | 1 | 14.0 | 562.494 | 442.114 | 7.684 | 7.684 | 1.720 | 9490.469 | 0.319557 |
| | 2 | 14.0 | 562.494 | 441.794 | 7.689 | 7.689 | 1.724 | 9489.616 | 0.00000225186 |
| | 3 | 14.0 | 562.494 | 441.794 | 7.689 | 7.689 | 1.724 | 9489.616 | $0.111427 \times 10^{-15}$ |
| 8 | 1 | 16.0 | 581.794 | 460.773 | 9.418 | 9.418 | 1.729 | 9164.910 | 0.324706 |
| | 2 | 16.0 | 581.794 | 460.449 | 9.422 | 9.422 | 1.734 | 9164.034 | 0.00000233752 |
| | 3 | 16.0 | 581.794 | 460.449 | 9.422 | 9.422 | 1.734 | 9164.034 | $0.121672 \times 10^{-15}$ |
| 9 | 1 | 18.0 | 600.449 | 478.777 | 11.160 | 11.160 | 1.738 | 8834.040 | 0.329994 |
| | 2 | 18.0 | 600.449 | 478.447 | 11.165 | 11.165 | 1.743 | 8833.140 | 0.00000242742 |
| | 3 | 18.0 | 600.449 | 478.447 | 11.165 | 11.165 | 1.743 | 8833.140 | $0.122334 \times 10^{-15}$ |
| 10 | 1 | 20.0 | 618.447 | 496.113 | 12.913 | 12.913 | 1.748 | 8497.712 | 0.335428 |
| | 2 | 20.0 | 618.447 | 495.778 | 12.917 | 12.917 | 1.752 | 8496.787 | 0.00000252181 |
| | 3 | 20.0 | 618.447 | 495.778 | 12.917 | 12.917 | 1.752 | 8496.787 | $0.134342 \times 10^{-15}$ |

Case: 10 increments, $t^{v}_{\text{end}} = 0.0000001$

of the variation of the elasticity modulus (cf. Eq. 2.94) is difficult in this scale and must be evaluated separately. The trend of the ideal plastic material, cf. Fig. 6.23, is the same as in the case of the hardening material. However, the absolute reduction of the stress, which results in the case of ideal plastic material behavior in a negative slope in the stress-strain diagram ('softening'), is clearly smaller.

The applied elongation of the specimen, i.e. displacement boundary condition, is shown in Fig. 6.24 for the case of the hardening material with damage. It can be clearly seen that the unloading requires much smaller displacements since the condition $\sigma = 0$ was chosen for the end of the unloading to avoid entering the compressive regime. These smaller values for the unloading displacement are a direct result of the plastic deformation. The simplest way to stay in the tensile regime ($\sigma \geq 0$) is to

**Table 6.20** Numerical results for monotonic loading with $0 \leq u \leq u_{max}$ for nonlinear hardening material

| inc $n$ | cycle $i$ | $\varepsilon$ | $\sigma^{trial}$ | $\sigma$ | $\varepsilon^{pl}$ | $\varepsilon^{pl}_{eff}$ | $\Delta\lambda$ | $E^{elpl}$ | $\Vert \boldsymbol{v}^{(i+1)} - \boldsymbol{v}^{(i)} \Vert$ |
|---|---|---|---|---|---|---|---|---|---|
| – | – | $10^{-3}$ | MPa | MPa | $10^{-3}$ | $10^{-3}$ | $10^{-3}$ | MPa | – |
| 1 | 1 | 1.0 | 70.0 | 70.0 | 0.0 | 0.0 | 0.0 | 0.0 | 0.0 |
| 2 | 1 | 2.0 | 140.0 | 140.0 | 0.0 | 0.0 | 0.0 | 0.0 | 0.0 |
| 3 | 1 | 3.0 | 210.0 | 210.0 | 0.0 | 0.0 | 0.0 | 0.0 | 0.0 |
| 4 | 1 | 4.0 | 280.0 | 280.0 | 0.0 | 0.0 | 0.0 | 0.0 | 0.0 |
| 5 | 1 | 5.0 | 350.0 | 350.0 | 0.0 | 0.0 | 0.0 | 0.0 | 0.0 |
| 6 | 1 | 6.0 | 420.0 | 360.893 | 0.844 | 0.844 | 0.844 | 10741.746 | 0.0754480 |
| 7 | 1 | 7.0 | 430.893 | 371.559 | 1.692 | 1.692 | 0.848 | 10589.495 | 0.0762223 |
| 8 | 1 | 8.0 | 441.559 | 382.072 | 2.542 | 2.542 | 0.850 | 10436.064 | 0.0768138 |
| 9 | 1 | 9.0 | 452.072 | 392.431 | 3.394 | 3.394 | 0.852 | 10281.438 | 0.0774126 |
| 10 | 1 | 10.0 | 462.431 | 402.635 | 4.248 | 4.248 | 0.854 | 10125.602 | 0.0780191 |
| 11 | 1 | 11.0 | 472.635 | 412.683 | 5.105 | 5.105 | 0.856 | 9968.540 | 0.0786336 |
| 12 | 1 | 12.0 | 482.683 | 422.573 | 5.963 | 5.963 | 0.859 | 9810.235 | 0.0792562 |
| 13 | 1 | 13.0 | 492.573 | 432.304 | 6.824 | 6.824 | 0.861 | 9650.671 | 0.0798871 |
| 14 | 1 | 14.0 | 502.304 | 441.875 | 7.688 | 7.688 | 0.863 | 9489.831 | 0.0805264 |
| 15 | 1 | 15.0 | 511.875 | 451.284 | 8.553 | 8.553 | 0.866 | 9327.698 | 0.0811742 |
| 16 | 1 | 16.0 | 521.284 | 460.531 | 9.421 | 9.421 | 0.868 | 9164.255 | 0.0818308 |
| 17 | 1 | 17.0 | 530.531 | 469.613 | 10.291 | 10.291 | 0.870 | 8999.484 | 0.0824963 |
| 18 | 1 | 18.0 | 539.613 | 478.530 | 11.164 | 11.164 | 0.873 | 8833.367 | 0.0831709 |
| 19 | 1 | 19.0 | 548.530 | 487.280 | 12.039 | 12.039 | 0.875 | 8665.885 | 0.0838547 |
| 20 | 1 | 20.0 | 557.280 | 495.862 | 12.916 | 12.916 | 0.877 | 8497.020 | 0.0845479 |

Case: 20 increments, $t^{v}_{end} = 0.1$

use small displacement increments and to check in each step if the condition is still fulfilled.

The variation of the elasticity modulus and the damage variable are shown in Figs. 6.25 and 6.26. Looking at the development of the elasticity modulus, i.e. Fig. 6.25, it can be seen that the modulus is deleasing with ongoing deformation. The small horizontal plateaus are related to the elastic unloading where the modulus is not changing. Comparing the early and late stages (Fig. 6.25a, b) of deformation, it can be seen that the difference between the hardening and the ideal plastic material is increasing. The trend of the damage variable shown in Fig. 6.26 is inverse compared to the elasticity modulus which is of course the result of the relation between both variable as given in Eq. (2.94).

The development of the plastic strain and the internal hardening variable is shown in Fig. 6.27. At an early stage of the deformation, both variables are practically the same since the influence of the damage is not very significant. In this context, it might be good to review the equations for the update of the internal variable, plastic and equivalent plastic strain under the influence of damage evolution:

**Table 6.21** Numerical results for monotonic loading with $0 \leq u \leq u_{\text{max}}$ for nonlinear hardening material

| inc $n$ | cycle $i$ | $\varepsilon$ | $\sigma^{\text{trial}}$ | $\sigma$ | $\varepsilon^{\text{pl}}$ | $\varepsilon^{\text{pl}}_{\text{eff}}$ | $\Delta\lambda$ | $E^{\text{elpl}}$ | $\lVert \boldsymbol{v}^{(i+1)} - \boldsymbol{v}^{(i)} \rVert$ |
|---|---|---|---|---|---|---|---|---|---|
| – | – | $10^{-3}$ | MPa | MPa | $10^{-3}$ | $10^{-3}$ | $10^{-3}$ | MPa | – |
| 1 | 1 | 1.0 | 70.0 | 70.0 | 0.0 | 0.0 | 0.0 | 0.0 | 0.0 |
| 2 | 1 | 2.0 | 140.0 | 140.0 | 0.0 | 0.0 | 0.0 | 0.0 | 0.0 |
| 3 | 1 | 3.0 | 210.0 | 210.0 | 0.0 | 0.0 | 0.0 | 0.0 | 0.0 |
| 4 | 1 | 4.0 | 280.0 | 280.0 | 0.0 | 0.0 | 0.0 | 0.0 | 0.0 |
| 5 | 1 | 5.0 | 350.0 | 350.0 | 0.0 | 0.0 | 0.0 | 0.0 | 0.0 |
| 6 | 1 | 6.0 | 420.0 | 360.893 | 0.844 | 0.844 | 0.844 | 10741.746 | 0.0754480 |
| | 2 | 6.0 | 420.0 | 360.817 | 0.845 | 0.845 | 0.845 | 10741.553 | 0.000000122931 |
| 7 | 1 | 7.0 | 430.817 | 371.559 | 1.692 | 1.692 | 0.847 | 10589.494 | 0.0760291 |
| | 2 | 7.0 | 430.817 | 371.483 | 1.693 | 1.693 | 0.848 | 10589.299 | 0.000000125153 |
| 8 | 1 | 8.0 | 441.483 | 382.072 | 2.542 | 2.542 | 0.849 | 10436.064 | 0.0766176 |
| | 2 | 8.0 | 441.483 | 381.995 | 2.543 | 2.543 | 0.850 | 10435.865 | 0.000000127426 |
| 9 | 1 | 9.0 | 451.995 | 392.431 | 3.394 | 3.394 | 0.851 | 10281.438 | 0.0772138 |
| | 2 | 9.0 | 451.995 | 392.354 | 3.395 | 3.395 | 0.852 | 10281.237 | 0.000000129753 |
| 10 | 1 | 10.0 | 462.354 | 402.635 | 4.248 | 4.248 | 0.853 | 10125.602 | 0.0778177 |
| | 2 | 10.0 | 462.354 | 402.557 | 4.249 | 4.249 | 0.854 | 10125.398 | 0.000000132134 |
| 11 | 1 | 11.0 | 472.557 | 412.683 | 5.105 | 5.105 | 0.855 | 9968.539 | 0.0784296 |
| | 2 | 11.0 | 472.557 | 412.604 | 5.106 | 5.106 | 0.856 | 9968.333 | 0.000000134573 |
| 12 | 1 | 12.0 | 482.604 | 422.573 | 5.963 | 5.963 | 0.858 | 9810.234 | 0.0790495 |
| | 2 | 12.0 | 482.604 | 422.494 | 5.964 | 5.964 | 0.859 | 9810.025 | 0.000000137069 |
| 13 | 1 | 13.0 | 492.494 | 432.304 | 6.824 | 6.824 | 0.860 | 9650.670 | 0.0796777 |
| | 2 | 13.0 | 492.494 | 432.224 | 6.825 | 6.825 | 0.861 | 9650.458 | 0.000000139625 |
| 14 | 1 | 14.0 | 502.224 | 441.875 | 7.688 | 7.688 | 0.862 | 9489.830 | 0.0803142 |
| | 2 | 14.0 | 502.224 | 441.794 | 7.689 | 7.689 | 0.863 | 9489.616 | 0.000000142243 |
| 15 | 1 | 15.0 | 511.794 | 451.284 | 8.553 | 8.553 | 0.864 | 9327.697 | 0.0809592 |
| | 2 | 15.0 | 511.794 | 451.203 | 8.554 | 8.554 | 0.866 | 9327.480 | 0.000000144924 |
| 16 | 1 | 16.0 | 521.203 | 460.530 | 9.421 | 9.421 | 0.867 | 9164.254 | 0.0816129 |
| | 2 | 16.0 | 521.203 | 460.449 | 9.421 | 9.421 | 0.868 | 9164.034 | 0.000000147671 |
| 17 | 1 | 17.0 | 530.449 | 469.613 | 10.291 | 10.291 | 0.869 | 8999.483 | 0.0822754 |
| | 2 | 17.0 | 530.449 | 469.530 | 10.292 | 10.292 | 0.870 | 8999.260 | 0.000000150484 |
| 18 | 1 | 18.0 | 539.530 | 478.530 | 11.164 | 11.164 | 0.871 | 8833.366 | 0.0829470 |
| | 2 | 18.0 | 539.530 | 478.447 | 11.165 | 11.165 | 0.873 | 8833.140 | 0.000000153368 |
| 19 | 1 | 19.0 | 548.447 | 487.280 | 12.039 | 12.039 | 0.874 | 8665.884 | 0.0836277 |
| | 2 | 19.0 | 548.447 | 487.196 | 12.040 | 12.040 | 0.875 | 8665.655 | 0.000000156322 |
| 20 | 1 | 20.0 | 557.196 | 495.862 | 12.916 | 12.916 | 0.876 | 8497.020 | 0.0843178 |
| | 2 | 20.0 | 557.196 | 495.778 | 12.917 | 12.917 | 0.877 | 8496.787 | 0.000000159350 |

Case: 20 increments, $t^{v}_{\text{end}} = 0.001$ and 0.00001

**Table 6.22** Numerical results for monotonic loading with $0 \le u \le u_{max}$ for nonlinear hardening material

| inc $n$ | cycle $i$ | $\varepsilon$ | $\sigma^{trial}$ | $\sigma$ | $\varepsilon^{pl}$ | $\varepsilon^{pl}_{eff}$ | $\Delta\lambda$ | $E^{elpl}$ | $\Vert \boldsymbol{v}^{(i+1)} - \boldsymbol{v}^{(i)} \Vert$ |
|---|---|---|---|---|---|---|---|---|---|
| – | – | $10^{-3}$ | MPa | MPa | $10^{-3}$ | $10^{-3}$ | $10^{-3}$ | MPa | – |
| 1 | 1 | 1.0 | 70.0 | 70.0 | 0.0 | 0.0 | 0.0 | 0.0 | 0.0 |
| 2 | 1 | 2.0 | 140.0 | 140.0 | 0.0 | 0.0 | 0.0 | 0.0 | 0.0 |
| 3 | 1 | 3.0 | 210.0 | 210.0 | 0.0 | 0.0 | 0.0 | 0.0 | 0.0 |
| 4 | 1 | 4.0 | 280.0 | 280.0 | 0.0 | 0.0 | 0.0 | 0.0 | 0.0 |
| 5 | 1 | 5.0 | 350.0 | 350.0 | 0.0 | 0.0 | 0.0 | 0.0 | 0.0 |
| 6 | 1 | 6.0 | 420.0 | 360.893 | 0.844 | 0.844 | 0.844 | 10741.746 | 0.0754480 |
| | 2 | 6.0 | 420.0 | 360.817 | 0.845 | 0.845 | 0.845 | 10741.553 | 0.000000122931 |
| | 3 | 6.0 | 420.0 | 360.817 | 0.845 | 0.845 | 0.845 | 10741.553 | $0.537078 \times 10^{-18}$ |
| 7 | 1 | 7.0 | 430.817 | 371.559 | 1.692 | 1.692 | 0.847 | 10589.494 | 0.0760291 |
| | 2 | 7.0 | 430.817 | 371.483 | 1.693 | 1.693 | 0.848 | 10589.299 | 0.000000125153 |
| | 3 | 7.0 | 430.817 | 371.483 | 1.693 | 1.693 | 0.848 | 10589.299 | $0.529465 \times 10^{-18}$ |
| 8 | 1 | 8.0 | 441.483 | 382.072 | 2.542 | 2.542 | 0.849 | 10436.064 | 0.0766176 |
| | 2 | 8.0 | 441.483 | 381.995 | 2.543 | 2.543 | 0.850 | 10435.865 | 0.000000127426 |
| | 3 | 8.0 | 441.483 | 381.995 | 2.543 | 2.543 | 0.850 | 10435.865 | $0.850916 \times 10^{-17}$ |
| 9 | 1 | 9.0 | 451.995 | 392.431 | 3.394 | 3.394 | 0.851 | 10281.438 | 0.0772138 |
| | 2 | 9.0 | 451.995 | 392.354 | 3.395 | 3.395 | 0.852 | 10281.237 | 0.000000129753 |
| | 3 | 9.0 | 451.995 | 392.354 | 3.395 | 3.395 | 0.852 | 10281.237 | $0.853125 \times 10^{-17}$ |
| 10 | 1 | 10.0 | 462.354 | 402.635 | 4.248 | 4.248 | 0.853 | 10125.602 | 0.0778177 |
| | 2 | 10.0 | 462.354 | 402.557 | 4.249 | 4.249 | 0.854 | 10125.398 | 0.000000132134 |
| | 3 | 10.0 | 462.354 | 402.557 | 4.249 | 4.249 | 0.854 | 10125.398 | $0.300000 \times 10^{-22}$ |
| 11 | 1 | 11.0 | 472.557 | 412.683 | 5.105 | 5.105 | 0.855 | 9968.539 | 0.0784296 |
| | 2 | 11.0 | 472.557 | 412.604 | 5.106 | 5.106 | 0.856 | 9968.333 | 0.000000134573 |
| | 3 | 11.0 | 472.557 | 412.604 | 5.106 | 5.106 | 0.856 | 9968.333 | $0.100000 \times 10^{-22}$ |
| 12 | 1 | 12.0 | 482.604 | 422.573 | 5.963 | 5.963 | 0.858 | 9810.234 | 0.0790495 |
| | 2 | 12.0 | 482.604 | 422.494 | 5.964 | 5.964 | 0.859 | 9810.025 | 0.000000137069 |
| | 3 | 12.0 | 482.604 | 422.494 | 5.964 | 5.964 | 0.859 | 9810.025 | $0.100000 \times 10^{-22}$ |
| 13 | 1 | 13.0 | 492.494 | 432.304 | 6.824 | 6.824 | 0.860 | 9650.670 | 0.0796777 |
| | 2 | 13.0 | 492.494 | 432.224 | 6.825 | 6.825 | 0.861 | 9650.458 | 0.000000139625 |
| | 3 | 13.0 | 492.494 | 432.224 | 6.825 | 6.825 | 0.861 | 9650.458 | $0.862136 \times 10^{-17}$ |
| 14 | 1 | 14.0 | 502.224 | 441.875 | 7.688 | 7.688 | 0.862 | 9489.830 | 0.0803142 |
| | 2 | 14.0 | 502.224 | 441.794 | 7.689 | 7.689 | 0.863 | 9489.616 | 0.000000142243 |
| | 3 | 14.0 | 502.224 | 441.794 | 7.689 | 7.689 | 0.863 | 9489.616 | $0.300000 \times 10^{-22}$ |
| 15 | 1 | 15.0 | 511.794 | 451.284 | 8.553 | 8.553 | 0.864 | 9327.697 | 0.0809592 |
| | 2 | 15.0 | 511.794 | 451.203 | 8.554 | 8.554 | 0.866 | 9327.480 | 0.000000144924 |
| | 3 | 15.0 | 511.794 | 451.203 | 8.554 | 8.554 | 0.866 | 9327.480 | $0.300000 \times 10^{-22}$ |

(continued)

**Table 6.22** (continued)

| inc $n$ | cycle $i$ | $\varepsilon$ | $\sigma^{\text{trial}}$ | $\sigma$ | $\varepsilon^{\text{pl}}$ | $\varepsilon^{\text{pl}}_{\text{eff}}$ | $\Delta\lambda$ | $E^{\text{elpl}}$ | $\Vert \boldsymbol{v}^{(i+1)} - \boldsymbol{v}^{(i)} \Vert$ |
|---|---|---|---|---|---|---|---|---|---|
| – | – | $10^{-3}$ | MPa | MPa | $10^{-3}$ | $10^{-3}$ | $10^{-3}$ | MPa | – |
| 16 | 1 | 16.0 | 521.203 | 460.530 | 9.421 | 9.421 | 0.867 | 9164.254 | 0.0816129 |
| | 2 | 16.0 | 521.203 | 460.449 | 9.421 | 9.421 | 0.868 | 9164.034 | 0.000000147671 |
| | 3 | 16.0 | 521.203 | 460.449 | 9.421 | 9.421 | 0.868 | 9164.034 | $0.869085 \times 10^{-17}$ |
| 17 | 1 | 17.0 | 530.449 | 469.613 | 10.291 | 10.291 | 0.869 | 8999.483 | 0.0822754 |
| | 2 | 17.0 | 530.449 | 469.530 | 10.292 | 10.292 | 0.870 | 8999.260 | 0.000000150484 |
| | 3 | 17.0 | 530.449 | 469.530 | 10.292 | 10.292 | 0.870 | 8999.260 | $0.359970 \times 10^{-17}$ |
| 18 | 1 | 18.0 | 539.530 | 478.530 | 11.164 | 11.164 | 0.871 | 8833.366 | 0.0829470 |
| | 2 | 18.0 | 539.530 | 478.447 | 11.165 | 11.165 | 0.873 | 8833.140 | 0.000000153368 |
| | 3 | 18.0 | 539.530 | 478.447 | 11.165 | 11.165 | 0.873 | 8833.140 | $0.608818 \times 10^{-17}$ |
| 19 | 1 | 19.0 | 548.447 | 487.280 | 12.039 | 12.039 | 0.874 | 8665.884 | 0.0836277 |
| | 2 | 19.0 | 548.447 | 487.196 | 12.040 | 12.040 | 0.875 | 8665.655 | 0.000000156322 |
| | 3 | 19.0 | 548.447 | 487.196 | 12.040 | 12.040 | 0.875 | 8665.655 | $0.529579 \times 10^{-17}$ |
| 20 | 1 | 20.0 | 557.196 | 495.862 | 12.916 | 12.916 | 0.876 | 8497.020 | 0.0843178 |
| | 2 | 20.0 | 557.196 | 495.778 | 12.917 | 12.917 | 0.877 | 8496.787 | 0.000000159350 |
| | 3 | 20.0 | 557.196 | 495.778 | 12.917 | 12.917 | 0.877 | 8496.787 | $0.169936 \times 10^{-17}$ |

Case: 20 increments, $t^{v}_{\text{end}} = 0.0000001$

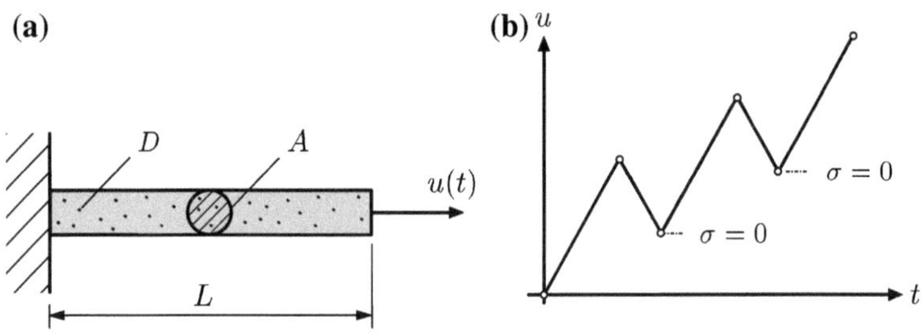


**Fig. 6.21** Reversed loading of an elastic-plastic rod element under damage effects in the pure tensile regime ($\sigma \geq 0$): **a** configuration and **b** displacement-time diagram

**Table 6.23** Material parameters used for the simulation, adapted from [11, 74]

| Quantity | Value |
|---|---|
| $E$ in MPa | 210000 |
| $\nu$ – | 0.3 |
| Hardening curve in MPa | $k(\kappa) = 620 + 3300\,[1 - \exp(-0.4\kappa)]$ |
| $s$ – | 1.0 |
| $r$ in MPa | 3.5 |

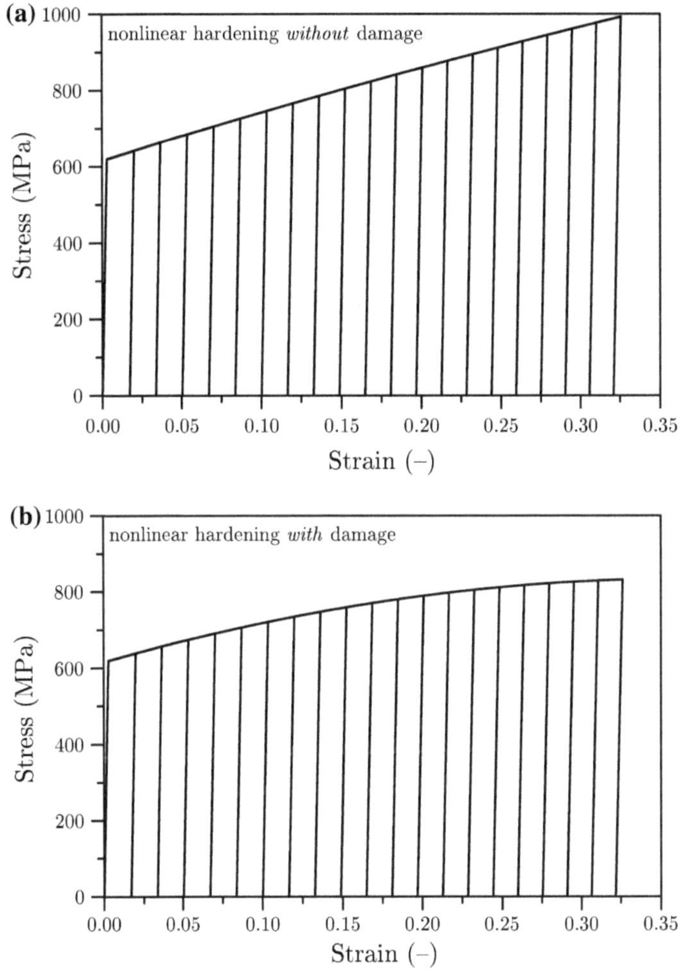


**Fig. 6.22** Stress-strain diagram for nonlinear hardening: **a** without damage and **b** with damage

$$\kappa_{n+1} = \kappa_n + \Delta\lambda_{n+1}\,, \tag{6.154}$$

$$\varepsilon^{\text{pl}}_{n+1} = \varepsilon^{\text{pl}}_n + \frac{\Delta\lambda_{n+1}}{1 - D_{n+1}}\text{sgn}(\sigma_{n+1})\,, \tag{6.155}$$

$$\varepsilon^{\text{pl}}_{\text{eff},n+1} = \varepsilon^{\text{pl}}_{\text{eff},n} + \frac{\Delta\lambda_{n+1}}{1 - D_{n+1}}\,. \tag{6.156}$$

Important is here if we compare this approach with the classical description without damage that the internal hardening variable and the equivalent plastic strain are different if $D \neq 0$. This can be clearly seen in a late stage of the deformation (cf. Fig. 6.27b) where the plastic strain is larger than the internal hardening variable.

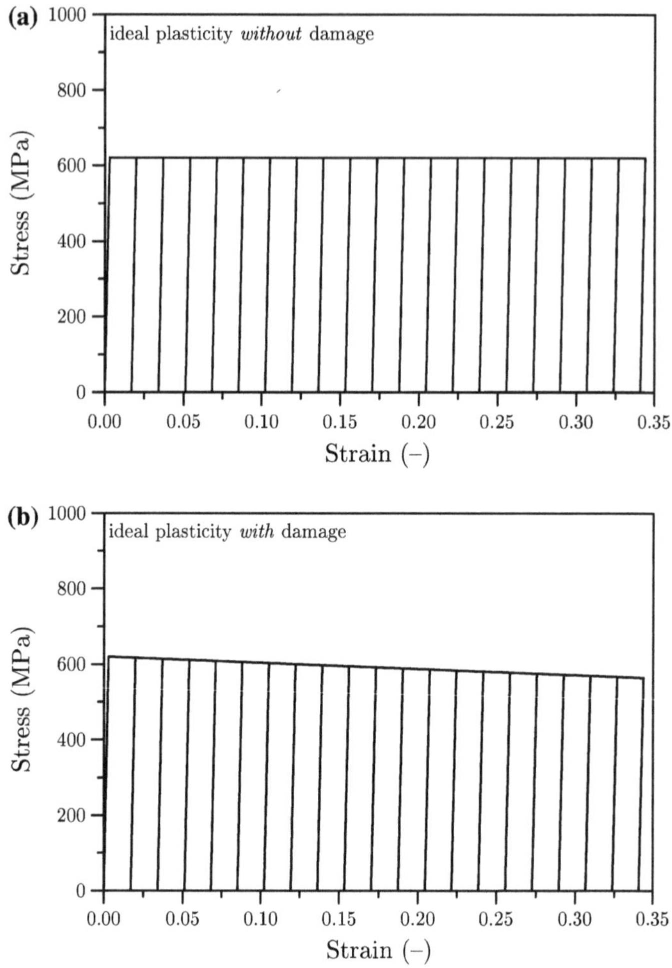


**Fig. 6.23** Stress-strain diagram for ideal plasticity: **a** without damage and **b** with damage

Comparing the hardening material with the ideal plastic behavior, it can be concluded that the increase is much higher for the ideal plastic case for the chosen axes.

The evolution of the damage variable with increasing plastic strain is shown in Fig. 6.28 where a quite different behavior is obtained for the hardening and the ideal plastic material. The ideal plastic material shows a practically linear increase of the damage variable whereas the hardening material tends to exponentially increase which clearly increases the difference for larger deformations. It should be noted here the functional dependency is given in terms of the plastic strain component. Fortunately, this plastic strain component is equal to the equivalent plastic strain and conclusions can be transferred to the general three-dimensional case at least to a certain extend.

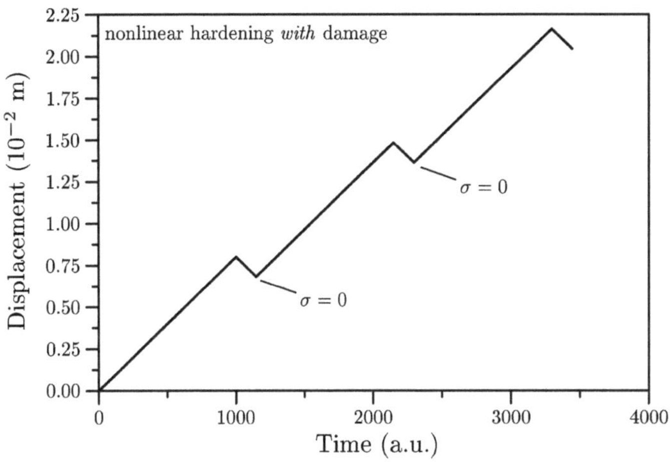


**Fig. 6.24** Displacement boundary condition as a function of time

## 6.8 Example: Monotonic tensile loading of an elastic-plastic rod element under damage effects (Gurson)

Consider a pure one-dimensional tensile test as schematically shown in Fig. 2.11 where no effects from contraction or clamping of the specimen are considered. Damage can develop according to the model by GURSON. Use the matrix solution scheme as given in Sect. 6.5.2 to consider a monotonic tensile load in the form of an end displacement boundary condition to simulate a tensile test with $\sigma \geq 0$. Staying in the tensile regime implies that any effects of kinematic hardening can be disregarded for this material. The elastic range is characterized by a YOUNG's modulus of $E = 70000$ MPa and the plastic range by a quadratic flow curve in the form of $k(\kappa) = \left(350 + 12900\kappa - 1.25 \times 10^5\kappa^2\right)$ MPa where the equivalent plastic strain $|\varepsilon^{\text{pl}}|$ should be used as the hardening variable. Determine for each iteration step and cycle the variables $\varepsilon$, $\sigma^{\text{trial}}$, $\sigma$, $\varepsilon^{\text{pl}}_{\text{eff}}$, $\Delta\lambda$, $D$, $\|\boldsymbol{v}^{(i+1)} - \boldsymbol{v}^{(i)}\|$ for the case of

- 20 equidistant increments with a strain inlement for each load step of $|\Delta\varepsilon_n| = 0.001$,
- a termination criterion $t^v_{\text{end}}$ of 0.00001 for the vector norm $\|\boldsymbol{v}^{(i+1)} - \boldsymbol{v}^{(i)}\|$ during the iteration
- and the three different cases of initial damage $D_0 = 0.0$, 0.001 and 0.01.

**6.8 Solution**

The results for the case $D_0 = 0.0$ can be basically taken from Table 6.21, at least for the last cycle in each increment. Using the iteration scheme for the GURSON model from Sect. 6.5.2 with initial damage $D_0 = 0.0$ gives an iteration for a VON MISES material where *no* damage is generated during the deformation. Since the basic definitions for the material in Example 6.6 and the here used formulations

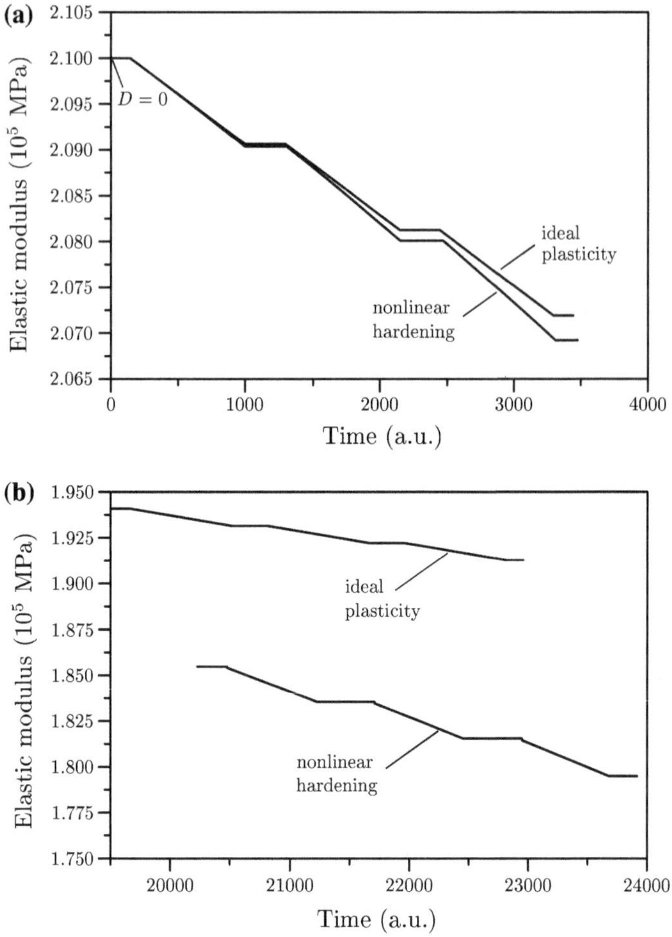


**Fig. 6.25** Elastic modulus as a function of time: **a** at the beginning of the load cycles; **b** at the end of the load cycles

are slightly different (i.e. the normalization), different values are obtained for the consistency parameter and the number of cycles till convergence is achieved.

The results for $D_0 = 0.001$ are presented in Table 6.24. The first four increments are pure elastic and the result is obtained with a single cycle for these steps. In the elasto-plastic range ($n \geq 5$), the chosen convergence criteria $t^v_{\text{end}} = 0.00001$ requires three ($n = 5$) or four ($n \geq 6$) cycles within each iteration. Furthermore, it can be seen that the damage is increasing with increasing deformation and the graphical representation shown in Fig. 6.29 indicates that the damage inleases practically in a linear manner for the chosen material parameters.

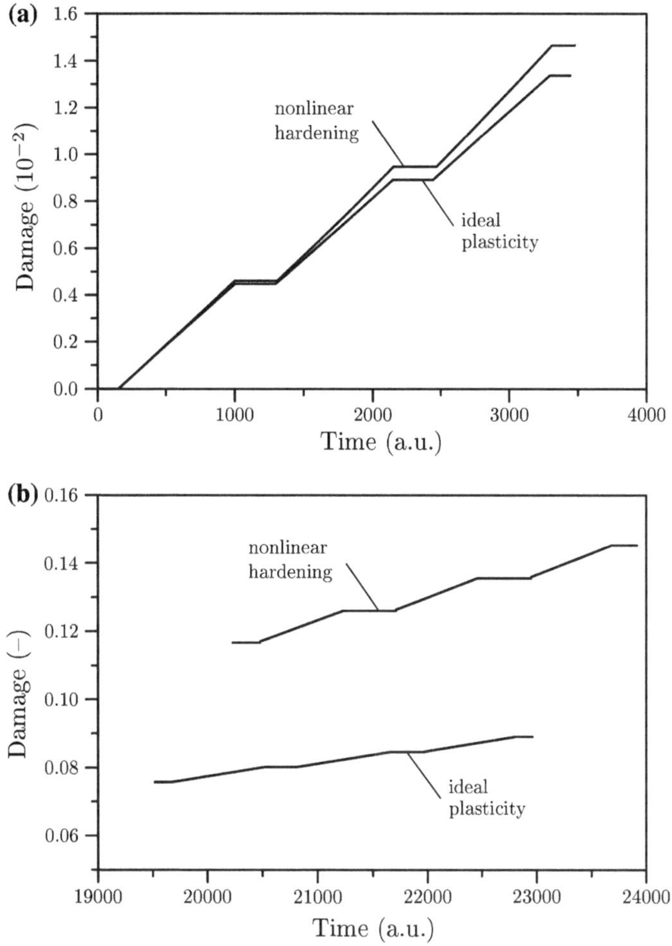


**Fig. 6.26** Damage variable as a function of time: **a** at the beginning of the load cycles; **b** at the end of the load cycles

The graphical representations of the stress-strain diagrams is given in Fig. 6.30 where it can be seen that damage results in lower stress values compared to the undamaged von Mises model. However, the difference is quite small under the chosen model parameters. The von Mises results and the Gurson approach for $D_0 = 0.001$ show practically no difference in the chosen scale of Fig. 6.30.

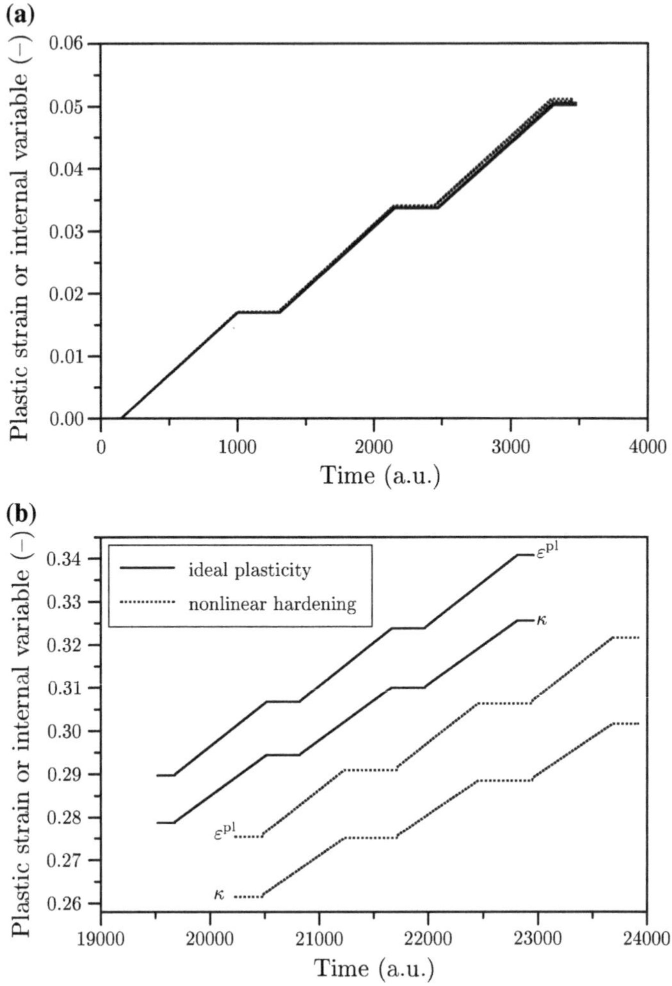


**Fig. 6.27** Plastic strain and internal variable as a function of time: **a** at the beginning of the load cycles; **b** at the end of the load cycles

## 6.7 Supplementary Problems

**6.9 Plastic modulus and elasto-plastic material modulus**
Discuss the case (a) $E^{pl} = E$ and (b) $E^{elpl} = E$.

**6.10 Back projection at linear hardening**
Calculate Example 6.1 for the following linear flow curve of a steal: $k(\kappa) = (690 + 21000\kappa)$ MPa. The modulus of elasticity equals 210,000 MPa. The geometric dimensions of Example 6.1 can be assumed.

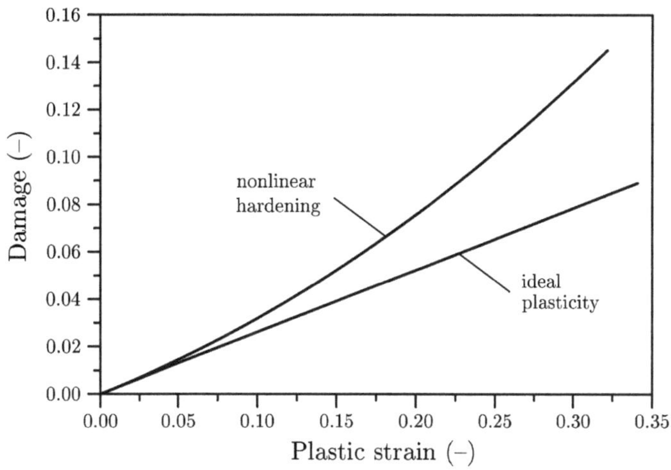


**Fig. 6.28** Damage variable as a function of plastic strain

**Table 6.24** Numerical results for monotonic loading with $|\Delta\varepsilon_n| = 0.001$ for a nonlinear hardening material and the GURSON damage model

| inc $n$ | cycle $i$ | $\varepsilon$ | $\sigma^{\text{trial}}$ | $\sigma$ | $\varepsilon^{\text{pl}}_{\text{eff}}$ | $\Delta\lambda$ | $D$ | $\Vert\boldsymbol{v}^{(i+1)} - \boldsymbol{v}^{(i)}\Vert$ |
|---|---|---|---|---|---|---|---|---|
| – | – | $10^{-3}$ | MPa | MPa | $10^{-3}$ | $10^{-3}$ | $10^{-2}$ | – |
| 1 | 1 | 1.0 | 70.0 | 70.0 | 0.0 | 0.0 | 0.1 | 0.0 |
| 2 | 1 | 2.0 | 140.0 | 140.0 | 0.0 | 0.0 | 0.1 | 0.0 |
| 3 | 1 | 3.0 | 210.0 | 210.0 | 0.0 | 0.0 | 0.1 | 0.0 |
| 4 | 1 | 4.0 | 280.0 | 280.0 | 0.0 | 0.0 | 0.1 | 0.0 |
| 5 | 1 | 5.0 | 350.0 | 349.988 | 0.000168 | 0.0294 | 0.1 | 0.321193 |
| | 2 | 5.0 | 350.0 | 349.667 | 0.00476 | 0.832 | 0.100000370 | 0.000285830 |
| | 3 | 5.0 | 350.0 | 349.666 | 0.00476 | 0.834 | 0.100000372 | 0.000000000137 |
| 6 | 1 | 6.0 | 419.667 | 417.752 | 0.0376 | 3.993 | 0.100000372 | 49.338 |
| | 2 | 6.0 | 419.667 | 368.413 | 0.874 | 108.338 | 0.10000944 | 7.825 |
| | 3 | 6.0 | 419.667 | 360.588 | 0.853 | 151.527 | 0.1000914 | 0.117 |
| | 4 | 6.0 | 419.667 | 360.471 | 0.850 | 152.720 | 0.1000723 | 0.00000537 |
| 7 | 1 | 7.0 | 430.471 | 428.577 | 0.883 | 4.092 | 0.1000723 | 49.734 |
| | 2 | 7.0 | 430.471 | 378.844 | 1.721 | 112.999 | 0.1000829 | 7.612 |
| | 3 | 7.0 | 430.471 | 371.231 | 1.701 | 156.506 | 0.100164 | 0.106 |
| | 4 | 7.0 | 430.471 | 371.124 | 1.698 | 157.637 | 0.100143 | 0.00000704 |
| 8 | 1 | 8.0 | 441.124 | 439.248 | 1.739 | 4.192 | 0.100143 | 50.114 |
| | 2 | 8.0 | 441.124 | 389.135 | 2.572 | 117.641 | 0.100156 | 7.413 |
| | 3 | 8.0 | 441.124 | 381.723 | 2.551 | 161.442 | 0.100236 | 0.0978 |
| | 4 | 8.0 | 441.124 | 381.625 | 2.548 | 162.514 | 0.100215 | 0.00000819 |

(continued)

**Table 6.24** (continued)

| inc $n$ | cycle $i$ | $\varepsilon$ | $\sigma^{\text{trial}}$ | $\sigma$ | $\varepsilon_{\text{eff}}^{\text{pl}}$ | $\Delta\lambda$ | $D$ | $\\|\boldsymbol{v}^{(i+1)} - \boldsymbol{v}^{(i)}\\|$ |
|---|---|---|---|---|---|---|---|---|
| – | – | $10^{-3}$ | MPa | MPa | $10^{-3}$ | $10^{-3}$ | $10^{-2}$ | – |
| 9 | 1 | 9.0 | 451.625 | 449.764 | 2.579 | 4.295 | 0.100215 | 50.480 |
| | 2 | 9.0 | 451.625 | 399.284 | 3.424 | 122.262 | 0.100229 | 7.223 |
| | 3 | 9.0 | 451.625 | 392.061 | 3.403 | 166.333 | 0.100308 | 0.0899 |
| | 4 | 9.0 | 451.625 | 391.971 | 3.400 | 167.353 | 0.100286 | 0.00000896 |
| 10 | 1 | 10.0 | 461.971 | 460.123 | 3.431 | 4.399 | 0.100286 | 50.834 |
| | 2 | 10.0 | 461.971 | 409.288 | 4.278 | 126.860 | 0.100301 | 7.043 |
| | 3 | 10.0 | 461.971 | 402.245 | 4.257 | 171.180 | 0.100380 | 0.0828 |
| | 4 | 10.0 | 461.971 | 402.163 | 5.254 | 172.150 | 0.100358 | 0.00000945 |
| 11 | 1 | 11.0 | 472.162 | 470.324 | 4.285 | 4.506 | 0.100358 | 51.178 |
| | 2 | 11.0 | 472.162 | 419.147 | 5.135 | 131.433 | 0.100374 | 6.873 |
| | 3 | 11.0 | 472.162 | 412.274 | 5.113 | 175.983 | 0.100451 | 0.0765 |
| | 4 | 11.0 | 472.162 | 412.197 | 5.111 | 176.907 | 0.100429 | 0.00000973 |
| 12 | 1 | 12.0 | 482.197 | 480.367 | 5.141 | 4.615 | 0.100429 | 51.511 |
| | 2 | 12.0 | 482.197 | 428.857 | 5.994 | 135.981 | 0.100446 | 6.711 |
| | 3 | 12.0 | 482.197 | 422.145 | 5.972 | 180.741 | 0.100523 | 0.0708 |
| | 4 | 12.0 | 482.197 | 422.075 | 5.970 | 181.623 | 0.100499 | 0.00000985 |
| 13 | 1 | 13.0 | 492.075 | 490.250 | 6.000 | 4.727 | 0.100499 | 51.835 |
| | 2 | 13.0 | 492.075 | 438.416 | 6.855 | 140.500 | 0.100518 | 6.557 |
| | 3 | 13.0 | 492.075 | 431.859 | 6.833 | 185.453 | 0.100594 | 0.0656 |
| | 4 | 13.0 | 492.075 | 431.793 | 6.831 | 186.296 | 0.100570 | 0.00000985 |
| 14 | 1 | 14.0 | 501.793 | 499.972 | 6.861 | 4.842 | 0.100570 | 52.150 |
| | 2 | 14.0 | 501.793 | 447.823 | 7.719 | 144.991 | 0.100590 | 6.411 |
| | 3 | 14.0 | 501.793 | 441.413 | 7.696 | 190.120 | 0.100665 | 0.0610 |
| | 4 | 14.0 | 501.793 | 441.352 | 7.694 | 190.926 | 0.100641 | 0.00000977 |
| 15 | 1 | 15.0 | 511.352 | 509.532 | 7.724 | 4.960 | 0.100641 | 52.456 |
| | 2 | 15.0 | 511.352 | 457.076 | 8.585 | 149.450 | 0.100662 | 6.271 |
| | 3 | 15.0 | 511.352 | 450.805 | 8.562 | 194.740 | 0.100737 | 0.0568 |
| | 4 | 15.0 | 511.352 | 450.749 | 8.560 | 195.512 | 0.100712 | 0.00000962 |
| 16 | 1 | 16.0 | 520.749 | 518.929 | 8.590 | 5.082 | 0.100712 | 52.756 |
| | 2 | 16.0 | 520.749 | 466.173 | 9.453 | 153.878 | 0.100734 | 6.137 |
| | 3 | 16.0 | 520.749 | 460.036 | 9.429 | 199.313 | 0.100808 | 0.0530 |
| | 4 | 16.0 | 520.749 | 459.983 | 9.428 | 200.053 | 0.100783 | 0.00000942 |
| 17 | 1 | 17.0 | 529.983 | 528.160 | 9.458 | 5.208 | 0.100783 | 53.048 |
| | 2 | 17.0 | 529.983 | 475.112 | 10.323 | 158.272 | 0.100806 | 6.010 |
| | 3 | 17.0 | 529.983 | 469.103 | 10.300 | 203.838 | 0.100879 | 0.0495 |
| | 4 | 17.0 | 529.983 | 469.053 | 10.298 | 204.549 | 0.100854 | 0.00000920 |

(continued)

**Table 6.24** (continued)

| inc $n$ | cycle $i$ | $\varepsilon$ | $\sigma^{\text{trial}}$ | $\sigma$ | $\varepsilon_{\text{eff}}^{\text{pl}}$ | $\Delta\lambda$ | $D$ | $\Vert\boldsymbol{v}^{(i+1)} - \boldsymbol{v}^{(i)}\Vert$ |
|---|---|---|---|---|---|---|---|---|
| – | – | $10^{-3}$ | MPa | MPa | $10^{-3}$ | $10^{-3}$ | $10^{-2}$ | – |
| 18 | 1 | 18.0 | 539.053 | 537.226 | 10.328 | 5.338 | 0.100854 | 53.334 |
| | 2 | 18.0 | 539.053 | 483.892 | 11.196 | 162.631 | 0.100878 | 5.887 |
| | 3 | 18.0 | 539.053 | 478.005 | 11.172 | 208.316 | 0.100950 | 0.0463 |
| | 4 | 18.0 | 539.053 | 477.958 | 11.171 | 208.998 | 0.100925 | 0.00000895 |
| 19 | 1 | 19.0 | 547.958 | 546.124 | 11.200 | 5.473 | 0.100925 | 53.614 |
| | 2 | 19.0 | 547.958 | 492.510 | 12.071 | 166.954 | 0.100949 | 5.770 |
| | 3 | 19.0 | 547.958 | 486.740 | 12.047 | 212.744 | 0.101021 | 0.0434 |
| | 4 | 19.0 | 547.958 | 486.697 | 12.045 | 213.400 | 0.100996 | 0.00000868 |
| 20 | 1 | 20.0 | 556.697 | 554.854 | 12.076 | 5.613 | 0.100996 | 53.888 |
| | 2 | 20.0 | 556.697 | 500.966 | 12.949 | 171.239 | 0.101022 | 5.658 |
| | 3 | 20.0 | 556.697 | 495.308 | 12.925 | 217.122 | 0.101093 | 0.0408 |
| | 4 | 20.0 | 556.697 | 495.267 | 12.923 | 217.754 | 0.101067 | 0.00000840 |

Case: 20 increments, $D_0 = 0.001$, $t_{\text{end}}^{v} = 0.00001$

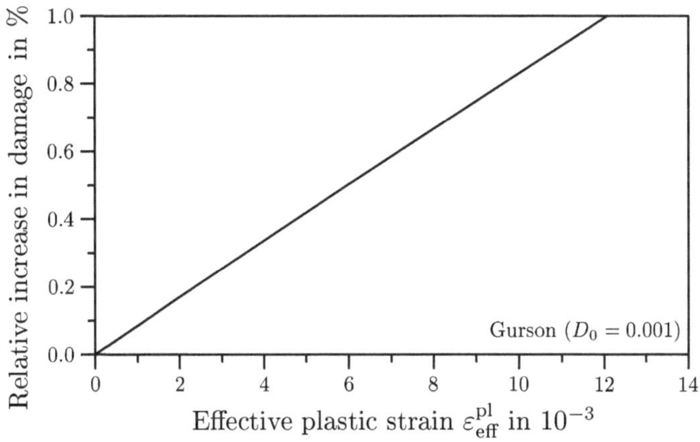


**Fig. 6.29** Relative increase of damage $\frac{D-D_0}{D_0} \times 100$ as a function of effective plastic strain

(a) For 10 increments with $\Delta\varepsilon = 0.001$,
(b) For 20 increments with $\Delta\varepsilon = 0.0005$,
(c) For 20 increments with $\Delta\varepsilon = 0.001$.

Compare and interpret all the results.

**6.11 Back projection at nonlinear hardening**

Calculate Example 6.1 for the following nonlinear flow curve: $k(\kappa) = (350 + 12900\kappa - 1.25 \times 10^5 \kappa^2)$ MPa. All other parameters can be taken as in Example 6.1.

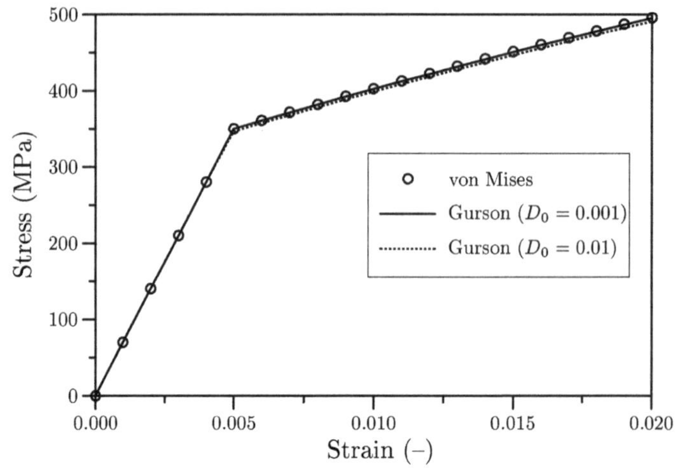


**Fig. 6.30** Stress-strain diagrams for different yield conditions, i.e. VON MISES and GURSON

**6.12 Back projection for rod at fixed support at both ends**

Calculate for the illustrated rod in Fig. 6.31 with fixed support at both ends, the displacement of the point of load application. The rod has an elasto-plastic material behavior ($E = 1 \times 10^5$ MPa; $E^{\text{elpl}} = 1 \times 10^3$ MPa; $k_{\text{t}}^{\text{init}} = 200$ MPa) and a force of $F = 6 \times 10^4$ N is applied equally in 3 increments. It can be assumed that the material behavior is identical under tensile and compression loading. Calculate the displacement of the point of load application and determine the stress and strain in both elements.

As convergence criteria, an absolute displacement difference at the point of load application of $10^{-5}$ mm can be given.

**6.13 Back projection for a finite element at ideal-plastic material behavior**

Discuss the case of one single finite element with force boundary condition at ideal-plastic material behavior. It can be assumed thereby that the applied force increases

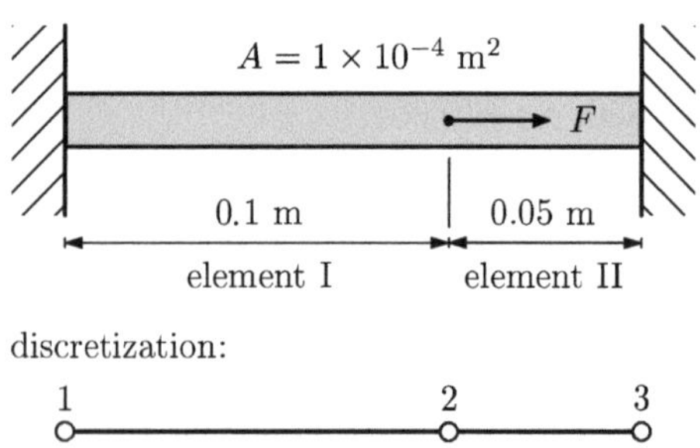


**Fig. 6.31** Sample problem back projection for a rod with fixed supports at both ends

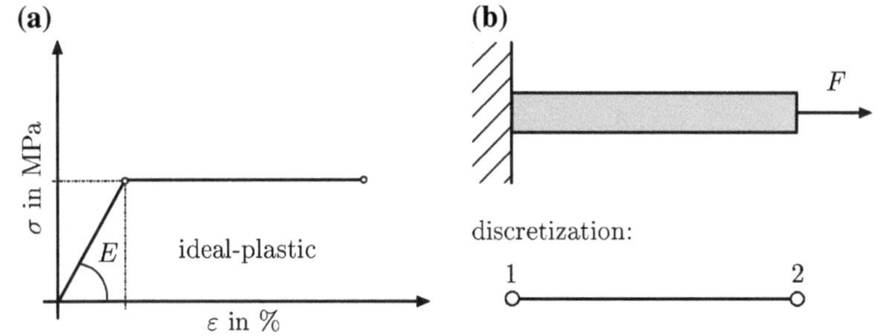


**Fig. 6.32** Sample problem back projection for a finite element in the case of ideal-plastic material behavior

linearly starting with zero. The problem and the material behavior are schematically illustrated in Fig. 6.32. Why is no convergence achieved in the plastic region at the force boundary condition? What changes if the force boundary condition is substituted by a displacement boundary condition?

### 6.14 Back projection for a von Mises material

Given is the following form of the one-dimensional VON MISES yield condition: $F = \left(\frac{|\sigma|}{k_t}\right)^2 - 1 = 0$.

Derive the fully implicit backward-EULER algorithm for isotropic (strain) hardening.

### 6.15 Back projection for a nonlinear hardening material

Given is a one-dimensional continuum rod of length $L = 0.4$ m and cross-sectional area $A = 1 \times 10^{-4}\,\text{m}^2$. The left-hand side is fixed and the right-hand side is displaced in 10 equal steps up to a value of $8 \times 10^{-3}$ m. YOUNG's modulus is equal to 70,000 MPa and the flow curve is described by $k(\varepsilon^{\text{pl}}) = c_1 \times \left(1 - c_2 \text{e}^{-n\varepsilon^{\text{pl}}}\right)$ where $c_1 = 800.0$ MPa, $c_2 = 9/16 = 0.5625$ and $n = 30.0$. Use the fully implicit backward-EULER algorithm for isotropic hardening to calculate all the state variables $(\sigma, \varepsilon, \varepsilon^{\text{pl}}, \varepsilon^{\text{pl}}_{\text{eff}}, E^{\text{elpl}})$ in each increment ($t^v_{\text{end}} = 0.001$).

### 6.16 Cyclic loading of an elastic-plastic continuum rod made of steel

Consider a pure one-dimensional tensile test as schematically shown in Fig. 6.33a. Use the matrix solution scheme as given in Sect. 6.2 to simulate a cycling load in the form of an end displacement boundary condition (cf. Fig. 6.33b) which is applied in 20 equidistant increments ($u_{\text{max}} = 0.002$ m). The cross-sectional area is $A = 10^{-4}$ $\text{m}^2$ and the length equals $L = 0.4$ m. The elastic range is characterized by YOUNG's modulus $E = 210000$ MPa and the plastic range should be modeled as

(a) an ideal plastic material with a yield stress of $k = 690$ MPa or
(b) a linear hardening material with $k^{\text{init}} = 690$ MPa and $E^{\text{pl}} = 21000$ MPa.

Assume that the material behavior is the same in the tensile and compressive regime of the isotropic hardening material to determine all the state variables $(\sigma, \varepsilon, \varepsilon^{\text{pl}}, \varepsilon^{\text{pl}}_{\text{eff}}, E^{\text{elpl}})$ in each inlement.

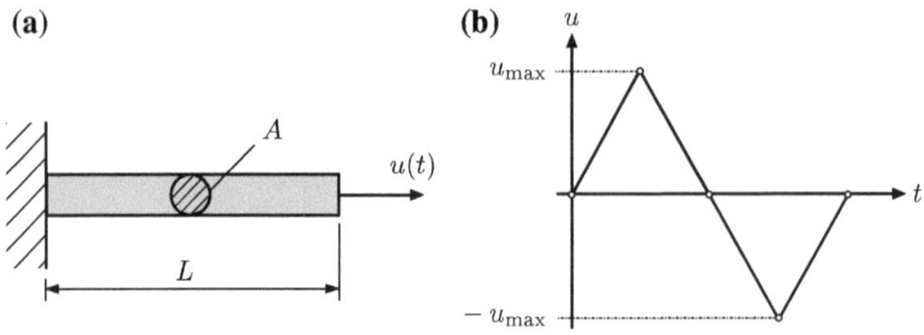


**Fig. 6.33** Cyclic loading of an elastic-plastic continuum rod: **a** configuration and **b** applied displacement boundary condition

**6.17 Back projection for a Gurson material with isotropic hardening**
Extend the derivation of the fully implicit backward-EULER scheme for the GURSON model as given in Sect. 6.5.2 to the case of isotropic hardening, i.e. $k_t = k_t(\kappa)$.

**6.18 Monotonic tensile loading of a linear-elastic ideal-plastic rod element under damage effects (Gurson)**
Consider a pure one-dimensional tensile test as schematically shown in Fig. 2.11 where no effects from contraction or clamping of the specimen are considered. Damage can develop according to the GURSON model. Use the matrix solution scheme as given in Sect. 6.5.2 to consider a monotonic tensile load in the form of an end displacement boundary condition to simulate a tensile test with $\sigma \geq 0$. Staying in the tensile regime implies that any effects of kinematic hardening can be disregarded for this material. The elastic range is characterized by a YOUNG's modulus of $E = 70000$ MPa and the ideal-plastic range by a yield stress of 350 MPa. Determine for each iteration step and cycle the variables $\varepsilon$, $\sigma^{\text{trial}}$, $\sigma$, $\varepsilon_{\text{eff}}^{\text{pl}}$, $\Delta\lambda$, $D$, $\|\boldsymbol{v}^{(i+1)} - \boldsymbol{v}^{(i)}\|$ for the case of 10 equidistant increments with a strain increment of $|\Delta\varepsilon_n| = 0.001$. The termination criterion is $t_{\text{end}}^{\boldsymbol{v}} = 0.00001$ for the vector norm $\|\boldsymbol{v}^{(i+1)} - \boldsymbol{v}^{(i)}\|$ during the iteration and the initial damage is $D_0 = 0.01$.

# Chapter 7
# Alternative Approach: The Finite Difference Method

**Abstract** This chapter introduces the finite difference method as an alternative approximation method. The method is derived based on Taylor's series approximations of the field variables as well as based on the weighted residual method. In the elastic range, rods as well as Bernoulli and Timoshenko beams are considered while in the plastic range only Bernoulli beams are treated.

## 7.1 Idea and Derivation of the Method

The first widely known approximation method for partial differential equations is the finite difference method [3, 100] which approximates the governing differential equations of a field problem using local expansions for the variables, generally truncated TAYLOR's series. Comprehensive descriptions of the method can be found, for example, in [25, 41, 70].

As a first step to solve a differential equation for a one-dimensional problem, it is assumed that the numerical solution $u(X)$ is to be determined only at a set of $n$ points within the domain $X \in [0, L]$ including the ends, cf. Fig. 7.1. These $n$ nodes or grid points will be used to derive approximations to the derivatives of the function $u(X)$. To simplify the approach, let us assume in the following that these $n$ nodes are equally spaced, at a distance equal to[1] $\Delta X = \frac{L}{n-1}$. As in the case of the finite element method, high gradients would require a smaller node spacing in order to meet the requirements on the accuracy of the approximate solution. However, the general idea of the finite difference method will be here introduced based on an equi-spaced division of the domain $0 \leq X \leq L$. A typical inner node of the domain is denoted in the following by $i$ and the two neighboring nodes are called $i - 1$ on the left-hand and $i + 1$ on the right-hand side. For sufficient smooth functions $u(X)$, a TAYLOR's series expansion around node $i$ gives:

[1] It should be noted here that the first node in this derivation is denoted by '1' and not as in some references as '0'.

 
A. Öchsner, *Elasto-Plasticity of Frame Structure Elements*,
DOI 10.1007/978-3-662-44225-8_7

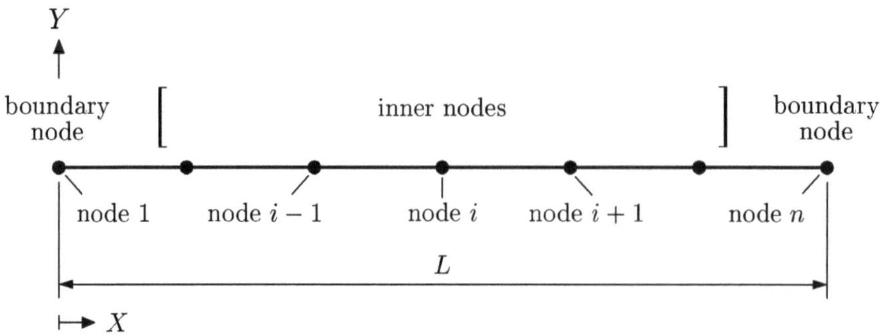


**Fig. 7.1** Finite difference model of a one-dimensional problem

$$u_{i+1} = u_i + \left(\frac{\mathrm{d}u}{\mathrm{d}X}\right)_i \Delta X + \left(\frac{\mathrm{d}^2u}{\mathrm{d}X^2}\right)_i \frac{\Delta X^2}{2} + \cdots + \left(\frac{\mathrm{d}^ku}{\mathrm{d}X^k}\right)_i \frac{\Delta X^k}{k!}, \tag{7.1}$$

$$u_{i-1} = u_i - \left(\frac{\mathrm{d}u}{\mathrm{d}X}\right)_i \Delta X + \left(\frac{\mathrm{d}^2u}{\mathrm{d}X^2}\right)_i \frac{\Delta X^2}{2} - \cdots + \left(\frac{\mathrm{d}^ku}{\mathrm{d}X^k}\right)_i \frac{\Delta X^k}{k!}. \tag{7.2}$$

The infinite series of Eqs. (7.1) and (7.2) are truncated for practical use after a certain number of terms. As a result of this approximation, the so-called truncation errors occurs. Summing up the expression of Eqs. (7.1) and (7.2) gives

$$u_{i+1} + u_{i-1} = 2u_i + \left(\frac{\mathrm{d}^2u}{\mathrm{d}X^2}\right)_i \Delta X^2 + \frac{1}{12}\left(\frac{\mathrm{d}^4u}{\mathrm{d}X^4}\right)_i \Delta X^4 + \cdots, \tag{7.3}$$

or rearranged for the second order derivative:

$$\left(\frac{\mathrm{d}^2u}{\mathrm{d}X^2}\right)_i = \frac{u_{i+1} - 2u_i + u_{i-1}}{\Delta X^2} \underbrace{- \frac{1}{12}\left(\frac{\mathrm{d}^4u}{\mathrm{d}X^4}\right)_i \Delta X^2 - \cdots}_{O(\Delta X^2)}. \tag{7.4}$$

The symbol '$O$' in Eq. (7.4) reads 'order of' and states that if the second order derivative of $u(X)$ is approximated by the first expression on the right-hand side of Eq. (7.4), then the truncation error is of order of $\Delta X^2$. This approximation is a second order accurate approximation because of the truncated terms and is called the centered difference scheme. In a similar way, other derivatives can be derived from Eqs. (7.1) and (7.2). Subtracting of Eq. (7.2) from (7.1) gives

$$u_{i+1} - u_{i-1} = 2\left(\frac{\mathrm{d}u}{\mathrm{d}X}\right)_i \Delta X + \frac{1}{3}\left(\frac{\mathrm{d}^3u}{\mathrm{d}X^3}\right)_i \Delta X^3 + \cdots \tag{7.5}$$

or rearranged for the first order derivative

$$\left(\frac{\mathrm{d}u}{\mathrm{d}X}\right)_i = \frac{u_{i+1} - u_{i-1}}{2\Delta X} \underbrace{- \frac{1}{6}\left(\frac{\mathrm{d}^3u}{\mathrm{d}X^3}\right)_i \Delta X^2 - \cdots}_{O(\Delta X^2)}. \tag{7.6}$$

This last approximation of the first order derivative is called centered difference or centered EULER and the truncation error is of order $\Delta X^2$. Rearranging Eq. (7.1), i.e.

$$u_{i+1} - u_i = \left(\frac{\mathrm{d}u}{\mathrm{d}X}\right)_i \Delta X + \left(\frac{\mathrm{d}^2u}{\mathrm{d}X^2}\right)_i \frac{\Delta X^2}{2} + \cdots, \tag{7.7}$$

or

$$\left(\frac{\mathrm{d}u}{\mathrm{d}X}\right)_i = \frac{u_{i+1} - u_i}{\Delta X} \underbrace{- \left(\frac{\mathrm{d}^2u}{\mathrm{d}X^2}\right)_i \Delta X - \cdots}_{O(\Delta X)}, \tag{7.8}$$

which gives an expression for the forward difference or forward EULER approximation of the first order derivative. In a similar way, Eq. (7.2) can be rearranged to obtain the backward difference or backward EULER approximation of the first order derivative as:

$$\left(\frac{\mathrm{d}u}{\mathrm{d}X}\right)_i = \frac{u_i - u_{i-1}}{\Delta X} \underbrace{+ \left(\frac{\mathrm{d}^2u}{\mathrm{d}X^2}\right)_i \Delta X - \cdots}_{O(\Delta X)}. \tag{7.9}$$

Graphical representations of the finite difference approximations of first order derivatives are shown in Fig. 7.2.

Further derivatives of $u(X)$ of different order can be derived if $u$ is evaluated at nodes $i+2, i-2$ etc. through a TAYLOR's series about the node $i$. Finite difference formulae of the first order derivative with truncation error of order $\Delta X^2$, based on a forward or backward difference approximation, can be obtained in the following way: Let us consider in Eq. (7.2) the second order derivative, i.e.

$$u_{i-1} - u_i = -\left(\frac{\mathrm{d}u}{\mathrm{d}X}\right)_i \Delta X + \frac{1}{2}\left(\frac{\mathrm{d}^2u}{\mathrm{d}X^2}\right)_i \Delta X^2 - \frac{1}{6}\left(\frac{\mathrm{d}^3u}{\mathrm{d}X^3}\right)_i \Delta X^3 + \cdots \tag{7.10}$$

or

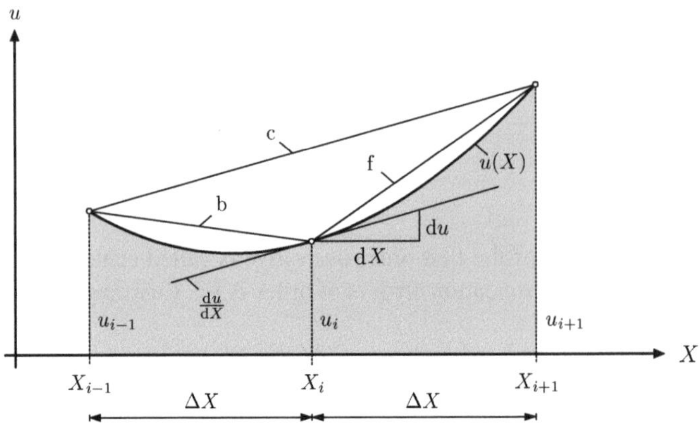


**Fig. 7.2** Graphical representation of first order derivative approximations: *c* centered; *f* forward; *b* backward difference

$$\left(\frac{\mathrm{d}u}{\mathrm{d}X}\right)_i = \frac{u_i - u_{i-1}}{\Delta X} + \frac{1}{2}\left(\frac{\mathrm{d}^2 u}{\mathrm{d}X^2}\right)_i \Delta X \underbrace{- \frac{1}{6}\left(\frac{\mathrm{d}^3 u}{\mathrm{d}X^3}\right)_i \Delta X^2 + \cdots}_{O(\Delta X^2)}, \tag{7.11}$$

where the second order derivative can be replaced by a backward approximation[2] with truncation error of order $\Delta X^2$, i.e.

$$\left(\frac{\mathrm{d}^2 u}{\mathrm{d}X^2}\right)_i = \frac{u_i - 2u_{i-1} + u_{i-2}}{\Delta X^2} + \left(\frac{\mathrm{d}^3 u}{\mathrm{d}X^3}\right)_i \Delta X - O(\Delta X^2). \tag{7.12}$$

Thus, the backward approximation of the first order derivative with truncation error of order $\Delta X^2$ is obtained as:

$$\begin{aligned}\left(\frac{\mathrm{d}u}{\mathrm{d}X}\right)_i &= \frac{u_i - u_{i-1}}{\Delta X} + \frac{u_i - 2u_{i-1} + u_{i-2}}{2\Delta X} + \frac{1}{2}\left(\left(\frac{\mathrm{d}^3 u}{\mathrm{d}X^3}\right)_i \Delta X^2 - O(\Delta X^3)\right) \\ &= \frac{3u_i - 4u_{i-1} + u_{i-2}}{2\Delta X} + O(\Delta X^2). \end{aligned} \tag{7.13}$$

Some common expressions for derivatives of different order and the respective truncation errors are summarized in Table 7.1.

[2] This formulation can be obtained based on a TAYLOR's series expansion for $i-2$ up to $\frac{\mathrm{d}^6 u}{\mathrm{d}X^6}$ and introducing a backward difference approximation of the first order derivative which contains the terms up to $\frac{\mathrm{d}^6 u}{\mathrm{d}X^6}$.

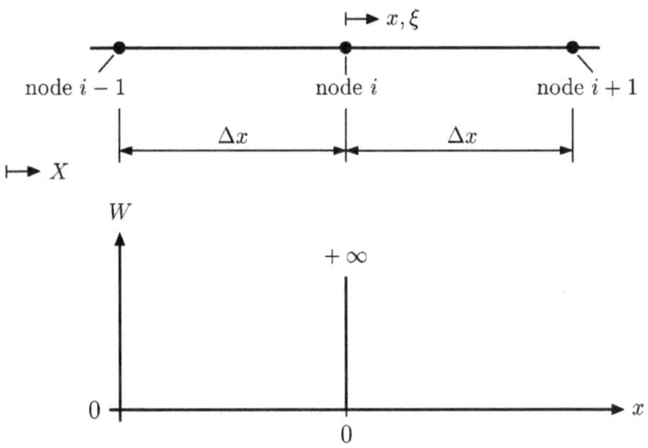


**Fig. 7.3** Typical cell for the cell collocation method and appropriate weight function

It is possible to derive the finite difference approximations based on a special type of collocations method [16], cf. Sect. 5.1. This so-called cell collocation method [57] considers a cell as shown in Fig. 7.3. Let us remind here that the point collocation method uses as weight function the DIRAC delta function.

To approximate, for example, a second order derivative $\frac{\mathrm{d}^2u}{\mathrm{d}x^2}$, a local approximate function can be written as

$$u = \sum_{k=i-1}^{i} N_k \times u_k = N_{i-1}u_{i-1} + N_i u_i + N_{i+1}u_{i+1}, \tag{7.14}$$

where $u_{i-1}, \ldots, u_{i+1}$ are the values of the function at the nodes. The shape functions $N_k$ are given in natural coordinates ($-1 \leq \xi \leq 1$) and can take the following quadratic form (cf. Fig. 7.4), [16]:

$$N_{i-1} = \frac{1}{2}\xi(\xi - 1), \quad N_i = (1-\xi)(1+\xi), \quad N_{i+1} = \frac{1}{2}\xi(1+\xi). \tag{7.15}$$

The second order derivative of the approximate function as given in Eq. (7.14) can be written under consideration of $\frac{\mathrm{d}\xi}{\mathrm{d}x} = \frac{1}{\Delta x}$ as:

$$\begin{aligned}\frac{\mathrm{d}^2u}{\mathrm{d}x^2} &= \frac{1}{\Delta x^2}\left(\frac{\mathrm{d}^2N_{i-1}}{\mathrm{d}\xi^2}u_{i-1} + \frac{\mathrm{d}^2N_i}{\mathrm{d}\xi^2}u_i + \frac{\mathrm{d}^2N_{i+1}}{\mathrm{d}\xi^2}u_{i+1}\right)\\ &= \frac{1}{\Delta x^2}\left(1u_{i-1} - 2u_i + 1u_{i+1}\right).\end{aligned} \tag{7.16}$$

**Table 7.1** Finite difference approximations for various differentiations, partly taken from [7, 25]

| Derivative | Finite difference approximation | Type | Error |
|---|---|---|---|
| $\left(\frac{\mathrm{d}u}{\mathrm{d}X}\right)_i$ | $\frac{u_{i+1}-u_i}{\Delta X}$ | Forward diff. | $O(\Delta X)$ |
| | $\frac{-3u_i+4u_{i+1}-u_{i+2}}{2\Delta X}$ | Forward diff. | $O(\Delta X^2)$ |
| | $\frac{u_i-u_{i-1}}{\Delta X}$ | Backward diff. | $O(\Delta X)$ |
| | $\frac{3u_i-4u_{i-1}+u_{i-2}}{2\Delta X}$ | Backward diff. | $O(\Delta X^2)$ |
| | $\frac{u_{i+1}-u_{i-1}}{2\Delta X}$ | Centered diff. | $O(\Delta X^2)$ |
| $\left(\frac{\mathrm{d}^2u}{\mathrm{d}X^2}\right)_i$ | $\frac{u_{i+2}-2u_{i+1}+u_i}{\Delta X^2}$ | Forward diff. | $O(\Delta X)$ |
| | $\frac{-u_{i+3}+4u_{i+2}-5u_{i+1}+2u_i}{\Delta X^2}$ | Forward diff. | $O(\Delta X^2)$ |
| | $\frac{u_i-2u_{i-1}+u_{i-2}}{\Delta X^2}$ | Backward diff. | $O(\Delta X)$ |
| | $\frac{2u_i-5u_{i-1}+4u_{i-2}-u_{i-3}}{\Delta X^2}$ | Backward diff. | $O(\Delta X^2)$ |
| | $\frac{u_{i+1}-2u_i+u_{i-1}}{\Delta X^2}$ | Centered diff. | $O(\Delta X^2)$ |
| $\left(\frac{\mathrm{d}^3u}{\mathrm{d}X^3}\right)_i$ | $\frac{u_{i+3}-3u_{i+2}+3u_{i+1}-u_i}{\Delta X^3}$ | Forward diff. | $O(\Delta X)$ |
| | $\frac{-3u_{i+4}+14u_{i+3}-24u_{i+2}+18u_{i+1}-5u_i}{2\Delta X^3}$ | Forward diff. | $O(\Delta X^2)$ |
| | $\frac{u_i-3u_{i-1}+3u_{i-2}-u_{i-3}}{\Delta X^3}$ | Backward diff. | $O(\Delta X)$ |
| | $\frac{5u_i-18u_{i-1}+24u_{i-2}-14u_{i-3}+3u_{i-4}}{2\Delta X^3}$ | Backward diff. | $O(\Delta X^2)$ |
| | $\frac{u_{i+2}-2u_{i+1}+2u_{i-1}-u_{i-2}}{2\Delta X^3}$ | Centered diff. | $O(\Delta X^2)$ |
| $\left(\frac{\mathrm{d}^4u}{\mathrm{d}X^4}\right)_i$ | $\frac{u_{i+4}-4u_{i+3}+6u_{i+2}-4u_{i+1}+u_i}{\Delta X^4}$ | Forward diff. | $O(\Delta X)$ |
| | $\frac{-2u_{i+5}+11u_{i+4}-24u_{i+3}+26u_{i+2}-14u_{i+1}+3u_i}{\Delta X^4}$ | Forward diff. | $O(\Delta X^2)$ |

(Continued)

**Table 7.1** (Continued)

| Derivative | Finite difference approximation | Type | Error |
|---|---|---|---|
| | $\dfrac{u_i - 4u_{i-1} + 6u_{i-2} - 4u_{i-3} + u_{i-4}}{\Delta X^4}$ | Backward diff. | $O(\Delta X)$ |
| | $\dfrac{3u_i - 14u_{i-1} + 26u_{i-2} - 24u_{i-3} + 11u_{i-4} - 2u_{i-5}}{\Delta X^4}$ | Backward diff. | $O(\Delta X^2)$ |
| | $\dfrac{u_{i+2} - 4u_{i+1} + 6u_i - 4u_{i-1} + u_{i-2}}{\Delta X^4}$ | Centered diff. | $O(\Delta X^2)$ |

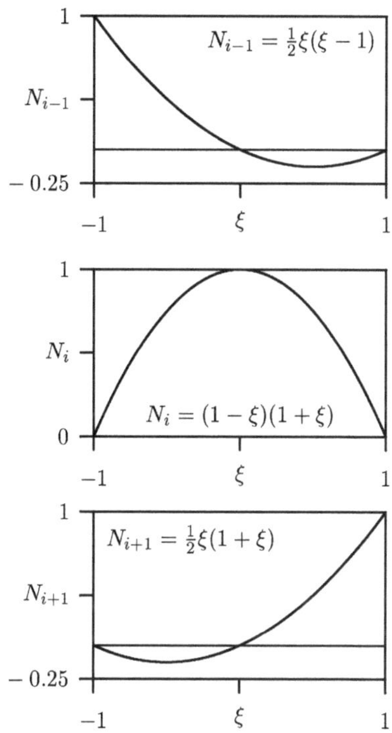


**Fig. 7.4** Shape functions for cell collocation to derive a finite difference scheme, adapted from [16]

Thus, the residuum

$$r = \frac{1}{\Delta x^2}(1u_{i-1} - 2u_i + 1u_{i+1}) \neq 0 \tag{7.17}$$

can be used to formulate the collocation statement at node $i$ as:

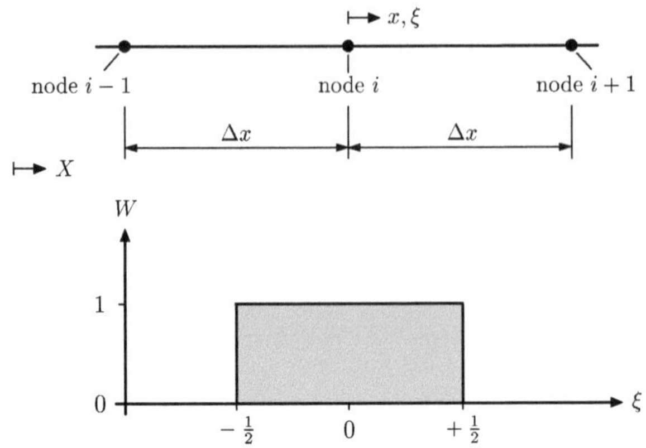


**Fig. 7.5** Typical subregion for the collocation by subregion method and appropriate weight function, adapted from [16]

$$\int_{i-1}^{i+1} rW\mathrm{d}x = \int_{i-1}^{i+1} \frac{1}{\Delta x^2}(1u_{i-1} - 2u_i + 1u_{i+1})\,\delta(x - x_i) \overset{!}{=} 0. \tag{7.18}$$

Under consideration of the properties of the Dirac delta function, cf. (Eq. 5.12), the last equation gives the statement

$$\frac{1}{\Delta x^2}(1u_{i-1} - 2u_i + 1u_{i+1}) = 0, \tag{7.19}$$

which is equivalent to the centered difference scheme as given in Table 7.1.

Instead of using the weighted residual method in the form of cell collocation, the method of collocation by subregions can be alternatively used to derive the finite difference method, cf. [16]. For this approach, the weight function is chosen as a step-type function as shown in Fig. 7.5.

Let us consider again a second order derivative $\frac{\mathrm{d}^2 u}{\mathrm{d}x^2}$ for which the weighted residual statement can be written as:

$$\int_{-\Delta x/2}^{\Delta x/2} \frac{\mathrm{d}^2 u}{\mathrm{d}x^2} W\mathrm{d}x = 0. \tag{7.20}$$

Integrating by parts gives the following expression

$$\int_{-\Delta x/2}^{\Delta x/2} \frac{\mathrm{d}^2 u}{\mathrm{d}x^2} W\mathrm{d}x = \left[\frac{\mathrm{d}u}{\mathrm{d}x}W\right]_{-\Delta x/2}^{\Delta x/2} - \int_{-\Delta x/2}^{\Delta x/2} \frac{\mathrm{d}u}{\mathrm{d}x}\frac{\mathrm{d}W}{\mathrm{d}x}\mathrm{d}x = 0, \tag{7.21}$$

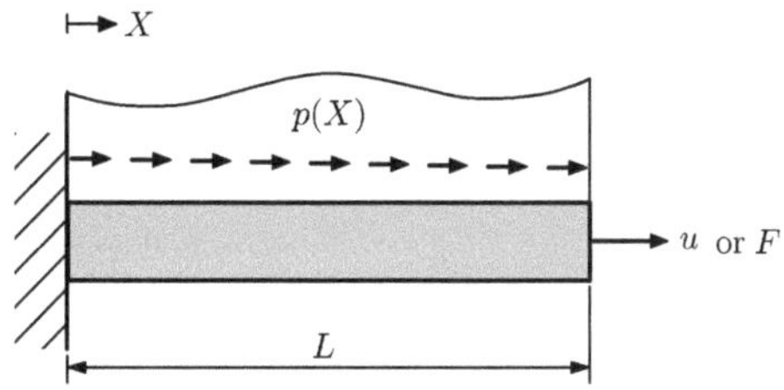


**Fig. 7.6** Rod loaded with a distributed load $p(X)$ and a displacement boundary condition $u$ or an end load $F$

or rearranged as:

$$\int_{-\Delta x/2}^{\Delta x/2} \frac{du}{dx}\frac{dW}{dx}dx = \left[\frac{du}{dx}W\right]_{-\Delta x/2}^{\Delta x/2}. \tag{7.22}$$

It holds for the step function that $W = 1$ and $\frac{dW}{dx} = 0$ and the last equation can be simplified to:

$$0 = \left.\frac{du}{dx}\right|_{+\Delta x/2} - \left.\frac{du}{dx}\right|_{-\Delta x/2}. \tag{7.23}$$

Replacing the differentials 'd' by incremental differences, Eq. (7.23) can be written as

$$0 = \frac{u_{i+1} - u_i}{\Delta x} - \frac{u_i - u_{i-1}}{\Delta x}, \tag{7.24}$$

which corresponds to the centered difference scheme as given in Table 7.1.

## 7.2 Investigation of Rods in the Elastic Range

### *7.2.1 Constant Material and Geometry Parameters*

Let us consider in the following a loaded rod of length $L$ as shown in Fig. 7.6. The tensile stiffness $EA$, i.e. the material and geometric parameters, are considered to be constant within this section. The left-hand end is fixed and the right-hand side is either loaded by a prescribed displacement $u$ or a single force $F$. Along the length of the rod is a distributed load $p(X)$ acting.

The problem is described according to Table 3.2 in the domain $X \in [0, L]$ by the following partial differential equation

$$EA\frac{\mathrm{d}^2 u}{\mathrm{d}X^2} = -p(X) \tag{7.25}$$

and the boundary conditions

$$u(X = 0) = 0 \text{ and } u(X = L) = u \tag{7.26}$$

or

$$u(X = 0) = 0 \text{ and } \left.\frac{\mathrm{d}u}{\mathrm{d}X}\right|_{X=L} = \frac{F}{EA}. \tag{7.27}$$

In a more formal way, Eq. (7.25) can be written as

$$EA\mathcal{L}\{u\} = -p(X), \tag{7.28}$$

where the differential operator is given by $\mathcal{L} = \frac{\mathrm{d}^2}{\mathrm{d}X^2}$. Taking from Table 7.1 the centered difference scheme of second order accuracy, a finite difference approximation of the partial differential equation (7.25) can be written for node $i$ as:

$$EA\left(\frac{u_{i+1} - 2u_i + u_{i-1}}{\Delta X^2}\right) = -p_i, \tag{7.29}$$

or as

$$\frac{EA}{\Delta X}(-u_{i+1} + 2u_i - u_{i-1}) = R_i, \tag{7.30}$$

where the equivalent nodal force $R_i$, resulting from a distributed load $p(X)$, is in general given for an *inner* node $i$ as: $R_i = \int_{-\Delta x/2}^{\Delta x/2} p(\hat{x})\mathrm{d}\hat{x}$. In this integral, the local coordinate $x$ has its origin at the location of node $i$. In the case of the boundary nodes, the equivalent nodal loads must be calculated as $R_1 = \int_0^{\Delta x/2} p(\hat{x})\mathrm{d}\hat{x}$ or $R_n = \int_{-\Delta x/2}^{0} p(\hat{x})\mathrm{d}\hat{x}$ in order to completely distribute the entire load $p(X)$ to the nodes.

Let us assume for simplicity that the distributed load shown in Fig. (7.6) is zero ($p(X) = 0$) and introduce $n = 5$ nodes, i.e. two boundary nodes and three inner nodes. The evaluation of the finite difference approximation according to Eq. (7.30) at the inner nodes $i = 2, \ldots, 4$ gives:

$$\text{node 2:}\ \frac{EA}{\Delta X}(-u_3 + 2u_2 - u_1) = 0, \tag{7.31}$$

$$\text{node 3:}\ \frac{EA}{\Delta X}(-u_4 + 2u_3 - u_2) = 0, \tag{7.32}$$

$$\text{node 4:}\ \frac{EA}{\Delta X}(-u_5 + 2u_4 - u_3) = 0. \tag{7.33}$$

Considering at both ends displacement boundary conditions, i.e. Eq. (7.26), the system of equations (7.31) till (7.33) can be written with $u_1 = 0$ and $u_5 = u$ as

$$\text{node 2:}\ \frac{EA}{\Delta X}(0 + 2u_2 - u_3) = 0, \tag{7.34}$$

$$\text{node 3:}\ \frac{EA}{\Delta X}(-u_2 + 2u_3 - u_4) = 0, \tag{7.35}$$

$$\text{node 4:}\ \frac{EA}{\Delta X}(-u_3 + 2u_4) = \frac{EA}{\Delta X}u, \tag{7.36}$$

or in matrix notation as

$$\frac{EA}{\Delta X}\begin{bmatrix} 2 & -1 & 0 \\ -1 & 2 & -1 \\ 0 & -1 & 2 \end{bmatrix}\begin{bmatrix} u_2 \\ u_3 \\ u_4 \end{bmatrix} = \begin{bmatrix} 0 \\ 0 \\ \frac{EA}{\Delta X}u \end{bmatrix}. \tag{7.37}$$

The last system of equations would be obtained the same way based on a finite element approximation with linear rod elements. Solution of the systems of equations given in Eq. (7.37) gives

$$\begin{bmatrix} u_2 \\ u_3 \\ u_4 \end{bmatrix} = \begin{bmatrix} \frac{1}{4}u_0 \\ \frac{2}{4}u_0 \\ \frac{3}{4}u_0 \end{bmatrix}, \tag{7.38}$$

which gives the same values as the analytical solution, i.e. linear increasing from 0 at $X = 0$ to $u_0$ at $X = L$.

The consideration of the force boundary condition at the right-hand end requires more efforts. Considering the equilibrium condition according to Eq. (3.6), i.e. $\frac{\mathrm{d}N}{\mathrm{d}X} = -p(X)$, together with the integrated form of the differential equation (3.10) gives the following expression for the internal normal force $N$ as a function of the derivative of $u(X)$:

$$EA\left(\frac{\mathrm{d}u(X)}{\mathrm{d}X}\right)_i = N_i(X). \tag{7.39}$$

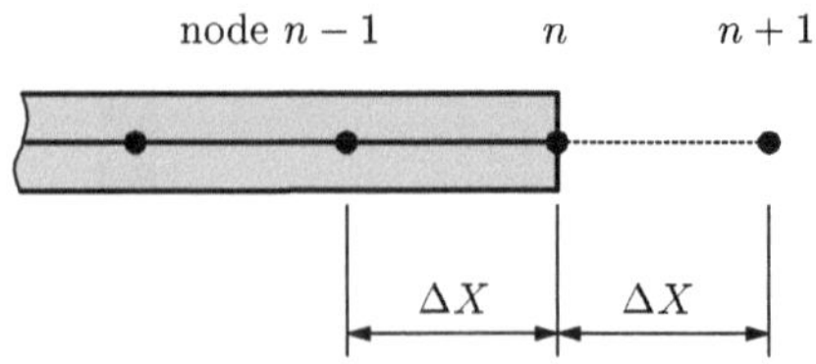


**Fig. 7.7** Fictitious finite difference node $n+1$ outside the rod, adapted from [7]

To evaluate the gradient in Eq. (7.39), the different formulations given in Table 7.1 can be used. If the higher accurate centered difference scheme should be used, it is necessary to introduce a fictitious node outside the rod as shown in Fig. 7.7.

Then, the finite difference approximation for the boundary node 5 can be written as:

$$\text{node 5: } \frac{EA}{\Delta X}(-u_6 + 2u_5 - u_4) = 0, \tag{7.40}$$

and the gradient at the boundary node together with the force equilibrium, i.e. $N_5 = F$, reads as

$$\left(\frac{\mathrm{d}u(X)}{\mathrm{d}X}\right)_5 = \frac{u_6 - u_4}{2\Delta X} = \frac{F}{EA}, \tag{7.41}$$

from which the expression for $u_6$ can be obtained as:

$$u_6 = u_4 + \frac{2F\Delta X}{EA}. \tag{7.42}$$

This result can be introduced into Eq. (7.40) to receive the required expression to substitute $u_5$ into Eq. (7.33) by given values:

$$u_5 = u_4 + \frac{F\Delta X}{EA}. \tag{7.43}$$

Thus, the final system of equations under consideration of the force boundary condition is given as:

$$\frac{EA}{\Delta X}\begin{bmatrix} 2 & -1 & 0 \\ -1 & 2 & -1 \\ 0 & -1 & 1 \end{bmatrix}\begin{bmatrix} u_2 \\ u_3 \\ u_4 \end{bmatrix} = \begin{bmatrix} 0 \\ 0 \\ F \end{bmatrix}. \tag{7.44}$$

It should be mentioned here that the same result can be obtained in this case based on the finite element method if the equation for node 5 is eliminated. Solution of this system of equations gives

$$\begin{bmatrix} u_2 \\ u_3 \\ u_4 \end{bmatrix} = \begin{bmatrix} \frac{\Delta X F}{EA} \\ \frac{2\Delta X F}{EA} \\ \frac{3\Delta X F}{EA} \end{bmatrix}, \tag{7.45}$$

which gives the same result as the analytical solution under consideration of $\Delta X = \frac{L}{4}$, cf. Table 3.3. Alternatively, we may write the system of equations under consideration of the displacement at node 5 as:

$$\frac{EA}{\Delta X} \begin{bmatrix} 2 & -1 & 0 & 0 \\ -1 & 2 & -1 & 0 \\ 0 & -1 & 2 & -1 \\ 0 & 0 & -1 & 1 \end{bmatrix} \begin{bmatrix} u_2 \\ u_3 \\ u_4 \\ u_5 \end{bmatrix} = \begin{bmatrix} 0 \\ 0 \\ 0 \\ F \end{bmatrix}, \tag{7.46}$$

or solved for the nodal unknowns:

$$\begin{bmatrix} u_2 \\ u_3 \\ u_4 \\ u_5 \end{bmatrix} = \begin{bmatrix} \frac{\Delta X F}{EA} \\ \frac{2\Delta X F}{EA} \\ \frac{3\Delta X F}{EA} \\ \frac{4\Delta X F}{EA} \end{bmatrix}. \tag{7.47}$$

Substituting $\Delta X = \frac{L}{4}$ gives the exact analytical solution as presented in Table 3.3.

### 7.2.2 Varying Material and Geometry Parameters

Let us consider in the following the case that the tensile stiffness is a function of the Cartesian coordinate $X$. Thus, the generalized problem shown in Fig. 7.6 can be described in the domain $X \in [0, L]$ by the following partial differential equation

$$\frac{\mathrm{d}}{\mathrm{d}X}\left( E(X)A(X)\frac{\mathrm{d}u}{\mathrm{d}X} \right) = -p(X) \tag{7.48}$$

and the boundary conditions

$$u(X=0) = 0 \text{ and } u(X=L) = u \tag{7.49}$$

or

$$u(X=0) = 0 \text{ and } \left.\frac{\mathrm{d}u}{\mathrm{d}X}\right|_{X=L} = \frac{F}{E(L)A(L)}. \tag{7.50}$$

The product of the varying modulus and cross section can be combined in an auxiliary function

$$k(X) = E(X)A(X), \tag{7.51}$$

and the differential equation (7.52) reads in a more general notation as:

$$\frac{\mathrm{d}}{\mathrm{d}X}\left(k(X)\frac{\mathrm{d}u}{\mathrm{d}X}\right) = -p(X). \tag{7.52}$$

A finite difference approximation of this equation can be introduced in different ways. Let us first neglect for simplicity the distributed load ($p(X) = 0$) since the treatment of a distributed load is discussed in Sect. 7.2.1.

The first approach is based on the introduction of an auxiliary function $v(x)$ of the form:

$$v(X) = k(X)\frac{\mathrm{d}u}{\mathrm{d}X}. \tag{7.53}$$

Thus, the differential equation (7.52) can be expressed in a much more simpler way under the simplification that the distributed load is neglected as:

$$\frac{\mathrm{d}v(X)}{\mathrm{d}X} = 0. \tag{7.54}$$

The first task is now to approximate the first order derivative $\frac{\mathrm{d}v}{\mathrm{d}X}$. To this end, let us consider the centered difference expression from Table 7.1 and replace $\Delta X$ by $\frac{\Delta X}{2}$. Thus, the centered difference approximation is given by

$$\left(\frac{\mathrm{d}v(X)}{\mathrm{d}X}\right)_i = \frac{v_{i+\frac{1}{2}} - v_{i-\frac{1}{2}}}{\Delta X}, \tag{7.55}$$

where, for example, the notation '$i + \frac{1}{2}$' refers to the value of the function $v$ in the middle of node $i$ and $i + 1$, cf. Fig. 7.8.

Based on the definition of the auxiliary function $v$ in Eq. (7.53), the functional values of $v$ in Eq. (7.55) at the intermediate locations can be stated as:

$$v_{i+\frac{1}{2}} = k_{i+\frac{1}{2}}\left(\frac{\mathrm{d}u}{\mathrm{d}X}\right)_{i+\frac{1}{2}} = k_{i+\frac{1}{2}}\frac{u_{i+1} - u_i}{\Delta X}, \tag{7.56}$$

$$v_{i-\frac{1}{2}} = k_{i-\frac{1}{2}}\left(\frac{\mathrm{d}u}{\mathrm{d}X}\right)_{i-\frac{1}{2}} = k_{i-\frac{1}{2}}\frac{u_i - u_{i-1}}{\Delta X}. \tag{7.57}$$

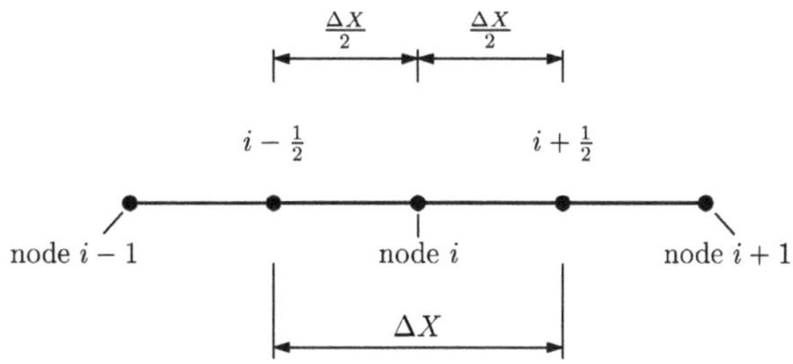


**Fig. 7.8** Definition of the intermediate position $i-\frac{1}{2}$ and $i+\frac{1}{2}$

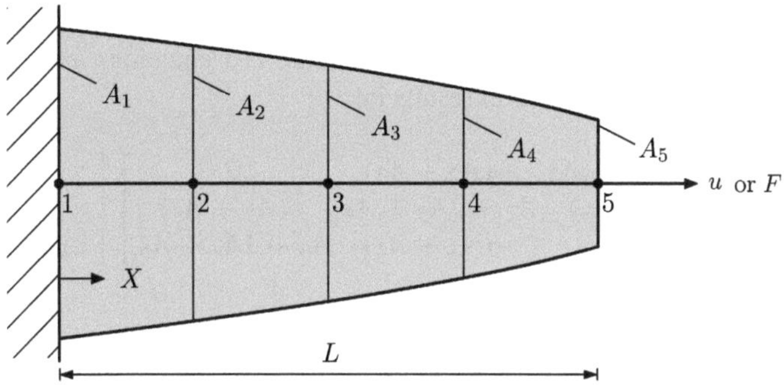


**Fig. 7.9** Rod with linear varying cross-section area discretized by five nodes in a finite difference approach

Thus, the finite difference approximation of the problem given in Eq. (7.54) can be written for node $i$ as:

$$\frac{1}{\Delta X}\left(-k_{i-\frac{1}{2}}u_{i-\frac{1}{2}} + (k_{i-\frac{1}{2}} + k_{i+\frac{1}{2}})u_i - k_{i+\frac{1}{2}}u_{i+\frac{1}{2}}\right) = 0. \tag{7.58}$$

It should be noted here that the last equation was multiplied by $\Delta X$. This is in general required to obtain from distributed loads the equivalent nodal force $R_i$, cf. Sect. 7.2.1. In addition, the last equation was multiplied by '$-1$' to make the comparison with the finite element approach easier.

In order to simplify the explanations of the different approaches and possibilities, let us consider in the following a rod with linear varying cross-sectional area as shown in Fig. 7.9. This structure should be discretized by five nodes and either a displacement $u$ or a force $F$ should be prescribed at the right-hand end.

Evaluation of the finite difference scheme given in Eq. (7.58) for the inner nodes shown in Fig. (7.9) gives:

$$\text{node 2:} \frac{E}{\Delta X}\left(-\frac{A_1+A_2}{2}u_1+\left(\frac{A_1+A_2}{2}+\frac{A_2+A_3}{2}\right)u_2-\frac{A_2+A_3}{2}u_3\right)=0, \tag{7.59}$$

$$\text{node 3:} \frac{E}{\Delta X}\left(-\frac{A_2+A_3}{2}u_2+\left(\frac{A_2+A_3}{2}+\frac{A_3+A_4}{2}\right)u_3-\frac{A_3+A_4}{2}u_4\right)=0, \tag{7.60}$$

$$\text{node 4:} \frac{E}{\Delta X}\left(-\frac{A_3+A_4}{2}u_3+\left(\frac{A_3+A_4}{2}+\frac{A_4+A_5}{2}\right)u_4-\frac{A_4+A_5}{2}u_5\right)=0. \tag{7.61}$$

The consideration of the displacement boundary conditions according to Eq. (7.49), i.e. $u_1 = 0$ and $u_5 = u$, in the three last equations and rearranging the set of equations in matrix form results in:

$$\frac{E}{2\Delta X}\begin{bmatrix} A_1+2A_2+A_3 & -(A_2+A_3) & 0 \\ -(A_2+A_3) & A_2+2A_3+A_4 & -(A_3+A_4) \\ 0 & -(A_3+A_4) & A_3+2A_4+A_5 \end{bmatrix}\begin{bmatrix} u_2 \\ u_3 \\ u_4 \end{bmatrix} = \begin{bmatrix} 0 \\ 0 \\ \frac{E}{2\Delta X}(A_4+A_5)u \end{bmatrix}. \tag{7.62}$$

The same system of equations would be obtained based on a finite element approach with linear elements as introduced in Sect. 5.7.2.

If the boundary condition at the right-hand end is given as a single force $F$, one may follow first of all the idea based on a fictitious node $n+1$ outside the structure as shown in the previous section in Fig. 7.7. The evaluation of node 5 based on the general scheme (7.58) gives

$$\text{node 5:} \frac{E}{\Delta X}\left(-\frac{A_4+A_5}{2}u_4+\left(\frac{A_4+A_5}{2}+\frac{A_5+A_6}{2}\right)u_5-\frac{A_5+A_6}{2}u_6\right)=0. \tag{7.63}$$

If the centered difference scheme from Table 7.1 is used, the gradient at the boundary node together with the force equilibrium, i.e. $N_5 = F$, reads as

$$\left(\frac{\mathrm{d}u}{\mathrm{d}X}\right)_5 = \frac{u_6-u_4}{2\Delta X} = \frac{F}{EA_5}. \tag{7.64}$$

The last equation can be rearranged for $u_6$ and this relationship can be introduced into the evaluation of node 5 according to Eq. (7.63). This gives finally an expression for the displacement $u_5$ in the form:

$$u_5 = u_4 + \frac{2F\Delta X}{E}\frac{A_5+A_6}{A_5(A_4+2A_5+A_6)}. \tag{7.65}$$

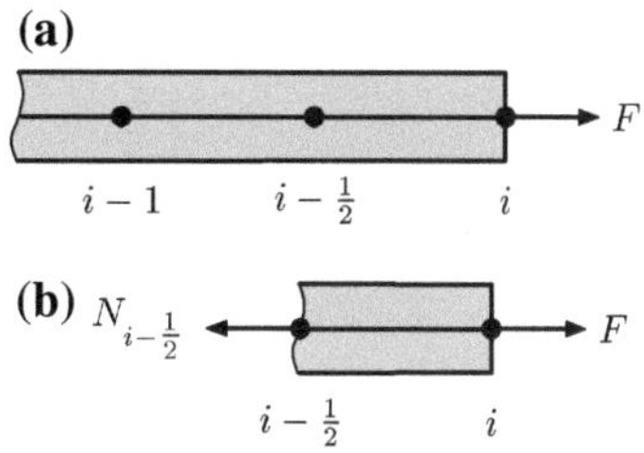


**Fig. 7.10** Definition of a centered difference scheme at the position $i-\frac{1}{2}$. Note that the distance between node $i-1$ and $i$ is equal to $\Delta X$

The last equation still contains the fictitious area $A_6$ and might be not the best approach to solve the problem.

An alternative approach can be based on the backward difference scheme as given in Table 7.1. Then, the gradient at the boundary node together with the force equilibrium reads as:

$$\left(\frac{\mathrm{d}u}{\mathrm{d}X}\right)_5 = \frac{u_5 - u_4}{\Delta X} = \frac{F}{EA_5}, \tag{7.66}$$

or rearranged for $u_5$ as:

$$u_5 = u_4 + \frac{F\Delta X}{EA_5}. \tag{7.67}$$

The last equation can be introduced into the evaluation of node 4 according to Eq. (7.61) to finally result in:

$$\frac{E}{2\Delta X}(-(A_3 + A_4)u_3 + (A_3 + A_4)u_4) = \frac{A_4 + A_5}{2A_5}F. \tag{7.68}$$

Thus, the final system of equations is given by Eqs. (7.59) and (7.60) under consideration of the boundary condition at node 1 ($u_1 = 0$) and Eq. (7.68) which considers the force boundary condition at node 5. This systems of equations reads in matrix from as:

$$\frac{E}{2\Delta X}\begin{bmatrix} A_1 + 2A_2 + A_3 & -(A_2 + A_3) & 0 \\ -(A_2 + A_3) & A_2 + 2A_3 + A_4 & -(A_3 + A_4) \\ 0 & -(A_3 + A_4) & A_3 + A_4 \end{bmatrix}\begin{bmatrix} u_2 \\ u_3 \\ u_4 \end{bmatrix} = \begin{bmatrix} 0 \\ 0 \\ \frac{A_4+A_5}{2A_5}F \end{bmatrix}. \tag{7.69}$$

The left-hand side of the last equation is identical to the finite element approach but the load vector differs from the finite element solution where only $F$ would appear.

An alternative approach can be based on the idea that the first order derivative is evaluated in the middle of node $i-1$ and $i$ due to a centered difference scheme, cf. Fig. 7.10a.

The corresponding centered difference scheme is given by

$$\left(\frac{\mathrm{d}u}{\mathrm{d}X}\right)_{i-\frac{1}{2}} = \frac{u_i - u_{i-1}}{\Delta X} = \frac{N_{i-\frac{1}{2}}}{k_{i-\frac{1}{2}}}, \tag{7.70}$$

whereas it follows from the force equilibrium (cf. Fig. 7.10b) that $N_{i-\frac{1}{2}} = F$ as long as there is no distributed load involved in the considered section of the rod. Applying the node numbering given in Fig. 7.9, Eq. (7.70) can be rearranged to obtain the following expression:

$$u_5 = u_4 + \frac{2F\Delta X}{E(A_4 + A_5)}, \tag{7.71}$$

which can be introduced into Eq. (7.61) to finally obtain:

$$\frac{E}{2\Delta X}(-(A_3 + A_4)u_3 + (A_3 + A_4)u_4) = F. \tag{7.72}$$

The last equation can be combined with Eqs. (7.59) and (7.60) to obtain the systems of equations as:

$$\frac{E}{2\Delta X}\begin{bmatrix} A_1 + 2A_2 + A_3 & -(A_2 + A_3) & 0 \\ -(A_2 + A_3) & A_2 + 2A_3 + A_4 & -(A_3 + A_4) \\ 0 & -(A_3 + A_4) & A_3 + A_4 \end{bmatrix}\begin{bmatrix} u_2 \\ u_3 \\ u_4 \end{bmatrix} = \begin{bmatrix} 0 \\ 0 \\ F \end{bmatrix}. \tag{7.73}$$

This formulation is identical to the finite element approach.

To investigate the difference between the system of equations given in Eq. (7.69) and Eq. (7.73), let us consider in the following different ratios $a$ between the area $A_5$ and $A_1$, cf. Example 5.5. The comparison with the analytical solution, cf. supplementary problem 5.31, will provide some understanding which is the better approach.

From Table 7.2 where the numerical errors between the different implementations of the force boundary condition and the exact solution are summarized, it can be concluded that the approach based on Eq. (7.73) gives a much more accurate approximation of the problem under consideration. This is a direct result of that fact that Eq. (7.73) is derived under consideration of a centered difference scheme (error $\sim O(\Delta X^2)$), whereas Eq. (7.69) is based on a backward difference scheme where the error is of order $O(\Delta X)$. Table 7.2 contains no numerical errors for the coordinate $X = 0$ since the result is exact (boundary condition) and the fraction would give $\frac{0}{0}$. Furthermore, it can be seen that the error is decreasing for increasing ratio $A_5/A_1$. For the limiting case $A_5 = A_1$, i.e. constant cross-sectional area, the solution would be exact.

A different approach than the presented strategies can be based on the product rule of differential calculus. Let us assume for simplicity that the YOUNG's modulus in

**Table 7.2** Relative error, $\frac{u_{\text{approx}}-u_{\text{exact}}}{u_{\text{exact}}} \times 100$, in percent for different implementations of a force boundary condition

| Equations | Node 1 $X = 0$ | Node 2 $X = \frac{L}{4}$ | Node 3 $X = \frac{2L}{4}$ | Node 4 $X = \frac{4L}{4}$ | Node 5 $X = L$ |
|---|---|---|---|---|---|
| $a = \frac{A_5}{A_1} = 0.1$ | | | | | |
| (7.69) | – | 111.3569 | 110.8318 | 109.3803 | 48.1861 |
| (7.73) | – | −0.5379 | −0.7850 | −1.4681 | −5.9208 |
| $a = \frac{A_5}{A_1} = 0.2$ | | | | | |
| (7.69) | – | 49.3807 | 49.1516 | 48.6324 | 26.0422 |
| (7.73) | – | −0.4129 | −0.5656 | −0.9117 | −2.1644 |
| $a = \frac{A_5}{A_1} = 0.3$ | | | | | |
| (7.69) | – | 28.7698 | 28.6528 | 28.4275 | 16.9195 |
| (7.73) | – | −0.3073 | −0.3979 | −0.5723 | −1.0116 |

the differential equation according to Eq. (7.52) is constant and consider the product of the functions $A(X)$ and $\frac{du}{dX}$. Then, the differential of the product is given by

$$E\left(\frac{dA}{dX}\frac{du}{dX} + A(X)\frac{d^2u}{dx^2}\right) = 0. \tag{7.74}$$

Introduction of centered difference schemes for the first and second order derivative according to Table 7.1 gives:

$$E\left(\left(\frac{dA}{dX}\right)_i \frac{u_{i+1} - u_{i-1}}{2\Delta X} + A_i \frac{u_{i+1} - 2u_i + u_{i-1}}{\Delta X^2}\right) = 0. \tag{7.75}$$

Let us assume as in the previous derivations a linear varying cross-sectional area. Thus, the gradient of the cross-sectional area function can be introduced similar to the centered difference scheme[3] as

$$\frac{dA}{dX} = \frac{A_{i+1} - A_{i-1}}{2\Delta X}, \tag{7.76}$$

[3] At this stage, many different ways are possible on how to introduce this gradient. A formulation similar to the centered difference scheme seems the natural choice and other approaches will be not discussed here.

which can be introduced into Eq. (7.75) to finally obtain the finite difference scheme:

$$\frac{E}{\Delta X}\left(-\left[\frac{1}{4}A_{i-1}+A_i-\frac{1}{4}A_{i+1}\right]u_{i-1}+2A_iu_i\right.$$
$$\left.-\left[-\frac{1}{4}A_{i-1}+A_i+\frac{1}{4}A_{i+1}\right]u_{i+1}\right). \tag{7.77}$$

It should be noted here that the last equation is multiplied by $\Delta X$ and $(-1)$. Evaluation of this finite difference scheme for the inner nodes ($i = 2, 3, 4$) shown in Fig. (7.9) gives:

$$\frac{E}{\Delta X}\left(-\left[\frac{1}{4}A_1+A_2-\frac{1}{4}A_3\right]u_1+2A_2u_2-\left[-\frac{1}{4}A_1+A_2+\frac{1}{4}A_3\right]u_3\right)=0, \tag{7.78}$$

$$\frac{E}{\Delta X}\left(-\left[\frac{1}{4}A_2+A_3-\frac{1}{4}A_4\right]u_2+2A_3u_3-\left[-\frac{1}{4}A_2+A_3+\frac{1}{4}A_4\right]u_4\right)=0, \tag{7.79}$$

$$\frac{E}{\Delta X}\left(-\left[\frac{1}{4}A_3+A_4-\frac{1}{4}A_5\right]u_3+2A_4u_4-\left[-\frac{1}{4}A_3+A_4+\frac{1}{4}A_5\right]u_4\right)=0. \tag{7.80}$$

The force boundary condition can be introduced based on the idea of the centered difference scheme as given in Eq. (7.71). Based on this expression, Eq. (7.80) can be rewritten to obtain the following expression:

$$\frac{E}{\Delta X}\left(-\left[\frac{1}{4}A_3+A_4-\frac{1}{4}A_5\right]u_3-\left[-\frac{1}{4}A_3-A_4+\frac{1}{4}A_5\right]u_4\right)$$
$$=F\frac{-\frac{1}{2}A_3+2A_4+\frac{1}{2}A_5}{A_4+A_5}. \tag{7.81}$$

Thus, the final system of equations for this approach, which is based on the product rule of differential calculus, is given by the following matrix scheme:

$$\frac{E}{2\Delta X}\begin{bmatrix} 4A_2 & \frac{A_1}{2}-2A_2-\frac{A_3}{2} & 0 \\ -\frac{A_2}{2}-2A_3-\frac{A_4}{2} & 4A_3 & \frac{A_2}{2}-2A_3-\frac{A_4}{2} \\ 0 & -\frac{A_3}{2}-2A_4+\frac{A_5}{2} & \frac{A_3}{2}+2A_4-\frac{A_5}{2} \end{bmatrix}\begin{bmatrix} u_2 \\ u_3 \\ u_4 \end{bmatrix}$$

$$
= \begin{bmatrix} 0 \\ 0 \\ F\frac{-\frac{1}{2}A_3+2A_4+\frac{1}{2}A_5}{A_4+A_5} \end{bmatrix} . \tag{7.82}
$$

The evaluation of the cases presented in Table 7.2 reveals that this derivation gives identical results as Eq. (7.73), i.e. an identical solution as a finite element approach.

# 7.3 Investigation of Bernoulli Beams in the Elastic Range

## *7.3.1 Approximation of the Differential Equation*

Let us consider in the following a loaded BERNOULLI beam of length $L$ as shown in Fig. 7.11. The bending stiffness $EI_Z$, i.e. the material and geometric parameters, are considered to be constant in the following derivations. The boundary conditions are not shown in the figure and will be chosen in different ways throughout this chapter. In the general case, a variable distributed load $q_Y(X)$ is acting along the length of the beam.

The problem under consideration can be described according to Table 4.5 by the following fourth-order differential equation

$$
EI_Z \frac{d^4 u_Y}{dX^4} = q_Y(X) \tag{7.83}
$$

and appropriate boundary conditions. Taking from Table 7.1 the centered difference scheme of second order accuracy, a finite difference approximation of the fourth-order differential equation can be written for node $i$ as:

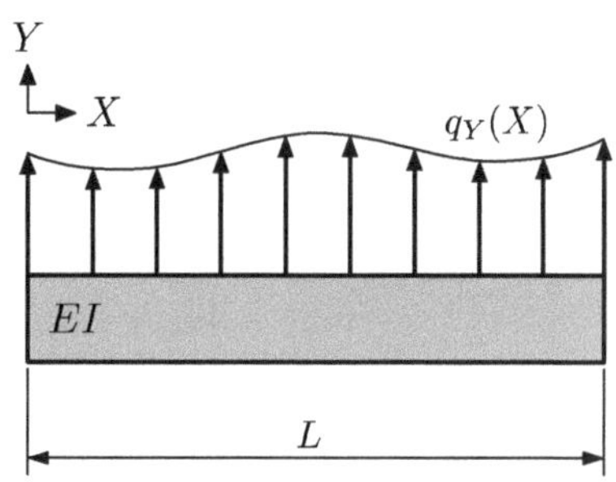


**Fig. 7.11** BERNOULLI beam loaded with a distributed load $q_Y(X)$

$$EI_Z\left(\frac{u_{i+2}-4u_{i+1}+6u_i-4u_{i-1}+u_{i-2}}{\Delta X^4}\right)=q_i, \tag{7.84}$$

or as

$$\frac{EI_Z}{\Delta X^3}(u_{i+2}-4u_{i+1}+6u_i-4u_{i-1}+u_{i-2})=R_i, \tag{7.85}$$

where the equivalent nodal force $R_i$, resulting from a distributed load $q(X)$, is in general given for an *inner* node $i$ as: $R_i=\int_{-\Delta x/2}^{\Delta x/2} q(\hat{x})\mathrm{d}\hat{x}$. In this integral, the local coordinate $x$ has its origin at the location of node $i$. In the case of the boundary nodes, the equivalent nodal loads must be calculated as $R_1=\int_0^{\Delta x/2} q(\hat{x})\mathrm{d}\hat{x}$ or $R_n=\int_{-\Delta x/2}^{0} q(\hat{x})\mathrm{d}\hat{x}$ in order to completely distribute the entire load $q(X)$ to the nodes. If a single force $F$ is acting at a node $i$, then this force can be summed up with the equivalent nodal force $R_i$. It should be noted here that—as in the case of the finite element method—the finite difference method allows the action of forces only at nodes.

The derivation of the second order derivative was demonstrated in Sect. 7.1 and can be used to calculate the centered difference scheme for the fourth order derivative as follows. The approximation is according to Table 7.1 given for node $i$ as:

$$\left(\frac{\mathrm{d}^2u}{\mathrm{d}X^2}\right)_i \approx \frac{u_{i+1}-2u_i+u_{i-1}}{\Delta X^2}. \tag{7.86}$$

In a similar way, we can write the approximation for node $i+1$ and $i-1$ as

$$\left(\frac{\mathrm{d}^2u}{\mathrm{d}X^2}\right)_{i+1} \approx \frac{u_{i+2}-2u_{i+1}+u_i}{\Delta X^2}, \tag{7.87}$$

$$\left(\frac{\mathrm{d}^2u}{\mathrm{d}X^2}\right)_{i-1} \approx \frac{u_i-2u_{i-1}+u_{i-2}}{\Delta X^2}. \tag{7.88}$$

If we consider the left-hand sides (i.e. the fractions for the second order derivatives) of the last three equations as functions $u$, we can introduce these three expressions in Eq. (7.86) to obtain the approximation for the second order derivative of the second order derivative, i.e. the fourth order derivative, as:

$$\left(\frac{\mathrm{d}^4u}{\mathrm{d}X^4}\right)_i \approx \frac{u_{i+2}-2u_{i+1}+u_i-2(u_{i+1}-2u_i+u_{i-1})+u_i-2u_{i-1}+u_{i-2}}{\Delta X^4}$$

$$=\frac{u_{i+2}-4u_{i+1}+6u_i-4u_{i-1}+u_{i-2}}{\Delta X^4}. \tag{7.89}$$

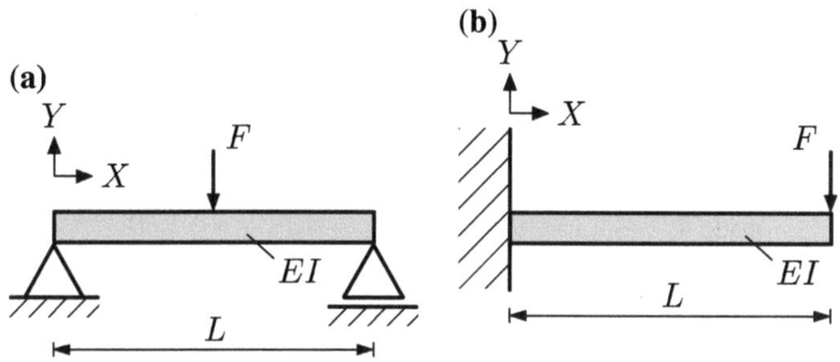


**Fig. 7.12** **a** Simply supported and **b** cantilevered BERNOULLI beam loaded by a single force

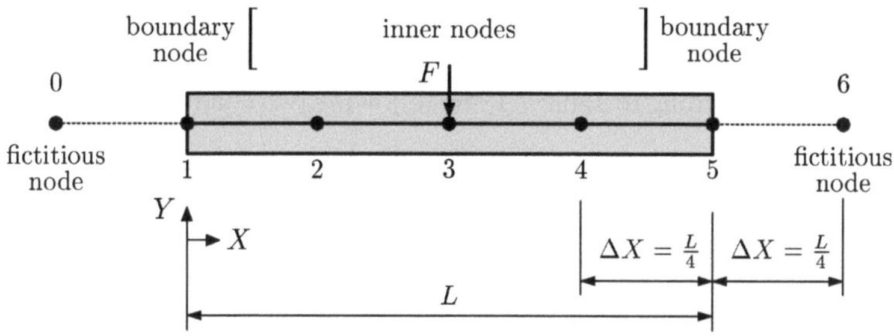


**Fig. 7.13** Finite difference discretization of the simply supported BERNOULLI beam

## *7.3.2 Solved Problems*

**7.1 Example: Finite difference approximation of a simply supported and cantilevered beam loaded by a single force**

Given is a BERNOULLI beam with different supports as shown in Fig. 7.12. The bending stiffness $EI_Z$ is constant and the length is equal to $L$. The simply supported beam (a) is loaded in the middle by a single force while the cantilevered beam (b) is loaded at its right-hand end by a single force $F$. Derive for both cases a finite difference approximation based on five grid points, i.e. an equidistant spacing of $\Delta X = \frac{L}{4}$.

Determine for both cases

- the displacement of the beam at the force application point,
- the analytical solution and
- calculate the relative error between the analytical and finite difference solution.

**7.1 Solution**

(a) The finite difference discretization of the simply supported beam is shown in Fig. 7.13 where at first only the inner nodes will be considered.

Evaluation of the finite difference approximation of the fourth-order differential equation according to Eq. (7.85) at the inner nodes $i = 2, \ldots, 4$ gives:

$$\text{node 2:} \quad \frac{EI_Z}{\Delta X^3}(u_4 - 4u_3 + 6u_2 - 4u_1 + u_0) = 0, \tag{7.90}$$

$$\text{node 3:} \quad \frac{EI_Z}{\Delta X^3}(u_5 - 4u_4 + 6u_3 - 4u_2 + u_1) = -F, \tag{7.91}$$

$$\text{node 4:} \quad \frac{EI_Z}{\Delta X^3}(u_6 - 4u_5 + 6u_4 - 4u_3 + u_2) = 0. \tag{7.92}$$

The vertical displacement is zero at both ends and it can be immediately concluded that $u_1 = u_5 = 0$. The fictitious nodes $i = 0$ and $i = 6$ outside the domain can be eliminated based on the boundary condition that the moment must be equal to zero at the supports, i.e. $M_Z(X = 0) = M_Z(X = L) = 0$. Application of the bending differential equation in the form with the bending moment according to Eq. (4.32), i.e. $EI_Z \frac{\mathrm{d}^2 u}{\mathrm{d}X^2} = M_Z$, and the centered finite difference approximation of the second order derivative according to Table 7.1, the following two conditions can be derived:

$$EI_Z \left. \frac{\mathrm{d}^2 u}{\mathrm{d}X^2} \right|_1 = EI_Z \frac{u_2 - 2u_1 + u_0}{\Delta X^2} = 0, \tag{7.93}$$

$$EI_Z \left. \frac{\mathrm{d}^2 u}{\mathrm{d}X^2} \right|_5 = EI_Z \frac{u_6 - 2u_5 + u_4}{\Delta X^2} = 0, \tag{7.94}$$

from which the two conditions $u_0 = -u_2$ and $u_6 = -u_4$ can be obtained. Introducing these relationships for the fictitious nodes and the condition of zero displacement at the supports into the system of equations according to (7.90)–(7.92) gives:

$$\text{node 2:} \quad \frac{EI_Z}{\Delta X^3}(5u_2 - 4u_3 + u_4) = 0, \tag{7.95}$$

$$\text{node 3:} \quad \frac{EI_Z}{\Delta X^3}(-4u_2 + 6u_3 - 4u_4) = -F, \tag{7.96}$$

$$\text{node 4:} \quad \frac{EI_Z}{\Delta X^3}(u_2 - 4u_3 + 5u_4) = 0, \tag{7.97}$$

or in matrix notation:

$$\frac{EI_Z}{\Delta X^3} \begin{bmatrix} 5 & -4 & 1 \\ -4 & 6 & -4 \\ 1 & -4 & 5 \end{bmatrix} \begin{bmatrix} u_2 \\ u_3 \\ u_4 \end{bmatrix} = \begin{bmatrix} 0 \\ -F \\ 0 \end{bmatrix}. \tag{7.98}$$

The solution of this linear system of equations gives the unknown nodal values as:

$$u_2 = -\frac{F\Delta X^3}{EI_Z}, \quad u_3 = -\frac{3F\Delta X^3}{4EI_Z}, \quad u_4 = -\frac{F\Delta X^3}{EI_Z}, \tag{7.99}$$

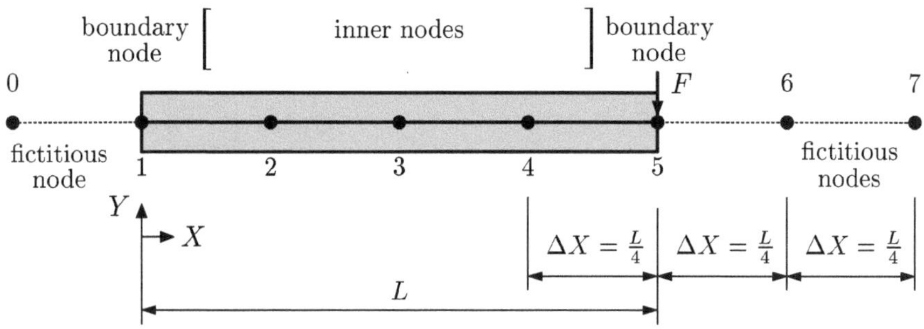


**Fig. 7.14** Finite difference discretization of the cantilevered BERNOULLI beam

or with $\Delta X = \frac{L}{4}$ as:

$$u_2 = -\frac{FL^3}{64EI_Z}, \quad u_3 = -\frac{3FL^3}{128EI_Z}, \quad u_4 = -\frac{FL^3}{64EI_Z}. \tag{7.100}$$

The analytical solution for the displacement at the force application point can be taken from Table 4.11 as $\frac{-FL^3}{48EI_Z}$ and the relative error is obtained as:

$$\text{relative error} = \frac{\frac{3}{128} - \frac{1}{48}}{\frac{1}{48}} \times 100 = 12.5\,\%. \tag{7.101}$$

(b) The finite difference discretization of the cantilevered beam is shown in Fig. 7.14. Evaluation of the finite difference approximation of the inner nodes gives similar to part (a) the following three equations:

$$\text{node 2:} \ \frac{EI_Z}{\Delta X^3}(u_4 - 4u_3 + 6u_2 - 4u_1 + u_0) = 0, \tag{7.102}$$

$$\text{node 3:} \ \frac{EI_Z}{\Delta X^3}(u_5 - 4u_4 + 6u_3 - 4u_2 + u_1) = 0, \tag{7.103}$$

$$\text{node 4:} \ \frac{EI_Z}{\Delta X^3}(u_6 - 4u_5 + 6u_4 - 4u_3 + u_2) = 0. \tag{7.104}$$

The vertical displacement is zero at the left-hand boundary because of the fixed support and it follows immediately that $u_1 = 0$ holds. In addition, the rotation is zero at the fixed support, i.e $\left.\frac{du}{dX}\right|_1 = 0$, and a centered finite difference approach according to Table 7.1 gives the condition $u_0 = u_2$.

The equilibrium between the internal reactions and the external load, cf. Fig. 7.15, and the application of the bending differential equations in the form of Eqs. (4.37) and (4.38) gives the following conditions at the right-hand boundary:

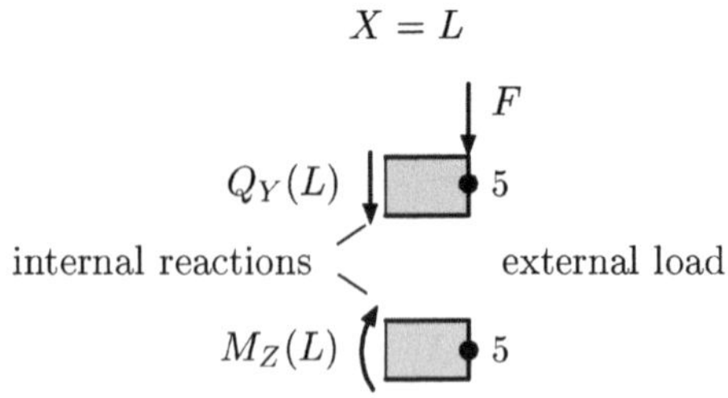


**Fig. 7.15** Equilibrium between internal reactions and external load at the right-hand boundary ($X = L$)

$$EI_Z \left.\frac{\mathrm{d}^2u}{\mathrm{d}X^2}\right|_5 = M_Z(X = L) = 0, \tag{7.105}$$

$$EI_Z \left.\frac{\mathrm{d}^3u}{\mathrm{d}X^3}\right|_5 = -Q_Y(X = L) = F. \tag{7.106}$$

Application of the centered difference approximations of the derivatives as given in Table 7.1, gives

$$EI_Z \times \frac{u_6 - 2u_5 + u_4}{\Delta X^2} = 0 \quad \text{or} \quad u_6 = 2u_5 - u_4. \tag{7.107}$$

The evaluation of the condition for the shear force gives in the case of the centered difference scheme

$$EI_Z \times \frac{u_7 - 2u_6 + 2u_4 - u_3}{2\Delta X^3} = F, \tag{7.108}$$

which introduces as $i = 7$ a second fictitious node at the right-hand boundary. To overcome this problem, the finite difference approximation for the boundary node $i = 5$ can be written as:

$$\text{node 5:} \ \frac{EI_Z}{\Delta X^3}(u_7 - 4u_6 + 6u_5 - 4u_4 + u_3) = -F. \tag{7.109}$$

Introduction of the relationships for $u_0, u_1, u_6$ and $u_7$ which are obtained from the boundary conditions in the finite difference approximations for nodes 2 till 5 gives the following system of equations:

$$\text{node 2:} \ \frac{EI_Z}{\Delta X^3}(7u_2 - 4u_3 + u_4) = 0, \tag{7.110}$$

$$\text{node 3:} \ \frac{EI_Z}{\Delta X^3}(-4u_2 + 6u_3 - 4u_4 + u_5) = 0, \tag{7.111}$$

$$\text{node 4:} \ \frac{EI_Z}{\Delta X^3}(u_2 - 4u_3 + 5u_4 - 2u_5) = 0, \tag{7.112}$$

$$\text{node 5: } \frac{EI_Z}{\Delta X^3}(2u_3 - 4u_4 + 2u_5) = -3F, \tag{7.113}$$

or in matrix form:

$$\frac{EI_Z}{\Delta X^3}\begin{bmatrix} 7 & -4 & 1 & 0 \\ -4 & 6 & -4 & 1 \\ 1 & -4 & 5 & -2 \\ 0 & 2 & -4 & 2 \end{bmatrix}\begin{bmatrix} u_2 \\ u_3 \\ u_4 \\ u_5 \end{bmatrix} = \begin{bmatrix} 0 \\ 0 \\ 0 \\ -3F \end{bmatrix}. \tag{7.114}$$

The solution of this linear system of equations gives the unknown nodal values as:

$$u_2 = -\frac{3F\Delta X^3}{EI_Z}, \quad u_3 = -\frac{21F\Delta X^3}{2EI_Z}, \quad u_4 = -\frac{21F\Delta X^3}{EI_Z}, \quad u_5 = -\frac{33F\Delta X^3}{EI_Z}, \tag{7.115}$$

or with $\Delta X = \frac{L}{4}$ as:

$$u_2 = -\frac{3FL^3}{64EI_Z}, \quad u_3 = -\frac{21FL^3}{128EI_Z}, \quad u_4 = -\frac{21FL^3}{64EI_Z}, \quad u_5 = -\frac{33FL^3}{64EI_Z}. \tag{7.116}$$

The analytical solution for the displacement at the force application point can be taken from Table 4.10 as $\frac{-FL^3}{3EI_Z}$ and the relative error is obtained as:

$$\text{relative error} = \frac{\frac{33}{64} - \frac{1}{3}}{\frac{1}{3}} \times 100 = 54.689\,\%. \tag{7.117}$$

A different way of solution can be chosen by avoiding the second ($i = 7$) fictitious node at the right-hand end. To this end, a backward finite difference approximation (cf. Table 7.1) can be introduced into the condition for the internal shear force at the right-hand end:

$$EI_Z \underbrace{\frac{5u_5 - 18u_4 + 24u_3 - 14u_2 + 3u_1}{2\Delta X^3}}_{\left.\frac{d^3u}{dX^3}\right|_5} = -Q_Y(X = L) = F. \tag{7.118}$$

The fifth equation reads now under consideration of the boundary condition at the left-hand end ($u_1 = 0$)

$$\text{node 5: } 5u_5 - 18u_4 + 24u_3 - 14u_2 = \frac{2\Delta X^3 F}{EI_Z}, \tag{7.119}$$

and the final system of equations is obtained as:

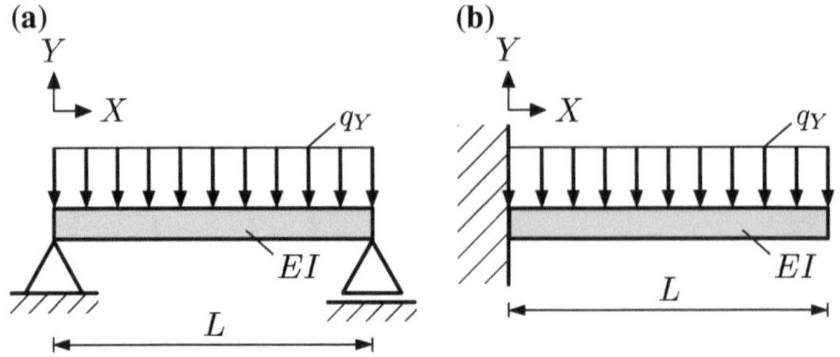


**Fig. 7.16** **a** Simply supported and **b** cantilevered BERNOULLI beam loaded by a distributed load

$$\frac{EI_Z}{\Delta X^3}\begin{bmatrix} 7 & -4 & 1 & 0 \\ -4 & 6 & -4 & 1 \\ 1 & -4 & 5 & -2 \\ -14 & 24 & -18 & 5 \end{bmatrix}\begin{bmatrix} u_2 \\ u_3 \\ u_4 \\ u_5 \end{bmatrix} = \begin{bmatrix} 0 \\ 0 \\ 0 \\ 2F \end{bmatrix}. \tag{7.120}$$

The solution of this linear system of equations gives with $\Delta X = \frac{L}{4}$ the unknown nodal values as:

$$u_2 = -\frac{FL^3}{32EI_Z}, \quad u_3 = -\frac{7FL^3}{64EI_Z}, \quad u_4 = -\frac{7FL^3}{32EI_Z}, \quad u_5 = -\frac{11FL^3}{32EI_Z}, \tag{7.121}$$

and the relative error is obtained as:

$$\text{relative error} = \frac{\frac{11}{32} - \frac{1}{3}}{\frac{1}{3}} \times 100 = 3.126\,\%. \tag{7.122}$$

**7.2 Example: Finite difference approximation of a simply supported and cantilevered beam loaded by a distributed load**

Given is a BERNOULLI beam with different supports as shown in Fig. 7.16. The bending stiffness $EI_Z$ is constant and the length is equal to $L$. The simply supported beam (a) and the cantilevered beam (b) are loaded by a constant distributed load $q_Y$. Derive for both cases a finite difference approximation based on five grid points, i.e. an equidistant spacing of $\Delta X = \frac{L}{4}$.
Determine for both cases

- the displacement in the middle of the beam for (a) and at the free tip for case (b),
- the analytical solution and
- calculate the relative error between the analytical and finite difference solution.

**7.2 Solution**

The problem of this example is very similar to the previous example 7.1 and major parts can be handled in a similar way.

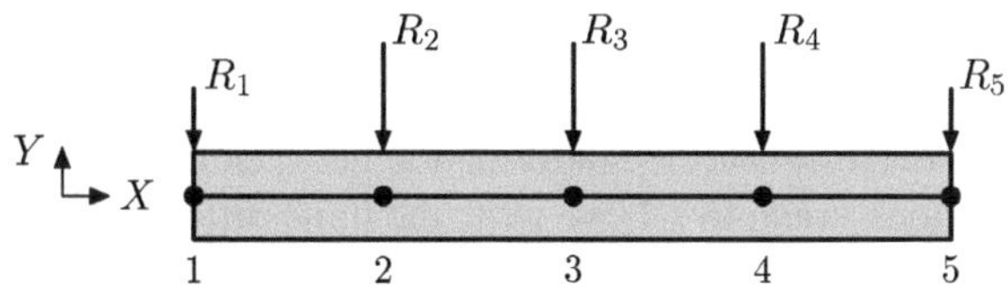


**Fig. 7.17** Equivalent force system for the distributed load problem shown in Fig. 7.16b: $R_1 = R_5 = -\frac{q_Y \Delta X}{2}$, $R_2 = R_3 = R_4 = q_Y \Delta X$

(a) For the case of the simply supported beam we write—as in the case of example 7.1—the finite difference approximation according to Eq. (7.85) at the inner nodes $i = 2, \ldots, 4$ under consideration of the distributed load as:

$$\text{node 2: } \frac{EI_Z}{\Delta X^3}(u_4 - 4u_3 + 6u_2 - 4u_1 + u_0) = -q_Y \Delta X, \tag{7.123}$$

$$\text{node 3: } \frac{EI_Z}{\Delta X^3}(u_5 - 4u_4 + 6u_3 - 4u_2 + u_1) = -q_Y \Delta X, \tag{7.124}$$

$$\text{node 4: } \frac{EI_Z}{\Delta X^3}(u_6 - 4u_5 + 6u_4 - 4u_3 + u_2) = -q_Y \Delta X. \tag{7.125}$$

Consideration of the boundary conditions, i.e. $u_1 = u_5 = 0$, $u_0 = -u_2$ and $u_6 = -u_4$, results in the following system of equations:

$$\frac{EI_Z}{\Delta X^3}\begin{bmatrix} 5 & -4 & 1 \\ -4 & 6 & -4 \\ 1 & -4 & 5 \end{bmatrix}\begin{bmatrix} u_2 \\ u_3 \\ u_4 \end{bmatrix} = \begin{bmatrix} -q_Y \Delta X \\ -q_Y \Delta X \\ -q_Y \Delta X \end{bmatrix}. \tag{7.126}$$

The solution of this linear system of equations gives the unknown nodal values as:

$$u_2 = -\frac{5 q_Y \Delta X^4}{2EI_Z}, \quad u_3 = -\frac{7 q_Y \Delta X^4}{2EI_Z}, \quad u_4 = -\frac{5 q_Y \Delta X^4}{2EI_Z}, \tag{7.127}$$

or with $\Delta X = \frac{L}{4}$ as:

$$u_2 = -\frac{5 q_Y L^4}{512 EI_Z}, \quad u_3 = -\frac{7 q_Y L^4}{512 EI_Z}, \quad u_4 = -\frac{5 q_Y L^4}{512 EI_Z}. \tag{7.128}$$

The analytical solution for the displacement in the middle of the beam can be taken from Table 4.11 as $\frac{-5 q_Y L^4}{384 EI_Z}$ and the relative error is obtained as:

$$\text{relative error} = \frac{\frac{7}{512} - \frac{5}{384}}{\frac{5}{384}} \times 100 = 5.0\,\%. \tag{7.129}$$

The graphical comparison between this finite difference approach and the analytical solution is presented in Fig. 7.18a where it can be seen that the finite difference approach slightly overestimates the deformation between the boundary supports.
(b) As in the previous example 7.1, we write the finite difference approximation for nodes $i = 2, \ldots, 5$ under consideration of the distributed load as[4]:

$$\text{node 2: } \frac{EI_Z}{\Delta X^3}(u_4 - 4u_3 + 6u_2 - 4u_1 + u_0) = -q_Y \Delta X, \tag{7.130}$$

$$\text{node 3: } \frac{EI_Z}{\Delta X^3}(u_5 - 4u_4 + 6u_3 - 4u_2 + u_1) = -q_Y \Delta X, \tag{7.131}$$

$$\text{node 4: } \frac{EI_Z}{\Delta X^3}(u_6 - 4u_5 + 6u_4 - 4u_3 + u_2) = -q_Y \Delta X, \tag{7.132}$$

$$\text{node 5: } \frac{EI_Z}{\Delta X^3}(u_7 - 4u_6 + 6u_5 - 4u_4 + u_3) = -q_Y \frac{\Delta X}{2}. \tag{7.133}$$

Consideration of the boundary conditions at node 1 gives as in the previous example $u_1 = 0$ and $u_0 = u_2$.

The equilibrium between the internal reactions and the external load based on the equivalent force system (cf. Fig. 7.17) gives here

$$EI_Z \left. \frac{\mathrm{d}^2 u}{\mathrm{d}X^2} \right|_5 = M_Z(X = L) = 0, \tag{7.134}$$

$$EI_Z \left. \frac{\mathrm{d}^3 u}{\mathrm{d}X^3} \right|_5 = -Q_Y(X = L) = q_Y \frac{\Delta X}{2}, \tag{7.135}$$

from which the following two conditions can be derived if a centered difference approximation is introduced:

$$u_6 = 2u_5 - u_4, \tag{7.136}$$

$$u_7 = 4u_5 - 4u_4 + u_3 + \frac{q_Y \Delta X^4}{EI_Z}. \tag{7.137}$$

Thus, the final system of equations is given as:

$$\frac{EI_Z}{\Delta X^3} \begin{bmatrix} 7 & -4 & 1 & 0 \\ -4 & 6 & -4 & 1 \\ 1 & -4 & 5 & -2 \\ 0 & 2 & -4 & 2 \end{bmatrix} \begin{bmatrix} u_2 \\ u_3 \\ u_4 \\ u_5 \end{bmatrix} = \begin{bmatrix} -q_Y \Delta X \\ -q_Y \Delta X \\ -q_Y \Delta X \\ -q_Y \frac{3\Delta X}{2} \end{bmatrix}. \tag{7.138}$$

The solution of this linear system of equations gives the unknown nodal values as:

---

[4] It is important to consider for the equivalent nodal force at the boundary node 5 only the effective length of $\frac{\Delta X}{2}$: $R_5 = q_Y \frac{\Delta X}{2}$.

$$u_2 = -\frac{9 q_Y \Delta X^4}{2 E I_Z}, \quad u_3 = -\frac{57 q_Y \Delta X^4}{4 E I_Z}, \quad u_4 = -\frac{53 q_Y \Delta X^4}{2 E I_Z}, \quad u_5 = -\frac{79 q_Y \Delta X^4}{2 E I_Z}, \tag{7.139}$$

or with $\Delta X = \frac{L}{4}$ as:

$$u_2 = -\frac{9 q_Y L^4}{512 E I_Z}, \quad u_3 = -\frac{57 q_Y L^4}{1024 E I_Z}, \quad u_4 = -\frac{53 q_Y L^4}{512 E I_Z}, \quad u_5 = -\frac{79 q_Y L^4}{512 E I_Z}. \tag{7.140}$$

The analytical solution for the displacement at the beam tip can be taken from Table 4.10 as $\frac{-q_Y L^4}{8 E I_Z}$ and the relative error is obtained as:

$$\text{relative error} = \frac{\frac{79}{512} - \frac{1}{8}}{\frac{1}{8}} \times 100 = 23.438\,\%. \tag{7.141}$$

A different approach to the solution can be chosen by replacing the boundary condition (7.135) by the analytical boundary condition which is not based on the equivalent system, i.e.

$$E I_Z \left.\frac{\mathrm{d}^3 u}{\mathrm{d}X^3}\right|_5 = -Q_Y(X = L) = 0, \tag{7.142}$$

and the condition for the fictitious node 7 is obtained as

$$u_7 = 4u_5 - 4u_4 + u_3. \tag{7.143}$$

Thus, the final system of equations is obtained for this case as:

$$\frac{E I_Z}{\Delta X^3}\begin{bmatrix} 7 & -4 & 1 & 0 \\ -4 & 6 & -4 & 1 \\ 1 & -4 & 5 & -2 \\ 0 & 2 & -4 & 2 \end{bmatrix}\begin{bmatrix} u_2 \\ u_3 \\ u_4 \\ u_5 \end{bmatrix} = \begin{bmatrix} -q_Y \Delta X \\ -q_Y \Delta X \\ -q_Y \Delta X \\ -q_Y \frac{\Delta X}{2} \end{bmatrix}. \tag{7.144}$$

The solution of this linear system of equations gives with $\Delta X = \frac{L}{4}$ the unknown nodal values as:

$$u_2 = -\frac{7 q_Y L^4}{512 E I_Z}, \quad u_3 = -\frac{43 q_Y L^4}{1024 E I_Z}, \quad u_4 = -\frac{39 q_Y L^4}{512 E I_Z}, \quad u_5 = -\frac{57 q_Y L^4}{512 E I_Z}, \tag{7.145}$$

and a much smaller relative error is obtained:

$$\text{relative error} = \frac{\frac{57}{512} - \frac{1}{8}}{\frac{1}{8}} \times 100 = 10.938\,\%. \tag{7.146}$$

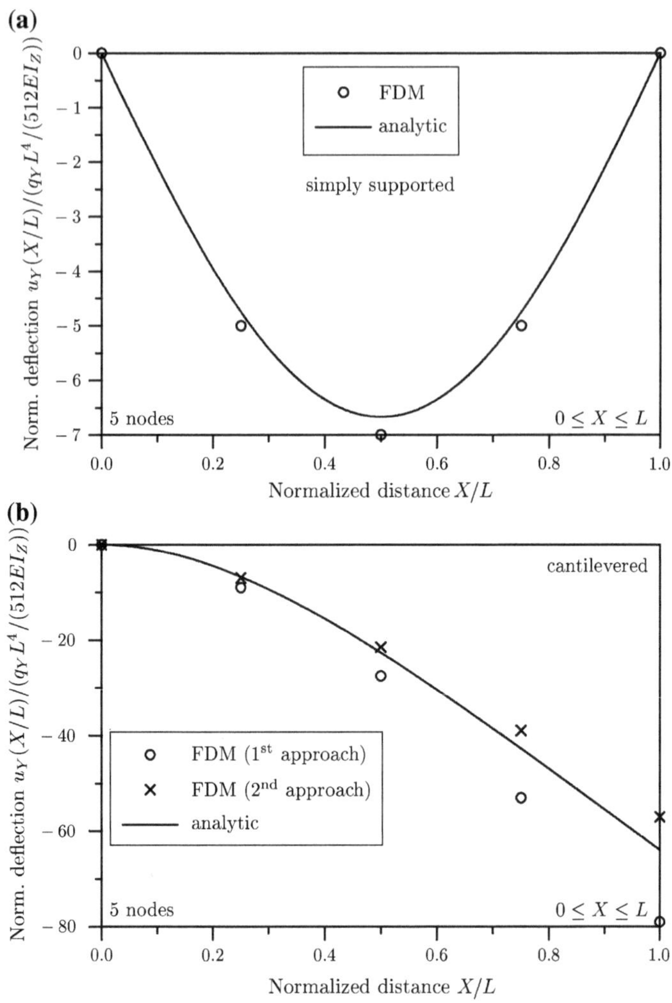


**Fig. 7.18** Displacement along the normalized beam axis obtained from a FDM approach compared to the analytical solution: **a** simply supported and **b** cantilevered beam, cf. Fig. 7.16

The graphical comparison between these two approaches and the analytical solution is given in Fig. 7.18. Depending on the approach, a different behavior, i.e. to over- or underestimate the analytical solution is obtained.

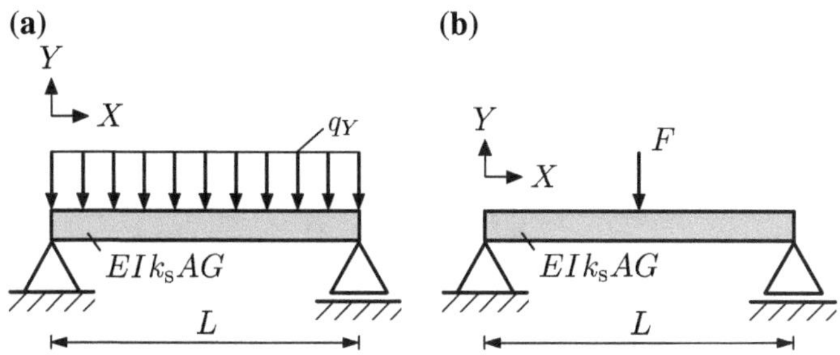


**Fig. 7.19** Simply supported TIMOSHENKO beam: **a** distributed load case; **b** single force case

## 7.4 Investigation of Timoshenko Beams in the Elastic Range

The consideration of TIMOSHENKO beams requires the simultaneous solution of a coupled system of differential equations as given in Table 4.19. In order to focus on the basic idea of the numerical approach, the derivations in the following are restricted to the case of constant material parameters.

**7.3 Example: Finite difference approximation of simply supported beams under different loading conditions**
Given is a simply supported TIMOSHENKO beam as shown in Fig. 7.19. The material parameters of the beam are constant and the length is equal to $L$. The simply supported beam is loaded by a constant distributed load $q_Y$ for case (a) and by a single force $F$ for case (b). Use first five domain nodes of equidistant spacing ($\Delta X = \frac{L}{4}$) and then 17 domain nodes of equidistant spacing ($\Delta X = \frac{L}{16}$) for the finite difference approximation.
Determine for both cases

- the general expression for the vertical displacement in the middle of the beam, i.e. $X = \frac{L}{2}$,
- the analytical solution and
- calculate the relative error between the analytical and finite difference solution for a squared cross-sectional area of length 0.5. Other values are: $E = 10000$; $\nu = 0.0$; $L = 10$; $q_y = 1$; $F = 0.1$.

**7.3 Solution**
The coupled differential equations which describe the problem can be extracted from Table 4.19 as:

$$EI_Z \frac{\mathrm{d}^2\phi_Z}{\mathrm{d}X^2} + k_s GA \left( \frac{\mathrm{d}u_Y}{\mathrm{d}X} - \phi_Z \right) = 0, \tag{7.147}$$

$$k_s GA \left( \frac{\mathrm{d}^2 u_Y}{\mathrm{d}X^2} - \frac{\mathrm{d}\phi_Z}{\mathrm{d}X} \right) = -q_Y. \tag{7.148}$$

As can be seen from these two equations, a finite difference approximation is required for first and second order derivatives. These approximations can be taken from Table 7.1 and are given for second order accurate centered difference schemes as[5]:

$$\frac{\mathrm{d}^2\phi_Z}{\mathrm{d}X^2} \approx \frac{\phi_{i+1} - 2\phi_i + \phi_{i-1}}{\Delta X^2}, \quad \frac{\mathrm{d}u_Y}{\mathrm{d}X} \approx \frac{u_{i+1} - u_{i-1}}{2\Delta X}, \tag{7.149}$$

$$\frac{\mathrm{d}^2 u_Y}{\mathrm{d}X^2} \approx \frac{u_{i+1} - 2u_i + u_{i-1}}{\Delta X^2}, \quad \frac{\mathrm{d}\phi_Z}{\mathrm{d}X} \approx \frac{\phi_{i+1} - \phi_{i-1}}{2\Delta X}. \tag{7.150}$$

Thus, the coupled differential equations given in Eqs. (7.147) and (7.148) can be approximated by the following finite difference scheme:

$$EI_Z \frac{\phi_{i+1} - 2\phi_i + \phi_{i-1}}{\Delta X^2} + k_s GA \left( \frac{u_{i+1} - u_{i-1}}{2\Delta X} - \phi_i \right) = 0, \tag{7.151}$$

$$k_s GA \left( \frac{u_{i+1} - 2u_i + u_{i-1}}{\Delta X^2} - \frac{\phi_{i+1} - \phi_{i-1}}{2\Delta X} \right) = -q_Y. \tag{7.152}$$

(a) Distributed load case

Considering five domain nodes (cf. Fig. 7.13), i.e. inner nodes 2, 4 and 5, the finite difference approximations for the three inner nodes can be written as:

$$\text{node 2:} \quad EI_Z \frac{\phi_3 - 2\phi_2 + \phi_1}{\Delta X^2} + k_s GA \left( \frac{u_3 - u_1}{2\Delta X} - \phi_2 \right) = 0, \tag{7.153}$$

$$k_s GA \left( \frac{u_3 - 2u_2 + u_1}{\Delta X^2} - \frac{\phi_3 - \phi_1}{2\Delta X} \right) = -q_Y, \tag{7.154}$$

$$\text{node 3:} \quad EI_Z \frac{\phi_4 - 2\phi_3 + \phi_2}{\Delta X^2} + k_s GA \left( \frac{u_4 - u_2}{2\Delta X} - \phi_3 \right) = 0, \tag{7.155}$$

$$k_s GA \left( \frac{u_4 - 2u_3 + u_2}{\Delta X^2} - \frac{\phi_4 - \phi_2}{2\Delta X} \right) = -q_Y, \tag{7.156}$$

[5] The indices $Y$ and $Z$ will be abandoned in the following to simplify the notation.

$$\text{node 4:}\quad EI_Z\frac{\phi_5-2\phi_4+\phi_3}{\Delta X^2}+k_sGA\left(\frac{u_5-u_3}{2\Delta X}-\phi_4\right)=0, \tag{7.157}$$

$$k_sGA\left(\frac{u_5-2u_4+u_3}{\Delta X^2}-\frac{\phi_5-\phi_3}{2\Delta X}\right)=-q_Y. \tag{7.158}$$

Consideration of the boundary conditions for the deflection gives $u(0)=u_1=0$ and $u(L)=u_5=0$. The unknown rotations at the ends, i.e. $\phi_1$ and $\phi_5$, can be eliminated from the system of equations by considering the condition for the moments at the ends: $M(0)=0$ and $M(L)=0$. From Eq. (4.172), the relationship between bending moment and rotation is given by $M_Z(X)=EI_Z\frac{\mathrm{d}\phi_Z}{\mathrm{d}X}$ and in order to avoid fictitious nodes, the following forward and backward difference approximations for the first order derivative with second order accuracy can be taken from Table 7.1:

$$\frac{\mathrm{d}\phi}{\mathrm{d}X}\approx\frac{-3u_i+4u_{i+1}-u_{i+2}}{2\Delta X}\quad\text{(forward)}, \tag{7.159}$$

$$\frac{\mathrm{d}\phi}{\mathrm{d}X}\approx\frac{3u_i-4u_{i-1}+u_{i-2}}{2\Delta X}\quad\text{(backward)}. \tag{7.160}$$

Application of the forward scheme to the left-hand boundary node ($i=1$) and the backward scheme to the right-hand boundary node ($i=1$) gives the following two relationships for the elimination of the rotations at the boundaries:

$$\phi_1=\frac{4}{3}\phi_2-\frac{1}{3}\phi_3, \tag{7.161}$$

$$\phi_5=\frac{4}{3}\phi_4-\frac{1}{3}\phi_3. \tag{7.162}$$

Introducing these relationships in the system of equations given by Eqs. (7.153) till (7.158) and considering that the load is opposed to the positive $Y$-direction, the system of equations can be written in matrix from as given in Eq. (7.166).
The solution of this linear system of equations gives in the middle of the beam under consideration of ($\Delta X=\frac{L}{4}$) the vertical displacement as:

$$u_3=u(\tfrac{L}{2})=-\frac{\left(32EI_Z+3\,L^2k_sAG\right)L^2q}{4k_sAG\left(64EI_Z+L^2k_sAG\right)}. \tag{7.163}$$

The analytical solution is given in Table 4.21 as $u_Y\left(\frac{L}{2}\right)=\frac{5q_YL^4}{384EI_Z}+\frac{q_YL^2}{8k_sAG}$ and the relative error results for the given numerical values as $-97.208\,\%$. Thus, the approximation is not very satisfying and a higher node density is required to improve the results.

$$\left[\begin{array}{cccccc}
0 & -\frac{2EI_Z}{3\Delta X^2} - k_s AG & \frac{k_s AG}{2\Delta X} & \frac{2EI_Z}{3\Delta X^2} & 0 & 0 \\
-\frac{2k_s AG}{\Delta X^2} & \frac{2k_s AG}{3\Delta X} & \frac{k_s AG}{\Delta X^2} & -\frac{2k_s AG}{3\Delta X} & 0 & 0 \\
\hline
-\frac{k_s AG}{2\Delta X} & \frac{EI_Z}{\Delta X^2} & 0 & -\frac{2EI_Z}{\Delta X^2} - k_s AG & \frac{k_s AG}{2\Delta X} & \frac{EI_Z}{\Delta X^2} \\
\frac{k_s AG}{\Delta X^2} & \frac{k_s AG}{2\Delta X} & -\frac{2k_s AG}{\Delta X^2} & 0 & \frac{k_s AG}{\Delta X^2} & -\frac{k_s AG}{2\Delta X} \\
\hline
0 & 0 & -\frac{k_s AG}{2\Delta X} & \frac{2EI_Z}{3\Delta X^2} & 0 & -\frac{2EI_Z}{3\Delta X^2} - k_s AG \\
0 & 0 & \frac{k_s AG}{\Delta X^2} & \frac{2k_s AG}{3\Delta X} & -\frac{2k_s AG}{\Delta X^2} & -\frac{2k_s AG}{3\Delta X}
\end{array}\right]
\left[\begin{array}{c} u_2 \\ \phi_2 \\ \hline u_3 \\ \phi_3 \\ \hline u_4 \\ \phi_4 \end{array}\right]
=
\left[\begin{array}{c} 0 \\ q \\ \hline 0 \\ q \\ \hline 0 \\ q \end{array}\right]. \tag{7.164}$$

Considering 17 domain nodes, a linear system of dimension $30 \times 30$ can be derived based on the 15 inner nodes. The vertical displacement is now as

$$u_9 = u(\tfrac{L}{2}) = -\frac{\left(512EI_Z + 53L^2 k_s AG\right) L^2 q}{4k_s GA \left(1024EI_Z + L^2 k_s AG\right)} \tag{7.165}$$

obtained and the relative error decreases to $-66.348\,\%$. As can be seen from this result, a larger number of nodes is required to achieve an acceptable error in the case of TIMOSHENKO beams.

(b) Single force case

In order to account for single forces which are acting on the structure, the system of equations given in Eqs. (7.151) and (7.152) must be multiplied by $\Delta X$ to obtain on the right-hand side the expression for the equivalent nodal force. Since this example involves only single *forces*,[6] it would be sufficient to multiply only the second equation by $\Delta X$. Thus, the linear system of equations is obtained for this case as:

[6] In the case of *single* moments, Eq. (7.151) should be multiplied by $\Delta X$.

$$\begin{bmatrix} 0 & -\frac{2EI_Z}{3\Delta X^2}-k_sAG & \frac{k_sAG}{2\Delta X} & \frac{2EI_Z}{3\Delta X^2} & 0 & 0 \\ -\frac{2k_sAG}{\Delta X} & \frac{2k_sAG}{3} & \frac{k_sAG}{\Delta X} & -\frac{2k_sAG}{3} & 0 & 0 \\ \hline -\frac{k_sAG}{2\Delta X} & \frac{EI_Z}{\Delta X^2} & 0 & -\frac{2EI_Z}{\Delta X^2}-k_sAG & \frac{k_sAG}{2\Delta X} & \frac{EI_Z}{\Delta X^2} \\ \frac{k_sAG}{\Delta X} & \frac{k_sAG}{2} & -\frac{2k_sAG}{\Delta X} & 0 & \frac{k_sAG}{\Delta X} & -\frac{k_sAG}{2} \\ \hline 0 & 0 & -\frac{k_sAG}{2\Delta X} & \frac{2EI_Z}{3\Delta X^2} & 0 & -\frac{2EI_Z}{3\Delta X^2}-k_sAG \\ 0 & 0 & \frac{k_sAG}{\Delta X} & \frac{2k_sAG}{3} & -\frac{2k_sAG}{\Delta X} & -\frac{2k_sAG}{3} \end{bmatrix} \begin{bmatrix} u_2 \\ \phi_2 \\ \hline u_3 \\ \phi_3 \\ \hline u_4 \\ \phi_4 \end{bmatrix} = \begin{bmatrix} 0 \\ 0 \\ \hline 0 \\ F \\ \hline 0 \\ 0 \end{bmatrix}. \tag{7.166}$$

The vertical displacement is for the single load case obtained as

$$u_3 = u(\tfrac{L}{2}) = -\frac{\left(32EI_Z + 3L^2k_sAG\right)LF}{2k_sAG\left(64EI_Z + L^2k_sAG\right)} \tag{7.167}$$

and the relative error for the given numerical values is equal to $-96.514\,\%$. Using again 17 domain nodes changes the result in a general form to

$$u_9 = u(\tfrac{L}{2}) = -\frac{\left(512EI_Z + 43k_sAGL^2\right)LF}{2k_sGA\left(1024EI_Z + L^2k_sAG\right)}, \tag{7.168}$$

which gives a decreased relative error of $-65.875\,\%$.

## 7.5 Consideration of Bernoulli Beams with Plastic Material Behavior

The derivations of the previous sections will be extended in the following to elasto-plastic material behavior. For simplicity reasons, a simply supported beam under constant moment loading as shown in Fig. 7.20a is considered to explain an approach to consider plasticity for beams. The material is assumed to be

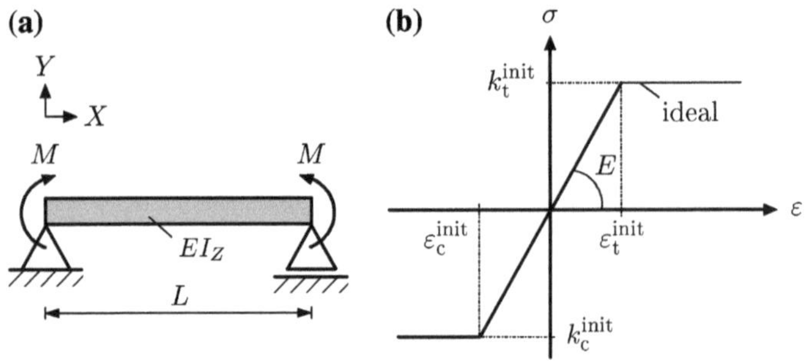


**Fig. 7.20** **a** Simply supported BERNOULLI beam under pure bending load; **b** linear-elastic/ideal-plastic material behavior

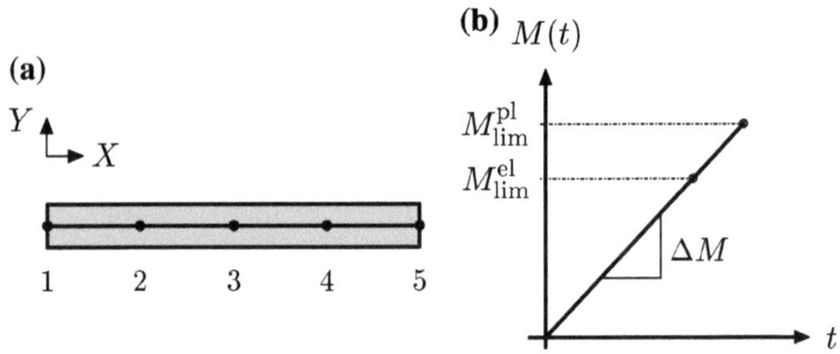


**Fig. 7.21** **a** Finite difference discretization of the beam and **b** moment-time function

linear-elastic/ideal-plastic as shown in Fig. 7.20b and this assumption allows to easily compare the numerical results with the analytical solutions of Sect. 4.2.6.1.

Let us use five domain nodes ($i = 1, \ldots, 5$) of equidistant spacing ($\Delta X = \frac{L}{4}$) for the finite difference approach as shown in Fig. 7.21a.

As in the case of the elasto-plastic finite *element* approach in Chap. 6, the load is now applied in incremental steps of magnitude $\Delta M$, see Fig. 7.21b. The indicated time $t$ is not important since we consider only time-independent material behavior and the expression 'time' can simply be replaced by 'step' or 'load step'. The following derivations do not consider the predictor-corrector scheme as in Chap. 6. A more simplified approach is introduced which is subject to the assumption of monotonic loading. Thus, the cases of unloading, load reversal and cyclic loading will be not covered in the framework of the following derivations.

The major challenge in the case of beams is that the stress state is changing over the height of the beam and a plastic layer is moving inwards with increasing load in the elasto-plastic range. One possible approach for this problem is the so-called layered approach[7] [2] where the cross section at the position of node $i$ is subdivided in a certain numbers of layers (see the grey cross section in Fig. 7.22).

[7] The layered approach can be applied in a similar manner in the framework of the finite element method, see Owen and Hinton [77].

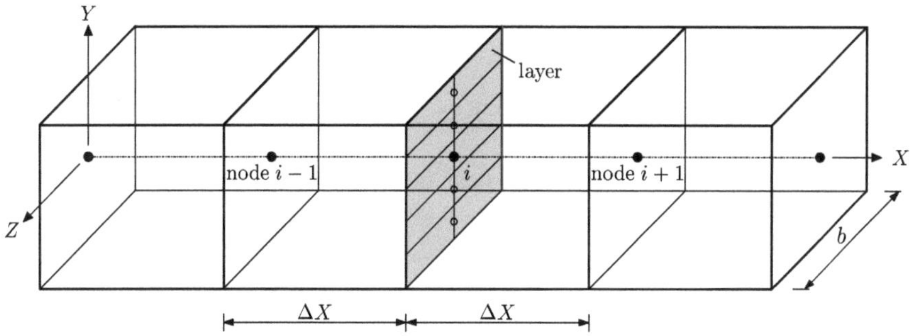


**Fig. 7.22** General configuration for the layered beam approach

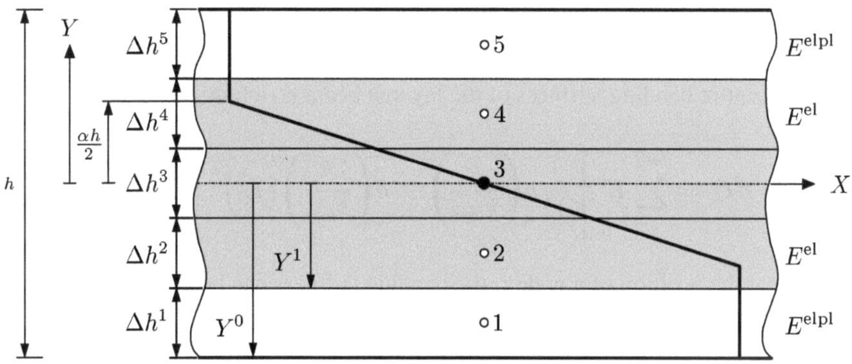


**Fig. 7.23** Example of a layered beam with five subdivisions ($k_{max} = 5$)

Let us have now a closer look at such a cross section as sown in Fig. 7.23.

The number of the layers ranges between $1 \leq k \leq k_{max} = 5$ and it is for simplicity assumed that each layer $k$ has the same height $\Delta h$. Furthermore, it is assumed that $k$ is an odd number which implies that the center (neural axis) of the beam is located in the middle of layer $\frac{k_{max}+1}{2}$. The coordinates of the top and bottom face of each layer, i.e. $Y^k$ and $Y^{k-1}$, can be expressed as

$$
\begin{aligned}
Y^0 &= -\frac{h}{2}, \\
Y^1 &= -\frac{h}{2} + \frac{h}{k_{max}}, \\
Y^2 &= -\frac{h}{2} + 2 \times \frac{h}{k_{max}}, \\
&\vdots \\
Y^k &= -\frac{h}{2} + k \times \frac{h}{k_{max}},
\end{aligned}
\tag{7.169}
$$

$$\vdots$$

$$Y^{k_{\max}} = +\frac{h}{2}.$$

The height of a layer $k$ ($1 \leq k \leq k_{\max}$) can be expressed as

$$\left|\Delta h^k\right| = Y^k - Y^{k-1}, \tag{7.170}$$

and the center coordinate of a layer $k$ is obtained as:

$$Y_{\mathrm{c}}^k = \frac{Y^k + Y^{k-1}}{2} = \frac{1}{2}\left(-h + \frac{(2k-1)h}{k_{\max}}\right). \tag{7.171}$$

Thus, the entire bending stiffness of the layered beam is obtained as (cf. Eq. B.6):

$$EI_Z = \sum_{k=1}^{k_{\max}} E^k \left[\frac{1}{12} b \left(\frac{h}{k_{\max}}\right)^3 + b\left(\frac{h}{k_{\max}}\right)\left(Y_{\mathrm{c}}^k\right)^2\right]. \tag{7.172}$$

Looking at the position of a node $i$, the bending stiffness can be expressed as

$$(EI_Z)_i = \sum_{k=1}^{k_{\max}} \frac{E_i^k}{E} E \underbrace{\frac{1}{12} bh^3}_{I_Z} \left[\frac{1}{(k_{\max})^3} + \frac{12}{k_{\max}}\left(\frac{Y_{\mathrm{c}}^k}{h}\right)^2\right]_i \tag{7.173}$$

$$= EI_Z \sum_{k=1}^{k_{\max}} \frac{E_i^k}{E}\left[\frac{1}{(k_{\max})^3} + \frac{12}{k_{\max}}\left(\frac{Y_{\mathrm{c}}^k}{h}\right)^2\right]_i, \tag{7.174}$$

or in dimensionless form and under consideration of Eq. (7.171) as:

$$\frac{(EI_Z)_i}{EI_Z} = \sum_{k=1}^{k_{\max}} \frac{E_i^k}{E}\left[\frac{1}{k_{\max}^3} + \frac{3}{k_{\max}}\left(\frac{2k-1}{k_{\max}} - 1\right)^2\right]_i = \alpha_i, \tag{7.175}$$

where it should be noted that $EI_Z$ is the bending stiffness of the entire cross section or beam in the pure elastic range and $E_i^k$ is the modulus of the $k$-th layer, i.e. $E_i^k = E$ in the elastic range and $E_i^k = 0$ in the plastic range. Thus, the dimensionless factor is in the pure elastic range $\alpha = 1$ and in the elastic-plastic range $0 \leq \alpha < 1$.

To decide if a layer is now considered as elastic ($E_i^k = E$) or plastic ($E_i^k = 0$), the following assumption is done: If the center of a layer—which is geometrically represented by the coordinate $Y_{\mathrm{c}}^k$—is in the elastic range, the *entire* layer $k$ is assumed

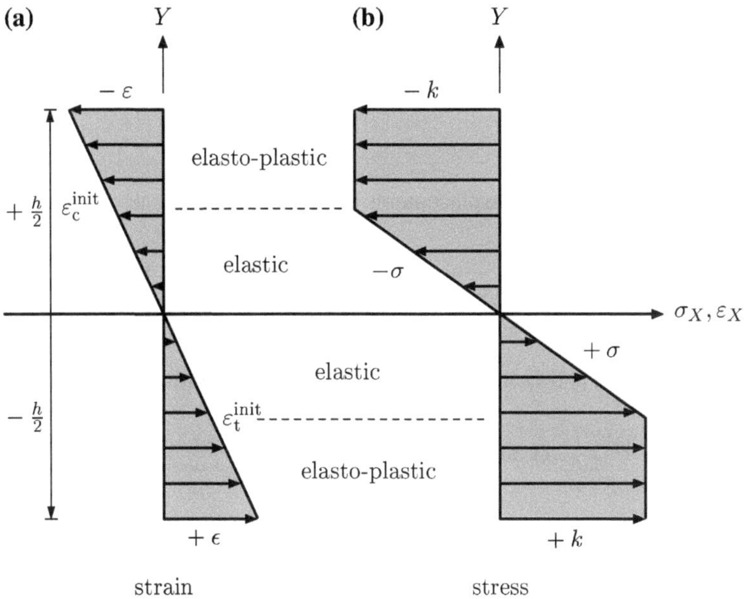


**Fig. 7.24** Difference between the **a** strain and **b** stress distribution in the elasto-plastic range for an ideal plastic material. The strain and stress state is uniaxial where only a single normal strain and stress ia acting

to be elastic. Vice versa, if the center of a layer is in the plastic range, the entire layer $k$ is assumed to be plastic. To develop now an appropriate strategy to decide if the center of a layer is in the pure elastic or elasto-plastic range, one may use the classical assumption that the strain is linearly distributed over the cross section even when the material is in the elasto-plastic range, see Fig. 7.24. Thus, the strain in the center of layer $k$ at node $i$ can be expressed according to Eq. (4.15) as:

$$(\varepsilon_X)_{k,i} = -\left( Y_{\mathrm{c}}^{k} \times \frac{\mathrm{d}^2 u_Y}{\mathrm{d}X^2} \right)_i , \tag{7.176}$$

where the second order derivative at node $i$ can be replaced by a centered difference scheme (truncation error of order $\Delta X^2$) as given in Table 7.1. Thus, the discretized form of Eq. (7.176) at node $i$ is given by:

$$(\varepsilon_X)_{k,i} = -\left( Y_{\mathrm{c}}^{k} \times \frac{u_{i+1} - 2u_i + u_{i-1}}{\Delta X^2} \right)_i . \tag{7.177}$$

It should be noted here that the displacements $u_i$ are calculated for the neutral fiber, i.e. for $Y_{\mathrm{c}}^{k} = 0$.

The analytical solutions in Sect. 4.2.6.1 are presented (see Fig. 4.22) in a normalized form as

$$\hat{u}_Y(X) = \frac{u_Y(X)}{\frac{M_{\text{lim}}^{\text{pl}} L^2}{E I_Z}}. \tag{7.178}$$

Introducing the normalized displacements $\hat{u}_Y$ in Eq. (7.177), ones obtains

$$(\varepsilon_X)_{k,i} = -\left( Y_{\text{c}}^k \times \frac{\hat{u}_{i+1} - 2\hat{u}_i + \hat{u}_{i-1}}{\Delta X^2} \times \frac{M_{\text{lim}}^{\text{pl}} L^2}{E I_Z} \right)_i. \tag{7.179}$$

Based on Eq. (4.78) and HOOKE's law, the normalizing factor can be expressed as:

$$\frac{M_{\text{lim}}^{\text{pl}} L^2}{E I_Z} \overset{(4.78)}{=} \frac{3 k I_Z}{E I_Z h} = \frac{3L^2}{h} \times \frac{k}{E} \overset{\text{HOOKE}}{=} \frac{3L^2}{h} \times \varepsilon^{\text{init}}. \tag{7.180}$$

Finally, one can state the condition for an elastic layer $k$ at node $i$ as:

$$\left| \frac{(\varepsilon_X)_k}{\varepsilon^{\text{init}}} \right|_i = \left| Y_{\text{c}}^k \times \frac{\hat{u}_{i+1} - 2\hat{u}_i + \hat{u}_{i-1}}{\Delta X^2} \times \frac{3L^2}{h} \right|_i \leq 1 \text{ (elastic)}. \tag{7.181}$$

If the fraction is larger than one, the layer is assumed to be in the plastic range. In the following, let us look at the entire finite difference solution of the problem. Since we assumed a constant moment loading of the beam (see Fig 7.20a), one may use the partial differential equation of the problem in the incremental form $E I_Z \frac{\text{d}^2 \Delta u}{\text{d}X^2} = \Delta M$ and a centered finite difference approximation to be evaluated at the inner nodes $i = 2, \ldots, 4$ to give:

$$\text{node 2: } \frac{(E I_Z)_2}{\Delta X^2} (\Delta u_3 - 2\Delta u_2 + \Delta u_1) = \Delta M_2, \tag{7.182}$$

$$\text{node 3: } \frac{(E I_Z)_3}{\Delta X^2} (\Delta u_4 - 2\Delta u_3 + \Delta u_2) = \Delta M_3, \tag{7.183}$$

$$\text{node 4: } \frac{(E I_Z)_4}{\Delta X^2} (\Delta u_5 - 2\Delta u_4 + \Delta u_3) = \Delta M_4. \tag{7.184}$$

The last three equations can be represented in matrix form under consideration of the boundary conditions $u_1 = u_5 = 0$ as:

$$\frac{1}{\Delta X^2} \begin{bmatrix} -2(E I_Z)_2 & 1(E I_Z)_2 & 0(E I_Z)_2 \\ 1(E I_Z)_3 & -2(E I_Z)_3 & 1(E I_Z)_3 \\ 0(E I_Z)_4 & 1(E I_Z)_4 & -2(E I_Z)_4 \end{bmatrix} \begin{bmatrix} \Delta u_2 \\ \Delta u_3 \\ \Delta u_4 \end{bmatrix} = \begin{bmatrix} \Delta M_2 \\ \Delta M_3 \\ \Delta M_4 \end{bmatrix}. \tag{7.185}$$

Since the coefficient matrix is now dependent on the solution (i.e. depending on the deformation, the $(EI_Z)_i$ may take different values), a residual form can be written as

$$\begin{bmatrix} r_2 \\ r_3 \\ r_4 \end{bmatrix} = \frac{1}{\Delta X^2} \begin{bmatrix} -2(EI_Z)_2 & 1(EI_Z)_2 & 0(EI_Z)_2 \\ 1(EI_Z)_3 & -2(EI_Z)_3 & 1(EI_Z)_3 \\ 0(EI_Z)_4 & 1(EI_Z)_4 & -2(EI_Z)_4 \end{bmatrix} \begin{bmatrix} \Delta u_2 \\ \Delta u_3 \\ \Delta u_4 \end{bmatrix} - \begin{bmatrix} \Delta M_2 \\ \Delta M_3 \\ \Delta M_4 \end{bmatrix}, \tag{7.186}$$

or in abbreviated form as:

$$\boldsymbol{r}(\Delta \boldsymbol{u}) = \boldsymbol{K}(\Delta \boldsymbol{u})\Delta \boldsymbol{u} - \Delta \boldsymbol{F}. \tag{7.187}$$

To solve this nonlinear system of equations, a complete NEWTON- RAPHSON iteration (iteration index $j$) can be applied in the following form [75]:

$$\Delta \boldsymbol{u}^{(j+1)} = \Delta \boldsymbol{u}^{(j)} - \left(\boldsymbol{K}_{\mathrm{T}}^{(j)}\right)^{-1} \boldsymbol{r}(\Delta \boldsymbol{u}^{(j)}), \tag{7.188}$$

where $\boldsymbol{K}_{\mathrm{T}}$ is the so-called tangent stiffness matrix. Equation (7.188) must be iterated as long as, for example, the normalized difference between two consecutive iteration steps is not below a certain threshold. Thus, the iteration can be stopped if the following condition is satisfied:

$$\sqrt{\frac{\left(\Delta u_2^{(j)} - \Delta u_2^{(j-1)}\right)^2 + \left(\Delta u_3^{(j)} - \Delta u_3^{(j-1)}\right)^2 + \left(\Delta u_4^{(j)} - \Delta u_4^{(j-1)}\right)^2}{\left(\Delta u_2^{(j)}\right)^2 + \left(\Delta u_3^{(j)}\right)^2 + \left(\Delta u_4^{(j)}\right)^2}} \leq t_{\mathrm{end}}, \tag{7.189}$$

where $t_{\mathrm{end}}$ is a small number. Before applying the iteration scheme of Eq. (7.188), it is appropriate to have a closer look at the tangent stiffness matrix $\boldsymbol{K}_{\mathrm{T}}$. In general, the tangent stiffness matrix can be expressed as [75]

$$\boldsymbol{K}_{\mathrm{T}} = \frac{\partial \boldsymbol{r}(\Delta \boldsymbol{u})}{\partial \Delta \boldsymbol{u}} = \boldsymbol{K} + \frac{\partial \boldsymbol{K}}{\partial \Delta \boldsymbol{u}} \Delta \boldsymbol{u}, \tag{7.190}$$

which can be written in our case of three unknowns as:

$$\boldsymbol{K}_{\mathrm{T}} = \underbrace{\boldsymbol{K}}_{3\times 3} + \underbrace{\left[ \frac{\partial \boldsymbol{K}}{\partial \Delta u_2} \Delta \boldsymbol{u} \,\middle|\, \frac{\partial \boldsymbol{K}}{\partial \Delta u_3} \Delta \boldsymbol{u} \,\middle|\, \frac{\partial \boldsymbol{K}}{\partial \Delta u_4} \Delta \boldsymbol{u} \right]}_{3\times 3 \text{ matrix}}, \tag{7.191}$$

or in components:

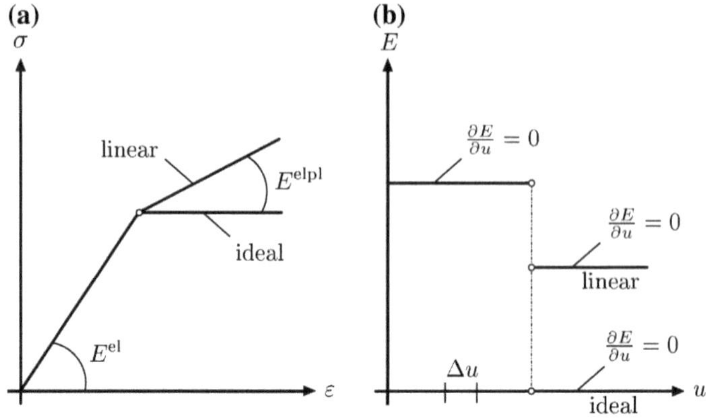


**Fig. 7.25** **a** Stress-strain diagram and **b** modulus as a function of displacement

$$\boldsymbol{K}_{\text{T}} = \begin{bmatrix} K_{11} & K_{12} & K_{13} \\ K_{21} & K_{22} & K_{23} \\ K_{31} & K_{32} & K_{33} \end{bmatrix} + \left( \left\| \begin{bmatrix} \frac{\partial K_{11}}{\partial \Delta u_2} & \frac{\partial K_{12}}{\partial \Delta u_2} & \frac{\partial K_{13}}{\partial \Delta u_2} \\ \frac{\partial K_{21}}{\partial \Delta u_2} & \frac{\partial K_{22}}{\partial \Delta u_2} & \frac{\partial K_{23}}{\partial \Delta u_2} \\ \frac{\partial K_{31}}{\partial \Delta u_2} & \frac{\partial K_{32}}{\partial \Delta u_2} & \frac{\partial K_{33}}{\partial \Delta u_2} \end{bmatrix} \begin{bmatrix} \Delta u_2 \\ \Delta u_3 \\ \Delta u_4 \end{bmatrix} \right\| \right.$$
$$\left. \begin{bmatrix} \frac{\partial K_{11}}{\partial \Delta u_3} & \frac{\partial K_{12}}{\partial \Delta u_3} & \frac{\partial K_{13}}{\partial \Delta u_3} \\ \frac{\partial K_{21}}{\partial \Delta u_3} & \frac{\partial K_{22}}{\partial \Delta u_3} & \frac{\partial K_{23}}{\partial \Delta u_3} \\ \frac{\partial K_{31}}{\partial \Delta u_3} & \frac{\partial K_{32}}{\partial \Delta u_3} & \frac{\partial K_{33}}{\partial \Delta u_3} \end{bmatrix} \begin{bmatrix} \Delta u_2 \\ \Delta u_3 \\ \Delta u_4 \end{bmatrix} \left\| \begin{bmatrix} \frac{\partial K_{11}}{\partial \Delta u_4} & \frac{\partial K_{12}}{\partial \Delta u_4} & \frac{\partial K_{13}}{\partial \Delta u_4} \\ \frac{\partial K_{21}}{\partial \Delta u_4} & \frac{\partial K_{22}}{\partial \Delta u_4} & \frac{\partial K_{23}}{\partial \Delta u_4} \\ \frac{\partial K_{31}}{\partial \Delta u_4} & \frac{\partial K_{32}}{\partial \Delta u_4} & \frac{\partial K_{33}}{\partial \Delta u_4} \end{bmatrix} \begin{bmatrix} \Delta u_2 \\ \Delta u_3 \\ \Delta u_4 \end{bmatrix} \right) . \tag{7.192}$$

The partial derivatives in Eq. (7.192) can be identified under consideration of Eq. (7.185) as partial derivatives of the bending stiffness $(EI_Z)_i$ with respect to one of the unknown displacement increments. Since the second moment of area can be considered in our specific case as a constant, the partial derivatives take the form:

$$\frac{\partial K_{kl}}{\partial \Delta u} \sim \frac{\partial E}{\partial \Delta u}. \tag{7.193}$$

Looking at Fig. 7.25, one can conclude that these partial derivatives are in the case of an ideal plastic—or even linear hardening—material equal to zero and the tangent stiffness matrix reduces for this special case to the stiffness matrix: $\boldsymbol{K}_{\text{T}} = \boldsymbol{K}$.

Thus, the iteration scheme of Eq. (7.188) reduces in this special case to:

$$\Delta \boldsymbol{u}^{(j+1)} = \Delta \boldsymbol{u}^{(j)} - \left(\boldsymbol{K}^{(j)}\right)^{-1} (\boldsymbol{K}(\Delta \boldsymbol{u})\Delta \boldsymbol{u} - \Delta \boldsymbol{F})^{(j)} \tag{7.194}$$

$$= \left(\boldsymbol{K}^{(j)}\right)^{-1} \Delta \boldsymbol{F}^{(j)}, \tag{7.195}$$

which is known under the expression direct or PICARD's iteration [87]. In components, we can write this scheme for our specific case as:

$$\begin{bmatrix} \Delta u_2 \\ \Delta u_3 \\ \Delta u_4 \end{bmatrix}^{(j+1)} = \frac{\Delta X^2}{4} \begin{bmatrix} \frac{-3}{(EI_Z)_2} & \frac{-2}{(EI_Z)_3} & \frac{-1}{(EI_Z)_4} \\ \frac{-2}{(EI_Z)_2} & \frac{-4}{(EI_Z)_3} & \frac{-2}{(EI_Z)_4} \\ \frac{-1}{(EI_Z)_2} & \frac{-2}{(EI_Z)_3} & \frac{-3}{(EI_Z)_4} \end{bmatrix}^{(j)} \begin{bmatrix} \Delta M_2 \\ \Delta M_3 \\ \Delta M_4 \end{bmatrix}^{(j)} \tag{7.196}$$

$$\overset{(7.176)}{=} \frac{M_{\text{lim}}^{\text{pl}} L^2}{64 E I_Z} \begin{bmatrix} \frac{-3}{\alpha_2} & \frac{-2}{\alpha_3} & \frac{-1}{\alpha_4} \\ \frac{-2}{\alpha_2} & \frac{-4}{\alpha_3} & \frac{-2}{\alpha_4} \\ \frac{-1}{\alpha_2} & \frac{-2}{\alpha_3} & \frac{-3}{\alpha_4} \end{bmatrix}^{(j)} \begin{bmatrix} \frac{\Delta M_2}{M_{\text{lim}}^{\text{pl}}} \\ \frac{\Delta M_3}{M_{\text{lim}}^{\text{pl}}} \\ \frac{\Delta M_4}{M_{\text{lim}}^{\text{pl}}} \end{bmatrix}^{(j)}, \tag{7.197}$$

or in normalized form as:

$$\begin{bmatrix} \Delta \hat{u}_2 \\ \Delta \hat{u}_3 \\ \Delta \hat{u}_4 \end{bmatrix}^{(j+1)} = \frac{1}{\frac{M_{\text{lim}}^{\text{pl}} L^2}{EI_Z}} \begin{bmatrix} \Delta u_2 \\ \Delta u_3 \\ \Delta u_4 \end{bmatrix}^{(j+1)} = \frac{1}{64} \begin{bmatrix} \frac{-3}{\alpha_2} & \frac{-2}{\alpha_3} & \frac{-1}{\alpha_4} \\ \frac{-2}{\alpha_2} & \frac{-4}{\alpha_3} & \frac{-2}{\alpha_4} \\ \frac{-1}{\alpha_2} & \frac{-2}{\alpha_3} & \frac{-3}{\alpha_4} \end{bmatrix}^{(j)} \begin{bmatrix} \frac{\Delta M_2}{M_{\text{lim}}^{\text{pl}}} \\ \frac{\Delta M_3}{M_{\text{lim}}^{\text{pl}}} \\ \frac{\Delta M_4}{M_{\text{lim}}^{\text{pl}}} \end{bmatrix}^{(j)}. \tag{7.198}$$

Let us now evaluate this iteration scheme for different ratios of the normalized moment in the pure elastic and the elasto-plastic range and compare these numerical results with the exact analytical solution. As shown in Fig. 7.21a, the beam is subdivided in five equidistant nodes and the number of layers is taken to be equal to 101 for the following calculations. Since the iteration scheme is quite sensible to the load step, the following increments should be assigned: Between $\frac{M}{M_{\text{lim}}^{\text{pl}}} = \frac{2}{3}$ and 0.8: 10 increments; between $\frac{M}{M_{\text{lim}}^{\text{pl}}} = 0.8$ and 0.9: 10 increments; between $\frac{M}{M_{\text{lim}}^{\text{pl}}} = 0.9$ and 0.95: 10 increments.

The results of the finite difference iterations are compared to the analytical solution in Fig. 7.26. As can be seen from this figure, the elastic finite difference solution is identical to the analytical solution and no incremental load application is required. The plastic range shows, however, an increasing difference between finite difference and analytical solution with increasing plastic deformation. To increase the accuracy in the plastic range, the following three parameters can be modified:

- the total number of nodes, i.e. a finer discretization,
- the number of the layers ($k_{\text{max}}$) over the height of the cross section and
- the size of the load increment $\Delta M$.

The influence of these parameters on the accuracy will be investigated in the scope of supplementary problems, see Sect. 7.6.

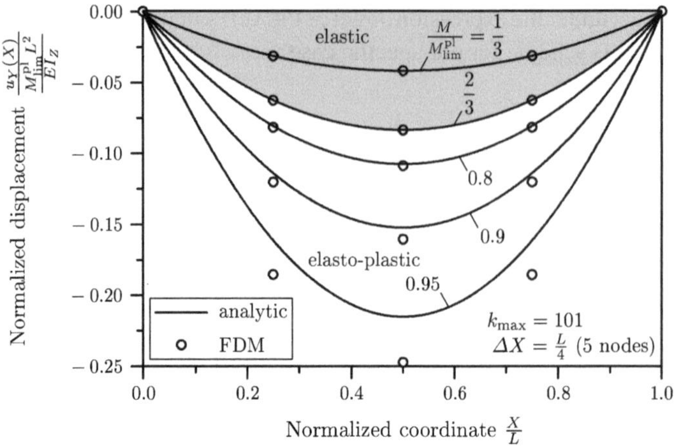


**Fig. 7.26** Normalized deformed shape of a simply supported BERNOULLI beam under pure bending load. Comparison between analytical and finite difference solution

**Fig. 7.27** Cantilevered rod loaded by a constant distributed load

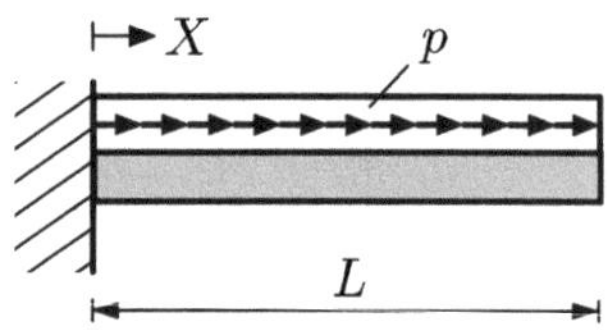


## 7.6 Supplementary Problems

**7.4 Forward difference approximation of the first order derivative**
Derive the finite difference approximation for the forward difference scheme of the first order derivative where the truncation error is of order $\Delta X^2$. The final result is given in Table 7.1.

**7.5 Centered difference approximation of the third order derivative**
Derive the finite difference approximation for the centered difference scheme of the third order derivative where the truncation error is of order $\Delta X^2$ based on the expressions for the first and second order derivatives. These expressions and the final result are given in Table 7.1.

**7.6 Finite difference approximation of a cantilevered rod with distributed load based on five domain nodes**
Given is a cantilevered rod of length $L$ with constant tensile stiffness $EA$ as shown in Fig. 7.27. The rod is loaded by a constant distributed load $p$. Use five domain nodes of equidistant spacing, i.e. $\Delta X = \frac{L}{4}$, for the finite difference approximation. Use only centered difference approximations of second order accuracy for the nodal evaluations and boundary conditions. Determine

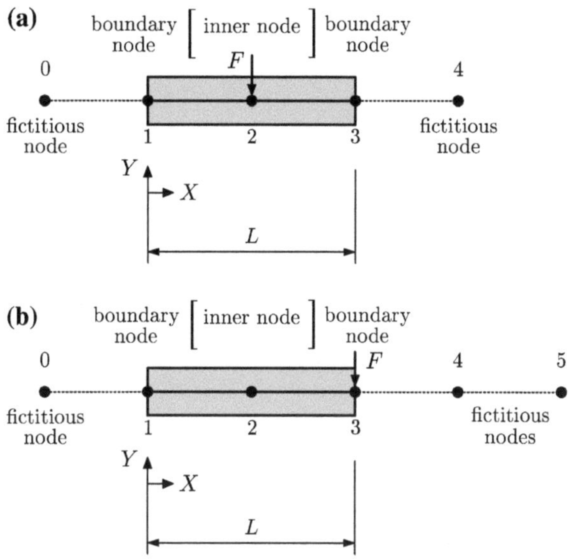


**Fig. 7.28** Finite difference approximation based on three domain nodes of **a** a simply supported and **b** a cantilevered BERNOULLI beam loaded by a single force. The schematic sketch of the problem is given in Fig. 7.12

- the horizontal displacement at the end of the rod, i.e. $X = L$,
- the analytical solution and
- calculate the relative error between the analytical and finite difference solution.

**7.7 Finite difference approximation of a simply supported and cantilevered beam based on three domain nodes**
Recalculate example 7.1 based on a discretization of the domain $0 \leq X \leq L$ with three nodes of equidistant spacing, cf. Fig. 7.28. Compare the results to the approach in example 7.1 which was based on five domain nodes.

**7.8 Centered difference approximation of the fourth order derivative**
Derive the finite difference approximation for the centered difference scheme of the fourth order derivative where the truncation error is of order $\Delta X^2$ based on TAYLOR's series expansions around node $i$ for locations $i+1, i-1, i+2$ and $i-2$. The final result is given in Table 7.1.

**7.9 Finite difference approximation of a cantilevered beam based on five domain nodes**
Given is a cantilevered BERNOULLI beam of length $L$ with constant bending stiffness $EI_Z$ as shown in Fig. 7.29. The beam is loaded by a single force $F$ at the position $X = \frac{3L}{4}$. Use five domain nodes of equidistant spacing, i.e. $\Delta X = \frac{L}{4}$, for the finite difference approximation. Use only centered difference approximations of second order accuracy for the nodal evaluations and boundary conditions. Determine

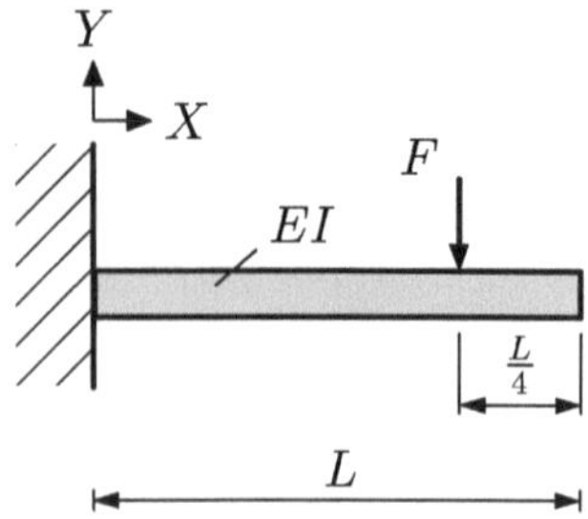


**Fig. 7.29** Cantilevered BERNOULLI beam loaded by a singe force at the position $X = \frac{3L}{4}$

- the vertical displacement at the end of the beam, i.e. $X = L$,
- the analytical solution and
- calculate the relative error between the analytical and finite difference solution.

**7.10 Finite difference approximation of a cantilevered beam based on five domain nodes—backward scheme**
Reconsider problem 7.9 and apply a backward finite difference approximation of the third order derivative where the truncation error is of order $\Delta X^2$ to formulate the equilibrium between the internal shear force and external load:

(a) at the free end, $X = L$,
(b) at the loading point, $X = \frac{3L}{4}$.

**7.11 Finite difference approximation of a cantilevered beam based on five domain nodes—conversion of tip load into distributed load**
Reconsider example 7.1 where a cantilevered beam loaded by a single force was investigated, see Fig. 7.30a. For five domain nodes and a centered difference scheme, a relative error of 54.689 % for the vertical deformation at $X = L$ was obtained. In order to improve the centered difference approach, the single force $F$ can be converted in an equivalent distributed load $q_Y = F/\Delta X$ between the two last nodes, see Fig. 7.30b. Calculate the relative error of the displacement at $X = L$.

**7.12 Finite difference approximation of a simply supported beam based on five domain nodes—bending moment approach**
Given is a simply supported BERNOULLI beam of length $L$ with constant bending stiffness $EI_Z$ as shown in Fig. 7.31. The beam is loaded by a constant distributed load $q_Y$. Use five domain nodes of equidistant spacing, i.e. $\Delta X = \frac{L}{4}$, for the finite difference approximation. Use the bending differential equation in the form of the moment, i.e. $EI_Z \frac{\mathrm{d}^2 u_Y}{\mathrm{d}X^2} = M_Z(X)$, to determine

- the vertical displacement in the middle of the beam, i.e. $X = \frac{L}{2}$,
- the analytical solution and
- calculate the relative error between the analytical and finite difference solution.

**7.13 Finite difference approximation of a simply supported beam under pure bending**

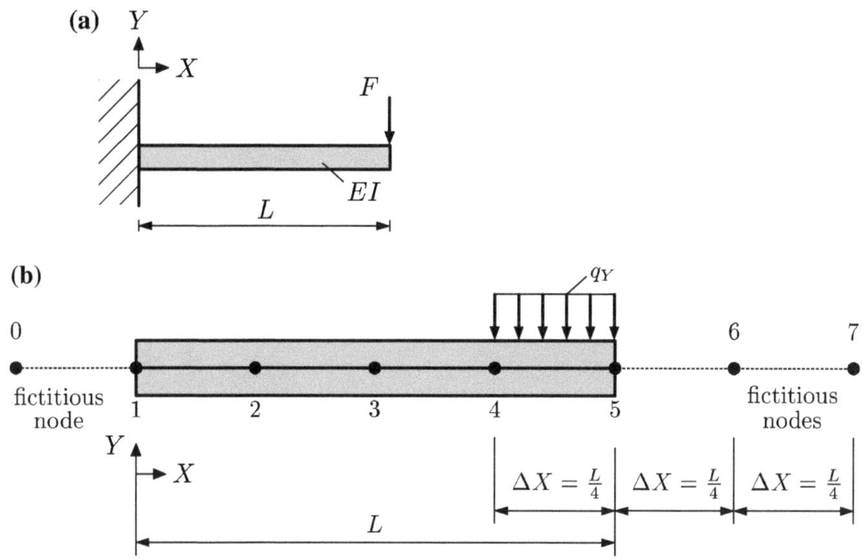


**Fig. 7.30** Finite difference approximation of a cantilevered beam based on five domain nodes—conversion of tip load into distributed load

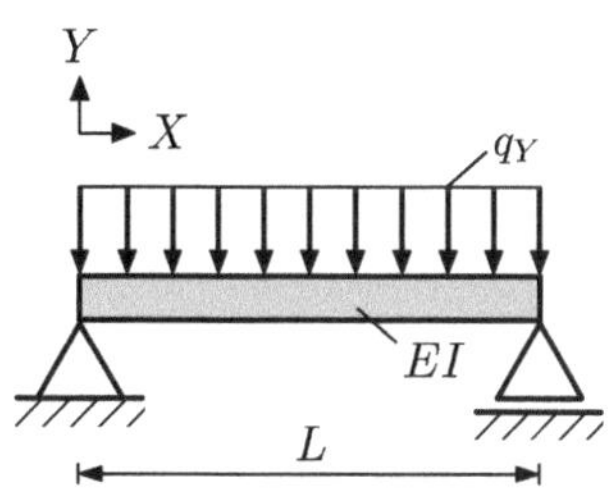


**Fig. 7.31** Simply supported BERNOULLI beam loaded by a distributed load

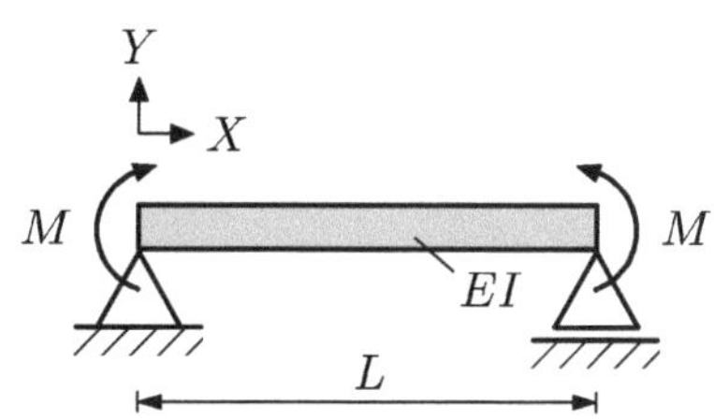


**Fig. 7.32** Simply supported BERNOULLI beam under pure bending

Given is a simply supported BERNOULLI beam of length $L$ with constant bending stiffness $EI_Z$ as shown in Fig. 7.32. The beam is loaded by single moments $M$ at its boundaries. Use (a) five and (b) eleven domain nodes of equidistant spacing, i.e. $\Delta X = \frac{L}{4}$ and $\Delta X = \frac{L}{10}$, for the finite difference approximation. Use the bending differential equation in the form of the distributed load and the moment, i.e. $EI_Z \frac{\mathrm{d}^4 u_Y}{\mathrm{d}X^4} = q_Y(X)$ and $EI_Z \frac{\mathrm{d}^2 u_Y}{\mathrm{d}X^2} = M_Z(X)$, to determine

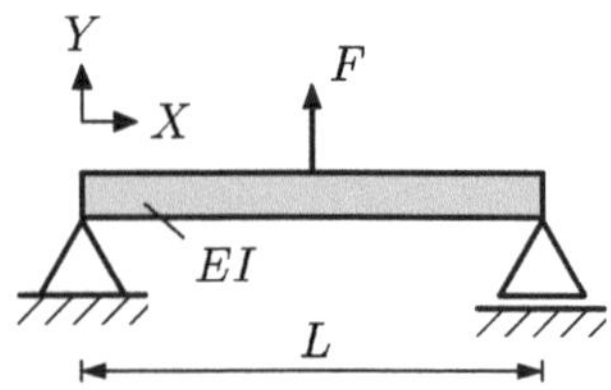


**Fig. 7.33** Simply supported BERNOULLI beam loaded by a centered single force

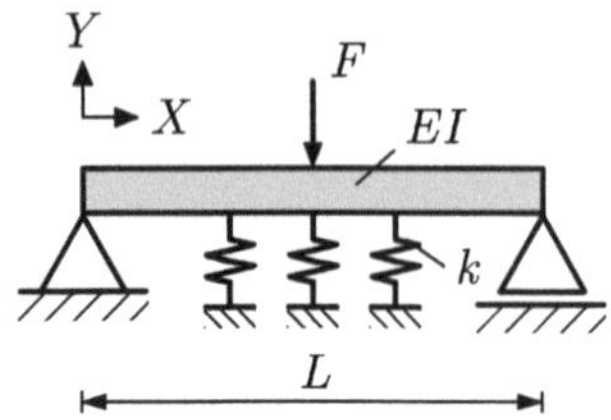


**Fig. 7.34** Simply supported BERNOULLI beam on elastic foundation loaded by a single force

- the vertical displacement at the nodes,
- the analytical solution and
- calculate the relative error between the analytical and finite difference solution for the displacement in the middle of the beam, i.e. $X = \frac{L}{2}$.

**7.14 Finite difference approximation of a simply supported beam—displacement, bending moment and shear force distribution**

Given is a simply supported BERNOULLI beam of length $L$ with constant bending stiffness $EI_z$ as shown in Fig. 7.33. The beam is loaded by a single force $F$ at $X = \frac{L}{2}$. Use five domain nodes of equivalent spacing, i.e. $\Delta X = \frac{L}{4}$, for the finite difference approximation. Use the fourth-order bending differential equation (7.83) under consideration of difference schemes of second order accuracy to determine

- the vertical displacement at each node,
- the bending moment at each node,
- the shear force at each node,
- the relative error between the FD and analytical solution at each node in regards to the displacement, moment and shear force.

**7.15 Finite difference approximation of a simply supported beam on elastic foundation based on five domain nodes**

Given is a simply supported BERNOULLI beam of length $L$ on an elastic foundation as shown in Fig. 7.34. The bending stiffness $EI_Z$ and the elastic foundation modulus $k$ are constant. Use five domain nodes of equidistant spacing, i.e. $\Delta X = \frac{L}{4}$, for the finite difference approximation to determine:

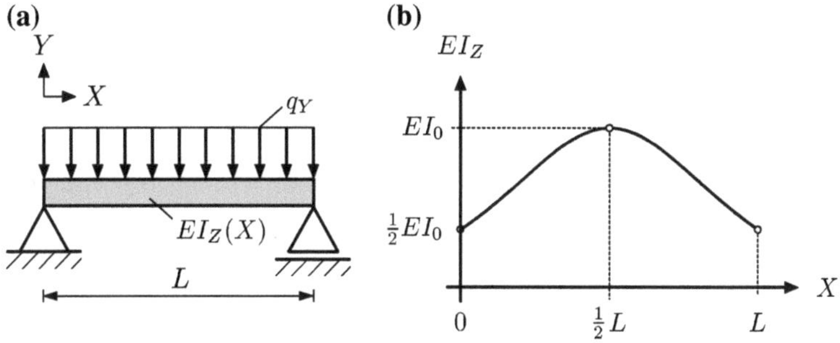


**Fig. 7.35** **a** Simply supported BERNOULLI beam loaded by a distributed load; **b** function of the varying bending stiffness $EI_Z = EI_Z(X)$

- The vertical displacement in the middle of the beam, i.e. $X = \frac{L}{2}$.
- Compare the finite difference approximation with the analytical solution for the case $k = 4$, $EI_Z = 1$ and $L = 1$.

**7.16 Finite difference approximation of a simply supported beam with varying bending stiffness—constant distributed load**

Given is a simply supported BERNOULLI beam of length $L$ with varying bending stiffness $EI_Z(X) = \frac{EI_0}{1+(2X/L-1)^2}$ as shown in Fig. 7.35. The beam is loaded by a constant distributed load $q_Y$. Use five domain nodes of equidistant spacing, i.e. $\Delta X = \frac{L}{4}$, for the finite difference approximation. Use the bending differential equation in the form of the moment, i.e. $EI_Z \frac{d^2 u_Y}{dX^2} = M_Z(X)$, to determine

- the vertical displacement in the middle of the beam, i.e. $X = \frac{L}{2}$,
- the analytical solution and
- calculate the relative error between the analytical and finite difference solution.

**7.17 Finite difference approximation of a simply supported beam with varying bending stiffness—constant bending moment**

Given is a simply supported BERNOULLI beam of length $L$ with varying bending stiffness $EI_Z(X) = \frac{EI_0}{1+(2X/L-1)^2}$ as shown in Fig. 7.36. The beam is loaded by single moments $M$ at its ends. Use five domain nodes of equidistant spacing, i.e. $\Delta X = \frac{L}{4}$, for the finite difference approximation. Use the bending differential equation in the form of the moment, i.e. $EI_Z \frac{d^2 u_Y}{dX^2} = M_Z$, to determine

- the vertical displacement in the middle of the beam, i.e. $X = \frac{L}{2}$,
- the analytical solution and
- calculate the relative error between the analytical and finite difference solution.

**7.18 Finite difference approximation of a beam fixed at both ends**

Given is a TIMOSHENKO beam of length $L$ which is fixed at both ends as shown in Fig. 7.37. The beam has constant material parameters and is loaded either by a single force $F$ or a constant distributed load $q_Y$. Use first five domain nodes of equidistant

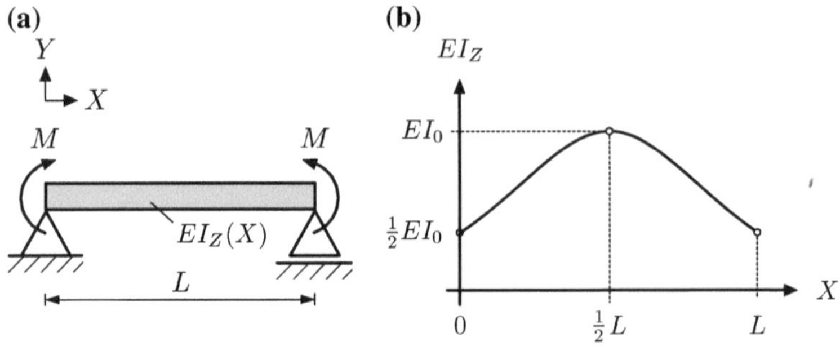


**Fig. 7.36** **a** Simply supported BERNOULLI beam loaded by single moments; **b** function of the varying bending stiffness $EI_Z = EI_Z(X)$

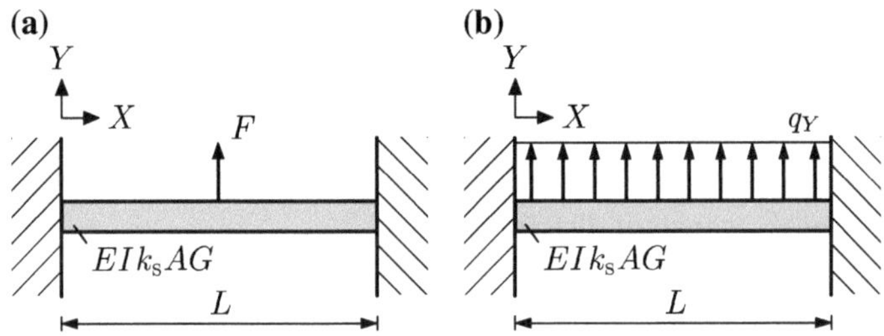


**Fig. 7.37** TIMOSHENKO beam fixed at both ends: **a** single force case; **b** distributed load case

spacing ($\Delta X = \frac{L}{4}$) and then 17 domain nodes of equidistant spacing ($\Delta X = \frac{L}{16}$) for the finite difference approximation to determine

- the general expression for the vertical displacement in the middle of the beam, i.e. $X = \frac{L}{2}$,
- the analytical solution and
- calculate the relative error between the analytical and finite difference solution for a squared cross-sectional area of length 0.5. Other values are: $E = 10000$; $\nu = 0.0$; $L = 10$; $q_y = 1$; $F = 0.1$.

**7.19 Convergence of finite difference approximation of a simply supported Timoshenko beam**

Given is a TIMOSHENKO beam of length $L$ which is simply supported as shown in Fig. 7.38. The beam has constant material parameters and is loaded by a constant distributed load $q_Y$. Use the following number of equidistant domain nodes to investigate the convergence of the solution: 5, 13, 23, 33, 53, 73 and 103. The analytical solution can be taken form Table 4.21. Determine

- the general expression for the vertical displacement in the middle of the beam, i.e. $X = \frac{L}{2}$,

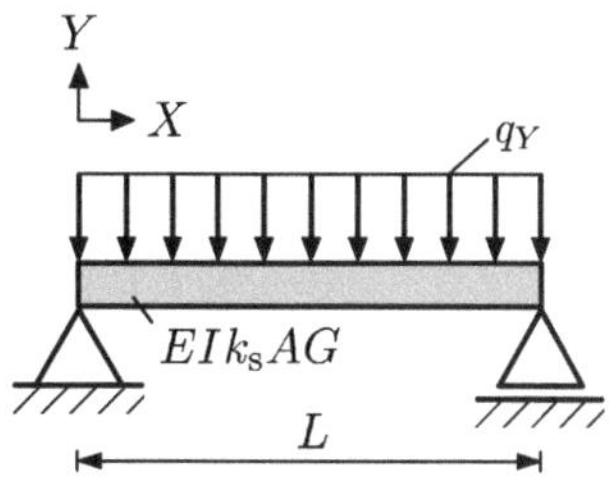


**Fig. 7.38** Simply supported TIMOSHENKO beam with distributed load

- a graphical representation of the relative error between the analytical and finite difference solution as a function of the domain node number for a squared cross-sectional area of length 0.5. Other values are: $E = 10000$; $\nu = 0.0$; $L = 10$; $q_y = 1$.

**7.20 Convergence of finite difference approximation of a cantilevered Timoshenko beam**

Given is a cantilevered TIMOSHENKO beam of length $L$ as shown in Fig. 7.39. The beam has constant material parameters and is either loaded in the negative $Y$-direction by a constant distributed load $q_Y$ or a single force $F$. Use the following number of equidistant domain nodes to investigate the convergence of the solution: 5, 13, 23, 33, 53, 73 and 103. The analytical solution can be taken form Table 4.21. Determine

- the general expression for the vertical displacement at the right-hand end of the beam, i.e. $X = L$,
- a graphical representation of the relative error between the analytical and finite difference solution as a function of the domain node number for a squared cross-sectional area of length 0.5. Other values are: $E = 10000$; $\nu = 0.0$; $L = 10$; $q_y = 1$; $F = 0.1$.

(a) In a first approach, a centered difference approximation should be written for the inner nodes. Thus, a system of equations with the dimension dim. $= 2 \times (\text{domain nodes} - 2)$ is obtained. The values of $u$ and $\phi$ at the right-hand boundary $X = L$ should be replaced by appropriate boundary conditions for $M_Z$ and $Q_Y$ based on backward difference schemes of $O(\Delta X^2)$.
(b) In a second approach, write a centered difference approximation which includes in addition the node at $X = L$. Thus, a system of equations with the dimension dim. $= 2 \times (\text{domain nodes} - 1)$ is obtained. Replace the values for $u$ and $\phi$ at the fictitious node based on appropriate boundary conditions for $M_Z$ and $Q_Y$ based on centered difference schemes of $O(\Delta X^2)$.

**7.21 Comparison between analytical and layer-wise integration of the bending stiffness**

Given is a segment of a BERNOULLI beam which is divided in $k_{\max}$ equidistant layers as shown in Fig. 7.40. The plastic zone extends to the coordinate $Y = \pm\frac{\alpha h}{2}$. Determine the bending stiffness $EI_z$ based on the exact approach and based on

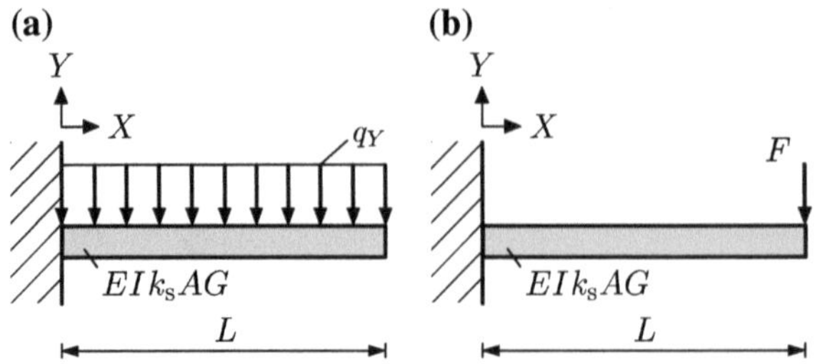


**Fig. 7.39** Cantilevered TIMOSHENKO beam with **a** distributed load and **b** single force

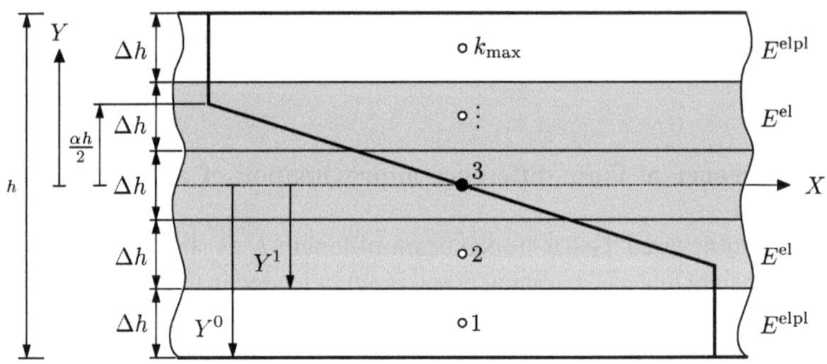


**Fig. 7.40** Schematic sketch for the layer-wise integration of the bending stiffness

layer-wise integration. Sketch the relative error between both approaches for the specific values $\alpha = \frac{11}{20}$ and $\frac{1}{2}$ in the range $0 \leq k \leq k_{max} = 151$.

**7.22 Investigation of the proportions of the relative bending stiffness**

The relative bending stiffness within the layered approach can be expressed according to Eq. (7.175) for the pure elastic case as

$$\frac{(EI_Z)_i}{EI_Z} = \sum_{k=1}^{k_{max}} 1 \times \left[\frac{1}{k_{max}^3} + \frac{3}{k_{max}}\left(\frac{2k-1}{k_{max}} - 1\right)^2\right]_i \tag{7.199}$$

$$= \sum_{k=1}^{k_{max}} 1 \times [\beta_1 + \beta_2]_i\,, \tag{7.200}$$

where the dimensionless factor $\beta_1$ is the proportion of the cross section related to the layer's own coordinate system and $\beta_2$ is the contribution from the offset between the coordinate system of the neutral axis and the layer's own coordinate system (see the parallel axis theorem). Some references neglect the contribution of $\beta_1$. Thus, determine the absolute difference between $\beta_1$, $\beta_2(k = k_{max})$ and $\beta_2\left(k = \frac{k_{max}+3}{2}\right)$ as well as the relative difference between $\beta_1$ and both functions of $\beta_2$ in the range $1 \leq k_{max} \leq 151$.

**7.23 Elasto-plastic finite difference solution: influence of layer number**
Consider again the problem of the simply supported BERNOULLI beam under constant bending load as shown in Figs. 7.20 and 7.26. The beam should be discretized by five equidistant nodes and the moment should be applied in the range $M/M_{\text{lim}}^{\text{pl}} = 2/3$ and 0.8 and in the range from 0.8 to 0.9 in ten equidistant steps. Calculate the deformation for $M/M_{\text{lim}}^{\text{pl}} = 0.9$ with $k_{\text{max}} = 11, 101, 1001, 10001$ and determine in addition the relative error to the analytical solution.

**7.24 Elasto-plastic finite difference solution: influence of load increment**
Consider again the problem of the simply supported BERNOULLI beam under constant bending load as shown in Figs. 7.20 and 7.26. The beam should be discretized by five equidistant nodes and the layers are $k_{\text{max}} = 101$. The moment should be applied in the range $M/M_{\text{lim}}^{\text{pl}} = 2/3$ and 0.8 and in the range from 0.8 to 0.9 in 5, 8 or 10 equidistant steps. Calculate the deformation for $M/M_{\text{lim}}^{\text{pl}} = 0.9$ and determine in addition the relative error to the analytical solution in dependence of the size of the load increment.

# Chapter 8
# Prelude to the General Three-Dimensional Case

**Abstract** This chapter introduces the general three-dimensional case. The basic equations of continuum mechanical modeling, i.e. the equilibrium, the constitutive and the kinematics relationships are used to derive the governing differential equation of static solids. The finite element approximation for linear-elastic material behavior is presented for hexahedron solid elements. The chapter closes with an introduction to the handling of elasto-plastic material behavior based on the fully implicit backward Euler algorithm.

## 8.1 Equilibrium

Figure 8.1 shows the normal and shear stresses which are acting on a differential volume element in the $x$-direction. All forces are drawn in their positive direction at each cut face. A positive cut face is obtained if the outward surface normal is directed in the positive direction of the corresponding coordinate axis. This means that the right-hand face in Fig. 8.1 is positive and the force $(\sigma_x + \frac{\partial \sigma_x}{\partial x}\mathrm{d}x)\mathrm{d}y\mathrm{d}z$ is oriented in the positive $x$-direction. In a similar way, the top face is positive, i.e. the outward surface normal is directed in the positive $y$-direction, and the shear force[1] is oriented in the positive $x$-direction. Since the volume element is assumed to be in equilibrium, forces resulting from stresses on the sides of the cuboid and from the body forces $f_i$ $(i = x, y, z)$ must be balanced. These body forces are defined as forces per unit volume which can be produced by gravity,[2] acceleration, magnetic fields, and so on.

The static equilibrium of forces in $x$-direction based on the seven force components—two normal forces, four shear forces and one body force—indicated in Fig. 8.1 gives after canceling with $\mathrm{d}V = \mathrm{d}x\mathrm{d}y\mathrm{d}z$:

$$\frac{\partial \sigma_x}{\partial x} + \frac{\partial \sigma_{yx}}{\partial y} + \frac{\partial \sigma_{zx}}{\partial z} + f_x = 0\,. \tag{8.1}$$

[1] In the case of a shear force $\sigma_{ij}$, the first index $i$ indicates that the stress acts on a plane normal to the $i$-axis and the second index $j$ denotes the direction in which the stress acts.

[2] If gravity is acting, the body force $f$ results as the product of density times standard gravity: $f = \frac{F}{V} = \frac{mg}{V} = \frac{m}{V}g = \varrho g$. The units can be checked by consideration of $1\,\mathrm{N} = 1\,\frac{\mathrm{mkg}}{\mathrm{s}^2}$.



A. Öchsner, *Elasto-Plasticity of Frame Structure Elements*,
DOI 10.1007/978-3-662-44225-8_8

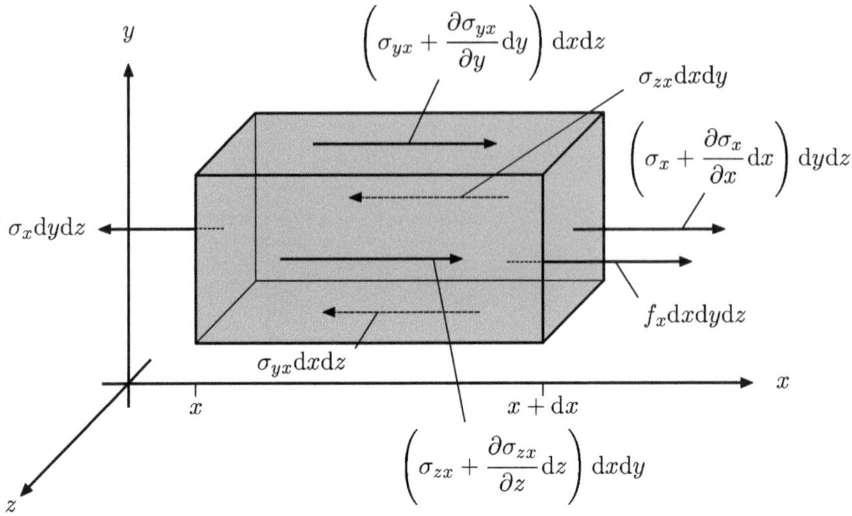


**Fig. 8.1** Stress and body forces which act on a differential volume element in $x$-direction

Based on the same approach, similar equations can be specified in the $y$- and $z$-direction:

$$\frac{\partial\sigma_y}{\partial y}+\frac{\partial\sigma_{yx}}{\partial x}+\frac{\partial\sigma_{yz}}{\partial z}+f_y=0\,, \tag{8.2}$$

$$\frac{\partial\sigma_z}{\partial z}+\frac{\partial\sigma_{xz}}{\partial x}+\frac{\partial\sigma_{yz}}{\partial y}+f_z=0\,. \tag{8.3}$$

These three balance equations can be written in matrix form as

$$\begin{bmatrix} \frac{\partial}{\partial x} & 0 & 0 & \frac{\partial}{\partial y} & 0 & \frac{\partial}{\partial z} \\ 0 & \frac{\partial}{\partial y} & 0 & \frac{\partial}{\partial x} & \frac{\partial}{\partial z} & 0 \\ 0 & 0 & \frac{\partial}{\partial z} & 0 & \frac{\partial}{\partial y} & \frac{\partial}{\partial x} \end{bmatrix} \begin{bmatrix} \sigma_x \\ \sigma_y \\ \sigma_z \\ \sigma_{xy} \\ \sigma_{yz} \\ \sigma_{xz} \end{bmatrix} + \begin{bmatrix} f_x \\ f_y \\ f_z \end{bmatrix} = \begin{bmatrix} 0 \\ 0 \\ 0 \end{bmatrix}, \tag{8.4}$$

or in symbolic notation:

$$\boldsymbol{\mathcal{L}}^{\mathrm{T}}\boldsymbol{\sigma}+\boldsymbol{b}=\boldsymbol{0}\,, \tag{8.5}$$

where $\boldsymbol{\mathcal{L}}$ is the differential operator matrix and $\boldsymbol{b}$ the column matrix of body forces.

## 8.2 Constitutive Equation

The generalized HOOKE's law for a linear-elastic isotropic material based on YOUNG's modulus $E$ and POISSON's ratio $\nu$ can be written for constant temperature with all components as

$$\begin{bmatrix} \sigma_x \\ \sigma_y \\ \sigma_z \\ \sigma_{xy} \\ \sigma_{yz} \\ \sigma_{xz} \end{bmatrix} = \frac{E}{(1+\nu)(1-2\nu)} \begin{bmatrix} 1-\nu & \nu & \nu & 0 & 0 & 0 \\ \nu & 1-\nu & \nu & 0 & 0 & 0 \\ \nu & \nu & 1-\nu & 0 & 0 & 0 \\ 0 & 0 & 0 & \frac{1-2\nu}{2} & 0 & 0 \\ 0 & 0 & 0 & 0 & \frac{1-2\nu}{2} & 0 \\ 0 & 0 & 0 & 0 & 0 & \frac{1-2\nu}{2} \end{bmatrix} \begin{bmatrix} \varepsilon_x \\ \varepsilon_y \\ \varepsilon_z \\ 2\,\varepsilon_{xy} \\ 2\,\varepsilon_{yz} \\ 2\,\varepsilon_{xz} \end{bmatrix}, \tag{8.6}$$

or in matrix form as

$$\boldsymbol{\sigma} = \boldsymbol{C}\boldsymbol{\epsilon}\,, \tag{8.7}$$

where $\boldsymbol{C}$ is the so-called elasticity matrix. It should be noted here that the engineering shear strain $\gamma_{ij} = 2\varepsilon_{ij}$ (for $i \neq j$) is used in the formulation of Eq. (8.6), see Sect. 4.3 for further details. Rearranging the elastic stiffness form given in Eq. (8.6) for the strains gives the elastic compliance form

$$\begin{bmatrix} \varepsilon_x \\ \varepsilon_y \\ \varepsilon_z \\ 2\,\varepsilon_{xy} \\ 2\,\varepsilon_{yz} \\ 2\,\varepsilon_{xz} \end{bmatrix} = \frac{1}{E} \begin{bmatrix} 1 & -\nu & -\nu & 0 & 0 & 0 \\ -\nu & 1 & -\nu & 0 & 0 & 0 \\ -\nu & -\nu & 1 & 0 & 0 & 0 \\ 0 & 0 & 0 & 2(1+\nu) & 0 & 0 \\ 0 & 0 & 0 & 0 & 2(1+\nu) & 0 \\ 0 & 0 & 0 & 0 & 0 & 2(1+\nu) \end{bmatrix} \begin{bmatrix} \sigma_x \\ \sigma_y \\ \sigma_z \\ \sigma_{xy} \\ \sigma_{yz} \\ \sigma_{xz} \end{bmatrix}, \tag{8.8}$$

or in matrix form as

$$\boldsymbol{\epsilon} = \boldsymbol{D}\boldsymbol{\sigma}\,, \tag{8.9}$$

where $\boldsymbol{D} = \boldsymbol{C}^{-1}$ is the so-called elastic compliance matrix. The general characteristic of HOOKE's law in the form of Eqs. (8.7) and (8.9) is that two independent material parameters are used. In addition to YOUNG's modulus $E$ and POISSON's ratio $\nu$, other elastic parameters can be used to form the set of two independent material parameters and the following Table 8.1 summarizes the conversion between the common material parameters.

**Table 8.1** Conversion of elastic constants: $\lambda$, $\mu$: LAMÉ's constants; $K$: bulk modulus; $G$: shear modulus; $E$: YOUNG's modulus; $\nu$: POISSON's ratio [21]

| | $\lambda, \mu$ | $E, \nu$ | $\mu, \nu$ | $E, \mu$ | $K, \nu$ | $G, \nu$ | $K, G$ |
|---|---|---|---|---|---|---|---|
| $\lambda$ | $\lambda$ | $\frac{\nu E}{(1+\nu)(1-2\nu)}$ | $\frac{2\mu\nu}{1-2\nu}$ | $\frac{\mu(E-2\mu)}{3\mu-E}$ | $\frac{3K\nu}{1+\nu}$ | $\frac{2G\nu}{1-2\nu}$ | $K-\frac{2G}{3}$ |
| $\mu$ | $\mu$ | $\frac{E}{2(1+\nu)}$ | $\mu$ | $\mu$ | $\frac{3K(1-2\nu)}{2(1+\nu)}$ | $\mu$ | $\mu$ |
| $K$ | $\lambda+\frac{2}{3}\mu$ | $\frac{E}{3(1-2\nu)}$ | $\frac{2\mu(1+\nu)}{3(1-2\nu)}$ | $\frac{\mu E}{3(3\mu-E)}$ | $K$ | $\frac{2G(1+\nu)}{3(1-2\nu)}$ | $K$ |
| $E$ | $\frac{\mu(3\lambda+2\mu)}{\lambda+\mu}$ | $E$ | $2\mu(1+\nu)$ | $E$ | $3K(1-2\nu)$ | $2G(1+\nu)$ | $\frac{9KG}{3K+G}$ |
| $\nu$ | $\frac{\lambda}{2(\lambda+\mu)}$ | $\nu$ | $\nu$ | $\frac{E}{2\mu}-1$ | $\nu$ | $\nu$ | $\frac{3K-2G}{2(3K+G)}$ |
| $G$ | $\mu$ | $\frac{E}{2(1+\nu)}$ | $\mu$ | $G$ | $\frac{3K(1-2\nu)}{2(1+\nu)}$ | $G$ | $G$ |

## 8.3 Kinematics

The kinematics or strain-displacement relations extract the stain field contained in a displacement field. Using engineering definitions of strain, the following relations can be obtained [22, 38]:

$$\varepsilon_x = \frac{\partial u_x}{\partial x}\,;\ \varepsilon_y = \frac{\partial u_y}{\partial y}\,;\ \varepsilon_z = \frac{\partial u_z}{\partial z}\,;\ \varepsilon_{xy} = \frac{1}{2}\left(\frac{\partial u_x}{\partial y} + \frac{\partial u_y}{\partial x}\right)\,; \tag{8.10}$$

$$\varepsilon_{xz} = \frac{1}{2}\left(\frac{\partial u_x}{\partial z} + \frac{\partial u_z}{\partial x}\right)\,;\ \varepsilon_{yz} = \frac{1}{2}\left(\frac{\partial u_y}{\partial z} + \frac{\partial u_z}{\partial y}\right). \tag{8.11}$$

In matrix format, these six relationships can be written as

$$\begin{bmatrix} \varepsilon_x \\ \varepsilon_y \\ \varepsilon_z \\ 2\varepsilon_{xy} \\ 2\varepsilon_{yz} \\ 2\varepsilon_{xz} \end{bmatrix} = \begin{bmatrix} \frac{\partial}{\partial x} & 0 & 0 \\ 0 & \frac{\partial}{\partial y} & 0 \\ 0 & 0 & \frac{\partial}{\partial z} \\ \frac{\partial}{\partial y} & \frac{\partial}{\partial x} & 0 \\ 0 & \frac{\partial}{\partial z} & \frac{\partial}{\partial y} \\ \frac{\partial}{\partial z} & 0 & \frac{\partial}{\partial x} \end{bmatrix} \begin{bmatrix} u_x \\ u_y \\ u_z \end{bmatrix}, \tag{8.12}$$

or symbolically as

$$\boldsymbol{\varepsilon} = \boldsymbol{\mathcal{L}} \boldsymbol{u}\,. \tag{8.13}$$

**Table 8.2** Fundamental governing equations of a continuum in the three-dimensional case

| Name | Matrix Notation | Tensor |
|---|---|---|
| Equilibrium | $\mathcal{L}^{\mathrm{T}}\boldsymbol{\sigma} + \boldsymbol{b} = \boldsymbol{0}$ | $\sigma_{ij,i} + b_i = 0$ |
| Constitution | $\boldsymbol{\sigma} = \boldsymbol{C}\boldsymbol{\epsilon}$ | $\sigma_{ij} = C_{ijkl}\varepsilon_{kl}$ |
| Kinematics | $\boldsymbol{\varepsilon} = \mathcal{L}\boldsymbol{u}$ | $\varepsilon_{ij} = \frac{1}{2}\left(u_{i,j} + u_{j,i}\right)$ |

## 8.4 Governing Differential Equation

The basic equations introduced in the previous three sections, i.e. the equilibrium, the constitutive and the kinematics equation are summarized in Table 8.2 where in addition the tensor notation[3] is given.

For the solution of the 15 unknown spatial functions (3 components of the displacement vector, 6 components of the symmetric strain tensor and 6 components of the symmetric stress tensor), a set of 15 scalar field equations is available:

- Equilibrium: 3,
- Constitution: 6,
- Kinematics: 6.

Furthermore, the boundary conditions are given:

$$\boldsymbol{u} \text{ on } \Gamma_u\,, \tag{8.14}$$

$$\boldsymbol{t} \text{ on } \Gamma_t\,, \tag{8.15}$$

where $\Gamma_u$ is the part of the boundary where a displacement boundary condition is prescribed and $\Gamma_t$ is the part of the boundary where a traction boundary condition, i.e. external force per unit area, is prescribed with $t_j = \sigma_{ij}n_j$, where $n_j$ are the components of the normal vector.

The 15 scalar field equations can be combined to eliminate the stress and strain fields. As a result, three scalar field equations for the three scalar displacement fields are obtained. These equations are called the LAMÉ-NAVIER[4] equations and can be derived as follows:

Introducing the constitutive equation according to (8.7) in the equilibrium equation (8.5) gives:

$$\mathcal{L}^{\mathrm{T}}\boldsymbol{C}\boldsymbol{\varepsilon} + \boldsymbol{b} = \boldsymbol{0}\,. \tag{8.16}$$

[3] A differentiation is there indicated by the use of a comma: The first index refers to the component and the comma indicates the partial derivative with respect to the second subscript corresponding to the relevant coordinate axis [22].

[4] Gabriel Léon Jean Baptiste LAMÉ (1795–1870), French mathematician.
Claude-Louis NAVIER (1785–1836), French engineer and physicist.

Introducing in the last equation the kinematics relations according to (8.13) gives finally the LAMÉ-NAVIER equations:

$$\mathcal{L}^{\mathrm{T}} \boldsymbol{C} \mathcal{L} \boldsymbol{u} + \boldsymbol{b} = \boldsymbol{0}\,. \tag{8.17}$$

Alternatively, the displacements may be substituted and the differential equations are obtained in terms of stresses. This formulation is known as the BELTRAMI-MICHELL[5] equations. If the body forces vanish ($\boldsymbol{b} = \boldsymbol{0}$), the partial differential equations in terms of stresses are called the BELTRAMI equations.

## 8.5 Derivation of the Principal Finite Element Equation

Let us assume in the following that the elasticity matrix in Eq. (8.17) is constant and that the exact solution is given by $\boldsymbol{u}^0$. Thus, the differential equation in terms of displacements can be written as:

$$\mathcal{L}^{\mathrm{T}} \boldsymbol{C} \mathcal{L} \boldsymbol{u}^0 + \boldsymbol{b} = \boldsymbol{0}\,. \tag{8.18}$$

Replacing the exact solution by an approximate solution $\boldsymbol{u}$, a residual $\boldsymbol{r}$ is obtained:

$$\boldsymbol{r} = \mathcal{L}^{\mathrm{T}} \boldsymbol{C} \mathcal{L} \boldsymbol{u} + \boldsymbol{b} \neq \boldsymbol{0}\,. \tag{8.19}$$

The inner product is obtained by weighting the residual and integration as

$$\int_V \boldsymbol{W}^{\mathrm{T}}(\boldsymbol{x}) \left( \mathcal{L}^{\mathrm{T}} \boldsymbol{C} \mathcal{L} \boldsymbol{u} + \boldsymbol{b} \right) \mathrm{d}V = 0\,, \tag{8.20}$$

where $\boldsymbol{W} = \begin{bmatrix} W_x \; W_y \; W_z \end{bmatrix}^{\mathrm{T}}$ is the column matrix of weight functions and $\boldsymbol{x} = \begin{bmatrix} x \; y \; z \end{bmatrix}^{\mathrm{T}}$ is the column matrix of Cartesian coordinates. Application of the GREEN-GAUSS theorem[6] (cf. Sect. A.9) gives the weak formulation as:

$$\int_V (\mathcal{L}\boldsymbol{W})^{\mathrm{T}} \boldsymbol{C} \, (\mathcal{L}\boldsymbol{u}) \, \mathrm{d}V = \int_A \boldsymbol{W}^{\mathrm{T}} \boldsymbol{t} \, \mathrm{d}A + \int_V \boldsymbol{W}^{\mathrm{T}} \boldsymbol{b} \, \mathrm{d}V\,, \tag{8.21}$$

---

[5] Eugenio BELTRAMI (1835–1900), Italian mathematician.
John Henry MICHELL (1863–1940), Australian mathematician.

[6] The supplementary problem (8.5) shows a possible way on how to use the theorem.

where the column matrix of traction forces $\boldsymbol{t} = \begin{bmatrix} t_x & t_y & t_z \end{bmatrix}^{\mathrm{T}}$ can be understood as the expression[7] $(\boldsymbol{C}\boldsymbol{\mathcal{L}}\boldsymbol{u})^{\mathrm{T}}\boldsymbol{n} = \boldsymbol{\sigma}^{\mathrm{T}}\boldsymbol{n}$.

Any further development of Eq. (8.21) requires now that the general expressions for the displacement and weight function, i.e. $\boldsymbol{u}$ and $\boldsymbol{W}$, are approximated by some functional representations. The nodal approach for the displacements[8] can be generally written for a three-dimensional element with $n$ nodes as:

$$u_x(\boldsymbol{x}) = N_1 u_{1x} + N_2 u_{2x} + N_3 u_{3x} + \cdots + N_n u_{nx}\,, \tag{8.22}$$

$$u_y(\boldsymbol{x}) = N_1 u_{1y} + N_2 u_{2y} + N_3 u_{3y} + \cdots + N_n u_{ny}\,, \tag{8.23}$$

$$u_z(\boldsymbol{x}) = N_1 u_{1z} + N_2 u_{2z} + N_3 u_{3z} + \cdots + N_n u_{nz}\,, \tag{8.24}$$

or in matrix notation as:

$$\begin{bmatrix} u_x \\ u_y \\ u_z \end{bmatrix} = \left[\begin{array}{ccc|ccc|ccc|c|ccc} N_1 & 0 & 0 & N_2 & 0 & 0 & N_3 & 0 & 0 & \cdots & N_n & 0 & 0 \\ 0 & N_1 & 0 & 0 & N_2 & 0 & 0 & N_3 & 0 & \cdots & 0 & N_n & 0 \\ 0 & 0 & N_1 & 0 & 0 & N_2 & 0 & 0 & N_3 & \cdots & 0 & 0 & N_n \end{array}\right] \left[\begin{array}{c} u_{1x} \\ u_{1y} \\ u_{1z} \\ \hline u_{2x} \\ u_{2y} \\ u_{2z} \\ \hline u_{3x} \\ u_{3y} \\ u_{3z} \\ \hline \vdots \\ \hline u_{nx} \\ u_{ny} \\ u_{nz} \end{array}\right]. \tag{8.25}$$

Introducing the notations

$$N_i \mathbf{I} = \begin{bmatrix} N_i & 0 & 0 \\ 0 & N_i & 0 \\ 0 & 0 & N_i \end{bmatrix} \quad \text{and} \quad \boldsymbol{u}_{\mathrm{p}i} = \begin{bmatrix} u_{ix} \\ u_{iy} \\ u_{iz} \end{bmatrix}, \tag{8.26}$$

Equation (8.25) can be written as

[7] Strictly speaking, the traction forces must be calculated based on the stress *tensor* as $t_i = \sigma_{ji} n_j$ and not based on the column matrix of stress components.

[8] The following derivations are written under the simplification that each node reveals only displacement DOF and no rotations.

$$\begin{bmatrix} u_x \\ u_y \\ u_z \end{bmatrix} = \begin{bmatrix} N_1\mathbf{I} \; N_2\mathbf{I} \; N_3\mathbf{I} \cdots N_n\mathbf{I} \end{bmatrix} \begin{bmatrix} \boldsymbol{u}_{\text{p}1} \\ \boldsymbol{u}_{\text{p}2} \\ \boldsymbol{u}_{\text{p}3} \\ \vdots \\ \boldsymbol{u}_{\text{p}n} \end{bmatrix}, \tag{8.27}$$

or with $\boldsymbol{N}_i = N_i\mathbf{I}$ as

$$\begin{bmatrix} u_x \\ u_y \\ u_z \end{bmatrix} = \begin{bmatrix} \boldsymbol{N}_1 \; \boldsymbol{N}_2 \; \boldsymbol{N}_3 \cdots \boldsymbol{N}_n \end{bmatrix} \begin{bmatrix} \boldsymbol{u}_{\text{p}1} \\ \boldsymbol{u}_{\text{p}2} \\ \boldsymbol{u}_{\text{p}3} \\ \vdots \\ \boldsymbol{u}_{\text{p}n} \end{bmatrix}. \tag{8.28}$$

The last equation can be written in abbreviated form as:

$$\boldsymbol{u}^{\text{e}}(\boldsymbol{x}) = \boldsymbol{N}^{\text{T}}(\boldsymbol{x})\boldsymbol{u}_{\text{p}}, \tag{8.29}$$

which is the same structure as in the case of the one-dimensional elements, cf. Eq. (5.63). The column matrix of the weight functions in Eq. (8.21) is approximated in a similar way as the unknown displacements as:

$$\boldsymbol{W}(\boldsymbol{x}) = \boldsymbol{N}^{\text{T}}(\boldsymbol{x})\delta\boldsymbol{u}_{\text{p}}. \tag{8.30}$$

Introducing the approximations for $\boldsymbol{u}^{\text{e}}$ and $\boldsymbol{W}$ according to Eqs. (8.29) and (8.30) in the weak formulation gives:

$$\int_V \left(\boldsymbol{\mathcal{L}}\boldsymbol{N}^{\text{T}}\delta\boldsymbol{u}_{\text{p}}\right)^{\text{T}} \boldsymbol{C} \left(\boldsymbol{\mathcal{L}}\boldsymbol{N}^{\text{T}}\boldsymbol{u}_{\text{p}}\right) \text{d}V = \int_A (\boldsymbol{N}^{\text{T}}\delta\boldsymbol{u}_{\text{p}})^{\text{T}}\boldsymbol{t}\,\text{d}A + \int_V (\boldsymbol{N}^{\text{T}}\delta\boldsymbol{u}_{\text{p}})^{\text{T}}\boldsymbol{b}\,\text{d}V, \tag{8.31}$$

which we can write under the consideration that the matrix of displacements and virtual displacements are not affected by the integration as:

$$\delta\boldsymbol{u}_{\text{p}}^{\text{T}} \int_V \left(\boldsymbol{\mathcal{L}}\boldsymbol{N}^{\text{T}}\right)^{\text{T}} \boldsymbol{C} \left(\boldsymbol{\mathcal{L}}\boldsymbol{N}^{\text{T}}\right) \text{d}V \boldsymbol{u}_{\text{p}} = \delta\boldsymbol{u}_{\text{p}}^{\text{T}} \int_A \boldsymbol{N}\boldsymbol{t}\,\text{d}A + \delta\boldsymbol{u}_{\text{p}}^{\text{T}} \int_V \boldsymbol{N}\boldsymbol{b}\,\text{d}V, \tag{8.32}$$

which gives after elimination of $\delta\boldsymbol{u}_{\text{p}}^{\text{T}}$ the following statement for the principal finite element equation on element level as:

$$\int_V \left(\boldsymbol{\mathcal{L}}\boldsymbol{N}^{\text{T}}\right)^{\text{T}} \boldsymbol{C} \left(\boldsymbol{\mathcal{L}}\boldsymbol{N}^{\text{T}}\right) \text{d}V \boldsymbol{u}_{\text{p}} = \int_A \boldsymbol{N}\boldsymbol{t}\,\text{d}A + \int_V \boldsymbol{N}\boldsymbol{b}\,\text{d}V. \tag{8.33}$$

Thus, we can identify the following three element matrices from the principal finite element equation:

$$\text{Stiffness matrix } (3n \times 3n)\text{:}\quad \boldsymbol{K}^{\mathrm{e}} = \int_V \underbrace{\left(\boldsymbol{\mathcal{L}} \boldsymbol{N}^{\mathrm{T}}\right)^{\mathrm{T}}}_{\boldsymbol{B}} \boldsymbol{C} \underbrace{\left(\boldsymbol{\mathcal{L}} \boldsymbol{N}^{\mathrm{T}}\right)}_{\boldsymbol{B}^{\mathrm{T}}} \mathrm{d}V, \tag{8.34}$$

$$\text{Boundary force matrix } (3n \times 1)\text{:}\quad \boldsymbol{f}_t^{\mathrm{e}} = \int_A \boldsymbol{N} \boldsymbol{t} \, \mathrm{d}A, \tag{8.35}$$

$$\text{Body force matrix } (3n \times 1)\text{:}\quad \boldsymbol{f}_b^{\mathrm{e}} = \int_V \boldsymbol{N} \boldsymbol{b} \, \mathrm{d}V. \tag{8.36}$$

Based on these abbreviations, the principal finite element equation for a single element can be written as:

$$\boldsymbol{K}^{\mathrm{e}} \boldsymbol{u}_{\mathrm{p}} = \boldsymbol{f}_t^{\mathrm{e}} + \boldsymbol{f}_b^{\mathrm{e}}. \tag{8.37}$$

In the following, let us look at the $\boldsymbol{B}$-matrix, i.e. the matrix which contains the derivatives of the shape functions. Application of the matrix of differential operators according to Eq. (8.12) to the matrix of shape functions gives:

$$\boldsymbol{\mathcal{L}} \boldsymbol{N}^{\mathrm{T}} = \begin{bmatrix} \frac{\partial}{\partial x} & 0 & 0 \\ 0 & \frac{\partial}{\partial y} & 0 \\ 0 & 0 & \frac{\partial}{\partial z} \\ \frac{\partial}{\partial y} & \frac{\partial}{\partial x} & 0 \\ 0 & \frac{\partial}{\partial z} & \frac{\partial}{\partial y} \\ \frac{\partial}{\partial z} & 0 & \frac{\partial}{\partial x} \end{bmatrix} \left[\begin{array}{ccc|ccc|c|ccc} N_1 & 0 & 0 & N_2 & 0 & 0 & \cdots & N_n & 0 & 0 \\ 0 & N_1 & 0 & 0 & N_2 & 0 & \cdots & 0 & N_n & 0 \\ 0 & 0 & N_1 & 0 & 0 & N_2 & \cdots & 0 & 0 & N_n \end{array}\right] \tag{8.38}$$

$$= \left[\begin{array}{ccc|ccc|c|ccc} \frac{\partial N_1}{\partial x} & 0 & 0 & \frac{\partial N_2}{\partial x} & 0 & 0 & & \frac{\partial N_n}{\partial x} & 0 & 0 \\ 0 & \frac{\partial N_1}{\partial y} & 0 & 0 & \frac{\partial N_2}{\partial y} & 0 & & 0 & \frac{\partial N_n}{\partial y} & 0 \\ 0 & 0 & \frac{\partial N_1}{\partial z} & 0 & 0 & \frac{\partial N_2}{\partial z} & \cdots & 0 & 0 & \frac{\partial N_n}{\partial z} \\ \frac{\partial N_1}{\partial y} & \frac{\partial N_1}{\partial x} & 0 & \frac{\partial N_2}{\partial y} & \frac{\partial N_2}{\partial x} & 0 & & \frac{\partial N_n}{\partial y} & \frac{\partial N_n}{\partial x} & 0 \\ 0 & \frac{\partial N_1}{\partial z} & \frac{\partial N_1}{\partial y} & 0 & \frac{\partial N_2}{\partial z} & \frac{\partial N_2}{\partial y} & & 0 & \frac{\partial N_n}{\partial z} & \frac{\partial N_n}{\partial y} \\ \frac{\partial N_1}{\partial z} & 0 & \frac{\partial N_1}{\partial x} & \frac{\partial N_2}{\partial z} & 0 & \frac{\partial N_2}{\partial x} & & \frac{\partial N_n}{\partial z} & 0 & \frac{\partial N_n}{\partial x} \end{array}\right] = \boldsymbol{B}^{\mathrm{T}}, \tag{8.39}$$

which is a $(6 \times 3n)$-matrix. The transposed, i.e. $(\boldsymbol{\mathcal{L}} \boldsymbol{N}^{\mathrm{T}})^{\mathrm{T}}$, is thus a $(3n \times 6)$-matrix. Multiplication with the elasticity matrix, i.e. a $(6 \times 6)$-matrix, results in $(\boldsymbol{\mathcal{L}} \boldsymbol{N}^{\mathrm{T}})^{\mathrm{T}} \boldsymbol{C}$, which is a $(3n \times 6)$-matrix. The final multiplication, i.e. $(\boldsymbol{\mathcal{L}}\boldsymbol{N}^{\mathrm{T}})^{\mathrm{T}}\boldsymbol{C}(\boldsymbol{\mathcal{L}}\boldsymbol{N}^{\mathrm{T}})$ gives after integration the stiffness matrix with a dimension of $(3n \times 3n)$.

The integrations of the element matrices given in Eqs. (8.34)–(8.36) are approximated by numerical integration. To this end, the coordinates $(x, y, z)$ are transformed to the natural coordinates (unit space) $(\xi, \eta, \zeta)$ where each coordinate ranges from $-1$ to 1. In the scope of the coordinate transformation, attention must be paid to the derivatives. For example, the derivative of the shape functions with respect to the $x$-coordinate is transformed in the following way:

$$\frac{\partial N_i}{\partial x} \rightarrow \frac{\partial N_i}{\partial \xi}\frac{\partial \xi}{\partial x} + \frac{\partial N_i}{\partial \eta}\frac{\partial \eta}{\partial x} + \frac{\partial N_i}{\partial \zeta}\frac{\partial \zeta}{\partial x}. \tag{8.40}$$

Furthermore, the coordinate transformation requires that $\mathrm{d}V = \mathrm{d}x\mathrm{d}y\mathrm{d}z \rightarrow \mathrm{d}V' = J\mathrm{d}\xi\mathrm{d}\eta\mathrm{d}\zeta$, where $J$ is the Jacobian as given in the Appendix A.10.

## 8.6 Hexahedron Solid Elements

A simple representative of a three-dimensional finite element is an eight-node hexahedron as shown in Fig. 8.2. This element uses trilinear interpolation functions and the strains tend to be constant throughout the element. The stiffness matrix of this element is normally—in the case of full integration—calculated based on an eight-point GAUSS- LEGENDRE quadrature formula.

Let us derive the element formulation from the assumption that a linear displacement field is given in parametric space. For the $x$-component, we can write

$$u_x^{\mathrm{e}}(\xi, \eta, \zeta) = a_0 + a_1\xi + a_2\eta + a_3\zeta + a_4\xi\eta + a_5\eta\zeta + a_6\xi\zeta + a_7\xi\eta\zeta\,, \tag{8.41}$$

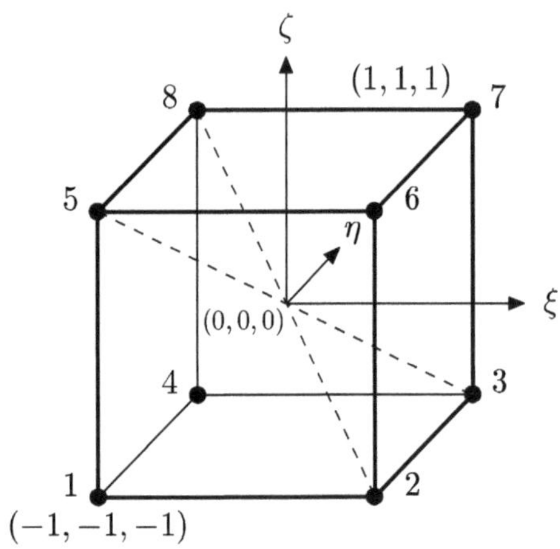


**Fig. 8.2** Three-dimensional eight-node hexahedron in parametric space

or in matrix notation

$$u_x^{\mathrm{e}}(\xi,\eta,\zeta) = \boldsymbol{\chi}^{\mathrm{T}}\boldsymbol{a} = \begin{bmatrix}1 & \xi & \eta & \zeta & \xi\eta & \eta\zeta & \xi\zeta & \xi\eta\zeta\end{bmatrix}\begin{bmatrix}a_0\\a_1\\a_2\\a_3\\a_4\\a_5\\a_6\\a_7\end{bmatrix}. \tag{8.42}$$

Evaluating Eq. (8.42) for all eight nodes of the hexahedron (cf. Fig. 8.2) gives:

$$\begin{aligned}
&\text{Node 1: } u_{1x} = u_x^{\mathrm{e}}(-1,-1,-1) = a_0 - a_1 - a_2 - a_3 + a_4 + a_5 + a_6 - a_7,\\
&\text{Node 2: } u_{2x} = u_x^{\mathrm{e}}(1,-1,-1) = a_0 + a_1 - a_2 - a_3 - a_4 + a_5 - a_6 + a_7,\\
&\qquad\vdots \qquad\qquad\qquad\qquad\qquad \vdots\\
&\text{Node 8: } u_{2x} = u_x^{\mathrm{e}}(-1,1,1) = a_0 - a_1 + a_2 + a_3 - a_4 + a_5 - a_6 - a_7,
\end{aligned}$$

or in matrix notation:

$$\begin{bmatrix}u_{1x}\\u_{2x}\\u_{3x}\\u_{4x}\\u_{5x}\\u_{6x}\\u_{7x}\\u_{8x}\end{bmatrix} = \begin{bmatrix}1 & -1 & -1 & -1 & 1 & 1 & 1 & -1\\ 1 & 1 & -1 & -1 & -1 & 1 & -1 & 1\\ 1 & 1 & 1 & -1 & 1 & -1 & -1 & -1\\ 1 & -1 & 1 & -1 & -1 & -1 & 1 & 1\\ 1 & -1 & -1 & 1 & 1 & -1 & -1 & 1\\ 1 & 1 & -1 & 1 & -1 & -1 & 1 & -1\\ 1 & 1 & 1 & 1 & 1 & 1 & 1 & 1\\ 1 & -1 & 1 & 1 & -1 & 1 & -1 & -1\end{bmatrix}\begin{bmatrix}a_0\\a_1\\a_2\\a_3\\a_4\\a_5\\a_6\\a_7\end{bmatrix}. \tag{8.43}$$

Solving for $\boldsymbol{a}$ gives

$$\begin{bmatrix}a_0\\a_1\\a_2\\a_3\\a_4\\a_5\\a_6\\a_7\end{bmatrix} = \frac{1}{8}\begin{bmatrix}1 & 1 & 1 & 1 & 1 & 1 & 1 & 1\\ -1 & 1 & 1 & -1 & -1 & 1 & 1 & -1\\ -1 & -1 & 1 & 1 & -1 & -1 & 1 & 1\\ -1 & -1 & -1 & -1 & 1 & 1 & 1 & 1\\ 1 & -1 & 1 & -1 & 1 & -1 & 1 & -1\\ 1 & 1 & -1 & -1 & -1 & -1 & 1 & 1\\ 1 & -1 & -1 & 1 & -1 & 1 & 1 & -1\\ -1 & 1 & -1 & 1 & 1 & -1 & 1 & -1\end{bmatrix}\begin{bmatrix}u_{1x}\\u_{2x}\\u_{3x}\\u_{4x}\\u_{5x}\\u_{6x}\\u_{7x}\\u_{8x}\end{bmatrix}, \tag{8.44}$$

or

$$\boldsymbol{a} = \boldsymbol{A}\boldsymbol{u}_{\mathrm{p}}. \tag{8.45}$$

The matrix of shape functions results as

$$N^{\mathrm{T}} = \left[N_1\ N_2\ N_3\ N_4\ N_5\ N_6\ N_7\ N_8\right] = \boldsymbol{\chi}^{\mathrm{T}}\boldsymbol{A} \tag{8.46}$$

or

$$N_1 = \frac{1}{8}(1-\xi)(1-\eta)(1-\zeta), \tag{8.47}$$

$$N_2 = \frac{1}{8}(1+\xi)(1-\eta)(1-\zeta), \tag{8.48}$$

$$N_3 = \frac{1}{8}(1+\xi)(1+\eta)(1-\zeta), \tag{8.49}$$

$$N_4 = \frac{1}{8}(1-\xi)(1+\eta)(1-\zeta), \tag{8.50}$$

$$N_5 = \frac{1}{8}(-1+\xi)(-1+\eta)(1+\zeta), \tag{8.51}$$

$$N_6 = \frac{1}{8}(1+\xi)(1-\eta)(1+\zeta), \tag{8.52}$$

$$N_7 = \frac{1}{8}(1+\xi)(1+\eta)(1+\zeta), \tag{8.53}$$

$$N_8 = \frac{1}{8}(1-\xi)(1+\eta)(1+\zeta), \tag{8.54}$$

or in a more compact form as

$$N_i = \frac{1}{8}(1+\xi\xi_i)(1+\eta\eta_i)(1+\zeta\zeta_i), \tag{8.55}$$

where $\xi_i$, $\eta_i$ and $\zeta_i$ are the coordinates of the nodes in parametric space ($i = 1, \ldots, 8$), cf. Fig. 8.2. The derivatives with respect to the parametric coordinates can easily be obtained as:

$$\frac{\partial N_i}{\partial \xi} = \frac{1}{8}(\xi_i)(1+\eta\eta_i)(1+\zeta\zeta_i)\,, \tag{8.56}$$

$$\frac{\partial N_i}{\partial \eta} = \frac{1}{8}(1+\xi\xi_i)(\eta_i)(1+\zeta\zeta_i)\,, \tag{8.57}$$

$$\frac{\partial N_i}{\partial \zeta} = \frac{1}{8}(1+\xi\xi_i)(1+\eta\eta_i)(\zeta_i)\,. \tag{8.58}$$

The geometrical derivatives in Eq. (8.40), e.g. $\frac{\partial \xi}{\partial x}$, $\frac{\partial \eta}{\partial x}$, $\frac{\partial \zeta}{\partial x}$, can be calculated on the basis of

$$\begin{bmatrix} \frac{\partial \xi}{\partial x} & \frac{\partial \xi}{\partial y} & \frac{\partial \xi}{\partial z} \\ \frac{\partial \eta}{\partial x} & \frac{\partial \eta}{\partial y} & \frac{\partial \eta}{\partial z} \\ \frac{\partial \zeta}{\partial x} & \frac{\partial \zeta}{\partial y} & \frac{\partial \zeta}{\partial z} \end{bmatrix} = \frac{1}{J} \begin{bmatrix} \frac{\partial y}{\partial \eta}\frac{\partial z}{\partial \zeta} - \frac{\partial y}{\partial \zeta}\frac{\partial z}{\partial \eta} & -\frac{\partial x}{\partial \eta}\frac{\partial z}{\partial \zeta} + \frac{\partial x}{\partial \zeta}\frac{\partial z}{\partial \eta} & \frac{\partial x}{\partial \eta}\frac{\partial y}{\partial \zeta} - \frac{\partial x}{\partial \zeta}\frac{\partial y}{\partial \eta} \\ -\frac{\partial y}{\partial \xi}\frac{\partial z}{\partial \zeta} + \frac{\partial y}{\partial \zeta}\frac{\partial z}{\partial \xi} & \frac{\partial x}{\partial \xi}\frac{\partial z}{\partial \zeta} - \frac{\partial x}{\partial \zeta}\frac{\partial z}{\partial \xi} & -\frac{\partial x}{\partial \xi}\frac{\partial y}{\partial \zeta} + \frac{\partial x}{\partial \zeta}\frac{\partial y}{\partial \xi} \\ \frac{\partial y}{\partial \xi}\frac{\partial z}{\partial \eta} - \frac{\partial y}{\partial \eta}\frac{\partial z}{\partial \xi} & -\frac{\partial x}{\partial \xi}\frac{\partial z}{\partial \eta} + \frac{\partial x}{\partial \eta}\frac{\partial z}{\partial \xi} & \frac{\partial x}{\partial \xi}\frac{\partial y}{\partial \eta} - \frac{\partial x}{\partial \eta}\frac{\partial y}{\partial \xi} \end{bmatrix}, \tag{8.59}$$

where the Jacobian $J$ is the determinant as given by

$$J = \left| \frac{\partial(x, y, z)}{\partial(\xi, \eta, \zeta)} \right| = x_\xi y_\eta z_\zeta + x_\eta y_\zeta z_\xi + x_\zeta y_\xi z_\eta - x_\eta y_\xi z_\zeta - x_\zeta y_\eta z_\xi - x_\xi y_\zeta z_\eta \,. \tag{8.60}$$

In Eq. (8.60), the abbreviation, e.g. $x_\xi$, stands for the partial derivative $\frac{\partial x}{\partial \xi}$ and so on. Let us assume the same interpolation for the global $x$, $y$ and $z$-coordinate as for the displacement field:

$$x(\xi, \eta, \zeta) = \sum_{i=1}^{8} N_i(\xi, \eta, \zeta) x_i, \tag{8.61}$$

$$y(\xi, \eta, \zeta) = \sum_{i=1}^{8} N_i(\xi, \eta, \zeta) y_i, \tag{8.62}$$

$$z(\xi, \eta, \zeta) = \sum_{i=1}^{8} N_i(\xi, \eta, \zeta) z_i, \tag{8.63}$$

where the global coordinates of the nodes $1, \ldots, 8$ can be used for $x_1, \ldots, x_8$, and so on. Thus, the derivatives can easily be obtained as:

$$\frac{\partial x}{\partial \xi} = \sum_{i=1}^{8} \frac{\partial N_i}{\partial \xi} x_i = \sum_{i=1}^{8} \frac{1}{8}(\xi_i)(1 + \eta\eta_i)(1 + \zeta\zeta_i) x_i, \tag{8.64}$$

$$\frac{\partial x}{\partial \eta} = \sum_{i=1}^{8} \frac{\partial N_i}{\partial \eta} x_i = \sum_{i=1}^{8} \frac{1}{8}(1 + \xi\xi_i)(\eta_i)(1 + \zeta\zeta_i) x_i, \tag{8.65}$$

$$\frac{\partial x}{\partial \zeta} = \sum_{i=1}^{8} \frac{\partial N_i}{\partial \zeta} x_i = \sum_{i=1}^{8} \frac{1}{8}(1 + \xi\xi_i)(1 + \eta\eta_i)(\zeta_i) x_i, \tag{8.66}$$

$$\frac{\partial y}{\partial \xi} = \sum_{i=1}^{8} \frac{\partial N_i}{\partial \xi} y_i = \sum_{i=1}^{8} \frac{1}{8}(\xi_i)(1 + \eta\eta_i)(1 + \zeta\zeta_i) y_i, \tag{8.67}$$

$$\frac{\partial y}{\partial \eta} = \sum_{i=1}^{8} \frac{\partial N_i}{\partial \eta} y_i = \sum_{i=1}^{8} \frac{1}{8}(1 + \xi\xi_i)(\eta_i)(1 + \zeta\zeta_i) y_i, \tag{8.68}$$

$$\frac{\partial y}{\partial \zeta} = \sum_{i=1}^{8} \frac{\partial N_i}{\partial \zeta} y_i = \sum_{i=1}^{8} \frac{1}{8}(1+\xi\xi_i)(1+\eta\eta_i)(\zeta_i)y_i, \tag{8.69}$$

$$\frac{\partial z}{\partial \xi} = \sum_{i=1}^{8} \frac{\partial N_i}{\partial \xi} z_i = \sum_{i=1}^{8} \frac{1}{8}(\xi_i)(1+\eta\eta_i)(1+\zeta\zeta_i)z_i, \tag{8.70}$$

$$\frac{\partial z}{\partial \eta} = \sum_{i=1}^{8} \frac{\partial N_i}{\partial \eta} z_i = \sum_{i=1}^{8} \frac{1}{8}(1+\xi\xi_i)(\eta_i)(1+\zeta\zeta_i)z_i, \tag{8.71}$$

$$\frac{\partial z}{\partial \zeta} = \sum_{i=1}^{8} \frac{\partial N_i}{\partial \zeta} z_i = \sum_{i=1}^{8} \frac{1}{8}(1+\xi\xi_i)(1+\eta\eta_i)(\zeta_i)z_i. \tag{8.72}$$

The derivatives of the shape functions with respect to the coordinates in parametric space are summarized in Table 8.3. Note that the derivatives (8.64)–(8.72) are simple constants and independent of $\xi$, $\eta$ and $\zeta$ for a cuboid or a parallelepiped (cf. Fig. 8.3).

Looking at the example of a cube with edge length $2a$ (cf. Fig. 8.3a), one can derive that $\frac{\partial x}{\partial \xi} = \frac{\partial y}{\partial \eta} = \frac{\partial z}{\partial \zeta} = a$, whereas all other geometrical derivatives are zero. For a cuboid with edge length $2a$, $2b$, and $2c$ (cf. Fig. 8.3b), similar derivation gives $\frac{\partial x}{\partial \xi} = a$, $\frac{\partial y}{\partial \eta} = b$, and $\frac{\partial z}{\partial \zeta} = c$. Considering the parallelepiped shown in Fig. 8.3c where the lower nodes are moved by $-d$ in the negative $x$-direction and the upper nodes by $+d$ in the positive $x$-direction, one obtains $\frac{\partial x}{\partial \xi} = a$, $\frac{\partial y}{\partial \eta} = b$, $\frac{\partial z}{\partial \zeta} = c$, and $\frac{\partial x}{\partial \zeta} = d$ (other derivatives are zero). For other cases, the geometrical derivatives become dependent on the parametric coordinates. For example, Fig. 8.3d shows a distorted cuboid where node 1 is translated by $d$ along the $x$-direction. For this case, one obtains:

$$\frac{\partial x}{\partial \xi} = a + \frac{1}{8}d - \frac{1}{8}\zeta d - \frac{1}{8}\eta d + \frac{1}{8}\eta\zeta d, \tag{8.73}$$

$$\frac{\partial x}{\partial \eta} = \frac{1}{8}d - \frac{1}{8}\zeta d - \frac{1}{8}\xi d + \frac{1}{8}\eta\zeta d, \tag{8.74}$$

$$\frac{\partial x}{\partial \zeta} = \frac{1}{8}d - \frac{1}{8}\eta d - \frac{1}{8}\xi d + \frac{1}{8}\eta\zeta d, \tag{8.75}$$

$$\frac{\partial y}{\partial \eta} = b, \tag{8.76}$$

$$\frac{\partial z}{\partial \zeta} = b. \tag{8.77}$$

On the basis of the derived equations, the element matrices given in Eqs. (8.34) till (8.36) can now be numerically evaluated. The integration is performed as in the case of the one-dimensional integrals based on GAUSS-LEGENDRE quadrature. For the

**Table 8.3** Derivatives of the shape functions in parametric space

| Node | $\frac{\partial N_i}{\partial \xi}$ | $\frac{\partial N_i}{\partial \eta}$ | $\frac{\partial N_i}{\partial \zeta}$ |
|---|---|---|---|
| 1 | $\frac{1}{8}(-1)(1-\eta)(1-\zeta)$ | $\frac{1}{8}(1-\xi)(-1)(1-\zeta)$ | $\frac{1}{8}(1-\xi)(1-\eta)(-1)$ |
| 2 | $\frac{1}{8}(1)(1-\eta)(1-\zeta)$ | $\frac{1}{8}(1+\xi)(-1)(1-\zeta)$ | $\frac{1}{8}(1+\xi)(1-\eta)(-1)$ |
| 3 | $\frac{1}{8}(1)(1+\eta)(1-\zeta)$ | $\frac{1}{8}(1+\xi)(1)(1-\zeta)$ | $\frac{1}{8}(1+\xi)(1+\eta)(-1)$ |
| 4 | $\frac{1}{8}(-1)(1+\eta)(1-\zeta)$ | $\frac{1}{8}(1-\xi)(1)(1-\zeta)$ | $\frac{1}{8}(1-\xi)(1+\eta)(-1)$ |
| 5 | $\frac{1}{8}(-1)(1-\eta)(1+\zeta)$ | $\frac{1}{8}(1-\xi)(-1)(1+\zeta)$ | $\frac{1}{8}(1-\xi)(1-\eta)(1)$ |
| 6 | $\frac{1}{8}(1)(1-\eta)(1+\zeta)$ | $\frac{1}{8}(1+\xi)(-1)(1+\zeta)$ | $\frac{1}{8}(1+\xi)(1-\eta)(1)$ |
| 7 | $\frac{1}{8}(1)(1+\eta)(1+\zeta)$ | $\frac{1}{8}(1+\xi)(1)(1+\zeta)$ | $\frac{1}{8}(1+\xi)(1+\eta)(1)$ |
| 8 | $\frac{1}{8}(-1)(1+\eta)(1+\zeta)$ | $\frac{1}{8}(1-\xi)(1)(1+\zeta)$ | $\frac{1}{8}(1-\xi)(1+\eta)(1)$ |

domain integrals, one can write that

$$\int_V f(x, y, z)\mathrm{d}V = \int_{V'} f'(\xi, \eta, \zeta)\mathrm{d}V' = \int_{-1}^{1}\int_{-1}^{1}\int_{-1}^{1} f'(\xi, \eta, \zeta) J \mathrm{d}\xi\mathrm{d}\eta\mathrm{d}\zeta \tag{8.78}$$

$$= \sum_{i=1}^{n} f'(\xi, \eta, \zeta)_i J_i w_i, \tag{8.79}$$

where the Jacobian is given in Eq. (8.60), $(\xi, \eta, \zeta)_i$ are the coordinates of the integration of GAUSS points and $w_i$ are the corresponding weight factors. The location of the integration points and values of associated weights are given in Table 8.4.

### *8.6.1 Evaluation of the Boundary Force Matrix*

The right-hand side of the weak statement contains the boundary force matrix according to Eq. (8.35). To evaluate this expression, it might be more advantageous to replace again the column matrix of traction forces by the stress tensor[9] and normal vector. Thus, we can write Eq. (8.35) in the following form:

[9] The stress *tensor* is used in the following since it makes the derivation easier, cf. Eq. (8.21) and the following comment and footnote.

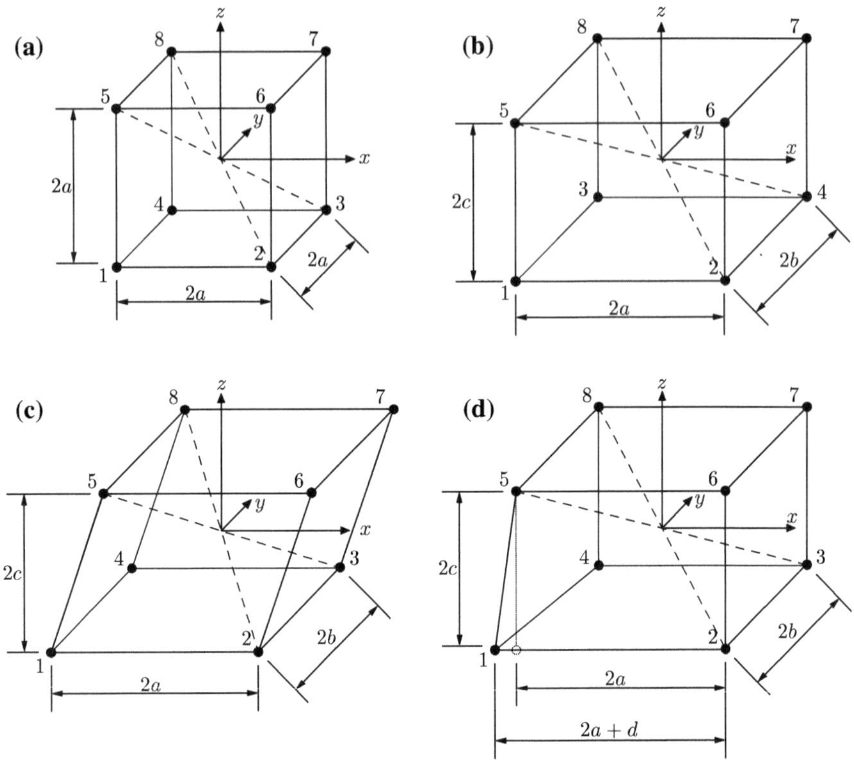


**Fig. 8.3** Different shapes of an eight-node hexahedron in the (x,y,z)-space: **a** cube; **b** cuboid; **c** parallelepiped; **d** distorted cuboid

$$\boldsymbol{f}_{\boldsymbol{t}}^{\mathrm{e}} = \int_A \underbrace{\begin{bmatrix} N_1 & 0 & 0 \\ 0 & N_1 & 0 \\ 0 & 0 & N_1 \\ \hline N_2 & 0 & 0 \\ 0 & N_2 & 0 \\ 0 & 0 & N_2 \\ \hline & \cdots & \\ & \cdots & \\ & \cdots & \\ \hline N_n & 0 & 0 \\ 0 & N_n & 0 \\ 0 & 0 & N_n \end{bmatrix}}_{\boldsymbol{N}} \underbrace{\begin{bmatrix} \sigma_{xx} & \sigma_{xy} & \sigma_{xz} \\ \sigma_{yx} & \sigma_{yy} & \sigma_{yz} \\ \sigma_{zx} & \sigma_{zy} & \sigma_{zz} \end{bmatrix} \begin{bmatrix} n_x \\ n_y \\ n_z \end{bmatrix}}_{\boldsymbol{t}} \mathrm{d}A, \tag{8.80}$$

or

**Table 8.4** Integration rules for hexahedral elements [67]

| Points | $\xi_i$ | $\eta_i$ | $\zeta_i$ | Weight $w_i$ | Error |
|---|---|---|---|---|---|
| 1 | 0 | 0 | 0 | 8 | $O(\xi^2)$ |
| 8 | $\pm 1/\sqrt{3}$ | $\pm 1/\sqrt{3}$ | $\pm 1/\sqrt{3}$ | 1 | $O(\xi^4)$ |
| 27 | 0 | 0 | 0 | $\left(\frac{8}{9}\right)^3$ | |
| | $\pm\sqrt{0.6}$ | 0 | 0 | $\left(\frac{5}{9}\right)\left(\frac{8}{9}\right)^2$ | |
| | 0 | $\pm\sqrt{0.6}$ | 0 | $\left(\frac{5}{9}\right)\left(\frac{8}{9}\right)^2$ | $O(\xi^6)$ |
| | 0 | 0 | $\pm\sqrt{0.6}$ | $\left(\frac{5}{9}\right)\left(\frac{8}{9}\right)^2$ | |
| | $\pm\sqrt{0.6}$ | $\pm\sqrt{0.6}$ | 0 | $\left(\frac{8}{9}\right)\left(\frac{5}{9}\right)^2$ | |
| | $\pm\sqrt{0.6}$ | 0 | $\pm\sqrt{0.6}$ | $\left(\frac{8}{9}\right)\left(\frac{5}{9}\right)^2$ | |
| | 0 | $\pm\sqrt{0.6}$ | $\pm\sqrt{0.6}$ | $\left(\frac{8}{9}\right)\left(\frac{5}{9}\right)^2$ | |
| | $\pm\sqrt{0.6}$ | $\pm\sqrt{0.6}$ | $\pm\sqrt{0.6}$ | $\left(\frac{5}{9}\right)^3$ | |

$$\boldsymbol{f}_t^{\mathrm{e}} = \int\limits_A \begin{bmatrix} N_1\sigma_{xx} & N_1\sigma_{xy} & N_1\sigma_{xz} \\ N_1\sigma_{yx} & N_1\sigma_{yy} & N_1\sigma_{yz} \\ N_1\sigma_{zx} & N_1\sigma_{zy} & N_1\sigma_{zz} \\ \hline N_2\sigma_{xx} & N_2\sigma_{xy} & N_2\sigma_{xz} \\ N_2\sigma_{yx} & N_2\sigma_{yy} & N_2\sigma_{yz} \\ N_2\sigma_{zx} & N_2\sigma_{zy} & N_2\sigma_{zz} \\ \hline & \cdots & \\ & \cdots & \\ & \cdots & \\ \hline N_n\sigma_{xx} & N_n\sigma_{xy} & N_n\sigma_{xz} \\ N_n\sigma_{yx} & N_n\sigma_{yy} & N_n\sigma_{yz} \\ N_n\sigma_{zx} & N_n\sigma_{zy} & N_n\sigma_{zz} \end{bmatrix} \begin{bmatrix} n_x \\ n_y \\ n_z \end{bmatrix} \mathrm{d}A. \tag{8.81}$$

The expression for the boundary force matrix given in Eq. (8.81) needs to be evaluated for each node along the element boundary. For node 1, the shape function $N_1$ is equal to one and identically zero for all other nodes. In addition, all other shape functions are identically zero at node 1. Since each node has three different normal vectors, cf. Fig. 8.4, one may calculate the expression for node 1 in $x$-direction by evaluating the first row of the following system:

Fig. 8.4 Normal vectors for the evaluation of the boundary force matrix at node 1

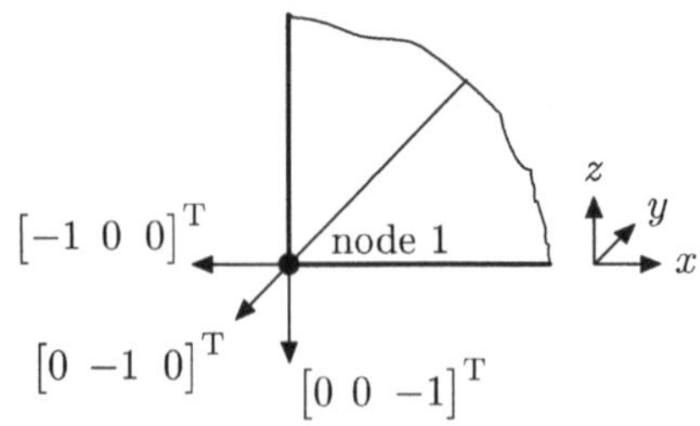


$$
\begin{bmatrix}
\sigma_{xx} & \sigma_{xy} & \sigma_{xz} \\
\sigma_{yx} & \sigma_{yy} & \sigma_{yz} \\
\sigma_{zx} & \sigma_{zy} & \sigma_{zz} \\ \hline
0 & 0 & 0 \\
0 & 0 & 0 \\
0 & 0 & 0 \\ \hline
 & \cdots & \\
 & \cdots & \\
 & \cdots & \\ \hline
0 & 0 & 0 \\
0 & 0 & 0 \\
0 & 0 & 0
\end{bmatrix}
\begin{bmatrix} -1 \\ 0 \\ 0 \end{bmatrix} A +
\begin{bmatrix}
\sigma_{xx} & \sigma_{xy} & \sigma_{xz} \\
\sigma_{yx} & \sigma_{yy} & \sigma_{yz} \\
\sigma_{zx} & \sigma_{zy} & \sigma_{zz} \\ \hline
0 & 0 & 0 \\
0 & 0 & 0 \\
0 & 0 & 0 \\ \hline
 & \cdots & \\
 & \cdots & \\
 & \cdots & \\ \hline
0 & 0 & 0 \\
0 & 0 & 0 \\
0 & 0 & 0
\end{bmatrix}
\begin{bmatrix} 0 \\ -1 \\ 0 \end{bmatrix} A +
\begin{bmatrix}
\sigma_{xx} & \sigma_{xy} & \sigma_{xz} \\
\sigma_{yx} & \sigma_{yy} & \sigma_{yz} \\
\sigma_{zx} & \sigma_{zy} & \sigma_{zz} \\ \hline
0 & 0 & 0 \\
0 & 0 & 0 \\
0 & 0 & 0 \\ \hline
 & \cdots & \\
 & \cdots & \\
 & \cdots & \\ \hline
0 & 0 & 0 \\
0 & 0 & 0 \\
0 & 0 & 0
\end{bmatrix}
\begin{bmatrix} 0 \\ 0 \\ -1 \end{bmatrix} A \,, \tag{8.82}
$$

or as

$$
-\sigma_{xx}A - \sigma_{xy}A - \sigma_{xz}A \,, \tag{8.83}
$$

which is balanced by the external force $F_{1x}$ at node 1. Similar results can be obtained for all other nodes and directions and the boundary force matrix can be written as

$$
\boldsymbol{f}_{t}^{\mathrm{e}} = \begin{bmatrix} F_{1x} & F_{1y} & F_{1z} & F_{2x} & F_{2y} & F_{2z} & \cdots & F_{8x} & F_{8y} & F_{8z} \end{bmatrix}^{\mathrm{T}} \,, \tag{8.84}
$$

where an external force is positive if directed to the positive coordinate direction.

## 8.6.2 Evaluation of the Body Force Matrix

The right-hand side of the weak statement contains the body force matrix according to Eq. (8.36). Let us consider in the following the special case that all the elements of the column matrix of body forces are constant, i.e. $\boldsymbol{b} = \begin{bmatrix} f & f & f \end{bmatrix}^{\mathrm{T}}$. Introducing the column matrix of the shape functions in Eq. (8.36) gives:

$$
\boldsymbol{f}_{\boldsymbol{b}}^{\mathrm{e}} = \int\limits_V \underbrace{\begin{bmatrix} N_1 & 0 & 0 \\ 0 & N_1 & 0 \\ 0 & 0 & N_1 \\ \hline N_2 & 0 & 0 \\ 0 & N_2 & 0 \\ 0 & 0 & N_2 \\ \hline & \cdots & \\ & \cdots & \\ & \cdots & \\ \hline N_8 & 0 & 0 \\ 0 & N_8 & 0 \\ 0 & 0 & N_8 \end{bmatrix}}_{\boldsymbol{N}} \underbrace{\begin{bmatrix} f \\ f \\ f \end{bmatrix}}_{\boldsymbol{b}} \mathrm{d}V = \int\limits_V \begin{bmatrix} N_1 f \\ N_1 f \\ N_1 f \\ \hline N_2 f \\ N_2 f \\ N_2 f \\ \hline \cdots \\ \cdots \\ \cdots \\ \hline N_8 f \\ N_8 f \\ N_8 f \end{bmatrix} \mathrm{d}V = f \int\limits_V \begin{bmatrix} N_1 \\ N_1 \\ N_1 \\ \hline N_2 \\ N_2 \\ N_2 \\ \hline \cdots \\ \cdots \\ \cdots \\ \hline N_8 \\ N_8 \\ N_8 \end{bmatrix} \mathrm{d}V. \tag{8.85}
$$

Thus, the integration over the single shape functions must be performed. It should be noted here that the shape functions in Eq. (8.85) are functions of the Cartesian coordinates, i.e. $N_i = N_i(x, y, z)$. In order to continue the derivation, let us assume that the hexahedron element is a cube with edge length $2a$ as shown in Fig. 8.3a. For this regular shape, the Jacobian simplifies to $J = a^3$. To numerically integrate the integral, a coordinate transformation must be performed which reads:

$$
\boldsymbol{f}_{\boldsymbol{b}}^{\mathrm{e}} = f \int\limits_{-1}^{1} \int\limits_{-1}^{1} \int\limits_{-1}^{1} \begin{bmatrix} N_1(\xi, \eta, \zeta) \\ N_1(\xi, \eta, \zeta) \\ N_1(\xi, \eta, \zeta) \\ \hline \cdots \\ \cdots \\ \cdots \\ \hline N_8(\xi, \eta, \zeta) \\ N_8(\xi, \eta, \zeta) \\ N_8(\xi, \eta, \zeta) \end{bmatrix} a^3 \mathrm{d}\xi \mathrm{d}\eta \mathrm{d}\zeta. \tag{8.86}
$$

Approximation of the integral based on an eight-point GAUSS- LEGENDRE quadrature formula (cf. Table 8.4) gives:

$$\boldsymbol{f}_{b}^{\mathrm{e}} \approx f \sum_{i=1}^{8} \begin{bmatrix} N_1(\xi,\eta,\zeta) \\ N_1(\xi,\eta,\zeta) \\ N_1(\xi,\eta,\zeta) \\ \hline \cdots \\ \cdots \\ \cdots \\ \hline N_8(\xi,\eta,\zeta) \\ N_8(\xi,\eta,\zeta) \\ N_8(\xi,\eta,\zeta) \end{bmatrix}_i a^3 w_i = f 8a^3 \begin{bmatrix} \frac{1}{8} \\ \frac{1}{8} \\ \frac{1}{8} \\ \hline \cdots \\ \cdots \\ \cdots \\ \hline \frac{1}{8} \\ \frac{1}{8} \\ \frac{1}{8} \end{bmatrix}. \tag{8.87}$$

It can be concluded from the last equation that the equivalent nodal loads in the case of a constant body force matrix are obtained by first calculating the resultant force, i.e. body force times volume ($f \times V$), and then equally distributing this force to all the eight nodes ($\frac{f \times V}{8}$). It shout be noted that the same result would be obtained based on analytical integration. The analytical integration for a cube with edge length $2a$ can be based on Eq. (8.85) and transforming the shape functions given in Eqs. (8.47)–(8.54) to the Cartesian space, for example, as $N_1(x,y,z) = \frac{1}{8}\left(1-\frac{x}{a}\right)\left(1-\frac{y}{a}\right)\left(1-\frac{z}{a}\right)$. Thus, analytical integration would require to evaluate the following expression:

$$\boldsymbol{f}_{b}^{\mathrm{e}} = f \int_{-a}^{a}\int_{-a}^{a}\int_{-a}^{a} \begin{bmatrix} N_1(x,y,z) \\ N_1(x,y,z) \\ N_1(x,y,z) \\ \hline \cdots \\ \cdots \\ \cdots \\ \hline N_8(x,y,z) \\ N_8(x,y,z) \\ N_8(x,y,z) \end{bmatrix} \mathrm{d}x\mathrm{d}y\mathrm{d}z. \tag{8.88}$$

## 8.7 Derivation of the Fully Implicit Backward Euler Algorithm for Isotropic Hardening

Let us first summarize the three ingredients of plastic analysis, i.e. the yield condition, the flow rule and the hardening rule for the general three-dimensional case. The yield condition can generally be expressed as

$$F(\boldsymbol{\sigma}, \boldsymbol{q}) \leq 0, \tag{8.89}$$

where $\boldsymbol{q} = \left[\kappa \ \boldsymbol{\alpha}\right]^{\mathrm{T}}$ is the column matrix of internal variables describing the hardening behavior of the material. Parameter $\kappa$ relates to isotropic hardening while the column matrix $\boldsymbol{\alpha}$ contains the kinematic hardening parameters.

The flow rule in its general form is given by

$$\mathrm{d}\varepsilon^{\mathrm{pl}} = \mathrm{d}\lambda \times \boldsymbol{r}(\boldsymbol{\sigma}, \boldsymbol{q}), \tag{8.90}$$

where $\mathrm{d}\lambda$ is a scalar called the plastic multiplier or consistency parameter and $\boldsymbol{r}(\boldsymbol{\sigma}, \boldsymbol{q})$ is the plastic flow direction. The plastic flow direction is often stated in terms of a plastic potential function $Q$ and the plastic strain increments are given by $\mathrm{d}\varepsilon^{\mathrm{pl}} = \mathrm{d}\lambda \frac{\partial Q}{\partial \sigma}$. The flow is said to be associated if $Q = F$, otherwise non-associated.

The evolution equation for the internal variables $\boldsymbol{q}$ can be specified in its general form as

$$\mathrm{d}\boldsymbol{q} = \mathrm{d}\lambda \times \boldsymbol{h}(\boldsymbol{\sigma}, \boldsymbol{q}), \tag{8.91}$$

where the function $\boldsymbol{h}$ describes the evolution of the hardening parameters.

The derivation of the fully implicit backward EULER algorithm for isotropic hardening ($\boldsymbol{q} \rightarrow q$) follows the steps presented in Sects. 6.1 and 6.2 where based on the previous increment ($n$) the field variables must be calculated for the new state ($n + 1$) at each integration point.

Under the assumption of pure linear-elastic material behavior, an elastic predictor calculates the trial stress as:

$$\boldsymbol{\sigma}_{n+1}^{\mathrm{trial}} = \boldsymbol{\sigma}_n + \underbrace{\boldsymbol{C}\Delta\boldsymbol{\varepsilon}_n}_{\text{predictor } \Delta\boldsymbol{\sigma}_n^{\mathrm{el}}} . \tag{8.92}$$

The corresponding hardening state corresponds to the state of the previous increment since the load step is assumed to be pure elastic:

$$q_{n+1}^{\mathrm{trial}} = q_n. \tag{8.93}$$

If the trial state is in the elastic range or on the yield surface ($F^{\mathrm{trial}} \leq 0$), the real state at $n + 1$ is equal to the calculated trial values of stress and hardening variables. However, if the stress state is outside the yield surface ($F^{\mathrm{trial}} > 0$), the second part of the algorithm must calculate based on the plastic corrector $\boldsymbol{\sigma}_n^{\mathrm{pl}}$ a valid state on the yield surface ($F_{n+1} = 0$), cf. Fig. 8.5.

The difference between the initial and target state (stress increment)

$$\Delta\boldsymbol{\sigma}_n = \boldsymbol{\sigma}_{n+1} - \boldsymbol{\sigma}_n \tag{8.94}$$

results according to HOOKE's law from the elastic part of the strain increment which can be obtained as the difference between the total strain increment and the plastic part as:

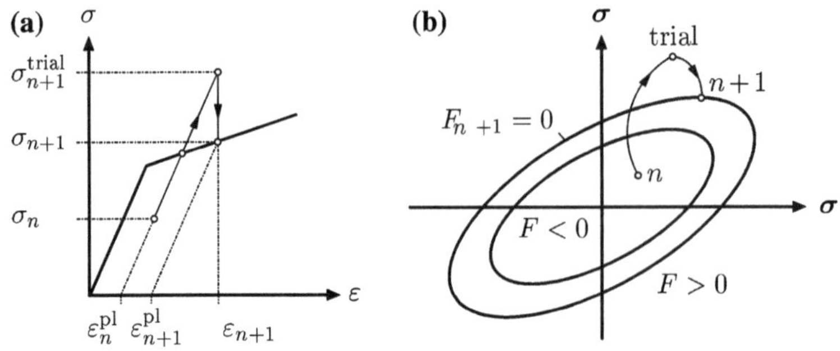


**Fig. 8.5** Schematic representation of the integration algorithm: **a** uniaxial stress-strain diagram; **b** multidimensional stress space. The $\boldsymbol{\sigma}$-$\boldsymbol{\sigma}$ coordinate system represents the $n$-dimensional stress space

$$\Delta\boldsymbol{\sigma}_n = \boldsymbol{C}\Delta\boldsymbol{\varepsilon}_n^{\text{el}} = \boldsymbol{C}\left(\Delta\boldsymbol{\varepsilon}_n - \Delta\boldsymbol{\varepsilon}_n^{\text{pl}}\right). \tag{8.95}$$

The total strain increment $\Delta\boldsymbol{\varepsilon}_n$ can be expressed with the trial stress state according to Fig. 8.5 as:

$$\Delta\boldsymbol{\varepsilon}_n = \boldsymbol{\varepsilon}_{n+1} - \boldsymbol{\varepsilon}_n = \boldsymbol{C}^{-1}\left(\boldsymbol{\sigma}_{n+1}^{\text{trial}} - \boldsymbol{\sigma}_n\right). \tag{8.96}$$

Introducing the last equation and the flow rule[10] in Eq. (8.95), the final state $\boldsymbol{\sigma}_{n+1}$ can be obtained in dependence of the trial stress state $\boldsymbol{\sigma}_{n+1}^{\text{trial}}$ as:

$$\boldsymbol{\sigma}_{n+1} = \boldsymbol{\sigma}_{n+1}^{\text{trial}} - \Delta\lambda_{n+1}\boldsymbol{C}\boldsymbol{r}(\boldsymbol{\sigma}). \tag{8.97}$$

Evaluating the function $\boldsymbol{r}(\boldsymbol{\sigma})$ during the iteration in the final state results in the implicit backward-EULER algorithm. Thus, the equations

$$\boldsymbol{\sigma}_{n+1} = \boldsymbol{\sigma}_{n+1}^{\text{trial}} - \Delta\lambda_{n+1}\boldsymbol{C}\boldsymbol{r}(\boldsymbol{\sigma}_{n+1}, q_{n+1}), \tag{8.98}$$

$$q_{n+1} = q_n + \Delta\lambda_{n+1}\boldsymbol{h}(\boldsymbol{\sigma}_{n+1}, q_{n+1}), \tag{8.99}$$

$$F = F(\boldsymbol{\sigma}_{n+1}, q_{n+1}), \tag{8.100}$$

are fulfilled in the final state $(n+1)$. However, a residual $\boldsymbol{r}$ remains for each of these equations outside the final state:

[10] At this point, we change again the notation from $\mathrm{d}\lambda$ to $\Delta\lambda$.

$$\begin{aligned}
\boldsymbol{r}_\sigma(\boldsymbol{\sigma}, q, \Delta\lambda) &= \boldsymbol{\sigma} - \boldsymbol{\sigma}_{n+1}^{\text{trial}} + \Delta\lambda \boldsymbol{C}\boldsymbol{r}(\boldsymbol{\sigma}, q) \neq 0 \ \text{ or}, \\
&= \boldsymbol{C}^{-1}\boldsymbol{\sigma} - \boldsymbol{C}^{-1}\boldsymbol{\sigma}_{n+1}^{\text{trial}} + \Delta\lambda \boldsymbol{r}(\boldsymbol{\sigma}, q) \neq 0, \qquad (8.101) \\
\boldsymbol{r}_q(\boldsymbol{\sigma}, q, \Delta\lambda) &= q - q_n - \Delta\lambda \boldsymbol{h}(\boldsymbol{\sigma}, q) \neq 0 \ \text{ or}, \\
&= -q + q_n + \Delta\lambda \boldsymbol{h}(\boldsymbol{\sigma}, q) \neq 0, \qquad (8.102) \\
r_F(\boldsymbol{\sigma}, q) &= F(\boldsymbol{\sigma}, q) \neq 0. \qquad (8.103)
\end{aligned}$$

Thus, the final stress and hardening state is the root of a matrix function $\boldsymbol{m}$ which is composed of the residual functions. Furthermore, it is useful to collect all variables in a single variable matrix $\boldsymbol{v}$:

$$\boldsymbol{m}(\boldsymbol{v}) \in \left(\mathbb{R}^8 \to \mathbb{R}^8\right) = \begin{bmatrix} \boldsymbol{r}_\sigma(\boldsymbol{v}) \\ \boldsymbol{r}_q(\boldsymbol{v}) \\ r_F(\boldsymbol{v}) \end{bmatrix}, \quad \boldsymbol{v} = \begin{bmatrix} \boldsymbol{\sigma} \\ q \\ \Delta\lambda \end{bmatrix}. \qquad (8.104)$$

The root is found by a NEWTON iteration (iteration index: $i$):

$$\boldsymbol{v}^{(i+1)} = \boldsymbol{v}^{(i)} - \left[\frac{\mathrm{d}\boldsymbol{m}}{\partial \boldsymbol{v}}\left(\boldsymbol{v}^{(i)}\right)\right]^{-1} \boldsymbol{m}\left(\boldsymbol{v}^{(i)}\right), \qquad (8.105)$$

where

$$\boldsymbol{v}^{(0)} = \begin{bmatrix} \boldsymbol{\sigma}^{(0)} \\ q^{(0)} \\ \Delta\lambda^{(0)} \end{bmatrix} = \begin{bmatrix} \boldsymbol{\sigma}_{n+1}^{\text{trial}} \\ q_n \\ 0 \end{bmatrix}, \qquad (8.106)$$

can be used as the initial value. The Jacobian matrix $\frac{\partial \boldsymbol{m}}{\partial \boldsymbol{v}}$ of the residual functions is obtained from the partial derivatives of Eqs. (8.101)–(8.103) as:

$$\frac{\partial \boldsymbol{m}}{\partial \boldsymbol{v}}(\boldsymbol{\sigma}, q, \Delta\lambda) = \begin{bmatrix} \frac{\partial \boldsymbol{r}_\sigma}{\partial \boldsymbol{\sigma}} & \frac{\partial \boldsymbol{r}_\sigma}{\partial q} & \frac{\partial \boldsymbol{r}_\sigma}{\partial \Delta\lambda} \\ \frac{\partial \boldsymbol{r}_q}{\partial \boldsymbol{\sigma}} & \frac{\partial \boldsymbol{r}_q}{\partial q} & \frac{\partial \boldsymbol{r}_q}{\partial \Delta\lambda} \\ \frac{\partial r_F}{\partial \boldsymbol{\sigma}} & \frac{\partial r_F}{\partial q} & \frac{\partial r_F}{\partial \Delta\lambda} \end{bmatrix} (\boldsymbol{\sigma}, q, \Delta\lambda) \qquad (8.107)$$

$$= \begin{bmatrix} \boldsymbol{C}^{-1} + \Delta\lambda \frac{\partial \boldsymbol{r}}{\partial \boldsymbol{\sigma}} & \Delta\lambda \frac{\partial \boldsymbol{r}}{\partial q} & \boldsymbol{r} \\ \Delta\lambda \frac{\partial \boldsymbol{h}}{\partial \boldsymbol{\sigma}} & -\mathbf{I} + \Delta\lambda \frac{\partial \boldsymbol{h}}{\partial q} & \boldsymbol{h} \\ \frac{\partial F}{\partial \boldsymbol{\sigma}} & \frac{\partial F}{\partial q} & 0 \end{bmatrix} (\boldsymbol{\sigma}, q, \Delta\lambda). \qquad (8.108)$$

In addition to the fulfillment of the plasticity equations (8.98)–(8.100) in each integration point, the global equilibrium must be fulfilled. To be able to apply the NEWTON iteration for this task, it is required even for small strains in the general three-dimensional case to derive the appropriate elasto-plastic tangent modulus,[11] cf. [115]. The consistent elasto-plastic modulus results for the three-dimensional case as:

$$\boldsymbol{C}_{n+1}^{\text{elpl}} = \frac{\partial \boldsymbol{\sigma}_{n+1}}{\partial \boldsymbol{\varepsilon}_{n+1}} = \frac{\partial \Delta \boldsymbol{\sigma}_{n}}{\partial \boldsymbol{\varepsilon}_{n+1}}. \tag{8.109}$$

The inverse of the Jacobian matrix $\frac{\partial \boldsymbol{m}}{\partial \boldsymbol{v}}$ which has to be evaluated in the converged state of the above mentioned NEWTON iteration gives

$$\left[\left(\frac{\partial \boldsymbol{m}}{\partial \boldsymbol{v}}\right)_{n+1}\right]^{-1} = \begin{bmatrix} \tilde{\boldsymbol{m}}_{11} & \tilde{\boldsymbol{m}}_{12} & \tilde{\boldsymbol{m}}_{13} \\ \tilde{\boldsymbol{m}}_{21} & \tilde{\boldsymbol{m}}_{22} & \tilde{\boldsymbol{m}}_{23} \\ \tilde{\boldsymbol{m}}_{31} & \tilde{\boldsymbol{m}}_{32} & \tilde{\boldsymbol{m}}_{33} \end{bmatrix}_{n+1}, \tag{8.110}$$

from which the elasto-plastic tangent modulus results as [115]:

$$\boldsymbol{C}_{n+1}^{\text{elpl}} = \tilde{\boldsymbol{m}}_{11}\,. \tag{8.111}$$

## 8.8 Supplementary Problems

**8.1 Hooke's law in terms of shear and bulk modulus**
Derive HOOKE's law for a linear isotropic material in terms of shear modulus $G$ and bulk modulus $K$ in the elastic stiffness form ($\boldsymbol{\sigma} = \boldsymbol{\sigma}(\boldsymbol{\varepsilon})$) and the elastic compliance form $\boldsymbol{\varepsilon} = \boldsymbol{\varepsilon}(\boldsymbol{\sigma})$.

**8.2 Hooke's law for the plane stress state**
Derive HOOKE's law for a two-dimensional plane stress state ($\sigma_z = \sigma_{yz} = \sigma_{xz} = 0$) in its elastic stiffness ($\boldsymbol{\sigma} = \boldsymbol{\sigma}(\boldsymbol{\epsilon})$) and elastic compliance ($\boldsymbol{\varepsilon} = \boldsymbol{\varepsilon}(\boldsymbol{\sigma})$) form in terms of YOUNG's modulus $E$ and POISSON's ratio $\nu$ for a linear isotropic material.

**8.3 Hooke's law for the plane strain state**
Derive HOOKE's law for a two-dimensional plane strain state ($\varepsilon_z = \varepsilon_{yz} = \varepsilon_{xz} = 0$) in its elastic stiffness ($\boldsymbol{\sigma} = \boldsymbol{\sigma}(\boldsymbol{\epsilon})$) and elastic compliance ($\boldsymbol{\varepsilon} = \boldsymbol{\varepsilon}(\boldsymbol{\sigma})$) form in terms of YOUNG's modulus $E$ and POISSON's ratio $\nu$ for a linear isotropic material.

**8.4 Beltrami-Michell equations**
Derive the mathematical expressions for the BELTRAMI-MICHELL equations.

**8.5 Green-Gauss theorem applied to equilibrium equation in $x$-direction**

[11] Alternative notation is consistent tangent modulus or algorithmic modulus.

Write the equilibrium equation as given in Eq. (8.1) in matrix form, i.e. $\boldsymbol{\mathcal{L}}_x^{\mathrm{T}}\boldsymbol{\sigma}_x + f_x = 0$, where $\boldsymbol{\mathcal{L}}_x^{\mathrm{T}} = \left[\frac{\partial}{\partial x}\; 0\; 0\; \frac{\partial}{\partial y}\; 0\; \frac{\partial}{\partial z}\right]$ and $\boldsymbol{\sigma}_x = \left[\sigma_x\; 0\; 0\; \sigma_{xy}\; 0\; \sigma_{xz}\right]^{\mathrm{T}}$. Under consideration of the constitutive equation and the strain-displacement relationship, replace the stress vector $\boldsymbol{\sigma}_x$ by the displacement vector $\boldsymbol{u} = \left[u_x\; u_y\; u_z\right]^{\mathrm{T}}$ and write the weighted residual statement for the $x$-axis in the form $\int_V W_x(\cdots)\mathrm{d}V = 0$. Use the GREEN- GAUSS theorem as given in Sect. A.9 to derive the weak form for the $x$-direction.

**8.6 Green-Gauss theorem applied to derive general 3D weak form**
How can the result from supplementary problem 8.5 be used to derive the general expression for the weak form as given in Eq. (8.21)?

**8.7 Body force matrix for gravity**
Simplify the body force matrix as given in Eq. (8.87) for the special case of a body force due to standard gravity $g$ acting in the negative $y$-direction.

# Appendix A
# Mathematics

## A.1 Greek Alphabet

See Table A.1.

## A.2 Frequently Used Constants

$$
\begin{aligned}
\pi &= 3.14159\,,\\
e &= 2.71828\,,\\
\sqrt{2} &= 1.41421\,,\\
\sqrt{3} &= 1.73205\,,\\
\sqrt{5} &= 2.23606\,,\\
\sqrt{e} &= 1.64872\,,\\
\sqrt{\pi} &= 1.77245\,.
\end{aligned}
$$

## A.3 Special Products

$$(x+y)^2 = x^2 + 2xy + y^2\,, \tag{A.1}$$

$$(x-y)^2 = x^2 - 2xy + y^2\,, \tag{A.2}$$

$$(x+y)^3 = x^3 + 3x^2y + 3xy^2 + y^3\,, \tag{A.3}$$

$$(x-y)^3 = x^3 - 3x^2y + 3xy^2 - y^3\,, \tag{A.4}$$

$$(x+y)^4 = x^4 + 4x^3y + 6x^2y^2 + 4xy^3 + y^4\,, \tag{A.5}$$

$$(x-y)^4 = x^4 - 4x^3y + 6x^2y^2 - 4xy^3 + y^4\,. \tag{A.6}$$



A. Öchsner, *Elasto-Plasticity of Frame Structure Elements*,
DOI 10.1007/978-3-662-44225-8

**Table A.1** The Greek alphabet

| Name | Small letters | Capital letters |
|---|---|---|
| Alpha | $\alpha$ | $A$ |
| Beta | $\beta$ | $B$ |
| Gamma | $\gamma$ | $\Gamma$ |
| Delta | $\delta$ | $\Delta$ |
| Epsilon | $\epsilon$ | $E$ |
| Zeta | $\zeta$ | $Z$ |
| Eta | $\eta$ | $H$ |
| Theta | $\theta, \vartheta$ | $\Theta$ |
| Iota | $\iota$ | $I$ |
| Kappa | $\kappa$ | $K$ |
| Lambda | $\lambda$ | $\Lambda$ |
| My | $\mu$ | $M$ |
| Ny | $\nu$ | $N$ |
| Xi | $\xi$ | $\Xi$ |
| Omikron | $o$ | $O$ |
| Pi | $\pi$ | $\Pi$ |
| Rho | $\rho, \varrho$ | $P$ |
| Sigma | $\sigma$ | $\Sigma$ |
| Tau | $\tau$ | $T$ |
| Ypsilon | $\upsilon$ | $\Upsilon$ |
| Phi | $\phi, \varphi$ | $\Phi$ |
| Chi | $\chi$ | $X$ |
| Psi | $\psi$ | $\Psi$ |
| Omega | $\omega$ | $\Omega$ |

## A.4 Trigonometric Functions

### Definition on a right-angled triangle

The triangle $ABC$ is in $C$ right-angled and has edges of length $a, b, c$. The trigonometric functions of the angle $\alpha$ are defined in the following manner (Fig. A.1):

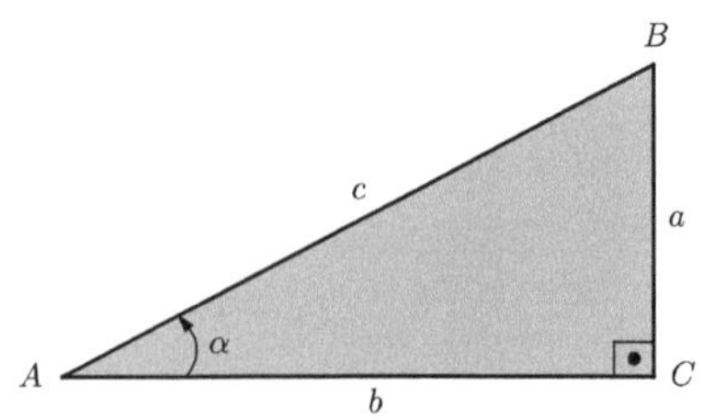


**Fig. A.1** Right-angled triangle

$$\text{sine of } \alpha = \sin\alpha = \frac{a}{c} = \frac{\text{opposite}}{\text{hypotenuse}}\,, \tag{A.7}$$

$$\text{cosine of } \alpha = \cos\alpha = \frac{b}{c} = \frac{\text{adjacent}}{\text{hypotenuse}}\,, \tag{A.8}$$

$$\text{tangent of } \alpha = \tan\alpha = \frac{a}{b} = \frac{\text{opposite}}{\text{adjacent}}\,, \tag{A.9}$$

$$\text{cotangent of } \alpha = \cot\alpha = \frac{b}{a} = \frac{\text{adjacent}}{\text{opposite}}\,, \tag{A.10}$$

$$\text{secant of } \alpha = \sec\alpha = \frac{c}{b} = \frac{\text{hypotenuse}}{\text{adjacent}}\,, \tag{A.11}$$

$$\text{cosecant of } \alpha = \csc\alpha = \frac{c}{a} = \frac{\text{hypotenuse}}{\text{opposite}}\,. \tag{A.12}$$

**Addition formulae**

$$\sin(\alpha \pm \beta) = \sin\alpha\cos\beta \pm \cos\alpha\sin\beta\,, \tag{A.13}$$

$$\cos(\alpha \pm \beta) = \cos\alpha\cos\beta \mp \sin\alpha\sin\beta\,, \tag{A.14}$$

$$\tan(\alpha \pm \beta) = \frac{\tan\alpha \pm \tan\beta}{1 \mp \tan\alpha\tan\beta}\,, \tag{A.15}$$

$$\cot(\alpha \pm \beta) = \frac{\cot\alpha\cot\beta \mp 1}{\cot\beta \pm \cot\beta}\,. \tag{A.16}$$

**Identity formula**

$$\sin^2\alpha + \cos^2\alpha = 1\,. \tag{A.17}$$

**Double-angle formulae**

$$\sin(2\alpha) = 2\sin\alpha \cdot \cos\alpha\,, \tag{A.18}$$

$$\cos(2\alpha) = \cos^2\alpha - \sin^2\alpha\,, \tag{A.19}$$

$$= 2\cos^2\alpha - 1\,, \tag{A.20}$$

$$= 1 - 2\sin^2\alpha \tag{A.21}$$

$$\tan(2\alpha) = \frac{2\tan\alpha}{1 - \tan^2\alpha}\,. \tag{A.22}$$

**Table A.2** Analytic values of sine, cosine, tangent and cotangent for different angles

| $\alpha$ in degree | $\alpha$ in radian | $\sin\alpha$ | $\cos\alpha$ | $\tan\alpha$ | $\cot\alpha$ |
|---|---|---|---|---|---|
| 0° | 0 | 0 | 1 | 0 | $\pm\infty$ |
| 30° | $\frac{1}{6}\pi$ | $\frac{1}{2}$ | $\frac{\sqrt{3}}{2}$ | $\frac{\sqrt{3}}{3}$ | $\sqrt{3}$ |
| 45° | $\frac{1}{4}\pi$ | $\frac{\sqrt{2}}{2}$ | $\frac{\sqrt{2}}{2}$ | 1 | 1 |
| 60° | $\frac{1}{3}\pi$ | $\frac{\sqrt{3}}{2}$ | $\frac{1}{2}$ | $\sqrt{3}$ | $\frac{\sqrt{3}}{3}$ |
| 90° | $\frac{1}{2}\pi$ | 1 | 0 | $\pm\infty$ | 0 |
| 120° | $\frac{2}{3}\pi$ | $\frac{\sqrt{3}}{2}$ | $-\frac{1}{2}$ | $-\sqrt{3}$ | $-\frac{\sqrt{3}}{3}$ |
| 135° | $\frac{3}{4}\pi$ | $\frac{\sqrt{2}}{2}$ | $-\frac{\sqrt{2}}{2}$ | 1 | 1 |
| 150° | $\frac{5}{6}\pi$ | $\frac{1}{2}$ | $-\frac{\sqrt{3}}{2}$ | $-\frac{\sqrt{3}}{3}$ | $-\sqrt{3}$ |
| 180° | $\pi$ | 0 | $-1$ | 0 | $\pm\infty$ |
| 210° | $\frac{7}{6}\pi$ | $-\frac{1}{2}$ | $-\frac{\sqrt{3}}{2}$ | $\frac{\sqrt{3}}{3}$ | $\sqrt{3}$ |
| 225° | $\frac{5}{4}\pi$ | $-\frac{\sqrt{2}}{2}$ | $-\frac{\sqrt{2}}{2}$ | 1 | 1 |
| 240° | $\frac{4}{3}\pi$ | $-\frac{\sqrt{3}}{2}$ | $-\frac{1}{2}$ | $\sqrt{3}$ | $\frac{\sqrt{3}}{3}$ |
| 270° | $\frac{3}{2}\pi$ | $-1$ | 0 | $\pm\infty$ | 0 |
| 300° | $\frac{5}{3}\pi$ | $-\frac{\sqrt{3}}{2}$ | $\frac{1}{2}$ | $-\sqrt{3}$ | $-\frac{\sqrt{3}}{3}$ |
| 315° | $\frac{7}{4}\pi$ | $-\frac{\sqrt{2}}{2}$ | $\frac{\sqrt{2}}{2}$ | $-1$ | $-1$ |
| 330° | $\frac{11}{6}\pi$ | $-\frac{1}{2}$ | $\frac{\sqrt{3}}{2}$ | $-\frac{\sqrt{3}}{3}$ | $-\sqrt{3}$ |
| 360° | $2\pi$ | 0 | 1 | 0 | $\pm\infty$ |

**Analytic values for different angles**
See Table A.2.

**Recursion formulae**
**Trigonometric functions in terms of exponential functions**

$$\sin x = \frac{\mathrm{e}^{ix} - \mathrm{e}^{-ix}}{2i}, \quad \cos x = \frac{\mathrm{e}^{ix} + \mathrm{e}^{-ix}}{2}, \tag{A.23}$$

$$\tan x = \frac{\mathrm{e}^{ix} - \mathrm{e}^{-ix}}{i\left(\mathrm{e}^{ix} + \mathrm{e}^{-ix}\right)}, \quad \cot x = \frac{i\left(\mathrm{e}^{ix} + \mathrm{e}^{-ix}\right)}{\mathrm{e}^{ix} - \mathrm{e}^{-ix}}. \tag{A.24}$$

**Relations between inverse trigonometric functions**
See Table A.3.

$$\sin^{-1}\alpha = \arcsin\alpha = -\arcsin(-\alpha) = \tfrac{\pi}{2} - \arccos\alpha = \arctan\tfrac{\alpha}{\sqrt{1-\alpha^2}}, \tag{A.25}$$

**Table A.3** Recursion formulae for trigonometric functions

| | $-\alpha$ | $90° \pm \alpha$<br>$\frac{\pi}{2} \pm \alpha$ | $180° \pm \alpha$<br>$\pi \pm \alpha$ | $270° \pm \alpha$<br>$\frac{3\pi}{2} \pm \alpha$ | $k(360°) \pm \alpha$<br>$2k\pi \pm \alpha$ |
|---|---|---|---|---|---|
| sin | $-\sin\alpha$ | $\cos\alpha$ | $\mp\sin\alpha$ | $-\cos\alpha$ | $\pm\sin\alpha$ |
| cos | $\cos\alpha$ | $\mp\sin\alpha$ | $-\cos\alpha$ | $\pm\sin\alpha$ | $\cos\alpha$ |
| tan | $-\tan\alpha$ | $\mp\cot\alpha$ | $\pm\tan\alpha$ | $\mp\cot\alpha$ | $\pm\tan\alpha$ |
| csc | $-\csc\alpha$ | $\sec\alpha$ | $\mp\csc\alpha$ | $-\sec\alpha$ | $\pm\csc\alpha$ |
| sec | $\sec\alpha$ | $\mp\csc\alpha$ | $-\sec\alpha$ | $\pm\csc\alpha$ | $\sec\alpha$ |
| cot | $-\cot\alpha$ | $\mp\tan\alpha$ | $\pm\cot\alpha$ | $\mp\tan\alpha$ | $\pm\cot\alpha$ |

$$\cos^{-1}\alpha = \arccos\alpha = \pi - \arccos(-\alpha) = \tfrac{\pi}{2} - \arcsin\alpha = \mathrm{arccot}\frac{\alpha}{\sqrt{1-\alpha^2}}, \quad (A.26)$$

$$\tan^{-1}\alpha = \arctan\alpha = -\arctan(-\alpha) = \tfrac{\pi}{2} - \mathrm{arccot}\alpha = \arcsin\frac{\alpha}{\sqrt{1+\alpha^2}}, \quad (A.27)$$

$$\cot^{-1}\alpha = \mathrm{arccot}\alpha = \pi - \mathrm{arccot}(-\alpha) = \tfrac{\pi}{2} - \arctan\alpha = \arccos\frac{\alpha}{\sqrt{1+\alpha^2}}. \quad (A.28)$$

## A.5 Exponential Functions

### Exponential functions versus trigonometric and hyperbolic functions

$$e^{ix} = \cos x + i\sin x, \quad e^{-ix} = \cos x - i\sin x, \quad (A.29)$$

$$e^{x} = \cosh x + \sinh x, \quad e^{-x} = \cosh x - \sinh x. \quad (A.30)$$

## A.6 Hyperbolic Functions

### Hyperbolic functions in terms of exponential functions

$$\sinh x = \frac{e^x - e^{-x}}{2}, \quad \cosh x = \frac{e^x + e^{-x}}{2}, \quad (A.31)$$

$$\tanh x = \frac{e^x - e^{-x}}{e^x + e^{-x}}, \quad \coth x = \frac{e^x + e^{-x}}{e^x - e^{-x}}. \quad (A.32)$$

### Functions of negative arguments

$$\sinh(-x) = -\sinh(x), \quad \cosh(-x) = \cosh(x), \quad (A.33)$$

$$\tanh(-x) = -\tanh(x), \quad \coth(-x) = -\coth(x). \quad (A.34)$$

### Inverse hyperbolic functions

$$\sinh^{-1}(\alpha) = \mathrm{arsinh}(\alpha) = \ln\left(\alpha + \sqrt{\alpha^2+1}\right), \quad (A.35)$$

$$\cosh^{-1}(\alpha) = \text{arcosh}(\alpha) = \ln\left(\alpha - \sqrt{\alpha^2 - 1}\right) \text{ for } -\infty < x \leq 0\,, \tag{A.36}$$

$$= \ln\left(\alpha + \sqrt{\alpha^2 - 1}\right) \text{ for } 0 \leq x < +\infty\,, \tag{A.37}$$

$$\tanh^{-1}(\alpha) = \text{artanh}(\alpha) = \ln\sqrt{\frac{1+\alpha}{1-\alpha}} = \frac{1}{2}\ln\frac{1+\alpha}{1-\alpha} \text{ for} |x| < 1\,, \tag{A.38}$$

$$\coth^{-1}(\alpha) = \text{arcoth}(\alpha) = \ln\sqrt{\frac{\alpha+1}{\alpha-1}} = \frac{1}{2}\ln\frac{\alpha+1}{\alpha-1} \text{ for} |x| > 1\,. \tag{A.39}$$

**Selected values**

$$\sinh(0) = 0\,, \quad \cosh(0) = 1\,, \quad \tanh(0) = 0\,, \quad \coth(0) = \pm\infty\,, \tag{A.40}$$

$$\tanh(\infty) = 1\,, \quad \tanh(-\infty) = -1\,, \quad \coth(\infty) = 1\,, \quad \coth(-\infty) = -1\,, \tag{A.41}$$

$$\sinh^{-1}(0) = 0\,, \quad \tanh^{-1}(0) = 0\,. \tag{A.42}$$

## A.7 Derivatives

- $\frac{\mathrm{d}}{\mathrm{d}x}\left(\frac{1}{x}\right) = -\frac{1}{x^2}$
- $\frac{\mathrm{d}}{\mathrm{d}x} x^n = n \times x^{n-1}$
- $\frac{\mathrm{d}}{\mathrm{d}x} \sqrt[n]{x} = \frac{1}{n \times \sqrt[n]{x^{n-1}}}$
- $\frac{\mathrm{d}}{\mathrm{d}x} \sin(x) = \cos(x)$
- $\frac{\mathrm{d}}{\mathrm{d}x} \cos(x) = -\sin(x)$
- $\frac{\mathrm{d}}{\mathrm{d}x} \ln(x) = \frac{1}{x}$
- $\frac{\mathrm{d}}{\mathrm{d}x} |x| = \begin{cases} -1 \text{ for } x < 0 \\ 1 \text{ for } x > 0 \end{cases}$
- $\frac{\mathrm{d}}{\mathrm{d}x} \sinh(x) = \cosh(x)$
- $\frac{\mathrm{d}}{\mathrm{d}x} \cosh(x) = \sinh(x)$
- $\frac{\mathrm{d}}{\mathrm{d}x} (f(x) \times g(x)) = \frac{\mathrm{d}f(x)}{\mathrm{d}x} g(x) + f(x)\frac{\mathrm{d}g(x)}{\mathrm{d}x}$ (product rule)
- $\frac{\mathrm{d}}{\mathrm{d}x}\left(\frac{f(x)}{g(x)}\right) = \frac{\mathrm{d}f(x)/\mathrm{d}x \times g(x) - f(x) \times \mathrm{d}g(x)/\mathrm{d}x}{[g(x)]^2}$ (quotient rule)

## A.8 Integrals

- $\int e^x \mathrm{d}x = e^x$
- $\int \sqrt{x}\mathrm{d}x = \frac{2}{3}x^{\frac{3}{2}}$
- $\int \frac{\mathrm{d}x}{\sqrt{a^2+x^2}} = \mathrm{arsinh}\left(\frac{x}{a}\right)$
- $\int \sin(x)\mathrm{d}x = -\cos(x)$
- $\int \cos(x)\mathrm{d}x = \sin(x)$
- $\int \sin(\alpha x)\cdot\cos(\alpha x)\mathrm{d}x = \frac{1}{2\alpha}\sin^2(\alpha x)$
- $\int \sin^2(\alpha x)\mathrm{d}x = \frac{1}{2}(x - \sin(\alpha x)\cos(\alpha x)) = \frac{1}{2}(x - \frac{1}{2\alpha}\sin(2\alpha x))$
- $\int \cos^2(\alpha x)\mathrm{d}x = \frac{1}{2}(x + \sin(\alpha x)\cos(\alpha x)) = \frac{1}{2}(x + \frac{1}{2\alpha}\sin(2\alpha x))$
- $\int \mathrm{arsinh}\left(\frac{x}{a}\right)\mathrm{d}x = x \times \mathrm{arsinh}\left(\frac{x}{a}\right) - \sqrt{x^2+a^2}$

## A.9 Integration by Parts

- One-dimensional case:

$$\int_a^b f(x)g'(x)\mathrm{d}x = f(x)g(x)|_a^b - \int_a^b f'(x)g(x)\mathrm{d}x$$

$$= f(x)g(x)|_b - f(x)g(x)|_a - \int_a^b f'(x)g(x)\mathrm{d}x\,. \tag{A.43}$$

- Two-dimensional case (plane): $\mathrm{d}\Omega = \mathrm{d}A = \mathrm{d}x\mathrm{d}y$

$$\int_\Omega fg'\mathrm{d}\Omega = \int_\Gamma fg\,n_x\,\mathrm{d}\Gamma - \int_\Omega f'g\,\mathrm{d}\Omega\,. \tag{A.44}$$

- Three-dimensional case (space): $\mathrm{d}\Omega = \mathrm{d}x\mathrm{d}y\mathrm{d}z$

$$\int_\Omega fg'\mathrm{d}\Omega = \int_\Gamma fg\,n_x\,\mathrm{d}\Gamma - \int_\Omega f'g\,\mathrm{d}\Omega\,. \tag{A.45}$$

Remark: $n_x$ is the cosine between the outward normal and the $x$-direction.

Or more general ($\Omega$ is the domain and $\Gamma$ refers to the boundary):

$$\int_{\Omega} f g_{,i} \mathrm{d}\Omega = \int_{\Gamma} f g\, n_i\, \mathrm{d}\Gamma - \int_{\Omega} f_{,i} g\, \mathrm{d}\Omega\,, \tag{A.46}$$

$$\int_{\Omega} f_{,i} g\, \mathrm{d}\Omega = \int_{\Gamma} f g\, n_i\, \mathrm{d}\Gamma - \int_{\Omega} f g_{,i}\, \mathrm{d}\Omega\,. \tag{A.47}$$

- GREEN-GAUSS theorem: $\alpha$: scalar; $\boldsymbol{b}$: vector

$$\int_{\Omega} \alpha \nabla^{\mathrm{T}} \boldsymbol{b}\, \mathrm{d}\Omega = \int_{\Gamma} \alpha \boldsymbol{b}^{\mathrm{T}} \boldsymbol{n} \mathrm{d}\Gamma - \int_{\Omega} (\nabla^{\mathrm{T}} \alpha) \boldsymbol{b}\, \mathrm{d}\Omega\,, \tag{A.48}$$

where $n$ is the unit outward vector acting on the boundary surface.

– Special case: $\boldsymbol{b} \rightarrow \beta \boldsymbol{b}$ ($\beta$: scalar)

$$\int_{\Omega} \alpha \nabla^{\mathrm{T}} (\beta \boldsymbol{b})\, \mathrm{d}\Omega = \int_{\Gamma} \alpha (\beta \boldsymbol{b})^{\mathrm{T}} \boldsymbol{n} \mathrm{d}\Gamma - \int_{\Omega} (\nabla^{\mathrm{T}} \alpha)(\beta \boldsymbol{b})\, \mathrm{d}\Omega\,. \tag{A.49}$$

– Special case: $\boldsymbol{b} \rightarrow \boldsymbol{K}\boldsymbol{b}$ ($\boldsymbol{K}$: matrix)

$$\int_{\Omega} \alpha \nabla^{\mathrm{T}} (\boldsymbol{K} \boldsymbol{b})\, \mathrm{d}\Omega = \int_{\Gamma} \alpha (\boldsymbol{K} \boldsymbol{b})^{\mathrm{T}} \boldsymbol{n} \mathrm{d}\Gamma - \int_{\Omega} (\nabla^{\mathrm{T}} \alpha)(\boldsymbol{K} \boldsymbol{b})\, \mathrm{d}\Omega\,. \tag{A.50}$$

## A.10 Integration and Coordinate Transformation

(a) One-dimensionale case:

Let $T : \mathbb{R} \rightarrow \mathbb{R}$ given by $x = g(u)$ be a one-dimensional transformation from $S$ to $R$. If $g$ has a continuous partial derivative such that the Jacobian is never zero, then

$$\int_{R} f(x) \mathrm{d}x = \int_{S} f(g(u)) \left| \frac{\mathrm{d}x}{\mathrm{d}u} \right| \mathrm{d}u\,, \tag{A.51}$$

where the Jacobian is $J = \left|\frac{\mathrm{d}x}{\mathrm{d}u}\right| = x_u$.

(b) Two-dimensionale case:

Let $T : \mathbb{R}^2 \rightarrow \mathbb{R}^2$ given by $x = g(u, v)$ and $y = h(u, v)$ be a transformation on the plane that is one from a region $S$ to a region $R$. If $g$ and $h$ have continuous partial derivatives such that the Jacobian is never zero, then

$$\iint_R f(x,y)\mathrm{d}y\mathrm{d}x = \iint_S f(g(u,v),h(u,v)) \left|\frac{\partial(x,y)}{\partial(u,v)}\right| \mathrm{d}u\,\mathrm{d}v\,, \tag{A.52}$$

where the Jacobian $J = \left|\frac{\partial(x,y)}{\partial(u,v)}\right| = \begin{vmatrix} x_u & x_v \\ y_u & y_v \end{vmatrix} = x_u \cdot y_v - x_v \cdot y_u = \frac{\partial x}{\partial u} \cdot \frac{\partial y}{\partial v} - \frac{\partial x}{\partial v} \cdot \frac{\partial y}{\partial u}$.

(c) Three-dimensionale case:

Let $T : \mathbb{R}^3 \to \mathbb{R}^3$ given by $x = g(u,v,w)$, $y = h(u,v,w)$ and $z = k(u,v,w)$ be a transformation on the space that is one from a space $S$ to a space $R$. If $g$, $h$ and $k$ have continuous partial derivatives such that the Jacobian is never zero, then

$$\begin{aligned} &\iiint_R f(x,y,z)\mathrm{d}z\mathrm{d}y\mathrm{d}x \\ &= \iiint_S f(g(u,v,w),h(u,v,w),k(u,v,w)) \left|\frac{\partial(x,y,z)}{\partial(u,v,w)}\right| \mathrm{d}w\,\mathrm{d}u\,\mathrm{d}v\,, \end{aligned} \tag{A.53}$$

where the Jacobian is $J = \left|\frac{\partial(x,y,z)}{\partial(u,v,w)}\right| = \begin{vmatrix} x_u & x_v & x_w \\ y_u & y_v & y_w \\ z_u & z_v & z_w \end{vmatrix}$.

Remark: A useful fact is that the Jacobian of the inverse transformation is the reciprocal of the Jacobian of the original transformation, e. g.

$$\left|\frac{\partial(x,y)}{\partial(u,v)}\right| = \left|\frac{\partial(u,v)}{\partial(x,y)}\right|^{-1}, \tag{A.54}$$

or

$$\left|\frac{\partial(u,v)}{\partial(x,y)}\right| = \left|\frac{\partial(x,y)}{\partial(u,v)}\right|^{-1}. \tag{A.55}$$

A useful relation in the scope of coordinate transformations holds for the inverse matrix together with its determinant in the following way:

$$\underbrace{\begin{bmatrix} u_x & u_y & u_z \\ v_x & v_y & v_z \\ w_x & w_y & w_z \end{bmatrix}}_{J} = \underbrace{\begin{bmatrix} x_u & x_v & x_w \\ y_u & y_v & y_w \\ z_u & z_v & z_w \end{bmatrix}^{-1}}_{J'^{-1}} = \frac{1}{J} \cdot \begin{bmatrix} y_v z_w - y_w z_v & -x_v z_w + x_w z_v & x_v y_w - x_w y_v \\ -y_u z_w + y_w z_u & x_u z_w - x_w z_u & -x_u y_w + x_w y_u \\ y_u z_v - y_v z_u & -x_u z_v + x_v z_u & x_u y_v - x_v y_u \end{bmatrix}, \tag{A.56}$$

where the Jacobian $J$ is the determinant as given by

$$J = \left| \frac{\partial(x, y, z)}{\partial(u, v, w)} \right| = x_u y_v z_w + x_v y_w z_u + x_w y_u z_v - x_v y_u z_w - x_w y_v z_u - x_u y_w z_v \,. \tag{A.57}$$

For the 2D case, Eq. (A.56) can be simplified to

$$\underbrace{\begin{bmatrix} u_x & u_y \\ v_x & v_y \end{bmatrix}}_{J} = \underbrace{\begin{bmatrix} x_u & x_v \\ y_u & y_v \end{bmatrix}^{-1}}_{J'^{-1}} = \frac{1}{J} \cdot \begin{bmatrix} y_v & -x_v \\ -y_u & x_u \end{bmatrix}, \tag{A.58}$$

where the Jacobian $J$ is the determinant as given by

$$J = \left| \frac{\partial(x, y)}{\partial(u, v)} \right| = \begin{vmatrix} x_u & x_v \\ y_u & y_v \end{vmatrix} = x_u y_v - x_v y_u \,. \tag{A.59}$$

# A.11 Numerical Integration

## *A.11.1 Simpson's Rule*

To derive the SIMPSON's[1] formula, three points $(\xi_1, f_1)$, $(\xi_2, f_2)$ and $(\xi_3, f_3)$ are selected (cf. Fig. A.2) and the general equation for a parabola, i.e.

$$f(\xi) = a + b \times \xi + c \times \xi^2 \,, \tag{A.60}$$

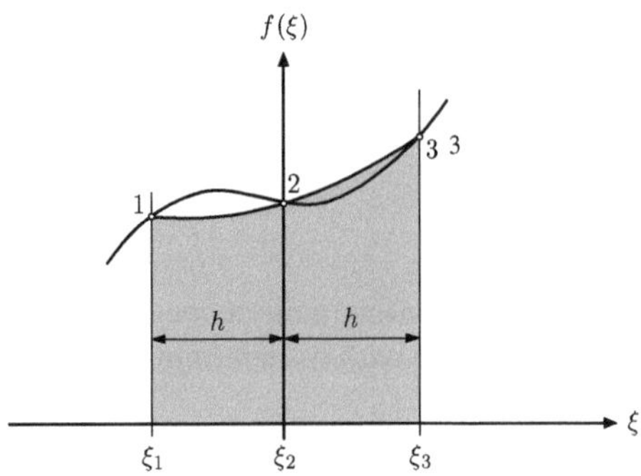


**Fig. A.2** Schematic sketch for the derivation of the SIMPSON's integration rule

[1] Thomas SIMPSON (1710–1761), English mathematician.

is separately evaluated for each of these three points:

$$f_1 = f(\xi_1) = a + b \times \xi_1 + c \times \xi_1^2 , \tag{A.61}$$

$$f_2 = f(\xi_2) = a + b \times \xi_2 + c \times \xi_2^2 , \tag{A.62}$$

$$f_3 = f(\xi_3) = a + b \times \xi_3 + c \times \xi_3^2 . \tag{A.63}$$

Solving the last system of equations for the unknown coefficients $a$, $b$ and $c$ yields:

$$a = \frac{\xi_1^2\xi_2 f_3 - \xi_2\xi_3^2 f_1 + \xi_2^2\xi_3 f_1 - \xi_1\xi_2^2 f_3 + \xi_1\xi_3^2 f_2 - \xi_1^2\xi_3 f_2}{-\xi_2\xi_3^2 - \xi_1\xi_2^2 + \xi_1\xi_3^2 + \xi_1^2\xi_2 - \xi_1^2\xi_3 + \xi_2^2\xi_3} , \tag{A.64}$$

$$b = \frac{\xi_1^2 f_2 - \xi_1^2 f_3 + \xi_2^2 f_3 + \xi_3 f_1^2 - \xi_3^2 f_2 - \xi_2^2 f_1}{-\xi_2\xi_3^2 - \xi_1\xi_2^2 + \xi_1\xi_3^2 + \xi_1^2\xi_2 - \xi_1^2\xi_3 + \xi_2^2\xi_3} , \tag{A.65}$$

$$c = -\frac{\xi_2 f_3 + \xi_3 f_1 + \xi_1 f_2 - \xi_1 f_3 - \xi_2 f_1 - \xi_3 f_2}{-\xi_2\xi_3^2 - \xi_1\xi_2^2 + \xi_1\xi_3^2 + \xi_1^2\xi_2 - \xi_1^2\xi_3 + \xi_2^2\xi_3} . \tag{A.66}$$

Introducing these values for $a$, $b$ and $c$ in the general form of the parabola according to Eq. (A.60), substituting $\xi_2 = \xi_1 + h,\ \xi_3 = \xi_1 + 2h$ and rearranging yields:

$$f(\xi) = f_1 + \frac{(\xi - \xi_1)(f_2 - f_1)}{h} + \frac{(\xi - \xi_1)(\xi - \xi_1 - h)(f_3 - 2f_2 + f_1)}{2h^2} . \tag{A.67}$$

After a brief calculation we get the area under the parabola as

$$\int\limits_{\xi_1}^{\xi_3} f(\xi)\,\mathrm{d}\xi = \frac{h}{3} \times (f_1 + 4f_2 + f_3) , \tag{A.68}$$

which is also called SIMPSON's one-third rule. For the natural coordinate range, i. e. $\xi_1 = -1 < \xi < \xi_3 = 1$, Eq. (A.68) can be written as:

$$\int\limits_{-1}^{1} f(\xi)\,\mathrm{d}\xi = \frac{1}{3} \times \Big(f(-1) + 4f(0) + f(1)\Big) . \tag{A.69}$$

When we add up a sequence of $n$ of these segments, we find

$$\int\limits_{\xi_1}^{\xi_n} f(\xi)\mathrm{d}\xi = \frac{h}{3} \times \Big(\underbrace{f_1 + 4f_2 + f_3}_{\text{1st segment}} + \underbrace{f_3 + 4f_4 + f_5}_{\text{2nd segment}} + \cdots + \underbrace{f_{n-2} + 4f_{n-1} + f_n}_{n\text{th segment}}\Big) , \tag{A.70}$$

$$= \frac{h}{3}\Big(f_1 + 4f_2 + 2f_3 + 4f_4 + 2f_5 + \cdots + 2f_{n-2} + 4f_{n-1} + f_n\Big) . \tag{A.71}$$

The last equation can be written in a general form as a function of the natural coordinates $\xi$ as

$$\int_{-1}^{1} f(\xi)\,\mathrm{d}\xi \approx \sum_{i=1}^{n} f(\xi_i) \times w(\xi_i)\,. \tag{A.72}$$

Values for the abscissae $\xi$ and weights $w$ are given in Table A.4.

**Table A.4** Values for the abscissae $\xi_i$ and weights $w_i$ for SIMPSON's integration rule, $\int_{-1}^{1} f(\xi)\mathrm{d}\xi \approx \sum_i f(\xi_i)w_i$

| No. points $n$ | Abscissae $\xi_i$ | Weights $w_i$ |
|---|---|---|
| 3 | 0 | $\frac{4}{3}$ |
| | $\pm 1$ | $\frac{1}{3}$ |
| 5 | 0 | $\frac{1}{3}$ |
| | $\pm 0.5$ | $\frac{2}{3}$ |
| | $\pm 1$ | $\frac{1}{6}$ |
| 9 | 0 | $\frac{1}{6}$ |
| | $\pm 0.25$ | $\frac{1}{3}$ |
| | $\pm 0.5$ | $\frac{1}{6}$ |
| | $\pm 0.75$ | $\frac{1}{3}$ |
| | $\pm 1$ | $\frac{1}{12}$ |
| 17 | 0 | $\frac{1}{12}$ |
| | $\pm 0.125$ | $\frac{1}{6}$ |
| | $\pm 0.25$ | $\frac{1}{12}$ |
| | $\pm 0.375$ | $\frac{1}{6}$ |
| | $\pm 0.5$ | $\frac{1}{12}$ |
| | $\pm 0.625$ | $\frac{1}{6}$ |
| | $\pm 0.75$ | $\frac{1}{12}$ |
| | $\pm 0.875$ | $\frac{1}{6}$ |
| | $\pm 1$ | $\frac{1}{24}$ |
| $n$ | 0 | ... |
| $= 2^m + 1$ | $\vdots$ | $\vdots$ |
| $(m = 1, 2, 3, \ldots)$ | ... | $2 \times \frac{2}{3(n-1)}$ |
| $(\wedge\, n > 3)$ | ... | $4 \times \frac{2}{3(n-1)}$ |
| | $\pm 1$ | $\frac{2}{3(n-1)}$ |

## *A.11.2 Gauss-Legendre Quadrature*

The principal idea of GAUSS-LEGENDRE quadrature can be introduced by taking not only the weights $w_i$, but also the location of the abscissae $\xi_i$ as unknowns (cf. Fig. A.3) and to require that an arbitrary polynomial of a certain order is exactly integrated.

Additionally, let us assume that the weights and abscissae are symmetric around the midpoint ($\xi = 0$) of the range of integration ($-1 \leq \xi \leq 1$). Similar to the previous section (cf. Fig. A.2), let us assume three functional evaluations of $f(\xi)$ in the scope of this derivation. Thus, the integral can be approximated by

$$\int_{-1}^{1} f(\xi)\,\mathrm{d}\xi \approx f(\xi_1) \times w_1 + f(\xi_2) \times w_2 + f(\xi_3) \times w_3\,, \tag{A.73}$$

and additionally requiring that the formula gives exact expressions for integrating a fifth order polynomial, i.e.

$$\int_{-1}^{1} (a_0 + a_1\xi + a_2\xi^2 + a_3\xi^3 + a_4\xi^4 + a_5\xi^5)\mathrm{d}\xi\,, \tag{A.74}$$

where $a_0, \ldots, a_5$ are arbitrary coefficients. Since integration is additive, it will be sufficient to require that Eq. (A.74) is exact for the single functions $f(\xi) = 1, \xi, \ldots, \xi^5$. Thus, to determine the six unknowns $\xi_1, \xi_2, \xi_3$ and $w_1, w_2, w_3$, the following six integral equations need to be evaluated:

$$\int_{-1}^{1} 1\mathrm{d}\xi = [\xi]_{-1}^{1} = 2 = w_1 + w_2 + w_3\,, \tag{A.75}$$

$$\int_{-1}^{1} \xi\mathrm{d}\xi = \left[\tfrac{1}{2}\xi^2\right]_{-1}^{1} = 0 = w_1\xi_1 + w_2\xi_2 + w_3\xi_3\,, \tag{A.76}$$

$$\int_{-1}^{1} \xi^2\mathrm{d}\xi = \left[\tfrac{1}{3}\xi^3\right]_{-1}^{1} = \tfrac{2}{3} = w_1\xi_1^2 + w_2\xi_2^2 + w_3\xi_3^2\,, \tag{A.77}$$

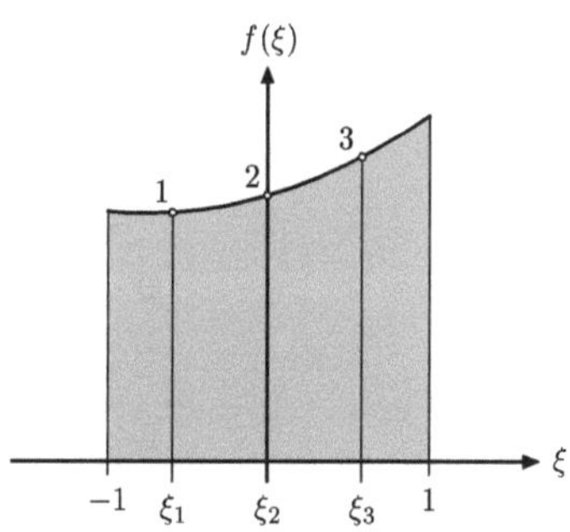


**Fig. A.3** Schematic sketch for the derivation of the GAUSS-LEGENDRE integration rule

$$\int_{-1}^{1} \xi^3 \mathrm{d}\xi = \left[\tfrac{1}{4}\xi^4\right]_{-1}^{1} = 0 = w_1\xi_1^3 + w_2\xi_2^3 + w_3\xi_3^3\,, \tag{A.78}$$

$$\int_{-1}^{1} \xi^4 \mathrm{d}\xi = \left[\tfrac{1}{5}\xi^5\right]_{-1}^{1} = \tfrac{2}{5} = w_1\xi_1^4 + w_2\xi_2^4 + w_3\xi_3^4\,, \tag{A.79}$$

$$\int_{-1}^{1} \xi^5 \mathrm{d}\xi = \left[\tfrac{1}{6}\xi^6\right]_{-1}^{1} = 0 = w_1\xi_1^5 + w_2\xi_2^5 + w_3\xi_3^5\,. \tag{A.80}$$

These six simultaneous nonlinear equations can be solved to obtain:

$$\xi_1 = -0.7745966692\,, \qquad w_1 = 0.5555555556\,, \tag{A.81}$$

$$\xi_2 = 0.0\,, \qquad w_2 = 0.8888888889\,, \tag{A.82}$$

$$\xi_3 = 0.7745966692\,, \qquad w_3 = 0.5555555556\,. \tag{A.83}$$

In Table A.5, coefficients and arguments for $n$-point GAUSS-LEGENDRE quadrature rules are given for the integral of the form

$$\int_{-1}^{1} f(\xi)\,\mathrm{d}\xi \approx \sum_{i=1}^{n} f(\xi_i) \times w(\xi_i)\,. \tag{A.84}$$

The general theory of GAUSS-LEGENDRE integration is based on so-called LEGENDRE polynomials. Without going into detail of this theory, useful formulae for the numerical determination of abscissas and weights can be specified.

**Table A.5** Values for the abscissae $\xi_i$ and weights $w_i$ for GAUSS-LEGENDRE integration rule, $\int_{-1}^{1} f(\xi)\mathrm{d}\xi \approx \sum_i f(\xi_i) w_i$

| No. points $n$ | Abscissae $\xi_i$ | Weights $w_i$ |
|---|---|---|
| 1 | 0 | 2 |
| 2 | $\pm 1/\sqrt{3}$ | 1 |
| 3 | $\pm\sqrt{0.6}$ | $\frac{5}{9}$ |
| | 0 | $\frac{8}{9}$ |
| 4 | $\pm\frac{1}{35}\sqrt{525+70\sqrt{30}}$ | $\frac{1}{36}\left(18-\sqrt{30}\right)$ |
| | $\pm\frac{1}{35}\sqrt{525-70\sqrt{30}}$ | $\frac{1}{36}\left(18+\sqrt{30}\right)$ |
| 5 | 0 | $\frac{128}{225}$ |
| | $\pm\frac{1}{21}\sqrt{245-14\sqrt{70}}$ | $\frac{1}{900}\left(322+13\sqrt{70}\right)$ |
| | $\pm\frac{1}{21}\sqrt{245+14\sqrt{70}}$ | $\frac{1}{900}\left(322-13\sqrt{70}\right)$ |

The best way to generate an explicit formulae for a LEGENDRE polynomial $P_n(\xi)$ is to use recurrence relations [1, 54]. These recurrence relationships $P_i(\xi)$ with $1 \leq i \leq n-1$ are particularly useful for computer evaluation of the roots of the LEGENDRE polynomials, their derivatives and thus the corresponding weights. Two such relations that are widely used are:

$$(i+1)P_{i+1}(\xi) = (2i+1)\xi P_i(\xi) - iP_{i-1}(\xi) \quad \text{for} \quad i \geq 1\,, \tag{A.85}$$

and for the derivative of the LEGENDRE polynomials

$$(\xi^2-1)\underbrace{\frac{\mathrm{d}P_i(\xi)}{\mathrm{d}\xi}}_{P_i'} = i\left(\xi P_i(\xi) - P_{i-1}(\xi)\right) \quad \text{for} \quad i \geq 1\,, \tag{A.86}$$

where the first two polynomials are given as $P_0 = 1$ and $P_1 = \xi$. The weights for a $n$-point integration rule are given by

$$w_i(\xi_i) = \frac{2}{n \cdot P_{n-1}(\xi_i)P_n'(\xi_i)} = \frac{2}{(1-\xi_i^2)(P_n'(\xi_i))^2} = \frac{2(1-\xi_i^2)}{(n+1)^2(P_{n+1}(\xi_i))^2}\,, \tag{A.87}$$

where the roots of $P_n$ are consecutively taken for the $\xi_i$ in Eq. (A.87). Constructing a LEGENDRE polynomial of degree $n$ based on Eq. (A.85) and numerically determining the roots (e.g. based on NEWTON's method) which are equal to the abscissas, gives together with Eqs. (A.86)–(A.87) the complete set for a $n$-point GAUSS-LEGENDRE integration rule. The general $n$-point GAUSS-LEGENDRE rule is exact for polynomial functions of degree $\leq 2n-1$. Let us return to the example given at the beginning of this section where the idea of GAUSS-LEGENDRE integration has been introduced based on a 3-point rule. The evaluation of the recurrence relation (A.85) gives:

$$\begin{aligned}
& P_0 = 1 \quad \text{(known)}\,,\\
& P_1 = \xi \quad \text{(known)}\,,\\
i=1: \quad & 2P_2 = 3\xi P_1 - 1P_0\\
\rightarrow \quad & P_2 = \tfrac{1}{2}\left(3\xi^2 - 1\right)\,,\\
i=2: \quad & 3P_3 = 5\xi P_2 - 2P_1\\
\rightarrow \quad & P_3 = \tfrac{1}{2}\left(5\xi^3 - 3\xi\right)\,.
\end{aligned}$$

The roots of $P_3$ can numerically be calculated as $\xi_1 = -\sqrt{15}/5$, $\xi_2 = 0$ and $\xi_3 = \sqrt{15}/5$. The respective derivatives follow from Eq. (A.86) as:

$$\begin{aligned}
& P_0' = 0 \quad \text{(known)}\,,\\
i=1: \quad & (\xi^2-1)P_1' = 1 \cdot (\xi \cdot P_1 - P_0)\\
\rightarrow \quad & P_1' = 1\,,\\
i=2: \quad & (\xi^2-1)P_2' = 2 \cdot (\xi \cdot P_2 - P_1)
\end{aligned}$$

$$
\begin{aligned}
&\rightarrow && P_2' = 3\xi\,, \\
i = 3: &\quad && (\xi^2 - 1)P_3' = 3 \cdot (\xi \cdot P_3 - P_2) \\
&\rightarrow && P_3' = \tfrac{3}{2(\xi^2-1)}\left(5\xi^4 - 6\xi^2 + 1\right) = \tfrac{1}{2}\left(15\xi^2 - 3\right).
\end{aligned}
$$

Finally, the weights can be expressed according to Eq. (A.87) as:

$$
w_3 = \frac{2}{3 \cdot P_2 \cdot P_3'} = \frac{2}{\frac{9}{2}\frac{3\xi^2-1}{\xi^2-1}\left(\frac{5}{2}\xi^3 - \frac{3}{2}\xi^2 - \frac{3}{2}\xi + \frac{1}{2}\right)}\,, \tag{A.88}
$$

which gives for the roots of the LEGENDRE polynomial (Tables A.6, A.7 and A.8):

$$
w_3\left(-\frac{\sqrt{15}}{5}\right) = \frac{5}{9},\; w_3(0) = \frac{8}{9} \text{ and } w_3\left(\frac{\sqrt{15}}{5}\right) = \frac{5}{9}\,.
$$

**Table A.6** LEGENDRE polynomials $P_{i+1}(\xi) = \frac{1}{i+1} \cdot ((2i+1)\xi P_i - iP_{i-1})$ with $i \geq 1$

| | |
|---|---|
| $P_0 = 1$ | (A.89) |
| $P_1 = \xi$ | (A.90) |
| $P_2 = \frac{1}{2}(3\xi^2 - 1)$ | (A.91) |
| $P_3 = \frac{1}{2}(5\xi^3 - 3\xi)$ | (A.92) |
| $P_4 = \frac{1}{8}(35\xi^4 - 30\xi^2 + 3)$ | (A.93) |
| $P_5 = \frac{1}{8}(63\xi^5 - 70\xi^3 + 15\xi)$ | (A.94) |
| $P_6 = \frac{1}{16}(231\xi^6 - 315\xi^4 + 105\xi^2 - 5)$ | (A.95) |
| $P_7 = \frac{1}{16}(429\xi^7 - 693\xi^5 + 315\,\xi^3 - 35\xi)$ | (A.96) |
| $P_8 = \frac{1}{128}(6435\xi^8 - 12012\xi^6 + 6930\xi^4 - 1260\xi^2 + 35)$ | (A.97) |
| $P_9 = \frac{1}{128}(12155\xi^9 - 25740\xi^7 + 18018\xi^5 - 4620\xi^3 + 315\xi)$ | (A.98) |
| $P_{10} = \frac{1}{256}(46189\xi^{10} - 109395\xi^8 + 90090\xi^6 - 30030\xi^4 + 3465\xi^2 - 63)$ | (A.99) |

**Table A.7** Derivatives of LEGENDRE polynomials $\frac{\mathrm{d}P_i(\xi)}{\mathrm{d}\xi} = P_i(\xi)' = \frac{i}{\xi^2-1} \cdot (\xi P_i - P_{i-1})$ with $i \geq 1$

| | |
|---|---|
| $P_0' = 0$ | (A.100) |
| $P_1' = 1$ | (A.101) |
| $P_2' = 3\xi$ | (A.102) |
| $P_3' = \frac{1}{2}(15\xi^2 - 3)$ | (A.103) |
| $P_4' = \frac{5}{2}\left(7\xi^2 - 3\right)\xi$ | (A.104) |
| $P_5' = \frac{1}{8}(315\xi^4 - 210\xi^2 + 15)$ | (A.105) |
| $P_6' = \frac{21}{8}\left(33\,\xi^4 - 30\,\xi^2 + 5\right)\xi$ | (A.106) |
| $P_7' = \frac{1}{16}(3003\xi^6 - 3465\xi^4 + 945\xi^2 - 35)$ | (A.107) |
| $P_8' = \frac{9}{16}\left(715\,\xi^6 - 1001\,\xi^4 + 385\,\xi^2 - 35\right)\xi$ | (A.108) |
| $P_9' = \frac{1}{128}(109395\xi^8 - 180180\xi^6 + 90090\xi^4 - 13860\xi^2 + 315)$ | (A.109) |
| $P_{10}' = \frac{55}{128}\left(4199\,\xi^8 - 7956\,\xi^6 + 4914\,\xi^4 - 1092\,\xi^2 + 63\right)\xi$ | (A.110) |

**Table A.8** Abscissae and weight factors for GAUSS-LEGENDRE integration: $\int_{-1}^{1} f(\xi)\mathrm{d}\xi \approx \sum_{i=1}^{n} w_i f(\xi_i)$

| $\pm\xi_i$ | $w_i$ | $\pm\xi_i$ | $w_i$ |
|---|---|---|---|
| $n = 1$ | | $n = 12$ | |
| 0.000000000000000 | 1.000000000000000 | 0.125233408511469 | 0.249147045813403 |
| $n = 2$ | | 0.367831498998180 | 0.233492536538355 |
| 0.577350269189626 | 1.000000000000000 | 0.587317954286618 | 0.203167426723066 |
| $n = 3$ | | 0.769902674194305 | 0.160078328543346 |
| 0.000000000000000 | 0.888888888888889 | 0.904117256370475 | 0.106939325995318 |
| 0.774596669241483 | 0.555555555555556 | 0.981560634246719 | 0.047175336386512 |
| $n = 4$ | | $n = 16$ | |
| 0.339981043584856 | 0.652145154862546 | 0.095012509837637 | 0.189450610455068 |
| 0.861136311594053 | 0.347854845137454 | 0.281603550779259 | 0.182603415044924 |
| $n = 5$ | | 0.458016777657227 | 0.169156519395002 |
| 0.000000000000000 | 0.568888888888889 | 0.617876244402644 | 0.149595988816577 |
| 0.538469310105683 | 0.478628670499367 | 0.755404408355003 | 0.124628971255534 |
| 0.906179845938664 | 0.236926885056189 | 0.865631202387832 | 0.095158511682493 |
| $n = 6$ | | 0.944575023073233 | 0.062253523938648 |
| 0.238619186083197 | 0.467913934572691 | 0.989400934991650 | 0.027152459411754 |
| 0.661209386466265 | 0.360761573048139 | $n = 20$ | |
| 0.932469514203152 | 0.171324492379171 | 0.076526521133497 | 0.152753387130726 |
| $n = 7$ | | 0.227785851141645 | 0.149172986472604 |
| 0.000000000000000 | 0.417959183673469 | 0.373706088715420 | 0.142096109318382 |
| 0.405845151377397 | 0.381830050505119 | 0.510867001950827 | 0.131688638449177 |
| 0.741531185599395 | 0.279705391489277 | 0.636053680726515 | 0.118194531961518 |
| 0.949107912342759 | 0.129484966168869 | 0.746331906460151 | 0.101930119817241 |
| $n = 8$ | | 0.839116971822219 | 0.083276741576705 |
| 0.183434642495650 | 0.362683783378362 | 0.912234428251326 | 0.062672048334109 |
| 0.525532409916329 | 0.313706645877887 | 0.963971927277914 | 0.040601429800387 |
| 0.796666477413627 | 0.222381034453375 | 0.993128599185095 | 0.017614007139152 |
| 0.960289856497536 | 0.101228536290377 | $n = 24$ | |
| $n = 9$ | | 0.064056892862606 | 0.127938195346752 |
| 0.000000000000000 | 0.330239355001260 | 0.191118867473616 | 0.125837456346828 |
| 0.324253423403809 | 0.312347077040003 | 0.315042679696163 | 0.121670472927803 |
| 0.613371432700591 | 0.260610696402936 | 0.433793507626045 | 0.115505668053726 |
| 0.836031107326636 | 0.180648160694858 | 0.545421471388840 | 0.107444270115966 |
| 0.968160239507626 | 0.081274388361575 | 0.648093651936976 | 0.097618652104114 |
| $n = 10$ | | 0.740124191578554 | 0.086190161531953 |
| 0.148874338981631 | 0.295524224714753 | 0.820001985973903 | 0.073346481411080 |
| 0.433395394129247 | 0.269266719309996 | 0.886415527004401 | 0.059298584915437 |
| 0.679409568299024 | 0.219086362515982 | 0.938274552002733 | 0.044277438817416 |
| 0.865063366688985 | 0.149451349150581 | 0.974728555971310 | 0.028531388628934 |
| 0.973906528517172 | 0.066671344308688 | 0.995187219997021 | 0.012341229799987 |

### A.11.3 Gauss-Lobatto Rule

Sometimes it is desirable to have a numerical integration rule that specifically includes the two end points, i.e. $\xi = -1$ and $\xi = 1$, in the abscissae list when $n \geq 2$. The LOBATTO[2] rule [1, 99] is such an alternate choice. Its $n$ points will exactly integrate a polynomial of order $(2n-3)$ for $n > 2$. It is usually less accurate than the GAUSS rule but it can be useful for example in the case of a bending beam in order to capture the maximum stresses at the top and the bottom. Its data are included in Table A.9 and the approximation is given by

$$\int_{-1}^{1} f(\xi)\,\mathrm{d}\xi \approx w(-1)f(-1) + \sum_{i=2}^{n-1} w_i f(\xi_i) + w(1)f(1)\,. \tag{A.111}$$

The abscissae for a $n$-point Lobatto rule, i.e. $n$ abscissae for the functional evaluation of $f(\xi)$, are equal to the roots of the polynomial $P'_{n-1}$, cf. Eq. (A.86). Let us remind here that the polynomial $P'_{n-1}$ is of order $n-2$ and thus reveals $n-2$ roots which give with the two end points of the interval, i.e. $\xi = -1$ and $\xi = 1$, a set of $n$ abscissae.

**Table A.9** Values for the abscissae $\xi_i$ and weights $w_i$ for LOBATTO integration rule, $\int_{-1}^{1} f(\xi)\mathrm{d}\xi \approx \sum_i f(\xi_i) w_i$

| No. points $n$ | Abscissae $\xi_i$ | Weights $w_i$ |
|---|---|---|
| 1 | 0 | 2 |
| 2 | $\pm 1$ | 1 |
| 3 | $\pm 1$ | $\frac{1}{3}$ |
| | 0 | $\frac{4}{3}$ |
| 4 | $\pm 1$ | $\frac{1}{6}$ |
| | $\pm\sqrt{0.2}$ | $\frac{5}{6}$ |
| 5 | $\pm 1$ | 0.1 |
| | $\pm\sqrt{3/7}$ | $\frac{49}{90}$ |
| | 0 | $\frac{64}{90}$ |
| 6 | $\pm 1$ | $\frac{1}{15}$ |
| | $\pm\sqrt{t_1}$ | $0.6/[t_1(1-t_0)^2]$ |
| | $\pm\sqrt{t_2}$ | $0.6/[t_2(1-t_0)^2]$ |

where $t_0 = \sqrt{7}$; $t_{1(2)} = (7 + (-)2t_0)/21$

[2] Rehuel LOBATTO (1797–1866), Dutch mathematician.

The weights are for a $n$-point integration rule

$$w_i(\xi_i) = \frac{2}{n(n-1)(P_{n-1}(\xi_i))^2}, \tag{A.112}$$

where the $n-2$ roots of $P_{n-1}$ and $\xi = \pm 1$ are consecutively taken for the $\xi_i$ in Eq. (A.112).

Let us illustrate in the following the derivation of a 5-point integration rule. To this end, one needs to consider the derivative of the LEGENDRE polynomial of order $5-1=4$, i.e.

$$P'_4 = \tfrac{5}{2}\left(7\,\xi^2 - 3\right)\xi\,. \tag{A.113}$$

The three roots of this polynomial are equal to

$$\xi_1 = -0.6546536707\,, \tag{A.114}$$
$$\xi_2 = 0.0000000000\,, \tag{A.115}$$
$$\xi_3 = 0.6546536707\,, \tag{A.116}$$

which gives together with the two end points of the interval, i.e.

$$\xi_4 = -1.0000000000\,, \tag{A.117}$$
$$\xi_5 = 1.0000000000\,, \tag{A.118}$$

a set of five abscissae for the functional evaluation. The weights for a 5-point rule are equal to

$$w_i(\xi_i) = \frac{2}{5\cdot 4\cdot (P_4(\xi_i))^2}, \tag{A.119}$$

where the LEGENDRE polynomial of order four is given by

$$P_4 = \tfrac{1}{8}(35\xi^4 - 30\xi^2 + 3)\,. \tag{A.120}$$

Consecutively taken the $\xi_i$ gives the weights for the 5-point Lobatto rule as:

$$w(\xi_1) = 0.5444444444444444\,, \tag{A.121}$$
$$w(\xi_2) = 0.7111111111111111\,, \tag{A.122}$$
$$w(\xi_3) = 0.5444444444444444\,, \tag{A.123}$$
$$w(\xi_4) = 0.1111111111111111\,, \tag{A.124}$$
$$w(\xi_5) = 0.1111111111111111\,. \tag{A.125}$$

It can easily be shown that the condition $\sum_i w(\xi_i) \stackrel{!}{=} 2$ is fulfilled (Table A.10).

**Table A.10** Abscissae and weight factors for LOBATTO integration: $\int_{-1}^{1} f(\xi)\mathrm{d}\xi \approx w(-1)f(-1) + \sum_{i=2}^{n-1} w_i f(\xi_i) + w(1)f(1)$

| $\pm\xi_i$ | $w_i$ | $\pm\xi_i$ | $w_i$ |
|---|---|---|---|
| $n=3$ | | $n=8$ | |
| 1.000000000000000 | 0.333333333333333 | 1.000000000000000 | 0.035714285714286 |
| 0.000000000000000 | 1.333333333333333 | 0.871740148509607 | 0.210704227143507 |
| $n=4$ | | 0.591700181433142 | 0.341122692483505 |
| 1.000000000000000 | 0.166666666666667 | 0.209299217902479 | 0.412458794658704 |
| 0.447213595499958 | 0.833333333333333 | $n=9$ | |
| $n=5$ | | 1.000000000000000 | 0.027777777777778 |
| 1.000000000000000 | 0.100000000000000 | 0.899757995411460 | 0.165495361560807 |
| 0.654653670707977 | 0.544444444444444 | 0.677186279510738 | 0.274538712500161 |
| 0.000000000000000 | 0.711111111111111 | 0.363117463826178 | 0.346428510973046 |
| $n=6$ | | 0.000000000000000 | 0.371519274376417 |
| 1.000000000000000 | 0.066666666666667 | $n=10$ | |
| 0.765055323929465 | 0.378474956297847 | 1.000000000000000 | 0.022222222222222 |
| 0.285231516480645 | 0.554858377035486 | 0.919533908166459 | 0.133305990851066 |
| $n=7$ | | 0.738773865105505 | 0.224889342063126 |
| 1.000000000000000 | 0.047619047619048 | 0.477924949810444 | 0.292042683679684 |
| 0.830223896278567 | 0.276826047361566 | 0.165278957666387 | 0.327539761183897 |
| 0.468848793470714 | 0.431745381209863 | | |
| 0.000000000000000 | 0.487619047619048 | | |

## A.12 Taylor's Series Expansion

A TAYLOR's series expansion of $f(x)$ with respect to $x_0$ is given by:

$$f(x) = f(x_0) + \left(\frac{\mathrm{d}f}{\mathrm{d}x}\right)_{x_0} (x - x_0) + \frac{1}{2!}\left(\frac{\mathrm{d}^2 f}{\mathrm{d}x^2}\right)_{x_0} (x - x_0)^2 + \ldots + \frac{1}{k!}\left(\frac{\mathrm{d}^k f}{\mathrm{d}x^k}\right)_{x_0} (x - x_0)^k . \tag{A.126}$$

The first order approximation takes the first two terms in the series and approximates the function as:

$$f(x) = f(x_0 + \mathrm{d}x) \approx f(x_0) + \left(\frac{\mathrm{d}f}{\mathrm{d}x}\right)_{x_0} (x - x_0) . \tag{A.127}$$

Recall from calculus that the derivative gives the slope of the tangent line at a given point and that the point-slope form is given by $f(x) - f(x_0) = m \times (x - x_0)$. Thus, we can conclude that the first order approximation gives us the equation of a straight line passing through the point $(x_0, f(x_0))$ with a slope of $m = f'(x_0) = (\mathrm{d}f/\mathrm{d}x)_{x_0}$, cf. Fig. A.4.

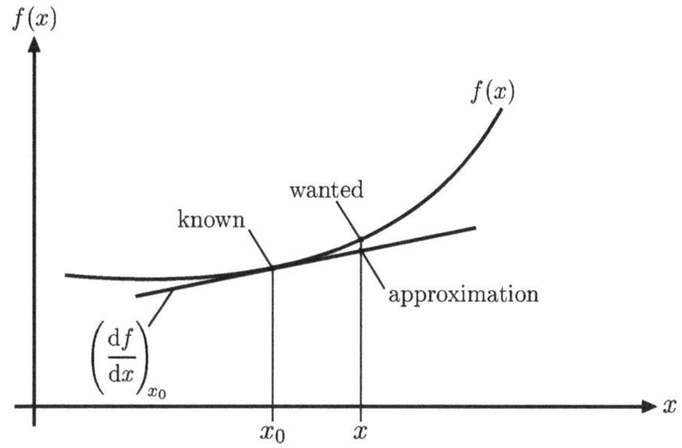


**Fig. A.4** Approximation of a function $f(x)$ by a first order TAYLOR's series

## A.13 Power Series

- $\frac{1}{1+x} = 1 - x + x^2 - x^3 + - \ldots$ for $|x| < 1$.

## A.14 Matrix Operations

A matrix $\boldsymbol{A}$ is defined as a set of quantities $a_{ij}$ ordered as follows

$$\boldsymbol{A} = \begin{matrix} \rightarrow j \\ \downarrow i \end{matrix} \begin{bmatrix} a_{11} & a_{12} & \ldots & a_{1n} \\ a_{21} & a_{22} & \ldots & a_{2n} \\ \vdots & \vdots & \ddots & \vdots \\ a_{m1} & a_{m2} & \ldots & a_{mn} \end{bmatrix} \tag{A.128}$$

and denoted by a bold upper case letter. As indicated, the position of a term $a_{ij}$ is defined by indices $i$ and $j$: the first index determines the row and the second index determines the column in the matrix. For example, $a_{23}$ is the term in the third row and in the second column. The matrix (A.128) is of order $m \times n$. A matrix is said to be square if $m = n$, and rectangular if $m \neq n$. An $m \times 1$ matrix is called a column matrix and denoted by a bold lower case variable. For example, the column matrix $\boldsymbol{b}$

$$\boldsymbol{b} = \begin{bmatrix} b_1 \\ b_2 \\ \vdots \\ b_m \end{bmatrix} \tag{A.129}$$

is defined as the matrix of order $m \times 1$.

The transposed matrix of a matrix $\boldsymbol{A}$ is denoted $\boldsymbol{A}^{\mathrm{T}}$ and represents the matrix with the columns and rows interchanged, i. e.

$$(\boldsymbol{A}^{\mathrm{T}})_{ij} = (\boldsymbol{A})_{ji} . \qquad \text{(A.130)}$$

If we use symmetric matrices for which the elements are symmetric with respect to the main diagonal, then

$$\boldsymbol{A}^{\mathrm{T}} = \boldsymbol{A} . \qquad \text{(A.131)}$$

The transposed of the matrix (A.129) results in a row matrix:

$$\boldsymbol{b}^{\mathrm{T}} = \begin{bmatrix} b_1 \; b_2 \; \ldots \; b_m \end{bmatrix} . \qquad \text{(A.132)}$$

The product of two transposes satisfies

$$(\boldsymbol{AB})^{\mathrm{T}} = \boldsymbol{B}^{\mathrm{T}}\boldsymbol{A}^{\mathrm{T}} . \qquad \text{(A.133)}$$

Proof:

$$(\boldsymbol{B}^{\mathrm{T}}\boldsymbol{A}^{\mathrm{T}})_{ij} = (b^{\mathrm{T}})_{ik}(a^{\mathrm{T}})_{kj} = b_{ki}a_{jk} = a_{jk}b_{ki} = (\boldsymbol{AB})_{ji} = (\boldsymbol{AB})^{\mathrm{T}}_{ij} .$$

The inverse of a square matrix $\boldsymbol{A}$ is a matrix $\boldsymbol{A}^{-1}$ such that

$$\boldsymbol{A}\boldsymbol{A}^{-1} = \mathbf{I} , \qquad \text{(A.134)}$$

where $\mathbf{I}$ is the identity matrix. A square matrix $\boldsymbol{A}$ has an inverse if the determinant $|\boldsymbol{A}| \neq 0$. A matrix possessing an inverse is called nonsingular, or invertible. The inverse of a product $\boldsymbol{AB}$ of matrices $\boldsymbol{A}$ and $\boldsymbol{B}$ can be expressed in terms of $\boldsymbol{A}^{-1}$ and $\boldsymbol{B}^{-1}$ as

$$(\boldsymbol{AB})^{-1} = \boldsymbol{B}^{-1}\boldsymbol{A}^{-1} . \qquad \text{(A.135)}$$

Proof:

$$\boldsymbol{C} \equiv \boldsymbol{AB} .$$

Then

$$\boldsymbol{B} = \boldsymbol{A}^{-1}\boldsymbol{AB} = \boldsymbol{A}^{-1}\boldsymbol{C}$$

and

$$\boldsymbol{A} = \boldsymbol{ABB}^{-1} = \boldsymbol{CB}^{-1} .$$

Therefore,

$$\boldsymbol{C} = \boldsymbol{AB} = (\boldsymbol{CB}^{-1})(\boldsymbol{A}^{-1}\boldsymbol{C}) = \boldsymbol{CB}^{-1}\boldsymbol{A}^{-1}\boldsymbol{C}$$

so

$$\boldsymbol{C}\boldsymbol{B}^{-1}\boldsymbol{A}^{-1} = \mathbf{I} \quad \text{and} \quad \boldsymbol{B}^{-1}\boldsymbol{A}^{-1} = \boldsymbol{C}^{-1} = (\boldsymbol{A}\boldsymbol{B})^{-1}. \tag{A.136}$$

## A.14.1 Matrix Multiplication

The product $\boldsymbol{C}$ of two matrices $\boldsymbol{A}$ and $\boldsymbol{B}$ is defined by

$$c_{ij} = a_{ik} \times b_{jk}, \tag{A.137}$$

where $k$ is summed over for all possible values of $i$ and $j$. Writing out the product explicitly,

$$\begin{bmatrix} c_{11} & c_{12} & \dots & c_{1p} \\ c_{21} & c_{22} & \dots & c_{2p} \\ \vdots & \vdots & \ddots & \vdots \\ c_{n1} & c_{n2} & \dots & c_{np} \end{bmatrix} = \begin{bmatrix} a_{11} & a_{12} & \dots & a_{1m} \\ a_{21} & a_{22} & \dots & a_{2m} \\ \vdots & \vdots & \ddots & \vdots \\ a_{n1} & a_{n2} & \dots & a_{nm} \end{bmatrix} \times \begin{bmatrix} b_{11} & b_{12} & \dots & b_{1p} \\ b_{21} & b_{22} & \dots & b_{2p} \\ \vdots & \vdots & \ddots & \vdots \\ b_{m1} & b_{m2} & \dots & b_{mp} \end{bmatrix}, \tag{A.138}$$

where

$$\begin{aligned} c_{11} &= a_{11}b_{11} + a_{12}b_{21} + \dots + a_{1m}b_{m1} \\ c_{12} &= a_{11}b_{12} + a_{12}b_{22} + \dots + a_{1m}b_{m2} \\ c_{1p} &= a_{11}b_{1p} + a_{12}b_{2p} + \dots + a_{1m}b_{mp} \\ c_{21} &= a_{21}b_{11} + a_{22}b_{21} + \dots + a_{2m}b_{m1} \\ &\vdots \\ c_{np} &= a_{n1}b_{1p} + a_{n2}b_{2p} + \dots + a_{nm}b_{mp}. \end{aligned} \tag{A.139}$$

It can be seen from the rule given in (A.139) that the element $c_{ij}$ of the matrix $\boldsymbol{C}$ is obtained by multiplying row $i$ of matrix $\boldsymbol{A}$ with column $j$ of matrix $\boldsymbol{B}$:

$$\begin{bmatrix} & & \\ & c_{ij} & \\ & & \end{bmatrix} = {}^{i}\begin{bmatrix} & & \\ — & — & — \\ & & \\ & & \end{bmatrix} \times \begin{bmatrix} & \overset{j}{|} & \\ & | & \\ & | & \\ & | & \end{bmatrix}. \tag{A.140}$$

The matrix multiplication is not commutative

$$\boldsymbol{A} \times \boldsymbol{B} \neq \boldsymbol{B} \times \boldsymbol{A} \tag{A.141}$$

but obeys the following laws:

$$\boldsymbol{A} \times (\boldsymbol{B} \times \boldsymbol{C}) = (\boldsymbol{A} \times \boldsymbol{B}) \times \boldsymbol{C} \qquad \text{associative law}, \tag{A.142}$$

$$(\boldsymbol{A} + \boldsymbol{B}) \times \boldsymbol{C} = \boldsymbol{A} \times \boldsymbol{C} + \boldsymbol{B} \times \boldsymbol{C} \qquad \text{distributive law}. \tag{A.143}$$

## A.14.2 Scalar Product

The scalar product[3] of two vectors $\boldsymbol{a}$ and $\boldsymbol{b}$ is given by:

$$\boldsymbol{a} : \boldsymbol{b} = \begin{bmatrix} a_1 \\ a_2 \\ \vdots \\ a_n \end{bmatrix} : \begin{bmatrix} b_1 \\ b_2 \\ \vdots \\ b_n \end{bmatrix} = a_1 b_1 + a_2 b_2 + \cdots + a_n b_n \,. \tag{A.144}$$

The same result is obtained by the following matrix multiplication:

$$\boldsymbol{a}^{\mathrm{T}} \times \boldsymbol{b} = \begin{bmatrix} a_1 \; a_2 \; \ldots \; a_n \end{bmatrix} \times \begin{bmatrix} b_1 \\ b_2 \\ \vdots \\ b_n \end{bmatrix} = a_1 b_1 + a_2 b_2 + \cdots + a_n b_n \,. \tag{A.145}$$

If vectors $\boldsymbol{a}$ and $\boldsymbol{b}$ are defined in a Cartesian coordinate system $(x, y, z)$, then

$$\boldsymbol{a} : \boldsymbol{b} = |\boldsymbol{a}| \times |\boldsymbol{b}| \cdot \cos\alpha \tag{A.146}$$

and it holds the geometric interpretation that the scalar product of the vectors $\boldsymbol{a}$ and $\boldsymbol{b}$ is equal to the product of the length of $\boldsymbol{a}$ projected onto $\boldsymbol{b}$, or visa versa. For example, the length of vector $\boldsymbol{a}$ is obtained as $|\boldsymbol{a}| = \sqrt{a_x^2 + a_y^2 + a_z^2}$.

## A.14.3 Dyadic Product

The dyadic product of two vectors $\boldsymbol{a}$ and $\boldsymbol{b}$ is given by

$$\boldsymbol{a} \otimes \boldsymbol{b} = \begin{bmatrix} a_1 \\ a_2 \\ \vdots \\ a_n \end{bmatrix} \otimes \begin{bmatrix} b_1 \\ b_2 \\ \vdots \\ b_n \end{bmatrix} = \begin{bmatrix} a_1 b_1 & a_1 b_2 & \ldots & a_1 b_n \\ a_2 b_1 & a_2 b_2 & \ldots & a_2 b_n \\ \vdots & \vdots & \ddots & \vdots \\ a_n b_1 & \vdots & \ldots & a_n b_n \end{bmatrix} . \tag{A.147}$$

[3] In the literature, the alternative designation *dot product* and *inner product* can be found.

The same result is obtained by the following matrix multiplication:

$$\boldsymbol{a} \times \boldsymbol{b}^{\mathrm{T}} = \begin{bmatrix} a_1 \\ a_2 \\ \vdots \\ a_n \end{bmatrix} \times \begin{bmatrix} b_1 \; b_2 \ldots b_n \end{bmatrix} = \begin{bmatrix} a_1 b_1 & a_1 b_2 & \ldots & a_1 b_n \\ a_2 b_1 & a_2 b_2 & \ldots & a_2 b_n \\ \vdots & \vdots & \ddots & \vdots \\ a_n b_1 & \vdots & \ldots & a_n b_n \end{bmatrix}. \tag{A.148}$$

## A.14.4 Inverse of Matrices

Equation for a 2 × 2 matrix:

$$\boldsymbol{A}^{-1} = \begin{bmatrix} a & b \\ c & d \end{bmatrix}^{-1} = \frac{1}{ad - bc} \times \begin{bmatrix} d & -b \\ -c & a \end{bmatrix}. \tag{A.149}$$

Equation for a 3 × 3 matrix:

$$\boldsymbol{A}^{-1} = \begin{bmatrix} a & b & c \\ d & e & f \\ g & h & i \end{bmatrix}^{-1} = \frac{1}{\det(\boldsymbol{A})} \begin{bmatrix} ei - fh & ch - bi & bf - ce \\ fg - di & ai - cg & cd - af \\ dh - eg & bg - ah & ae - bd \end{bmatrix}, \tag{A.150}$$

where

$$\det(\boldsymbol{A}) = aei + bfg + cdh - ceg - afh - bdi. \tag{A.151}$$

Equation for a 4 × 4 matrix:

$$\boldsymbol{A}^{-1} = \begin{bmatrix} a & b & c & d \\ e & f & g & h \\ i & j & k & l \\ m & n & o & p \end{bmatrix}^{-1} = \frac{1}{\det(\boldsymbol{A})} \begin{bmatrix} b_{11} & b_{12} & b_{13} & b_{14} \\ b_{21} & b_{22} & b_{23} & b_{24} \\ b_{31} & b_{32} & b_{33} & b_{34} \\ b_{41} & b_{42} & b_{43} & b_{44} \end{bmatrix}, \tag{A.152}$$

where

$$\begin{aligned} \det(\boldsymbol{A}) = {} & afkp + agln + ahjo + belo + bgip + bhkm \\ & + cejp + cflm + chin + dekn + dfio + dgjm \\ & - aflo - agjp - ahkn - bekp - bglm - bhio \\ & - celn - cfip - chjm - dejo - dfkm - dgin, \end{aligned} \tag{A.153}$$

and

$$b_{11} = fkp + gln + hjo - flo - gjp - hkn, \tag{A.154}$$

$$
\begin{aligned}
b_{12} &= blo + cjp + dkn - bkp - cln - djo\,,\\
b_{13} &= bgp + chn + dfo - bho - cfp - dgn\,,\\
b_{14} &= bhk + cfl + dgj - bgl - chj - dfk\,,\\
b_{21} &= elo + gip + hkm - ekp - glm - hio\,,\\
b_{22} &= akp + clm + dio - alo - cip - dkm\,,\\
b_{23} &= aho + cep + dgm - agp - chm - deo\,,\\
b_{24} &= agl + chi + dek - ahk - cel - dgi\,,\\
b_{31} &= ejp + flm + hin - eln - fip - hjm\,,\\
b_{32} &= aln + bip + djm - ajp - blm - din\,,\\
b_{33} &= afp + bhm + den - ahn - bep - dfm\,,\\
b_{34} &= ahj + bel + dfi - afl - bhi - dej\,,\\
b_{41} &= ekn + fio + gjm - ejo - fkm - gin\,,\\
b_{42} &= ajo + bkm + cin - akn - bio - cjm\,,\\
b_{43} &= agn + beo + cfm - afo - bgm - cen\,,\\
b_{44} &= afk + bgi + cej - agj - bek - cfi\,.
\end{aligned}
$$

Blockwise inversion

$$
\begin{bmatrix} \boldsymbol{A} & \boldsymbol{B} \\ \boldsymbol{C} & \boldsymbol{D} \end{bmatrix}^{-1} = \begin{bmatrix} \boldsymbol{A}^{-1} + \boldsymbol{A}^{-1}\boldsymbol{B}(\boldsymbol{D} - \boldsymbol{C}\boldsymbol{A}^{-1}\boldsymbol{B})^{-1}\boldsymbol{C}\boldsymbol{A}^{-1} & -\boldsymbol{A}^{-1}\boldsymbol{B}(\boldsymbol{D} - \boldsymbol{C}\boldsymbol{A}^{-1}\boldsymbol{B})^{-1} \\ -(\boldsymbol{D} - \boldsymbol{C}\boldsymbol{A}^{-1}\boldsymbol{B})^{-1}\boldsymbol{C}\boldsymbol{A}^{-1} & (\boldsymbol{D} - \boldsymbol{C}\boldsymbol{A}^{-1}\boldsymbol{B})^{-1} \end{bmatrix}, \tag{A.155}
$$

where $\boldsymbol{A}$ and $\boldsymbol{D}$ must be square, so that they can be inverted. This is the case if and only if $\boldsymbol{A}$ and $\boldsymbol{D} - \boldsymbol{C}\boldsymbol{A}^{-1}\boldsymbol{B}$ are nonsingular. Alternatively, the inversion procedure can be expressed if $\boldsymbol{D}$ and $\boldsymbol{A} - \boldsymbol{B}\boldsymbol{D}^{-1}\boldsymbol{C}$ are nonsingular as:

$$
\begin{bmatrix} \boldsymbol{A} & \boldsymbol{B} \\ \boldsymbol{C} & \boldsymbol{D} \end{bmatrix}^{-1} = \begin{bmatrix} (\boldsymbol{A} - \boldsymbol{B}\boldsymbol{D}^{-1}\boldsymbol{C})^{-1} & -(\boldsymbol{A} - \boldsymbol{B}\boldsymbol{D}^{-1}\boldsymbol{C})^{-1}\boldsymbol{B}\boldsymbol{D}^{-1} \\ -\boldsymbol{D}^{-1}\boldsymbol{C}(\boldsymbol{A} - \boldsymbol{B}\boldsymbol{D}^{-1}\boldsymbol{C})^{-1} & \boldsymbol{D}^{-1} + \boldsymbol{D}^{-1}\boldsymbol{C}(\boldsymbol{A} - \boldsymbol{B}\boldsymbol{D}^{-1}\boldsymbol{C})^{-1}\boldsymbol{B}\boldsymbol{D}^{-1} \end{bmatrix}. \tag{A.156}
$$

## A.15 Elementary Geometry

- Intercept theorem Fig. A.5

$$
\overline{SA} : \overline{SA'} = \overline{SB} : \overline{SB'}\,, \tag{A.157}
$$

$$
\overline{AB} : \overline{A'B'} = \overline{SA} : \overline{SA'}\,. \tag{A.158}
$$

**Fig. A.5** Intercept theorem

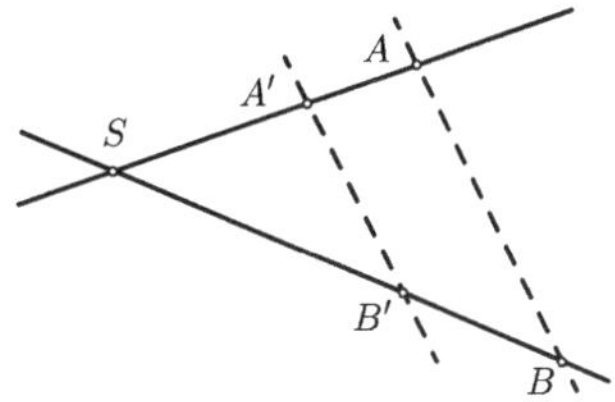


# A.16 Analytical Geometry

## *A.16.1 Straight-Line Equations*

• Point-slope form:

Given is a point $(x_1, y_1)$ and a slope $m$:

$$y - y_1 = m(x - x_1)\,, \tag{A.159}$$

$$y = (y_1 - mx_1) + m \times x\,. \tag{A.160}$$

• Slope-intercept form:

Given is a slope $m$ and the $y$-intercept $b$:

$$y = b + m \times x\,. \tag{A.161}$$

• Two-point form:

Given are the points $(x_1, y_1)$ and $(x_2, y_2)$:

$$\frac{y - y_1}{x - x_1} = \frac{y_1 - y_2}{x_1 - x_2}\,, \tag{A.162}$$

$$y = \left(y_1 - \frac{y_1 - y_2}{x_1 - x_2} x_1\right) + \frac{y_1 - y_2}{x_1 - x_2} \times x\,. \tag{A.163}$$

# Appendix B
# Mechanics

## B.1 Centroids

The coordinates $(z_S, y_S)$ of the centroid $S$ of the plane surface shown in Fig. B.1 can be expressed as

$$z_S = \frac{\int z \, \mathrm{d}A}{\int \mathrm{d}A}, \tag{B.1}$$

$$y_S = \frac{\int y \, \mathrm{d}A}{\int \mathrm{d}A}, \tag{B.2}$$

where the integrals $\int z \, \mathrm{d}A$ and $\int y \, \mathrm{d}A$ are known as the first moments of area.[4] In the case of surfaces composed of $n$ simple shapes, the integrals can be replaced by summations to obtain:

$$z_S = \frac{\sum\limits_{i=1}^{n} z_i A_i}{\sum\limits_{i=1}^{n} A_i}, \tag{B.3}$$

$$y_S = \frac{\sum\limits_{i=1}^{n} y_i A_i}{\sum\limits_{i=1}^{n} A_i}. \tag{B.4}$$

[4] A better expression would be moment of surface since area means strictly speaking the measure of the size of the surface which is different to the surface itself.



A. Öchsner, *Elasto-Plasticity of Frame Structure Elements*,
DOI 10.1007/978-3-662-44225-8

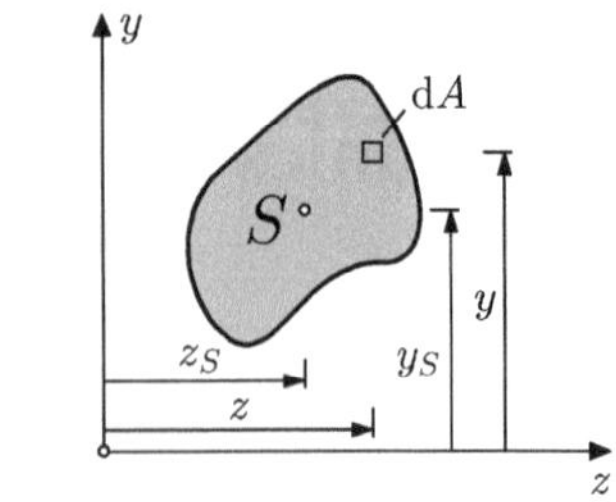


Fig. B.1 Plane surface with centroid $S$

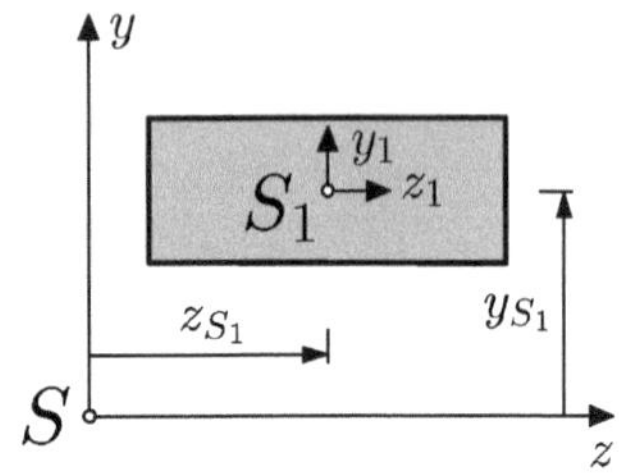


Fig. B.2 Configuration for the parallel-axis theorem

## B.2 Parallel-Axis Theorem

The parallel-axis theorem gives the relationship between the second moment of area[5] with respect to a centroidal axis $(z_1, y_1)$ and the second moment of area with respect to any parallel axis[6] $(z, y)$. For the rectangular shown in Fig. B.2, the relations can be expressed as:

$$I_y = I_{y_1} + z_{S_1}^2 \times A_1 \,, \tag{B.5}$$

$$I_z = I_{z_1} + y_{S_1}^2 \times A_1 \,, \tag{B.6}$$

$$I_{zy} = I_{z_1 y_1} - z_{S_1} y_{S_1} \times A_1 \,. \tag{B.7}$$

## B.3 Bending of Bernoulli Beams in the $x$–$z$ Plane

The presented derivation follows the line of reasoning given in Sects. 4.2.1–4.2.4. A beam with length $L$ under constant moment loading $M_y(x) = \text{const.}$, meaning under *pure* bending, is considered, see Fig. B.3.

---

[5] The second moment of area is also called in the literature the second moment of inertia. However, the expression moment of inertia is in the context of properties of surfaces misleading since no mass or movement is involved.

[6] This arbitrary axis can be for example the axis trough the common centroid $S$ of a composed surface.

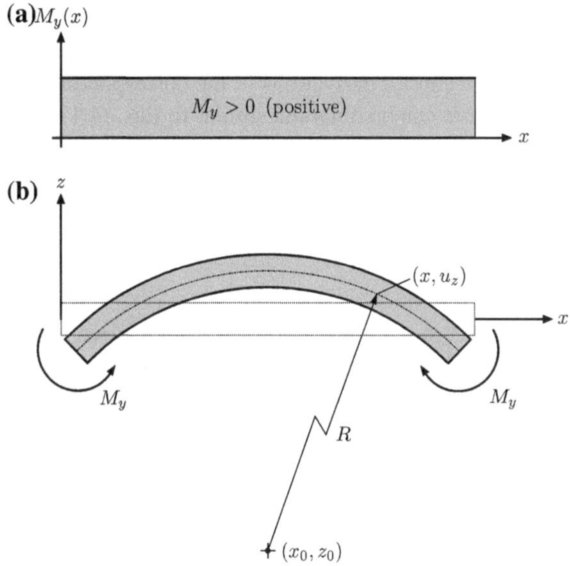


**Fig. B.3** Beam under pure bending in the $x$–$z$ plane: **a** moment distribution; **b** deformed beam. Note that the deformation is overdrawn for better illustration

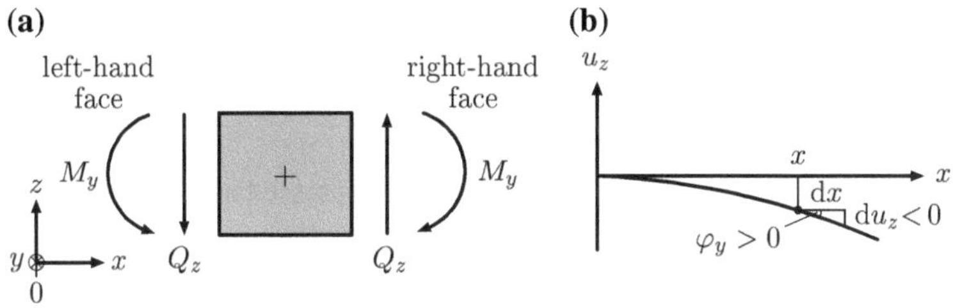


**Fig. B.4** Positive definition of **a** internal reactions and **b** rotation (but negative slope)

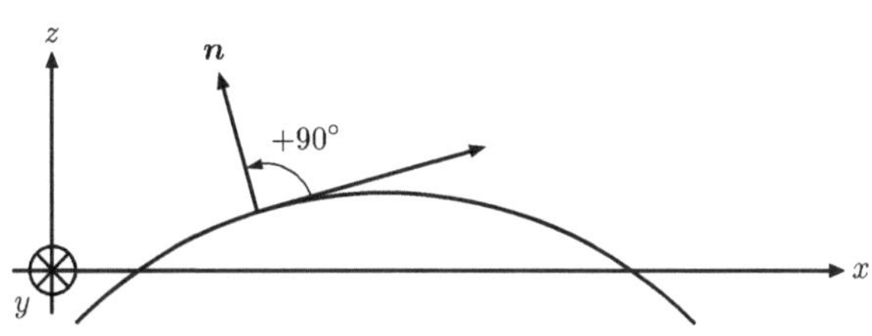


**Fig. B.5** On the definition of a negative curvature in the $x$–$z$ plane

In the case of a deformation in the $x$–$z$ plane, it is more important to preciously distinguish between the positive orientation of the internal reactions and the positive rotational angle, see Fig. B.4. The internal reactions at a right-hand boundary are directed in the positive directions of the coordinate axes. Thus, a positive moment at a right-hand boundary is clockwise oriented, see Fig. B.4. However, a negative slope results in a positive rotational angle which is clockwise oriented, see Fig. B.4. This difference requires some careful derivations of the corresponding equations.

The radius of curvature can be obtained similar to Eqs. (4.1)–(4.9) as:

$$|R| = \frac{\left(1 + \left(\frac{\mathrm{d}u_z}{\mathrm{d}x}\right)^2\right)^{3/2}}{\left|\frac{\mathrm{d}^2 u_z}{\mathrm{d}x^2}\right|}. \tag{B.8}$$

To decide if the radius of curvature is positive of negative, let us have a look at Fig. B.5 where a curve with its tangential and normal vectors is shown. Since the curve in this configuration is bending away from the normal vector, it holds that $\frac{\mathrm{d}^2 u_z}{\mathrm{d}x^2} < 0$ and the radius of curvature is obtained for a positive bending moment as:

$$R = -\frac{\left(1 + \left(\frac{\mathrm{d}u_z}{\mathrm{d}x}\right)^2\right)^{3/2}}{\left(\frac{\mathrm{d}^2 u_z}{\mathrm{d}x^2}\right)}. \tag{B.9}$$

Thus, we can write for small deflections:

$$R = -\frac{1}{\frac{\mathrm{d}^2 u_z}{\mathrm{d}x^2}} \quad \text{or} \quad \kappa = -\frac{\mathrm{d}^2 u_z}{\mathrm{d}x^2}. \tag{B.10}$$

The strain can be based on Fig. B.6 generally expressed as

$$\varepsilon_x = \frac{\mathrm{d}s - \mathrm{d}x}{\mathrm{d}x}, \tag{B.11}$$

where the lengths of the circular arcs are expressed as:

$$\mathrm{d}x = R\mathrm{d}\varphi_y, \tag{B.12}$$

$$\mathrm{d}s = (R + z)\mathrm{d}\varphi_y, \tag{B.13}$$

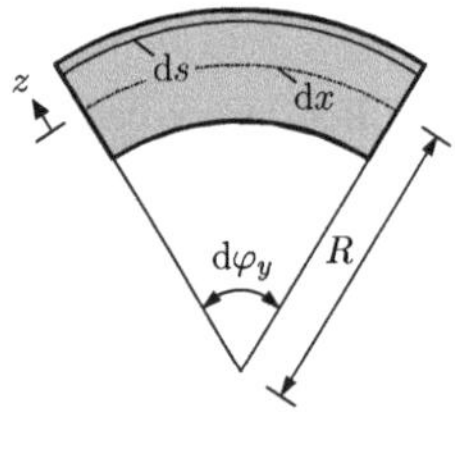


**Fig. B.6** Segment of a beam under pure bending in the $x$–$z$ plane. Note that the deformation is overdrawn for better illustration

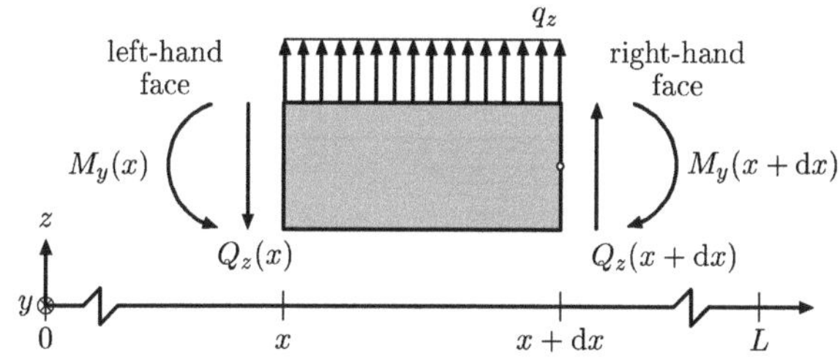


**Fig. B.7** Infinitesimal beam element in the $x$–$z$ plane with internal reactions and constant distributed load

and the strain can be expressed as

$$\varepsilon_x = \frac{(R+z)\mathrm{d}\varphi_y - R\mathrm{d}\varphi_y}{\mathrm{d}x} = z\frac{\mathrm{d}\varphi_y}{\mathrm{d}x} \overset{(B.12)}{=} z\frac{1}{R} \overset{(B.10)}{=} -z\frac{\mathrm{d}^2 u_z(x)}{\mathrm{d}x^2}\,. \tag{B.14}$$

Looking at Fig. B.7, the equilibrium of the vertical forces can be stated as:

$$-\,Q_z(x) + Q_z(x+\mathrm{d}x) + q_z\mathrm{d}x = 0\,, \tag{B.15}$$

which can be written after a TAYLOR's series expansion as:

$$\frac{\mathrm{d}Q_z(x)}{\mathrm{d}x} = -q_z\,. \tag{B.16}$$

The equilibrium of moments around the reference point at $x + \mathrm{d}x$ gives finally:

$$\frac{\mathrm{d}M_y(x)}{\mathrm{d}x} = Q_z(x)\,, \tag{B.17}$$

or under consideration of Eq. (B.16)

$$\frac{\mathrm{d}^2 M_y(x)}{\mathrm{d}x^2} = \frac{\mathrm{d}Q_z(x)}{\mathrm{d}x} = -q_z(x)\,. \tag{B.18}$$

The one-dimensional HOOKE's law according to Eq. (3.3) can be written with Eq. (B.14) as:

$$\sigma_x(x,y) = -Ez\frac{\mathrm{d}^2 u_z(x)}{\mathrm{d}x^2}\,. \tag{B.19}$$

The internal moment which results from the stress distribution shown in Fig. B.8 can be expressed as:

$$M_y = \int_A (+z)(\sigma_x)\mathrm{d}A = -\int_A zEz\frac{\mathrm{d}^2 u_z(x)}{\mathrm{d}x^2}\,\mathrm{d}A = -EI_y\frac{\mathrm{d}^2 u_z(x)}{\mathrm{d}x^2} = \frac{I_y\sigma_z}{z}\,. \tag{B.20}$$

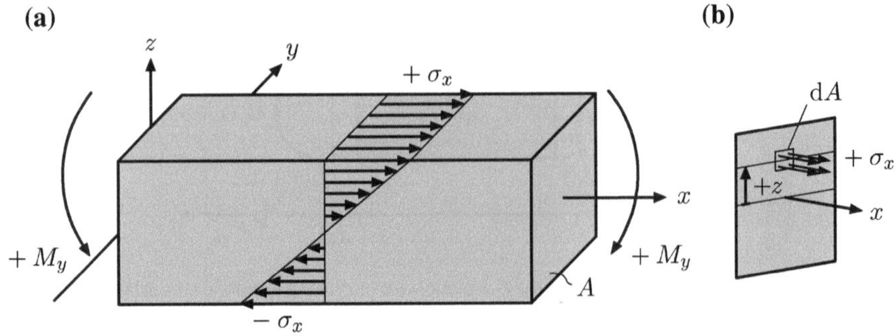


**Fig. B.8** **a** Schematic representation of the normal stress distribution $\sigma_x = \sigma_x(z)$ of a bending beam; **b** Definition and position of an infinitesimal surface element for the derivation of the resulting moment action of the normal stress distribution

**Table B.1** Elementary basic equations for the bending of a BERNOULLI beam in the $x$–$z$ plane

| Name | Equation |
|---|---|
| Kinematics | $\varepsilon_x(x,z) = -z\frac{\mathrm{d}^2u_z(x)}{\mathrm{d}x^2}$ |
| Equilibrium | $\frac{\mathrm{d}Q_z(x)}{\mathrm{d}x} = -q_z(x)\,;\ \frac{\mathrm{d}M_y(x)}{\mathrm{d}x} = Q_z(x)$ |
| Constitutive equation | $\sigma_x(x,z) = E\varepsilon_x(x,z)$ |
| Stress | $\sigma_x(x,z) = \frac{M_y(x)}{I_y}\,z(x)$ |
| Diff'equation | $EI_y\frac{\mathrm{d}^2u_z(x)}{\mathrm{d}x^2} = -M_y(x)$ |
| | $EI_y\frac{\mathrm{d}^3u_z(x)}{\mathrm{d}x^3} = -Q_z(x)$ |
| | $EI_y\frac{\mathrm{d}^4u_z(x)}{\mathrm{d}x^4} = q_z(x)$ |

The differential equations are given under the assumption of constant bending stiffness $EI_y$

Finally, the basic equations for the bending of a beam in the $x$–$z$ plane are summarized in Table B.1. It should be noted that the derivations are independent of the direction of the $z$-axis, i.e. upward or downwards does not change the final equations.

## B.4 Shear Stress Distribution in an Elastic-Plastic Beam

The following derivations are adapted from [79] and consider a configuration as schematically shown in Fig. 4.18. We assume in the following that a plastic zone has partially developed in an upper and lower band close to the free surface. The derivation will show—under the assumptions of the beam theory—that in the elastic zone a normal stress $\sigma_x = \sigma$ and a shear stress $\tau_{xy} = \tau$ is acting while the shear stress $\tau$ is zero in the plastic zones.

Let us first consider the derivation of the stress distributions $\sigma = \sigma(x, y)$ and $\tau = \tau(x, y)$ for $y > 0$, i.e. in the upper part of the beam. The equilibrium equation (cf. Eq. (8.1))

$$\frac{\partial \sigma}{\partial x} + \frac{\partial \tau}{\partial y} = 0\,, \tag{B.21}$$

the VON MISES yield condition (cf. Eq. (2.16))

$$\sigma^2 + 3\tau^2 = 3k_s^2 \tag{B.22}$$

and the boundary condition

$$\tau(x, +\tfrac{h}{2}) = 0 \tag{B.23}$$

serve as the initial equations for the derivation. At the upper free surface, i.e. at $y = +\frac{h}{2}$, the yield condition (B.22) reduces due to the boundary condition (B.23) to:

$$\sigma(x, +\tfrac{h}{2}) = \sqrt{3}k_s\,. \tag{B.24}$$

Since the yield stress is assumed to be constant, i.e. $k_s =$ const., Eq. (B.21) reads at the free surface as:

$$\frac{\partial \tau}{\partial y}(x, +\tfrac{h}{2}) = 0\,. \tag{B.25}$$

Differentiation of the yield condition (B.22) with respect to the $y$-coordinate gives:

$$\frac{\partial}{\partial y}\left(\sigma^2 + 3\tau^2\right) = \frac{\partial}{\partial y}\left(3k_s^2\right)\,,$$

$$2\sigma\frac{\partial \sigma}{\partial y} + 6\tau\frac{\partial \tau}{\partial y} = 0\,,$$

or

$$\sigma\frac{\partial \sigma}{\partial y} + 3\tau\frac{\partial \tau}{\partial y} = 0\,. \tag{B.26}$$

Combining the last expression with Eq. (B.25) gives at the free surface:

$$\frac{\partial \sigma}{\partial y}(x, +\tfrac{h}{2}) = 0\,. \tag{B.27}$$

Differentiation of the equilibrium equation (B.21) with respect to the $y$-coordinate, i.e.

$$\frac{\partial}{\partial y}\left(\frac{\partial \sigma}{\partial x}+\frac{\partial \tau}{\partial y}\right)=0\,, \tag{B.28}$$

gives under consideration of the SCHWARZ's theorem,[7] i.e. the equality of mixed partial derivatives:

$$\frac{\partial^2 \tau}{\partial y^2}(x,+\tfrac{h}{2})=0\,. \tag{B.29}$$

Differentiation of Eq. (B.26) with respect to the $y$-coordinate and considerations of Eqs. (B.25), (B.27) and (B.29) gives:

$$\frac{\partial}{\partial y}\left(\sigma\frac{\partial \sigma}{\partial y}+3\tau\frac{\partial \tau}{\partial y}\right)=0\,,$$

$$\underbrace{\frac{\partial \sigma}{\partial y}\frac{\partial \sigma}{\partial y}}_{=0,(B.27)}+\sigma\frac{\partial^2 \sigma}{\partial y^2}+3\underbrace{\frac{\partial \tau}{\partial y}\frac{\partial \tau}{\partial y}}_{=0,(B.25)}+3\tau\underbrace{\frac{\partial^2 \tau}{\partial y^2}}_{=0,(B.29)}=0\,,$$

or

$$\frac{\partial^2 \sigma}{\partial y^2}(x,+\tfrac{h}{2})=0\,. \tag{B.30}$$

Following this line of reasoning (or see [56]), it can be shown that higher order derivatives of the normal stress $\sigma$ and the shear stress $\tau$ with respect to the $y$-coordinate vanish at the upper free surface, i.e. for $y=\frac{h}{2}$. This allows the conclusion that $\sigma=\sqrt{3}k_s$ and $\tau=0$ holds not only along $y=\frac{h}{2}$ as stated in Eqs. (B.23) and (B.24), but throughout the entire upper plastic region of the beam. To understand this conclusion, consider that the stress values at $y=+\frac{h}{2}$ are given as $\sigma=\sqrt{3}k_s=\text{const.}$ and $\tau=0$. Since the first order derivative represents the slope and the second order derivative the the curvature, the constant values of $\sigma=\sqrt{3}k_s$ and $\tau=0$ at $y=+\frac{h}{2}$ do not change inside the plastic region. The whole shear force in the elasto-plastic region ($0\leq x\leq\varsigma$, cf. Fig. 4.25b) is therefore taken by the central elastic core ($|y|\leq\frac{\alpha h}{2}$, cf. Fig. 4.25c). At the elasto-plastic boundaries, i.e. $|y|=\frac{\alpha h}{2}$, the shear force must be zero ($\tau=0$) for continuity reasons. The distribution of the shear force itself for $|y|\leq\frac{\alpha h}{2}$ is parabolic as shown in Fig. B.9.

[7] Hermann Amandus SCHWARZ (1843–1921), German mathematician: $\frac{\partial^2 f}{\partial x_j \partial x_i}=\frac{\partial^2 f}{\partial x_i \partial x_j}$.

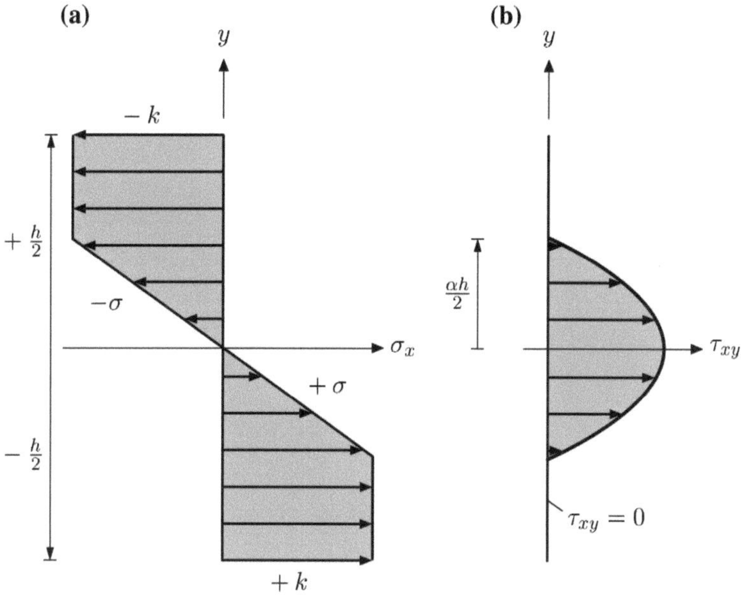


**Fig. B.9** Stress distribution in an elasto-plastic beam: **a** normal stress and **b** shear stress

# Appendix C
# Units and Conversion

## C.1 SI Base Units

The International System of Units (SI)[8] must be used in scientific publications to express physical units. This system consists of the seven base quantities—length, mass, time, electric current, thermodynamic temperature, amount of substance, and luminous intensity—and their respective base units are the meter, kilogram, second, ampere, kelvin, mole, and candela.[9]

## C.2 Coherent SI derived Units

A coherent SI derived unit is defined uniquely as a product of powers of base units that include no numerical factor other than 1. Table C.1 gives some examples of derived units and their expression in terms of base units.

## C.3 Consistent Units

The application of a finite element code does normally not require that a specific system of units is selected. A finite element code keeps through an analysis *consistent* units and requires only that a user assigns the absolute measure without specifying a specific unit. Thus, the units considered by the user during the pre-processing phase are maintained for the post-processing phase. The user must assure that the

---

[8] The original name is known in French as: Système International d'Unités.

[9] More information on units can be found in the brochures of the Bureau International des Poids et Mesures (BIPM): www.bipm.org/en/si.



A. Öchsner, *Elasto-Plasticity of Frame Structure Elements*,
DOI 10.1007/978-3-662-44225-8

**Table C.1** Example of coherent SI derived units

| Quantity | Coherent derived unit | | | |
|---|---|---|---|---|
| | Name | Symbol | In terms of other SI units | In terms of SI base units |
| Celsius temperature | Degree celsius | °C | | K |
| Energy, work | Joule | J | N m | $m^2\,kg\,s^{-2}$ |
| Force | Newton | N | | $m\,kg\,s^{-2}$ |
| Plane angle | Radian | rad | 1 | m/m |
| Power | Watt | W | J/s | $m^2\,kg\,s^{-3}$ |
| Pressure, stress | Pascal | Pa | $N/m^2$ | $m^{-1}\,kg\,s^{-2}$ |

**Table C.2** Example of consistent units

| Property | Unit |
|---|---|
| Length | mm |
| Area | $mm^2$ |
| Force | N |
| Pressure | $MPa = \frac{N}{mm^2}$ |
| Moment | Nmm |
| Moment of inertia | $mm^4$ |
| E-Modulus | $MPa = \frac{N}{mm^2}$ |
| Density | $\frac{Ns^2}{mm^4}$ |
| Time | s |
| Mass | $10^3$kg |

considered units are consistent, i.e. they fit each other. The following Table C.2 shows an example of consistent units

Pay attention to the unit of the density. The following example shows the conversion of the density of steel:

$$\varrho_{St} = 7.8\,\frac{kg}{dm^3} = 7.8 \times 10^3\,\frac{kg}{m^3} = 7.8 \times 10^{-6}\,\frac{kg}{mm^3}\,. \tag{C.1}$$

With

$$1\,N = 1\,\frac{m\,kg}{s^2} = 1 \times 10^3\,\frac{mm\,kg}{s^2} \quad \text{und} \quad 1\,kg = 1 \times 10^{-3}\,\frac{Ns^2}{mm} \tag{C.2}$$

follows the consistent density to:

$$\varrho_{St} = 7.8 \times 10^{-9}\,\frac{Ns^2}{mm^4}\,. \tag{C.3}$$

**Table C.3** Example of consistent English units

| Property | Unit |
|---|---|
| Length | in |
| Area | $in^2$ |
| Force | lbf |
| Pressure | $psi = \frac{lbf}{in^2}$ |
| Moment | lbf in |
| Moment of inertia | $in^4$ |
| E-Modulus | $psi = \frac{lbf}{in^2}$ |
| Density | $\frac{lbf\,sec^2}{in^4}$ |
| Time | sec |

Since literature reports time by time also other units, the following Table C.3 shows an example of consistent English units:

Pay attention to the conversion of the density:

$$\varrho_{St} = 0.282\,\frac{lb}{in^3} = 0.282\,\frac{1}{in^3} \times 0.00259\,\frac{lbf\,sec^2}{in} = 0.73038 \times 10^{-3}\,\frac{lbf\,sec^2}{in^4}\,. \tag{C.4}$$

## C.4 Conversion of Important English Units to the Metric System

See Table C.4.

**Table C.4** Conversion of important U.S. customary units and British Imperial units ('English units') to Metric units (m: meter; cm: centimeter; g: gram; N: newton; J: joule; W: watt)

| Type | English unit | Conversion |
|---|---|---|
| Length | Inch | 1 in = 0.025400 m |
| | Foot | 1 ft = 0.304800 m |
| | Yard | 1 yd = 0.914400 m |
| | Mile (statute) | 1 mi = 1609.344 m |
| | Mile (nautical) | 1 nm = 1852.216 m |
| Area | Square inch | 1 sq in = 1 $in^2$ = 6.45160 $cm^2$ |
| | Square foot | 1 sq ft = 1 $ft^2$ = 0.092903040 $m^2$ |
| | Square yard | 1 sq yd = 1 $yd^2$ = 0.836127360 $m^2$ |
| | Square mile | 1 sq mi = 1 $mi^2$ = 2589988.110336 $m^2$ |
| | Acre | 1 ac = 4046.856422400 $m^2$ |

(Continued)

**Table C.4** (Continued)

| | | |
|---|---|---|
| Volume | Cubic inch | 1 cu in = 1 $in^3$ = 0.000016387064 $m^3$ |
| | Cubic foot | 1 cu ft = 1 $ft^3$ = 0.028316846592 $m^3$ |
| | Cubic yard | 1 cu yd = 1 $yd^3$ = 0.764554857984 $m^3$ |
| Mass | Ounce | 1 oz = 28.349523125 g |
| | Pound (mass) | 1 $lb_m$ = 453.592370 g |
| | Short ton | 1 sh to = 907184.74 g |
| | Long ton | 1 lg to = 1016046.9088 g |
| Force | Pound-force | 1 lbf = 1 $lb_F$ = 4.448221615260500 N |
| | Poundal | 1 pdl = 0.138254954376 N |
| Stress | Pound-force per square inch | 1 psi = 1 $\frac{lbf}{in^2}$ = 6894.75729316837 $\frac{N}{m^2}$ |
| | Pound-force per square foot | 1 $\frac{lbf}{ft^2}$ = 47.880258980336 $\frac{N}{m^2}$ |
| Energy | British thermal unit | 1 Btu = 1055.056 J |
| | Calorie | 1 cal = 4185.5 J |
| Power | Horsepower | 1 hp = 745.699871582270 W |

# Answers to Supplementary Problems

## Problems from Chap. 2

**2.1 Derivation of basic equations for von Mises' yield condition**
See Table C.5.

**2.2 Derivation of the elasto-plastic modulus for the one-dimensional case with isotropic hardening**
$\boldsymbol{h} \rightarrow 1, \boldsymbol{q} \rightarrow \kappa, r \rightarrow \text{sgn}(\sigma), \frac{\partial F}{\partial \sigma} \rightarrow \text{sgn}(\sigma), \frac{\partial F}{\partial \sigma} \times r \rightarrow 1$.

$$E^{\text{elpl}} = \frac{E \times \frac{\partial F}{\partial \kappa}}{\frac{\partial F}{\partial \kappa} - E}\,. \tag{C.5}$$

**2.3 Derivation of the elasto-plastic modulus matrix for Tresca yield condition**

$$\boldsymbol{C}^{\text{elpl}} = \begin{bmatrix} E & 0 \\ 0 & G \end{bmatrix} - \frac{1}{(E + H + E^{\text{pl}})(\sigma - \alpha_\sigma)^2 + (16G + 16H + 4E^{\text{pl}})(\tau - \alpha_\tau)^2} \times \begin{bmatrix} E^2(\sigma - \alpha_\sigma)^2 & 4EG(\sigma - \alpha_\sigma)(\tau - \alpha_\tau) \\ 4EG(\sigma - \alpha_\sigma)(\tau - \alpha_\tau) & 16G^2(\tau - \alpha_\tau)^2 \end{bmatrix}\,. \tag{C.6}$$

**2.4 Derivation of basic equations for Gurson's yield condition**
See Table C.6.

## Problems from Chap. 3

**3.5 Differential equation for a rod with variable distributed load**

$$-N_x(x) + p_x\left(x + \frac{\text{d}x}{2}\right)\text{d}x + N_x(x + \text{d}x) = 0\,. \tag{C.7}$$


A. Öchsner, *Elasto-Plasticity of Frame Structure Elements*,
DOI 10.1007/978-3-662-44225-8

**Table C.5** Basic equations of a formulation of the 1D VON MISES model

| 1D VON MISES model | |
|---|---|
| *Yield condition* | |
| $F = \left(\frac{\|\sigma\|}{k_t}\right)^2 - 1 = 0$ | $F = \frac{\|\sigma\|}{k_t} - 1 = 0$ |
| *Flow rule* | |
| $\mathrm{d}\varepsilon^{\mathrm{pl}} = \mathrm{d}\lambda \frac{2\sigma}{k_t^2}$ | $\mathrm{d}\varepsilon^{\mathrm{pl}} = \mathrm{d}\lambda \frac{\mathrm{sgn}(\sigma)}{k_t}$ |
| *Evolution of hardening variable* | |
| $\mathrm{d}\kappa = \mathrm{d}\lambda \frac{2\sigma^2}{k_t^3}$ | $\mathrm{d}\kappa = \mathrm{d}\lambda \frac{\|\sigma\|}{k_t^2}$ |

**Table C.6** Basic equations of a formulation of the 1D GURSON model

| 1D GURSON model |
|---|
| *Yield condition* |
| $F = \|\sigma\|^2 + 2D\cosh\left(\frac{\sigma}{2k_t(\kappa)}\right) k_t(\kappa)^2 - (1 + D^2)k_t(\kappa)^2 = 0$ |
| *Flow rule* |
| $\mathrm{d}\varepsilon^{\mathrm{pl}} = \mathrm{d}\lambda\left(2\sigma + Dk_t(\kappa)\sinh\left(\frac{\sigma}{2k_t(\kappa)}\right)\right)$ |
| *Damage evolution equation* |
| $\mathrm{d}D = \mathrm{d}\lambda 3(D - D^2)k_t(\kappa)\sinh\left(\frac{\sigma}{2k_t(\kappa)}\right)$ |
| *Evolution of hardening variable* |
| $\mathrm{d}\kappa = \frac{\mathrm{d}\lambda}{(1-D)}\left(\frac{2\sigma^2}{k_t(\kappa)} + D\sigma\sinh\left(\frac{\sigma}{2k_t(\kappa)}\right)\right)$ |

### 3.6 Differential equation for a rod with variable cross-sectional area

$$-\sigma_(x)A(x) + \sigma_x(x + \mathrm{d}x)A(x + \mathrm{d}x) + p_x \mathrm{d}x = 0\,. \tag{C.8}$$

### 3.7 Deflection of a rod due to distributed load

$$u_x(L) = \frac{p_0}{EA}\left((a_2 - a_1)L - \frac{1}{2}(L - a_1)^2\right). \tag{C.9}$$

### 3.8 Deformation of a bi-material rod

$$k_{\mathrm{I}}u(x) = -\frac{p_0 x^2}{2} + c_1 x + c_2 \quad \text{for} \quad 0 \le x \le L\,, \tag{C.10}$$

$$k_{\mathrm{II}}u(x) = c_3 x + c_4 \quad \text{for} \quad L \le x \le 2L\,. \tag{C.11}$$

Boundary and transmission conditions are as follows:

$$u(0) = 0\,, \quad u(2L) = u_0\,, \quad u(L^-) = u(L^+)\,, \quad k_{\mathrm{I}}\frac{\mathrm{d}u(L^-)}{\mathrm{d}x} = k_{\mathrm{II}}\frac{\mathrm{d}u(L^+)}{\mathrm{d}x}\,. \tag{C.12}$$

The last condition states that the force at the interface is the same. If the geometry on both sides of the interface is the same, the traction continuity is obtained.

$$c_1 = \frac{2k_{\mathrm{I}} + k_{\mathrm{II}}}{2(k_{\mathrm{I}} + k_{\mathrm{II}})} p_0 L + \frac{k_{\mathrm{I}} k_{\mathrm{II}}}{k_{\mathrm{I}} + k_{\mathrm{II}}} \frac{u_0}{L}, \tag{C.13}$$

$$c_2 = 0, \tag{C.14}$$

$$c_3 = -\frac{k_{\mathrm{II}}}{2(k_{\mathrm{I}} + k_{\mathrm{II}})} p_0 L + \frac{k_{\mathrm{I}} k_{\mathrm{II}}}{k_{\mathrm{I}} + k_{\mathrm{II}}} \frac{u_0}{L}, \tag{C.15}$$

$$c_4 = \frac{k_{\mathrm{II}}}{k_{\mathrm{I}} + k_{\mathrm{II}}} p_0 L^2 + \frac{k_{\mathrm{II}}(k_{\mathrm{II}} - k_{\mathrm{I}})}{k_{\mathrm{I}} + k_{\mathrm{II}}} u_0 . \tag{C.16}$$

Section $0 \leq x \leq L$:

$$u(x) = \frac{1}{k_{\mathrm{I}}} \left( -\frac{p_0 x^2}{2} + \left[ \frac{2k_{\mathrm{I}} + k_{\mathrm{II}}}{2(k_{\mathrm{I}} + k_{\mathrm{II}})} p_0 L + \frac{k_{\mathrm{I}} k_{\mathrm{II}}}{k_{\mathrm{I}} + k_{\mathrm{II}}} \frac{u_0}{L} \right] x \right), \tag{C.17}$$

$$\varepsilon(x) = \frac{1}{k_{\mathrm{I}}} \left( -p_0 x + \frac{2k_{\mathrm{I}} + k_{\mathrm{II}}}{2(k_{\mathrm{I}} + k_{\mathrm{II}})} p_0 L + \frac{k_{\mathrm{I}} k_{\mathrm{II}}}{k_{\mathrm{I}} + k_{\mathrm{II}}} \frac{u_0}{L} \right), \tag{C.18}$$

$$\sigma(x) = \frac{E_{\mathrm{I}}}{k_{\mathrm{I}}} \left( -p_0 x + \frac{2k_{\mathrm{I}} + k_{\mathrm{II}}}{2(k_{\mathrm{I}} + k_{\mathrm{II}})} p_0 L + \frac{k_{\mathrm{I}} k_{\mathrm{II}}}{k_{\mathrm{I}} + k_{\mathrm{II}}} \frac{u_0}{L} \right). \tag{C.19}$$

Section $L \leq x \leq 2L$:

$$\begin{aligned} u(x) =& \frac{1}{k_{\mathrm{II}}} \left( \left[ -\frac{k_{\mathrm{II}}}{2(k_{\mathrm{I}} + k_{\mathrm{II}})} p_0 L + \frac{k_{\mathrm{I}} k_{\mathrm{II}}}{k_{\mathrm{I}} + k_{\mathrm{II}}} \frac{u_0}{L} \right] x \right. \\ & \left. + \frac{k_{\mathrm{II}}}{k_{\mathrm{I}} + k_{\mathrm{II}}} p_0 L^2 + \frac{k_{\mathrm{II}}(k_{\mathrm{II}} - k_{\mathrm{I}})}{k_{\mathrm{I}} + k_{\mathrm{II}}} u_0 \right), \end{aligned} \tag{C.20}$$

$$\varepsilon(x) = -\frac{1}{2(k_{\mathrm{I}} + k_{\mathrm{II}})} p_0 L + \frac{k_{\mathrm{I}}}{k_{\mathrm{I}} + k_{\mathrm{II}}} \frac{u_0}{L} = \text{const.}, \tag{C.21}$$

$$\sigma(x) = -\frac{E_{\mathrm{II}}}{2(k_{\mathrm{I}} + k_{\mathrm{II}})} p_0 L + \frac{E_{\mathrm{II}} k_{\mathrm{I}}}{k_{\mathrm{I}} + k_{\mathrm{II}}} \frac{u_0}{L} = \text{const.} \tag{C.22}$$

The distribution of stress, strain and displacement is shown in Fig. C.1. The special case $u_0 = 0$ is shown in Fig. C.2.

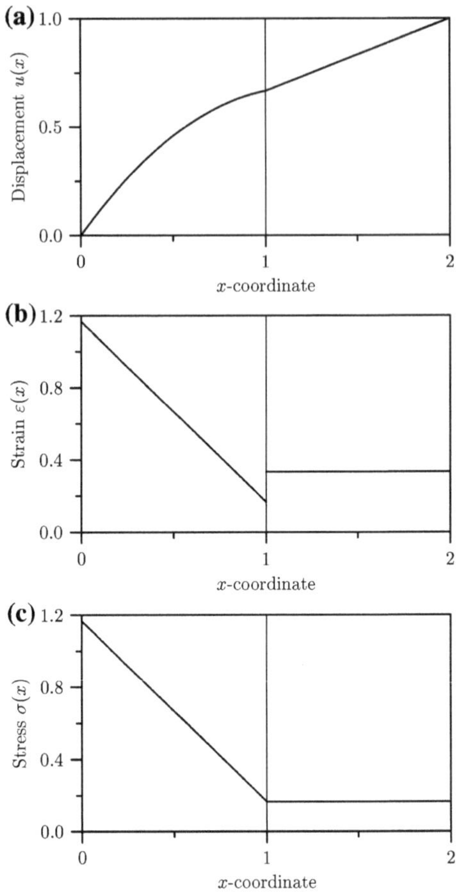


**Fig. C.1** Bi-material rod: **a** displacement $u(x)$; **b** strain $\varepsilon(x)$; **c** stress $\sigma(x)$

**Fig. C.2** Bi-material rod with fixed end displacement $u_0 = 0$

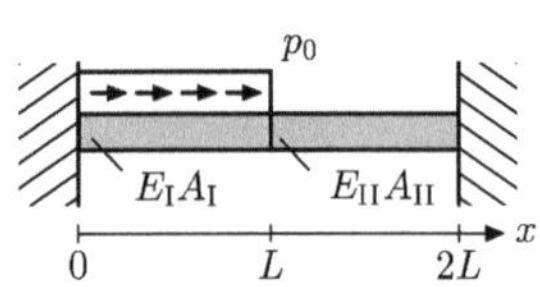


## 3.9 Elongation of a rod made of a functionally graded material

$$A\frac{\mathrm{d}}{\mathrm{d}x}\left(E(x)\frac{\mathrm{d}u_x(x)}{\mathrm{d}x}\right) = 0\,, \tag{C.23}$$

$$u_x(x) = \frac{c_1 L}{A(E_1 - E_0)}\ \ln\left(E_0 + \frac{x}{L}\,(E_1 - E_0)\right) + c_2\,. \tag{C.24}$$

Evaluation of boundary conditions gives $c_1 = F$ and $c_2 = -\frac{FL \ln E_0}{A(E_1 - E_0)}$.

$$u_x(x) = \frac{FL}{A(E_1 - E_0)} \ln\left(1 + \frac{x}{L}\left(\frac{E_1}{E_0} - 1\right)\right), \tag{C.25}$$

$$\lim_{E_1 \to E_0} u_x(x) = \frac{Fx}{E_0 A}. \tag{C.26}$$

**3.10 Elongation of a rod with elastic embedding**
Partial differential equation (PDE):

$$\frac{1}{\beta^2} \times \frac{\mathrm{d}^2 u_x(x)}{\mathrm{d}x^2} - u = 0 \text{ with } \beta = \sqrt{\frac{k}{EA}}. \tag{C.27}$$

General solution of the PDE:

$$u_x(x) = c_1 \mathrm{e}^{-\beta x} + c_2 \mathrm{e}^{\beta x} \tag{C.28}$$

$$= a \cosh(\beta x) + b \sinh(\beta x). \tag{C.29}$$

Considering of the boundary conditions, i.e. $u_x(0) = 0$ and $EA\frac{\mathrm{d}u_x(L)}{\mathrm{d}x} = F$, allows the determination of the constants as:

$$a = 0 \text{ and } b = \frac{F}{\beta \cosh(\beta L) EA}. \tag{C.30}$$

$$u_x(x) = \frac{\tanh(\beta x)}{\beta} \times \frac{F}{EA}, \tag{C.31}$$

$$u_x(L) = \frac{\tanh(\beta L)}{\beta} \times \frac{F}{EA}. \tag{C.32}$$

Limiting cases:

$$u_x(x)|_{\beta \to 0} = \frac{Fx}{EA}, \quad u_x(x)\,|_{\beta \to \infty} = 0. \tag{C.33}$$

**3.11 Elongation of a rod due to constant distributed load in the elasto-plastic range**
Normal force distribution: $N_x(x) = p_0(L - x)$.
Elasto-plastic range $0 \leq x \leq \varsigma = L - \frac{Ak_t^{\text{init}}}{p_0}$:

$$AE^{\text{elpl}} u_x(x) = -\frac{p_0 x^2}{2} + \left(c_1 - Ak_t^{\text{init}}\left(1 - \frac{E^{\text{elpl}}}{E}\right)\right) x + c_2. \tag{C.34}$$

Elastic range $\varsigma \leq x \leq L$:

$$AE u_x(x) = -\frac{p_0 x^2}{2} + c_3 x + c_4 \,. \tag{C.35}$$

With the following constants of integration:

$$c_1 = c_3 = p_0 L \,, \; c_2 = 0 \,, \tag{C.36}$$

$$c_4 = \left(\frac{E}{E^{\text{elpl}}} - 1\right)\left(-\frac{p_0 \varsigma^2}{2} + p_0 L \varsigma - A k_{\text{t}}^{\text{init}} \varsigma\right) \tag{C.37}$$

$$= \frac{1}{2}\left(\frac{E}{E^{\text{elpl}}} - 1\right)\left(-L\sqrt{p_0} + \frac{A k_{\text{t}}^{\text{init}}}{\sqrt{p_0}}\right)^2 . \tag{C.38}$$

**3.12 Elongation of a rod with changing cross section in the elasto-plastic range**

Normal stress distribution: $\sigma_x(x) = F/A(x)$.
Elasto-plastic range $0 \leq x \leq \varsigma = \frac{FL - A_1 k_{\text{t}}^{\text{init}} L}{k_{\text{t}}^{\text{init}}(A_2 - A_1)}$:

$$u_x(x) = \frac{c_1 L}{E^{\text{elpl}}(A_2 - A_1)} \ln\left(1 + \frac{x}{L}\left[\frac{A_2}{A_1} - 1\right]\right) - \frac{k_{\text{t}}^{\text{init}}}{E^{\text{elpl}}}\left[1 - \frac{E^{\text{elpl}}}{E}\right] x + c_2 \,. \tag{C.39}$$

Elastic range $\varsigma \leq x \leq L$:

$$u_x(x) = \frac{c_3 L}{E(A_2 - A_1)} \ln\left(1 + \frac{x}{L}\left[\frac{A_2}{A_1} - 1\right]\right) + c_4 \,. \tag{C.40}$$

With the following constants of integration:

$$c_1 = c_3 = F \,, \; c_2 = 0 \,, \tag{C.41}$$

$$c_4 = \left[1 - \frac{E^{\text{elpl}}}{E}\right]\left\{\frac{FL}{E^{\text{elpl}}(A_2 - A_1)} \ln\left(1 + \frac{\varsigma}{L}\left[\frac{A_2}{A_1} - 1\right]\right) - \frac{k_{\text{t}}^{\text{init}} \varsigma}{E^{\text{elpl}}}\right\} \tag{C.42}$$

$$= \frac{1 - \frac{E^{\text{elpl}}}{E}}{E^{\text{elpl}}(A_2 - A_1)}\left(FL \ln\left(\frac{F}{k_{\text{t}}^{\text{init}} A_1}\right) - FL + A_1 k_{\text{t}}^{\text{init}} L\right) . \tag{C.43}$$

**Problems from Chap.** 4

**4.4 Equilibrium for a differential beam element with variable distributed load**
The variable distributed load is evaluated in the middle in order to formulate the equilibrium relationship:

$$-Q_y(x)+Q_y(x+\mathrm{d}x)+q_y(x+\tfrac{1}{2}\,\mathrm{d}x)\mathrm{d}x=0\,. \tag{C.44}$$

$$M_z(x+\mathrm{d}x)-M_z(x)+Q_y(x)\mathrm{d}x-\frac{1}{2}q_y(x+\tfrac{1}{2}\,\mathrm{d}x)\mathrm{d}x^2=0\,. \tag{C.45}$$

**4.5 Bending line derived from stress distribution over cross section**

$$\sigma(y,\sigma_{\mathrm{max}})=-\frac{\sigma_{\mathrm{max}}}{\frac{h}{2}}\times y\,. \tag{C.46}$$

$$M(\sigma_{\mathrm{max}})=-2b\int\limits_0^{h/2}y\left(-\frac{\sigma_{\mathrm{max}}}{\frac{h}{2}}\times y\right)\mathrm{d}y=\frac{b\sigma_{\mathrm{max}}h^2}{6}=\frac{\sigma_{\mathrm{max}}I_z}{\frac{h}{2}}\,, \tag{C.47}$$

$$\kappa=\frac{\sigma_{\mathrm{max}}}{E\frac{h}{2}}=\frac{M}{EI_z}\,. \tag{C.48}$$

**4.6 Simply supported beam with centered single force**
The general solution can be written in the range $0\le x\le \frac{L}{2}$ as:

$$u_y(x)=c_1x^3+c_2x^2+c_3x+c_4\,. \tag{C.49}$$

Consideration of the boundary conditions, i.e. $u_y(0)=0$, $EI_z\frac{\mathrm{d}^2u_y}{\mathrm{d}x^2}(0)=M_z(0)=0$, $\frac{\mathrm{d}u_y}{\mathrm{d}x}\left(\frac{L}{2}\right)=0$ and $EI_z\frac{\mathrm{d}^3u_y}{\mathrm{d}x^3}\left(\frac{L}{2}\right)=-Q_y\left(\frac{L}{2}\right)=\frac{F}{2}$, allows the determination of the constants of integration as:

$$c_1=\frac{F}{12EI_z}\,,\quad c_2=0\,, \tag{C.50}$$

$$c_3=-\frac{L^2F}{16EI_z}\,,\quad c_4=0\,. \tag{C.51}$$

Thus, the following function for the vertical deflection is obtained:

$$u_y(x)=-\frac{Fx}{48EI_z}\left(3L^2-4x^2\right)\,. \tag{C.52}$$

**4.7 Simply supported beam under pure bending load**
The general solution can be written in the range $0\le x\le L$ as:

$$EI_zu_y(x)=\frac{1}{2}Mx^2+c_1x+c_2\,. \tag{C.53}$$

Consideration of the boundary conditions, i.e. $u_y(0) = 0$ and $u_y(L) = 0$, gives $c_1 = -\frac{1}{2}ML$ and $c_2 = 0$. Thus, the following function for the vertical deflection is obtained:

$$u_y(x) = \frac{M}{2EI_z}\left(x^2 - Lx\right) \quad \text{and} \quad u_y\left(\frac{L}{2}\right) = -\frac{ML^2}{8EI_z}. \tag{C.54}$$

**4.8 Bernoulli beam fixed at both ends**

The general solution is given by:

$$u_y(x) = \frac{1}{EI_z}\left(\frac{q_y x^4}{24} + \frac{c_1 x^3}{6} + \frac{c_2 x^2}{2} + c_3 x + c_4\right). \tag{C.55}$$

(a) Single force case ($0 \leq x \leq \frac{L}{2}$)

Boundary conditions: $u_y(0) = 0$, $\varphi_z(0) = 0$, $Q_y(0) = \frac{F}{2}$, $\varphi_z(\frac{L}{2}) = 0$.

$$u_y(x) = \frac{1}{EI_z}\left(-\frac{Fx^3}{12} + \frac{FLx^2}{16}\right), \varphi_z(x) = \frac{1}{EI_z}\left(-\frac{Fx^2}{4} + \frac{FLx}{8}\right). \tag{C.56}$$

$$u_{y,\text{max}} = \frac{FL^3}{192EI_z}, \varphi_{z,\text{max}} = \varphi_z(\tfrac{L}{4}) = \frac{FL^2}{64EI_z}. \tag{C.57}$$

(b) Distributed load case

Boundary conditions: $u_y(0) = 0$, $\varphi_z(0) = 0$, $Q_y(0) = \frac{q_y L}{2}$, $\varphi_z(\frac{L}{2}) = 0$.

$$u_y(x) = \frac{1}{EI_z}\left(\frac{q_y x^4}{24} - \frac{q_y L x^3}{12} + \frac{q_y L^2 x^2}{24}\right), \tag{C.58}$$

$$\varphi_z(x) = \frac{1}{EI_z}\left(\frac{q_y x^3}{6} - \frac{q_y L x^2}{4} + \frac{q_y L^2 x}{12}\right). \tag{C.59}$$

$$u_{y,\text{max}} = \frac{q_y L^4}{384EI_z}, \varphi_{z,\text{max}} = \varphi_z(\tfrac{3-\sqrt{3}}{6}L) = \frac{\sqrt{3}q_y L^3}{216EI_z}. \tag{C.60}$$

**4.9 Bernoulli beam fixed at both ends and loaded by a single moment**

One possibility to solve the problem is to release the right-hand boundary and to consider the reactions $R_2$ and $M_2$ as shown in Fig. C.3. The solutions for the single cases, i.e. where only $M$, $R_2$ or $M_2$ is acting, can be taken from Table 4.10 as:

External moment $M$: $u_y(L) = \frac{Ma^2}{2EI_z} + \underbrace{\frac{Ma}{EI_z}}_{\text{slope}} \underbrace{(L-a)}_{\text{length}}$,

$$\varphi_z(L) = \frac{Ma}{EI_z},$$

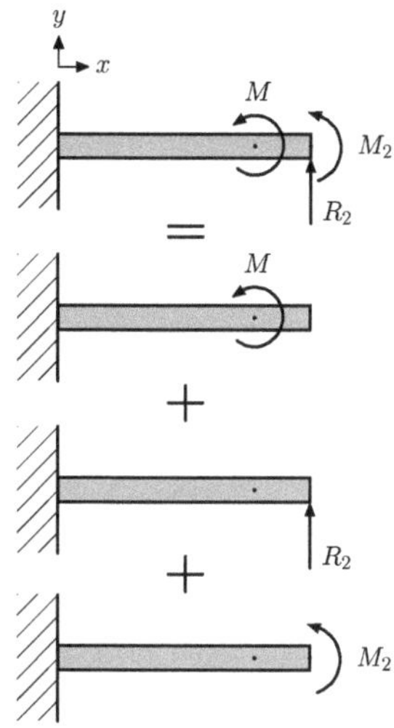


**Fig. C.3** Solution strategy for the BERNOULLI beam of second-order indetermination

Force $R_2$:

$$u_y(L) = \frac{R_2 L^3}{3EI_z},$$

$$\varphi_z(L) = \frac{R_2 L^2}{2EI_z},$$

Moment $M_2$:

$$u_y(L) = \frac{M_2 L^2}{2EI_z},$$

$$\varphi_z(L) = \frac{M_2 L}{EI_z}.$$

Superposition of the expressions for $u_y$ and $\varphi_z$ and consideration of the boundary conditions at $x = L$, i.e. $u_y(L) = 0$ and $\varphi_z(L) = 0$, gives the reactions at the right-hand boundary as:

$$R_2 = \frac{6M}{L^3}\left(a^2 - aL\right) \text{ and } M_2 = \frac{M}{L^2}\left(2aL - 3a^2\right). \tag{C.61}$$

The describing differential equation based on the moment distribution is given under consideration of a discontinuous functions as:

$$EI_z \frac{d^2 u_y(x)}{dx^2} = \frac{M}{L^2}\left(L^2 - 4aL + 3a^2\right) - \frac{6M}{L^3}\left(a^2 - aL\right)x - M\langle x - a\rangle^0. \tag{C.62}$$

Integration and consideration of the boundary conditions at the left-hand boundary gives the integration constants ($c_1 = c_2 = 0$) and the expression for the displacement as:

$$u_y(x) = \frac{-M}{2EI_z}\left[-\left(L^2 - 4aL + 3a^2\right)\left(\frac{x}{L}\right)^2 + 2\left(a^2 - aL\right)\left(\frac{x}{L}\right)^3 + \langle x-a\rangle^2\right]. \tag{C.63}$$

In the special case of $a = \frac{L}{2}$, the maxima for the deflection are obtained at $x = \frac{L}{3}$ and $x = \frac{2L}{3}$ while the maxima of the slope are obtained for $x = \frac{L}{6}$ and $x = \frac{5L}{6}$:

$$u_{y,\max} = \mp\frac{ML^2}{216EI_z} \text{ and } \varphi_{z,\max} = -\frac{ML}{48EI_z}. \tag{C.64}$$

**4.10 Deflection of a simply supported functionally graded beam**
General solution:

$$u_y(x) = \frac{M\ln(1+\alpha x)}{EI_0\alpha^2} + \frac{M\ln(1+\alpha x)x}{EI_0\alpha} - \frac{M}{EI_0\alpha^2} + c_1 x + c_2. \tag{C.65}$$

Consideration of the boundary conditions, i.e. $u_y(0) = 0$ and $u_y(L) = 0$, gives $c_2 = \frac{M}{EI_0\alpha^2}$ and $c_1 = -\frac{M\ln(1+\alpha L)}{EI_0\alpha^2 L} - \frac{M\ln(1+\alpha L)}{EI_0\alpha} + \frac{M}{EI_0\alpha}$. Thus, the normalized deflection is obtained as:

$$\frac{u_y(x)EI_0}{M} = \frac{1}{\alpha}\left[\frac{\ln(1+\alpha x)}{\alpha} - \frac{\ln(1+\alpha L)x}{\alpha L} + \ln\left(\frac{1+\alpha x}{1+\alpha L}\right)x\right]. \tag{C.66}$$

The graphical representation is given in Fig. C.4.

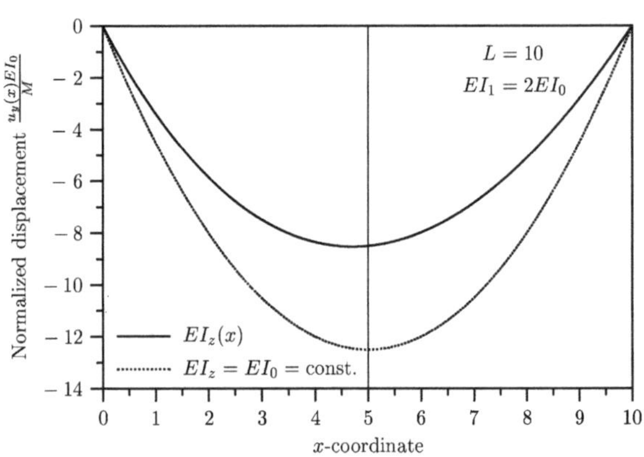


**Fig. C.4** Deflection of a simply supported BERNOULLI beam with variable and constant bending stiffness

**4.11 Equilibrium for a differential beam element on elastic foundation with distributed load**

$$-Q_y(x)+Q_y(x+\mathrm{d}x)+q_0\mathrm{d}x-ku_y(x+\tfrac{1}{2}\,\mathrm{d}x)\mathrm{d}x=0\,, \tag{C.67}$$

$$M_z(x+\mathrm{d}x)-M_z(x)-q_0\mathrm{d}x(\tfrac{1}{2}\mathrm{d}x)+ku_y(x+\tfrac{1}{2}\,\mathrm{d}x)\mathrm{d}x(\tfrac{1}{2}\mathrm{d}x)+Q_y(x)\mathrm{d}x=0\,. \tag{C.68}$$

It follows with

$$Q_y(x+\mathrm{d}x)\approx Q_y(x)+\frac{\mathrm{d}Q_y(x)}{\mathrm{d}x}\,\mathrm{d}x\,, \tag{C.69}$$

$$M_z(x+\mathrm{d}x)\approx M_z(x)+\frac{\mathrm{d}M_z(x)}{\mathrm{d}x}\,\mathrm{d}x\,, \tag{C.70}$$

$$u_y(x+\tfrac{1}{2}\mathrm{d}x)\approx u_y(x)+\frac{\mathrm{d}u_y(x)}{\mathrm{d}x}\,\tfrac{1}{2}\mathrm{d}x\,, \tag{C.71}$$

and under consideration that the second order infinitely small quantity $(\mathrm{d}x)^2$ may be ignored that:

$$\frac{\mathrm{d}Q_y(x)}{\mathrm{d}x}=-q_0+ku_y(x)\ \text{ and }\ \frac{\mathrm{d}M_z(x)}{\mathrm{d}x}=-Q_y(x)\,. \tag{C.72}$$

**4.12 Cantilevered beam on elastic foundation: single force case**
Partial differential equation (PDE):

$$EI_z\frac{\mathrm{d}^4u_y(x)}{\mathrm{d}x^4}+ku_y(x)=0\,, \tag{C.73}$$

or with the abbreviation $\beta=\sqrt[4]{\frac{k}{4EI_z}}$ as:

$$\frac{1}{4\beta^4}\frac{\mathrm{d}^4u_y(x)}{\mathrm{d}x^4}+u_y(x)=0\,. \tag{C.74}$$

The solution of the PDE can be found in reference [104] or simply obtained from a computer system for advanced mathematics in the following different forms:

$$u_y(x)=\mathrm{e}^{\beta x}(A\cos(\beta x)+B\sin(\beta x))+\mathrm{e}^{-\beta x}(C\cos(\beta x)+D\sin(\beta x))\,, \tag{C.75}$$

$$u_y(x)=c_1\sin(\beta x)\sinh(\beta x)+c_2\sin(\beta x)\cosh(\beta x)+c_3\cos(\beta x)\sinh(\beta x)+c_4\cos(\beta x)\cosh(\beta x)\,, \tag{C.76}$$

$$u_y(x)=a\mathrm{e}^{-\beta x}\sin(\beta x)+b\mathrm{e}^{-\beta x}\cos(\beta x)+c\mathrm{e}^{\beta x}\sin(\beta x)+d\mathrm{e}^{\beta x}\cos(\beta x)\,, \tag{C.77}$$

which can easily be transformed into each other based on the relations in the appendix. Consideration of the boundary conditions, i.e. $u_y(0) = 0$, $\frac{\mathrm{d}u_y(0)}{\mathrm{d}x} = 0$, $\frac{\mathrm{d}^2u_y(L)}{\mathrm{d}^2x} = 0$ and $EI_z\frac{\mathrm{d}^3u_y(L)}{\mathrm{d}x^3} = F$, allows the determination of the constants as:

$$c_1 = -\frac{(\sin(\beta L)\cosh(\beta L) + \cos(\beta L)\sinh(\beta L))\,F}{2\beta^3\left(\cos^2(\beta L) + \cosh^2(\beta L)\right)EI_z}, \tag{C.78}$$

$$c_2 = \frac{\cos(\beta L)\cosh(\beta L)\,F}{2\beta^3\left(\cos^2(\beta L) + \cosh^2(\beta L)\right)EI_z}, \tag{C.79}$$

$$c_3 = -\frac{\cos(\beta L)\cosh(\beta L)\,F}{2\beta^3\left(\cos^2(\beta L) + \cosh^2(\beta L)\right)EI_z}, \tag{C.80}$$

$$c_4 = 0\,. \tag{C.81}$$

Deflection at $x = L$:

$$u_y(L) = \frac{(\sin(\beta L)\cos(\beta L) - \cosh(\beta L)\sinh(\beta L))\,F}{2\beta^3\left(\cos^2(\beta L) + \cosh^2(\beta L)\right)EI_z}\,. \tag{C.82}$$

Limiting case $\beta \to 0$:

$$u_y(L) = -\frac{FL^3}{3EI_z}\,. \tag{C.83}$$

**4.13 Simply supported beam on elastic foundation: single force case**
The solution of the PDE can be found in reference [104] or simply obtained from a computer system for advanced mathematics in the following form:

$$\begin{aligned} u_y(x) = {} & c_1 \sin(\beta x)\sinh(\beta x) + c_2 \sin(\beta x)\cosh(\beta x) \\ & + c_3\cos(\beta x)\sinh(\beta x) + c_4\cos(\beta x)\cosh(\beta x)\,. \end{aligned} \tag{C.84}$$

Based on this formulation, the derivatives can be calculated as:

$$\begin{aligned} \frac{\mathrm{d}u_y(x)}{\mathrm{d}x} = {} & \beta\,(c_1\cos(\beta x)\sinh(\beta x) + c_1\sin(\beta x)\cosh(\beta x) \\ & + c_2\cos(\beta x)\cosh(\beta x) + c_2\sin(\beta x)\sinh(\beta x) \\ & - c_3\sin(\beta x)\sinh(\beta x) + c_3\cos(\beta x)\cosh(\beta x) \\ & - c_4\sin(\beta x)\cosh(\beta x) + c_4\cos(\beta x)\sinh(\beta x))\,. \end{aligned} \tag{C.85}$$

$$\begin{aligned} \frac{\mathrm{d}^2u_y}{\mathrm{d}x^2} = {} & 2\beta^2\,(c_1\cos(\beta x)\cosh(\beta x) + c_2\cos(\beta x)\sinh(\beta x) \\ & - c_3\sin(\beta x)\cosh(\beta x) - c_4\sin(\beta x)\sinh(\beta x))\,. \end{aligned} \tag{C.86}$$

$$\frac{d^3 u_y(x)}{dx^3} = 2\beta^3 \left(-c_1 \sin(\beta x)\cosh(\beta x) + c_1 \cos(\beta x)\sinh(\beta x)\right.$$
$$- c_2 \sin(\beta x)\sinh(\beta x) + c_2 \cos(\beta x)\cosh(\beta x)$$
$$- c_3 \cos(\beta x)\cosh(\beta x) - c_3 \sin(\beta x)\sinh(\beta x)$$
$$\left.-c_4 \cos(\beta x)\sinh(\beta x) - c_4 \sin(\beta x)\cosh(\beta x)\right) . \tag{C.87}$$

The evaluation of the boundary conditions, i.e. $u_y(0) = 0$, $M_z(0) = 0$, $\frac{du_y}{dx}(\frac{L}{2}) = 0$ and $EI_z \frac{d^3 u_y}{dx^3}(\frac{L}{2}) = -Q_y(\frac{L}{2}) = \frac{F}{2}$, allows to determine the constants of integration as:

$$c_1 = 0 \ , \ \ c_4 = 0 \,, \tag{C.88}$$

$$c_2 = \frac{\left(-\sin\left(\beta \frac{L}{2}\right)\sinh\left(\beta \frac{L}{2}\right) + \cos\left(\beta \frac{L}{2}\right)\cosh\left(\beta \frac{L}{2}\right)\right) F}{8\left(\left(\cosh\left(\beta \frac{L}{2}\right)\right)^2 - 1 + \left(\cos\left(\beta \frac{L}{2}\right)\right)^2\right) EI_z \beta^3} \,, \tag{C.89}$$

$$c_3 = -\frac{\left(\cos\left(\beta \frac{L}{2}\right)\cosh\left(\beta \frac{L}{2}\right) + \sin\left(\beta \frac{L}{2}\right)\sinh\left(\beta \frac{L}{2}\right)\right) F}{8\left(\left(\cosh\left(\beta \frac{L}{2}\right)\right)^2 - 1 + \left(\cos\left(\beta \frac{L}{2}\right)\right)^2\right) EI_z \beta^3} \,. \tag{C.90}$$

$$u_y\left(\tfrac{L}{2}\right) = -\frac{\left(\cosh\left(\beta \frac{L}{2}\right)\sinh\left(\beta \frac{L}{2}\right) - \sin\left(\beta \frac{L}{2}\right)\cos\left(\beta \frac{L}{2}\right)\right) F}{8\left(\left(\cosh\left(\beta \frac{L}{2}\right)\right)^2 - 1 + \left(\cos\left(\beta \frac{L}{2}\right)\right)^2\right) EI_z \beta^3} \,. \tag{C.91}$$

The limiting case $\beta \to 0$ gives exactly the expression given in Table 4.11.

### 4.14 Beam on elastic foundation: distributed load case

$$EI_z \frac{d^4 u_y(x)}{dx^4} + k u_y(x) - \frac{q_0}{k} = 0 \,, \tag{C.92}$$

$$u_y(x) = \frac{q_0}{k} + c_1 \sin(\beta x)\sinh(\beta x) + c_2 \sin(\beta x)\cosh(\beta x)$$
$$+ c_3 \cos(\beta x)\sinh(\beta x) + c_4 \cos(\beta x)\cosh(\beta x) \,. \tag{C.93}$$

### 4.15 Integration of a discontinuity function

$$\int (x - a_1)^2 \langle x - a_2 \rangle^0 dx = (x - a_1)^2 \langle x - a_2 \rangle^1 -$$
$$- (x - a_1)^1 \langle x - a_2 \rangle^2 + \frac{1}{3} \langle x - a_2 \rangle^3 + c \,. \tag{C.94}$$

**4.16 Cantilevered beam in the elasto-plastic range loaded by a single moment**
The differential equations in the pure elastic and elasto-plastic range are given as:

$$EI_z \frac{d^2 u_y}{dx^2} = M \quad \text{and} \quad \frac{d^2 u_y}{dx^2} = \frac{2k}{Eh\sqrt{3}\sqrt{1-\frac{M}{M_{\text{lim}}^{\text{pl}}}}} \,. \tag{C.95}$$

Integration and consideration of the boundary conditions, i.e. $u_y(0) = 0$ and $\frac{du_y(0)}{dx} = 0$, gives:

$$u_y(x) = \frac{1}{2}\frac{M_{\text{lim}}^{\text{pl}} L^2}{EI_z}\frac{M}{M_{\text{lim}}^{\text{pl}}} \times \left(\frac{x}{L}\right)^2 \quad \text{(elastic)}\,, \tag{C.96}$$

$$u_y(x) = \frac{k}{Eh\sqrt{3}\sqrt{1-\frac{M}{M_{\text{lim}}^{\text{pl}}}}} \times x^2 \quad \text{(elasto-plastic)}\,. \tag{C.97}$$

**4.17 Cantilevered beam in the elasto-plastic range loaded by a distributed load**
The pure elastic solution is given in Table 4.10. In the elasto-plastic range, the part $L \leq x \leq \varsigma$ will remain elastic while the rest of the beam is in the elasto-plastic range. Both regions can be described by the differential equations:

$$EI_z \frac{d^2 u_y}{dx^2} = \frac{qx^2}{2} \quad \text{and} \quad \frac{d^2 u_y}{dx^2} = \frac{2k}{Eh\sqrt{3}\sqrt{1-\frac{qL^2}{2M_{\text{lim}}^{\text{pl}}}\left(\frac{x}{L}\right)^2}} \,. \tag{C.98}$$

Integration twice gives for the regions $L \leq x \leq \varsigma$ and $0 \leq x \leq \varsigma$:

$$EI_z \frac{du_y}{dx} = \frac{qx^3}{6} + c_1 \,, \tag{C.99}$$

$$EI_z u_y = \frac{qx^4}{24} + c_1 x + c_2 \,, \tag{C.100}$$

$$\frac{du_y}{dx} = \frac{2\sqrt{6}}{3}\frac{k}{hE}\sqrt{\frac{M_{\text{lim}}^{\text{pl}}}{q}} \arcsin\sqrt{\frac{qx^2}{2M_{\text{lim}}^{\text{pl}}}} + c_3 \,, \tag{C.101}$$

and under consideration of, for example,

$$\iint \frac{b}{\sqrt{1-ax^2}}\,dx = \frac{bx}{\sqrt{a}} \arcsin\sqrt{ax^2} + \frac{b}{a}\sqrt{1-ax^2} + c_3 x + c_4 \,,$$

$$u_y = \frac{2k}{\sqrt{3}Eh}\sqrt{\frac{2M^{\text{pl}}_{\text{lim}}}{q}}\, x \arcsin\sqrt{\frac{qx^2}{2M^{\text{pl}}_{\text{lim}}}} + \frac{2k}{\sqrt{3}Eh}\left(\frac{2M^{\text{pl}}_{\text{lim}}}{q}\right)\sqrt{1-\frac{qx^2}{2M^{\text{pl}}_{\text{lim}}}} + c_3 x + c_4 \,. \tag{C.102}$$

Consideration of the boundary conditions, i.e. $\frac{\mathrm{d}u_y(L)}{\mathrm{d}x} = 0$ and $u_y(L) = 0$, and the continuity conditions, i.e. $\frac{\mathrm{d}u_y(\varsigma^-)}{\mathrm{d}x} = \frac{\mathrm{d}u_y(\varsigma^+)}{\mathrm{d}x}$ and $u_y(\varsigma^-) = u_y(\varsigma^+)$, and the identities $qL^3 = M^{\text{pl}}_{\text{lim}} L \left(\frac{qL^2}{M^{\text{pl}}_{\text{lim}}}\right)$ and $\varsigma = L\sqrt{\frac{4}{3}\frac{M^{\text{pl}}_{\text{lim}}}{pL^2}}$ allows to determine the constants of integration:

$$c_3 = -\frac{2\sqrt{6}k}{3hE}\sqrt{\frac{M^{\text{pl}}_{\text{lim}}}{q}}\arcsin\sqrt{\frac{qL^2}{2M^{\text{pl}}_{\text{lim}}}}\,, \tag{C.103}$$

$$c_4 = -\frac{2k}{\sqrt{3}hE}\left(\frac{2M^{\text{pl}}_{\text{lim}}}{q}\right)\sqrt{1-\frac{qL^2}{2M^{\text{pl}}_{\text{lim}}}}\,, \tag{C.104}$$

$$c_1 = \frac{\sqrt{6}M^{\text{pl}}_{\text{lim}}}{3}\sqrt{\frac{M^{\text{pl}}_{\text{lim}}}{q}}\left(\arcsin\sqrt{\frac{2}{3}} - \arcsin\sqrt{\frac{qL^2}{2M^{\text{pl}}_{\text{lim}}}} - \frac{2\sqrt{2}}{9}\right), \tag{C.105}$$

$$c_2 = \frac{M^{\text{pl}}_{\text{lim}}}{3}\left(\frac{2M^{\text{pl}}_{\text{lim}}}{q}\right)\left(\frac{4}{3} - \sqrt{3}\sqrt{1-\frac{qL^2}{2M^{\text{pl}}_{\text{lim}}}}\right). \tag{C.106}$$

Thus, the deflection in the regions $0 \le x \le \varsigma$ and $\varsigma \le x \le L$ is given as:

$$u_y = \frac{1}{EI_z}\left(\frac{qx^4}{24} + \frac{\sqrt{6}M^{\text{pl}}_{\text{lim}}}{3}\sqrt{\frac{M^{\text{pl}}_{\text{lim}}}{q}}\left(\arcsin\sqrt{\frac{2}{3}} - \arcsin\sqrt{\frac{qL^2}{2M^{\text{pl}}_{\text{lim}}}} - \frac{2\sqrt{2}}{9}\right)x + \frac{M^{\text{pl}}_{\text{lim}}}{3}\left(\frac{2M^{\text{pl}}_{\text{lim}}}{q}\right)\left(\frac{4}{3} - \sqrt{3}\sqrt{1-\frac{qL^2}{2M^{\text{pl}}_{\text{lim}}}}\right)\right), \tag{C.107}$$

$$u_y = \frac{2k}{\sqrt{3}Eh}\sqrt{\frac{2M_{\text{lim}}^{\text{pl}}}{q}}\,x \arcsin\sqrt{\frac{qx^2}{2M_{\text{lim}}^{\text{pl}}}}$$
$$+\frac{2k}{\sqrt{3}Eh}\left(\frac{2M_{\text{lim}}^{\text{pl}}}{q}\right)\sqrt{1-\frac{qx^2}{2M_{\text{lim}}^{\text{pl}}}}$$
$$-\frac{2\sqrt{6}k}{3hE}\sqrt{\frac{M_{\text{lim}}^{\text{pl}}}{q}}\arcsin\sqrt{\frac{qL^2}{2M_{\text{lim}}^{\text{pl}}}}\times x$$
$$-\frac{2k}{\sqrt{3}hE}\left(\frac{2M_{\text{lim}}^{\text{pl}}}{q}\right)\sqrt{1-\frac{qL^2}{2M_{\text{lim}}^{\text{pl}}}}\,. \tag{C.108}$$

Comment: If the solution is required which is based on a coordinate system originated at the fixed support, the transformation $\tilde{x} = L - x$ can be used. However, the actual coordinate system with origin at the free end is more advantageous for the mathematical derivation.

Alternatively, the derivation can be based in the plastic range on the following differential equation

$$\frac{\mathrm{d}^2 u_y}{\mathrm{d}x^2} = \frac{M_{\text{lim}}^{\text{pl}}}{\sqrt{3}EI_z\sqrt{1-\frac{M_z(L)}{M_{\text{lim}}^{\text{pl}}}\left(\frac{x}{L}\right)^2}}\,, \tag{C.109}$$

which results in the following constants of integration:

$$c_1 = \frac{\sqrt{3}}{3}LM_{\text{lim}}^{\text{pl}}\sqrt{\frac{M_{\text{lim}}^{\text{pl}}}{M_z(L)}}\left(\arcsin\sqrt{\frac{2}{3}} - \arcsin\sqrt{\frac{M_z(L)}{M_{\text{lim}}^{\text{pl}}}} - \frac{2\sqrt{2}}{9}\right), \tag{C.110}$$

$$c_2 = \frac{M_{\text{lim}}^{\text{pl}}L^2}{3}\left(\frac{M_{\text{lim}}^{\text{pl}}}{M_z(L)}\right)\left(\frac{4}{3} - \sqrt{3}\sqrt{1-\frac{M_z(L)}{M_{\text{lim}}^{\text{pl}}}}\right), \tag{C.111}$$

$$c_3 = -\frac{\sqrt{3}}{3}\frac{M_{\text{lim}}^{\text{pl}}L}{EI_z}\sqrt{\frac{M_{\text{lim}}^{\text{pl}}}{M_z(L)}}\arcsin\sqrt{\frac{M_z(L)}{M_{\text{lim}}^{\text{pl}}}}\,, \tag{C.112}$$

$$c_4 = -\frac{M_{\text{lim}}^{\text{pl}}L^2}{\sqrt{3}EI_z}\frac{M_{\text{lim}}^{\text{pl}}}{M_z(L)}\sqrt{1-\frac{M_z(L)}{M_{\text{lim}}^{\text{pl}}}}\,. \tag{C.113}$$

**4.18 Cantilevered beam in the elasto-plastic range loaded by a distributed load—normalized displacement representation**

The normalized deflection in the region $0 \leq x \leq \varsigma$ is given under consideration of $qL^4 = M_{\text{lim}}^{\text{pl}} L^2 \times 2\left(\frac{M_z(L)}{M_{\text{lim}}^{\text{pl}}}\right)$ as:

$$\frac{u_y(x)}{\frac{M_{\text{lim}}^{\text{pl}} L^2}{EI_z}} = \frac{1}{12}\left(\frac{M_z(L)}{M_{\text{lim}}^{\text{pl}}}\right) \times \left(\frac{x}{L}\right)^4 + \frac{\sqrt{3}}{3}\sqrt{\frac{M_{\text{lim}}^{\text{pl}}}{M_z(L)}}\left(\arcsin\sqrt{\frac{2}{3}} - \arcsin\sqrt{\frac{M_z(L)}{M_{\text{lim}}^{\text{pl}}}} - \frac{2\sqrt{2}}{9}\right)\left(\frac{x}{L}\right) + \frac{1}{3}\left(\frac{M_{\text{lim}}^{\text{pl}}}{M_z(L)}\right)\left(\frac{4}{3} - \sqrt{3}\sqrt{1 - \frac{M_z(L)}{M_{\text{lim}}^{\text{pl}}}}\right). \tag{C.114}$$

And for $\varsigma \leq x \leq L$ as:

$$\frac{u_y(x)}{\frac{M_{\text{lim}}^{\text{pl}} L^2}{EI_z}} = \frac{1}{\sqrt{3}}\sqrt{\frac{M_{\text{lim}}^{\text{pl}}}{M_z(L)}}\left(\frac{x}{L}\right)\arcsin\sqrt{\frac{M_z(L)}{M_{\text{lim}}^{\text{pl}}}\left(\frac{x}{L}\right)^2} + \frac{1}{\sqrt{3}}\frac{M_{\text{lim}}^{\text{pl}}}{M_z(L)}\sqrt{1 - \frac{M_z(L)}{M_{\text{lim}}^{\text{pl}}}\left(\frac{x}{L}\right)^2} - \frac{1}{\sqrt{3}}\sqrt{\frac{M_{\text{lim}}^{\text{pl}}}{M_z(L)}}\left(\frac{x}{L}\right)\arcsin\sqrt{\frac{M_z(L)}{M_{\text{lim}}^{\text{pl}}}} - \frac{1}{\sqrt{3}}\frac{M_{\text{lim}}^{\text{pl}}}{M_z(L)}\sqrt{1 - \frac{M_z(L)}{M_{\text{lim}}^{\text{pl}}}}. \tag{C.115}$$

**4.19 Cantilevered beam in the elasto-plastic range loaded by a distributed load—separation between pure elastic and elasto-plastic zone**

Balancing the external and internal bending moment for $M_z(L) > M_{\text{lim}}^{\text{el}}$ gives:

$$\frac{1}{3}M_{\text{lim}}^{\text{pl}}\left(3 - \alpha^2\right) = \frac{2M_{\text{lim}}^{\text{pl}}}{3\sqrt{3}\sqrt{1 - \frac{M_z(L)}{M_{\text{lim}}^{\text{pl}}}\left(\frac{x}{L}\right)^2}}. \tag{C.116}$$

For $\alpha = 1$ and $x = \varsigma$, one obtains:

$$\frac{\varsigma}{L} = \sqrt{\frac{2}{3}\frac{M_{\text{lim}}^{\text{pl}}}{M_z(L)}}. \tag{C.117}$$

Limiting cases:

$$M_z(L) = M_{\text{lim}}^{\text{el}} = \frac{2}{3} M_{\text{lim}}^{\text{pl}} \quad \rightarrow \quad \varsigma = L \,, \tag{C.118}$$

$$M_z(L) = M_{\text{lim}}^{\text{pl}} \quad \rightarrow \quad \varsigma = L\sqrt{\frac{2}{3}} \,. \tag{C.119}$$

**4.20 Simply supported beam in the elasto-plastic range loaded by a distributed load**

Bending moment distribution:

$$M_z(x) = \frac{q}{2}\left(\frac{L^2}{4} - x^2\right) . \tag{C.120}$$

Parameter $\alpha$:

$$\alpha = \sqrt{3}\sqrt{1 - \frac{q}{q_{\text{lim}}^{\text{pl}}}\left(1 - 4\left(\frac{x}{L}\right)^2\right)} \quad \text{with} \quad q_{\text{lim}}^{\text{pl}} = \frac{2bh^2k}{L^2} \,. \tag{C.121}$$

Pure elastic range, i.e. $\varsigma = \frac{L}{2}\sqrt{1 - \frac{2}{3}\frac{q_{\text{lim}}^{\text{pl}}}{q}} \le x \le \frac{L}{2}$:

$$u_y(x) = \frac{1}{EI_z}\left[\frac{q}{2}\left(\frac{L^2x^2}{8} - \frac{x^4}{12}\right) + c_1x + c_2\right] . \tag{C.122}$$

Elasto-plastic range, i.e. $0 \le x \le \varsigma$:

$$u_y(x) = \frac{kL}{\sqrt{3}Eh}\sqrt{\frac{q_{\text{lim}}^{\text{pl}}}{q}}\left[x \times \text{arsinh}\left(\frac{x}{\frac{L}{2}\sqrt{\frac{q_{\text{lim}}^{\text{pl}}}{q} - 1}}\right) - \sqrt{x^2 + \frac{L^2}{4}\left(\frac{q_{\text{lim}}^{\text{pl}}}{q} - 1\right)}\right]$$
$$+ c_3x + c_4 \,. \tag{C.123}$$

Constants of integration based on boundary conditions:

$$c_1 = \frac{kI_zL}{\sqrt{3}h}\sqrt{\frac{q_{\text{lim}}^{\text{pl}}}{q}} \times \text{arsinh}\left(\frac{\sqrt{1 - \frac{2}{3}\frac{q_{\text{lim}}^{\text{pl}}}{q}}}{\sqrt{\frac{q_{\text{lim}}^{\text{pl}}}{q} - 1}}\right)$$
$$- \frac{qL^3}{16}\left(\left[1 - \frac{2}{3}\frac{q_{\text{lim}}^{\text{pl}}}{q}\right]^{1/2} - \frac{1}{3}\left[1 - \frac{2}{3}\frac{q_{\text{lim}}^{\text{pl}}}{q}\right]^{3/2}\right) , \tag{C.124}$$

$$c_2 = -c_1 \times \frac{L}{2} - \frac{5}{384} \times qL^4 \,, \qquad c_3 = 0 \,, \tag{C.125}$$

$$c_4 = \frac{1}{4}\frac{kL^2}{Eh}\frac{q}{q_{\text{lim}}^{\text{pl}}}\left(\left[1-\frac{2}{3}\frac{q_{\text{lim}}^{\text{pl}}}{q}\right]-\frac{1}{6}\left[1-\frac{2}{3}\frac{q_{\text{lim}}^{\text{pl}}}{q}\right]^2+\frac{2}{3}\left[\frac{q_{\text{lim}}^{\text{pl}}}{q}\right]^2-\frac{5}{6}\right)$$
$$+\frac{kL^2}{2Eh}\frac{q}{q_{\text{lim}}^{\text{pl}}}\left(1-\sqrt{1-\frac{2}{3}\frac{q_{\text{lim}}^{\text{pl}}}{q}}\right)\times\left(\left[1-\frac{2}{3}\frac{q_{\text{lim}}^{\text{pl}}}{q}\right]^{1/2}-\frac{1}{3}\left[1-\frac{2}{3}\frac{q_{\text{lim}}^{\text{pl}}}{q}\right]^{3/2}\right)$$
$$-\frac{kL^2}{2\sqrt{3}Eh}\sqrt{\frac{q_{\text{lim}}^{\text{pl}}}{q}}\times\text{arsinh}\left(\frac{\sqrt{1-\frac{2}{3}\frac{q_{\text{lim}}^{\text{pl}}}{q}}}{\sqrt{\frac{q_{\text{lim}}^{\text{pl}}}{q}-1}}\right). \qquad \text{(C.126)}$$

**4.21 Calculation of the shear stress distribution in a rectangular cross section**
Starting point is the infinitesimal beam element given in Fig. 4.8. Under the assumption that no distributed load is acting, it follows from Eq. (4.20) that $Q_y(x) \approx Q_y(x+\text{d}x)$. The next step is to replace the internal reactions, i.e. the bending moment and the shear force, by the corresponding normal and shear stresses. To this end, a part is cut off at $y = y'$, see Fig. C.5.

Force equilibrium in the $x$-direction gives:

$$\int_{y'}^{y=+\frac{h}{2}} \sigma_x(x) b\,\text{d}y - \int_{y'}^{y=+\frac{h}{2}} \left(\sigma_x(x) + \frac{\text{d}\sigma_x(x)}{\text{d}x}\text{d}x\right) b\,\text{d}y + \tau_{xy} b\,\text{d}x = 0\,. \qquad \text{(C.127)}$$

It results from Eq. (4.33) after differentiation with respect to the $x$-coordinate:

$$\frac{\text{d}\sigma_x(x)}{\text{d}x} = -\frac{y}{I_z}\frac{\text{d}M_z(x)}{\text{d}x} = \frac{Q_y(x)\times y}{I_z}\,. \qquad \text{(C.128)}$$

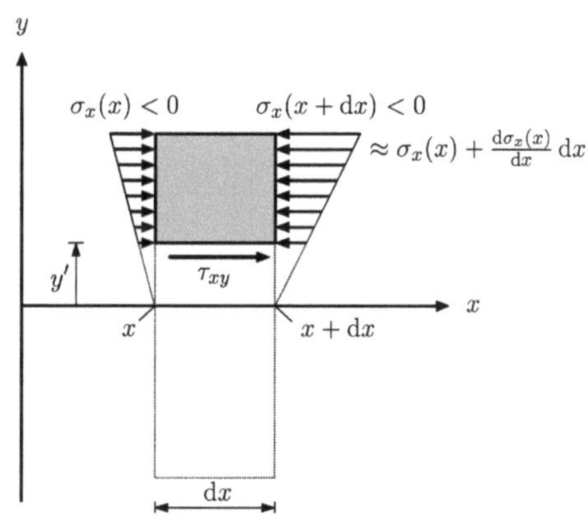


**Fig. C.5** Infinitesimal beam element of size $\text{d}x \times \left(\frac{h}{2} - y'\right)$ in the $x$–$y$ plane with normal and shear stresses

Thus,

$$\tau_{xy} = \frac{Q_y(x)}{I_z} \int\limits_{y'}^{y=+\frac{h}{2}} y\,\mathrm{d}y = \frac{Q_y(x)}{I_z}\left[\frac{1}{2}y^2\right]_{y'}^{\frac{h}{2}} = \frac{Q_y(x)}{2I_z}\left(\left(\frac{h}{2}\right)^2 - (y')^2\right). \quad (C.129)$$

Maximum shear stress for $y' = 0$:

$$\tau_{xy,\max} = \frac{Q_y(x)h^2}{8I_z} = \frac{3Q_y(x)}{2bh} = \frac{3Q_y(x)}{2A}. \quad (C.130)$$

**4.22 Calculation of the shear correction factor for rectangular cross section**

$$\int\limits_{\Omega} \frac{1}{2G}\tau_{xy}^2 \mathrm{d}\Omega \stackrel{!}{=} \int\limits_{\Omega_\mathrm{s}} \frac{1}{2G}\left(\frac{Q_y}{A_\mathrm{s}}\right)^2 \mathrm{d}\Omega_\mathrm{s}, \quad (C.131)$$

$$k_\mathrm{s} = \frac{Q_y}{A\int\limits_A \tau_{xy}^2 \mathrm{d}A} = \frac{5}{6}. \quad (C.132)$$

**4.23 Differential equation under consideration of distributed moment**
Shear force: no difference, meaning $\frac{\mathrm{d}Q_y(x)}{\mathrm{d}x} = -q_y(x)$.

Bending moment:

$$M_z(x+\mathrm{d}x) - M_z(x) + Q_y(x)\mathrm{d}x - \frac{1}{2}q_y\mathrm{d}x^2 + m_z\mathrm{d}x = 0. \quad (C.133)$$

$$\frac{\mathrm{d}M_z(x)}{\mathrm{d}x} = -Q_y(x) - m_z, \quad (C.134)$$

$$\frac{\mathrm{d}^2M_z(x)}{\mathrm{d}x^2} + \frac{\mathrm{d}m_z(x)}{\mathrm{d}x} = q_y(x). \quad (C.135)$$

Differential equations:

$$\frac{\mathrm{d}}{\mathrm{d}x}\left(EI_z\frac{\mathrm{d}\phi_z}{\mathrm{d}x}\right) + k_\mathrm{s}AG\left(\frac{\mathrm{d}u_y}{\mathrm{d}x} - \phi_z\right) = -m_z(x), \quad (C.136)$$

$$\frac{\mathrm{d}}{\mathrm{d}x}\left[k_\mathrm{s}AG\left(\frac{\mathrm{d}u_y}{\mathrm{d}x} - \phi_z\right)\right] = -q_y(x). \quad (C.137)$$

**4.24 Analytical calculation of the distribution of the deflection and rotation for a cantilevered beam under point load**

Boundary conditions:

$$u_y(x=0)=0\ ,\quad \phi_z(x=0)=0\,, \tag{C.138}$$

$$M_z(x=0)=FL\ ,\quad Q_y(x=0)=F\,. \tag{C.139}$$

Integration constants:

$$c_1=-F\ ;\ c_2=FL\ ;\ c_3=\frac{EI_z}{k_sAG}F\ ;\ c_4=0\,. \tag{C.140}$$

Course of the displacement:

$$u_y(x)=\frac{1}{EI_z}\left(-F\frac{x^3}{6}+FL\frac{x^2}{2}+\frac{EI_zF}{k_sAG}x\right)\,. \tag{C.141}$$

Course of the rotation:

$$\phi_z(x)=\frac{1}{EI_z}\left(-F\frac{x^2}{2}+FLx\right)\,. \tag{C.142}$$

Maximal bending:

$$u_y(x=L)=\frac{1}{EI_z}\left(\frac{FL^3}{3}+\frac{EI_zFL}{k_sAG}\right)\,. \tag{C.143}$$

Bending at the loading point:

$$\phi_z(x=L)=\frac{FL^2}{2EI_z}\,. \tag{C.144}$$

Limit value:

$$u_y(x=L)=\frac{4F}{b}\left(\frac{L}{h}\right)^3+\frac{F}{k_sbG}\left(\frac{L}{h}\right)\,. \tag{C.145}$$

$$u_y(L)\big|_{h\ll L}\rightarrow\frac{4F}{b}\left(\frac{L}{h}\right)^3=\frac{FL^3}{3EI_z}\,, \tag{C.146}$$

$$u_y(L)\big|_{h\gg L}\rightarrow\frac{F}{k_sbG}\left(\frac{L}{h}\right)=\frac{FL}{k_sAG}\,. \tag{C.147}$$

**4.25 Analytical calculation of the normalized deflection for beams with shear contribution**

$$I_z = \frac{bh^3}{12}\,,\; A = hb\,,\; k_s = \frac{5}{6}\,, G = \frac{E}{2(1+\nu)}\,. \tag{C.148}$$

$$u_{y,\,\text{norm}} = \frac{1}{3} + \frac{1+\nu}{5}\left(\frac{h}{L}\right)^2\,, \tag{C.149}$$

$$u_{y,\,\text{norm}} = \frac{1}{8} + \frac{1+\nu}{10}\left(\frac{h}{L}\right)^2\,, \tag{C.150}$$

$$u_{y,\,\text{norm}} = \frac{1}{48} + \frac{1+\nu}{5}\left(\frac{h}{L}\right)^2\,. \tag{C.151}$$

**4.26 Cantilevered beam loaded by a single force**
Boundary conditions:

$$u_y(x=L) = 0\,,\quad \phi_z(x=L) = 0\,, \tag{C.152}$$

$$M_z(x=0) = 0\,,\quad Q_y(x=0) = F\,. \tag{C.153}$$

Integration constants:

$$c_1 = -F\,;\; c_2 = 0\,;\; c_3 = \frac{EI_zF}{k_sAG} + \frac{FL^2}{2}\,;\; c_4 = -\frac{EI_zFL}{k_sAG} - \frac{FL^3}{3}\,. \tag{C.154}$$

Displacement distribution:

$$u_y(x) = \frac{FL^3}{6EI_z}\left(-\frac{x^3}{L^3} + \frac{3x}{L} - 2\right) + \frac{FL^3(1+\nu)}{6k_sEI_z}\left(\frac{h^2}{L^2}\right)\left(\frac{x}{L} - 1\right)\,. \tag{C.155}$$

**4.27 Simply supported beam in the elastic range loaded by a distributed load**
Boundary conditions:

$$u_y(x=-L) = 0\,,\quad u_y(x=+L) = 0\,, \tag{C.156}$$

$$M_z(x=-L) = 0\,,\quad M_z(x=+L) = 0\,. \tag{C.157}$$

Integration constants:

$$c_1 = 0\,;\; c_2 = -\frac{EI_zq}{k_sAG} - \frac{qL^2}{2}\,;\; c_3 = 0\,;\; c_4 = \frac{EI_zqL^2}{2k_sAG} + \frac{5qL^4}{24}\,. \tag{C.158}$$

Displacement distribution:

$$u_y(x) = \frac{qL^4}{24EI_z}\left(\frac{x^4}{L^4} - \frac{6x^2}{L^2} + 5\right) + \frac{qL^4(1+\nu)}{12k_sEI_z}\left(\frac{h^2}{L^2}\right)\left(1 - \frac{x^2}{L^2}\right)\,. \tag{C.159}$$

**4.28 Cantilevered beam loaded by a single force**
Boundary conditions:

$$u_y(x=L)=0\ ,\ \ \phi_z(x=L)=0\,, \tag{C.160}$$
$$M_z(x=0)=0\ ,\ \ Q_y(x=0)=F\,. \tag{C.161}$$

Integration constants:

$$c_1=-F\ ;\ c_2=0\ ;\ c_3=\frac{3EI_zF}{2AG}+\frac{FL^2}{2}\ ;\ c_4=-\frac{3EI_zFL}{2AG}-\frac{FL^3}{3}\,. \tag{C.162}$$

Displacement distribution:

$$u_y(x)=\frac{FL^3}{6EI_z}\left(-\frac{x^3}{L^3}+\frac{3x}{L}-2\right)+\frac{FL^3(1+\nu)}{4EI_z}\left(\frac{h^2}{L^2}\right)\left(-1+\frac{x}{L}\right)\,. \tag{C.163}$$

**4.29 Shear stress distribution for a cantilevered beam loaded by a single force**

$$\tau_{xy}=G\left(\frac{\partial u_y}{\partial x}+\frac{\partial u_x}{\partial y}\right)\,. \tag{C.164}$$

With:

$$\frac{\partial u_x}{\partial y}=-\phi_z-\frac{4y^2}{h^2}\left(-\phi_z+\frac{\partial u_y}{\partial x}\right)\,, \tag{C.165}$$

$$\frac{\partial u_y}{\partial x}=\frac{FL^2}{6EI_z}\left(-\frac{3x^2}{L^2}+\frac{6x}{L}\right)+\frac{FL^2(1+\nu)}{4EI_z}\left(\frac{h^2}{L^2}\right)\,. \tag{C.166}$$

$$\tau_{xy}=\frac{F}{2I_z}\left[\left(\frac{h}{2}\right)^2-y^2\right]\,. \tag{C.167}$$

**4.30 Simply supported beam in the elastic range loaded by a distributed load**
Boundary conditions:

$$u_y(x=-L)=0\ ,\ \ u_y(x=+L)=0\,, \tag{C.168}$$
$$M_z(x=-L)=0\ ,\ \ M_z(x=+L)=0\,. \tag{C.169}$$

Integration constants:

$$c_1=0\ ;\ c_2=-\frac{6EI_zq_y}{5AG}-\frac{qL^2}{2}\ ;\ c_3=0\ ;\ c_4=\frac{3EI_zL^2q}{5AG}+\frac{5q_yL^4}{24}\,. \tag{C.170}$$

Displacement distribution:

$$u_y(x) = \frac{q_y L^4}{24EI_z}\left(\frac{x^4}{L^4} - \frac{6x^2}{L^2} + 5\right) + \frac{q_y L^4(1+\nu)}{10EI_z}\left(\frac{h^2}{L^2}\right)\left(1 - \frac{x^2}{L^2}\right). \quad \text{(C.171)}$$

**4.31 Simply supported beam—shear stress distribution in the elasto-plastic range**

$$\frac{\mathrm{d}\left(\frac{\alpha h}{2}\right)}{\mathrm{d}x} = \frac{3qx}{2bk\left(\frac{\alpha h}{2}\right)}, \quad \text{(C.172)}$$

$$\tau_{xy}(x, y) = \frac{3}{4}\frac{qx}{b\left(\frac{\alpha h}{2}\right)}\left[1 - \left(\frac{y}{\left(\frac{\alpha h}{2}\right)}\right)^2\right]. \quad \text{(C.173)}$$

**Problems from Chap. 5**

**5.15 Method of least squares**
The approach might be not the best since positive and negative errors are canceled.

**5.16 Weighted residual method with arbitrary distributed load for a rod**

$$\int_0^L \left(EA\frac{\mathrm{d}^2 u_x(x)}{\mathrm{d}x^2} + p_x(x)\right) W(x)\mathrm{d}x = 0 \quad \text{(C.174)}$$

$$\int_0^L EA\frac{\mathrm{d}u_x}{\mathrm{d}x}\frac{\mathrm{d}W}{\mathrm{d}x}\,\mathrm{d}x = \int_0^L Wp_x(x)\mathrm{d}x + EA\left[W\frac{\mathrm{d}u_x}{\mathrm{d}x}\right]_0^L \quad \text{(C.175)}$$

$$\ldots = \delta \boldsymbol{u}_{\mathrm{p}}^{\mathrm{T}} \int_0^L \boldsymbol{N} p_x(x)\mathrm{d}x + \cdots \quad \text{(C.176)}$$

$$\ldots = \int_0^L \begin{bmatrix} N_1 \\ N_2 \end{bmatrix} q_y(x)\mathrm{d}x + \cdots \quad \text{(C.177)}$$

The additional expression on the right-hand side gives the equivalent nodal loads for a distributed load according to Eq. (5.255).

**5.17 Finite element approximation with a single linear rod element**
The analytical solutions are given in Table 3.3. The finite element solutions are obtained as:

$$\text{(a)}\;\; u_x(x) = \frac{F_0 x}{EA}\,, \qquad u_x(L) = \frac{F_0 L}{EA}\,, \tag{C.178}$$

$$\text{(b)}\;\; u_x(x) = \frac{p_0}{EA}\left(\frac{xL}{2}\right), \qquad u_x(L) = \frac{p_0 L^2}{2EA}\,, \tag{C.179}$$

$$\text{(c)}\;\; u_x(x) = \frac{p_0}{EA}\left(\frac{xL}{3}\right), \qquad u_x(L) = \frac{p_0 L^2}{3EA}\,. \tag{C.180}$$

The finite element approximation is equal to the exact solution only in the case of the single force. In the case of distributed loads, the same solution is only obtained at the nodes but the distribution between the nodes is different.

**5.18 Finite element approximation with a single quadratic rod element**
The analytical solutions are given in Table 3.3. The finite element solutions are obtained as:

$$\text{(a)}\;\; u_{2x} = \frac{F_0 L}{2EA}\,, \qquad u_{3x} = \frac{F_0 L}{EA}\,, \qquad u_x(x) = \frac{F_0 x}{EA}\,, \tag{C.181}$$

$$\text{(b)}\;\; u_{2x} = \frac{3p_0 L^2}{8EA}\,, \qquad u_{3x} = \frac{p_0 L^2}{2EA}\,, \qquad u_x(x) = \frac{p_0}{EA}\left(xL - \frac{x^2}{2}\right), \tag{C.182}$$

$$\text{(c)}\;\; u_{2x} = \frac{11 p_0 L^2}{48EA}\,, \qquad u_{3x} = \frac{p_0 L^2}{3EA}\,, \qquad u_x(x) = \frac{p_0}{EA}\left(\frac{7}{12}xL - \frac{x^2}{4}\right). \tag{C.183}$$

The finite element solution at the nodes is in all cases equal to the analytical solution. However, only in case (a) and (b) both are the same between the nodes.

**5.19 Equivalent nodal loads for a quadratic distribution (linear rod element)**

$$\text{(a)}\;\; F_{1x} = \frac{p_0^* L^3}{12}\,, \quad F_{2x} = \frac{p_0^* L^3}{4}\,, \tag{C.184}$$

$$\text{(b)}\;\; F_{1x} = \frac{p_0 L}{12}\,, \quad F_{2x} = \frac{p_0 L}{4}\,. \tag{C.185}$$

**5.20 Derivation of shape functions for a quadratic rod element**

$$\boldsymbol{N}^{\mathrm{T}} = \boldsymbol{\chi}^{\mathrm{T}}\boldsymbol{A} = \begin{bmatrix} 1 & \xi & \xi^2 \end{bmatrix} \frac{1}{2} \begin{bmatrix} 0 & 1 & 0 \\ -\frac{1}{2} & 0 & \frac{1}{2} \\ \frac{1}{2} & -1 & \frac{1}{2} \end{bmatrix}. \tag{C.186}$$

**5.21 Derivation of shape functions and stiffness matrix for a quadratic rod element with unevenly distributed nodes**

$$N_1(\xi) = \frac{b}{2(b-1)} - \frac{\xi}{2} - \frac{\xi^2}{2(b-1)}\,, \tag{C.187}$$

$$N_2(\xi) = -\frac{1}{b^2-1} + \frac{\xi^2}{b^2-1}\,, \tag{C.188}$$

$$N_3(\xi) = \frac{b}{2(1+b)} + \frac{\xi}{2} + \frac{\xi^2}{2(1+b)}\,. \tag{C.189}$$

$$\boldsymbol{K} = \frac{EA}{L}\begin{bmatrix} \frac{7-6b+3b^2}{3(b-1)^2} & -\frac{8}{3(b^2-1)(b-1)} & -\frac{3b^2+1}{3(b^2-1)} \\ -\frac{8}{3(b^2-1)(b-1)} & \frac{16}{3(b^2-1)^2} & \frac{8}{3(1+b)(b^2-1)} \\ -\frac{3b^2+1}{3(b^2-1)} & \frac{8}{3(1+b)(b^2-1)} & \frac{7+6b+3b^2}{3(1+b)^2} \end{bmatrix}. \tag{C.190}$$

For a value of $b = 0.9$ ($\xi = -0.9$) , the following matrix is obtained:

$$\boldsymbol{K}_{b=0.9} = \frac{EA}{L}\begin{bmatrix} 134.3333333 & -140.3508772 & 6.017543860 \\ -140.3508772 & 147.7377654 & -7.386888275 \\ 6.017543860 & -7.386888275 & 1.369344414 \end{bmatrix}. \tag{C.191}$$

If the inner node is close to the left or right-hand boundary, the stiffness matrix will contain very small and very large values. This can result in numerical problems ($\rightarrow$ inversion of the matrix). As a rule of thumb, the inner node should be at least $\frac{1}{4}$ from the outer nodes.

**5.22 Derivation of shape functions for a cubic rod element**

$$\boldsymbol{N}^{\mathrm{T}} = \boldsymbol{\chi}^{\mathrm{T}}\boldsymbol{A} = \begin{bmatrix}1 & \xi & \xi^2 & \xi^3\end{bmatrix}\frac{1}{2}\begin{bmatrix} -\frac{1}{16} & \frac{9}{16} & \frac{9}{16} & -\frac{1}{16} \\ \frac{1}{16} & -\frac{27}{16} & \frac{27}{16} & -\frac{1}{16} \\ \frac{9}{16} & -\frac{9}{16} & -\frac{9}{16} & \frac{9}{16} \\ -\frac{9}{16} & \frac{27}{16} & -\frac{27}{16} & \frac{9}{16} \end{bmatrix}. \tag{C.192}$$

**5.23 Structure composed of three linear rod elements**
Displacement vector:

$$\text{(a)}\ \boldsymbol{u} = \frac{LF_0}{EA}\begin{bmatrix} 0 & \frac{1}{3} & \frac{5}{6} & \frac{11}{6}\end{bmatrix}^{\mathrm{T}}, \tag{C.193}$$

$$\text{(b)}\ \boldsymbol{u} = u_0\begin{bmatrix} 0 & \frac{2}{11} & \frac{5}{11} & 1\end{bmatrix}^{\mathrm{T}}. \tag{C.194}$$

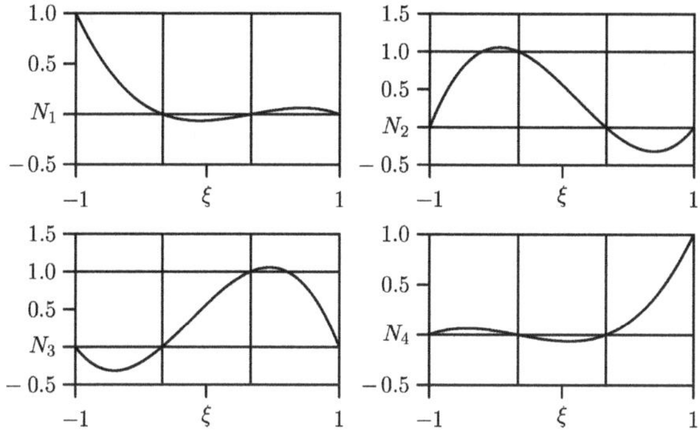


**Fig. C.6** Shape functions for a cubic rod element with equidistant nodes plotted over the natural coordinate ($\xi$)

Reaction forces:

$$\text{(a)}\ F_{\text{R1}} = -F_0\,, \tag{C.195}$$

$$\text{(b)}\ F_{\text{R1}} = -\frac{6EAu_0}{11L}\,,\quad F_{\text{R4}} = +\frac{6EAu_0}{11L}\,. \tag{C.196}$$

See Fig. C.6.

**5.24 Finite element approximation of a rod with four elements: comparison of displacement, strain and stress distribution with analytical solution**
Analytical solution:

$$EAu(X) = -\frac{p_0X^2}{2} + c_1X + c_2\,. \tag{C.197}$$

Boundary conditions: $u(0) = 0$ and $N_X(L) = 0$.

$$u(X) = \frac{p_0L^2}{EA}\left(\frac{X}{L} - \frac{1}{2}\left(\frac{X}{L}\right)^2\right), \tag{C.198}$$

$$\varepsilon(X) = \frac{p_0L}{EA}\left(1 - \frac{X}{L}\right), \tag{C.199}$$

$$\sigma(X) = \frac{p_0L}{A}\left(1 - \frac{X}{L}\right). \tag{C.200}$$

Finite element solution:

$$u_2 = \frac{7p_0L^2}{32EA}\,,\quad u_3 = \frac{3p_0L^2}{8EA}\,,\quad u_4 = \frac{15p_0L^2}{32EA}\,,\quad u_5 = \frac{p_0L^2}{2EA}\,, \tag{C.201}$$

$$\varepsilon_{\text{I}} = \frac{7p_0L}{8EA}\,,\ \varepsilon_{\text{II}} = \frac{5p_0L}{8EA}\,,\ \varepsilon_{\text{III}} = \frac{3p_0L}{8EA}\,,\ \varepsilon_{\text{IV}} = \frac{p_0L}{8EA}\,, \tag{C.202}$$

$$\varepsilon_1 = \varepsilon_{\text{I}}\,,\ \varepsilon_2 = \frac{\varepsilon_{\text{I}}+\varepsilon_{\text{II}}}{2}\,,\ \varepsilon_3 = \frac{\varepsilon_{\text{II}}+\varepsilon_{\text{III}}}{2}\,,\ \varepsilon_4 = \frac{\varepsilon_{\text{III}}+\varepsilon_{\text{IV}}}{2}\,,\ \varepsilon_5 = \varepsilon_{\text{IV}}\,, \tag{C.203}$$

$$\sigma_{\text{I}} = \frac{7p_0L}{8A}\,,\ \sigma_{\text{II}} = \frac{5p_0L}{8A}\,,\ \sigma_{\text{III}} = \frac{3p_0L}{8A}\,,\ \sigma_{\text{IV}} = \frac{p_0L}{8A}\,, \tag{C.204}$$

$$\sigma_1 = \sigma_{\text{I}}\,,\ \sigma_2 = \frac{\sigma_{\text{I}}+\sigma_{\text{II}}}{2}\,,\ \sigma_3 = \frac{\sigma_{\text{II}}+\sigma_{\text{III}}}{2}\,,\ \sigma_4 = \frac{\sigma_{\text{III}}+\sigma_{\text{IV}}}{2}\,,\ \sigma_5 = \sigma_{\text{IV}}\,. \tag{C.205}$$

See Fig. C.7.

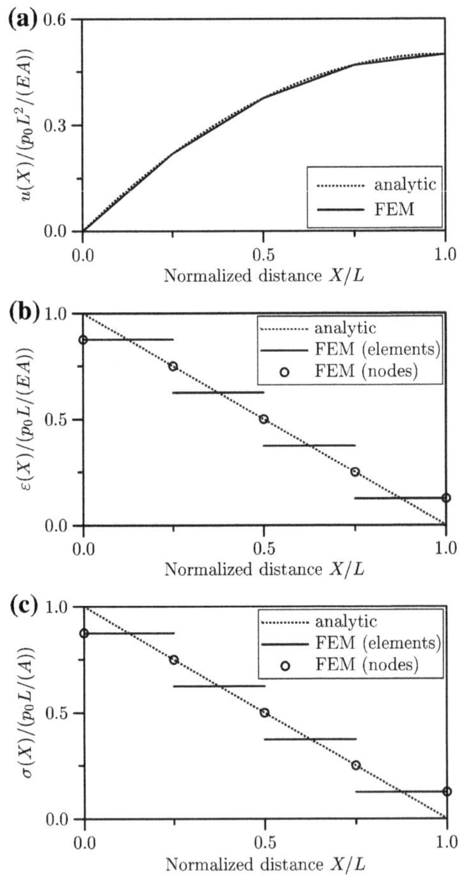


**Fig. C.7** Rod element discretized by four elements: **a** displacement; **b** strain and **c** stress

**5.25 Elongation of a bi-material rod: finite element solution and comparison with analytical solution**
The analytical solution is discussed in supplementary problem 3.8.
Finite element solution:

$$u_1 = 0\,, \tag{C.206}$$

$$u_2 = \frac{3k_\mathrm{I}p_0L^2 + k_\mathrm{II}p_0L^2 + 4k_\mathrm{I}k_\mathrm{II}u_0}{8k_\mathrm{I}(k_\mathrm{I}+k_\mathrm{II})}\,, \tag{C.207}$$

$$u_3 = \frac{p_0L^2 + 2k_\mathrm{II}u_0}{2(k_\mathrm{I}+k_\mathrm{II})}\,, \tag{C.208}$$

$$u_4 = \frac{p_0L^2 + 2k_\mathrm{I}u_0 + 4k_\mathrm{II}u_0}{4(k_\mathrm{I}+k_\mathrm{II})}\,, \tag{C.209}$$

$$u_5 = u_0\,. \tag{C.210}$$

$$\varepsilon_\mathrm{I} = \frac{3k_\mathrm{I}p_0L^2 + k_\mathrm{II}p_0L^2 + 4k_\mathrm{I}k_\mathrm{II}u_0}{4Lk_\mathrm{I}(k_\mathrm{I}+k_\mathrm{II})}\,, \tag{C.211}$$

$$\varepsilon_\mathrm{II} = \frac{k_\mathrm{I}p_0L^2 - k_\mathrm{II}p_0L^2 + 4k_\mathrm{I}k_\mathrm{II}u_0}{4Lk_\mathrm{I}(k_\mathrm{I}+k_\mathrm{II})}\,, \tag{C.212}$$

$$\varepsilon_\mathrm{III} = \frac{-p_0L^2 + 2k_\mathrm{I}u_0}{2L(k_\mathrm{I}+k_\mathrm{II})}\,, \tag{C.213}$$

$$\varepsilon_\mathrm{IV} = \frac{-p_0L^2 + 2k_\mathrm{I}u_0}{2L(k_\mathrm{I}+k_\mathrm{II})}\,. \tag{C.214}$$

$$\sigma_\mathrm{I} = \frac{E_\mathrm{I}(3k_\mathrm{I}p_0L^2 + k_\mathrm{II}p_0L^2 + 4k_\mathrm{I}k_\mathrm{II}u_0)}{4Lk_\mathrm{I}(k_\mathrm{I}+k_\mathrm{II})}\,, \tag{C.215}$$

$$\sigma_\mathrm{II} = \frac{E_\mathrm{I}(k_\mathrm{I}p_0L^2 - k_\mathrm{II}p_0L^2 + 4k_\mathrm{I}k_\mathrm{II}u_0)}{4Lk_\mathrm{I}(k_\mathrm{I}+k_\mathrm{II})}\,, \tag{C.216}$$

$$\sigma_\mathrm{III} = \frac{E_\mathrm{II}(-p_0L^2 + 2k_\mathrm{I}u_0)}{2L(k_\mathrm{I}+k_\mathrm{II})}\,, \tag{C.217}$$

$$\sigma_\mathrm{IV} = \frac{E_\mathrm{II}(-p_0L^2 + 2k_\mathrm{I}u_0)}{2L(k_\mathrm{I}+k_\mathrm{II})}\,. \tag{C.218}$$

The values of strain and stress at inner nodes are obtained by averaging the elemental values. The distribution of stress, strain and displacement is shown in Fig. C.8.

**5.26 Stress distribution for a fixed-fixed rod structure**
Solution vector:

$$\boldsymbol{u} = \frac{LF}{EA}\begin{bmatrix} \frac{1}{6} & \frac{1}{3} & \frac{1}{2} & \frac{1}{3} & \frac{1}{6} \end{bmatrix}^\mathrm{T}\,. \tag{C.219}$$

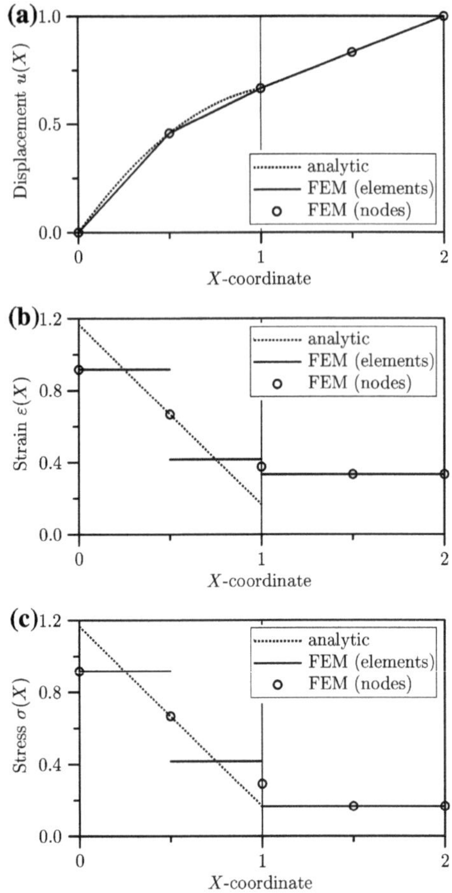


**Fig. C.8** Bi-material rod discretized by four elements: **a** displacement; **b** strain and **c** stress

The difference between the elemental stress values and the averaged nodal values is shown in Fig. C.9.

**5.27 Linear rod element with variable cross section: derivation of stiffness matrix**

(a) $d = d(x)$ linear

$$\boldsymbol{K}^{\mathrm{e}} = \frac{E}{L} \times \frac{\pi(d_1^2 + d_1 d_2 + d_2^2)}{12} \begin{bmatrix} 1 & -1 \\ -1 & 1 \end{bmatrix} \tag{C.220}$$

or by replacing the diameters with the cross-section areas, i.e. $d_i = 2\sqrt{\frac{A_i}{\pi}}$,

$$\boldsymbol{K}^{\mathrm{e}} = \frac{E}{L} \times \frac{\pi(A_1 + A_2 + \sqrt{A_1 A_2})}{3} \begin{bmatrix} 1 & -1 \\ -1 & 1 \end{bmatrix}. \tag{C.221}$$

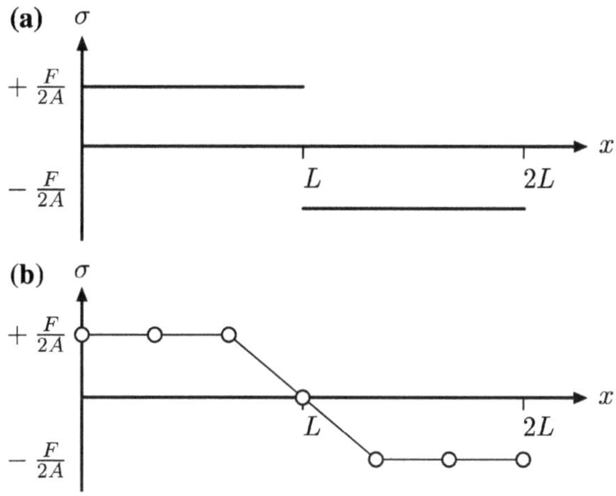


**Fig. C.9** **a** Stress distribution based on elemental values and **b** based on nodal averaging

(b) $A = A(x)$ linear

$$\boldsymbol{K}^{\mathrm{e}} = \frac{E}{L} \times \frac{A_1 + A_2}{2} \begin{bmatrix} 1 & -1 \\ -1 & 1 \end{bmatrix}. \tag{C.222}$$

Two-point GAUSS integration gives for both cases the same result as the analytical integration!

### 5.28 Quadratic rod element with variable cross section: derivation of stiffness matrix

(a) $d = d(x)$ linear

$$\boldsymbol{K}^{\mathrm{e}} = \frac{\pi E}{60L} \times \begin{bmatrix} 23d_1^2 + 9d_1 d_2 + 3d_2^2 & -2(13d_1^2 + 4d_1 d_2 + 3d_2^2) & 3d_1^2 - d_1 d_2 + 3d_2^2 \\ -2(13d_1^2 + 4d_1 d_2 + 3d_2^2) & 8(4d_1^2 + 2d_1 d_2 + 4d_2^2) & -2(3d_1^2 + 4d_1 d_2 + 13d_2^2) \\ 3d_1^2 - d_1 d_2 + 3d_2^2 & -2(3d_1^2 + 4d_1 d_2 + 13d_2^2) & 3d_1^2 + 9d_1 d_2 + 23d_2^2 \end{bmatrix}. \tag{C.223}$$

(b) $A = A(x)$ linear

$$\boldsymbol{K}^{\mathrm{e}} = \frac{E}{3L} \begin{bmatrix} \frac{11A_1+3A_2}{2} & -2(3A_1 + A_2) & \frac{A_1+A_2}{2} \\ -2(3A_1 + A_2) & 8(A_1 + A_2) & -2(A_1 + 3A_2) \\ \frac{A_1+A_2}{2} & -2(A_1 + 3A_2) & \frac{3A_1+11A_2}{2} \end{bmatrix}. \tag{C.224}$$

**5.29 Linear rod element with variable cross section: comparison of displacements between FE and analytical solution for a single element**
(a) Force boundary condition

- Finite element solution:

$$u_2 = \frac{2LF_0}{E(A_1 + A_2)}. \tag{C.225}$$

For the case $A_2 = 2A_1 = 2A$ we obtain:

$$u_2 = \frac{2LF_0}{E(A + 2A)} = \frac{2}{3}\frac{LF_0}{EA} = 0.\bar{6}\,\frac{LF_0}{EA}. \tag{C.226}$$

For the case $A_2 = A_1 = A$ we obtain:

$$u_2 = \frac{LF_0}{EA}. \tag{C.227}$$

- Analytical solution:

$$u(x) = \frac{F_0 L}{E(A_2 - A_1)} \times \ln\left(1 + \frac{x}{L}\left(\frac{A_2}{A_1} - 1\right)\right), \tag{C.228}$$

or at the right-hand end:

$$u(L) = \frac{F_0 L}{E(A_2 - A_1)} \ln\left(\frac{A_2}{A_1}\right). \tag{C.229}$$

For the case $A_2 = 2A_1 = 2A$ we obtain:

$$u(L) = \frac{F_0 L}{EA} \ln(2) \approx 0.693\,\frac{LF_0}{EA}. \tag{C.230}$$

For the case $A_2 = A_1 = A$ we obtain under consideration of $\lim \left.\frac{\ln x}{x-1}\right|_{x \to 1} = 1$ where $x = \frac{A_2}{A_1}$:

$$u(L) = \frac{LF_0}{EA}. \tag{C.231}$$

The finite element and the analytical solution is different which comes from the fact that $EA$ is not constant!

(b) Displacement boundary condition

- Finite element solution:

$$u(x) = \frac{x}{L} \times u_0 \,, \tag{C.232}$$

$$u(L) = u_0 \,. \tag{C.233}$$

The finite element solution is independent of the cross section ratio.

- Analytical solution:

$$u(x) = \frac{u_0}{\ln\left(\frac{A_2}{A_1}\right)} \times \ln\left(1 + \frac{x}{L}\left(\frac{A_2}{A_1} - 1\right)\right) \tag{C.234}$$

or at the right-hand end:

$$u(L) = u_0 \,. \tag{C.235}$$

**5.30 Quadratic rod element with variable cross section: comparison of end displacement between FE and analytical solution for single element**
Finite element solution:

$$u_{3x} = \frac{A_1 + A_2}{A_1^2 + 4A_1A_2 + A_2^2} \times \frac{3LF_0}{E} \,. \tag{C.236}$$

For the case $A_2 = 2A_1 = 2A$ we obtain:

$$u_{3x} = \frac{9}{13} \times \frac{LF_0}{EA} \approx 0.6923 \times \frac{LF_0}{EA} \,. \tag{C.237}$$

**5.31 Subdivided structure with variable cross section: comparison of displacements and stresses between FE and analytical solution for four elements**
- Finite element solution:

$$\boldsymbol{u} = \frac{F_0L}{EA_1} \times \left[0 \; \tfrac{5}{18} \; \tfrac{40}{63} \; \tfrac{143}{126} \; \tfrac{124}{63}\right]^{\mathrm{T}} . \tag{C.238}$$

The following vector states the constant stresses in each element and should not be confused with the averaged nodal values:

$$\boldsymbol{\sigma} = \frac{F_0}{A_1} \times \left[\tfrac{10}{9} \; \tfrac{10}{7} \; \tfrac{2}{1} \; \tfrac{10}{3}\right]^{\mathrm{T}} . \tag{C.239}$$

• Analytical solutions:

$$\frac{u(X)}{\frac{F_0 L}{EA_1}} = \frac{1}{\left(\frac{A_5}{A_1} - 1\right)} \times \ln\left(1 + \frac{X}{L}\left(\frac{A_5}{A_1} - 1\right)\right). \tag{C.240}$$

$$\boldsymbol{u} = \frac{F_0 L}{EA_1} \times \left[0 \;\; \tfrac{5}{4} \times \ln\left(\tfrac{4}{5}\right) \;\; \tfrac{5}{4} \times \ln\left(\tfrac{3}{5}\right) \;\; \tfrac{5}{4} \times \ln\left(\tfrac{2}{5}\right) \;\; \tfrac{5}{4} \times \ln(5)\right]^{\mathrm{T}}. \tag{C.241}$$

$$\frac{\varepsilon(X)}{\frac{F_0}{EA_1}} = \frac{1}{1 + \frac{X}{L}\left(\frac{A_5}{A_1} - 1\right)}. \tag{C.242}$$

$$\frac{\sigma(X)}{\frac{F_0}{A_1}} = \frac{1}{1 + \frac{X}{L}\left(\frac{A_5}{A_1} - 1\right)}. \tag{C.243}$$

A comparison between the finite element and the analytical solution is given in Fig. C.10.

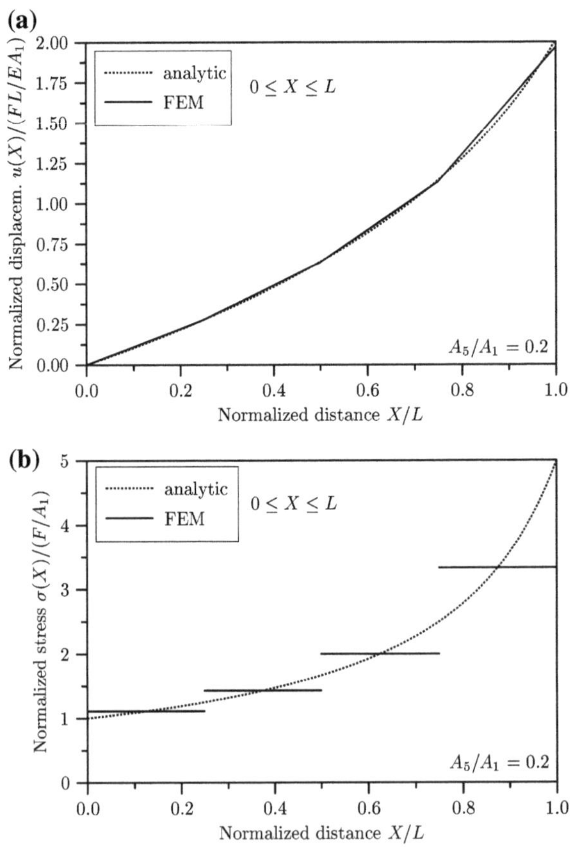


**Fig. C.10** Comparison between FE and analytical solution: **a** displacements; **b** stresses

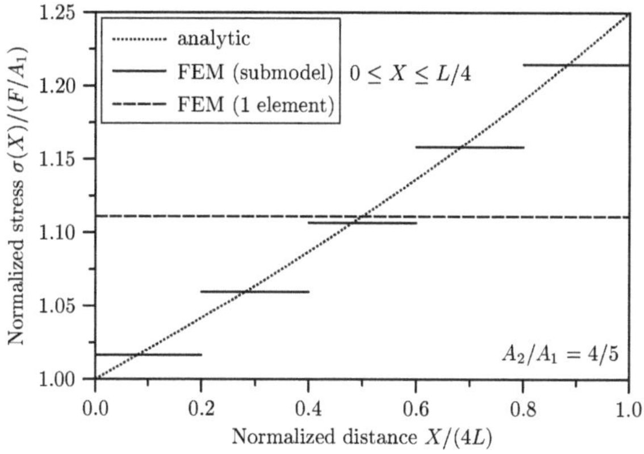


**Fig. C.11** Stress distribution based on the submodel and comparison with the analytical solution and the result from the coarse mesh

### 5.32 Submodel of a structure with variable cross section

The following vector states the constant stresses in each element and should not be confused with the averaged nodal values:

$$\sigma = \frac{F_0}{A_1} \times \left[1.016366\ 1.059615\ 1.106709\ 1.158184\ 1.214681\right]^{\mathrm{T}} . \tag{C.244}$$

The stress distribution based on the submodel and a comparison with the analytical solution is presented in Fig. C.11.

### 5.34 Rod with elastic embedding: stiffness matrix

$$\int_0^L W(x) \left( EA \frac{\mathrm{d}^2 u_x}{\mathrm{d}x^2} - k u_x \right) \mathrm{d}x = 0 , \tag{C.245}$$

$$\boldsymbol{K}^{\mathrm{e}} = \cdots + k \int_0^L \boldsymbol{N}(x) \boldsymbol{N}(x)^{\mathrm{T}} \mathrm{d}x . \tag{C.246}$$

(a) Linear shape functions:

$$\boldsymbol{K}^{\mathrm{e}} = \frac{EA}{L} \begin{bmatrix} 1 & -1 \\ -1 & 1 \end{bmatrix} + \frac{kL}{6} \begin{bmatrix} 2 & 1 \\ 1 & 2 \end{bmatrix} . \tag{C.247}$$

(b) Quadratic shape functions:

$$\boldsymbol{K}^{\mathrm{e}} = \frac{EA}{3L}\begin{bmatrix} 7 & -8 & 1 \\ -8 & 16 & -8 \\ 1 & -8 & 7 \end{bmatrix} + \frac{kL}{30}\begin{bmatrix} 4 & 2 & -1 \\ 2 & 16 & 2 \\ -1 & 2 & 4 \end{bmatrix}. \tag{C.248}$$

**5.34 Rod with elastic embedding: single force case**
(a) Linear shape functions:

$$\left(\frac{EA}{L} + \frac{kL}{3}\right) u_2 = F\,, \tag{C.249}$$

$$u_2 = \frac{F}{\frac{EA}{L} + \frac{kL}{3}}\,, \tag{C.250}$$

$$u_2|_{k\to 0} = \frac{FL}{EA}\,, \quad u_2|_{EA\to 0} = \frac{3F}{kL}\,. \tag{C.251}$$

Comparison between FE and analytical solution:

$$u_x(L)_{\mathrm{FE}} = 0.50F\,, \tag{C.252}$$
$$u_x(L)_{\mathrm{analyt}} = 0.542304F\,. \tag{C.253}$$

(b) Quadratic shape functions:

$$\begin{bmatrix} \frac{16EA}{3L} + \frac{16kL}{30} & -\frac{8EA}{3L} + \frac{2kL}{30} \\ -\frac{8EA}{3L} + \frac{2kL}{30} & \frac{7EA}{3L} + \frac{4kL}{30} \end{bmatrix} \begin{bmatrix} u_2 \\ u_3 \end{bmatrix} = \begin{bmatrix} 0 \\ F \end{bmatrix}, \tag{C.254}$$

$$u_2 = \frac{24\left(10EA + kL^2\right)LF}{240E^2A^2 + 104EAkL^2 + 3k^2L^4}\,, \tag{C.255}$$

$$u_2|_{k\to 0} = \frac{FL}{EA}\,, \quad u_2|_{EA\to 0} = \frac{8F}{kL}\,. \tag{C.256}$$

Comparison between FE and analytical solution:

$$u_x(L)_{\mathrm{FE}} = 0.538860F\,, \tag{C.257}$$
$$u_x(L)_{\mathrm{analyt}} = 0.542304F\,. \tag{C.258}$$

**5.35 Weighted residual method with arbitrary distributed load for a beam**

$$\int_0^L \left(EI_z \frac{\mathrm{d}^4 u_y(x)}{\mathrm{d}x^4} - q_y(x)\right) W(x)\mathrm{d}x = 0 \tag{C.259}$$

$$\int_0^L EI_z \frac{\mathrm{d}^2 u_y}{\mathrm{d}x^2}\frac{\mathrm{d}^2 W}{\mathrm{d}x^2}\,\mathrm{d}x = \int_0^L W q_y(x)\mathrm{d}x + \left[ -W\frac{\mathrm{d}^3 u_y}{\mathrm{d}x^3} + \frac{\mathrm{d}W}{\mathrm{d}x}\frac{\mathrm{d}^2 u_y}{\mathrm{d}x^2} \right]_0^L \tag{C.260}$$

$$\cdots = \delta \boldsymbol{u}_{\mathrm{p}}^{\mathrm{T}} \int_0^L \boldsymbol{N}^{\mathrm{T}} q_y(x)\mathrm{d}x + \cdots \tag{C.261}$$

$$\cdots = \int_0^L \begin{bmatrix} N_{1u} \\ N_{1\varphi} \\ N_{2u} \\ N_{2\varphi} \end{bmatrix} q_y(x)\mathrm{d}x + \cdots \tag{C.262}$$

The additional expression on the right-hand side results in the equivalent nodal loads for a distributed load according to Eq. (5.255).

**5.36 Stiffness matrix for bending in the $x$–$z$ plane**

For bending in the $x$–$z$ plane it must be considered that the rotation is defined by $\varphi_y(x) = -\frac{\mathrm{d}u_z(x)}{\mathrm{d}x}$. Thus, the following shape functions can be derived:

$$N_{1u}^{xz} = 1 - 3\left(\frac{x}{L}\right)^2 + 2\left(\frac{x}{L}\right)^3, \tag{C.263}$$

$$N_{1\varphi}^{xz} = -x + 2\frac{x^2}{L} - \frac{x^3}{L^2}, \tag{C.264}$$

$$N_{2u}^{xz} = 3\left(\frac{x}{L}\right)^2 - 2\left(\frac{x}{L}\right)^3, \tag{C.265}$$

$$N_{2\varphi}^{xz} = \frac{x^2}{L} - \frac{x^3}{L^2}. \tag{C.266}$$

A comparison with the shape functions for bending in the $x$–$y$ plane according to Eqs. (5.222) till (5.225) yields that the shape functions for the rotations are multiplied by $(-1)$.

**5.37 Beam with variable cross-sectional area**

The second moments of area are obtained as:

$$I_z(x) = \frac{\pi}{64}\left(d_1 + (d_2 - d_1)\frac{x}{L}\right)^4 \text{ (circle)}, \tag{C.267}$$

$$I_z(x) = \frac{b}{12}\left(d_1 + (d_2 - d_1)\frac{x}{L}\right)^3 \text{ (rectangle)}. \tag{C.268}$$

For the circular and rectangular cross section one obtains:

$$
\boldsymbol{k}^{\mathrm{e}} = \frac{\pi E}{64L^3}\begin{bmatrix}
\frac{12(11d_2^4+11d_1^4+5d_2^3d_1+3d_2^2d_1^2+5d_2\,d_1^3)}{35} & \frac{2(19d_2^4+47\,d_1^4+8d_2^3d_1+9d_2^2d_1^2+22d_2\,d_1^3)L}{35} & -\frac{12(11d_2^4+11d_1^4+5d_2^3d_1+3d_2^2d_1^2+5d_2\,d_1^3)}{35} & \frac{2(47d_2^4+19d_1^4+22d_2^3d_1+9d_2^2d_1^2+8d_2\,d_1^3)L}{35} \\
\frac{2(19\,d_2^4+47\,d_1^4+8\,d_2^3d_1+9\,d_2^2d_1^2+22\,d_2\,d_1^3)L}{35} & \frac{4(3\,d_2^4+17\,d_1^4+2\,d_2^3d_1+4\,d_2^2d_1^2+9\,d_2\,d_1^3)L^2}{35} & -\frac{2(19\,d_2^4+47\,d_1^4+8\,d_2^3d_1+9\,d_2^2d_1^2+22\,d_2\,d_1^3)L}{35} & \frac{2(13\,d_2^4+13\,d_1^4+4\,d_2^3d_1+d_2^2d_1^2+4\,d_2\,d_1^3)L^2}{35} \\
-\frac{12(11\,d_2^4+11\,d_1^4+5\,d_2^3d_1+3\,d_2^2d_1^2+5\,d_2\,d_1^3)}{35} & -\frac{2(19\,d_2^4+47\,d_1^4+8\,d_2^3d_1+9\,d_2^2d_1^2+22\,d_2\,d_1^3)L}{35} & \frac{12(11\,d_2^4+11\,d_1^4+5\,d_2^3d_1+3\,d_2^2d_1^2+5\,d_2\,d_1^3)}{35} & -\frac{2(47\,d_2^4+19\,d_1^4+22\,d_2^3d_1+9\,d_2^2d_1^2+8\,d_2\,d_1^3)L}{L} \\
\frac{2(47\,d_2^4+19\,d_1^4+22\,d_2^3d_1+9\,d_2^2d_1^2+8\,d_2\,d_1^3)L}{35} & \frac{2(13\,d_2^4+13\,d_1^4+4\,d_2^3d_1+d_2^2d_1^2+4\,d_2\,d_1^3)L^2}{35} & -\frac{2(47\,d_2^4+19\,d_1^4+22\,d_2^3d_1+9\,d_2^2d_1^2+8\,d_2\,d_1^3)L}{35} & \frac{4(17\,d_2^4+3\,d_1^4+9\,d_2^3d_1+4\,d_2^2d_1^2+2\,d_2\,d_1^3)L^2}{35}
\end{bmatrix} \tag{C.269}
$$

$$
\boldsymbol{k}^{\mathrm{e}} = \frac{bE}{12L^3}\begin{bmatrix}
\frac{3(7\,d_2^3+3\,d_2^2d_1+3\,d_2\,d_1^2+7\,d_1^3)}{5} & \frac{3(2\,d_2^3+d_2^2d_1+2\,d_2\,d_1^2+5\,d_1^3)L}{5} & -\frac{3(7\,d_2^3+3\,d_2^2d_1+3\,d_2\,d_1^2+7\,d_1^3)}{5} & \frac{3(5\,d_2^3+2\,d_2^2d_1+d_2\,d_1^2+2\,d_1^3)L}{5} \\
\frac{3(2\,d_2^3+d_2^2d_1+2\,d_2\,d_1^2+5\,d_1^3)L}{5} & \frac{(2\,d_2^3+2\,d_2^2d_1+5\,d_2\,d_1^2+11\,d_1^3)L^2}{5} & -\frac{3(2\,d_2^3+d_2^2d_1+2\,d_2\,d_1^2+5\,d_1^3)L}{5} & \frac{(4\,d_2^3+d_2^2d_1+d_2\,d_1^2+4\,d_1^3)L^2}{5} \\
-\frac{3(7\,d_2^3+3\,d_2^2d_1+3\,d_2\,d_1^2+7\,d_1^3)}{5} & -\frac{3(2\,d_2^3+d_2^2d_1+2\,d_2\,d_1^2+5\,d_1^3)L}{5} & \frac{3(7\,d_2^3+3\,d_2^2d_1+3\,d_2\,d_1^2+7\,d_1^3)}{5} & -\frac{3(5\,d_2^3+2\,d_2^2d_1+d_2\,d_1^2+2\,d_1^3)L}{5} \\
\frac{3(5\,d_2^3+2\,d_2^2d_1+d_2\,d_1^2+2\,d_1^3)L}{5} & \frac{(4\,d_2^3+d_2^2d_1+d_2\,d_1^2+4\,d_1^3)L^2}{5} & -\frac{3(5\,d_2^3+2\,d_2^2d_1+d_2\,d_1^2+2\,d_1^3)L}{5} & \frac{(11\,d_2^3+5\,d_2^2d_1+2\,d_2\,d_1^2+2\,d_1^3)L^2}{5}
\end{bmatrix} \tag{C.270}
$$

## 5.38 Equivalent nodal loads for quadratic distributed load

| $q(x) = q_0x^2$ | $q(x) = q_0\left(\frac{x}{L}\right)^2$ |
|---|---|
| $F_{1y} = -\frac{q_0L^3}{15}$ | $F_{1y} = -\frac{q_0L}{15}$ |
| $M_{1z} = -\frac{q_0L^4}{60}$ | $M_{1z} = -\frac{q_0L^2}{60}$ |
| $F_{2y} = -\frac{4q_0L^3}{15}$ | $F_{2y} = -\frac{4q_0L}{15}$ |
| $M_{1z} = \frac{q_0L^4}{30}$ | $M_{1z} = \frac{q_0L^2}{30}$ |

**5.39 Beam with variable cross section loaded by a single force**
Analytical solution:

$$EI_z(x)\frac{\mathrm{d}^2u_y(x)}{\mathrm{d}x^2} = M_z(x)\,, \tag{C.271}$$

$$\frac{E\pi h^4}{64}\left(2-\frac{x}{L}\right)^4\frac{\mathrm{d}^2u_y(x)}{\mathrm{d}x^2} = -F(L-x)\,. \tag{C.272}$$

$$u_y(x) = \frac{FL}{E\pi h^4}\left(\frac{64L^3}{2(-2L+x)}+\frac{64L^4}{6(-2L+x)}\right)+\frac{16L}{3}x+\frac{40L^2}{3}\,. \tag{C.273}$$

$$u_y(L) = -\frac{8}{3}\frac{FL^3}{E\pi h^4} \approx -2.666667\frac{FL^3}{E\pi h^4}\,. \tag{C.274}$$

Finite element solution:

$$u_y(L) = -\frac{7360}{2817}\frac{FL^3}{E\pi h^4} \approx -2.612709\frac{FL^3}{E\pi h^4}\,. \tag{C.275}$$

**5.40 Beam on elastic foundation: stiffness matrix**

$$\int_0^L W(x)\left(EI_z\frac{\mathrm{d}^4u_y}{\mathrm{d}x^4}+ku_y\right)\mathrm{d}x = 0\,, \tag{C.276}$$

$$\boldsymbol{K}^{\mathrm{e}} = \dots + k\int_0^L \boldsymbol{N}(x)\boldsymbol{N}(x)^{\mathrm{T}}\mathrm{d}x\,, \tag{C.277}$$

$$\boldsymbol{K}^{\mathrm{e}} = \frac{EI_z}{L^3}\begin{bmatrix} 12 & 6L & -12 & 6L \\ 6L & 4L^2 & -6L & 2L^2 \\ -12 & -6L & 12 & -6L \\ 6L & 2L^2 & -6L & 4L^2 \end{bmatrix} + \frac{kL}{420}\begin{bmatrix} 156 & 22L & 54 & -13L \\ 22L & 4L^2 & 13L & -3L^2 \\ 54 & 13L & 156 & -22L \\ -13L & -3L^2 & -22L & 4L^2 \end{bmatrix}\,. \tag{C.278}$$

**5.41 Beam on elastic foundation: single force case**

The reduced system of equations:

$$\begin{bmatrix} 12\dfrac{EI_z}{L^3}+\dfrac{156}{420}kL & -6\dfrac{EI_zL}{L^3}-\dfrac{22}{420}kL^2 \\ -6\dfrac{EI_zL}{L^3}-\dfrac{22}{420}kL^2 & 4\dfrac{EI_zL^2}{L^3}+\dfrac{4}{420}kL^3 \end{bmatrix}\begin{bmatrix} u_{2y} \\ \varphi_{2z} \end{bmatrix}=\begin{bmatrix} -F \\ 0 \end{bmatrix}. \tag{C.279}$$

Deflection and rotation of the beam at $x = L$:

$$u_{2y}=-\frac{12\left(420EI_z+kL^4\right)L^3F}{15120E^2{I_z}^2+1224EI_zkL^4+k^2L^8}, \tag{C.280}$$

$$\varphi_{2z}=-\frac{6\left(1260EI_z+11\,kL^4\right)L^2F}{15120E^2{I_z}^2+1224EI_zkL^4+k^2L^8}. \tag{C.281}$$

Special case 1, $k = 0$:

$$u_{2y}|_{k=0}=-\frac{FL^3}{3EI_z},\quad \varphi_{2z}|_{k=0}=-\frac{FL^2}{2EI_z}. \tag{C.282}$$

Special case 2, $EI_z = 0$:

$$u_{2y}|_{EI_z=0}=-\frac{12F}{kL},\quad \varphi_{2z}|_{EI_z=0}=-\frac{66F}{kL^2}. \tag{C.283}$$

Comparison between FE and analytical solution:

$$u_y(L)_{\text{FE}}=-0.253994\times F, \tag{C.284}$$

$$u_y(L)_{\text{analyt}}=-0.254166\times F. \tag{C.285}$$

**5.42 Beam on nonlinear elastic foundation: stiffness matrix**

$$k(u_y)=k_0-\frac{u_y}{u_1}(k_0-k_1)=k_0\Bigg(1-u_y\times\underbrace{\frac{1-k_1/k_0}{u_1}}_{\alpha_{01}}\Bigg), \tag{C.286}$$

$$\boldsymbol{K}^{\text{e}}=\cdots+k_0\int\limits_0^L\left(1-u_y(x)\times\alpha_{01}\right)\boldsymbol{N}(x)\boldsymbol{N}(x)^{\text{T}}\text{d}x, \tag{C.287}$$

where $u_y(x)$ can be taken from Table 5.10.

$$
\boldsymbol{K}^{\mathrm{e}} = \frac{EI_z}{L^3}\begin{bmatrix} 12 & 6L & -12 & 6L \\ 6L & 4L^2 & -6L & 2L^2 \\ -12 & -6L & 12 & -6L \\ 6L & 2L^2 & -6L & 4L^2 \end{bmatrix} + \frac{k_0 L}{420}\begin{bmatrix} 156 & 22L & 54 & -13L \\ 22L & 4L^2 & 13L & -3L^2 \\ 54 & 13L & 156 & -22L \\ -13L & -3L^2 & -22L & 4L^2 \end{bmatrix}
$$

$$
- \frac{k_0 L \times \alpha_{01}}{2520}\left[\begin{matrix} -43\varphi_{2z}L + 162u_{2y} + 774u_{1y} + 97\varphi_{1z}L & L\left(-9\varphi_{2z}L + 35u_{2y} + 97u_{1y} + 16\varphi_{1z}L\right) \\ L\left(-9\varphi_2 zL + 35u_{2y} + 97u_{1y} + 16\varphi_{1z}L\right) & L^2\left(-2\varphi_{2z}L + 8u_{2y} + 16u_{1y} + 3\varphi_{1z}L\right) \\ -35\varphi_{2z}L + 162u_{2y} + 162u_{1y} + 35\varphi_{1z}L & L\left(8\varphi_{1z}L + 35u_{1y} + 43u_{2y} - 9\varphi_{2z}L\right) \\ -L\left(-8\varphi_{2z}L + 35u_{2y} + 43u_{1y} + 9\varphi_{1z}L\right) & -L^2\left(2\varphi_{1z}L + 9u_{1y} + 9u_{2y} - 2\varphi_{2z}L\right) \end{matrix}\right.
$$

$$
\left.\begin{matrix} -35\varphi_{2z}L + 162u_{2y} + 162u_{1y} + 35\varphi_{1z}L & -L\left(-8\varphi_{2z}L + 35u_{2y} + 43u_{1y} + 9\varphi_{1z}L\right) \\ L\left(8\varphi_{1z}L + 35u_{1y} + 43u_{2y} - 9\varphi_{2z}L\right) & -L^2\left(2\varphi_{1z}L + 9u_{1y} + 9u_{2y} - 2\varphi_{2z}L\right) \\ 43\varphi_{1z}L + 162u_{1y} + 774u_{2y} - 97\varphi_{2z}L & -L\left(9\varphi_{1z}L + 35u_{1y} + 97u_{2y} - 16\varphi_{2z}L\right) \\ -L\left(9\varphi_{1z}L + 35u_{1y} + 97u_{2y} - 16\varphi_{2z}L\right) & L^2\left(2\varphi_{1z}L + 8u_{1y} + 16u_{2y} - 3\varphi_{2z}L\right) \end{matrix}\right] . \tag{C.288}
$$

**5.43 Cantilevered beam with triangular shaped distributed load**
Finite element solution:

$$
\frac{u_y\left(\frac{x}{L}\right)}{\left(\frac{q}{24EI_z}\right)} = -\left(\frac{18}{5}L^4\left(\frac{x}{L}\right)^2 - \frac{7}{5}L^4\left(\frac{x}{L}\right)^3\right). \tag{C.289}
$$

Analytical solution:

$$
\frac{u_y\left(\frac{x}{L}\right)}{\left(\frac{q}{24EI_z}\right)} = -\left(4L^4\left(\frac{x}{L}\right)^2 - 2L^4\left(\frac{x}{L}\right)^3 + \frac{L^4}{5}\left(\frac{x}{L}\right)^5\right). \tag{C.290}
$$

A comparison between the finite element and the analytical solution is shown in Fig. C.12.

**5.44 Finite element approximation with a single beam element**
(a) FE solution for simply supported beam

$$
u_y(0) = 0\,, \quad u_y(L) = 0\,, \tag{C.291}
$$

$$
u_y(a) = -\frac{F}{6EI} \times \frac{2a^2b^2(a^2+ab+b^2)}{(a+b)^3}\,. \tag{C.292}
$$

The analytical solution gives at $x = 0$ and $x = L$ the same solution as the FE approach but at the location of the force: $u_y(a) = -\frac{Fa^2b^2}{3EI(a+b)}$.

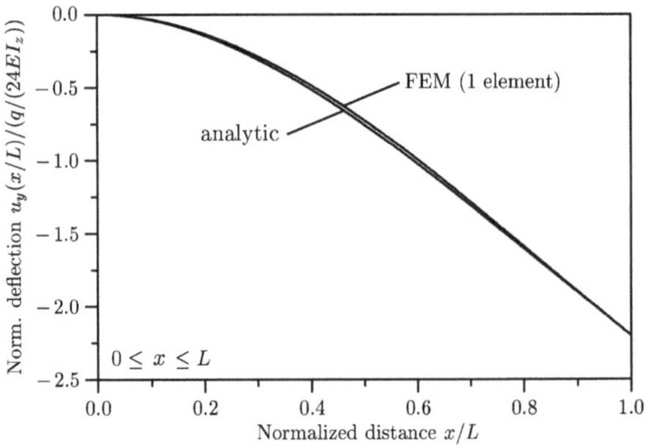


**Fig. C.12** Comparison of finite element and analytical solution for a cantilevered beam with triangular shaped distributed load

(b) FE solution for cantilevered beam

$$u_y(0) = 0\,, \quad u_y(L) = -\frac{F}{12EI} \times (4a^3 + 6a^2 b)\,, \tag{C.293}$$

$$u_y(a) = -\frac{F}{EI} \times \frac{a^4(a^2 + 3ab + 3b^2)}{3(a+b)^3}\,. \tag{C.294}$$

The analytical solution gives at $x = 0$ and $x = L$ the same solution as the FE approach but at the location of the force: $u_y(a) = -\frac{Fa^3}{3EI}$.

**5.45 Cantilevered beam: moment curvature relationship**

$$u_y(0) = 0\,, \quad u_y(L) = -\frac{5FL^3}{6EI_z}\,, \quad u_y(2L) = -\frac{8FL^3}{3EI_z}\,, \tag{C.295}$$

$$\varphi_z(0) = 0\,, \quad \varphi_z(L) = -\frac{3FL^2}{2EI_z}\,, \quad \varphi_z(2L) = -\frac{2FL^2}{EI_z}\,, \tag{C.296}$$

$$\kappa_z(0) = -\frac{2FL}{EI_z}\,, \quad \kappa_z(L) = -\frac{FL}{EI_z}\,, \quad \kappa_z(2L) = 0\,, \tag{C.297}$$

$$M_z(0) = -2FL\,, \quad M_z(L) = -FL\,, \quad M_z(2L) = 0\,, \tag{C.298}$$

$$\frac{M_z}{\kappa_z}(0) = EI_z\,, \quad \frac{M_z}{\kappa_z}(L) = EI_z\,, \quad \frac{M_z}{\kappa_z}(2L) = \frac{0}{0}\,. \tag{C.299}$$

The finite element solution is identical with the analytical solution.

**5.46 Cantilevered stepped beam**

(a) $I_{\text{I}} = 2I$ and $I_{\text{II}} = I$:

$$u_{2y} = -\frac{5FL^3}{12EI}\,, \qquad u_{3y} = -\frac{3FL^3}{2EI}\,, \tag{C.300}$$

$$\varphi_{2z} = -\frac{3FL^2}{4EI}\,, \qquad \varphi_{3z} = -\frac{5FL^2}{4EI}\,, \tag{C.301}$$

$$R_{1y} = F\,,\ M_{1z} = 2LF\,, \tag{C.302}$$

$$M_1 = -2LF\,,\ M_2 = -LF\,,\ M_3 = 0\,, \tag{C.303}$$

$$M_{\text{I}} = 0.5FL(\xi - 3)\,,\ M_{\text{II}} = 0.5FL(\xi - 1)\,, \tag{C.304}$$

$$Q_{\text{I}} = F\,,\ Q_{\text{II}} = F\,, \tag{C.305}$$

$$\sigma_{\text{I}}(0) = -\frac{LF}{J}\,y\,,\ \sigma_{\text{I}}(0.5L) = -\frac{3LF}{4J}\,y\,,\ \sigma_{\text{I}}(L) = -\frac{LF}{2J}\,y\,, \tag{C.306}$$

$$\sigma_{\text{I}}(L) = -\frac{LF}{J}\,y\,,\ \sigma_{\text{I}}(1.5L) = -\frac{LF}{2J}\,y\,,\ \sigma_{\text{I}}(2L) = 0\,y\,. \tag{C.307}$$

(b) $I_{\text{I}} = I_{\text{II}} = I$:

$$u_{2y} = -\frac{5FL^3}{6EI}\,, \qquad u_{3y} = -\frac{8FL^3}{3EI}\,, \tag{C.308}$$

$$\varphi_{2z} = -\frac{3FL^2}{2EI}\,, \qquad \varphi_{3z} = -\frac{2FL^2}{EI}\,, \tag{C.309}$$

$$R_{1y} = F\,,\ M_{1z} = 2LF\,, \tag{C.310}$$

$$M_1 = -2LF\,,\ M_2 = -LF\,,\ M_3 = 0\,, \tag{C.311}$$

$$M_{\text{I}} = 0.5FL(\xi - 3)\,,\ M_{\text{II}} = 0.5FL(\xi - 1)\,, \tag{C.312}$$

$$Q_{\text{I}} = F\,,\ Q_{\text{II}} = F\,, \tag{C.313}$$

$$\sigma_{\text{I}}(0) = -\frac{2LF}{J}\,y\,,\ \sigma_{\text{I}}(0.5L) = -\frac{3LF}{2J}\,y\,,\ \sigma_{\text{I}}(L) = -\frac{LF}{J}\,y\,, \tag{C.314}$$

$$\sigma_{\text{I}}(L) = -\frac{LF}{J}\,y\,,\ \sigma_{\text{I}}(1.5L) = -\frac{LF}{2J}\,y\,,\ \sigma_{\text{I}}(2L) = 0\,y\,. \tag{C.315}$$

## 5.47 Overhang beam with distributed load and single force

$$\varphi_{1Z} = \frac{L_{\text{I}}(-L_{\text{I}}^2 q + 4L_{\text{II}}F)}{24EI} \; ; \; \varphi_{2Z} = -\frac{L_{\text{I}}(-L_{\text{I}}^2 q + 8L_{\text{II}}F)}{24EI}, \tag{C.316}$$

$$u_{3Y} = -\frac{L_{\text{II}}(-L_{\text{I}}^3 q + 8L_{\text{I}}L_{\text{II}}F + 8L_{\text{II}}^2 F)}{24EI}, \tag{C.317}$$

$$\varphi_{3Z} = -\frac{-L_{\text{I}}^3 q + 12L_{\text{II}}^2 F + 8L_{\text{I}}L_{\text{II}}F}{24EI}, \tag{C.318}$$

$$R_{1Y} = -\frac{L_{\text{II}}F}{L_{\text{I}}} + \frac{L_{\text{I}}q}{2} \; ; \; R_{2Y} = \frac{(L_{\text{I}} + L_{\text{II}})F}{L_{\text{I}}} + \frac{L_{\text{I}}q}{2}, \tag{C.319}$$

$$u_Y(X = \tfrac{1}{2}L_{\text{I}}) = \frac{L_{\text{I}}^2(-L_{\text{I}}^2 q + 6L_{\text{II}}F)}{96EI}, \tag{C.320}$$

$$M_Z(X = \tfrac{1}{2}L_{\text{I}}) = \frac{L_{\text{I}}^2 q}{12} - \frac{L_2 F}{2} \; ; \; Q_Y(X = \tfrac{1}{2}L_{\text{I}}) = -\frac{L_{\text{II}}F}{L_{\text{I}}}, \tag{C.321}$$

$$u_Y(\tfrac{1}{2}L_{\text{I}}) = \frac{L_{\text{I}}L_{\text{II}}(-L_{\text{I}}^3 q - 4L_{\text{I}}^2 L_{\text{II}} q - 4L_{\text{I}}L_{\text{II}}^2 q - L_{\text{II}}^3 q + 8L_{\text{I}}L_{\text{III}}F + 4L_{\text{II}}L_{\text{III}}F)}{24(L_{\text{I}} + L_{\text{II}})EI}. \tag{C.322}$$

## 5.48 Beam structure with a gap

Force to close the gap:

$$F = \frac{6EI\delta}{5L^3}. \tag{C.323}$$

Deflection and rotation at the free end:

- $u_{2y}(x = L) < \delta$:

$$u_{3y} = -\frac{8L^3 F}{3EI}, \; \varphi_{3z} = -\frac{2L^2 F}{EI}. \tag{C.324}$$

- $u_{2y}(x = L) \geq \delta$:

$$u_{3y} = -\frac{30\delta EI + 7L^3 F}{12EI}, \; \varphi_{3z} = -\frac{3(2\delta EI + L^3 F)}{4LEI}. \tag{C.325}$$

Maximum normalized stress $\frac{\sigma_x(y_{\text{max}})}{y_{\text{max}}}$ at the nodes 1, 2 and 3:

- $u_{2y}(x = L) < \delta$:

$$\frac{2LF}{I}, \; \frac{LF}{I}, \; 0. \tag{C.326}$$

- $u_{2y}(x = L) \geq \delta$:

$$-\frac{-6\delta EI + L^3F}{2L^2I}\ ,\ \frac{LF}{I}\ ,\ 0\,. \tag{C.327}$$

## 5.49 Advanced example: beam element with nonlinear bending stiffness

$$a = EI_0\,,\, b = \frac{EI_0}{\kappa_1}\,(4\beta_{05} - \beta_1 - 3)\ ,\, c = -\frac{4EI_0}{\kappa_1^2}\left(\beta_{05} - \frac{1}{2}\,\beta_1 - \frac{1}{2}\right). \tag{C.328}$$

$$EI(\varepsilon) = EI_0\Bigg(1 - \underbrace{\frac{(3 + \beta_1 - 4\beta_{05})}{\kappa_1}}_{\alpha_1}\cdot\kappa - \underbrace{\frac{4(-\frac{1}{2} - \frac{1}{2}\beta_1 + \beta_{05})}{\kappa_1^2}}_{\alpha_2}\cdot\kappa^2\Bigg). \tag{C.329}$$

$$\boldsymbol{K}^{\mathrm{e}} = (5.347) - \frac{4EI_0 \times \alpha_2}{5L^7}\begin{bmatrix} K_{11} & K_{12} & K_{13} & K_{14} \\ & K_{22} & K_{23} & K_{24} \\ & & K_{33} & K_{34} \\ \text{sym.} & & & K_{44} \end{bmatrix}. \tag{C.330}$$

$$\begin{aligned} K_{11} = {} & 324\,u_1\,\varphi_1\,L + 324\,u_1\,\varphi_2\,L + 132\,\varphi_1\varphi_2L^2 - 324\,u_2\,\varphi_2\,L - 324\,u_2\varphi_1L \\ & + 324\,u_1^2 - 648\,u_1\,u_2 + 96\,\varphi_1^2L^2 + 324\,u_2{}^2 + 96\,\varphi_2^2L^2\,, \end{aligned}$$

$$\begin{aligned} K_{12} = {} & 3\,(64\,u_1\,\varphi_1\,L + 44\,u_1\,\varphi_2\,L + 22\,\varphi_1\varphi_2L^2 - 44\,u_2\,\varphi_2\,L - 64\,u_2\varphi_1\,L \\ & + 54\,u_1^2 - 108\,u_1\,u_2 + 21\,\varphi_1^2L^2 + 54\,u_2{}^2 + 11\,\varphi_2^2L^2)L\,, \end{aligned}$$

$$\begin{aligned} K_{13} = {} & -324\,u_1\,\varphi_1\,L - 324\,u_1\,\varphi_2\,L - 132\,\varphi_1\varphi_2L^2 + 324\,u_2\,\varphi_2\,L + 324\,u_2\varphi_1\,L \\ & - 324\,u_1^2 + 648\,u_1\,u_2 - 96\,\varphi_1^2L^2 - 324\,u_2{}^2 - 96\,\varphi_2^2L^2\,, \end{aligned}$$

$$\begin{aligned} K_{14} = {} & 3\,(44\,u_1\,\varphi_1\,L + 64\,u_1\,\varphi_2\,L + 22\,\varphi_1\varphi_2L^2 - 64\,u_2\,\varphi_2\,L - 44\,u_2\varphi_1\,L \\ & + 54\,u_1^2 - 108\,u_1\,u_2 + 11\,\varphi_1^2L^2 + 54\,u_2{}^2 + 21\,\varphi_2^2L^2)L\,, \end{aligned}$$

$$\begin{aligned} K_{22} = {} & 2\,(63\,u_1\,\varphi_1\,L + 33\,u_1\,\varphi_2\,L + 19\,\varphi_1\varphi_2L^2 - 33\,u_2\,\varphi_2\,L - 63\,u_2\varphi_1\,L \\ & + 48\,u_1^2 - 96\,u_1\,u_2 + 22\,\varphi_1^2L^2 + 48\,u_2{}^2 + 7\,\varphi_2^2L^2)L^2\,, \end{aligned}$$

$$\begin{aligned} K_{23} = {} & -3\,(64\,u_1\,\varphi_1\,L + 44\,u_1\,\varphi_2\,L + 22\,\varphi_1\varphi_2L^2 - 44\,u_2\,\varphi_2\,L - 64\,u_2\varphi_1\,L \\ & + 54\,u_1^2 - 108\,u_1\,u_2 + 21\,\varphi_1^2L^2 + 54\,u_2{}^2 + 11\,\varphi_2^2L^2)L\,, \end{aligned}$$

$$\begin{aligned} K_{24} = {} & (66\,u_1\,\varphi_1\,L + 66\,u_1\,\varphi_2\,L + 28\,\varphi_1\varphi_2L^2 - 66\,u_2\,\varphi_2\,L - 66\,u_2\varphi_1\,L \\ & + 66\,u_1^2 - 132\,u_1\,u_2 + 19\,\varphi_1^2L^2 + 66\,u_2{}^2 + 19\,\varphi_2^2L^2)L^2\,, \end{aligned}$$

$$K_{33} = 324\,u_1\,\varphi_1\,L + 324\,u_1\,\varphi_2\,L + 132\,\varphi_1\varphi_2 L^2 - 324\,u_2\,\varphi_2\,L - 324\,u_2\varphi_1\,L \\ + 324\,u_1^2 - 648\,u_1\,u_2 + 96\,\varphi_1^2 L^2 + 324\,u_2{}^2 + 96\,\varphi_2^2 L^2\,,$$

$$K_{34} = -3\,(44\,u_1\,\varphi_1\,L + 64\,u_1\,\varphi_2\,L + 22\,\varphi_1\varphi_2 L^2 - 64\,u_2\,\varphi_2\,L - 44\,u_2\varphi_1\,L \\ + 54\,u_1^2 - 108\,u_1\,u_2 + 11\,\varphi_1^2 L^2 + 54\,u_2{}^2 + 21\,\varphi_2^2 L^2)L\,,$$

$$K_{44} = 2\,(33\,u_1\,\varphi_1\,L + 63\,u_1\,\varphi_2\,L + 19\,\varphi_1\varphi_2 L^2 - 63\,u_2\,\varphi_2\,L - 33\,u_2\varphi_1\,L \\ + 48\,u_1^2 - 96\,u_1\,u_2 + 7\,\varphi_1^2 L^2 + 48\,u_2{}^2 + 22\,\varphi_2^2 L^2)L^2\,.$$

**5.50 Timoshenko beam element with quadratic shape functions for the displacement and linear shape functions for the rotation**

The nodal displacement at the middle node as a function of the other unknowns results in:

$$u_{2y} = \frac{u_{1y} + u_{3y}}{2} + \frac{\phi_{1z} - \phi_{3z}}{8} L + \frac{1}{32} \frac{6L}{k_s AG} \int_0^L q_y(x) N_{2u}(x) \mathrm{d}x\,. \tag{C.331}$$

The additional load vector on the right-hand side results in:

$$\cdots = \cdots + \begin{bmatrix} \int_0^L q_y(x) N_{1u} \mathrm{d}x + \frac{1}{2} \int_0^L q_y(x) N_{2u} \mathrm{d}x \\ +\frac{1}{8} L \int_0^L q_y(x) N_{2u} \mathrm{d}x \\ \int_0^L q_y(x) N_{3u} \mathrm{d}x + \frac{1}{2} \int_0^L q_y(x) N_{2u} \mathrm{d}x \\ -\frac{1}{8} L \int_0^L q_y(x) N_{2u} \mathrm{d}x \end{bmatrix}\,. \tag{C.332}$$

With $\int_0^L N_{1u} \mathrm{d}x = \frac{L}{6}$, $\int_0^L N_{2u} \mathrm{d}x = \frac{2L}{3}$ and $\int_0^L N_{3u} \mathrm{d}x = \frac{L}{6}$ the following results for a constant distributed load $q_y$:

$$\cdots = \cdots + \begin{bmatrix} \frac{1}{2} q_y L \\ +\frac{1}{12} q_y L^2 \\ \frac{1}{2} q_y L \\ -\frac{1}{12} q_y L^2 \end{bmatrix}\,. \tag{C.333}$$

This result is identical with the equivalent distributed load for a BERNOULLI beam. For this see Table 5.9.

**5.51 Timoshenko beam element with cubic shape functions for the displacement and quadratic shape functions for the rotation**
The element is exact!
Deformation in the $x$–$y$ plane:

$$\frac{2EI_z}{L^3(1+12\Lambda)}\begin{bmatrix} 6 & 3L & -6 & 3L \\ 3L & 2L^2(1+3\Lambda) & -3L & L^2(1-6\Lambda) \\ -6 & -3L & 6 & -3L \\ 3L & L^2(1-6\Lambda) & -3L & 2L^2(1+3\Lambda) \end{bmatrix}\begin{bmatrix} u_{1y} \\ \phi_{1z} \\ u_{2y} \\ \phi_{2z} \end{bmatrix} = \begin{bmatrix} F_{1y} \\ M_{1z} \\ F_{2y} \\ M_{2z} \end{bmatrix}. \tag{C.334}$$

Deformation in the $x$–$z$ plane:

$$\frac{2EI_y}{L^3(1+12\Lambda)}\begin{bmatrix} 6 & -3L & -6 & -3L \\ -3L & 2L^2(1+3\Lambda) & 3L & L^2(1-6\Lambda) \\ -6 & 3L & 6 & 3L \\ -3L & L^2(1-6\Lambda) & 3L & 2L^2(1+3\Lambda) \end{bmatrix}\begin{bmatrix} u_{1z} \\ \phi_{1y} \\ u_{2z} \\ \phi_{2y} \end{bmatrix} = \begin{bmatrix} F_{1z} \\ M_{1y} \\ F_{2z} \\ M_{2y} \end{bmatrix}. \tag{C.335}$$

**5.52 Plane truss structure arranged in a square**
(a) Force boundary condition

$$u_{2X} = \frac{(4+\sqrt{2})LF}{2(2+\sqrt{2})EA}\ ,\quad u_{2Y} = -\frac{\sqrt{2}\,LF}{2(2+\sqrt{2})EA}\,, \tag{C.336}$$

$$R_{1X} = -\frac{(4+\sqrt{2})F}{2(2+\sqrt{2})}\ ,\quad R_{1Y} = 0\ ,\quad R_{3X} = -\frac{\sqrt{2}\,F}{2(2+\sqrt{2})}\,, \tag{C.337}$$

$$R_{3Y} = -\frac{\sqrt{2}\,F}{2(2+\sqrt{2})}\ ,\quad R_{4X} = 0\ ,\quad R_{4Y} = \frac{\sqrt{2}\,F}{2(2+2\sqrt{2})}\,, \tag{C.338}$$

$$F_{\mathrm{I}} = \frac{(4+\sqrt{2})F}{2(2+\sqrt{2})}\ ,\quad F_{\mathrm{II}} = \frac{F}{2+\sqrt{2}}\ ,\quad F_{\mathrm{III}} = -\frac{\sqrt{2}\,F}{2(2+\sqrt{2})}\,. \tag{C.339}$$

(b) Displacement boundary condition

$$u_{2X} = u\ ,\quad u_{2Y} = -\frac{\sqrt{2}\,u}{4+\sqrt{2}}\,, \tag{C.340}$$

$$R_{1X} = -\frac{EAu}{L}\ ,\quad R_{1Y} = 0\ ,\quad R_{2X} = \frac{2(2+\sqrt{2})EAu}{(4+\sqrt{2})L}\,, \tag{C.341}$$

$$R_{2Y} = 0 \ , \ R_{3X} = -\frac{\sqrt{2}\,EAu}{(4+\sqrt{2})L} \ , \ R_{3Y} = -\frac{\sqrt{2}\,EAu}{(4+\sqrt{2})L} \, , \tag{C.342}$$

$$R_{4X} = 0 \ , \ R_{4Y} = \frac{\sqrt{2}\,EAu}{(4+\sqrt{2})L} \, , \tag{C.343}$$

$$F_{\mathrm{I}} = \frac{EAu}{L} \ , \ F_{\mathrm{II}} = \frac{2EAu}{(4+\sqrt{2})L} \ , \ F_{\mathrm{III}} = -\frac{\sqrt{2}EAu}{(4+\sqrt{2})L} \, . \tag{C.344}$$

**5.53 Plane truss structure arranged in a triangle**
(a) Force boundary condition

$$u_{4X} = \frac{LF_X}{2EA} \, , \ u_{4Y} = \frac{LF_Y}{EA} \, , \tag{C.345}$$

$$R_{1X} = -\frac{F_X}{4} + \frac{F_Y}{2} \, , \ R_{1Y} = \frac{F_X}{4} - \frac{F_Y}{2} \, , \ R_{2X} = -\frac{F_X}{2} \, , \ R_{2Y} = 0 \, , \tag{C.346}$$

$$R_{3X} = -\frac{F_X}{4} - \frac{F_Y}{2} \, , \ R_{3Y} = -\frac{F_X}{4} - \frac{F_Y}{2} \, , \tag{C.347}$$

$$F_I = \frac{F_X}{2} \, , \ F_{II} = \frac{\sqrt{2}(F_X - 2F_Y)}{4} \, , \ F_{III} = \frac{\sqrt{2}(F_X + 2F_Y)}{4} \, . \tag{C.348}$$

(b) Displacement boundary condition

$$R_{1X} = -\frac{EA(u_X - u_Y)}{2L} \, , \ R_{1Y} = \frac{EA(u_X - u_Y)}{2L} \, , \tag{C.349}$$

$$R_{2X} = -\frac{EAu_X}{L} \, , R_{2Y} = 0 \, , \tag{C.350}$$

$$R_{3X} = -\frac{EA(u_X + u_Y)}{2L} \, , \ R_{3Y} = -\frac{EA(u_X + u_Y)}{2L} \, , \tag{C.351}$$

$$R_{4X} = \frac{2EAu_X}{L} \, , \ R_{4Y} = \frac{EAu_Y}{L} \, , \tag{C.352}$$

$$F_I = \frac{EAu_X}{L} \, , \ F_{II} = \frac{EA(u_X - u_Y)}{\sqrt{2}L} \, , \ F_{III} = \frac{EA(u_X + u_Y)}{\sqrt{2}L} \, . \tag{C.353}$$

### 5.54 Plane beam rod structure

$$\begin{bmatrix} u_{2X} \\ u_{2Y} \\ \varphi_{2Z} \end{bmatrix} = \begin{bmatrix} -\dfrac{4L^3F}{E\left(2AL^2+3I\right)} \\ -\dfrac{8L^3F}{E\left(2AL^2+3I\right)} \\ -\dfrac{6L^2F}{E\left(2AL^2+3I\right)} \end{bmatrix}, \tag{C.354}$$

$$\begin{bmatrix} R_{1X} \\ R_{1Y} \\ M_{1Z} \end{bmatrix} = \begin{bmatrix} \dfrac{2AL^2F}{2AL^2+3I} \\ \dfrac{3IF}{2AL^2+3I} \\ \dfrac{6ILF}{2AL^2+3I} \end{bmatrix} ; \quad \begin{bmatrix} R_{3X} \\ R_{3Y} \end{bmatrix} = \begin{bmatrix} -\dfrac{2AL^2F}{2AL^2+3I} \\ \dfrac{2AL^2F}{2AL^2+3I} \end{bmatrix}. \tag{C.355}$$

### 5.55 Stiffness matrix for a generalized beam element for different rotation angles at the nodes

$$\boldsymbol{K}^{\mathrm{e}}_{XY}(\alpha_1,\alpha_2) = \begin{bmatrix} \boldsymbol{K}_{11} & \boldsymbol{K}_{12} \\ \boldsymbol{K}_{21} & \boldsymbol{K}_{22} \end{bmatrix}, \tag{C.356}$$

where the submatrices are given under consideration of the abbreviations s = sin and c = cos as:

$$\boldsymbol{K}_{11} = \begin{bmatrix} \dfrac{12I}{L^3}\mathrm{s}^2\alpha_1 + \dfrac{A}{L}\mathrm{c}^2\alpha_1 & \left(-\dfrac{12I}{L^3}+\dfrac{A}{L}\right)\mathrm{s}\alpha_1\mathrm{c}\alpha_1 & -\dfrac{6I}{L^2}\mathrm{s}\alpha_1 \\ \left(-\dfrac{12I}{L^3}+\dfrac{A}{L}\right)\mathrm{s}\alpha_1\mathrm{c}\alpha_1 & \dfrac{12I}{L^3}\mathrm{c}^2\alpha_1 + \dfrac{A}{L}\mathrm{s}^2\alpha_1 & \dfrac{6I}{L^2}\mathrm{c}\alpha_1 \\ -\dfrac{6I}{L^2}\mathrm{s}\alpha_1 & \dfrac{6I}{L^2}\mathrm{c}\alpha_1 & \dfrac{4I}{L} \end{bmatrix},$$

$$\boldsymbol{K}_{12} = \begin{bmatrix} -\dfrac{12I}{L^3}\mathrm{s}\alpha_1\mathrm{s}\alpha_2 - \dfrac{A}{L}\mathrm{c}\alpha_1\mathrm{c}\alpha_2 & \dfrac{12I}{L^3}\mathrm{s}\alpha_1\mathrm{c}\alpha_2 - \dfrac{A}{L}\mathrm{c}\alpha_1\mathrm{s}\alpha_2 & -\dfrac{6I}{L^2}\mathrm{s}\alpha_1 \\ \dfrac{12I}{L^3}\mathrm{c}\alpha_1\mathrm{s}\alpha_2 - \dfrac{A}{L}\mathrm{s}\alpha_1\mathrm{c}\alpha_2 & -\dfrac{12I}{L^3}\mathrm{c}\alpha_1\mathrm{c}\alpha_2 - \dfrac{A}{L}\mathrm{s}\alpha_1\mathrm{s}\alpha_2 & \dfrac{6I}{L^2}\mathrm{c}\alpha_1 \\ \dfrac{6I}{L^2}\mathrm{s}\alpha_2 & -\dfrac{6I}{L^2}\mathrm{c}\alpha_2 & \dfrac{2I}{L} \end{bmatrix},$$

$$\boldsymbol{K}_{21} = \begin{bmatrix} -\frac{12I}{L^3}\mathrm{s}\alpha_1\mathrm{s}\alpha_2 - \frac{A}{L}\mathrm{c}\alpha_1\mathrm{c}\alpha_2 & \frac{12I}{L^3}\mathrm{c}\alpha_1\mathrm{s}\alpha_2 - \frac{A}{L}\mathrm{s}\alpha_1\mathrm{c}\alpha_2 & \frac{6I}{L^2}\mathrm{s}\alpha_2 \\ \frac{12I}{L^3}\mathrm{s}\alpha_1\mathrm{c}\alpha_2 - \frac{A}{L}\mathrm{c}\alpha_1\mathrm{s}\alpha_2 & -\frac{12I}{L^3}\mathrm{c}\alpha_1\mathrm{c}\alpha_2 - \frac{A}{L}\mathrm{s}\alpha_1\mathrm{s}\alpha_2 & -\frac{6I}{L^2}\mathrm{c}\alpha_2 \\ -\frac{6I}{L^2}\mathrm{s}\alpha_1 & \frac{6I}{L^2}\mathrm{c}\alpha_1 & \frac{2I}{L} \end{bmatrix},$$

$$\boldsymbol{K}_{22} = \begin{bmatrix} \frac{12I}{L^3}\mathrm{s}^2\alpha_2 + \frac{A}{L}\mathrm{c}^2\alpha_2 & \left(-\frac{12I}{L^3} + \frac{A}{L}\right)\mathrm{s}\alpha_2\mathrm{c}\alpha_2 & \frac{6I}{L^2}\mathrm{s}\alpha_2 \\ \left(-\frac{12I}{L^3} + \frac{A}{L}\right)\mathrm{s}\alpha_2\mathrm{c}\alpha_2 & \frac{12I}{L^3}\mathrm{c}^2\alpha_2 + \frac{A}{L}\mathrm{s}^2\alpha_2 & -\frac{6I}{L^2}\mathrm{c}\alpha_2 \\ \frac{6I}{L^2}\mathrm{s}\alpha_2 & -\frac{I}{L^2}\mathrm{c}\alpha_2 & \frac{4I}{L} \end{bmatrix}.$$

**5.56 Mechanical properties of squared frame structure**

(a) [quarter model, BC: $\frac{F}{2}$]

$$u_{3Y} = \frac{1}{384}\frac{\left(5AL^2 + 96\,I\right)LF}{EIA}, \tag{C.357}$$

$$E_{\text{struct}} = \frac{\sigma}{\varepsilon} = \frac{192EIA}{\left(5AL^2 + 96I\right)Ld}. \tag{C.358}$$

For $A = \frac{\pi d^2}{4}$ and $I = \frac{\pi d^4}{64}$:

$$E_{\text{struct}} = \frac{3Ed^3\pi}{\left(6d^2 + 5L^2\right)L}. \tag{C.359}$$

(b) [quarter model, BC: $\frac{F}{2}$]

$$u_{2Y} = \frac{1}{48}\frac{\left(AL^2 + 12\,I\right)LF}{EIA}, \tag{C.360}$$

$$E_{\text{struct}} = \frac{\sigma}{\varepsilon} = \frac{24EIA}{\left(AL^2 + 12I\right)Ld}. \tag{C.361}$$

For $A = \frac{\pi d^2}{4}$ and $I = \frac{\pi d^4}{64}$:

$$E_{\text{struct}} = \frac{3Ed^3\pi}{2\left(3d^2 + 4L^2\right)L}. \tag{C.362}$$

**5.57 Squared frame structure: different ways of load application**
(a) [quarter model]

$$u_{2Y} = \frac{1}{4}\frac{LF}{EA}, \tag{C.363}$$

$$E_{\text{struct}} = \frac{\sigma}{\varepsilon} = \frac{2EA}{Ld}. \tag{C.364}$$

For $A = \frac{\pi d^2}{4}$ and $I = \frac{\pi d^4}{64}$:

$$E_{\text{struct}} = \frac{dE\pi}{2L}. \tag{C.365}$$

**5.58 Mechanical properties of idealized honeycomb structure**
The problem under consideration of the symmetry is shown in Fig. C.13.
(a)

$$u_{2Y} = \frac{1}{24}\frac{\left(AL^2 + 18\,I\right)LF}{EIA}, \tag{C.366}$$

$$E_{\text{struct}} = \frac{\sigma}{\varepsilon} = \frac{\frac{F}{\frac{1}{2}(\sqrt{3}+1)Ld}}{\frac{\frac{1}{24}\frac{(AL^2+18\,I)LF}{EIA}}{\frac{1}{2}\sqrt{3}L}} = \frac{24EIA}{\left(1+\sqrt{3}\right)\left(AL^2+18\,I\right)Ld}. \tag{C.367}$$

For $A = \frac{\pi d^2}{4}$ and $I = \frac{\pi d^4}{64}$:

$$E_{\text{struct}} = \frac{3Ed^3\pi}{\left(1+\sqrt{3}\right)\left(8L^2+9d^2\right)L}. \tag{C.368}$$

(b)

$$u_{3Y} = \frac{1}{16}\frac{\left(AL^2 + 6\,I\right)LF}{EIA}, \tag{C.369}$$

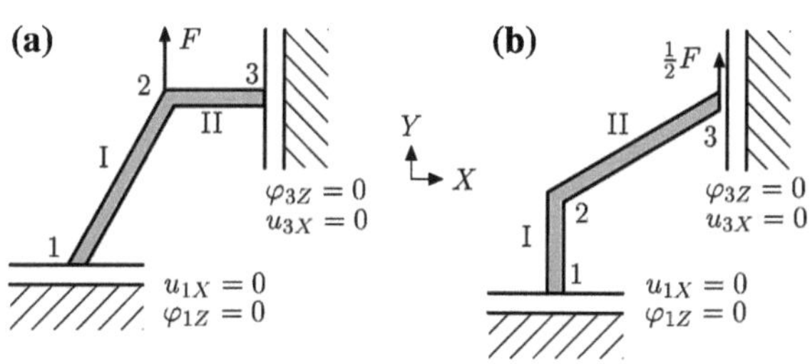


**Fig. C.13** Honeycomb structure approximated under consideration of symmetry: **a** flat orientation and **b** pointy or angled orientation

$$E_{\text{struct}} = \frac{\sigma}{\varepsilon} = \frac{\dfrac{F/2}{\frac{1}{2}\sqrt{3}Ld}}{\dfrac{\frac{1}{16}\frac{(AL^2+6I)LF}{EIA}}{L}} = \frac{16EIA}{\sqrt{3}\left(AL^2+6I\right)Ld}\,. \tag{C.370}$$

For $A = \frac{\pi d^2}{4}$ and $I = \frac{\pi d^4}{64}$:

$$E_{\text{struct}} = \frac{2}{\sqrt{3}} \times \frac{Ed^3\pi}{\left(8L^2+3d^2\right)L}\,. \tag{C.371}$$

**5.59 Bridge structure (computational problem)**
Since the solution is obtained numerically,[10] the obtained numbers must be carefully interpreted. Looking at case a) load 1), i.e. the symmetrical loading, the displacement at node 3 in the horizontal direction must be zero. However, there is a small value calculated, cf. Table C.7. This value must be related to the vertical displacement at node 3 and one can see that $u_{3X}$ is several orders of magnitude smaller than $u_{3Y}$. Thus, from a practical point of view, $u_{3X}$ can be considered as equal to zero.

Some rotations in Table C.8 could not be evaluated (indicated by '—') since the rod elements do not have a rotational degree of freedom (Tables C.9).
General expressions for the nodal unknowns for case (c) load (1):

$$u_{2X}=0\,,\ u_{2Y}=-2.414214\,\frac{FL}{EA}\,,\ u_{3X}=0\,,\ u_{3Y}=-4.828427\,\frac{FL}{EA}\,,$$
$$u_{4X}=0\,,\ u_{4Y}=-2.414214\,\frac{FL}{EA}\,,\ u_{6X}=-\frac{FL}{EA}\,,\ u_{6Y}=u_{2Y}\,, \tag{C.372}$$
$$u_{7X}=0\,,\ u_{7Y}=-4.828427\,\frac{FL}{EA}\,,\ u_{8X}=-\frac{FL}{EA}\,,\ u_{8Y}=u_{6Y}\,.$$

General expressions for the nodal unknowns for case (c) load (2):

$$u_{2X}=0.25\,\frac{FL}{EA}\,,\ u_{2Y}=-3.871320\,\frac{FL}{EA}\,,\ u_{3X}=0.5\,\frac{FL}{EA}\,,\ u_{3Y}=-2.414214\,\frac{FL}{EA}\,,$$
$$u_{4X}=u_{2X}\,,\ u_{4Y}=-0.957107\,\frac{FL}{EA}\,,\ u_{6X}=-u_{4X}\,,\ u_{6Y}=u_{4Y}\,, \tag{C.373}$$
$$u_{7X}=u_{2X}\,,\ u_{7Y}=u_{3Y}\,,\ u_{8X}=0.75\,\frac{FL}{EA}\,,\ u_{8Y}=-2.871320\,\frac{FL}{EA}\,.$$

See Table C.10.

[10] The actual results were obtained with Maple 8.0 and the accuracy setting 'Digits := 25'.

**Table C.7** Numerical results (case a) for the bridge structure problem with $L = 4{,}000\,\text{mm}$, $E = 200{,}000\,\text{MPa}$, $I = 80\,\text{mm}^4$ and $A = 10\,\text{mm}^2$

| Node Nr. | $u_X$ | $u_Y$ | $\varphi_Z$ |
|---|---|---|---|
| *Case (a) Load (1)* | | | |
| 1 | 0.0 | 0.0 | $-0.105623 \times 10^{-5}F$ |
| 2 | $-0.110022 \times 10^{-7}F$ | $-0.004828F$ | $-0.111306 \times 10^{-5}F$ |
| 3 | $-0.855248 \times 10^{-8}F$ | $-0.009657F$ | $-0.141264 \times 10^{-11}F$ |
| 4 | $0.244971 \times 10^{-8}F$ | $-0.004828F$ | $0.111306 \times 10^{-5}F$ |
| 5 | 0.0 | 0.0 | $0.105623 \times 10^{-5}F$ |
| 6 | $-0.002000F$ | $-0.004828F$ | $0.100792 \times 10^{-5}F$ |
| 7 | $-0.427622 \times 10^{-8}F$ | $-0.009657F$ | $-0.126408 \times 10^{-11}F$ |
| 8 | $0.002000F$ | $-0.004828F$ | $-0.100792 \times 10^{-5}F$ |
| *Case (a) Load (2)* | | | |
| 1 | 0.0 | 0.0 | $-0.210602 \times 10^{-5}F$ |
| 2 | $0.000500F$ | $-0.007743F$ | $-0.460909 \times 10^{-6}F$ |
| 3 | $0.001000F$ | $-0.004828F$ | $0.607748 \times 10^{-6}F$ |
| 4 | $0.000500F$ | $-0.001914F$ | $0.488052 \times 10^{-6}F$ |
| 5 | 0.0 | 0.0 | $0.368358 \times 10^{-6}F$ |
| 6 | $-0.000500F$ | $-0.001914F$ | $0.466897 \times 10^{-6}F$ |
| 7 | $0.0005F$ | $-0.004828F$ | $0.379865 \times 10^{-6}F$ |
| 8 | $0.001500F$ | $-0.005743F$ | $-0.107569 \times 10^{-6}F$ |
| *Case (a) Load (3)* | | | |
| 1 | 0.0 | 0.0 | $-0.022606F$ |
| 2 | $0.000435F$ | $-0.004693F$ | $0.014409F$ |
| 3 | $0.000693F$ | $-0.002386F$ | $-0.001727F$ |
| 4 | $0.000349F$ | $-0.000898F$ | $0.000287F$ |
| 5 | 0.0 | 0.0 | $-0.000108F$ |
| 6 | $-0.000165F$ | $-0.000878F$ | $0.000118F$ |
| 7 | $0.000315F$ | $-0.002383F$ | $0.000205F$ |
| 8 | $0.000777F$ | $-0.003443F$ | $0.000380F$ |
| *Case (a) Load (4)* | | | |
| 1 | 0.0 | 0.0 | $0.003528F$ |
| 2 | $0.000265F$ | $-0.006822F$ | $-0.012608F$ |
| 3 | $0.000671F$ | $-0.007799F$ | $0.008826F$ |
| 4 | $0.000323F$ | $-0.003309F$ | $-0.001471F$ |
| 5 | 0.0 | 0.0 | $0.000542F$ |
| 6 | $-0.001319F$ | $-0.003409F$ | $-0.000532F$ |
| 7 | $0.000280F$ | $-0.007783F$ | $-0.001513F$ |
| 8 | $0.001967F$ | $-0.005759F$ | $0.000788F$ |

**Table C.8** Numerical results (case b) for the bridge structure problem with $L = 4{,}000$ mm, $E = 200{,}000$ MPa, $I = 80$ mm$^4$ and $A = 10$ mm$^2$

| Node Nr. | $u_X$ | $u_Y$ | $\varphi_Z$ |
|---|---|---|---|
| *Case (b) Load (1)* | | | |
| 1 | 0.0 | 0.0 | $-0.103467 \times 10^{-5}F$ |
| 2 | 0.0 | $-0.004828F$ | $-0.155198 \times 10^{-5}F$ |
| 3 | 0.0 | $-0.009657F$ | 0.0 |
| 4 | 0.0 | $-0.004828F$ | $0.155198 \times 10^{-5}F$ |
| 5 | 0.0 | 0.0 | $0.103467 \times 10^{-5}F$ |
| 6 | $-0.002000F$ | $-0.004828F$ | – |
| 7 | 0.0 | $-0.009657F$ | – |
| 8 | $0.002000F$ | $-0.004828F$ | – |
| *Case (b) Load (2)* | | | |
| 1 | 0.0 | 0.0 | $-0.264480 \times 10^{-5}F$ |
| 2 | $0.000500F$ | $-0.007743F$ | $-0.517336 \times 10^{-6}F$ |
| 3 | $0.001000F$ | $-0.004828F$ | $0.109282 \times 10^{-5}F$ |
| 4 | $0.000500F$ | $-0.001914F$ | $0.517336 \times 10^{-6}F$ |
| 5 | 0.0 | 0.0 | $0.459164 \times 10^{-6}F$ |
| 6 | $-0.000500F$ | $-0.001914F$ | – |
| 7 | $0.000500F$ | $-0.004828F$ | – |
| 8 | $0.001500F$ | $-0.005743F$ | – |
| *Case (b) Load (3)* | | | |
| 1 | 0.0 | 0.0 | $-0.045761F$ |
| 2 | $0.000344F$ | $-0.004935F$ | $0.02901F$ |
| 3 | $0.000687F$ | $-0.002156F$ | $-0.007812F$ |
| 4 | $0.000344F$ | $-0.000928F$ | $0.002232F$ |
| 5 | 0.0 | 0.0 | $-0.001116F$ |
| 6 | $-0.000103F$ | $-0.000848F$ | – |
| 7 | $0.000344F$ | $-0.002156F$ | – |
| 8 | $0.000790F$ | $-0.003480F$ | – |
| *Case (b) Load (4)* | | | |
| 1 | 0.0 | 0.0 | $0.012275F$ |
| 2 | $0.000344F$ | $-0.007091F$ | $-0.024555F$ |
| 3 | $0.000687F$ | $-0.008019F$ | $0.023438F$ |
| 4 | $0.000344F$ | $-0.003084F$ | $-0.006695F$ |
| 5 | 0.0 | 0.0 | $0.003349F$ |
| 6 | $-0.001317F$ | $-0.003325F$ | – |
| 7 | $0.000344F$ | $-0.008019F$ | – |
| 8 | $0.002004F$ | $-0.005957F$ | – |

**Table C.9** Numerical results (case c) for the bridge structure problem with $L = 4{,}000\,\text{mm}$, $E = 200{,}000\,\text{MPa}$, $I = 80\,\text{mm}^4$ and $A = 10\,\text{mm}^2$

| Node Nr. | $u_X$ | $u_Y$ | $\varphi_Z$ |
|---|---|---|---|
| *Case (c) Load (1)* | | | |
| 1 | 0.0 | 0.0 | — |
| 2 | 0.0 | $-0.004828F$ | — |
| 3 | 0.0 | $-0.009657F$ | — |
| 4 | 0.0 | $-0.004828F$ | — |
| 5 | 0.0 | 0.0 | — |
| 6 | $-0.002000F$ | $-0.004828F$ | — |
| 7 | 0.0 | $-0.009657F$ | — |
| 8 | $0.002000F$ | $-0.004828F$ | — |
| *Case (c) Load (2)* | | | |
| 1 | 0.0 | 0.0 | — |
| 2 | $0.000500F$ | $-0.007743F$ | — |
| 3 | $0.001000F$ | $-0.004828F$ | — |
| 4 | $0.000500F$ | $-0.001914F$ | — |
| 5 | 0.0 | 0.0 | — |
| 6 | $-0.000500F$ | $-0.001914F$ | — |
| 7 | $0.000500F$ | $-0.004828F$ | — |
| 8 | $0.001500F$ | $-0.005743F$ | — |

General expressions for the nodal unknowns for case (d) load (1):

$$\begin{aligned}
&\varphi_{1Z} = -\frac{FL^2}{EI}\,,\; u_{2X} = 0\,,\; u_{2Y} = -0.916667\,\frac{FL^3}{EI}\,,\; \varphi_{2Z} = -0.75\,\frac{FL^2}{EI}\,,\\
&u_{3X} = 0\,,\; u_{3Y} = -1.333333\,\frac{FL^3}{EI}\,,\; \varphi_{3Z} = 0\,,\; u_{4X} = 0\,,\\
&u_{4Y} = -0.916667\,\frac{FL^3}{EI}\,,\; \varphi_{4Z} = 0.75\,\frac{FL^2}{EI}\,,\; \varphi_{5Z} = \frac{FL^2}{EI}\,.
\end{aligned} \tag{C.374}$$

It should be noted here that the case (d) corresponds to a simply supported beam under bending.

## Problems from Chap. 6

### 6.9 Plastic modulus and elasto-plastic modulus

(a) $E^{\text{elpl}} = \frac{E\times E^{\text{pl}}}{E+E^{\text{pl}}} = E \;\Rightarrow E = 0.$

Pure linear-plastic behavior without elastic part, meaning pure elastic behavior on macro-level.
(b) $E^{\text{pl}} = E \;\Rightarrow\; E^{\text{elpl}} = \frac{E\times E}{E+E} = \frac{1}{2}\,E.$

Linear hardening, whereupon the elasto-plastic modulus is half of the elastic modulus $E$.

**Table C.10** Numerical results (case d) for the bridge structure problem with $L = 4{,}000$ mm, $E = 200{,}000\,\text{MPa}$, $I = 80\,\text{mm}^4$ and $A = 10\,\text{mm}^2$

| Node Nr. | $u_X$ | $u_Y$ | $\varphi_Z$ |
|---|---|---|---|
| *Case (d) Load (1)* | | | |
| 1 | 0.0 | 0.0 | $-1.000000F$ |
| 2 | 0.0 | $-3666.666667F$ | $-0.750000F$ |
| 3 | 0.0 | $-5333.333333F$ | 0.0 |
| 4 | 0.0 | $-3666.666667F$ | $0.750000F$ |
| 5 | 0.0 | 0.0 | $1.000000F$ |
| *Case (d) Load (2)* | | | |
| 1 | 0.0 | 0.0 | $-0.875000F$ |
| 2 | 0.0 | $-3000F$ | $-0.500000F$ |
| 3 | 0.0 | $-3666.666667F$ | $0.125000F$ |
| 4 | 0.0 | $-2333.333333F$ | $0.500000F$ |
| 5 | 0.0 | 0.0 | $0.625000F$ |
| *Case (d) Load (3)* | | | |
| 1 | 0.0 | 0.0 | $-0.546875F$ |
| 2 | 0.0 | $-1687.500000F$ | $-0.234375F$ |
| 3 | 0.0 | $-1958.333333F$ | $0.078125F$ |
| 4 | 0.0 | $-1229.166667F$ | $0.265625F$ |
| 5 | 0.0 | 0.0 | $0.328125F$ |
| *Case (d) Load (4)* | | | |
| 1 | 0.0 | 0.0 | $-1.015625F$ |
| 2 | 0.0 | $-3645.833333F$ | $-0.703125F$ |
| 3 | 0.0 | $-4875.000000F$ | $0.109375F$ |
| 4 | 0.0 | $-3187.500000F$ | $0.671875F$ |
| 5 | 0.0 | 0.0 | $0.859375F$ |

**6.10 Back projection at linear hardening**
See Tables C.11, C.12, C.13.

**6.11 Back projection at nonlinear hardening**
See Table C.14

**6.12 Back projection for a rod with fixed support on both sides**
The flow curve results in:

$$k(\kappa) = 200\ \text{MPa} + 1010.\bar{1}0\ \text{MPa} \times \kappa\,. \tag{C.375}$$

The iteration scheme of the NEWTON-RAPHSON method can be used as follows:

$$\Delta u_2^{(i+1)} = \frac{\Delta F^{(i)}}{A\left(\frac{\tilde{E}^{\text{I}}}{L^{\text{I}}} + \frac{\tilde{E}^{\text{II}}}{L^{\text{II}}}\right)}\,. \tag{C.376}$$

**Table C.11** Numerical values of the back projection for a continuum rod in the case of linear hardening (10 increments; $\Delta\varepsilon = 0.001$)

| Inc | $\varepsilon$ | $\sigma^{\text{trial}}$ | $\sigma$ | $\kappa$ | $\mathrm{d}\lambda$ | $E^{\text{elpl}}$ |
|---|---|---|---|---|---|---|
| – | – | MPa | MPa | $10^{-3}$ | $10^{-3}$ | MPa |
| 1 | 0.001 | 210.0 | 210.0 | 0.0 | 0.0 | 0.0 |
| 2 | 0.002 | 420.0 | 420.0 | 0.0 | 0.0 | 0.0 |
| 3 | 0.003 | 630.0 | 630.0 | 0.0 | 0.0 | 0.0 |
| 4 | 0.004 | 840.0 | 703.636 | 0.649 | 0.649 | 19090.909 |
| 5 | 0.005 | 913.636 | 722.727 | 1.558 | 0.909 | 19090.909 |
| 6 | 0.006 | 932.727 | 741.818 | 2.468 | 0.909 | 19090.909 |
| 7 | 0.007 | 951.818 | 760.909 | 3.377 | 0.909 | 19090.909 |
| 8 | 0.008 | 970.909 | 780.000 | 4.286 | 0.909 | 19090.909 |
| 9 | 0.009 | 990.000 | 799.091 | 5.195 | 0.909 | 19090.909 |
| 10 | 0.010 | 1009.091 | 818.182 | 6.104 | 0.909 | 19090.909 |

For the fulfillment of the convergence criteria nine cycles are necessary for the second increment and four cycles for the third increment (Table C.15).

**6.13 Back projection for a finite element at ideal-plastic material behavior**
In the elastic area, the calculation of the displacement on the node can take place. As soon as plastic material behavior occurs, no convergence can be achieved since no unique relationship between stress and strain exists. If the boundary force condition is being substituted by a boundary displacement condition, the strain in the rod is known and the stress can be calculated.

**6.14 Back projection for a von Mises material**
The derivation is similar to the approach in Sect. 6.2. In the final state $(n+1)$, the following three equations must be fulfilled:

$$\sigma_{n+1} = \sigma_{n+1}^{\text{trial}} + \Delta\lambda_{n+1} E \frac{2\sigma_{n+1}}{(k_{\text{t}}(\kappa_{n+1}))^2}\,, \tag{C.377}$$

$$\kappa_{n+1} = \kappa_n + \Delta\lambda_{n+1} \frac{2\sigma_{n+1}^2}{(k_{\text{t}}(\kappa_{n+1}))^3}\,, \tag{C.378}$$

$$F_{n+1} = \left(\frac{|\sigma_{n+1}|}{k_{\text{t}}(\kappa_{n+1}}\right)^2 - 1 = 0\,. \tag{C.379}$$

$$\frac{\partial \boldsymbol{m}}{\partial \boldsymbol{v}} = \begin{bmatrix} 1 + \dfrac{2\Delta\lambda E}{k_{\text{t}}^2} & -\dfrac{4\Delta\lambda E\sigma}{k_{\text{t}}^3}\dfrac{\mathrm{d}k_{\text{t}}}{\mathrm{d}\kappa} & \dfrac{2\sigma E}{k_{\text{t}}^2} \\ -\dfrac{4\Delta\lambda\sigma}{k_{\text{t}}^3} & 1 + \dfrac{6\Delta\lambda\sigma^2}{{}_{\text{t}}k^4}\dfrac{\mathrm{d}k_{\text{t}}}{\mathrm{d}\kappa} & -\dfrac{2\sigma^2}{k_{\text{t}}^3} \\ \dfrac{2\sigma}{k_{\text{t}}^2} & -\dfrac{2\sigma^2}{k_{\text{t}}^3}\dfrac{\mathrm{d}k_{\text{t}}}{\mathrm{d}\kappa} & 0 \end{bmatrix}. \tag{C.380}$$

**Table C.12** Numerical values of the back projection for a continuum rod in the case of linear hardening (20 increments; $\Delta\varepsilon = 0.0005$)

| Inc | $\varepsilon$ | $\sigma^{\text{trial}}$ | $\sigma$ | $\kappa$ | $\mathrm{d}\lambda$ | $E^{\text{elpl}}$ |
|---|---|---|---|---|---|---|
| – | – | MPa | MPa | $10^{-3}$ | $10^{-3}$ | MPa |
| 1 | 0.0005 | 105.0 | 105.0 | 0.0 | 0.0 | 0.0 |
| 2 | 0.0010 | 210.0 | 210.0 | 0.0 | 0.0 | 0.0 |
| 3 | 0.0015 | 315.0 | 315.0 | 0.0 | 0.0 | 0.0 |
| 4 | 0.0020 | 420.0 | 420.0 | 0.0 | 0.0 | 0.0 |
| 5 | 0.0025 | 525.0 | 525.0 | 0.0 | 0.0 | 0.0 |
| 6 | 0.0030 | 630.0 | 630.0 | 0.0 | 0.0 | 0.0 |
| 7 | 0.0035 | 735.0 | 694.091 | 0.195 | 0.195 | 19090.909 |
| 8 | 0.0040 | 799.091 | 703.636 | 0.649 | 0.455 | 19090.909 |
| 9 | 0.0045 | 808.636 | 713.182 | 1.104 | 0.455 | 19090.909 |
| 10 | 0.0050 | 818.182 | 722.727 | 1.558 | 0.455 | 19090.909 |
| 11 | 0.0055 | 827.727 | 732.273 | 2.013 | 0.455 | 19090.909 |
| 12 | 0.0060 | 837.273 | 741.818 | 2.468 | 0.455 | 19090.909 |
| 13 | 0.0065 | 846.818 | 751.364 | 2.922 | 0.455 | 19090.909 |
| 14 | 0.0070 | 856.364 | 760.909 | 3.377 | 0.455 | 19090.909 |
| 15 | 0.0075 | 865.909 | 770.455 | 3.831 | 0.455 | 19090.909 |
| 16 | 0.0080 | 875.455 | 780.000 | 4.286 | 0.455 | 19090.909 |
| 17 | 0.0085 | 885.000 | 789.545 | 4.740 | 0.455 | 19090.909 |
| 18 | 0.0090 | 894.545 | 799.091 | 5.195 | 0.455 | 19090.909 |
| 19 | 0.0095 | 904.091 | 808.636 | 5.649 | 0.455 | 19090.909 |
| 20 | 0.0100 | 913.636 | 818.182 | 6.104 | 0.455 | 19090.909 |

**Table C.13** Numerical values of the back projection for continuum rod at linear hardening (20 increments; $\Delta\varepsilon = 0.001$)

| Inc | $\varepsilon$ | $\sigma^{\text{trial}}$ | $\sigma$ | $\kappa$ | $\mathrm{d}\lambda$ | $E^{\text{elpl}}$ |
|---|---|---|---|---|---|---|
| – | – | MPa | MPa | $10^{-3}$ | $10^{-3}$ | MPa |
| 1 | 0.001 | 210.0 | 210.0 | 0.0 | 0.0 | 0.0 |
| 2 | 0.002 | 420.0 | 420.0 | 0.0 | 0.0 | 0.0 |
| 3 | 0.003 | 630.0 | 630.0 | 0.0 | 0.0 | 0.0 |
| 4 | 0.004 | 840.0 | 703.636 | 0.649 | 0.649 | 19090.909 |
| 5 | 0.005 | 913.636 | 722.727 | 1.558 | 0.909 | 19090.909 |
| 6 | 0.006 | 932.727 | 741.818 | 2.468 | 0.909 | 19090.909 |
| 7 | 0.007 | 951.818 | 760.909 | 3.377 | 0.909 | 19090.909 |
| 8 | 0.008 | 970.909 | 780.000 | 4.286 | 0.909 | 19090.909 |
| 9 | 0.009 | 990.000 | 799.091 | 5.195 | 0.909 | 19090.909 |
| 10 | 0.010 | 1009.091 | 818.182 | 6.104 | 0.909 | 19090.909 |

(Continued)

**Table C.13** (Continued)

| | | | | | | |
|---|---|---|---|---|---|---|
| 11 | 0.011 | 1028.182 | 837.273 | 7.013 | 0.909 | 19090.909 |
| 12 | 0.012 | 1047.273 | 856.364 | 7.922 | 0.909 | 19090.909 |
| 13 | 0.013 | 1066.364 | 875.455 | 8.831 | 0.909 | 19090.909 |
| 14 | 0.014 | 1085.455 | 894.545 | 9.740 | 0.909 | 19090.909 |
| 15 | 0.015 | 1104.545 | 913.636 | 10.649 | 0.909 | 19090.909 |
| 16 | 0.016 | 1123.636 | 932.727 | 11.558 | 0.909 | 19090.909 |
| 17 | 0.017 | 1142.727 | 951.818 | 12.468 | 0.909 | 19090.909 |
| 18 | 0.018 | 1161.818 | 970.909 | 13.377 | 0.909 | 19090.909 |
| 19 | 0.019 | 1180.909 | 990.000 | 14.286 | 0.909 | 19090.909 |
| 20 | 0.020 | 1200.000 | 1009.091 | 15.195 | 0.909 | 19090.909 |

**Table C.14** Numerical values of the back projection for a continuum rod in the case of nonlinear hardening hardening (10 increments; $\Delta\varepsilon = 0.002$)

| Inc | $\varepsilon$ | $\sigma^{\text{trial}}$ | $\sigma$ | $\kappa$ | $\text{d}\lambda$ | $E^{\text{elpl}}$ |
|---|---|---|---|---|---|---|
| – | – | MPa | MPa | $10^{-3}$ | $10^{-3}$ | MPa |
| 1 | 0.002 | 140.0 | 140.0 | 0.0 | 0.0 | 0.0 |
| 2 | 0.004 | 280.0 | 280.0 | 0.0 | 0.0 | 0.0 |
| 3 | 0.006 | 420.0 | 360.817 | 0.845469 | 0.845469 | 10741.553 |
| 4 | 0.008 | 500.817 | 381.995 | 2.542923 | 1.697454 | 10435.865 |
| 5 | 0.010 | 521.995 | 402.557 | 4.249179 | 1.706256 | 10125.398 |
| 6 | 0.012 | 542.557 | 422.494 | 5.964375 | 1.715196 | 9810.025 |
| 7 | 0.014 | 562.494 | 441.794 | 7.688654 | 1.724279 | 9489.616 |
| 8 | 0.016 | 581.794 | 460.449 | 9.422161 | 1.733507 | 9164.034 |
| 9 | 0.018 | 600.449 | 478.447 | 11.165046 | 1.742885 | 8833.140 |
| 10 | 0.020 | 618.447 | 495.778 | 12.917462 | 1.752416 | 8496.787 |

**Table C.15** Numerical values for a rod with fixed support on both sides (3 increments; $\Delta F_2 = 2 \times 10^4$ N)

| Inc | $u_2$ | $\varepsilon^{\text{I}}$ | $\varepsilon^{\text{II}}$ | $\sigma^{\text{I}}$ | $\sigma^{\text{II}}$ | $\varepsilon^{\text{pl,I}}$ | $\varepsilon^{\text{pl,II}}$ |
|---|---|---|---|---|---|---|---|
| – | mm | $10^{-3}$ | $10^{-3}$ | MPa | MPa | $10^{-3}$ | $10^{-3}$ |
| 1 | 0.0666667 | 0.666667 | −1.33333 | 66.6667 | −133.333 | 0.0 | 0.0 |
| 2 | 0.19806 | 1.9806 | −3.9612 | 198.060 | −201.938 | 0.0 | −1.94182 |
| 3 | 6.88003 | 68.8003 | −137.601 | 266.008 | −333.992 | 66.1402 | −134.261 |

**Table C.16** Numerical results for monotonic loading with $0 \leq u \leq u_{\max}$ for nonlinear hardening material. Case: 10 increments, $t_{\mathrm{end}}^{v} = 0.001$

| Inc $n$ | cycle $i$ | $\varepsilon$ | $\sigma^{\mathrm{trial}}$ | $\sigma$ | $\varepsilon^{\mathrm{pl}}$ | $\varepsilon_{\mathrm{eff}}^{\mathrm{pl}}$ | $\Delta\lambda$ | $E^{\mathrm{elpl}}$ | $\lVert \boldsymbol{v}^{(i+1)} - \boldsymbol{v}^{(i)} \rVert$ |
|---|---|---|---|---|---|---|---|---|---|
| – | – | $10^{-3}$ | MPa | MPa | $10^{-3}$ | $10^{-3}$ | $10^{-3}$ | MPa | – |
| 1 | 1 | 2.0 | 140.0 | 140.0 | 0.0 | 0.0 | 0.0 | 0.0 | 0.0 |
| 2 | 1 | 4.0 | 280.0 | 280.0 | 0.0 | 0.0 | 0.0 | 0.0 | 0.0 |
| 3 | 1 | 6.0 | 420.0 | 361.317 | 0.838 | 0.838 | 0.838 | 11080.780 | 0.118788 |
| | 2 | 6.0 | 420.0 | 361.199 | 0.840 | 0.840 | 0.840 | 11080.305 | $0.478641 \times 10^{-6}$ |
| 4 | 1 | 8.0 | 501.199 | 383.359 | 2.523 | 2.523 | 1.683 | 10617.385 | 0.466817 |
| | 2 | 8.0 | 501.199 | 382.892 | 2.530 | 2.530 | 1.690 | 10615.583 | $0.708255 \times 10^{-5}$ |
| 5 | 1 | 10.0 | 522.892 | 404.124 | 4.227 | 4.227 | 1.697 | 10165.266 | 0.454138 |
| | 2 | 10.0 | 522.892 | 403.669 | 4.233 | 4.233 | 1.703 | 10163.575 | $0.641760 \times 10^{-5}$ |
| 6 | 1 | 12.0 | 543.669 | 423.997 | 5.943 | 5.943 | 1.710 | 9726.045 | 0.441270 |
| | 2 | 12.0 | 543.669 | 423.555 | 5.949 | 5.949 | 1.716 | 9724.461 | $0.579727 \times 10^{-5}$ |
| 7 | 1 | 14.0 | 563.555 | 443.004 | 7.671 | 7.671 | 1.722 | 9299.860 | 0.428257 |
| | 2 | 14.0 | 563.555 | 442.576 | 7.677 | 7.677 | 1.728 | 9298.380 | $0.522113 \times 10^{-5}$ |
| 8 | 1 | 16.0 | 582.576 | 461.173 | 9.412 | 9.412 | 1.734 | 8886.806 | 0.415144 |
| | 2 | 16.0 | 582.576 | 460.758 | 9.418 | 9.418 | 1.740 | 8885.425 | $0.468836 \times 10^{-5}$ |
| 9 | 1 | 18.0 | 600.758 | 478.528 | 11.164 | 11.164 | 1.746 | 8486.933 | 0.401972 |
| | 2 | 18.0 | 600.758 | 478.126 | 11.170 | 11.170 | 1.752 | 8485.648 | $0.419779 \times 10^{-5}$ |
| 10 | 1 | 20.0 | 618.126 | 495.098 | 12.927 | 12.927 | 1.758 | 8100.253 | 0.388782 |
| | 2 | 20.0 | 618.126 | 494.709 | 12.933 | 12.933 | 1.763 | 8099.059 | $0.374792 \times 10^{-5}$ |

**6.15 Back projection for a nonlinear hardening material**
See Table C.16.

**6.16 Cyclic loading of an elastic-plastic continuum rod made of steel**
See Tables C.17 and C.18.

**6.17 Back projection for a Gurson material with isotropic hardening**
The derivation results in a similar matrix as given in Eq. (6.117) where $k_{\mathrm{t}}$ must be replaced by $k_{\mathrm{t}}(\kappa)$. In addition, the second column must be replaced by the following expressions:

$$\frac{\partial r_\sigma}{\partial \kappa} = -\frac{\Delta\lambda E}{(k_{\mathrm{t}}(\kappa))^3}\left[4\sigma + D k_{\mathrm{t}}(\kappa)\sinh\left(\frac{\sigma}{2k_{\mathrm{t}}(\kappa)}\right) + \frac{D\sigma}{2}\cosh\left(\frac{\sigma}{2k_{\mathrm{t}}(\kappa)}\right)\right]\frac{\mathrm{d}k_{\mathrm{t}}(\kappa)}{\mathrm{d}\kappa}, \tag{C.381}$$

$$\begin{aligned}\frac{\partial r_\kappa}{\partial \kappa} = 1 - \frac{\Delta\lambda}{1-D}&\left[-\frac{6\sigma^2}{(k_{\mathrm{t}}(\kappa))^4} - \frac{2D\sigma}{(k_{\mathrm{t}}(\kappa))^3}\sinh\left(\frac{\sigma}{2k_{\mathrm{t}}(\kappa)}\right)\right.\\ &\left.-\frac{D\sigma^2}{2(k_{\mathrm{t}}(\kappa))^4}\cosh\left(\frac{\sigma}{2k_{\mathrm{t}}(\kappa)}\right)\right]\frac{\mathrm{d}k_{\mathrm{t}}(\kappa)}{\mathrm{d}\kappa},\end{aligned} \tag{C.382}$$

**Table C.17** Numerical results for cyclic loading with $u_{max} \leq u \leq -u_{max}$ for ideal plastic material

| Inc | $u$ | $\varepsilon$ | $\sigma^{trial}$ | $\sigma$ | $\varepsilon^{pl}$ | $\varepsilon^{pl}_{eff}$ | $\Delta\lambda$ | $E^{elpl}$ |
|---|---|---|---|---|---|---|---|---|
| – | mm | $10^{-3}$ | MPa | MPa | $10^{-3}$ | $10^{-3}$ | $10^{-3}$ | MPa |
| 1 | 0.4 | 1.0 | 210.0 | 210.0 | 0.0 | 0.0 | 0.0 | 0.0 |
| 2 | 0.8 | 2.0 | 420.0 | 420.0 | 0.0 | 0.0 | 0.0 | 0.0 |
| 3 | 1.2 | 3.0 | 630.0 | 630.0 | 0.0 | 0.0 | 0.0 | 0.0 |
| 4 | 1.6 | 4.0 | 840.0 | 690.0 | 0.714 | 0.714 | 0.714 | 0.0 |
| 5 | 2.0 | 5.0 | 900.0 | 690.0 | 1.714 | 1.714 | 1.0 | 0.0 |
| 6 | 1.6 | 4.0 | 480.0 | 480.0 | 1.714 | 1.714 | 0.0 | 0.0 |
| 7 | 1.2 | 3.0 | 270.0 | 270.0 | 1.714 | 1.714 | 0.0 | 0.0 |
| 8 | 0.8 | 2.0 | 60.0 | 60.0 | 1.714 | 1.714 | 0.0 | 0.0 |
| 9 | 0.4 | 2.0 | −150.0 | −150.0 | 1.714 | 1.714 | 0.0 | 0.0 |
| 10 | 0.0 | 0.0 | −360.0 | −360.0 | 1.714 | 1.714 | 0.0 | 0.0 |
| 11 | −0.4 | −1.0 | −570.0 | −570.0 | 1.714 | 1.714 | 0.0 | 0.0 |
| 12 | −0.8 | −2.0 | −780.0 | −690.0 | 1.286 | 2.143 | 0.429 | 0.0 |
| 13 | −1.2 | −3.0 | −900.0 | −690.0 | 0.286 | 3.143 | 1.0 | 0.0 |
| 14 | −1.6 | −4.0 | −900.0 | −690.0 | −0.714 | 4.143 | 1.0 | 0.0 |
| 15 | −2.0 | −5.0 | −900.0 | −690.0 | −1.714 | 5.143 | 1.0 | 0.0 |
| 16 | −1.6 | −4.0 | −480.0 | −480.0 | −1.714 | 5.143 | 0.0 | 0.0 |
| 17 | −1.2 | −3.0 | −270.0 | −270.0 | −1.714 | 5.143 | 0.0 | 0.0 |
| 18 | −0.8 | −2.0 | −60.0 | −60.0 | 1.714 | 5.143 | 0.0 | 0.0 |
| 19 | −0.4 | −1.0 | 150.0 | 150.0 | 1.714 | 5.143 | 0.0 | 0.0 |
| 20 | 0.0 | 0.0 | 369.0 | 360.0 | 1.714 | 5.143 | 0.0 | 0.0 |

$$\frac{\partial r_F}{\partial \kappa} = -\frac{1}{(k_t(\kappa))^3}\left[2|\sigma|^2 + D\sigma k_t(\kappa)\sinh\left(\frac{\sigma}{2k_t(\kappa)}\right)\right]\frac{\mathrm{d}k_t(\kappa)}{\mathrm{d}\kappa}, \tag{C.383}$$

$$\frac{\partial r_D}{\partial \kappa} = \frac{3\Delta\lambda(D - D^2)}{(k_t(\kappa))^2}\left[\sinh\left(\frac{\sigma}{2k_t(\kappa)}\right) + \frac{\sigma}{2k_t(\kappa)}\cosh\left(\frac{\sigma}{2k_t(\kappa)}\right)\right]\frac{\mathrm{d}k_t(\kappa)}{\mathrm{d}\kappa}. \tag{C.384}$$

**6.18 Monotonic tensile loading of a linear-elastic ideal-plastic rod element under damage effects (Gurson)**
See Table C.19.

**Problems from Chap.** 7

**7.4 Forward difference approximation of the first order derivative**
Intermediate result:

$$\left(\frac{\mathrm{d}u}{\mathrm{d}X}\right)_i = \frac{u_{i+1} - u_i}{\Delta X} - \frac{1}{2}\left(\frac{\mathrm{d}^2u}{\mathrm{d}X^2}\right)_i \Delta X \underbrace{-\frac{1}{6}\left(\frac{\mathrm{d}^3u}{\mathrm{d}X^3}\right)_i \Delta X^2 - \cdots}_{O(\Delta X^2)}. \tag{C.385}$$

**Table C.18** Numerical results for cyclic loading with $u_{max} \leq u \leq -u_{max}$ for linear hardening steel

| Inc | $u$ | $\varepsilon$ | $\sigma^{trial}$ | $\sigma$ | $\varepsilon^{pl}$ | $\varepsilon^{pl}_{eff}$ | $\Delta\lambda$ | $E^{elpl}$ |
|---|---|---|---|---|---|---|---|---|
| – | mm | $10^{-3}$ | MPa | MPa | $10^{-3}$ | $10^{-3}$ | $10^{-3}$ | MPa |
| 1 | 0.4 | 1.0 | 210.0 | 210.0 | 0.0 | 0.0 | 0.0 | 0.0 |
| 2 | 0.8 | 2.0 | 420.0 | 420.0 | 0.0 | 0.0 | 0.0 | 0.0 |
| 3 | 1.2 | 3.0 | 630.0 | 630.364 | 0.0 | 0.0 | 0.0 | 0.0 |
| 4 | 1.6 | 4.0 | 840.0 | 703.636 | 0.649 | 0.649 | 0.649 | 19090.909 |
| 5 | 2.0 | 5.0 | 913.636 | 722.727 | 1.558 | 1.558 | 0.909 | 19090.909 |
| 6 | 1.6 | 4.0 | 512.727 | 512.727 | 1.558 | 1.558 | 0.0 | 0.0 |
| 7 | 1.2 | 3.0 | 302.727 | 302.727 | 1.558 | 1.558 | 0.0 | 0.0 |
| 8 | 0.8 | 2.0 | 92.727 | 92.727 | 1.558 | 1.558 | 0.0 | 0.0 |
| 9 | 0.4 | 2.0 | −117.273 | −117.273 | 1.558 | 1.558 | 0.0 | 0.0 |
| 10 | 0.0 | 0.0 | −327.273 | −327.273 | 1.558 | 1.558 | 0.0 | 0.0 |
| 11 | −0.4 | −1.0 | −537.273 | −537.273 | 1.558 | 1.558 | 0.0 | 0.0 |
| 12 | −0.8 | −2.0 | −747.273 | −724.959 | 1.452 | 1.665 | 0.106 | 19090.909 |
| 13 | −1.2 | −3.0 | −934.959 | −744.050 | 0.543 | 2.574 | 0.909 | 19090.909 |
| 14 | −1.6 | −4.0 | −954.050 | −763.140 | −0.366 | 3.483 | 0.909 | 19090.909 |
| 15 | −2.0 | −5.0 | −973.140 | −782.231 | −1.275 | 4.392 | 0.909 | 19090.909 |
| 16 | −1.6 | −4.0 | −572.231 | −572.231 | −1.275 | 4.392 | 0.0 | 0.0 |
| 17 | −1.2 | −3.0 | −362.231 | −362.231 | −1.275 | 4.392 | 0.0 | 0.0 |
| 18 | −0.8 | −2.0 | −152.231 | −152.231 | −1.275 | 4.392 | 0.0 | 0.0 |
| 19 | −0.4 | −1.0 | 57.769 | 57.769 | −1.275 | 4.392 | 0.0 | 0.0 |
| 20 | 0.0 | 0.0 | 267.769 | 267.769 | −1.275 | 4.392 | 0.0 | 0.0 |

**7.5 Centered difference approximation of the third order derivative**
Starting point:

$$\left(\frac{du}{dX}\right)_i = \frac{u_{i+1} - u_{i-1}}{2\Delta X}\,. \tag{C.386}$$

Introducing in the above equation the expressions for the second order derivatives at nodes $i+1$ and $i-1$, i.e.

$$\left(\frac{d^2u}{dX^2}\right)_{i+1} \approx \frac{u_{i+2} - 2u_{i+1} + u_i}{\Delta X^2} \to u_{i+1}\,, \tag{C.387}$$

$$\left(\frac{d^2u}{dX^2}\right)_{i-1} \approx \frac{u_i - 2u_{i-1} + u_{i-2}}{\Delta X^2} \to u_{i-1}\,, \tag{C.388}$$

gives the derivative of the second order derivative, i.e. the third order derivative, as stated in Table 7.1.

**Table C.19** Numerical results for monotonic loading with $|\Delta\varepsilon_n| = 0.002$ for a linear-elastic ideal-plastic material and the GURSON damage model. Case: 10 increments, $D_0 = 0.01$, $t^{v}_{\text{end}} = 0.00001$

| Inc $n$ | cycle $i$ | $\varepsilon$ | $\sigma^{\text{trial}}$ | $\sigma$ | $\varepsilon^{\text{pl}}_{\text{eff}}$ | $\Delta\lambda$ | $D$ | $\Vert \boldsymbol{v}^{(i+1)} - \boldsymbol{v}^{(i)} \Vert$ |
|---|---|---|---|---|---|---|---|---|
| – | – | $10^{-3}$ | MPa | MPa | $10^{-3}$ | $10^{-3}$ | $10^{-2}$ | – |
| 1 | 1 | 2.0 | 140.0 | 140.0 | 0.0 | 0.0 | 1.0 | 0.0 |
| 2 | 1 | 4.0 | 280.0 | 280.0 | 0.0 | 0.0 | 1.0 | 0.0 |
| 3 | 1 | 6.0 | 420.0 | 352.569 | 1.168 | 140.109 | 1.0 | 6.454 |
| | 2 | 6.0 | 420.0 | 346.116 | 1.059 | 185.453 | 1.000815 | 0.0602 |
| | 3 | 6.0 | 420.0 | 346.056 | 1.055 | 186.482 | 1.000815 | 0.00000523 |
| | 4 | 6.0 | 420.0 | 346.056 | 1.055 | 186.482 | 1.000815 | 0.00000000 |
| 4 | 1 | 8.0 | 486.056 | 366.224 | 3.456 | 215.139 | 1.000815 | 19.619 |
| | 2 | 8.0 | 486.056 | 346.605 | 3.088 | 343.838 | 1.002357 | 0.555 |
| | 3 | 8.0 | 486.056 | 346.050 | 3.053 | 353.073 | 1.002361 | 0.000446 |
| | 4 | 8.0 | 486.056 | 346.050 | 3.053 | 353.090 | 1.002361 | 0.00000000 |
| 5 | 1 | 10.0 | 486.050 | 366.218 | 5.454 | 215.140 | 1.00236 | 19.619 |
| | 2 | 10.0 | 486.050 | 346.599 | 5.085 | 343.843 | 1.00390 | 0.555 |
| | 3 | 10.0 | 486.050 | 346.044 | 5.050 | 353.078 | 1.00391 | 0.000446 |
| | 4 | 10.0 | 486.050 | 346.043 | 5.050 | 353.094 | 1.00391 | 0.00000000 |
| 6 | 1 | 12.0 | 486.043 | 366.218 | 7.451 | 215.141 | 1.00391 | 19.620 |
| | 2 | 12.0 | 486.043 | 346.593 | 7.083 | 343.847 | 1.00546 | 0.555 |
| | 3 | 12.0 | 486.043 | 346.038 | 7.048 | 353.083 | 1.00546 | 0.000446 |
| | 4 | 12.0 | 486.043 | 346.037 | 7.048 | 353.099 | 1.00546 | 0.00000000 |
| 7 | 1 | 14.0 | 486.037 | 366.207 | 9.449 | 215.143 | 1.00546 | 19.620 |
| | 2 | 14.0 | 486.037 | 346.587 | 9.080 | 343.851 | 1.00701 | 0.555 |
| | 3 | 14.0 | 486.037 | 346.032 | 9.045 | 353.088 | 1.00701 | 0.000446 |
| | 4 | 14.0 | 486.037 | 346.031 | 9.045 | 353.104 | 1.00701 | 0.00000000 |
| 8 | 1 | 16.0 | 486.031 | 366.201 | 11.446 | 215.144 | 1.00701 | 19.620 |
| | 2 | 16.0 | 486.031 | 346.581 | 11.078 | 343.856 | 1.00856 | 0.555 |
| | 3 | 16.0 | 486.031 | 346.026 | 11.043 | 353.093 | 1.00857 | 0.000446 |
| | 4 | 16.0 | 486.031 | 346.025 | 11.043 | 353.108 | 1.00857 | 0.00000000 |
| 9 | 1 | 18.0 | 486.025 | 366.195 | 13.444 | 215.145 | 1.00857 | 19.620 |
| | 2 | 18.0 | 486.025 | 346.575 | 13.075 | 343.860 | 1.01012 | 0.555 |
| | 3 | 18.0 | 486.025 | 346.019 | 13.040 | 353.097 | 1.01013 | 0.000446 |
| | 4 | 18.0 | 486.025 | 346.019 | 13.040 | 353.114 | 1.01013 | 0.00000000 |
| 10 | 1 | 20.0 | 486.019 | 366.189 | 15.442 | 215.147 | 1.0101 | 19.621 |
| | 2 | 20.0 | 486.019 | 346.569 | 15.073 | 343.864 | 1.0117 | 0.555 |
| | 3 | 20.0 | 486.019 | 346.013 | 15.038 | 353.102 | 1.0117 | 0.000446 |
| | 4 | 20.0 | 486.019 | 346.013 | 15.038 | 353.119 | 1.0117 | 0.00000000 |

**7.6 Finite difference approximation of a cantilevered rod with distributed load based on five domain nodes**
The linear system of equations reads as:

$$\frac{EA}{\Delta X}\begin{bmatrix} 2 & -1 & 0 & 0 \\ -1 & 2 & -1 & 0 \\ 0 & -1 & 2 & -1 \\ 0 & 0 & -1 & 1 \end{bmatrix}\begin{bmatrix} u_2 \\ u_3 \\ u_4 \\ u_5 \end{bmatrix} = \begin{bmatrix} \Delta X p \\ \Delta X p \\ \Delta X p \\ \Delta X p \\ \frac{\Delta X p}{4} \end{bmatrix}, \tag{C.389}$$

or solved for the nodal unknowns:

$$\begin{bmatrix} u_2 \\ u_3 \\ u_4 \\ u_5 \end{bmatrix} = \begin{bmatrix} \frac{13}{4}\frac{\Delta X^2 p}{EA} \\ \frac{11}{2}\frac{\Delta X^2 p}{EA} \\ \frac{27}{4}\frac{\Delta X^2 p}{EA} \\ \frac{7}{1}\frac{\Delta X^2 p}{EA} \end{bmatrix}. \tag{C.390}$$

The analytical solution can be taken from Table 3.3 as $u_5 = \frac{pL^2}{2EA}$ whereas the finite difference solution gives $u_5 = \frac{7}{16}\frac{pL^2}{EA} \approx 0.4375 \times \frac{pL^2}{EA}$.

**7.7 Finite difference approximation of a simply supported and cantilevered beam based on three domain nodes**
(a) Simply supported beam:

$$\text{node 2:} \quad \frac{EI_Z}{\Delta X^3}(u_4 - 4u_3 + 6u_2 - 4u_1 + u_0) = -F. \tag{C.391}$$

Boundary conditions: $u(X=0) = u_1 = 0$, $u(X=L) = u_3 = 0$, $M_Z(X=0) = 0 \rightarrow u_0 = -u_2$, $M_Z(X=L) = 0 \rightarrow u_4 = -u_2$.

$$u_2 = -\frac{\Delta X^3 F}{4EI_Z} = -\underbrace{\frac{1}{32}}_{0.03125}\frac{FL^3}{EI_Z}. \tag{C.392}$$

Relative error to analytical solution: 50 %.

(b) Cantilevered beam:

$$\text{node 2:} \quad \frac{EI_Z}{\Delta X^3}(u_4 - 4u_3 + 6u_2 - 4u_1 + u_0) = 0, \tag{C.393}$$

$$\text{node 3:} \quad \frac{EI_Z}{\Delta X^3}(u_5 - 4u_4 + 6u_3 - 4u_2 + u_1) = -F. \tag{C.394}$$

Boundary conditions: $u(X=0)=u_1=0$, $\frac{\mathrm{d}u_1}{\mathrm{d}X}=0 \rightarrow u_0=u_2$, $EI_Z \left.\frac{\mathrm{d}^2u}{\mathrm{d}X^2}\right|_3 = M_Z=0 \rightarrow u_4=-u_2+2u_3$, $EI_Z \left.\frac{\mathrm{d}^3u}{\mathrm{d}X^3}\right|_3 = F \rightarrow u_5=4u_3-4u_2+\frac{2\Delta X^3 F}{EI_Z}$.
Linear system of equations:

$$\frac{EI_Z}{\Delta X^3}\begin{bmatrix} 3 & -1 \\ -4 & 2 \end{bmatrix}\begin{bmatrix} u_2 \\ u_3 \end{bmatrix} = \begin{bmatrix} 0 \\ -3F \end{bmatrix}. \tag{C.395}$$

Solution:

$$u_2 = -\frac{3}{2}\frac{\Delta X^3 F}{EI_Z} = -\frac{3}{16}\frac{FL^3}{EI_Z}, \tag{C.396}$$

$$u_3 = -\frac{9}{2}\frac{\Delta X^3 F}{EI_Z} = -\underbrace{\frac{9}{16}}_{0.5625}\frac{FL^3}{EI_Z}. \tag{C.397}$$

Relative error at node 3 compared to analytical solution: 68.752 %.

**7.8 Centered difference approximation of the fourth order derivative**

$$\begin{aligned} u_{i+1} = u_i &+ \left(\frac{\mathrm{d}^1u}{\mathrm{d}X^1}\right)_i \frac{\Delta X}{1!} + \left(\frac{\mathrm{d}^2u}{\mathrm{d}X^2}\right)_i \frac{\Delta X^2}{2!} + \left(\frac{\mathrm{d}^3u}{\mathrm{d}X^3}\right)_i \frac{\Delta X^3}{3!} + \\ &+ \left(\frac{\mathrm{d}^4u}{\mathrm{d}X^4}\right)_i \frac{\Delta X^4}{4!} + \left(\frac{\mathrm{d}^5u}{\mathrm{d}X^5}\right)_i \frac{\Delta X^5}{5!} + \left(\frac{\mathrm{d}^6u}{\mathrm{d}X^6}\right)_i \frac{\Delta X^6}{6!} + \cdots, \end{aligned} \tag{C.398}$$

$$\begin{aligned} u_{i-1} = u_i &- \left(\frac{\mathrm{d}^1u}{\mathrm{d}X^1}\right)_i \frac{\Delta X}{1!} + \left(\frac{\mathrm{d}^2u}{\mathrm{d}X^2}\right)_i \frac{\Delta X^2}{2!} - \left(\frac{\mathrm{d}^3u}{\mathrm{d}X^3}\right)_i \frac{\Delta X^3}{3!} \\ &+ \left(\frac{\mathrm{d}^4u}{\mathrm{d}X^4}\right)_i \frac{\Delta X^4}{4!} - \left(\frac{\mathrm{d}^5u}{\mathrm{d}X^5}\right)_i \frac{\Delta X^5}{5!} + \left(\frac{\mathrm{d}^6u}{\mathrm{d}X^6}\right)_i \frac{\Delta X^6}{6!} - \cdots. \end{aligned} \tag{C.399}$$

Summing up the expressions (C.398) and (C.399) and rearranging for the second order derivative gives:

$$\left(\frac{\mathrm{d}^2u}{\mathrm{d}X^2}\right)_i = \frac{u_{i+1}-2u_1+u_{i-1}}{\Delta X^2} - \left(\frac{\mathrm{d}^4u}{\mathrm{d}X^4}\right)_i \frac{\Delta X^2}{12} - \left(\frac{\mathrm{d}^6u}{\mathrm{d}X^6}\right)_i \frac{\Delta X^4}{360} - \cdots. \tag{C.400}$$

$$u_{i+2} = u_i + \left(\frac{\mathrm{d}^1u}{\mathrm{d}X^1}\right)_i \frac{(2\Delta X)}{1!} + \left(\frac{\mathrm{d}^2u}{\mathrm{d}X^2}\right)_i \frac{(2\Delta X)^2}{2!} + \left(\frac{\mathrm{d}^3u}{\mathrm{d}X^3}\right)_i \frac{(2\Delta X)^3}{3!}$$

$$+\left(\frac{\mathrm{d}^4u}{\mathrm{d}X^4}\right)_i\frac{(2\Delta X)^4}{4!}+\left(\frac{\mathrm{d}^5u}{\mathrm{d}X^5}\right)_i\frac{(2\Delta X)^5}{5!}+\left(\frac{\mathrm{d}^6u}{\mathrm{d}X^6}\right)_i\frac{(2\Delta X)^6}{6!}+\cdots, \tag{C.401}$$

$$u_{i-2}=u_i-\left(\frac{\mathrm{d}^1u}{\mathrm{d}X^1}\right)_i\frac{(2\Delta X)}{1!}+\left(\frac{\mathrm{d}^2u}{\mathrm{d}X^2}\right)_i\frac{(2\Delta X)^2}{2!}-\left(\frac{\mathrm{d}^3u}{\mathrm{d}X^3}\right)_i\frac{(2\Delta X)^3}{3!}$$
$$+\left(\frac{\mathrm{d}^4u}{\mathrm{d}X^4}\right)_i\frac{(2\Delta X)^4}{4!}-\left(\frac{\mathrm{d}^5u}{\mathrm{d}X^5}\right)_i\frac{(2\Delta X)^5}{5!}+\left(\frac{\mathrm{d}^6u}{\mathrm{d}X^6}\right)_i\frac{(2\Delta X)^6}{6!}-\cdots. \tag{C.402}$$

Summing up the expressions (C.401) and (C.402) and considering the expression for the second order derivative given in Eq. (C.400) allows to express the fourth order derivative finally as:

$$\left(\frac{\mathrm{d}^4u}{\mathrm{d}X^4}\right)_i=\frac{u_{i+2}-4u_{i+1}+6u_i-4u_{i-1}+u_{i-2}}{(\Delta X)^4}\underbrace{-\left(\frac{\mathrm{d}^6u}{\mathrm{d}X^6}\right)_i\frac{(\Delta X)^2}{6}}_{O(\Delta X^2)}. \tag{C.403}$$

**7.9 Finite difference approximation of a cantilevered beam based on five domain nodes**

Centered finite difference approximation at nodes $i=2,\ldots,5$:

$$\text{node 2: } \frac{EI_Z}{\Delta X^3}(u_4-4u_3+6u_2-4u_1+u_0)=0, \tag{C.404}$$
$$\text{node 3: } \frac{EI_Z}{\Delta X^3}(u_5-4u_4+6u_3-4u_2+u_1)=0, \tag{C.405}$$
$$\text{node 4: } \frac{EI_Z}{\Delta X^3}(u_6-4u_5+6u_4-4u_3+u_2)=-F, \tag{C.406}$$
$$\text{node 5: } \frac{EI_Z}{\Delta X^3}(u_7-4u_6+6u_5-4u_4+u_3)=0. \tag{C.407}$$

Boundary conditions:

$$u_1=0\ ,\quad \left.\frac{\mathrm{d}u}{\mathrm{d}X}\right|_1=0\ \rightarrow\ u_o=u_2. \tag{C.408}$$

$$M_Z(L)=0\ \rightarrow\ u_6=2u_5-u_4\ ,\quad Q_Y(L)=0\ \rightarrow\ u_7=2u_6-2u_4+u_3\,. \tag{C.409}$$

Linear system of equations:

$$\frac{EI_Z}{\Delta X^3}\begin{bmatrix}7&-4&1&0\\-4&6&-4&1\\1&-4&5&-2\\0&1&-2&1\end{bmatrix}\begin{bmatrix}u_2\\u_3\\u_4\\u_5\end{bmatrix}=\begin{bmatrix}0\\0\\-F\\0\end{bmatrix}. \tag{C.410}$$

Solution:

$$u_2 = -\frac{3FL^3}{128EI_Z}, \ u_3 = -\frac{5FL^3}{64EI_Z}, \ u_4 = -\frac{19FL^3}{128EI_Z}, \ u_5 = -\frac{7FL^3}{32EI_Z}. \qquad \text{(C.411)}$$

The analytical solution can be extracted from Table 4.8 as $u(L) = -\frac{81FL^3}{384EI_Z}$ and the relative error is obtained as 3.704 %.

**7.10 Finite difference approximation of a cantilevered beam based on five domain nodes—backward scheme**

(a)

$$EI_Z \left.\frac{\mathrm{d}^3 u}{\mathrm{d}X^3}\right|_5 = -Q_Y(X = L) = 0. \qquad \text{(C.412)}$$

$$\frac{EI_Z}{\Delta X^3}\begin{bmatrix} 7 & -4 & 1 & 0 \\ -4 & 6 & -4 & 1 \\ 1 & -4 & 5 & -2 \\ -14 & 24 & -18 & 5 \end{bmatrix}\begin{bmatrix} u_2 \\ u_3 \\ u_4 \\ u_5 \end{bmatrix} = \begin{bmatrix} 0 \\ 0 \\ -F \\ 0 \end{bmatrix}. \qquad \text{(C.413)}$$

$$\text{relative error} = \left|\frac{\frac{1}{8} + \frac{81}{384}}{-\frac{81}{384}}\right| \times 100 = 159.259\,\%. \qquad \text{(C.414)}$$

(b)

$$EI_Z \left.\frac{\mathrm{d}^3 u}{\mathrm{d}X^3}\right|_4 = -Q_Y\left(X = \frac{3L}{4}\right) = F. \qquad \text{(C.415)}$$

$$\frac{EI_Z}{\Delta X^3}\begin{bmatrix} 7 & -4 & 1 & 0 \\ -4 & 6 & -4 & 1 \\ 1 & -4 & 5 & -2 \\ 27 & -18 & 5 & 0 \end{bmatrix}\begin{bmatrix} u_2 \\ u_3 \\ u_4 \\ u_5 \end{bmatrix} = \begin{bmatrix} 0 \\ 0 \\ -F \\ 2F \end{bmatrix}. \qquad \text{(C.416)}$$

$$\text{relative error} = \frac{\frac{7}{32} - \frac{81}{384}}{\frac{81}{384}} \times 100 = 3.704\,\%. \qquad \text{(C.417)}$$

**7.11 Finite difference approximation of a cantilevered beam based on five domain nodes—conversion of tip load into distributed load**

The final system of equations is obtained for this case as:

$$\frac{EI_Z}{\Delta X^3}\begin{bmatrix} 7 & -4 & 1 & 0 \\ -4 & 6 & -4 & 1 \\ 1 & -4 & 5 & -2 \\ 0 & 2 & -4 & 2 \end{bmatrix}\begin{bmatrix} u_2 \\ u_3 \\ u_4 \\ u_5 \end{bmatrix} = \begin{bmatrix} 0 \\ 0 \\ -q_Y\frac{\Delta X}{2} \\ -q_Y\frac{\Delta X}{2} \end{bmatrix}, \qquad \text{(C.418)}$$

from which we get $u_5 = -\frac{25}{128}\frac{FL^3}{EI_Z}$ and a relative error of 41.406 %.

**7.12 Finite difference approximation of a simply supported beam based on five domain nodes—bending moment approach**

The bending moment distribution is obtained as $M_Z(X) = \frac{q_y X}{2}(L - X)$. Thus, the finite difference approximation is given by:

$$EI_Z \times \frac{u_{i+1} - 2u_i + u_{i-1}}{\Delta X^2} = \frac{q_Y X_i}{2}(L - X_i) \,. \tag{C.419}$$

Evaluation for nodes 2, 3 and 4 and consideration of the boundary conditions, i.e. $u_1 = u_5 = 0$, gives:

$$\begin{bmatrix} -2 & 1 & 0 \\ 1 & -2 & 1 \\ 0 & 1 & -2 \end{bmatrix} \begin{bmatrix} u_2 \\ u_3 \\ u_4 \end{bmatrix} = \begin{bmatrix} \frac{q_Y \Delta X^3}{2EI_Z}(L - \Delta X) \\ \frac{q_Y \Delta X^3}{EI_Z}(L - 2\Delta X) \\ \frac{3q_Y \Delta X^3}{2EI_Z}(L - 3\Delta X) \end{bmatrix} . \tag{C.420}$$

Solution:

$$u_2 = -\frac{5q_Y L^4}{512EI_Z} \,, \quad u_3 = -\frac{7q_Y L^4}{512EI_Z} \,, \quad u_4 = -\frac{5q_Y L^4}{512EI_Z} \,. \tag{C.421}$$

**7.13 Finite difference approximation of a simply supported beam under pure bending**

The finite difference approximations of the the 2nd and 4th order partial differential equation can be written as:

$$EI_Z \left( \frac{u_{i+1} - 2u_i + u_{i-1}}{\Delta X^2} \right) = M_i \,, \tag{C.422}$$

$$\frac{EI_Z}{\Delta X^3}(u_{i+2} - 4u_{i+1} + 6u_i - 4u_{i-1} + u_{i-2}) = R_i = 0 \,. \tag{C.423}$$

(a) Five domain nodes

The system of linear equation based on the 2nd order PDE is given by:

$$\frac{EI_Z}{\Delta X^2} \begin{bmatrix} -2 & 1 & 0 \\ 1 & -2 & 1 \\ 0 & 1 & -2 \end{bmatrix} \begin{bmatrix} u_2 \\ u_3 \\ u_4 \end{bmatrix} = \begin{bmatrix} M \\ M \\ M \end{bmatrix} , \tag{C.424}$$

from which its solution is obtained as:

$$u_2 = u_4 = -\frac{3ML^2}{32EI_Z} \,, \quad u_3 = -\frac{ML^2}{8EI_Z} \,. \tag{C.425}$$

The system of linear equation based on the 4th order PDE is given by:

$$\frac{EI_Z}{\Delta X^3}\begin{bmatrix} 5 & -4 & 1 \\ -4 & 6 & -4 \\ 1 & -4 & 5 \end{bmatrix}\begin{bmatrix} u_2 \\ u_3 \\ u_4 \end{bmatrix} = \begin{bmatrix} -M/\Delta X \\ 0 \\ -M/\Delta X \end{bmatrix}, \tag{C.426}$$

which gives the same results as in the case of the 2nd order PDE. Furthermore, the finite difference and analytical solutions are identical, cf. supplementary problem 4.7.

(b) Ten domain nodes
The system of linear equation based on the 2nd order PDE is given by:

$$\frac{EI_Z}{\Delta X^2}\begin{bmatrix} -2 & 1 & 0 & 0 & 0 & 0 & 0 & 0 & 0 \\ 1 & -2 & 1 & 0 & 0 & 0 & 0 & 0 & 0 \\ 0 & 1 & -2 & 1 & 0 & 0 & 0 & 0 & 0 \\ 0 & 0 & 1 & -2 & 1 & 0 & 0 & 0 & 0 \\ 0 & 0 & 0 & 1 & -2 & 1 & 0 & 0 & 0 \\ 0 & 0 & 0 & 0 & 1 & -2 & 1 & 0 & 0 \\ 0 & 0 & 0 & 0 & 0 & 1 & -2 & 1 & 0 \\ 0 & 0 & 0 & 0 & 0 & 0 & 1 & -2 & 1 \\ 0 & 0 & 0 & 0 & 0 & 0 & 0 & 1 & -2 \end{bmatrix}\begin{bmatrix} u_2 \\ u_3 \\ u_4 \\ u_5 \\ u_6 \\ u_7 \\ u_8 \\ u_9 \\ u_{10} \end{bmatrix} = \begin{bmatrix} M \\ M \\ M \\ M \\ M \\ M \\ M \\ M \\ M \end{bmatrix}, \tag{C.427}$$

from which its solution is obtained as:

$$u_2 = -\frac{9ML^2}{200EI_Z}, \quad u_3 = -\frac{2ML^2}{25EI_Z}, \quad u_4 = -\frac{21ML^2}{200EI_Z}, \tag{C.428}$$

$$u_5 = -\frac{3ML^2}{25EI_Z}, \quad u_6 = -\frac{ML^2}{8EI_Z}. \tag{C.429}$$

The system of linear equation based on the 4th order PDE is given by:

$$\frac{EI_Z}{\Delta X^3}\begin{bmatrix} 5 & -4 & 1 & 0 & 0 & 0 & 0 & 0 & 0 \\ -4 & 6 & -4 & 1 & 0 & 0 & 0 & 0 & 0 \\ 1 & -4 & 6 & -4 & 1 & 0 & 0 & 0 & 0 \\ 0 & 1 & -4 & 6 & -4 & 1 & 0 & 0 & 0 \\ 0 & 0 & 1 & -4 & 6 & -4 & 1 & 0 & 0 \\ 0 & 0 & 0 & 1 & -4 & 6 & -4 & 1 & 0 \\ 0 & 0 & 0 & 0 & 1 & -4 & 6 & -4 & 1 \\ 0 & 0 & 0 & 0 & 0 & 1 & -4 & 6 & -4 \\ 0 & 0 & 0 & 0 & 0 & 0 & 1 & -4 & 5 \end{bmatrix}\begin{bmatrix} u_2 \\ u_3 \\ u_4 \\ u_5 \\ u_6 \\ u_7 \\ u_8 \\ u_9 \\ u_{10} \end{bmatrix} = \begin{bmatrix} -M/\Delta X \\ 0 \\ 0 \\ 0 \\ 0 \\ 0 \\ 0 \\ 0 \\ -M/\Delta X \end{bmatrix}, \tag{C.430}$$

which gives the same results as in the case of the 2nd order PDE. Furthermore, the finite difference and analytical solutions are again identical, cf. supplementary problem 4.7.

**7.14 Finite difference approximation of a simply supported beam—displacement, bending moment and shear force distribution**

The FD solution for the displacement can be taken from example 7.1a as:

$$u_1 = u_5 = 0\,,\, u_2 = u_4 = \frac{FL^3}{64EI_z}\,,\, u_3 = \frac{3FL^3}{128EI_z}\,. \tag{C.431}$$

FD moment distribution:

$$EI_z \frac{\mathrm{d}^2 u}{\mathrm{d}X^2} = M_z \;\rightarrow\; (EI_z)_i \frac{u_{i+1} - 2u_i + u_{i-1}}{\Delta X^2} = M_{z,i}\,. \tag{C.432}$$

$$\rightarrow M_{z,2} = M_{z,4} = -\frac{FL}{8}\,,\, M_{z,3} = -\frac{FL}{4}\,. \tag{C.433}$$

The boundary values, i.e. $M_{z,1} = M_{z,5} = 0$ result from the BC or from a forward ($i = 1$) or backward ($i = 5$) scheme.
FD shear force distribution:

$$EI_z \frac{\mathrm{d}^3 u}{\mathrm{d}X^3} = -Q_y \;\rightarrow\; (EI_z)_i \frac{u_{i+2} - 2u_{i+1} + 2u_{i-1} - u_{i-2}}{2\Delta X^3} = -Q_{y,i}\,. \tag{C.434}$$

$$Q_{y,3} = -\frac{64EI_z}{2L^3}\Big(\underbrace{u_5}_{0} - 2u_4 + 2u_2 - \underbrace{u_1}_{0}\Big) = 0\,. \tag{C.435}$$

The boundary values, i.e. $Q_{y,1} = Q_{y,5} = -2F$ result from a forward ($i = 1$) or backward ($i = 5$) scheme. The values $Q_{y,3}$ and $Q_{y,4}$ cannot be calculated based on the actual subdivision.
Analytical solution ($0 \leq X \leq \frac{L}{2}$):

$$u(X) = \frac{F}{48EI_z}\left(3L^2X - 4X^3\right)\,,\, M_z(X) = -\frac{FX}{2}\,,\, Q_y(X) = -\frac{F}{2}\,. \tag{C.436}$$

**7.15 Finite difference approximation of a simply supported beam on elastic foundation based on five domain nodes**

Finite difference approximation of the partial differential equation:

$$EI_Z \times \frac{u_{i+2} - 4u_{i+1} + 6u_i - 4u_{i-1} + u_{i-2}}{\Delta X^3} + ku_i\Delta X = R_i\,. \tag{C.437}$$

Linear system of equations:

$$\begin{bmatrix} 5 + \frac{k\Delta X^4}{EI_Z} & -4 & 1 \\ -4 & 6 + \frac{k\Delta X^4}{EI_Z} & -4 \\ 1 & -4 & 5 + \frac{k\Delta X^4}{EI_Z} \end{bmatrix} \begin{bmatrix} u_2 \\ u_3 \\ u_4 \end{bmatrix} = \begin{bmatrix} 0 \\ -\frac{F\Delta X^3}{EI_Z} \\ 0 \end{bmatrix}\,. \tag{C.438}$$

Solution:

$$u_3 = u\left(\tfrac{L}{2}\right) = -\frac{\left(6EI_Z + \frac{kL^4}{256}\right)L^3 F}{64\left(4(EI_Z)^2 + \frac{3EI_Z kL^4}{64} + \frac{k^2L^8}{65536}\right)}. \tag{C.439}$$

Special case $k = 4$, $EI_Z = 1$ and $L = 1$:

$$u_3 = u\left(\tfrac{L}{2}\right) = -0.0224451F. \tag{C.440}$$

Relative error (analytical solution cf. supplementary problem 4.13):

$$\text{relative error} = \frac{-0.0224451F + 0.0200233F}{-0.0200233F} \times 100 = 12.094\,\%. \tag{C.441}$$

**7.16 Finite difference approximation of a simply supported beam with varying bending stiffness—constant distributed load**

With the bending stiffness and bending moment distribution, the following differential equation can be written:

$$\frac{EI_0}{1+\left(\frac{2X}{L}-1\right)^2} \times \frac{\mathrm{d}^2 u_Y(X)}{\mathrm{d}X^2} = \frac{q_Y X}{2}(L - X). \tag{C.442}$$

The following finite difference scheme can be derived:

$$u_{i+1} - 2u_i + u_{i-1} = \Delta X^2 \times \frac{q_Y X_i}{2}(L - X_i) \times \frac{1+\left(\frac{2X_i}{L}-1\right)^2}{EI_0}. \tag{C.443}$$

Under consideration of the boundary conditions, i.e. $u_1 = u_5 = 0$, the following system of equations is obtained:

$$\begin{bmatrix} -2 & 1 & 0 \\ 1 & -2 & 1 \\ 0 & 1 & -2 \end{bmatrix} \begin{bmatrix} u_2 \\ u_3 \\ u_4 \end{bmatrix} = \frac{q_Y L^4}{EI_0} \begin{bmatrix} \frac{15}{2048} \\ \frac{1}{128} \\ \frac{15}{2048} \end{bmatrix}. \tag{C.444}$$

Solution:

$$u_2 = -\frac{23 q_Y L^4}{2048 EI_0}, \quad u_3 = -\frac{31 q_Y L^4}{2048 EI_0}, \quad u_4 = -\frac{23 q_Y L^4}{2048 EI_0}. \tag{C.445}$$

Analytical solution:

$$u_Y(X) = -\frac{q_Y}{EI_0 L^2}\left(\frac{1}{15}X^6 - \frac{1}{12}X^4L^2 - \frac{1}{6}X^3L^3\right) + c_1 X + C_2. \tag{C.446}$$

Consideration of the boundary conditions gives $c_2 = 0$ and $c_1 = -\frac{11 q_Y L^3}{60 E I_0}$. Based on the analytical solution, the displacement in the middle of the beam is obtained as:

$$u_Y\left(\tfrac{L}{2}\right) = -\frac{q_Y L^4}{15 E I_0}. \tag{C.447}$$

$$\text{relative error} = \frac{-\frac{31}{2048} + \frac{1}{15}}{-\frac{1}{15}} \times 100 = -77.295\,\%. \tag{C.448}$$

**7.17 Finite difference approximation of a simply supported beam with varying bending stiffness—constant bending moment**

With the bending stiffness and bending moment distribution, the following differential equation can be written:

$$\frac{EI_0}{1 + \left(\frac{2X}{L} - 1\right)^2} \times \frac{\mathrm{d}^2 u_Y(X)}{\mathrm{d}X^2} = M. \tag{C.449}$$

The following finite difference scheme can be derived:

$$u_{i+1} - 2u_i + u_{i-1} = \Delta X^2 \times M \times \frac{1 + \left(\frac{2X_i}{L} - 1\right)^2}{EI_0}. \tag{C.450}$$

Under consideration of the boundary conditions, i.e. $u_1 = u_5 = 0$, the following system of equations is obtained:

$$\begin{bmatrix} -2 & 1 & 0 \\ 1 & -2 & 1 \\ 0 & 1 & -2 \end{bmatrix} \begin{bmatrix} u_2 \\ u_3 \\ u_4 \end{bmatrix} = \frac{ML^2}{EI_0} \begin{bmatrix} \frac{5}{64} \\ \frac{1}{16} \\ \frac{5}{64} \end{bmatrix}. \tag{C.451}$$

Solution:

$$u_2 = -\frac{7ML^2}{64EI_0}, \quad u_3 = -\frac{9ML^2}{64EI_0}, \quad u_4 = -\frac{7ML^2}{64EI_0}. \tag{C.452}$$

Analytical solution:

$$u_Y(X) = \frac{2M}{EI_0 L^2}\left(\frac{1}{2}X^2L^2 - \frac{1}{3}X^3L + \frac{1}{6}X^4\right) + c_1 X + C_2. \tag{C.453}$$

Consideration of the boundary conditions gives $c_2 = 0$ and $c_1 = -\frac{2ML}{EI_0}$. Based on the analytical solution, the displacement in the middle of the beam is obtained as:

$$u_Y\left(\tfrac{L}{2}\right) = -\frac{31ML^2}{48EI_0}. \tag{C.454}$$

$$\text{relative error} = \frac{-\frac{9}{64}+\frac{31}{48}}{-\frac{31}{48}} \times 100 = -78.226\,\% \,. \tag{C.455}$$

**7.18 Finite difference approximation of a beam fixed at both ends**
(a) Single force case

$$u_Y(0.5L) = \frac{\left(32EI_Y + k_s AGL^2\right) LF}{2k_s AG\left(k_s AGL^2 + 64EI_Y\right)} \quad (5 \text{ nodes}), \tag{C.456}$$

$$u_Y(0.5L) = \frac{\left(512EI_Y + 11k_s AGL^2\right) LF}{2k_s AG\left(k_s AGL^2 + 1024EI_Y\right)} \quad (17 \text{ nodes}). \tag{C.457}$$

Relative error is $-95.385$ % for 5 nodes and $-65.104$ % for 17 nodes.

(b) Distributed load case

$$u_Y(0.5L) = \frac{\left(32EI_Y + k_s AGL^2\right) L^2 q}{4k_s AG\left(k_s AGL^2 + 64EI_Y\right)} \quad (5 \text{ nodes}), \tag{C.458}$$

$$u_Y(0.5L) = \frac{\left(512EI_Y + 11k_s AGL^2\right) L^2 q}{4k_s AG\left(k_s AGL^2 + 1024EI_Y\right)} \quad (17 \text{ nodes}). \tag{C.459}$$

Relative error is $-95.274$ % for 5 nodes and $-64.267$ % for 17 nodes.

**7.19 Convergence of finite difference approximation of a simply supported Timoshenko beam**
Only the inner nodes are used to derive the finite difference approximation. The following general solutions consider that the load is opposed to the positive $Y$-direction. To calculate the numerical values, it is only required to insert the absolute value of $q$.

$$u_Y(0.5L) = -\frac{\left(32EI_Z + 3k_s AGL^2\right) L^2 q}{4k_s AG\left(k_s AGL^2 + 64EI_Z\right)} \quad (5 \text{ nodes}), \tag{C.460}$$

$$u_Y(0.5L) = -\frac{\left(512EI_Z + 53k_s AGL^2\right) L^2 q}{4k_s AG\left(k_s AGL^2 + 1024EI_Z\right)} \quad (13 \text{ nodes}), \tag{C.461}$$

$$u_Y(0.5L) = -\frac{\left(1936EI_Z + 201k_s AGL^2\right) L^2 q}{8k_s AG\left(k_s AGL^2 + 1936EI_s\right)} \quad (23 \text{ nodes}), \tag{C.462}$$

$$u_Y(0.5L) = -\frac{\left(2048EI_Z + 213k_s AGL^2\right) L^2 q}{4k_s AG\left(k_s AGL^2 + 4096EI_Z\right)} \quad (33 \text{ nodes}), \tag{C.463}$$

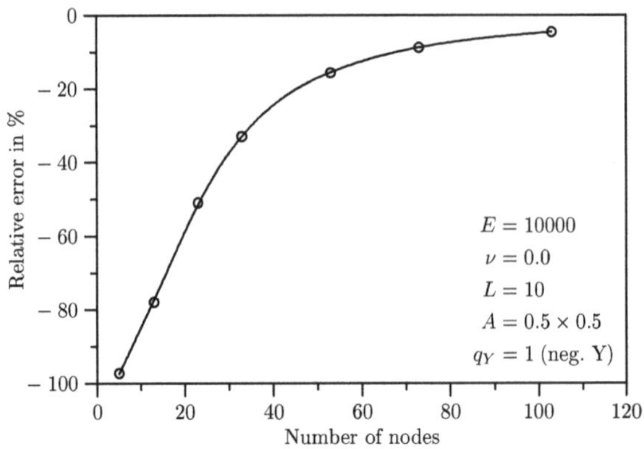


**Fig. C.14** Convergence rate of the finite difference approximation for a simply supported TIMOSHENKO beam with distributed load

$$u_Y(0.5L) = -\frac{\left(5408EI_Z + 563k_sAGL^2\right)L^2q}{4k_sAG\left(k_sAGL^2 + 10816EI_Z\right)} \quad (53 \text{ nodes}), \tag{C.464}$$

$$u_Y(0.5L) = -\frac{\left(31104EI_Z + 3239k_sAGL^2\right)L^2q}{12k_sAG\left(k_sAGL^2 + 20736EI_Z\right)} \quad (73 \text{ nodes}), \tag{C.465}$$

$$u_Y(0.5L) = -\frac{\left(124848EI_Z + 13003k_sAGL^2\right)L^2q}{24k_sAG\left(k_sAGL^2 + 41616EI_Z\right)} \quad (103 \text{ nodes}). \tag{C.466}$$

The convergence rate is shown in Fig. C.14.

**7.20 Convergence of finite difference approximation of a cantilevered Timoshenko beam**

(a) Solutions for dim. = 2 × (domain nodes – 2). The following general solutions consider that the load is opposed to the positive $Y$-direction.

$$u_Y(L) = -\frac{\left(4096E^2I_Z^2 + 1248EI_Zk_sAGL^2 + 25k_s^2A^2G^2L^4\right)L^2q_Y}{4k_sAG\left(32EI_Z + k_sAGL^2\right)\left(k_sAGL^2 + 64EI_Z\right)} \quad (5 \text{ nodes}), \tag{C.467}$$

$$u_Y(L) = -\frac{\left(32EI_Z + 11k_sAGL^2\right)LF}{k_sAG\left(32EI_Z + k_sAGL^2\right)} \quad (5 \text{ nodes}). \tag{C.468}$$

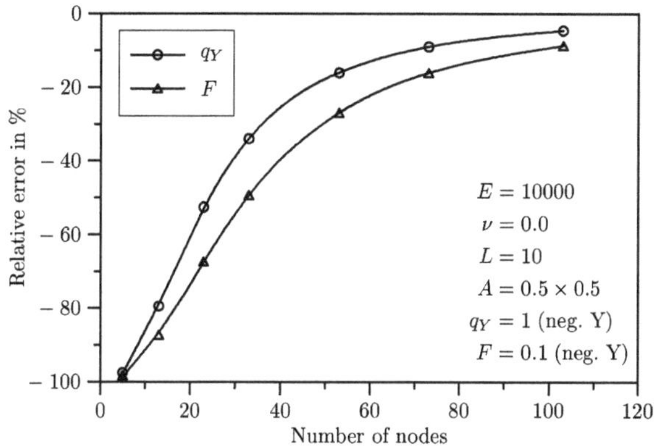


**Fig. C.15** Convergence rate of the finite difference approximation for a cantilevered TIMOSHENKO beam with distributed or point load; dim. = 2 × (domain nodes − 2)

The convergence rate is shown in Fig. C.15.

$$\begin{bmatrix} 0 & -\frac{2EI_Z}{\Delta X} - k_sAG\Delta X & \frac{k_sAG}{2} & \frac{EI_Z}{\Delta X} & 0 & 0 \\ -\frac{2k_sAG}{\Delta X} & 0 & \frac{k_sAG}{\Delta X} & -\frac{k_sAG}{2} & 0 & 0 \\ \hline -\frac{k_sAG}{2} & \frac{EI_Z}{\Delta X} & 0 & -\frac{2EI_Z}{\Delta X} - k_sAG\Delta X & \frac{k_sAG}{2} & \frac{EI_Z}{\Delta X} \\ \frac{k_sAG}{\Delta X} & \frac{k_sAG}{2} & -\frac{2k_sAG}{\Delta X} & 0 & \frac{k_sAG}{\Delta X} & -\frac{k_sAG}{2} \\ \hline 0 & 0 & -\frac{2k_sAG}{3} & \frac{2EI_Z}{3\Delta X} - \frac{k_sAG\Delta X}{9} & \frac{2k_sAG}{3} & -\frac{2EI_Z}{3\Delta X} - \frac{5k_sAG\Delta X}{9} \\ 0 & 0 & \frac{2k_sAG}{3\Delta X} & \frac{4k_sAG}{9} & -\frac{2k_sAG}{3\Delta X} & -\frac{2k_sAG}{9} \end{bmatrix} \begin{bmatrix} u_2 \\ \phi_2 \\ \hline u_3 \\ \phi_3 \\ \hline u_4 \\ \phi_4 \end{bmatrix} = \begin{bmatrix} 0 \\ q_Y\Delta X \\ \hline 0 \\ q_Y\Delta X \\ \hline 0 \\ q_Y\Delta X \end{bmatrix}. \tag{C.469}$$

In the case of the single force, the right-hand side of Eq. (C.469) must be replaced by the following column matrix:

$$\left[0\;0\;0\;0\;\tfrac{F\Delta X}{3}\;\tfrac{2F}{3}\right]^{\mathrm{T}}. \tag{C.470}$$

The value of the deflection at the right-hand end can be obtained from the solution of the linear system of equations as:

$$u(X = L) = -\frac{u_3}{3} + \frac{4u_4}{3} - \frac{2\Delta X\phi_3}{9} + \frac{8\Delta X\phi_4}{9} + \frac{2\Delta XQ_5}{3k_sAG}. \tag{C.471}$$

(b) Solutions for dim. = 2 × (domain nodes − 1).
The fictitious node at the right-hand boundary can be eliminated for example in the case of five domain nodes by the kinematic relationship

$$\phi_5 = \frac{\mathrm{d}u_Y(L)}{\mathrm{d}X} - \gamma_{XY}(L) \approx \frac{u_6 - u_4}{2\Delta X} - \frac{\tau_{XY}(L)}{G} = \frac{u_6 - u_4}{2\Delta X} - \frac{Q_Y(L)}{k_s AG}, \tag{C.472}$$

or

$$u_6 = u_4 + \phi_5 2\Delta X + \frac{Q_Y(L)}{k_s AG}. \tag{C.473}$$

The condition for the bending moment at the free boundary gives: $\phi_6 = \phi_4$ (See Fig. C.16).

In the case of this approach, the system of equations given in Eq. (C.469) is extended by two columns and rows whereas the last two rows reads as:

$$\begin{bmatrix} \cdots & 0 & \frac{EI_Z}{\Delta X} & 0 & -\frac{2EI_Z}{\Delta X} \\ \cdots & \frac{2k_s AG}{\Delta X} & 0 & -\frac{2k_s AG}{\Delta X} & 2k_s AG \end{bmatrix} \begin{bmatrix} u_5 \\ \phi_5 \end{bmatrix} = \begin{bmatrix} 0 \\ \frac{q_Y \Delta X}{2} \end{bmatrix}, \tag{C.474}$$

or in the case of the single force as:

$$\begin{bmatrix} \cdots & 0 & \frac{EI_Z}{\Delta X} & 0 & -\frac{2EI_Z}{\Delta X} \\ \cdots & \frac{2k_s AG}{\Delta X} & 0 & -\frac{2k_s AG}{\Delta X} & 2k_s AG \end{bmatrix} \begin{bmatrix} u_5 \\ \phi_5 \end{bmatrix} = \begin{bmatrix} -Q_Y(L)\Delta X \\ F - 2Q_Y(L) \end{bmatrix}. \tag{C.475}$$

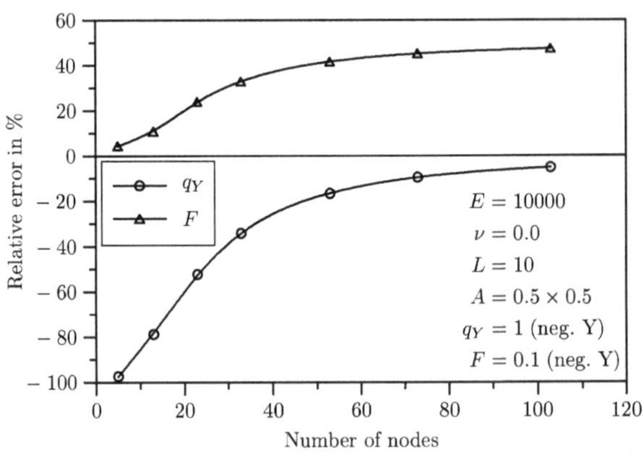


**Fig. C.16** Convergence rate of the finite difference approximation for a cantilevered TIMOSHENKO beam with distributed or point load; dim. = 2 × (domain nodes − 1)

Let us finally remind that the equilibrium between the internal reaction (shear force) and external load at the right boundary gives: $Q_Y(L) = -F$.

$$u_Y(L) = -\frac{\left(224EI_Z + 55k_sAGL^2\right)L^2q_Y}{8k_sAG\left(64EI_Z + k_sAGL^2\right)} \quad \text{(5 nodes)}, \tag{C.476}$$

$$u_Y(L) = -\frac{\left(3072E^2I_Z^2 + 1024EI_Zk_sAGL^2 + 11k_s^2A^2G^2L^4\right)LF}{32k_sAGEI_Z\left(64EI_Z + k_sAGL^2\right)} \quad \text{(5 nodes)}. \tag{C.477}$$

**7.21 Comparison between analytical and layer-wise integration of the bending stiffness**

See Fig. C.17.

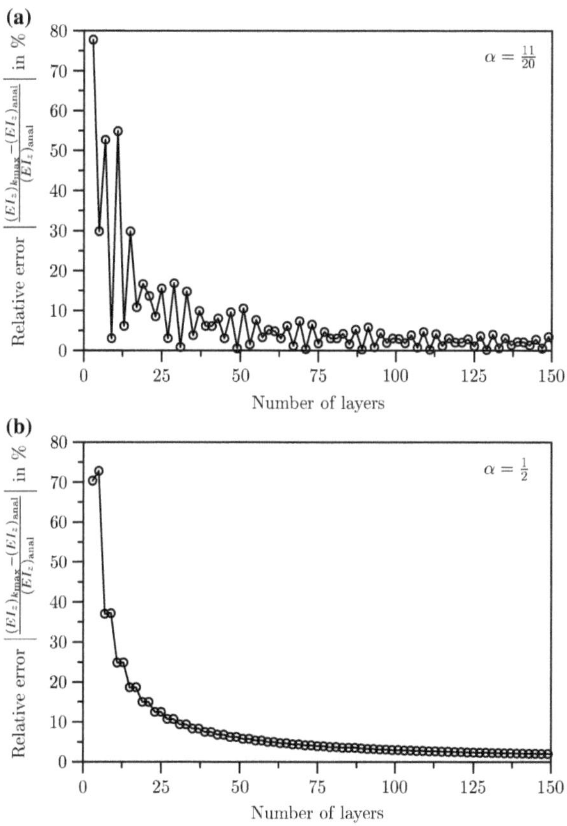


**Fig. C.17** Relative error as a function of the layer number for different size of the plastic zone

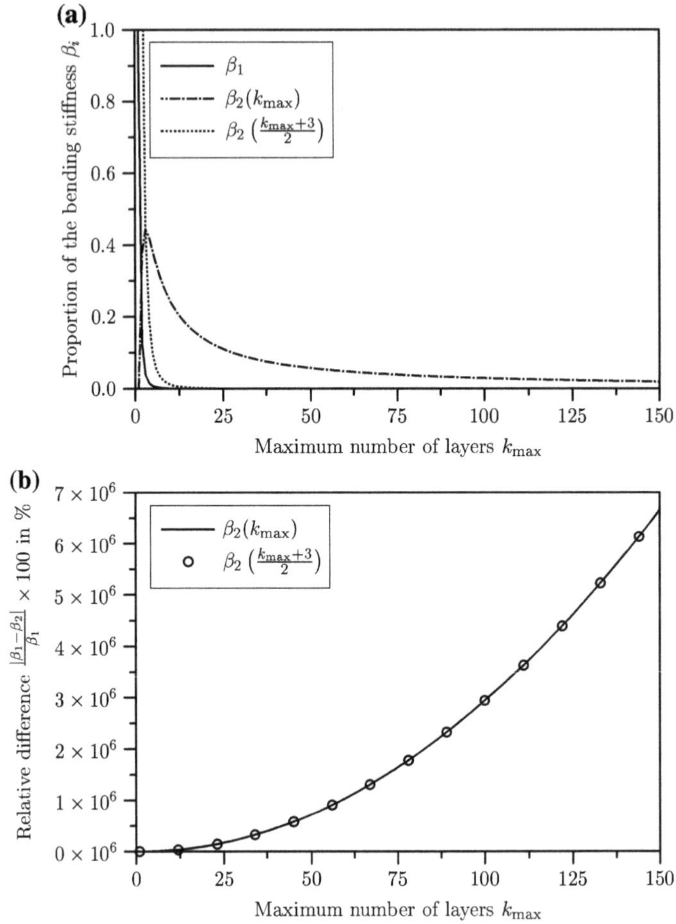


**Fig. C.18** **a** Absolute difference and **b** relative difference between the proportions of the relative bending stiffness in the elastic range within the layered approach

**7.22 Investigation of the proportions of the relative bending stiffness**
See Fig. C.18.

**7.23 Elasto-plastic finite difference solution: influence of layer number**
See Table C.20.

**7.24 Elasto-plastic finite difference solution: influence of load increment**
See Table C.21.

**Table C.20** Numerical results for deformed shape and comparison with analytical solution: influence of $k_{max}$

| | $k_{max} = 11$ | $k_{max} = 101$ | $k_{max} = 1001$ | $k_{max} = 10001$ |
|---|---|---|---|---|
| *Displacement* $u_Y(X)$ | | | | |
| $\frac{X}{L} = 0.0$ | 0.0 | 0.0 | 0.0 | 0.0 |
| $\frac{X}{L} = 0.25$ | −0.110762 | −0.120338 | −0.121357 | −0.121390 |
| $\frac{X}{L} = 0.5$ | −0.147682 | −0.160451 | −0.161810 | −0.161853 |
| *Relative error in %* | | | | |
| $\frac{X}{L} = 0.0$ | – | – | – | – |
| $\frac{X}{L} = 0.25$ | 2.933360 | 5.459040 | 6.352147 | 6.380597 |
| $\frac{X}{L} = 0.5$ | 2.933148 | 5.459271 | 6.352380 | 6.380830 |

**Table C.21** Numerical results for deformed shape and comparison with analytical solution: influence of load increment

| | $\Delta M = 1/5$ | $\Delta M = 1/8$ | $\Delta M = 1/10$ |
|---|---|---|---|
| *Displacement* $u_Y(X)$ | | | |
| $\frac{X}{L} = 0.0$ | 0.0 | 0.0 | 0.0 |
| $\frac{X}{L} = 0.25$ | −0.128549 | −0.119502 | −0.120338 |
| $\frac{X}{L} = 0.5$ | −0.171399 | −0.159336 | −0.160451 |
| *Relative error in %* | | | |
| $\frac{X}{L} = 0.0$ | – | – | – |
| $\frac{X}{L} = 0.25$ | 12.654530 | 4.726031 | 5.459040 |
| $\frac{X}{L} = 0.5$ | 12.654777 | 4.726261 | 5.459271 |

## Problems from Chap. 8

### 8.1 Hooke's law in terms of shear and bulk modulus

Elastic stiffness form:

$$\begin{bmatrix} \sigma_x \\ \sigma_y \\ \sigma_z \\ \sigma_{xy} \\ \sigma_{yz} \\ \sigma_{xz} \end{bmatrix} = \begin{bmatrix} K+\frac{4}{3}G & K-\frac{2}{3}G & K-\frac{2}{3}G & 0 & 0 & 0 \\ K-\frac{2}{3}G & K+\frac{4}{3}G & K-\frac{2}{3}G & 0 & 0 & 0 \\ K-\frac{2}{3}G & K-\frac{2}{3}G & K+\frac{4}{3}G & 0 & 0 & 0 \\ 0 & 0 & 0 & G & 0 & 0 \\ 0 & 0 & 0 & 0 & G & 0 \\ 0 & 0 & 0 & 0 & 0 & G \end{bmatrix} \begin{bmatrix} \varepsilon_x \\ \varepsilon_y \\ \varepsilon_z \\ 2\,\varepsilon_{xy} \\ 2\,\varepsilon_{yz} \\ 2\,\varepsilon_{xz} \end{bmatrix}. \quad \text{(C.478)}$$

Elastic compliance form:

$$\begin{bmatrix} \varepsilon_x \\ \varepsilon_y \\ \varepsilon_z \\ 2\,\varepsilon_{xy} \\ 2\,\varepsilon_{yz} \\ 2\,\varepsilon_{xz} \end{bmatrix} = \frac{1}{18KG} \begin{bmatrix} 6K+2G & -3K+2G & -3K+2G & 0 & 0 & 0 \\ -3K+2G & 6K+2G & -3K+2G & 0 & 0 & 0 \\ -3K+2G & -3K+2G & 6K+2G & 0 & 0 & 0 \\ 0 & 0 & 0 & 18K & 0 & 0 \\ 0 & 0 & 0 & 0 & 18K & 0 \\ 0 & 0 & 0 & 0 & 0 & 18K \end{bmatrix} \begin{bmatrix} \sigma_x \\ \sigma_y \\ \sigma_z \\ \sigma_{xy} \\ \sigma_{yz} \\ \sigma_{xz} \end{bmatrix}. \tag{C.479}$$

**8.2 Hooke's law for the plane stress state**
Elastic stiffness form:

$$\begin{bmatrix} \sigma_x \\ \sigma_y \\ \sigma_{xy} \end{bmatrix} = \frac{E}{1-\nu^2} \begin{bmatrix} 1 & \nu & 0 \\ \nu & 1 & 0 \\ 0 & 0 & \frac{1-\nu}{2} \end{bmatrix} \begin{bmatrix} \varepsilon_x \\ \varepsilon_y \\ 2\,\varepsilon_{xy} \end{bmatrix}. \tag{C.480}$$

Elastic compliance form:

$$\begin{bmatrix} \varepsilon_x \\ \varepsilon_y \\ 2\,\varepsilon_{xy} \end{bmatrix} = \frac{1}{E} \begin{bmatrix} 1 & -\nu & 0 \\ -\nu & 1 & 0 \\ 0 & 0 & 2(\nu+1) \end{bmatrix} \begin{bmatrix} \sigma_x \\ \sigma_y \\ \sigma_{xy} \end{bmatrix}. \tag{C.481}$$

Furthermore:

$$\varepsilon_z = -\frac{\nu}{1-\nu}\,(\varepsilon_x + \varepsilon_y)\,. \tag{C.482}$$

**8.3 Hooke's law for the plane strain state**
Elastic stiffness form:

$$\begin{bmatrix} \sigma_x \\ \sigma_y \\ \sigma_{xy} \end{bmatrix} = \frac{E}{(1+\nu)(1-2\nu)} \begin{bmatrix} 1-\nu & \nu & 0 \\ \nu & 1-\nu & 0 \\ 0 & 0 & \frac{1-2\nu}{2} \end{bmatrix} \begin{bmatrix} \varepsilon_x \\ \varepsilon_y \\ 2\,\varepsilon_{xy} \end{bmatrix}. \tag{C.483}$$

Elastic compliance form:

$$\begin{bmatrix} \varepsilon_x \\ \varepsilon_y \\ 2\,\varepsilon_{xy} \end{bmatrix} = \frac{1-\nu^2}{E} \begin{bmatrix} 1 & -\frac{\nu}{1-\nu} & 0 \\ -\frac{\nu}{1-\nu} & 1 & 0 \\ 0 & 0 & \frac{2}{1-\nu} \end{bmatrix} \begin{bmatrix} \sigma_x \\ \sigma_y \\ \sigma_{xy} \end{bmatrix}. \tag{C.484}$$

Furthermore:

$$\sigma_z = \nu\,(\sigma_x + \sigma_y)\,. \tag{C.485}$$

**8.4 Beltrami-Michell equations**

$$\boldsymbol{\mathcal{L}}^{\mathrm{T}}\boldsymbol{\sigma} + \boldsymbol{b} = \boldsymbol{0}\,. \tag{C.486}$$

**8.5 Green-Gauss theorem applied to equilibrium equation in $x$-direction**

$$\underbrace{\begin{bmatrix} \sigma_x \\ 0 \\ 0 \\ \sigma_{xy} \\ 0 \\ \sigma_{xy} \end{bmatrix}}_{\boldsymbol{\sigma}_x} = \underbrace{\frac{E}{(1+\nu)(1-2\nu)}\begin{bmatrix} 1-\nu & \nu & \nu & 0 & 0 & 0 \\ 0 & 0 & 0 & 0 & 0 & 0 \\ \nu & \nu & 1-\nu & 0 & 0 & 0 \\ 0 & 0 & 0 & \frac{1-2\nu}{2} & 0 & 0 \\ 0 & 0 & 0 & 0 & 0 & 0 \\ 0 & 0 & 0 & 0 & 0 & \frac{1-2\nu}{2} \end{bmatrix}}_{\boldsymbol{C}_x} \underbrace{\begin{bmatrix} \varepsilon_x \\ \varepsilon_y \\ \varepsilon_z \\ 2\,\varepsilon_{xy} \\ 0 \\ 2\,\varepsilon_{xz} \end{bmatrix}}_{\boldsymbol{\varepsilon}_x}, \tag{C.487}$$

$$\underbrace{\begin{bmatrix} \varepsilon_x \\ \varepsilon_y \\ \varepsilon_z \\ 2\varepsilon_{xy} \\ 0 \\ 2\varepsilon_{xz} \end{bmatrix}}_{\boldsymbol{\varepsilon}_x} = \underbrace{\begin{bmatrix} \frac{\partial}{\partial x} & 0 & 0 \\ 0 & \frac{\partial}{\partial y} & 0 \\ 0 & 0 & \frac{\partial}{\partial z} \\ \frac{\partial}{\partial y} & \frac{\partial}{\partial x} & 0 \\ 0 & 0 & 0 \\ \frac{\partial}{\partial z} & 0 & \frac{\partial}{\partial x} \end{bmatrix}}_{\boldsymbol{\mathcal{L}}_{\mathrm{m},x}} \cdot \underbrace{\begin{bmatrix} u_x \\ u_y \\ u_z \end{bmatrix}}_{\boldsymbol{u}}, \tag{C.488}$$

Weighted residual statement in $x$-direction:

$$\int_V \left( W_x \boldsymbol{\mathcal{L}}_x^{\mathrm{T}} \left( \boldsymbol{C}_x \boldsymbol{\mathcal{L}}_{\mathrm{m},x} \boldsymbol{u} \right) + W_x f_x \right) \mathrm{d}V = 0\,. \tag{C.489}$$

Application of the GREEN-GAUSS theorem gives:

$$\int_V \left( \boldsymbol{\mathcal{L}}_x^{\mathrm{T}} W_x \right) \boldsymbol{C}_x \left( \boldsymbol{\mathcal{L}}_{\mathrm{m},x} \boldsymbol{u} \right) \mathrm{d}V = \int_A W_x \left( \boldsymbol{C}_x \boldsymbol{\mathcal{L}}_{\mathrm{m},x} \boldsymbol{u} \right)^{\mathrm{T}} \boldsymbol{n}_x \mathrm{d}A + \int_V W_x f_x \mathrm{d}V\,. \tag{C.490}$$

**8.6 Green-Gauss theorem applied to derive general 3D weak form**
Derive the weak statement for all three coordinate directions separately and sum up these three equations.

**8.7 Body force matrix for gravity**
$\boldsymbol{f}_b^{\mathrm{e}} = \varrho g \times 8a^3 \left[ 0\ -\tfrac{1}{8}\ 0 \,|\, \cdots \,|\, 0\ -\tfrac{1}{8}\ 0 \right]^{\mathrm{T}}$.

# References

1. Abramowitz M, Stegun IA (1965) Hanbook of mathematical functions. Dover Publications, New York
2. Al-Amery RIM, Roberts TM (1990) Nonlinear finite difference analysis of composite beams with partial interaction. Comput Struct 35:81–87
3. de Allen DNG (1955) Relaxation methods. McGraw-Hill, New York
4. Altenbach H, Altenbach J, Naumenko K (1998) Ebene Flächentragwerke: Grundlagen der Modellierung und Berechnung von Scheiben und Platten. Springer, Berlin
5. Anandarajah A (2010) Computational methods in elasticity and plasticity: solids and porous media. Springer, New York
6. Backhaus G (1968) Zur Fließgrenze bei allgemeiner Verfestigung (On the yield condition for general workhardening). Z Angew Math Mech 48:99–108
7. Bathe K-J (1996) Finite element procedures. Prentice-Hall, Upper Saddle River
8. Bauschinger J (1886) Über die Veränderung der Elastizitätsgrenze und die Festigkeit des Eisens und Stahls durch Strecken und Quetschen, durch Erwärmen und Abkühlen und durch oftmals wiederholte Beanspruchungen (On the change of the elastic limit and strength of iron and steel by elongation, shortening, heating, cooling and repeated loading). Mitt Mech-Tech Lab Königl Tech Hochsch München 13:1–116
9. Beer FP, Johnston ER Jr, DeWolf JT, Mazurek DF (2009) Mechanics of materials. McGraw-Hill, New York
10. Belytschko T, Liu WK, Moran B (2000) Nonlinear finite elements for continua and structures. John Wiley & Sons, Chichester
11. Benallal A, Billardon R, Doghri I, Moret-Bailly L (1987) Crack initiation and propagation analysis taking into account initial strain hardening and damage fields. In: Luxmoore AR, Owen DRJ, Rajapakse YPS, Kanninen MF (eds) Numerical methods in fracture mechanics. Pineridge Press, Swansea
12. Betten J (2001) Kontinuumsmechanik. Springer, Berlin
13. de Borst R (1986) Non-linear analysis of frictional materials. Dissertation, Delft University of Technology.
14. Boresi AP, Schmidt RJ (2003) Advanced mechanics of materials. John Wiley & Sons, New York
15. Borja RI (2013) Plasticity: modeling and computation. Springer, Berlin
16. Brebbia CA, Telles JFC, Wrobel LC (1984) Boundary element techniques: theory and applications. Springer-Verlag, Berlin
17. Buchanan GR (1995) Schaum's outline of theory and problems of finite element analysis. McGraw-Hill, New York



A. Öchsner, *Elasto-Plasticity of Frame Structure Elements*,
DOI 10.1007/978-3-662-44225-8

18. Budynas RG (1999) Advanced strength and applied stress analysis. McGraw-Hill Book, Singapore
19. Chakrabarty J (2006) Theory of plasticity. Elsevier Butterworth-Heinemann, Oxford
20. Chakrabarty J (2010) Applied plasticity. Springer-Verlag, New York
21. Chen WF, Saleeb AF (1982) Constitutive equations for engineering materials. Volume 1: elasticity and modelling. Wiley, New York.
22. Chen WF, Han DJ (1988) Plasticity for structural engineers. Springer-Verlag, New York
23. Clebsch RFA (1862) Theorie der Elasticität fester Körper. Teubner, Leipzig, B.G.
24. Clough RW (1960) The finite element in plane strain analysis. In: Proceedings of the 2nd ASCE conference on electronic computation, Pittsburgh, PA.
25. Collatz L (1966) The numerical treatment of differential equations. Springer, Berlin
26. Cook RD, Malkus DS, Plesha ME, Witt RJ (2002) Concepts and applications of finite element analysis. Wiley, New York
27. Courant R (1943) Variational methods for the solution of problems of equilibrium and vibrations. B Am Math Soc 49:1–61
28. Cowper GR (1966) The shear coefficient in Timoshenko's beam theory. J Appl Mech 33:335–340
29. Crisfield MA (2000) Non-linear finite element analysis of solids and structures. Volume 2: advanced topics. Wiley, Chichester.
30. Crisfield MA (2001) Non-linear finite element analysis of solids and structures. Volume 1: essentials. Wiley, Chichester.
31. Czichos H, Hennecke M (eds) (2007)Hütte. Das Ingenieurwissen, Springer, Berlin
32. Doltsinis I (2000) Elements of plasticity: theory and computation. WIT Press, Southampton
33. Drucker DC (1952) A more fundamental approach to plastic stress-strain relations. In: Sternberg E et al (eds) Proceedings of the 1st US national congress on applied mechanics. Edward Brothers Inc, Michigan, pp 487–491
34. Drucker DC (1956) The effect of shear on the plastic bending of beams. J Appl MEch 23:509–514
35. Dubbel H, Grote K-H, Feldhusen J (eds) (2004) Dubbel. Taschenbuch für den Maschinenbau, Springer-Verlag, Berlin
36. Dunne F, Petrinic N (2005) Introduction to computational plasticity. Oxford University Press, Oxford
37. Elishakoff I, Irretier H (eds) (1987) Refined dynamical theories of beams, plates and shells and their applications. In: Proceedings of the Euromech-colloquium 219. Springer, Berlin.
38. Eschenauer H, Olhoff N, Schnell W (1997) Applied structural mechanics: Fundamentals of elasticity, load-bearing structures, structural optimization. Springer, Berlin
39. Esmaeilli M, Öchsner A (2011) A one-dimensional implemetation of a coupled elasto-plastic model for ductile damage. Materialwiss Werkst 42:444–451
40. Felippa CA (2001) A historical outline of matrix structural analysis: a play in three acts. Comput Struct 79:1313–1324
41. Forsythe GE, Wasow WR (1960) Finite-difference methods for partial differential equations. Wiley, New York
42. Galerkin BG (1915) Series solution of some problems of elastic equilibrium of rods and plates (Russion). Vestn Inzh Tech 19:897–908
43. Gere JM, Timoshenko SP (1991) Mechanics of materials. PWS-KENT Publishing Company, Boston
44. Ghugal YM, Shimpi RP (2001) A review of refined shear deformation theories for isotropic and anisotropic laminated beams. J Reinf Plast Comp 20:255–272
45. Gould PL (1988) Analysis of shells and plates. Springer-Verlag, New York
46. Gross D, Hauger W, Schröder J, Wall WA (2009) Technische Mechanik 2: Elastostatik. Springer, Berlin
47. Gruttmann F, Wagner W (2001) Shear correction factors in Timoshenko's beam theory for arbitrary shaped cross-sections. Comput Mech 27:199–207

48. Gurson AL (1977) Continuum theory of ductile rupture by void nucleation and growth: part I—yield criteria and flow rules for ductile media. J Eng Mater T ASME 99:2–15
49. Hartmann F, Katz C (2007) Structural analysis with finite elements. Springer, Berlin
50. Hencky H (1924) Zur Theorie plastischer deformation und der hierdurch im material hervorgerufenen Nachspannungen. Z Angew Math Mech 4:323–334
51. Heyman J (1998) Structural analysis: a historical approach. Cambridge University Press, Cambridge
52. Hibbeler RC (2008) Mechanics of materials. Prentice Hall, Singapore
53. Hill R (1950) The mathematical theory of plasticity. Oxford University Press, New York
54. Hildebrand FB (1956) Introduction to numerical analysis. McGraw-Hill, New York
55. Hodge PG (1959) Plastic analysis of structures. McGraw-Hill, New York
56. Horne MR (1951) The plastic theory of bending of mild steel beams with particular reference to the effect of shear forces. Proc R Soc Lon Ser-A 207:216–228
57. Lau PCM, Brebbia CA (1978) The cell collocation method in continuum mechanics. Int J Mech Sci 20:83–95
58. Johnson W, Yu TX (1981) On springback after the pure bending of beams and plates of elastic work-hardening material - III. Int J Mech Sci 23:687–696
59. Kassimali A (2012) Matrix analysis of structures. Cengage Learning, Stamford
60. Lehmann Th (1972) Einige Bemerkungen zu einer allgemeinen Klasse von Stoffgesetzen für große elasto-plastische Formänderungen (Some remarks on a general class of yield conditions for large elasto-plastic deformations). Ing Arch 41:297–310
61. Lemaitre J (1985) A continuous damage mechanics model for ductile fracture. J Eng Mater-T ASME 107:83–89
62. Lemaitre J (1985) Coupled elasto-plasticity and damage constitutive equations. Comput Method Appl 51:31–49
63. Lemaitre J (1996) A course on damage mechanics. Springer, Berlin
64. Lemaitre J, Desmorat R (2005) Engineering damage mechanics: ductile, creep, fatigue and brittle failures. Springer, Berlin
65. Levinson M (1981) A new rectangular beam theory. J Sound Vib 74:81–87
66. MacNeal RH (1978) A simple quadrilateral shell element. Comput Struct 8:175–183
67. MacNeal RH (1994) Finite elements: their design and performance. Marcel Dekker, New York
68. Melan E (1938) Zur Plastizität des räumlichen Kontinuums (On the plasticity of the spatial continuum). Ing Arch 9:116–126
69. von Mises R (1913) Mechanik des festen Körpers im plastisch deformablen Zustand. Nachr Königl Ges Wiss Göttingen. Math Phy Kl 1:582–592
70. Mitchell AR, Griffiths DF (1980) The finite difference method in partial differential equations. Wiley, New York
71. Moran B, Ortiz M, Shih CF (1990) Formulation of implicit finite element methods for multiplicative finite deformation plasticity. Int J Num Meth Eng 29:483–514
72. Nádai AL (1950) Theory of flow and fracture of solids. McGraw-Hill, New York
73. Neal BG (1977) The plastic methods of structural analysis. Halsted Press, New York
74. de Neto EA, Souza Perić D, Owen DRJ (2008) Computational methods for plasticity: theory and applications. Wiley, Chichester
75. Öchsner A, Merkel M (2013) One-dimensional finite elements: an introduction to the FE method. Springer, Berlin
76. Ortiz M, Popov EP (1985) Accuracy and stability of integration algorithms for elastoplastic constitutive equations. Int J Num Meth Eng 21:1561–1576
77. Owen DRJ, Hinton E (1980) Finite elements in plasticity: theory and practice. Pineridge Press Limited, Swansea
78. Pook L (2007) Metal fatigue—what it is, why it matters. Springer, Dordrecht
79. Prager W, Hodge PG (1951) Theory of perfectly plastic solids. Wiley, New York
80. Prager W (1955) The theory of plasticity; a survey of recent achievements (James Clayton Lecture). Proc Inst Mech Eng 169:41–57

81. Prager W (1956) A new method of analyzing stress and strain in work-hardening plastic solids. J Appl Mech 23:493–496
82. Reddy JN (1984) A simple higher-order theory for laminated composite plate. J Appl Mech 51:745–752
83. Reddy JN (1990) A general non-linear third-order theory of plates with moderate thickness. Int J Nonlinear Mech 25:677–686
84. Reddy JN (1997) Mechanics of laminated composite plates: theory and analysis. CRC Press, Boca Raton
85. Reddy JN (1997) On locking-free shear deformable beam finite elements. Comput Method Appl M 149:113–132
86. Reddy JN (1999) On the dynamic behaviour of the Timoshenko beam finite elements. Sadhana-Acad P Eng S 24:175–198
87. Reddy JN (2004) An introduction to nonlinear finite element analysis. Oxford University Press, Oxford
88. Reddy JN (2006) An introduction to the finite element method. McGraw Hill, Singapore
89. Ritz W (1909) Über eine neue Methode zur Lösung gewisser Variationsprobleme der mathematischen Physik. J Reine Angew Math 135:1–61
90. Russel WT, MacNeal RH (1953) An improved electrical analogy for the analysis of beams in bending. J Appl Mech 20:349
91. Samuelsson A, Zienkiewicz OC (2006) History of the stiffness method. Int J Numer Meth Eng 67:149–157
92. Sayyad AS (2011) Comparison of various refined beam theories for the bending and free vibration analysis of thick beams. Appl Comput Mech 5:217–230
93. Schijve J (2001) Fatigue of structures and materials. Kluwer Academic Pulishers, Dordrecht
94. Shield RT, Ziegler H (1958) On Prager's hardening rule. Z Angew Math Phys 9a:260–276.
95. Simo JC, Ortiz M (1985) A unified approach to finite deformation elastoplasticity based on the use of hyperelastic constitutive equations. Comput Method Appl M 49:221–245
96. Simo JC, Hughes TJR (1998) Computational inelasticity. Springer, New York
97. Stein E (2014) History of the finite element method—mathematics meets mechanics—part I: engineering developments. In: Stein E (ed) The history of theoretical, material and computational mechanics. Springer-Verlag, Berlin, pp 399–442
98. Steinke P (2010) Finite-Elemente-Methode - Rechnergestützte Einführung. Springer, Berlin
99. Stroud AH, Secrest D (1966) Gaussian quadrature formulas. Prentice Hall, Englewood Cliffs
100. Southwell RV (1946) Relaxation methods in theoretical physics. Clarendon Press, Oxford
101. Szabó I (2003) Einführung in die Technische Mechanik: Nach Vorlesungen István Szabó. Springer-Verlag, Berlin
102. Timoshenko SP (1921) On the correction for shear of the differential equation for transverse vibrations of prismatic bars. Philos Mag 41:744–746
103. Timoshenko S (1940) Strength of materials-part I elementary theory and problems. D. Van Nostrand Company, New York
104. Timoshenko S (1947) Strength of materials—part II advanced theory and problems. D. Van Nostrand Company, New York
105. Timoshenko S, Woinowsky-Krieger S (1959) Theory of plates and shells. McGraw-Hill Book Company, New York
106. Timoshenko SP, Goodier JN (1970) Theory of elasticity. McGraw-Hill, New York
107. Tresca H (1864) Mémoire sur l'écoulement des corps solides soumis à de fortes pressions. CR Hebd Acad Sci 59:754–758
108. Turner MJ, Clough RW, Martin HC, Topp LJ (1956) Stiffness and deflection analysis of complex structures. J Aeronaut Sci 23:805–823
109. Tvergaard V (1991) Mechanical modelling of ductile fracture. Meccanica 26:11–16
110. Twiss RJ, Moores EM (1992) Structural geology. WH Freeman & Co, New York
111. Underwood EE (1970) Quantitative stereology. Addison-Wesley, London
112. Wang CM (1995) Timoshenko beam-bending solutions in terms of Euler-Bernoulli solutions. J Eng Mech ASCE 121:763–765

113. Wang CM, Reddy JN, Lee KH (2000) Shear deformable beams and plates: relationships with classical solution. Elsevier, Oxford
114. Weaver W Jr, Gere JM (1980) Matrix analysis of framed structures. Van Nostrand Reinhold Company, New York
115. Wriggers P (2008) Nonlinear finite element methods. Springer, Berlin
116. Winkler E (1867) Die Lehre von der Elasticität und Festigkeit mit besonderer Rücksicht auf ihre Anwendung in der Technik. H. Dominicus, Prag
117. Yu TX, Zhang LC (1996) Plastic bending: theory and applications. World Scientific, Singapore
118. Ziegler H (1959) A modification of Prager's hardening rule. Q Appl Math 17:55–65
119. Zienkiewicz OC (1995) Origins, milestones and directions of the finite element method—a personal view. Arch Comput Method E 2:1–48
120. Zienkiewicz OC, Taylor RL (2000) The finite element method. Volume 1: the basis. Butterworth-Heinemann, Oxford.
121. Zienkiewicz OC (2004) The birth of the finite element method and of computational mechanics. Int J Numer Meth Eng 60:3–10

# Index




A. Öchsner, *Elasto-Plasticity of Frame Structure Elements*,
DOI 10.1007/978-3-662-44225-8

MIX
Papier aus verantwortungsvollen Quellen
Paper from responsible sources
FSC® C105338

If you have any concerns about our products, you can contact us on
**ProductSafety@springernature.com**

In case Publisher is established outside the EU, the EU authorized representative is:
**Springer Nature Customer Service Center GmbH**
**Europaplatz 3, 69115 Heidelberg, Germany**

Printed by Libri Plureos GmbH
in Hamburg, Germany